The chemistry of the **carbon–carbon triple bond** Part 2

THE CHEMISTRY OF FUNCTIONAL GROUPS

A series of advanced treatises under the general editorship of Professor Saul Patai

The chemistry of alkenes (2 volumes)
The chemistry of the carbonyl group (2 volumes)
The chemistry of the ether linkage
The chemistry of the amino group
The chemistry of the nitro and nitroso groups (2 parts)
The chemistry of carboxylic acids and esters
The chemistry of the carbon–nitrogen double bond
The chemistry of amides
The chemistry of the cyano group
The chemistry of the hydroxyl group (2 parts)
The chemistry of the azido group
The chemistry of acyl halides
The chemistry of the carbon–halogen bond (2 parts)
The chemistry of the quinonoid compounds (2 parts)
The chemistry of the thiol group (2 parts)
The chemistry of the hydrazo, azo and azoxy groups (2 parts)
The chemistry of amidines and imidates
The chemistry of cyanates and their thio derivatives (2 parts)
The chemistry of diazonium and diazo groups (2 parts)
The chemistry of the carbon–carbon triple bond (2 parts)
Supplement A: The chemistry of double-bonded functional groups (2 parts)

—C≡C—

The chemistry of the **carbon–carbon triple bond**

Part 2

Edited by

SAUL PATAI

The Hebrew University, Jerusalem

1978

JOHN WILEY & SONS

CHICHESTER — NEW YORK — BRISBANE — TORONTO

An Interscience ® Publication

Library of Congress Catalog Card No. 75–6913

ISBN 0 471 99497 9 (Pt. 1)
ISBN 0 471 99498 7 (Pt. 2)
ISBN 0 471 99496 0 (Set)

Printed in Great Britain by John Wright and Sons Ltd., at the Stonebridge Press, Bristol.

Contributing authors

J. Bastide	Centre Universitaire, Perpignan, France
D. A. Ben-Efraim	The Weizmann Institute of Science, Rehovot, Israel
K. A. Connors	School of Pharmacy, University of Wisconsin, Madison, Wisconsin, USA
J. D. Coyle	The Polytechnic, Wolverhampton, England
J. I. Dickstein	College of Du Page, Glen Ellyn, Illinois, USA
A. Gavezzotti	Istituto di Chimica Fisica e Centro CNR, Università di Milano, Milan, Italy
J. L. Hencher	University of Windsor, Windsor, Ontario, Canada
O. Henri-Rousseau	Centre Universitaire, Perpignan, France
A. C. Hopkinson	York University, Downsview, Ontario, Canada
A. M. Hudrlik	Rutgers University, New Brunswick, New Jersey, USA
P. F. Hudrlik	Rutgers University, New Brunswick, New Jersey, USA
W. D. Huntsman	Ohio University, Athens, Ohio, USA
Sir Ewart R. H. Jones	The Dyson Perrins Laboratory, Oxford University, Oxford, England
T. Kaneda	ISIR, Osaka University, Suita, Osaka, Japan
J. Klein	The Hebrew University, Jerusalem, Israel
J. C. Lavalley	U.E.R. de Sciences, Université de Caen, 14032 Caen Cedex, France
C. Lifshitz	The Hebrew University, Jerusalem, Israel
R. Lines	Chemical Centre, University of Lund, Lund, Sweden
A. Mandelbaum	Technion-Israel Institute of Technology, Haifa, Israel
S. I. Miller	Illinois Institute of Technology, Chicago, Illinois, USA
S. Misumi	ISIR, Osaka University, Suita, Osaka, Japan
M. Nakagawa	Osaka University, Toyonaka, Osaka 560, Japan
J. Saussey	U.E.R. de Sciences, Université de Caen, 14032 Caen Cedex, France
G. H. Schmid	University of Toronto, Toronto, Ontario, Canada
R. Shaw	Sunnyvale, California 94087, USA

M. Simonetta	Istituto di Chimica Fisica e Centro CNR, Università di Milano, Milan, Italy
V. Thaller	The Dyson Perrins Laboratory, Oxford University, Oxford, England
F. Théron	Université de Clermont-Ferrand, France
J. H. P. Utley	Queen Mary College, London, England
M. Verny	Université de Clermont-Ferrand, France
R. Vessière	Université de Clermont-Ferrand, France

Foreword

The present volume deals with the chemistry of the carbon–carbon triple bond. This is presented and organized again on the same general lines as described in the 'Preface to the series' printed on the following pages.

Some chapters originally planned for this volume did not materialize. These include a chapter on 'Free radical attacks involving carbon–carbon triple bonds', and a chapter on 'Arynes and hetarynes'. Tragically, the chapter on 'Directing and activating effects' is missing from this book owing to the untimely death of Professor Pentti Salomaa, a good friend, an excellent chemist and a devoted teacher, missed by all who knew him. It is hoped to include chapters on these subjects in 'Supplement C: The Chemistry of Triple-bonded Functional Groups', which is planned to be published in several years' time.

Jerusalem, October 1977 SAUL PATAI

The Chemistry of Functional Groups
Preface to the series

The series 'The Chemistry of Functional Groups' is planned to cover in each volume all aspects of the chemistry of one of the important functional groups in organic chemistry. The emphasis is laid on the functional group tested and on the effects which it exerts on the chemical and physical properties, primarily in the immediate vicinity of the group in question, and secondarily on the behaviour of the whole molecule. For instance, the volume *The Chemistry of the Ether Linkage* deals with reactions in which the C—O—C group is involved, as well as with the effects of the C—O—C group on the reactions of alkyl or aryl groups connected to the ether oxygen. It is the purpose of the volume to give a complete coverage of all properties and reactions of ethers in as far as these depend on the presence of the ether group but the primary subject matter is not the whole molecule, but the C—O—C functional group.

A further restriction in the treatment of the various functional groups in these volumes is that material included in easily and generally available secondary or tertiary sources, such as Chemical Reviews, Quarterly Reviews, Organic Reactions, various 'Advances' and 'Progress' series as well as textbooks (i.e. in books which are usually found in the chemical libraries of universities and research institutes) should not, as a rule, be repeated in detail, unless it is necessary for the balanced treatment of the subject. Therefore each of the authors is asked *not* to give an encyclopaedic coverage of his subject, but to concentrate on the most important recent developments and mainly on material that has not been adequately covered by reviews or other secondary sources by the time of writing of the chapter, and to address himself to a reader who is assumed to be at a fairly advanced post-graduate level.

With these restrictions, it is realized that no plan can be devised for a volume that would give a *complete* coverage of the subject with *no* overlap between chapters, while at the same time preserving the readability of the text. The Editor set himself the goal of attaining *reasonable* coverage with *moderate* overlap, with a minimum of cross-references between the chapters of each volume. In this manner, sufficient freedom is given to each author to produce readable quasi-monographic chapters.

The general plan of each volume includes the following main sections:

(a) An introductory chapter dealing with the general and theoretical aspects of the group.

(b) One or more chapters dealing with the formation of the functional group in question, either from groups present in the molecule, or by introducing the new group directly or indirectly.

(c) Chapters describing the characterization and characteristics of the functional groups, i.e. a chapter dealing with qualitative and quantitative methods of determination including chemical and physical methods, ultraviolet, infrared, nuclear magnetic resonance and mass spectra: a chapter dealing with activating and directive effects exerted by the group and/or a chapter on the basicity, acidity or complex-forming ability of the group (if applicable).

(d) Chapters on the reactions, transformations and rearrangements which the functional group can undergo, either alone or in conjunction with other reagents.

(e) Special topics which do not fit any of the above sections, such as photochemistry, radiation chemistry, biochemical formations and reactions. Depending on the nature of each functional group treated, these special topics may include short monographs on related functional groups on which no separate volume is planned (e.g. a chapter on 'Thioketones' is included in the volume *The Chemistry of the Carbonyl Group*, and a chapter on 'Ketenes' is included in the volume *The Chemistry of Alkenes*). In other cases certain compounds, though containing only the functional group of the title, may have special features so as to be best treated in a separate chapter, as e.g. 'Polyethers' in *The Chemistry of the Ether Linkage*, or 'Tetraaminoethylenes' in *The Chemistry of the Amino Group*.

This plan entails that the breadth, depth and thought-provoking nature of each chapter will differ with the views and inclinations of the author and the presentation will necessarily be somewhat uneven. Moreover, a serious problem is caused by authors who deliver their manuscript late or not at all. In order to overcome this problem at least to some extent, it was decided to publish certain volumes in several parts, without giving consideration to the originally planned logical order of the chapters. If after the appearance of the originally planned parts of a volume it is found that either owing to non-delivery of chapters, or to new developments in the subject, sufficient material has accumulated for publication of a supplementary volume, containing material on related functional groups, this will be done as soon as possible.

The overall plan of the volumes in the series 'The Chemistry of Functional Groups' includes the titles listed below:

The Chemistry of Alkenes (*two volumes*)
The Chemistry of the Carbonyl Group (*two volumes*)
The Chemistry of the Ether Linkage
The Chemistry of the Amino Group
The Chemistry of the Nitro and Nitroso Group (*two parts*)
The Chemistry of Carboxylic Acids and Esters
The Chemistry of the Carbon–Nitrogen Double Bond
The Chemistry of the Cyano Group
The Chemistry of Amides
The Chemistry of the Hydroxyl Group (*two parts*)
The Chemistry of the Azido Group
The Chemistry of Acyl Halides
The Chemistry of the Carbon–Halogen Bond (*two parts*)
The Chemistry of Quinonoid Compounds (*two parts*)
The Chemistry of the Thiol Group (*two parts*)
The Chemistry of Amidines and Imidates
The Chemistry of the Hydrazo, Azo and Azoxy Groups
The Chemistry of Cyanates and their Thio Derivatives
The Chemistry of Diazonium and Diazo Groups
The Chemistry of the Carbon–Carbon Triple Bond (*two parts*)
Supplement A: The Chemistry of Double-bonded Functional Groups (*two parts*)

Titles in press:

The Chemistry of Ketenes, Allenes and Related Compounds
Supplement B: The Chemistry of Acid Derivatives

Future volumes planned include:

The Chemistry of Cumulenes and Heterocumulenes
The Chemistry of Organometallic Compounds
The Chemistry of Sulphur-containing Compounds
Supplement C: The Chemistry of Triple-bonded Functional Groups
Supplement D: The Chemistry of Halides and Pseudo-halides
Supplement E: The Chemistry of $-NH_2$, $-OH$, *and* $-SH$ *Groups and their Derivatives*

Advice or criticism regarding the plan and execution of this series will be welcomed by the Editor.

The publication of this series would never have started, let alone continued, without the support of many persons. First and foremost among these is Dr Arnold Weissberger, whose reassurance and trust encouraged me to tackle this task, and who continues to help and advise me. The efficient and patient cooperation of several staff-members of the Publisher also rendered me invaluable aid (but unfortunately their code of ethics does not allow me to thank them by name). Many of my friends and colleagues in Israel and overseas helped me in the solution of various major and minor matters, and my thanks are due to all of them, especially to Professor Z. Rappoport. Carrying out such a long-range project would be quite impossible without the non-professional but none the less essential participation and partnership of my wife.

The Hebrew University
Jerusalem, Israel

SAUL PATAI

Contents

CHAPTER 12

Photochemistry of the C≡C bond

J. D. Coyle
The Polytechnic, Wolverhampton, England

I. INTRODUCTION

The electronic structure of alkynes is related to that of alkenes, and the photochemistry of the two classes of compound reflects this similarity. Because the photochemistry of alkenes has received greater attention and has already been described in systematic form[1, 2], it is not unexpected that the present account should point out the ways in which alkyne photochemistry parallels, or is markedly different from, that of alkenes. There is a considerable difference, however, in the range of compounds which has been studied in each class. Reports of photochemical reactions of alkynes very often refer to mono- or disubstituted acetylenes in which the substituents are alkyl, aryl or alkoxycarbonyl. There have been studies on diyne and enyne systems, but as yet there has emerged nothing in alkyne chemistry to match the wealth of photochemistry reported for dienes and polyenes. This reflects in part the greater tendency of the compounds containing the C≡C bond to undergo photopolymerization rather than any other reaction on irradiation. Within this limitation there is a wide variety of reactions open to the excited states of alkynes, and quite a number of the processes have synthetic application or potential.

From spectroscopic data it seems likely that the excited states involved in alkyne photochemistry are either (π, π^*) states, in which an electron from a bonding π molecular orbital has been promoted to an antibonding π^* molecular orbital, or Rydberg states, in which a π electron has been promoted to an extended σ-type orbital covering more than one nucleus. The spectra of acetylene, propyne and but-1-yne all show features characteristic of both types of electronic transition[3]. In the region 110–160 nm there are intense sharp bands, many of which can be assigned to two or three different Rydberg progressions, and in the region 160–210 nm there is a weaker and more diffuse band with a maximum around 170–180 nm. The longer wavelength region probably consists of two or three overlapping bands, of which that at longest wavelength can be attributed to a $\pi \rightarrow \pi^*$ transition. The lowest energy singlet state is (π, π^*) in nature; note that for some simple alkenes the Rydberg singlet seems to be lowest in energy. The energy of the lowest state is not easily assigned from the absorption spectrum since there is a long, weak absorption tail (210–240 nm) associated with the changed geometry of the (π, π^*) singlet; a value of 505 kJ mol^{-1} (121 kcal mol^{-1}, equivalent to a wavelength of 237 nm) has been given[4] for the singlet state energy of acetylene.

The Rydberg excited states of acetylene, like the ground state, are linear[5], but the preferred geometry of the lowest (π, π^*) singlet state is non-linear and transoid, with ∠HCC angles of about 120°. The (π, π^*) state is sometimes crudely represented as a biradical species (**1**), although a zwitterionic canonical pair (**2**) might be more appropriate for a singlet state.

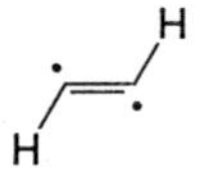

(**1**)

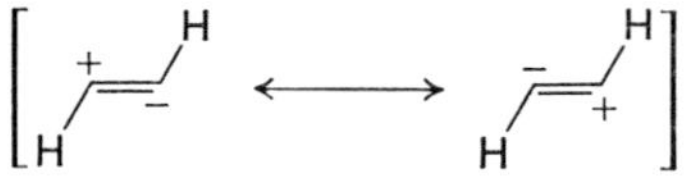

(**2**)

The energy of the lowest triplet excited state of simple alkynes is not known with certainty; these compounds do not phosphoresce, and the singlet → triplet absorption has not been characterized. A value as low as 190 kJ mol^{-1} (46 kcal mol^{-1}) has been suggested[6] for acetylene on the basis of electron impact studies. The triplet energies of phenyl-substituted acetylenes and of di- and poly-ynes can be assigned either from the enhanced singlet → triplet absorption spectrum under a high pressure of dissolved oxygen[7] or from phosphorescence data. Phenylacetylene has a triplet energy of ~300 kJ mol^{-1} (72 kcal mol^{-1}), diphenylacetylene 260 (62), butadiyne 330 (79·5) and octa-2,4,6-triyne 265 (63). Diphenylacetylene luminesces from both the singlet state (at room temperature) and the triplet state (at 77 K), and it also exhibits excimer fluorescence ($\lambda_{max} \sim 390$ nm)[8].

II. PHOTOFRAGMENTATION

Many simple alkynes undergo efficient homolytic bond cleavage on (vacuum) ultraviolet irradiation, particularly in the vapour phase (in solution the situation is often more complex because extensive polymerization occurs). The bond which breaks can be either that adjacent to the triple bond or the next one, i.e. the 'propargylic' bond (this term will be used here in an analogous way to the use of 'allylic' for C=C compounds). In the first category come the reactions of acetylene itself[9], which produces diacetylene (butadiyne), hydrogen and ethylene (equation 1) together with vinylacetylene, benzene and other polymers, and the reactions of alkynes with a halogen substituent on the triple bond (equation 2)[10]. Other groups which form a particularly stable free radical (i.e. a particularly weak ≡C—X bond) can be broken off in this way, such as trifluoromethyl[11] which then reacts with the original alkyne (equation 3) or can be trapped with added aliphatic hydrocarbon (equation 4).

$$CH{\equiv}CH \xrightarrow[\text{vapour}]{h\nu\ (185\ \text{nm})} CH{\equiv}C{-}C{\equiv}CH + H_2 + C_2H_4 \qquad (1)$$

$$\varphi = 0{\cdot}10 \qquad 0{\cdot}02 \qquad 0{\cdot}05$$

$$CH{\equiv}C{-}Br \xrightarrow{h\nu} C_2H^{\bullet}+Br^{\bullet} \quad (2)$$

$$CH{\equiv}C{-}CF_3 \xrightarrow{h\nu\ (190\ nm)} C_2H^{\bullet} + {}^{\bullet}CF_3$$

$$^{\bullet}CF_3 \xrightarrow{CH\equiv CCF_3} CF_3CH{=}\dot{C}CF_3 \longrightarrow CF_3CH{=}CHCF_3 \quad (3)$$

$$^{\bullet}CF_3 \xrightarrow{C_3H_8} CHF_3 \quad (4)$$

With shorter wavelength radiation acetylene also produces carbon and molecular hydrogen (equation 5) in a different primary process[12].

$$CH{\equiv}CH \xrightarrow[(124/147\ nm)]{h\nu} C_2+H_2 \quad (5)$$

Propargylic cleavage occurs for higher alkynes, and the propargyl radical (**3**) generated from propyne has been observed by e.s.r. spectroscopy[13]. With long wavelength radiation (206 nm) the major gaseous products from the photolysis of propyne in the gaseous phase[14] are hydrogen and hexa-1,5-diyne (**4**) formed by dimerization of the radical **3**.

$$CH_3C{\equiv}CH \xrightarrow[77\ K]{h\nu\ (250\text{–}380\ nm)} [\dot{C}H_2{-}C{\equiv}CH \leftrightarrow CH_2{=}C{=}\dot{C}H]$$

(3)

$$CH{\equiv}C{-}CH_2CH_2C{\equiv}CH$$

(4)

However, at shorter wavelengths (but still below the ionization threshold) molecular extrusion of hydrogen occurs, as evidenced by the fact that some of the hydrogen which is produced cannot be quenched by added free radical inhibitors. The main source of this molecular hydrogen seems to be[15] two-stage breakdown *via* a carbene, as suggested for the reaction of 3,3,3-trideuteriopropyne (equation 6).

$$CD_3C{\equiv}CH \xrightarrow{h\nu\ (124\ nm)} D_2+\ddot{C}D{-}C{\equiv}CH \longrightarrow C_3+HD \quad (6)$$

But-2-yne undergoes a similar reaction (equation 7) to give a diyne; but-2-ene is also formed as a reduction product[16]. The stable products arising from but-1-yne irradiation (equation 8) are hydrogen and vinylacetylene, together with smaller amounts of two- and three-carbon compounds, and this suggests that C—H cleavage is preferred over C—C cleavage in the excited state of this alkyne.

$$CH_3C{\equiv}CCH_3 \xrightarrow{h\nu} CH_3C{\equiv}CCH_2CH_2C{\equiv}CCH_3+CH_3CH{=}CHCH_3 \quad (7)$$

$$CH{\equiv}CCH_2CH_3 \xrightarrow{h\nu} CH{\equiv}C{-}CH{=}CH_2+H_2 \quad (8)$$

The production of acetylene as the major gaseous product in the reaction of butadiyne (equation 9) occurs in part via $C_2H^{\bullet}$ radicals when short wavelength radiation is used[17], but with 254 nm radiation it is suggested that a molecular process occurs in a non-linear excited state. The evidence for this additional process is that no C_2HD is formed when perdeuteriopropyne is present as a source of deuterium to trap the radicals.

$$CH{\equiv}C{-}C{\equiv}CH \xrightarrow{h\nu} 2\ CH{\equiv}C^{\bullet} \longrightarrow CH{\equiv}CH+C_2 \quad (9)$$

Products arising from propargylic C—C cleavage are not usually formed in solution, and the dialkyne **5** is photochemically inert to radiation of wavelength 254 nm [18]. However, cyclo-octa-1,5-diyne does cleave to give butatriene (equation 10), and the extra driving force here comes from the strain in the cyclic compound[19].

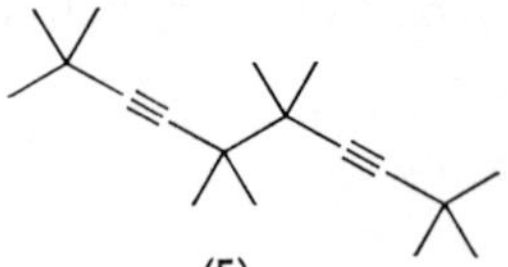

(5)

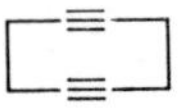 $\xrightarrow{h\nu\ (206\ \text{nm})}$ $2\ CH_2{=}C{=}C{=}CH_2$ (10)

The β,γ-acetylenic ketone **6** undergoes photochemical C—C cleavage at a position which is propargylic to the triple bond (equation 11)[20], but this is more likely to be an α-cleavage reaction of the excited state of the ketone group[21] rather than a reaction of the alkyne excited state.

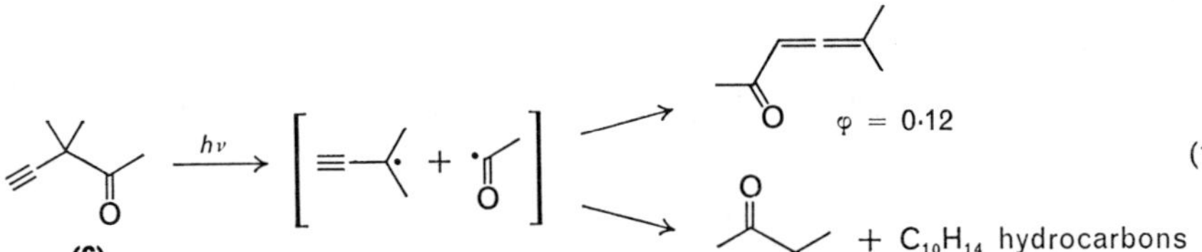

(6) (11)

III. PHOTOREARRANGEMENT

Propargylic cleavage of a $N{-}CH_3$ bond similar to that described for C—H and C—C bonds in the previous section, followed by recombination of the radicals produced, may be responsible for the very inefficient photorearrangement observed for the ynamine **7** to give 2-phenylisobutyronitrile[22]. Alternatively the reaction may be considered as a sequence of two consecutive [1,3] sigmatropic shifts rather than as a radical reaction.

$$PhC{\equiv}CNMe_2 \xrightarrow{h\nu\ (254\ \text{nm})} PhC(Me){=}C{=}NMe \xrightarrow{h\nu} Ph{-}C(Me)_2{-}C{\equiv}N$$

(7) 2·5%

There are few reports of rearrangement reactions for alkynes analogous to the extensive array of electrocyclic processes, sigmatropic shifts and di-π-methane reactions which are documented for alkenes. A major deterrent to 4π- or 6π-electron ring closures involving alkynes is the strain in the product (e.g. in reaction 12). Another difficulty may be the linear geometry of the alkyne unit—although the relaxed excited state with *trans* geometry may be suitably oriented for ring closure, the results from alkenes (notably the stereospecificity of the processes) suggest that a concerted reaction normally occurs *before* relaxation to the equilibrium geometry of the excited state.

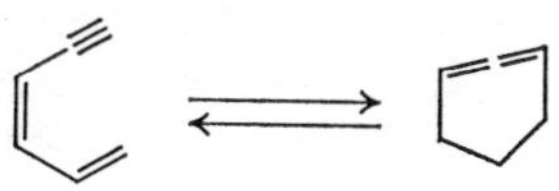

(12)

Aryl-substituted butenynes do appear to undergo a photochemical electrocyclic ring closure, which is followed by a hydrogen shift in the initially formed cyclohexa-1,2,4-triene to produce a compound with a new fused aromatic ring (equation 13), but the process has been shown to occur by way of non-concerted radical or ionic pathways[23].

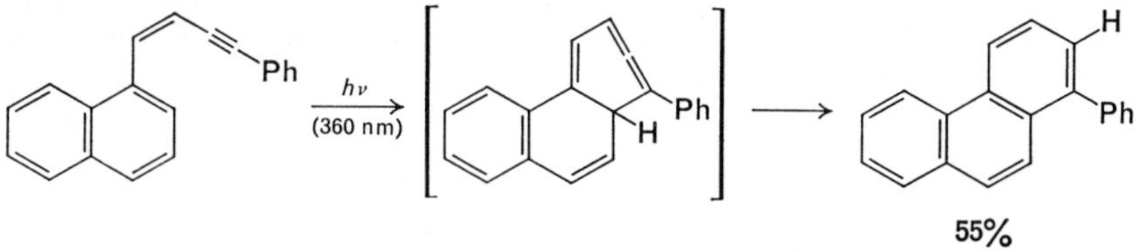

(13)

3,3-Dimethyl-1,5-diphenylpenta-1,4-diyne (**8**) does not undergo a photochemical di-π-methane rearrangement[24] like that of the corresponding diene (**9**). However, the enyne **10** does give products of the di-π-methane type, and the reaction is stereospecific[25], as can be seen from the reaction of **10** (equation 14) and of its *cis* isomer **11** (equation 15).

PhC≡C—CMe_2—C≡CPh
(**8**)

PhCH=CH—CMe_2—CH=CHPh
(**9**)

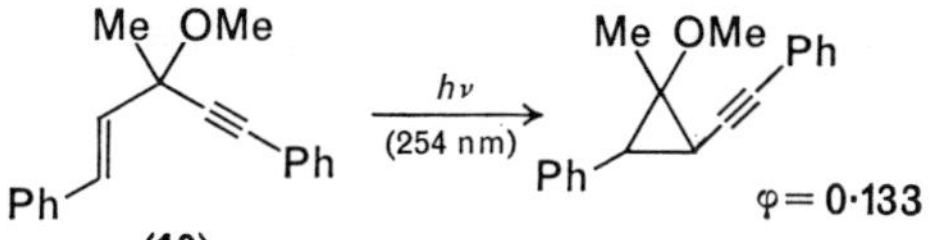

(**10**)

(14)

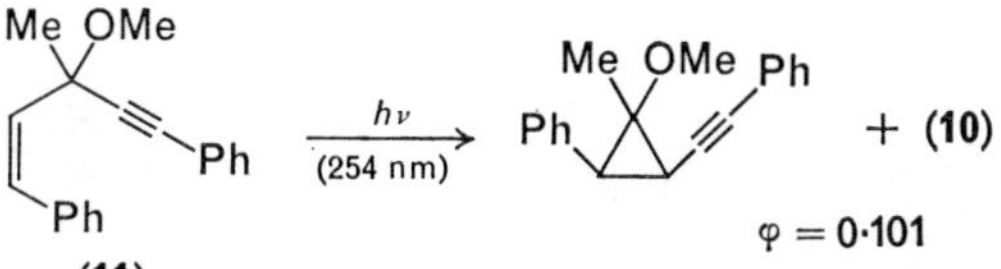

(**11**)

(15)

If the conclusions from Zimmerman's extensive studies on 1,4-dienes[26] can be applied to 1,4-enynes, then the triple bond is necessary for efficient reaction to occur and it plays an important role in the reaction pathway (equation 16).

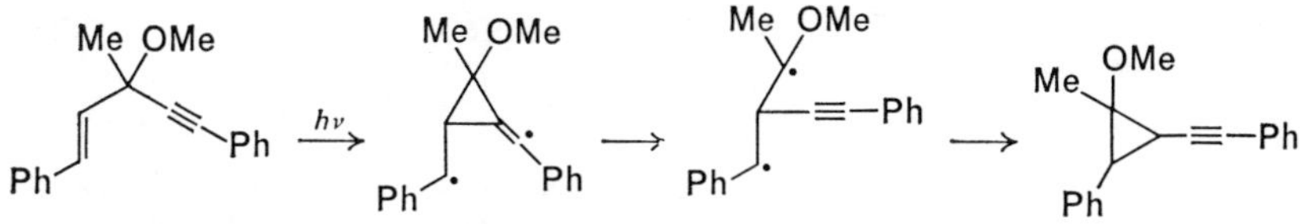

(16)

The rearrangement[27] of tetraphenylpropyne to tetraphenylcyclopropene (equation 17) is formally analogous to the di-π-methane reactions of 3-phenylalkenes, and it is likely that the mechanism is similar.

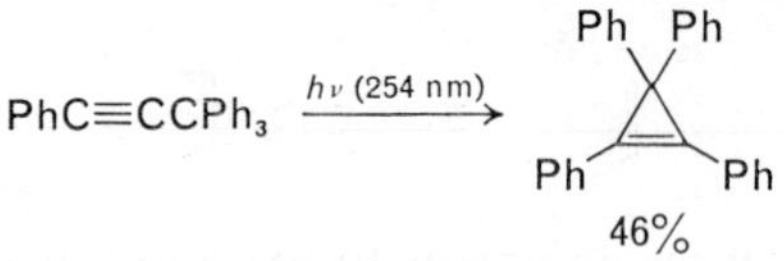

(17)

In this 3-phenylalkyne system the triple bond becomes incorporated into a cyclopropene ring, whereas from the enynes **10** and **11** there is no sign of a vinylcyclopropene product. The preference for reaction to take place at the alkene unit rather than at the alkyne unit in the photochemistry of enynes is seen again in the photochemistry of enepoly-yne chlorides[28], where only cyclopropyl chlorides are produced (equation 18). This preference is probably a reflection of the lower strain energy in a cyclopropane than in a cyclopropene ring.

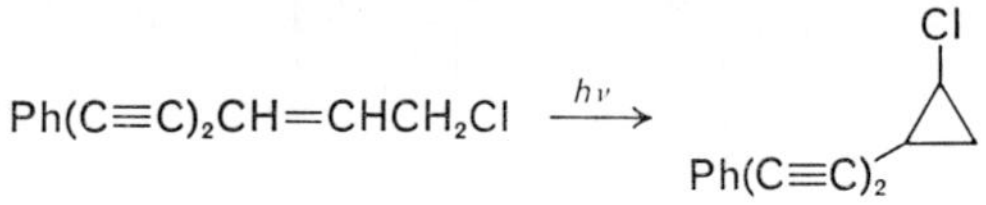

(18)

IV. PHOTOADDITION

A. Photoaddition Involving the Alkyne Excited State

In the photoaddition of a saturated hydrocarbon to ethyl propiolate (equation 19) it is likely that the excited state of the acetylenic ester initiates reaction by abstracting a hydrogen atom from the hydrocarbon[29]. The addition of cyclic ethers to an alkyne seems similar (equation 20), although a ketone sensitizer is required for addition of tetrahydropyran or dioxan[30]. When reactions of this type involve a conjugated acetylenic ester, the first-formed α,β-unsaturated ester can normally undergo further photochemical reaction to produce the β,γ isomer (see equations 19 and 20).

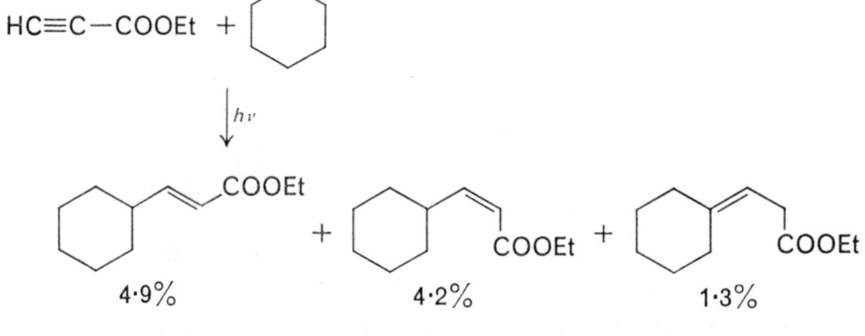

(19)

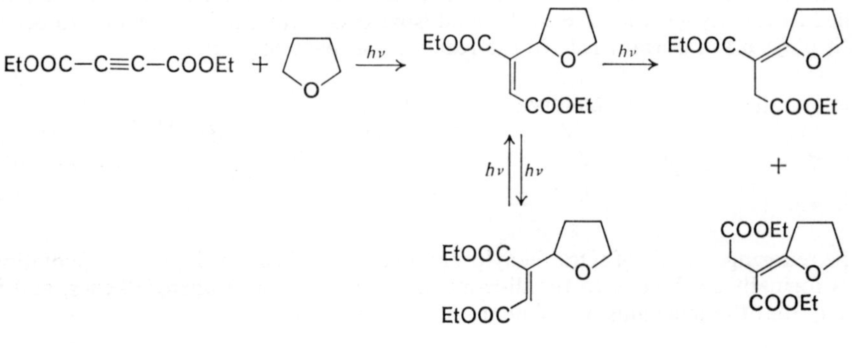

(20)

Alcohols can undergo photoaddition to alkynes, and the products from acetylenic esters[29] are γ-hydroxy-*trans*-α,β-unsaturated esters or the unsaturated lactones derived from the *cis* isomers by cyclization (equation 21). Reaction with acetylenic ketones provides a route to furans (equation 22)[31], and there is n.m.r. evidence for

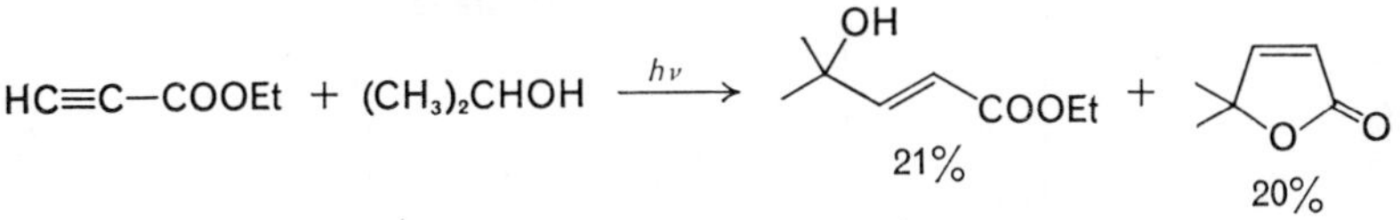
(21)

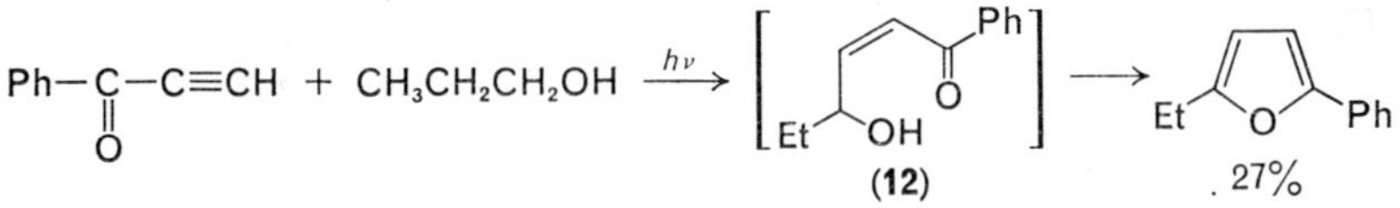
(22)

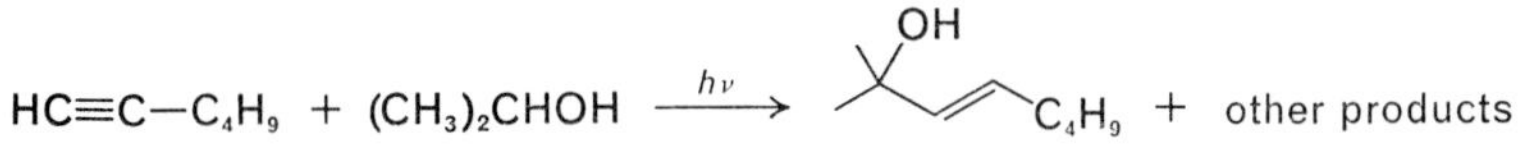
(23)

the suggested intermediate **12**. The carbonyl group is not essential to the reaction[29], and hex-1-yne gives an analogous addition product with propan-2-ol (equation 23).

In some systems an alcohol gives rise to photoreduction products of the alkyne rather than to products by photoaddition of the whole alcohol molecule. The acetylenic ester **13** gives the corresponding ethylenic ester on irradiation in methanol (equation 24). Under the same conditions the ketone $(CH_3)_3C{-}C{\equiv}C{-}COMe$

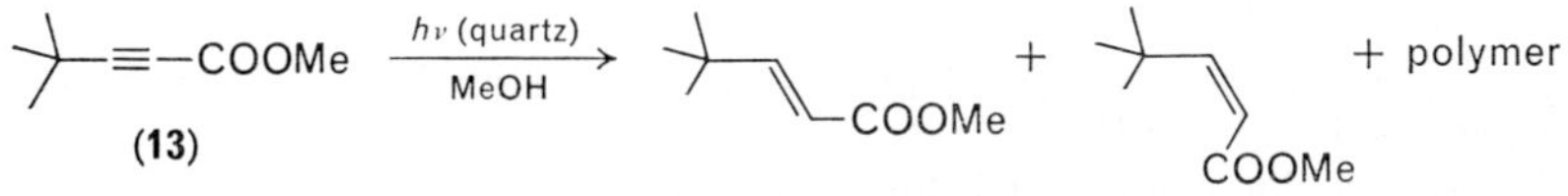
(24)

gives only pinacol products in which the triple bond is preserved[32]. The phenyl-substituted alkyne **14** reacts in a similar manner, with the *cis* alkene being the major product (equation 25). In the latter reaction deuterium-labelling studies indicate that

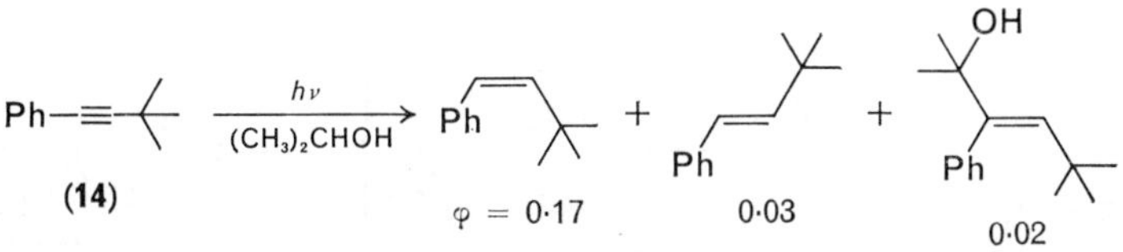
(25)

both of the 'new' hydrogen atoms in the product come mainly from the carbinolic position of propan-2-ol [24], and this suggests that the alkyne excited state abstracts a hydrogen atom in a radical-like process.

Dodeca-5,7-diyne undergoes photoreduction at one triple bond on irradiation in pentane or an alcohol (equation 26)[33], as does the 1,4-diyne **8**, although this second dialkyne also gives rise to a cyclopentadiene product[24], which may be formed by initial ring closure of the diyne to give a cyclopentadienyl biradical (equation 27).

Amines, too, undergo photoaddition to alkynes. From diphenylacetylene and a secondary amine are obtained an enamine (which is hydrolysed during work-up) and products which arise by further reaction of stilbene, the photoreduction product of diphenylacetylene (equation 28). The products can be rationalized in terms of an initial hydrogen transfer to the excited state of the alkyne; the fact that the N—H

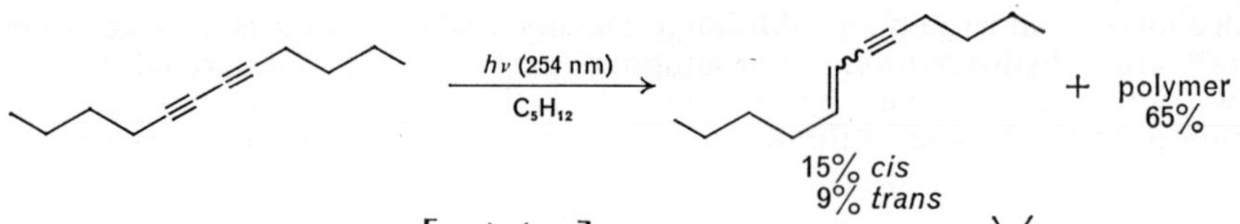

(26)

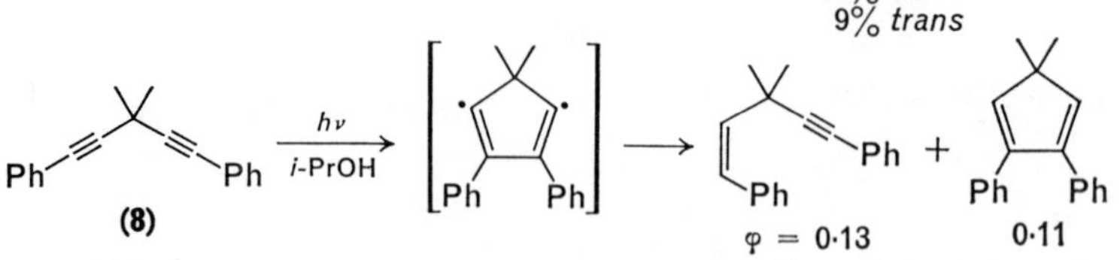

(27)

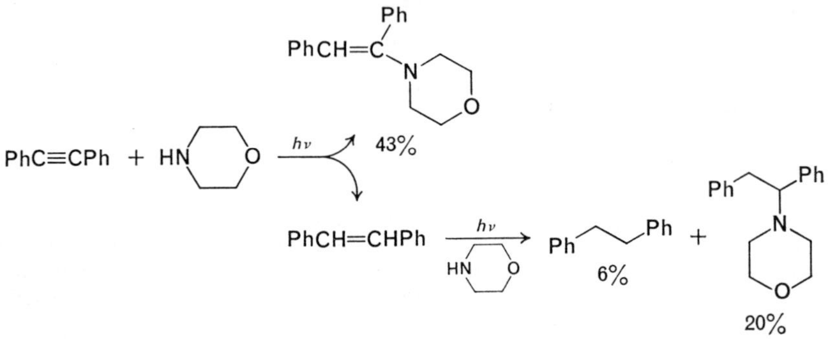

(28)

hydrogen seems to be transferred may indicate that an ionic rather than a free radical process is involved[34].

An intramolecular example of this type of addition, involving an amide and a C≡C triple bond, is afforded by the reaction of *o*-acetamidophenylacetylenes (equation 29). The product may arise by transfer of a hydrogen from the amide nitrogen to the acetylenic carbon, followed by ring closure to the amide oxygen[35].

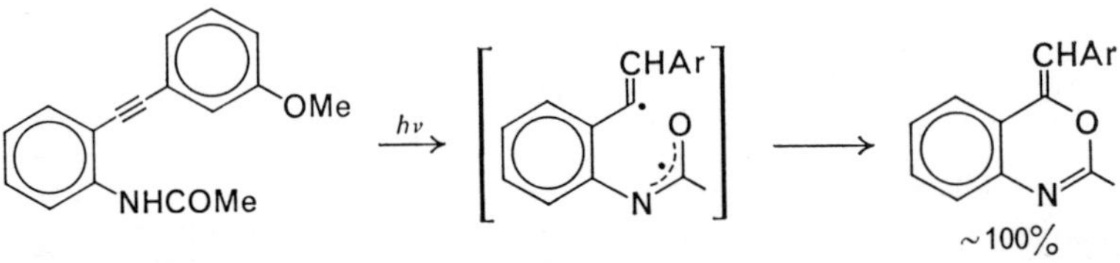

(29)

In the photoaddition of water (equation 30)[36] or acetic acid (equation 31)[37] to alkynes it seems likely that ionic addition occurs by protonation of the alkyne excited state, particularly in view of the observation that the hydration reaction is speeded up by acid and retarded by base. The sensitized addition of acetic acid to medium-ring cycloalkynes[38] to give enol acetates (equation 32) is strongly reminiscent of the analogous addition to cycloalkenes, which has been shown to go by way of protonation of the highly strained *trans*-cycloalkene.

An intramolecular addition of a hydroxylic group to a triple bond results in the formation of a benzofuran from *o*-hydroxyphenylacetylene when irradiated in basic solution (equation 33). Without the base the sole product is *o*-hydroxyacetophenone (**15**) formed by addition of water to the triple bond[39].

$$PhC{\equiv}CPh + H_2O \xrightarrow[MeCN]{h\nu\ (254\ nm)} PhCCH_2Ph\ (\text{C=O})\quad 100\% \tag{30}$$

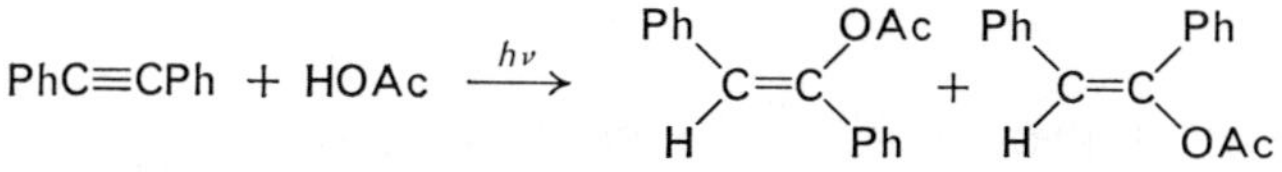

(31)

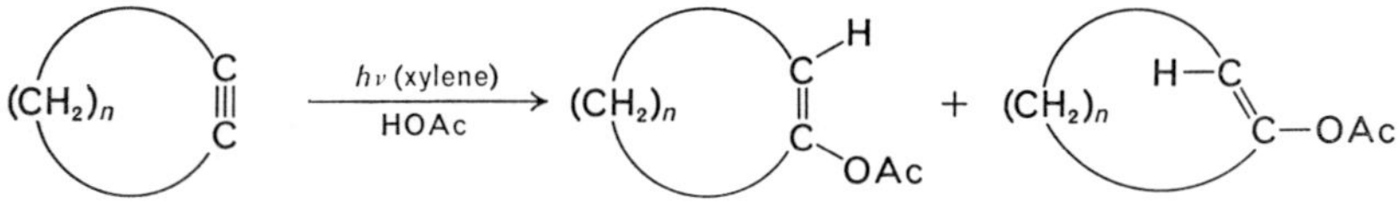

(32)

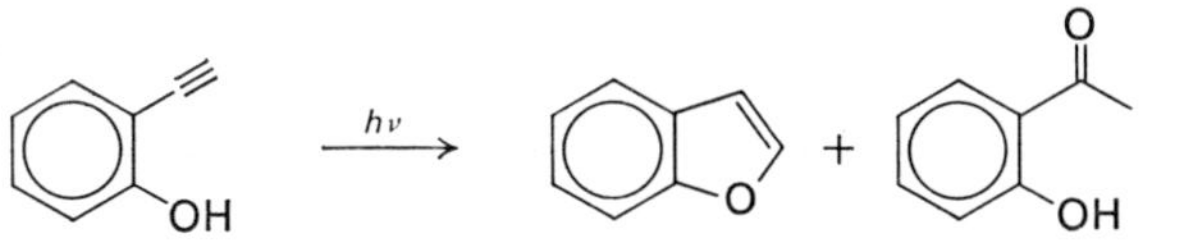

60% (15) 20%

(33)

Photodimerization of acetylene to give vinylacetylene (butenyne) and formation of polymers in the photolysis of alkynes generally are examples of photoaddition to alkynes. Photopolymerization of di- and poly-ynes has been studied, and for both conjugated diynes[40] or triynes[41] the polymerization process is a 1,4-addition reaction (equation 34). The products are highly unsaturated, and they tend to contain a high proportion of oxygen after exposure to the atmosphere.

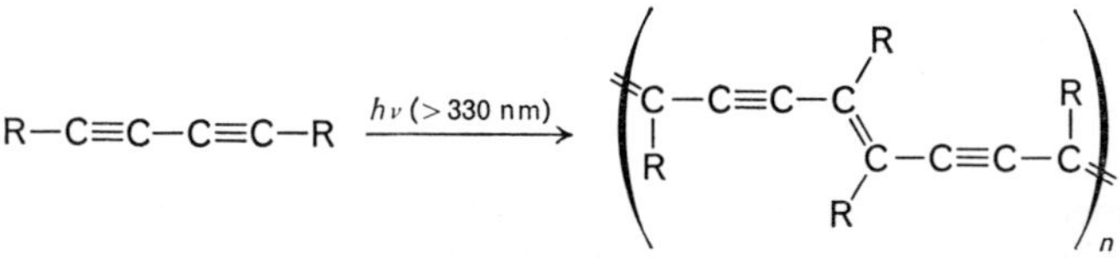

(34)

B. Photoaddition not Involving the Alkyne Excited State

A number of the photoaddition reactions of alkynes are not reactions of the alkyne excited state but involve the formation of free radicals by photocleavage of the other compound. The photochemical *anti*-Markownikoff addition of hydrogen bromide to alkynes (equation 35) under conditions where the ionic addition is very slow is an example of this[42].

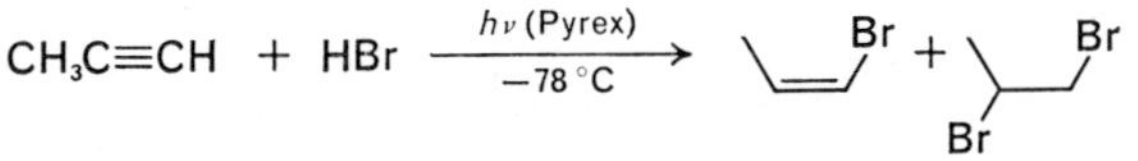

70% 30%

(35)

1 : 1 Cyclized adducts have been reported[43] for alkynes with benzaldehyde (equation 36), although a 1 : 2 alkyne : benzaldehyde adduct can be obtained

(equation 37)[44] and offers a closer analogy to the reactions of alkynes with aliphatic aldehydes or formamide (equation 38)[45]. In the formation of these 1 : 2 addition compounds very little of the intermediate α,β-unsaturated carbonyl compound is isolated.

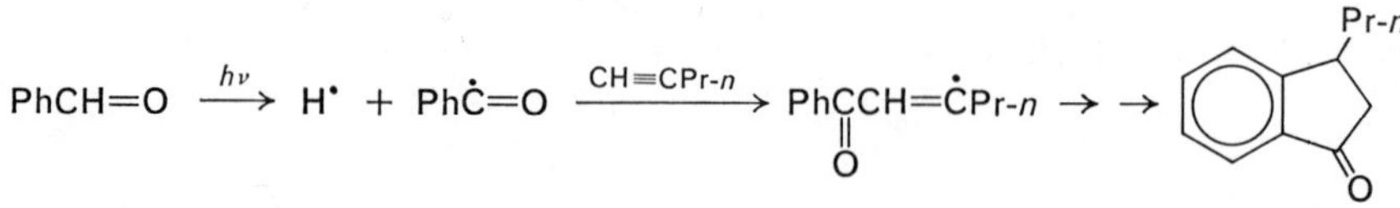

(36)

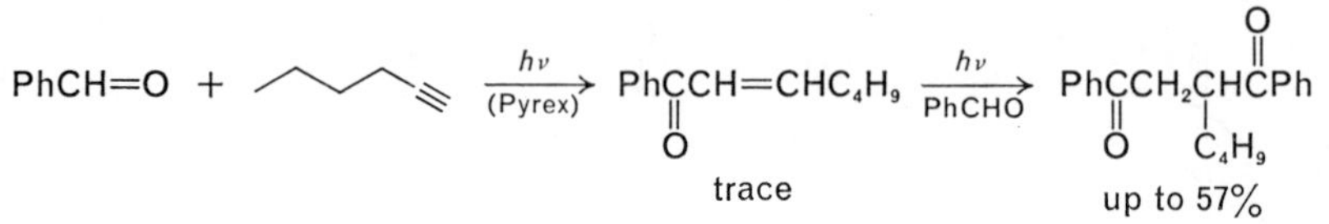

(37)

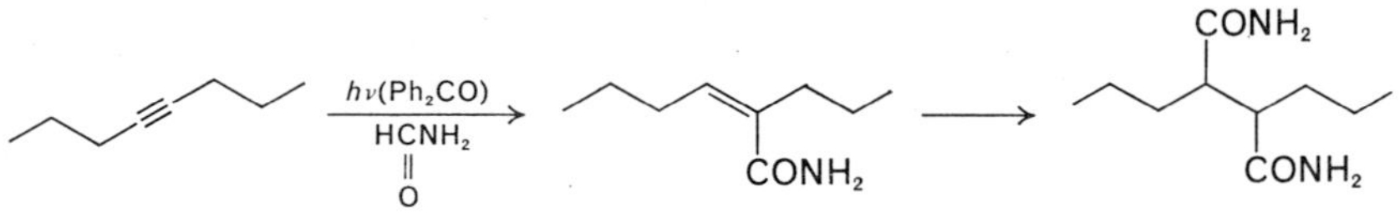

(38)

The reaction with formamide requires a sensitizer such as benzophenone, and $\cdot CONH_2$ radicals are produced by hydrogen abstraction from the amide by the excited state of the ketone. With bromotrichloromethane 2 : 1 alkyne : haloalkane adducts are produced as well as 1 : 1 adducts (equation 39). Extensive polymerization occurs, and peroxide-initiated reaction often gives better yields of simple products[46].

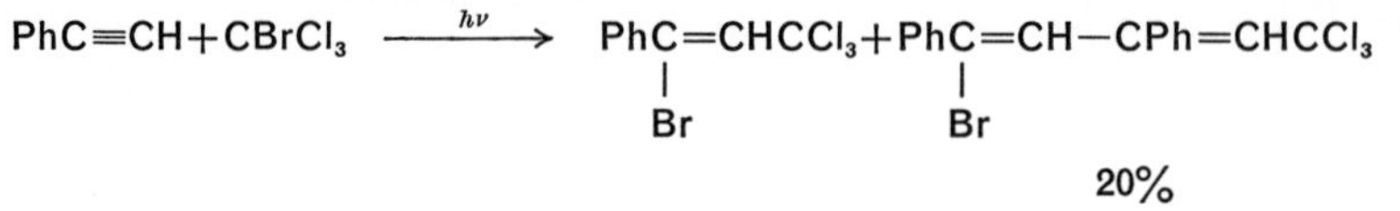

(39)

The distinction between those reactions which do involve the alkyne excited state and those which do not becomes blurred in the case of intramolecular processes of compounds where the alkyne is part of an extended conjugated system. For example, radical production as a result of homolytic cleavage of a P—C bond in the phosphine **16** leads ultimately to an intramolecular addition product involving one of the acetylenic C≡C bonds (equation 40)[47].

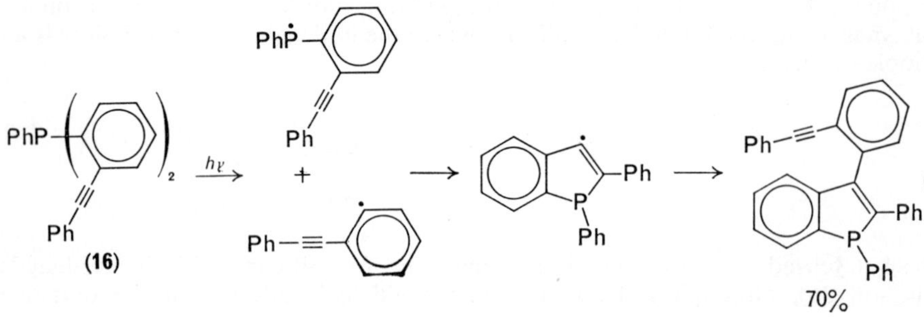

(40)

V. PHOTOCYCLOADDITION

A. Photocycloaddition to C=C Compounds

The areas in which work on alkyne photochemistry has been most prolific are those involving cycloaddition to carbon–carbon double bonds in alkenes, aromatics and related compounds. The simplest type of reaction involves formation of a cyclobutene from an alkene and an alkyne (equation 41)[48]. The cyclobutene product may itself be photolabile, and if radiation is used which is absorbed more strongly by the cyclobutene, the product isolated may be the 1,3-diene derived from it by electrocyclic ring opening (equation 42)[49].

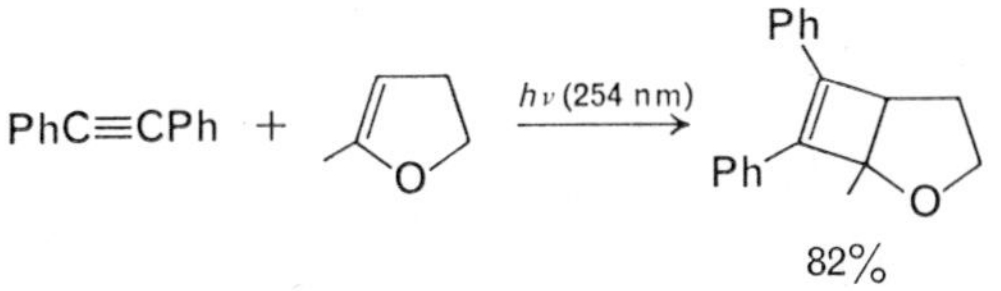

(41)

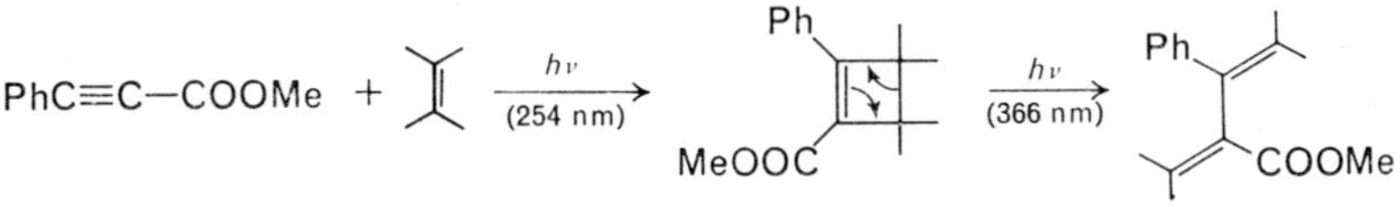

(42)

In both of these examples it is clear that the alkyne absorbs the radiation initially, and its excited state must be involved in reaction. The diphenylacetylene reaction (equation 41) is quenched by added pyrene and can be sensitized by triphenylene, and these observations suggest that it is a reaction of the (lowest) triplet state of the alkyne. If the alkyne is less strongly absorbing than the phenylacetylenes, as is dimethyl acetylenedicarboxylate for instance, it may be difficult to isolate any cyclobutene product at all in such a reaction, because the cyclobutene can undergo photochemical ring-opening or photochemical cycloaddition with a second molecule of alkene. If a second cycloaddition occurs, the product is a bicyclo [2.2.0] hexane or the 1,5-diene which this gives on thermal ring-opening (equation 43)[50].

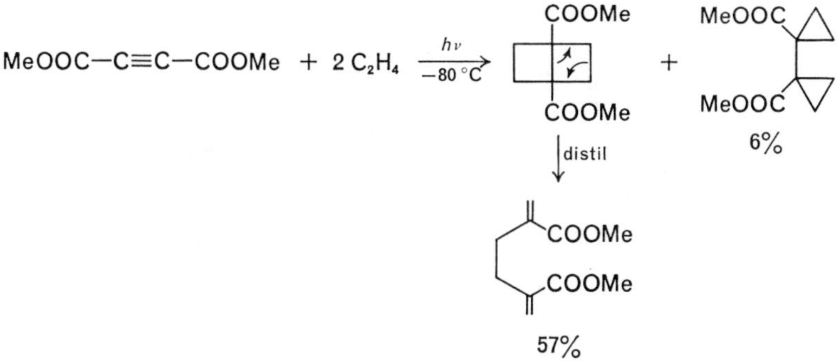

(43)

It is proposed that these cycloadditions are also reactions of the alkyne triplet state, and this seems reasonable in the light of the non-concerted nature of the reaction indicated by the fact that dimethyl acetylenedicarboxylate gives a mixture of the same four stereoisomers of the bicyclohexane product with either *cis*- or *trans*-but-2-ene (equation 44)[51a]. This is in contrast to the stereospecific nature of the Lewis acid promoted cycloaddition of alkenes and alkynes to give cyclobutenes[51b].

Further support for a photochemical triplet mechanism in a rather different system is that the compound **17** gives a vinylcyclopentenone (by way of intramolecular cycloaddition followed by electrocyclic ring-opening) only on triplet sensitization (equation 45)[20].

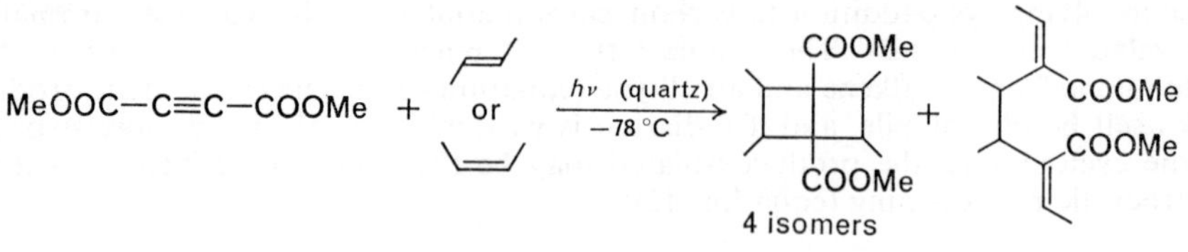

(44)

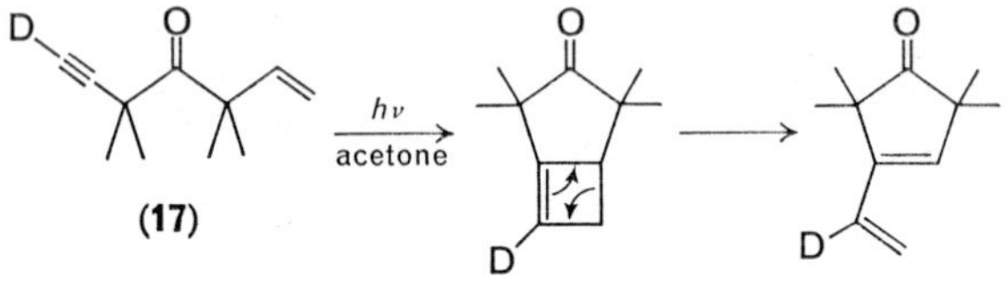

(45)

Cyclo-octa-1,5-diene reacts with diphenylacetylene to give a high yield of intramolecular reaction product of this 1 : 2 alkyne : alkene type (equation 46)[52].

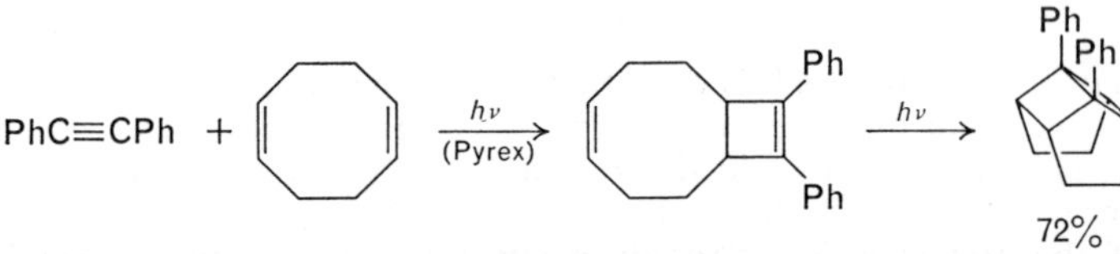

(46)

The photocycloaddition reaction seems to be equally successful in systems where the alkene absorbs the radiation initially, and this must happen when simple alkyl-substituted alkynes react with enones (equation 47)[53] or with enediones (equation 48)[54]. The reactions of cyclopentenone probably occur through a triplet state of the enone and perhaps through a complex between this excited state and the alkyne[55].

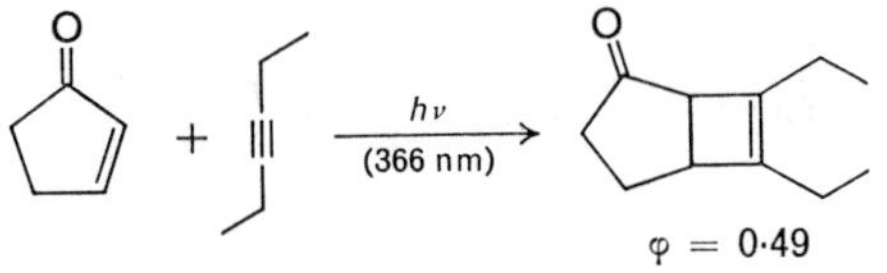

(47)

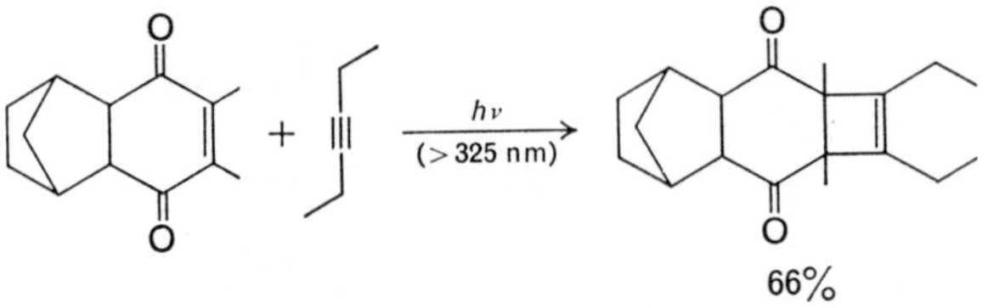

(48)

The preferred orientation of addition (equation 49) is the opposite of that found for alkene addition to cyclic enones, and this difference is not easily accounted for.

In some reports it is not possible to identify whether the alkyne or the alkene absorbs first, and it is possible that an excited state complex (an exciplex) can be produced irrespective of which excited state is formed first. However, the fashion for invoking an exciplex intermediate without evidence for it is a regrettable one.

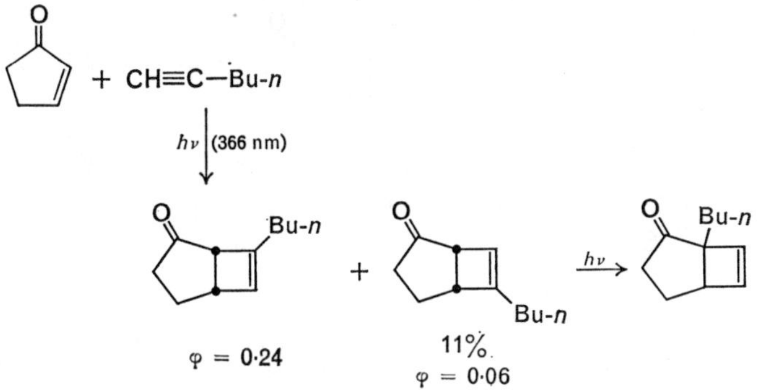

(49)

In the photoaddition reactions of some alkyne/alkene pairs an alternative 1 : 2 alkyne : alkene product arises[56] which contains a bicyclopropyl unit (equation 50, and see equation 43 above). The formation of the alternative product is dependent

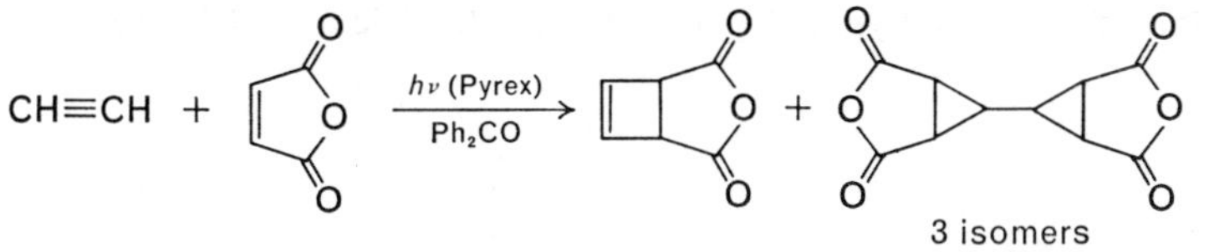

(50)

on alkyne structure (maleic anhydride and but-2-yne give only a cyclobutene[57]) and on concentration and temperature (maleic anhydride and acetylene give from 65% cyclobutene product to 84% bicyclopropyl product[56]).

The bicyclopropyl product is the major product when the two alkene units are in the same molecule, particularly with cyclohexa-1,4-diene systems which are fairly rigid (equation 51)[58]. Once again the reaction is effective whether the alkene (equation 51) or the alkyne (equation 52)[59a] absorbs the light initially.

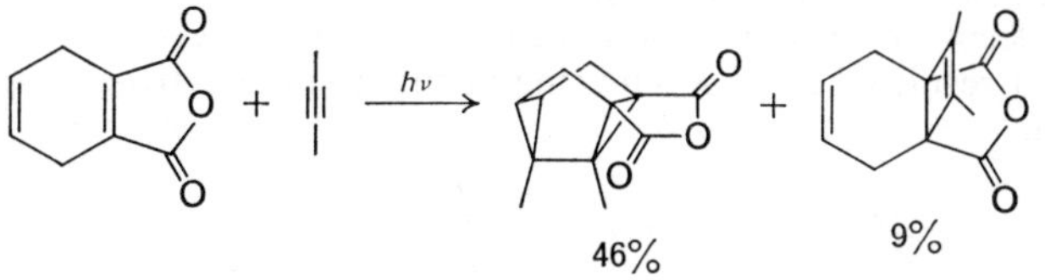

(51)

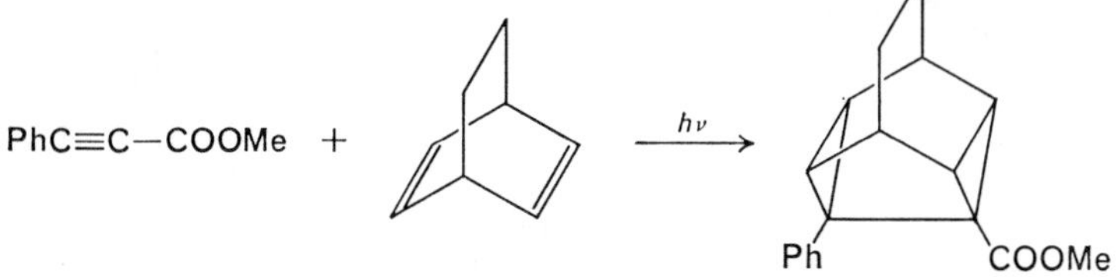

(52)

It has been suggested on the basis of kinetic results[56] that the mechanism involves the trapping of the first-formed biradical by the second alkene unit (equation 53), although another possibility is that the biradical rearranges to a cyclopropylcarbene before it reacts further.

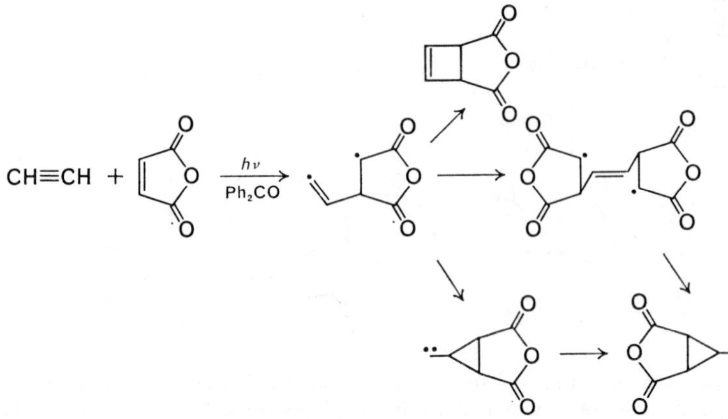

(53)

That a biradical species can be involved in the cycloaddition reactions is strongly supported by the isolation[59b] of an acyclic hydrogen-shifted product from the irradiation of 3,3-dimethylbut-1-yne and methylmaleic anhydride (equation 54).

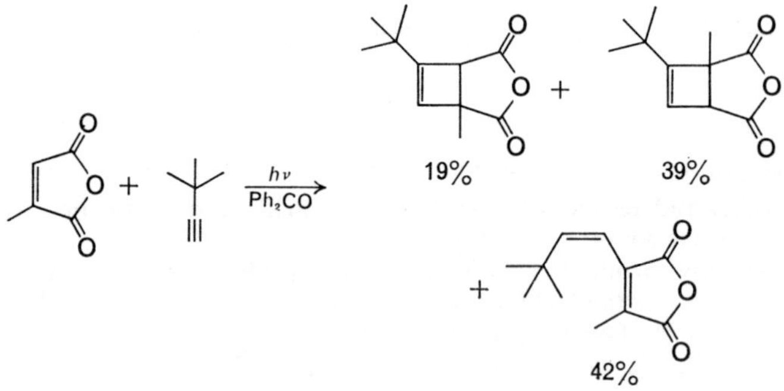

(54)

One report has appeared[60] of a system in which the first-formed cyclobutene adds a second molecule of alkyne photochemically, leading eventually to a benzene derivative as a result of electrocyclic ring-opening and a retro-Diels–Alder process (equation 55).

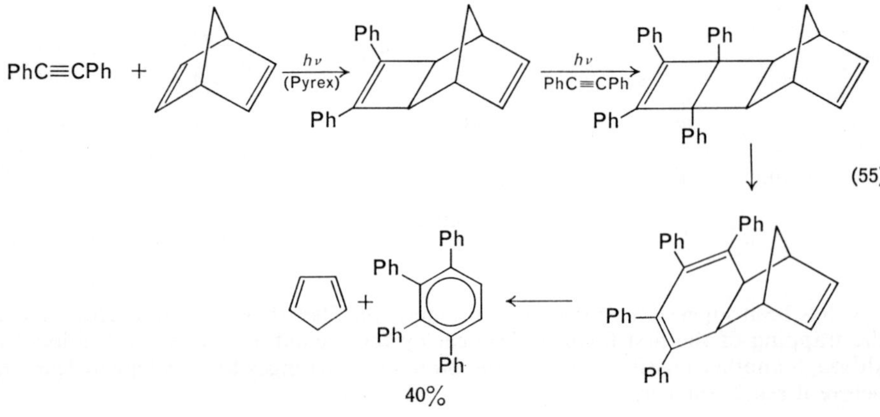

(55)

As in other photochemical reactions, alkenes seem to react in preference to alkynes, and in the sensitized photolysis of 2-methylbutenyne the product isolated is a dialkynylcyclobutane (equation 56) rather than a cyclobutene[61].

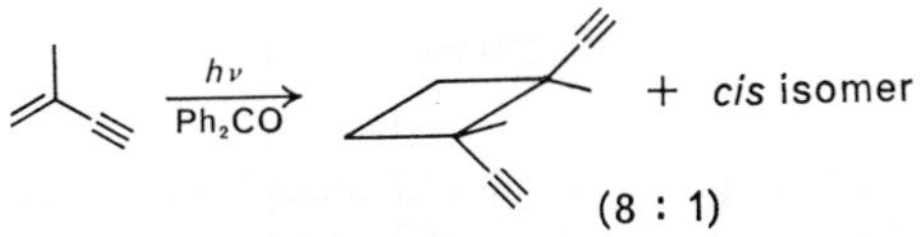

(56)

Photodimerization of an alkyne to give a cyclobutadiene has not been reported, although the tricyclic compound **18** formed as one product of γ-radiolysis of but-2-yne may well be produced by dimerization of tetramethylcyclobutadiene[16].

Photochemical trimerization of alkynes to give benzenes is, however, well documented. Benzene (and butadiyne) are the most abundant volatile products in

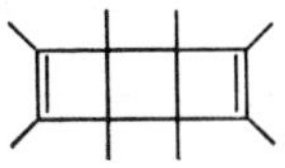

(18)

the 185 nm photolysis of acetylene (equation 1)[9, 62], whilst the Cd($^3P_1$)-sensitized process[63] gives mainly benzene (ϕ = 0·094) and vinylacetylene (ϕ = 0·02). Cd($^3P_1$) is not sufficiently energetic to abstract a hydrogen atom from acetylene, and hence butadiyne and hydrogen are not formed, but only oligomers of acetylene via the alkyne excited state. This cyclotrimerization process can be efficient and synthetically useful with substituted alkynes (equation 57)[64]. The mechanism is thought to involve a radical-initiated polymerization process (equation 58), though it is claimed that

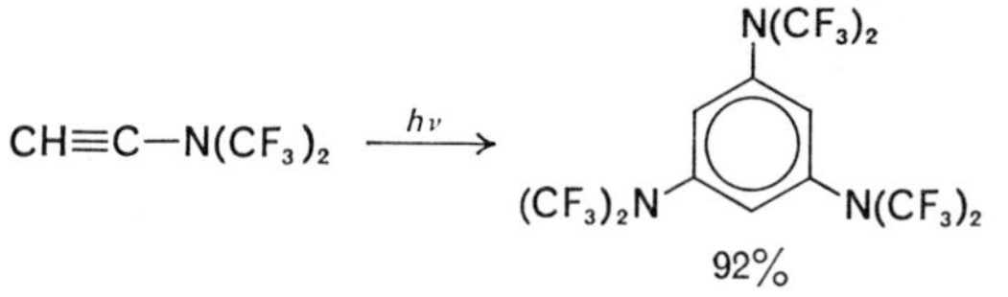

(57)

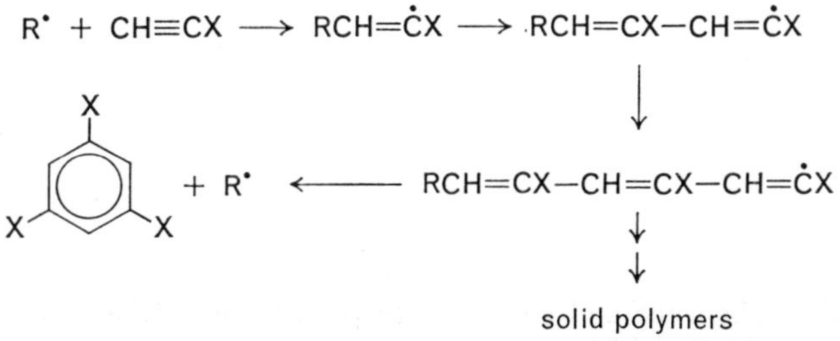

(58)

the results of deuterium-labelling studies in the Hg ($^3P_1$)-sensitized reaction[65] and on direct irradiation[62] are better explained by a molecular process. The evidence is not unambiguous.

If hydrogen sulphide is present the reaction can be diverted to give thiophen as a major product after incorporation of only two acetylene units (equation 59)[66]. This occurs whether the H_2S or the acetylene is excited first, and the thiophen yield increases at the expense of benzene as the H_2S concentration is increased. The mechanism probably involves internal trapping in the radical **19**.

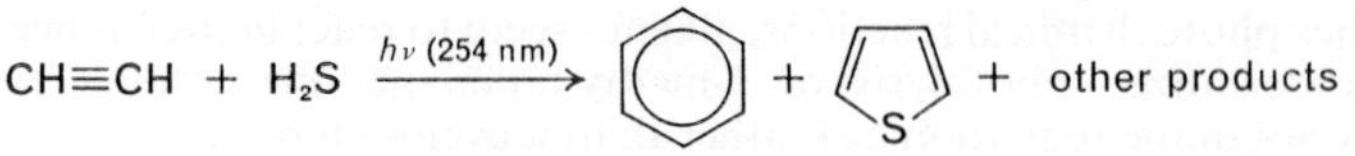

(59)

HS—CH=CH—CH=ĊH

(19)

Under certain conditions (−78 °C, liquid phase) ethenethiol (CH_2=CH—SH) is the major product of the photoaddition of hydrogen sulphide to acetylene[67].

There are few reports of photocycloaddition reactions of di- and poly-ynes which involve more than one of the alkyne groups, but the production of toluene or other alkylbenzenes in the irradiation of butadiyne with propene or other terminal alkenes (equation 60) is one such process[17].

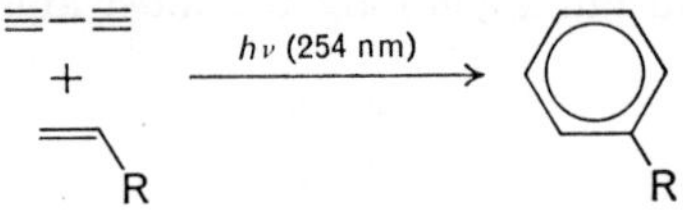

(60)

B. Photocycloaddition to Aromatic Compounds

Alkynes undergo cycloaddition on irradiation with benzene or naphthalene derivatives or with other aromatic compounds. With a benzene derivative the product is usually a cyclo-octatetraene which results from thermal electrocyclic ring-opening of the bicyclo-octatriene formed initially by 1,2-addition of the alkyne to the benzene ring (equation 61)[68, 69]. The intermediate can be trapped using a dienophile such as tetracyanoethylene (equation 62)[70]. The first step of the photo-addition process involves excitation of the alkyne[70], and orbital symmetry considerations suggest that concerted 1,2-addition is 'allowed' if the alkyne is excited but not if the benzene is excited[71].

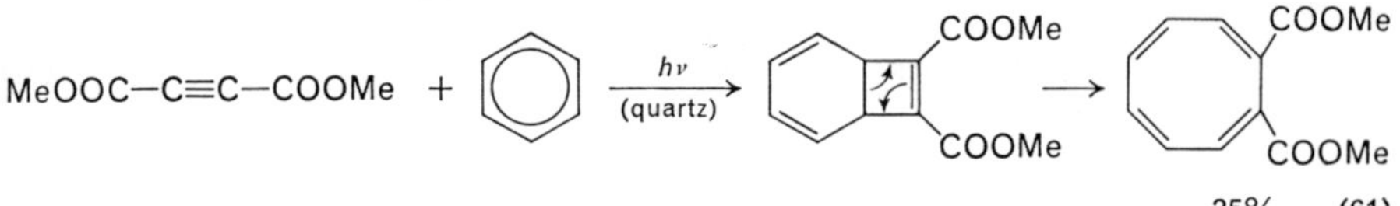

35% (61)

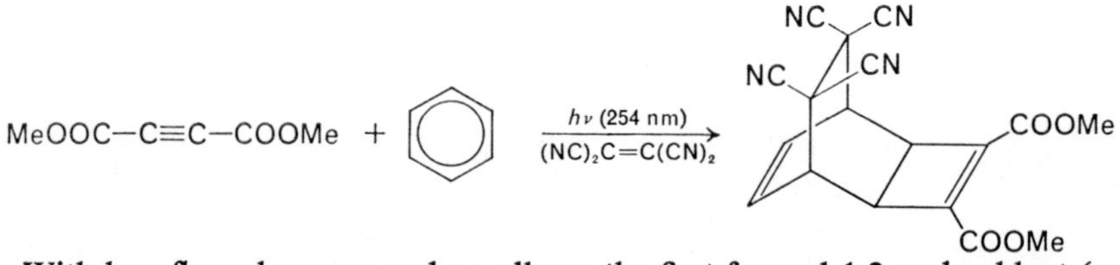

(62)

With hexafluorobenzene and an alkyne the first-formed 1,2-cycloadduct (e.g. **20**) can be isolated in high yield[72].

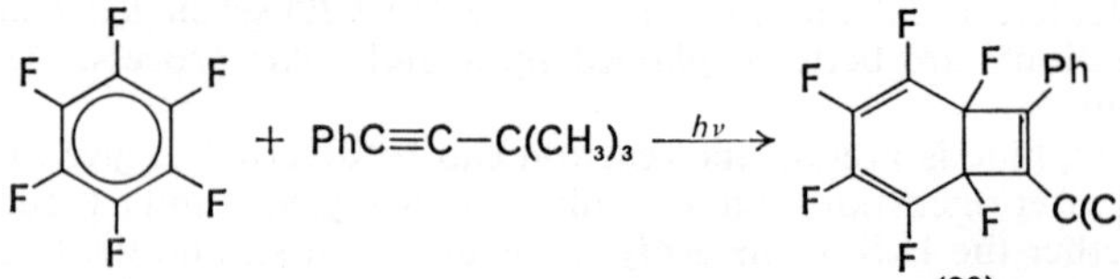

(20)

86%

The 1,2-cycloaddition reaction can take place in an intramolecular manner (equation 63), although in this example the initial excitation involves the aromatic group[73]. A reaction of a different type is thought to be involved in the first stage of the formation of azulene or naphthalene photodimers from diphenylacetylene (equation 64), though here it is claimed that an intermediate benzocyclobutadiene species has been detected[74]. The intermediate isomer of diphenylacetylene is formed via the triplet state and is relatively long-lived at −10 °C. The major dimers formed are 1,2,3-triphenylazulene and 1,2,3-triphenylnaphthalene; hexaphenylbenzene and octaphenylcubane are also produced[75].

Similar azulene dimers arise on irradiation of 1,2-diethynylbenzenes (equation 65)[76], and an intramolecular napthalene-type adduct from a 1,8-diethynylnaphthalene (equation 66)[77]. Interestingly, the latter reaction can also be brought about thermally.

Naphthalenes also undergo 1,2-photocycloaddition with diphenylacetylene, although the products isolated are not benzocyclo-octatetraenes, but rather adducts

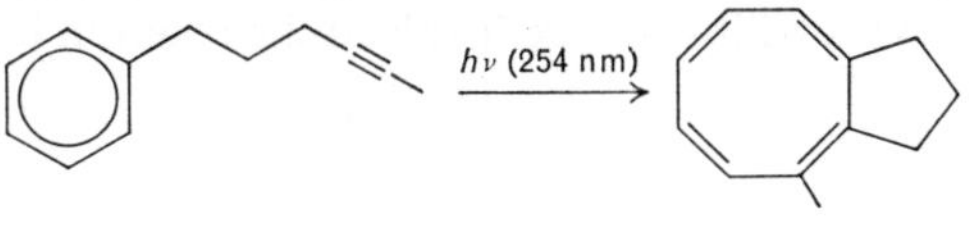

(63)

$\varphi = 0{\cdot}0026$

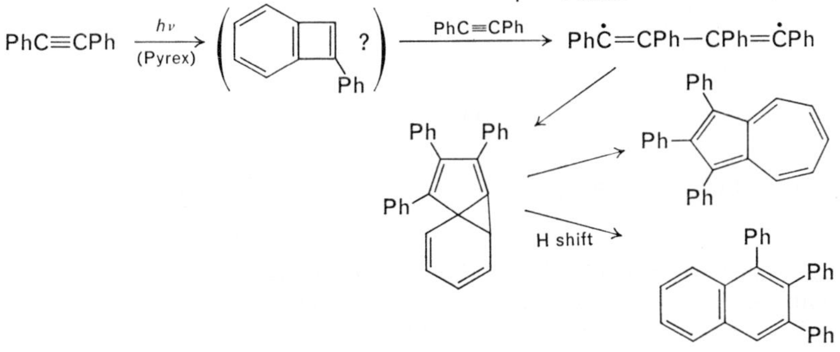

(64)

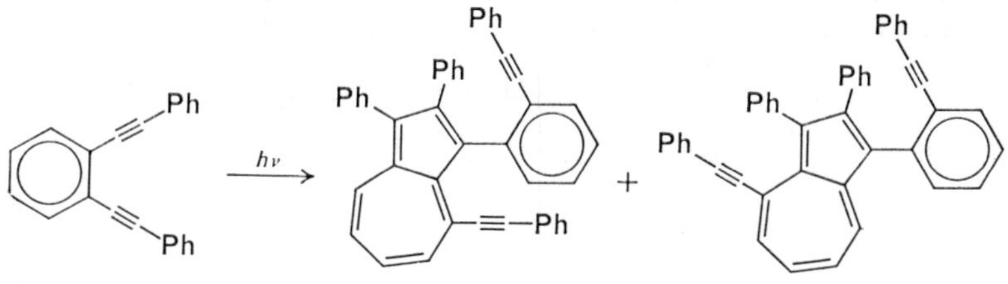

(65)

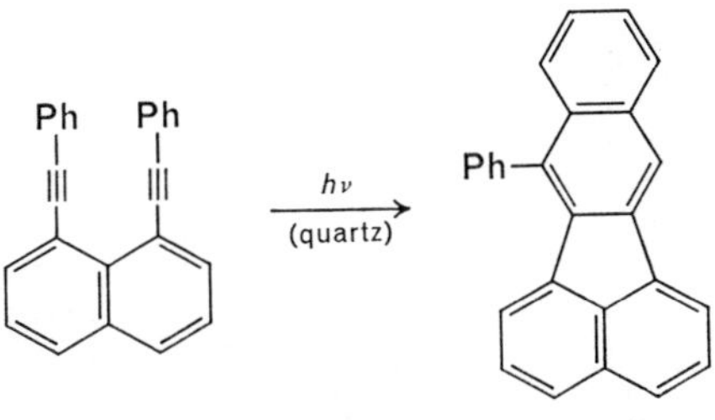

(66)

62%

produced by a second intramolecular photocycloaddition (equation 67)[78]. The intermediate cyclobutenes can be prepared (~100%) by heating the tetracyclic adducts, and they have been shown to undergo rapid photochemical reaction to

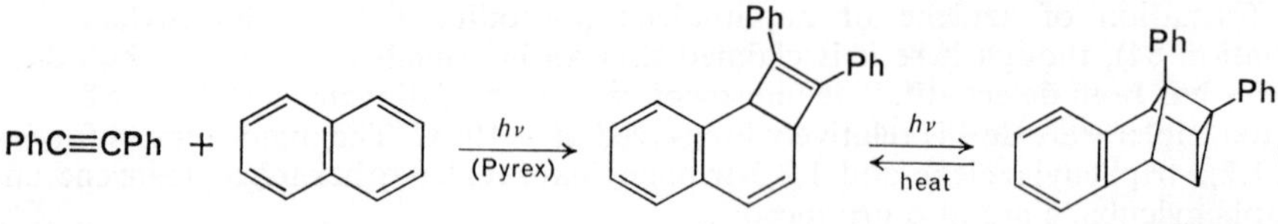

(67)

regenerate these adducts (80–90%); they are relatively inert to thermal conversion to benzocyclo-octatetraenes[79].

When the naphthalene ring is substituted, two products are generally formed[80], with production of an adduct involving the substituted ring being preferred (equation 68). Sometimes the reaction is very efficient and highly regioselective (equation 69)[81]. When amino, halo or acyl substituents are involved the reaction is very inefficient.

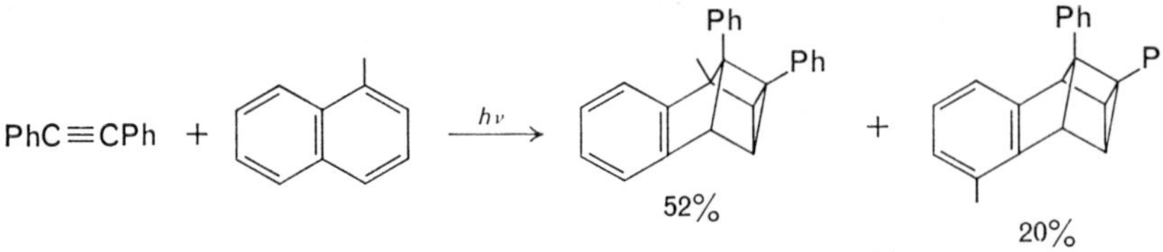

(68)

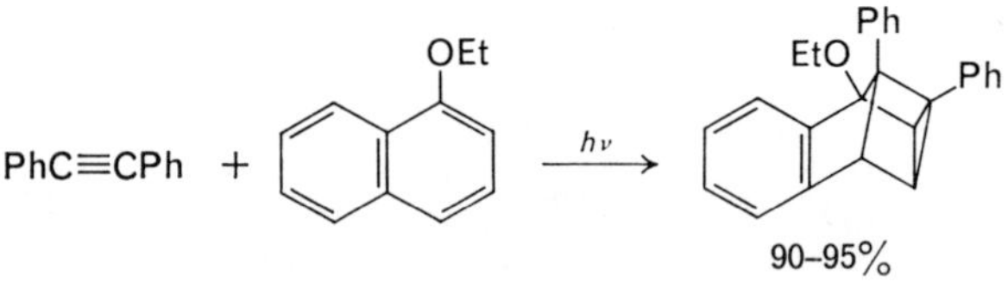

(69)

The mechanism of this reaction is thought to involve the singlet excited state of the naphthalene, which forms an exciplex with the alkyne. Earlier it had been suggested that the reaction involved diphenylacetylene excitation, in view of the fact that dimethyl acetylenedicarboxylate gives different products (equation 70) and in the latter case the alkyne is definitely not excited first[69]. However, selective excitation

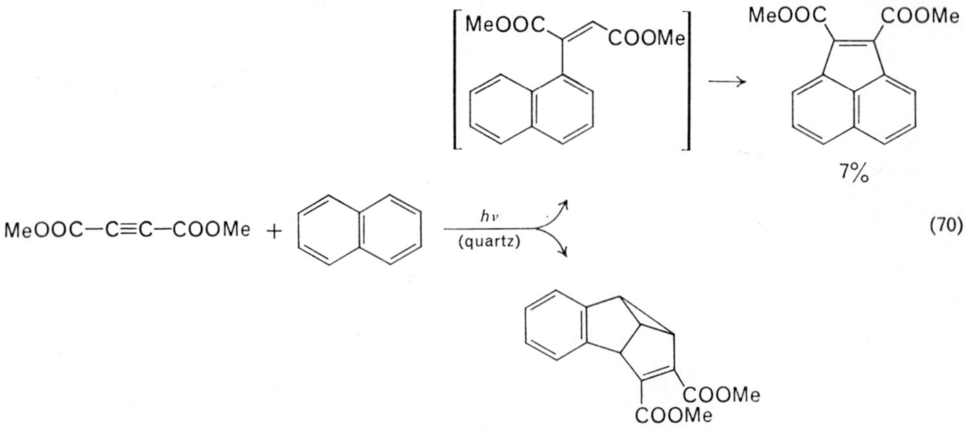

(70)

(21)

of the alkyne reduces the yield of adduct, and a study of the effect of each reagent on the fluorescence of the other suggests that interaction between the diphenylacetylene ground state and the naphthalene excited singlet state occurs and is responsible for the reaction[82].

The second product (**21**) in the reaction of the diester (equation 70) results from 1,3-addition to the aromatic ring. Such addition is the major mode of reaction for benzene/alkene photocycloadditions, but with alkynes it is less common. One of the few reported examples[83] is the addition of diphenylacetylene to esters of trimesic acid (equation 71).

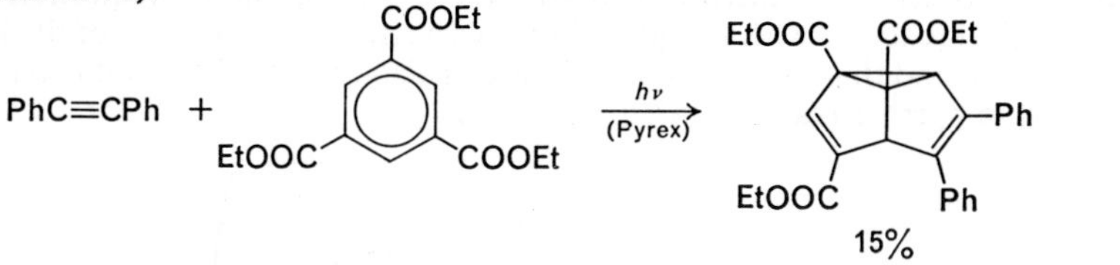

(71)

The 1,3-cycloadduct **22** which is isolated (25% yield) together with the 'expected' 1,2-bis(trifluoromethyl)cyclo-octatetraene (40%) from the photoreaction of benzene and perfluorobut-2-yne[84] is a valence isomer of the 'normal' 1,3-adduct **23** and may arise by a simple isomerization from it, although it is thought more likely that product **22** arises by sensitized photoisomerization of the 1,4-cycloadduct (**24**).

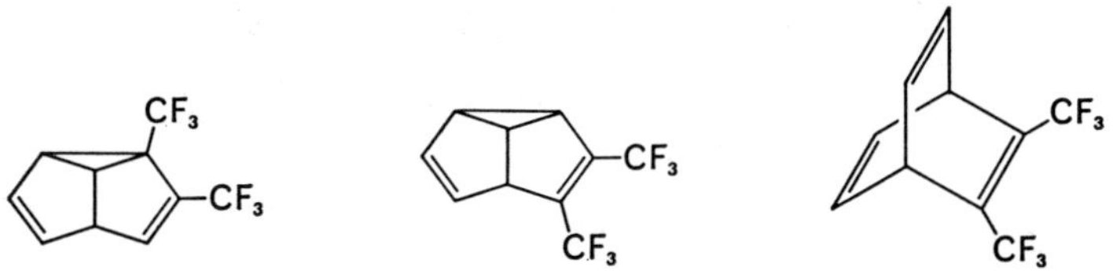

(**22**) (**23**) (**24**)

A pair of products (**25a**, **25b**) derived from methyl 2-naphthoate and diphenylacetylene appear at a glance to be 1,3-cycloaddition products, but the pattern of substituents is incompatible with this[85], and the compounds may arise by photochemical reaction of the normal tetracyclic adduct (equation 72).

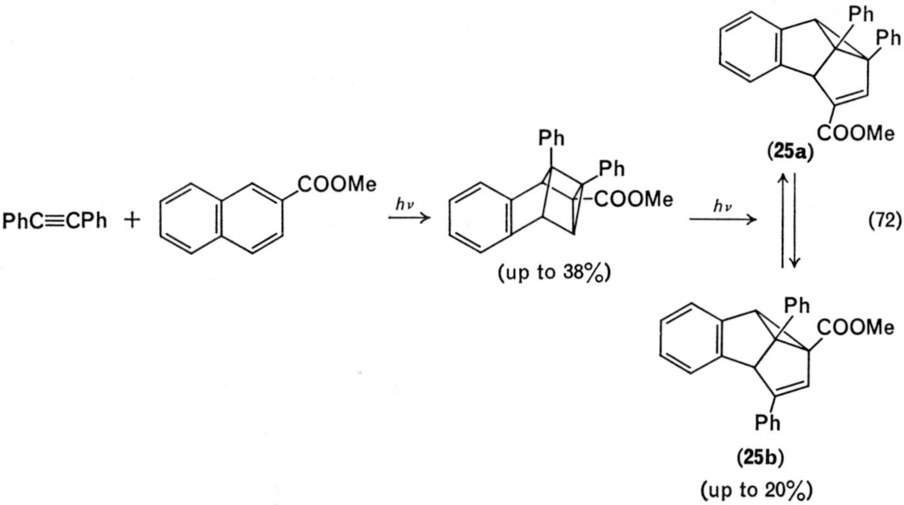

(72)

Of the 5-membered heteroaromatic systems, pyrrole reacts most like benzene in alkyne photocycloaddition, giving a 3,4-disubstituted azepine by 2,3-cycloaddition followed by electrocyclic ring-opening (equation 73)[86]. Azepines with a different substitution pattern have been made by thermal 2,5-addition of an alkyne to a pyrrole, followed by photochemical ring-closure and thermal ring-opening of the tetracyclic photoproduct (equation 74)[87].

Furans give 2,5-cycloadducts on irradiation with alkynes (equation 75)[88]. Thiophens probably behave similarly[89], but the product isolated is a substituted benzene which arises by extrusion of sulphur from the adduct (equation 76). The photochemical reaction with thiophen involves a triplet excited state of the thiophen, but both furan and thiophen cycloadditions can also be brought about thermally[89, 90], (compare the pyrrole reaction in equation 74).

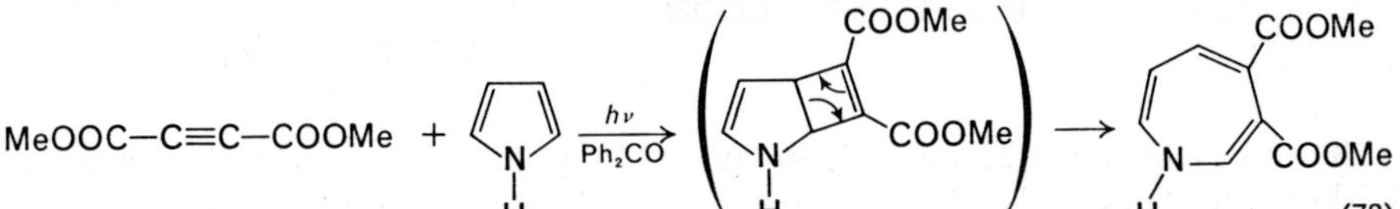

(73)

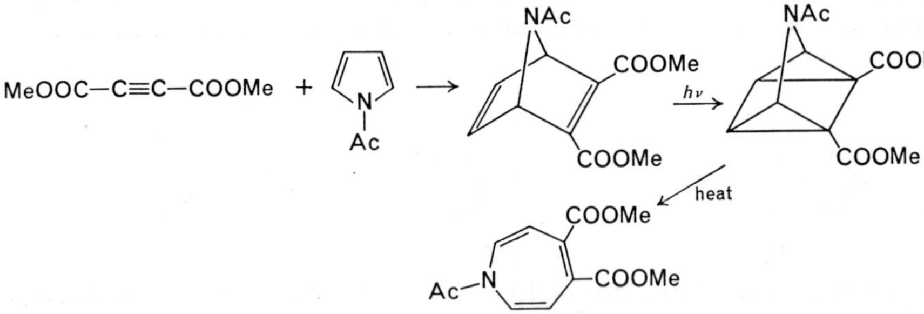

(74)

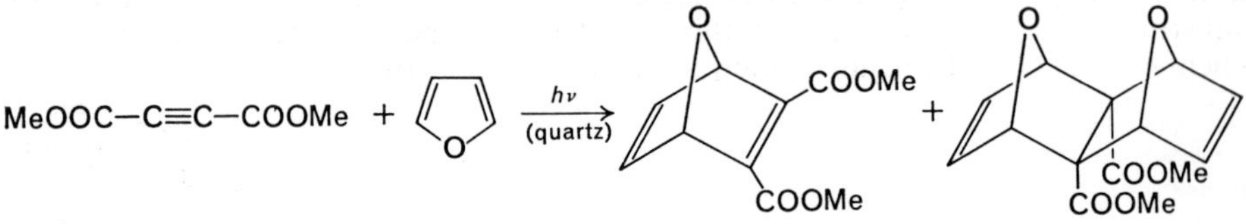

(75)

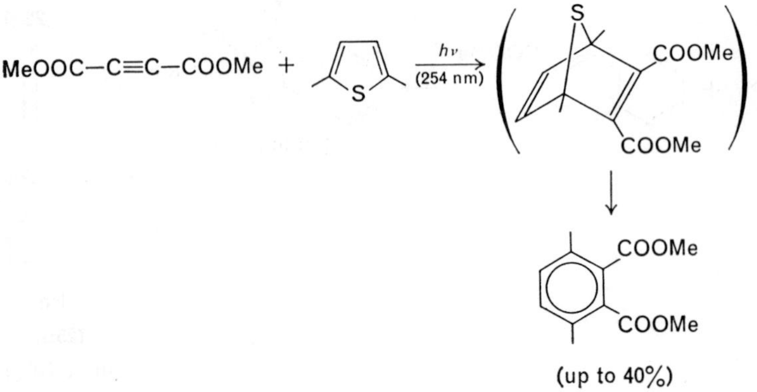

(76)

Indene behaves like an alkene in its photoreactions with alkynes and gives a cyclobutene (equation 77). The reaction can be triplet-sensitized, and the orientation of addition is that expected on the basis of the most stable biradical intermediate[91].

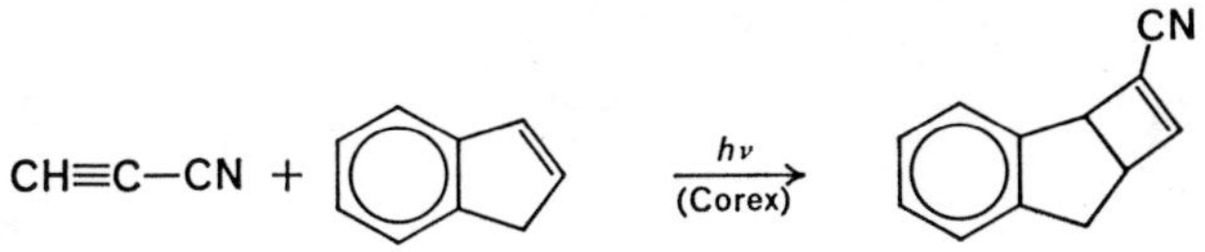

(77)

Benzothiophen reacts in a similar way, though the adduct isolated is usually not the one expected on the basis of straightforward cycloaddition. This is attributed to rearrangement of the first-formed adduct, and the 'normal' adduct (**26**) from diphenylacetylene has been shown to undergo rapid and efficient photochemical conversion to the major isolated product (equation 78)[92]. This rearrangement may occur by ring-opening and reclosure of the dihydrothiophen ring (i.e. *via* **27**), or by formation and ring-opening of a polycyclic intermediate (**28**). More recently, however, it has been suggested[93] that this rearrangement does not occur with adducts containing alkoxycarbonyl substituents, but that the initial addition occurs in a different mode (equation 79).

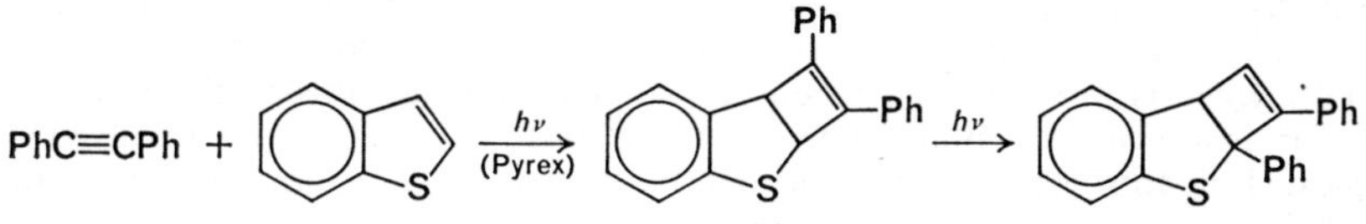

(78)

3% 30%

(**26**)

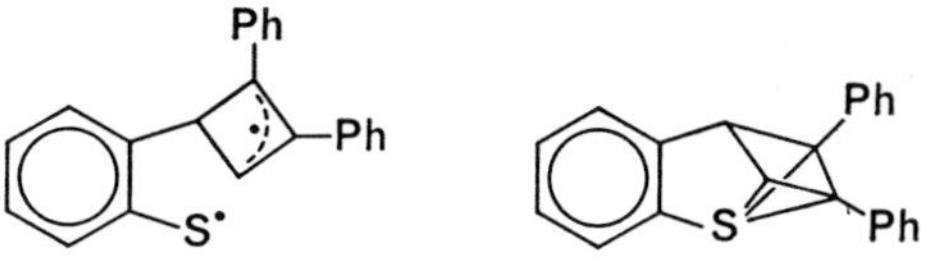

(**27**) (**28**)

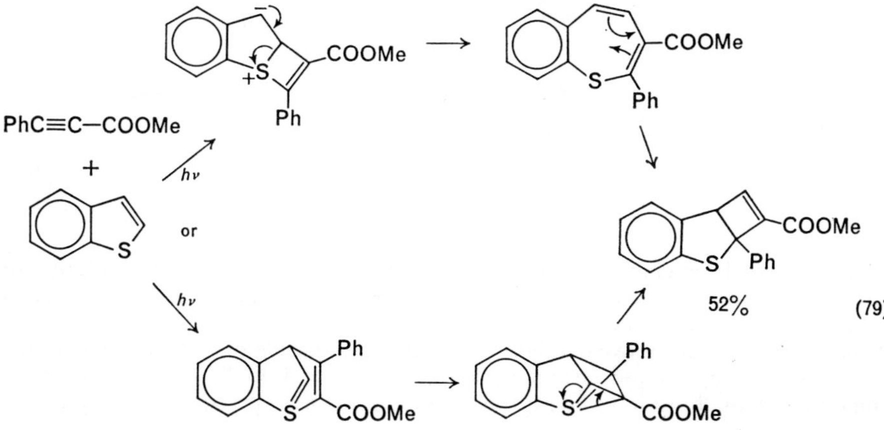

(79)

The reaction with benzothiophens provides a route to substituted naphthalenes by thermal ring-opening and sulphur extrusion of the photoadducts (equation 80)[93, 94].

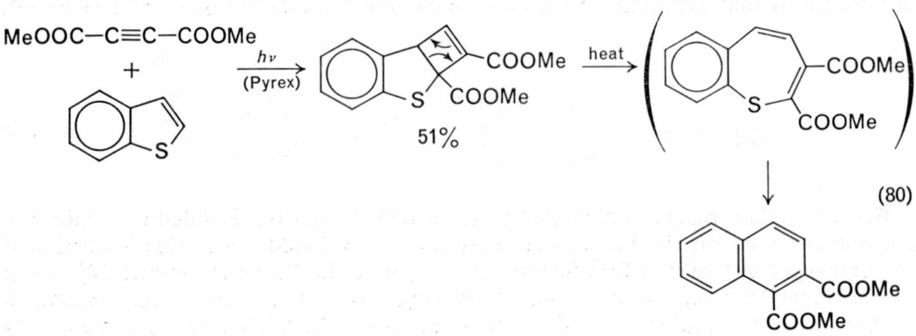

(80)

C. Photocycloaddition to Other Multiple Bonds (C=O, C=S, NO_2)

Photoreactions of alkynes with compounds such as ketones, thioketones or nitro-compounds often involve initial excitation of the non-alkyne addend, and because of this they do not necessarily involve an electronically excited state of the alkyne. However, we have noted in previous sections that it is not always clear whether or not the alkyne is the first species to be excited, nor does it follow that electronically excited alkyne is not involved when the alkyne does not absorb the radiation—it may, for instance, be obtained by energy transfer or it may be involved as an exciplex. For this reason, and for completeness, the account in this section is included.

Alkynes react photochemically with aromatic aldehydes or ketones[95] to give α,β-unsaturated carbonyl compounds (equation 81). This occurs by way of cyclo-addition to give an oxete, followed by thermal ring-opening of this intermediate, and the orientation of addition is in accord with a two-step cycloaddition via the more stable biradical intermediate (equation 82).

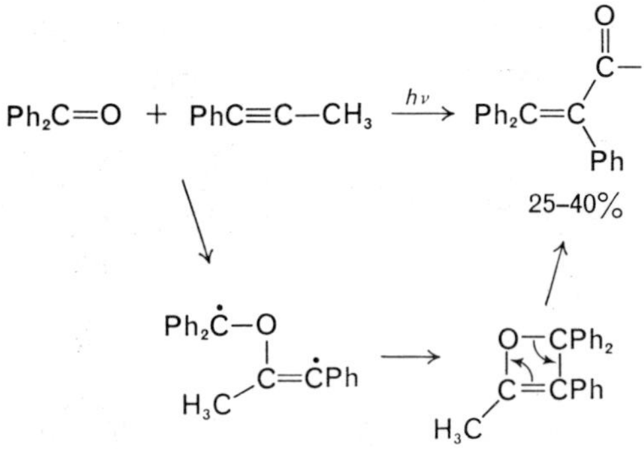

(81)

(82)

An oxete has been characterized as the product of low-temperature irradiation of benzaldehyde with but-2-yne (equation 83), and on warming it gives the normal unsaturated ketone[96]. A second product (**29**) appears on prolonged low-temperature irradiation as a result of cycloaddition of a second molecule of benzaldehyde to the oxete. With aldehydes and terminal alkynes the photocycloaddition reaction is in competition with photo-induced radical addition processes (see equations 36 and 37).

A non-photochemical process of this type has been reported[97] for a strongly electron-deficient ketone and an electron-rich alkyne (equation 84).

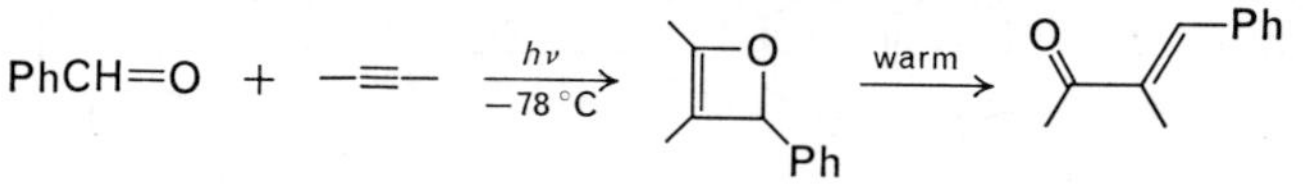

(83)

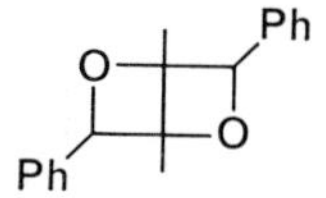

(29)

With alkoxyalkynes and aromatic ketones, in addition to the expected acrylate esters (equation 85) alkylidenecycloheptatrienes are also formed in the photoreaction[98], probably by a reaction from the biradical intermediate involving radical attack on an aromatic ring (equation 86).

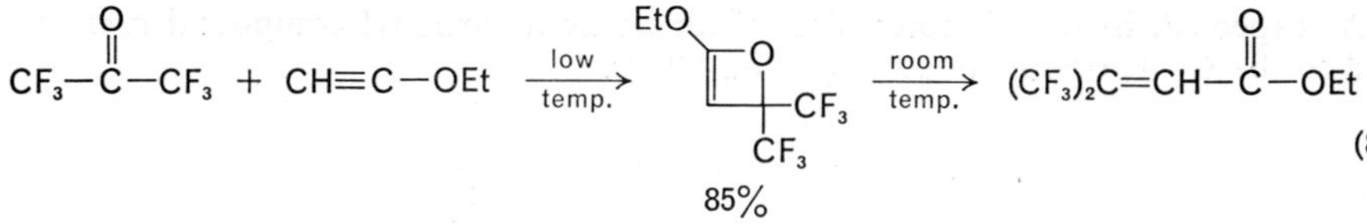

(84)

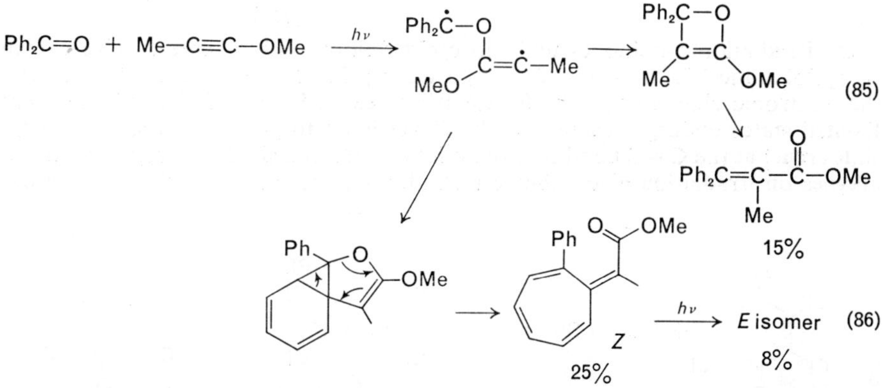

(85)

(86)

1,2-Cycloaddition of alkynes to 1,2-diketones[99] (or to *o*-quinones[100]) offers a route to unsaturated 1,4-diketones (e.g. **30**) and hence to furans (equation 87). 1,4-Cycloaddition to the dicarbonyl moiety occurs in the first stage of the photoreaction of phenanthraquinone and methoxyacetylene (equation 88), although

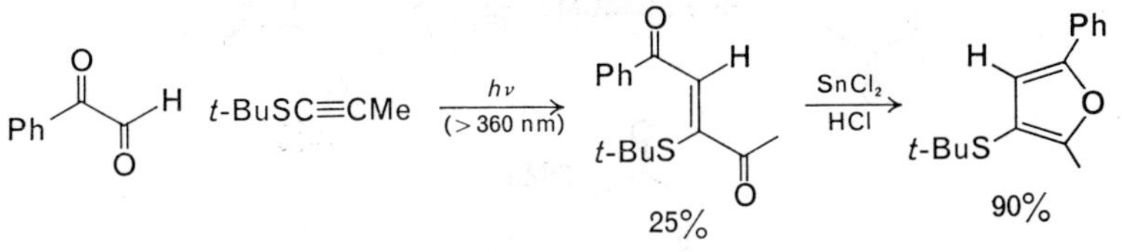

(87)

(**30**)

methoxypropyne gives a dioxole as a result of a 1,2-hydrogen shift in the biradical intermediate (equation 89)[101].

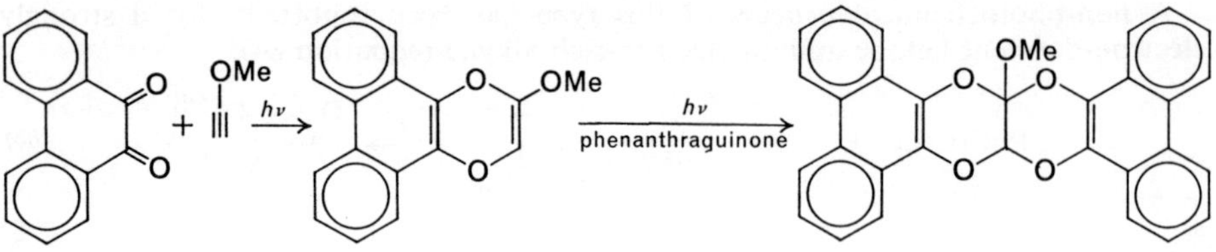

(88)

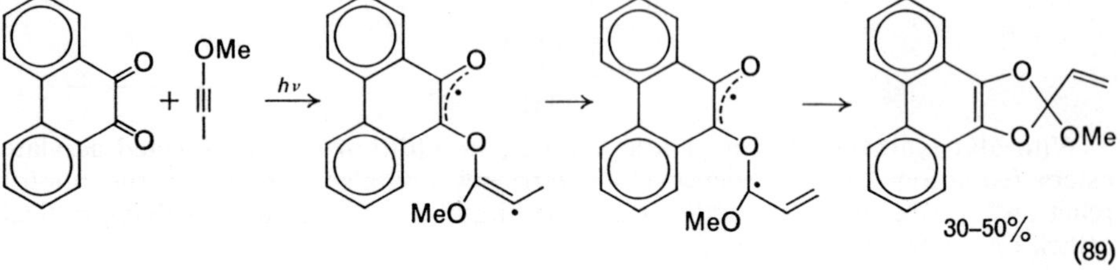

(89)

As expected, in the photoreaction of an enyne a carbonyl compound prefers to add to the C=C double bond (equation 90)[102].

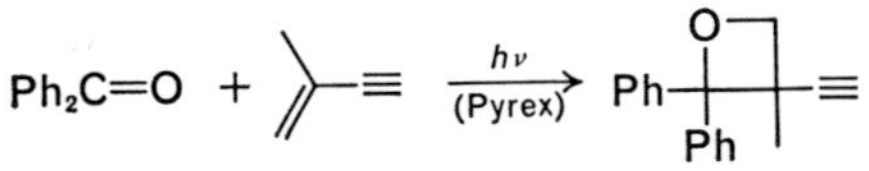

(90)

58%

On irradiation *p*-quinones undergo cycloaddition with alkenes to give oxetanes or cyclobutanes[103]; the major factor governing the choice of product seems to be the electronic character of the lowest triplet excited state of the quinone. (n,π^*) Excited states undergo reaction at the C=O bond to give oxetanes, whilst (π,π^*) states react at the C=C bond to give cyclobutanes. In a similar way, *p*-quinones and alkynes on irradiation give quinone methides (equation 91)[104] if the quinone has a

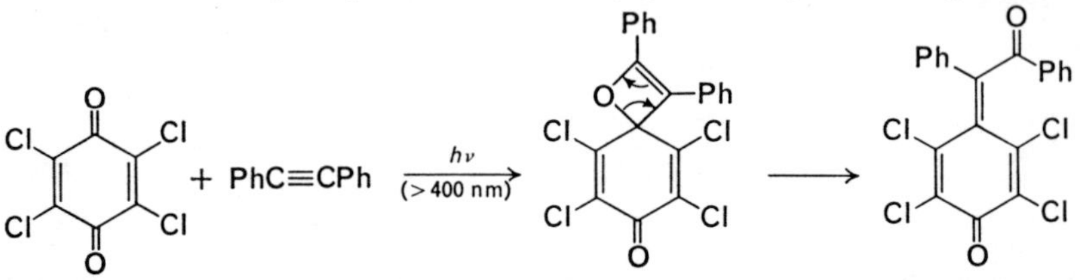

(91)

75%

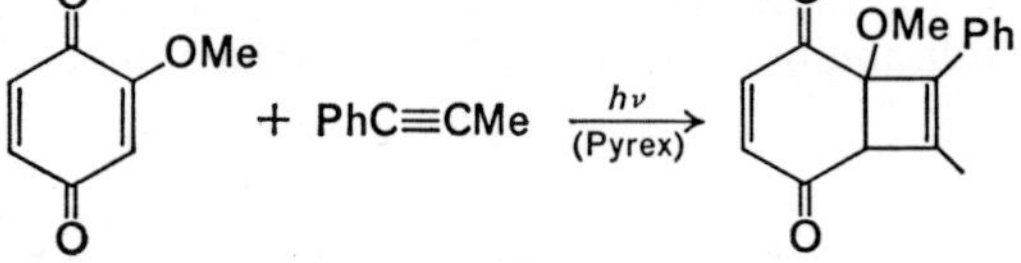

(92)

70%

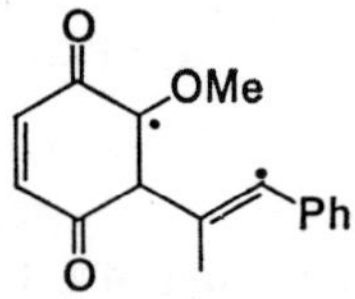

(31)

lowest (n,π^*) excited state, or cyclobutenes (equation 92)[105] if the quinone has a lowest (π,π^*) excited state. The quinone methides are formed by thermal ring-opening of oxetes. Sometimes a mixture of both types of product arises (equation 93)[106].

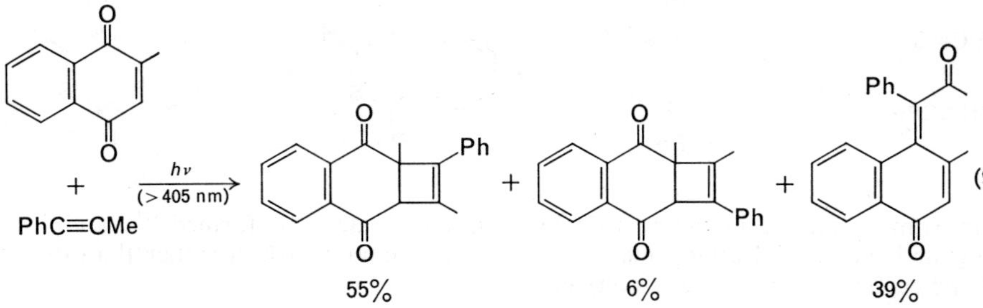

(93)

In the case of chloranil (equation 91) the product is formed when the wavelength of irradiation is such that the charge-transfer complex between the two reactants is the absorbing species. The orientation of cycloaddition to form the cyclobutene is that expected[107] on the basis of the most stable biradical intermediate (e.g. **31** for the reaction in equation 92; see also the product ratio in equation 93). In some cases the first-formed adduct may undergo subsequent photorearrangement by ring-opening and reclosure (equation 94)[108].

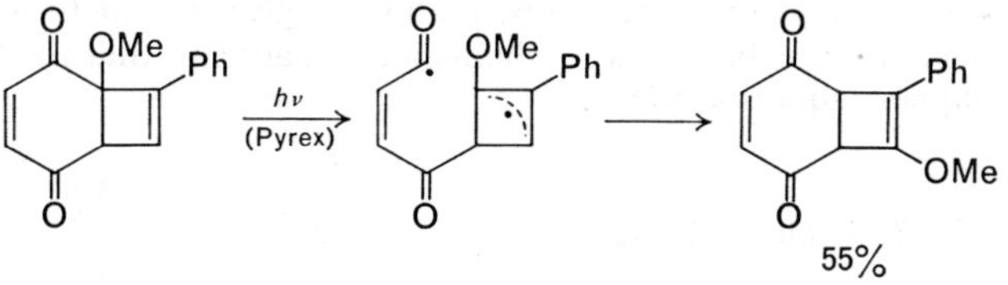

(94)

It is possible to form cyclobutene adducts from a quinone which has a lowest (n,π^*) excited state by a sequence involving 'protection' of the quinone as a Diels–Alder adduct with anthracene (equation 95)[109]. The enedione which undergoes photocycloaddition with alkyne reacts at the C=C bond rather than at the C=O bond (see also equation 48).

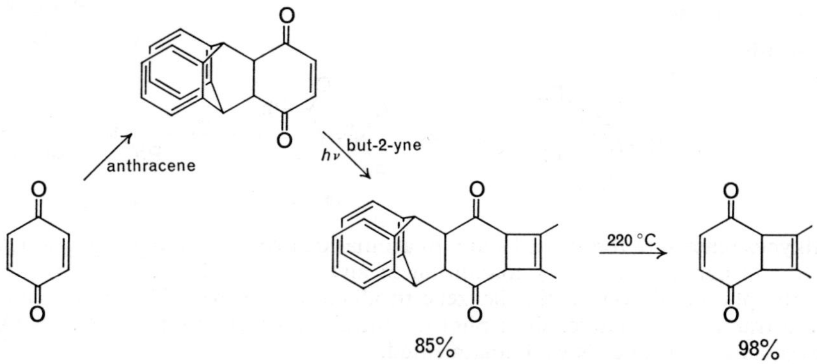

(95)

Aromatic thioketones react photochemically with alkynes to give isothiochromenes[110a]. This occurs by way of intramolecular attack in the first-formed biradical (equation 96).

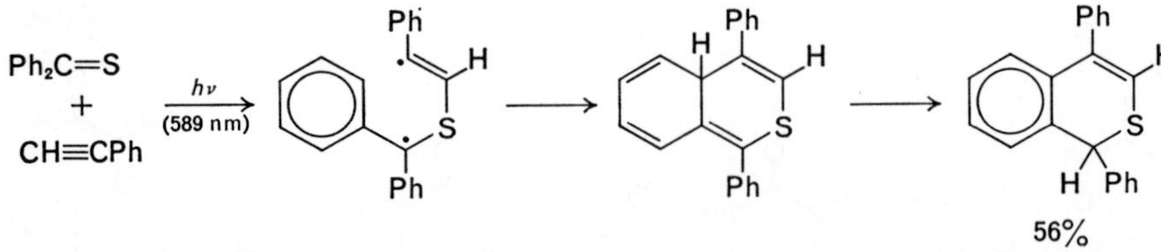
(96)

In some systems, however, thiochromenes (e.g. **32**) are formed[110b], and it is suggested that initial attack may occur on an aromatic carbon rather than on an alkyne carbon of the diphenylacetylene.

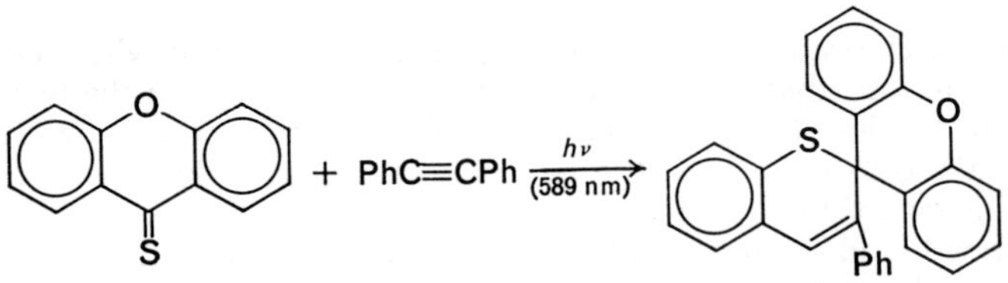
(32)
91%

Alkylthio- or dialkylamino-substituted alkynes give products which probably arise via a thiete intermediate in a mechanism analogous to that for the addition of ketones to alkynes (equation 97)[111].

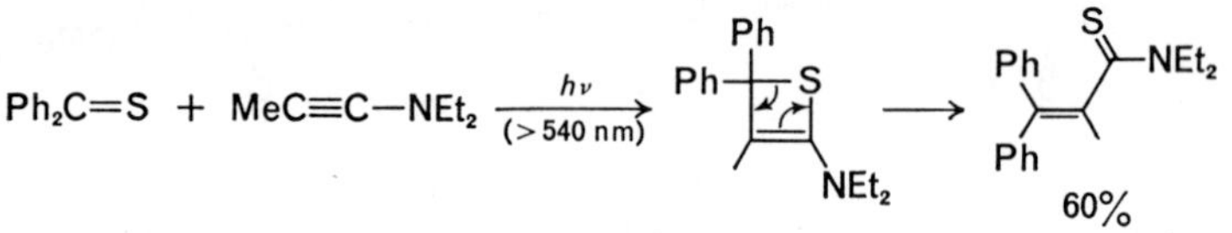
(97)

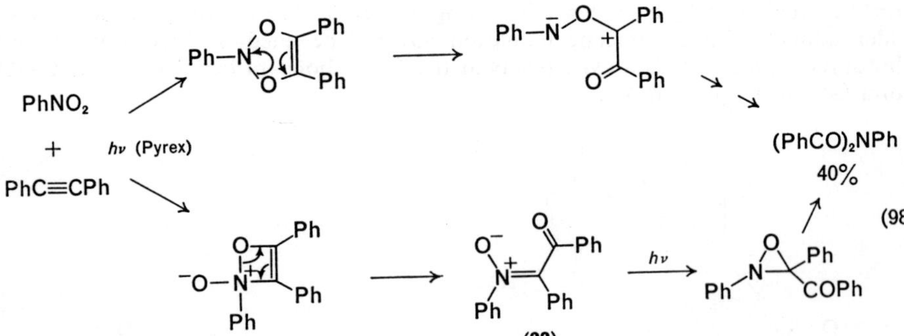
(33)
(98)

Alkynes react with the excited states of aromatic nitro compounds (equation 98)[112], perhaps in part by way of cycloaddition to give an unstable dioxazole by analogy with the photoaddition of nitrobenzene to alkenes. However, [2+2] cycloaddition is an attractive alternative, since photochemical transformation of nitrones (**33**) to amides via oxaziridines is well documented.

In the intramolecular photoreaction of 1-nitro-8-alkynylnaphthalenes (equation 99) the product (**34**) is formally analogous to that obtained by cycloaddition of an

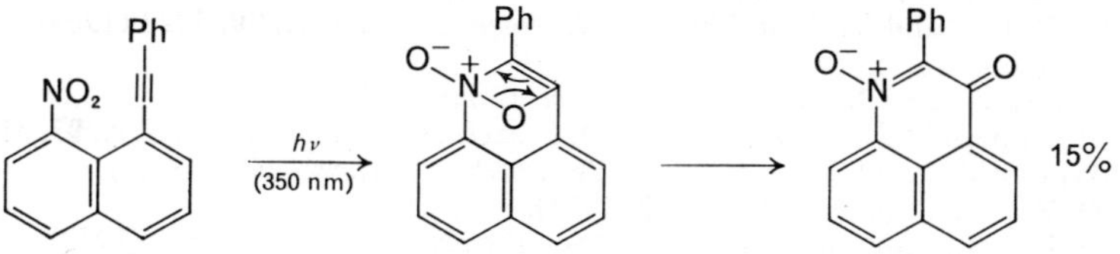

(99)

(**34**)

alkyne to a ketone followed by ring-opening, and it is possible that the reaction occurs through an oxazete[113].

There have been few reports of the cycloaddition of oxygen to alkynes, although the sensitized reaction of the ynamine **35** to give an α-ketoamide probably involves such a process (equation 100).

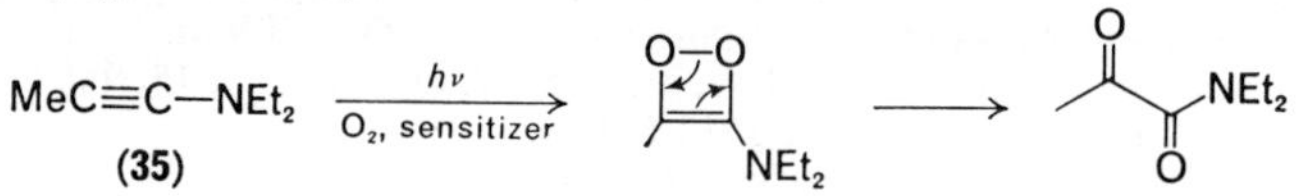

(**35**) (100)

VI. REFERENCES

1. J. D. Coyle, *Chem. Soc. Rev.*, **3**, 329 (1974).
2. J. A. Barltrop and J. D. Coyle, *Excited States in Organic Chemistry*, John Wiley and Sons, London, 1975, Chap. 8.
3. T. Nakayama and K. Watanabe, *J. Chem. Phys.*, **40**, 558 (1964).
4. A. D. Walsh, *J. Chem. Soc.*, 2288 (1953).
5. D. Demoulin, *Chem. Phys.*, **11**, 329 (1975).
6. C. R. Bowman and W. D. Miller, *J. Chem. Phys.*, **42**, 681 (1965).
7. D. F. Evans, *J. Chem. Soc.*, 1735 (1960).
8. B. Brocklehurst, D. C. Bull, M. Evans, P. M. Scott and G. Stanney, *J. Amer. Chem. Soc.*, **97**, 2977 (1975).
9. J. R. McNesby and H. Okabe, *Adv. Photochem.*, **3**, 157 (1964).
10. K. Evans, D. Heller, S. A. Rice and R. Scheps, *J. Chem. Soc. Faraday II*, **69**, 856 (1973).
11. D. F. Howarth and A. G. Sherwood, *Canad. J. Chem.*, **51**, 1655 (1973).
12. H. Okabe, *J. Chem. Phys.*, **62**, 2782 (1975).
13. C. P. Poole and R. S. Anderson, *J. Chem. Phys.*, **31**, 346 (1959).
14. A. Galli, P. Harteck and R. R. Reeves, *J. Phys. Chem.*, **71**, 2719 (1967).
15. W. A. Payne and L. J. Stief, *J. Chem. Phys.*, **56**, 3333(1972).
16. D. G. Whitten and W. Berngruber, *J. Amer. Chem. Soc.*, **93**, 3204 (1971).
17. G. J. Pontrelli, *J. Chem. Phys.*, **43**, 2571 (1965).
18. R. S. Macomber, *J. Org Chem.*, **38**, 816 (1973).
19. E. Kloster-Jensen and J. Wirz, *Helv. Chim. Acta*, **58**, 162 (1975).
20. M. A. Schexnayder and P. S. Engel, *J. Amer. Chem. Soc.*, **97**, 4825 (1975).
21. J. A. Barltrop and J. D. Coyle, *Excited States in Organic Chemistry*, John Wiley and Sons, London, 1975, Chap. 7, Section 7.3.3.
22. R. Selvarajan and J. H. Boyer, *Chem. Comm.*, 889 (1970).
23. A. H. A. Tinnemans and W. H. Laarhoven, *J. Chem. Soc., Perkin II*, 1111, 1115 (1976).
24. H. E. Zimmerman and J. A. Pincock, *J. Amer. Chem. Soc.*, **95**, 3246 (1973).
25. J. Perreten, D. M. Chihal, G. W. Griffin and N. S. Bhacca, *J. Amer. Chem. Soc.*, **95**, 3427 (1973); *J. Org. Chem.*, **41**, 3931 (1976).
26. H. E. Zimmerman and R. D. Little, *J. Amer. Chem. Soc.*, **94**, 8256 (1972).
27. B. Halton, M. Kulig, M. A. Battiste, J. Perreten, D. M. Gibson and G. W. Griffin, *J. Amer. Chem. Soc.*, **93**, 2327 (1971).
28. F. Bohlmann and W. Skuballa, *Tetrahedron Letters*, 2109 (1970).

29. G. Büchi and S. H. Feairheller, *J. Org. Chem.*, **34**, 609 (1969).
30. G. Ahlgren, *J. Org. Chem.*, **38**, 1369 (1973); P. Singh, *J. Org. Chem.*, **37**, 836 (1972).
31. T. Nishio and Y. Omote, *Chem. Letters.*, 103 (1976).
32. J. W. Wilson and V. S. Stubblefield, *J. Amer. Chem. Soc.*, **90**, 3423 (1968).
33. D. A. Ben-Efraim, *Tetrahedron Letters*, 957 (1967).
34. M. Kawanisi and K. Matsunaga, *Chem. Comm.*, 313 (1972).
35. T. D. Roberts, L. Munchausen and H. Schechter, *J. Amer. Chem. Soc.*, **97**, 3112 (1975).
36. T. Wooldridge and T. D. Roberts, *Tetrahedron Letters*, 4007 (1973.)
37. T. D. Roberts, *Chem. Comm.*, 362 (1971).
38. K. Fujita, K. Yamamoto and T. Shono, *Tetrahedron Letters*, 3865 (1973).
39. J. P. Ferris and F. R. Antonucci, *Chem. Comm.*, 126 (1972).
40. G. Wegner, *Makromol. Chem.*, **154**, 35 (1972).
41. J. Kiji, J. Kaiser, G. Wegner and R. C. Schulz, *Polymer*, **14**, 433 (1973).
42. P. S. Skell and R. G. Allen, *J. Amer Chem. Soc.*, **80**, 5997 (1958).
43. K. Fujita, K. Yamamoto and T. Shono, *Nippon Kagaku Kaishi*, 1933 (1973); *Chem. Abstr.*, **79**, 145647 (1973).
44. J. S. Bradshaw, R. D. Knudsen and W. W. Parish, *J. Org. Chem.*, **40**, 529 (1975).
45. G. Friedman and A. Komem, *Tetrahedron Letters*, 3357 (1968).
46. M. S. Kharasch, J. J. Jerome and W. H. Urry, *J. Org. Chem.*, **15**, 966 (1950).
47. W. Winter, *Tetrahedron Letters*, 3913 (1975).
48. M. P. Servé and H. M. Rosenberg, *J. Org. Chem.*, **35**, 1237 (1970).
49. D. R. Arnold and Y. C. Chang, *J. Heterocyclic Chem.*, **8**, 1097 (1971).
50. D. C. Owsley and J. J. Bloomfield, *J. Amer. Chem. Soc.*, **93**, 782 (1971).
51. (a) S. Majeti, V. A. Majeti and C. S. Foote, *Tetrahedron Letters*, 1177 (1975); (b) B. B. Snider, *J. Org. Chem.*, **41**, 3061 (1976).
52. T. Kubota and H. Sakurai, *J. Org. Chem.*, **38**, 1762 (1973).
53. P. de Mayo and M. C. Usselman, *Anal. Quim.*, **68**, 779 (1972).
54. J. A. Barltrop and D. Giles, *J. Chem. Soc. (C)*, 105 (1969).
55. E. P. Serebryakov and S. D. Kulomzina, *Izv. Akad. Nauk. SSSR, Ser. Khim.*, **12**, 2739 (1975); *Chem. Abstr.*, **84**, 073354 (1976).
56. W. Hartmann, *Chem. Ber.*, **104**, 2864 (1971).
57. R. Criegee, U. Zirngibl, H. Furrer, D. Seebach and G. Freund, *Chem. Ber.*, **97**, 2942 (1964).
58. R. Askani, *Chem. Ber.*, **98**, 3618 (1965).
59. (a) K. Fujita, K. Matsui and T. Shono, *Nippon Kagaku Kaishi*, 1024 (1975); *Chem. Abstr.*, **83**, 146784 (1975); (b) W. Mayer, D. Wendisch and W. Hartmann, *Tetrahedron Letters* 1277 (1977).
60. T. Kubota, K. Shima and H. Sakurai, *Chem. Comm.*, 360 (1971).
61. G. T. Kwiatkowski and D. B. Selley, *Tetrahedron Letters*, 3471 (1968).
62. M. Tsukada and S. Shida, *Bull. Chem. Soc. Japan*, **43**, 3621 (1970).
63. S. Tsunashima and S. Sato, *Bull. Chem. Soc. Japan*, **41**, 2281 (1968).
64. J. Freear and A. E. Tipping, *J. Chem. Soc., Perkin I*, 1074 (1975).
65. S. Shida and M. Tsukada, *Bull. Chem. Soc. Japan*, **43**, 2740 (1970).
66. M. Tsukada, T. Oka and S. Shida, *Chem. Letters*, **6**, 437 (1972).
67. O. P. Strausz, T. Hikida and H. E. Gunning, *Canad. J. Chem.*, **43**, 717 (1965).
68. D. Bryce-Smith and J. E. Lodge, *J. Chem. Soc.*, 695 (1963); E. Grovenstein, T. C. Campbell and T. Shibita, *J. Org. Chem.*, **34**, 2418 (1969).
69. L. A. Paquette, *Tetrahedron*, **31**, 2855 (1975).
70. D. Bryce-Smith, A. Gilbert and J. Grzonka, *Chem. Comm.*, 498 (1970).
71. D. Bryce-Smith, *Chem. Comm.*, 806 (1969).
72. B. Sket and M. Zupan, *J. Amer. Chem. Soc.*, **99**, 3504 (1977).
73. W. Lippke, W. Ferree and H. Morrison, *J. Amer. Chem. Soc.*, **96**, 2134 (1974); H. Morrison, T. Nylund and F. Palensky, *Chem. Comm.*, 4 (1976).
74. K. Ota, K. Murofushi, T. Hoshi and H. Inoue, *Tetrahedron Letters*, 1431 (1974); see Ref. 68 for a similar dimerization of phenylacetylene.
75. G. Büchi, C. W. Perry and E. W. Robb, *J. Org. Chem.*, **27**, 4106 (1962).
76. E. Müller, M. Sauerbier, D. Streichfuss, R. Thomas, W. Winter, G. Zountsas and J. Heiss, *Ann. Chem.* **750**, 63 (1971).

77. B. Bossenbroek, D. C. Sanders, H. M. Curry and H. Shechter, *J. Amer. Chem. Soc.*, **91,** 371 (1969).
78. P. J. Collin and W. H. F. Sasse, *Austr. J. Chem.*, **24,** 2325 (1971).
79. W. H. F. Sasse, *Austr. J. Chem.*, **22,** 1257 (1969).
80. W. H. F. Sasse, P. J. Collin, D. B. Roberts and G. Sugowdz, *Austr. J. Chem.*, **24,** 2151 (1971).
81. W. H. F. Sasse, P. J. Collin, D. B. Roberts and G. Sugowdz, *Austr. J. Chem.*, **24,** 2339 (1971).
82. R. J. McDonald and B. K. Selinger, *Z. Phys. Chem.*, **69,** 132 (1970).
83. T. Teitei and D. Wells, *Tetrahedron Letters*, 2299 (1975).
84. R. S. H. Liu and C. G. Krespan, *J. Org. Chem.*, **34,** 1271 (1969).
85. G. Sugowdz, P. J. Collin and W. H. F. Sasse, *Austr. J. Chem.*, **26,** 147 (1973).
86. R. P. Gandhi and V. K. Chadha, *Indian J. Chem.*, **9,** 305 (1971).
87. H. Prinzbach, R. Fuchs and R. Kitzing, *Angew. Chem. Int. Ed.*, **7,** 67 (1968).
88. R. P. Gandhi and V. K. Chadha, *Chem. Comm.*, 552 (1968).
89. H. J. Kuhn and K. Gollnick, *Chem. Ber.*, **106,** 674 (1973).
90. H. Prinzbach, M. Arguëlles and E. Druckrey, *Angew. Chem. Int. Ed.*, **5,** 1039 (1966).
91. R. M. Bowman, J. J. McCullough and J. S. Swenton, *Canad. J. Chem.*, **47,** 4503 (1969).
92. W. H. F. Sasse, P. J. Collin and D. B. Roberts, *Tetrahedron Letters*, 4791 (1969).
93. J. H. Dopper and D. C Neckers, *J. Org. Chem.*, **36,** 3755 (1971).
94. D. C. Neckers, J. H. Dopper and H. Wynberg, *Tetrahedron Letters*, 2913 (1969).
95. H. Polman, A. Mosterd and H. J. T. Bos, *Rec. Trav. Chim.*, **92,** 845 (1973).
96. L. E. Friedrich and J. D. Bower, *J. Amer. Chem. Soc.*, **95,** 6869 (1973).
97. W. J. Middleton, *J. Org. Chem.*, 1307 (1965).
98. H. Polman, J. S. M. Boleij and H. J. T. Bos, *Rec. Trav. Chim.*, **91,** 1088 (1972).
99. A. Mosterd, H. J. Matser and H. J. T. Bos, *Tetrahedron Letters*, 4179 (1974)
100. A. Mosterd, R. E. L. J. Lecluijze and H. J. T. Bos, *Rec. Trav. Chim.*, **94,** 72 (1975); A. Mosterd, L. J. De Noten and H. J. T. Bos, *Rec. Trav. Chim.*, **96**, 16 (1977).
101. H. J. T. Bos, H. Polman and P. F. E. van Montfort, *Chem. Comm.*, 188 (1973).
102. H. A. J. Carless, *Tetrahedron Letters*, 2265 (1972); R. J. C. Koster, D. G. Streefkerk, J. Oudshoorn-Van Veen and H. J. T. Bos, *Rec Trav. Chim.*, **93,** 157 (1974).
103. J. M. Bruce in *The Chemistry of the Quinonoid Compounds*, Part 1 (Ed. S. Patai), John Wiley and Sons, London, 1974, Chap. 9.
104. J. A. Barltrop and B. Hesp, *J. Chem. Soc.* (*C*), 1625 (1967).
105. S. P. Pappas, B. C. Pappas, and N. A. Portnoy, *J. Org. Chem.*, **34,** 520 (1969).
106. S. Farid, W. Kothe and G. Pfundt, *Tetrahedron Letters*, 4147 (1968).
107. S. P. Pappas and N. A. Portnoy, *Chem. Comm.*, 597 (1969).
108. F. A. L. Anet and D. P. Mullis, *Tetrahedron Letters*, 737 (1969).
109. P. Yates and G. V. Nair, *Synth. Comm.*, **3,** 337 (1973).
110. (a) A. Ohno, *Intern. J. Sulfur Chem. B*, **6,** 183 (1971); (b) H. Gotthardt and S. Nieberl, *Tetrahedron Letters*, 3563 (1976).
111. A. C. Brouwer and H. J. T. Bos, *Tetrahedron Letters*, 209 (1976).
112. M. L. Scheinbaum, *J. Org. Chem.*, **29,** 2200 (1964).
113. C. C. Leznoff and R. J. Hayward, *Canad. J. Chem.*, **49,** 3596 (1971).

CHAPTER **13**

Synthetic acyclic polyacetylenes

W. D. HUNTSMAN

Ohio University, Athens, Ohio

I. INTRODUCTION

The amount of work devoted to synthetic polyacetylenes has remained at a high level during the past two decades, following the period of intense activity in the 1950s. Much of the impetus has been provided by the discovery of large numbers of polyacetylenes in nature[1], but the challenge of synthesizing and studying molecules with large numbers of conjugated triple bonds has also been a major factor. To date, the longest synthetic polyacetylene seems to be $Et_3Si(C{\equiv}C)_{16}SiEt_3$. The synthesis of this and related long-chain polyynes was achieved through clever use of trialkylsilyl groups as protective groups[2] (see Section III.A).

One of the current areas of intense activity is the solid-state polymerization of certain diacetylene derivatives, a reaction in which conjugated polyenepolyyne chains are created with effectively infinite length (see Section IV.F).

The chemistry of acyclic, synthetic polyacetylenes will be considered in this chapter. Cyclic and naturally occuring polyacetylenes are covered in separate chapters. Although most of the discussion is directed toward conjugated polyynes, reactions of non-conjugated derivatives are included in which interactions between triple bonds play an important role. Earlier work in the area has been reviewed thoroughly[3–6] and consequently in this chapter attention is directed mainly toward some of the more recent advances.

II. PROPERTIES

A. Stability

Polyacetylenes generally exhibit low thermal stability, and the stability decreases with increasing number of triple bonds. Although it has been reported that butadiyne can be distilled at 10 °C without decomposition[3], other workers have found that it polymerizes rapidly above 0 °C [7]. Reports of detonations and extreme shock sensitivity of derivatives of diacetylene have appeared[8, 9] and emphasize the importance of taking adequate safety precautions when working with polyacetylenes in general. Triacetylene, $H(C{\equiv}C)_3H$, is extremely unstable, and even in the absence of air, it turns black and often explodes violently[10]. It is possible to isolate the polyynes $H(C{\equiv}C)_nH$ with $n = 3$, 4 and 5 as solids at low temperatures[11], but customarily these and higher polyynes are handled in dilute solutions in which the stability is much greater. Thus $H(C{\equiv}C)_{12}H$ can be obtained and handled at room temperature as a dilute solution in methanol[2].

The stability of disubstituted polyynes **1** is greater than that of the unsubstituted derivatives, and the increase is remarkable in cases where the substituent is a large bulky group such as *t*-butyl[12–14]. For the dimethyl derivatives **1a**, the tetrayne

$$R(C{\equiv}C)_nR$$

(1a) $R = CH_3$; **(1b)** $R = Ph$; **(1c)** $R = t\text{-}Bu$

($n = 4$) decomposes at 80 °C, whereas the hexayne ($n = 6$) decomposes at 5 °C [15]. With the diphenyl derivatives **1b**, the practical limit of stability is reached at $n = 8$ [16]; the decayne, $n = 10$, has been synthesized but it decomposes below room temperature[14]. The di-*t*-butyldodecayne **1c**, $n = 12$, can be heated nearly to 50 °C [14]! Presumably the bulky *t*-butyl groups provide stabilization by hindering the close approach of polyyne chains to each other[12]. Trialkylsilyl groups also increase the stability of terminal polyynes, and can serve as convenient derivatives for storage[2, 17]. Most conjugated polyynes are photosensitive and give brightly coloured polymers upon exposure to light[18]. The photosensitivity roughly parallels the thermal sensitivity.

B. Geometry

The chain in conjugated polyynes is linear, and bond lengths show clear alternation between triple and single bonds. Thus in diynes and triynes lengths of 1·20 Å and 1·38 Å are found for the triple and single bonds[18]. In the tetrayne **2** the same length (1·20 Å) is found for all of the triple bonds, and all single bonds are 1·38 Å except the central one which is shortened to 1·33 Å [19]. In the crystalline state, the carbon chain in **2** is slightly bowed with the central carbons being approximately

$$Me_3Si(C{\equiv}C)_4SiMe_3$$

(2)

0·5 Å away from the line joining the silicon atoms, but the deviation from linearity is ascribed to crystal packing forces.

C. Electronic Effects

Butadiyne is a stronger acid than acetylene as might be anticipated on the basis of enhanced electronegativity of *sp*-hybridized carbon[18]. As can be seen from Table 1, the dissociation of carboxylic acids is greatly enhanced by a triple bond in the α, β position, and is increased further by a second conjugated triple bond, but to a smaller extent. A third triple bond has a still smaller acid-strengthening effect[20].

TABLE 1. pK_a values for acetylenic acids[20]

Acid	pK_a
$C_3H_7CO_2H$	4·8
$C_2H_5C{\equiv}CCO_2H$	2·60
$C_2H_5(C{\equiv}C)_2CO_2H$	1·90
$C_2H_5(C{\equiv}C)_3CO_2H$	1·67

Studies of the alkaline cleavage of silyl derivatives **3** also point to enhanced inductive withdrawal by additional alkynyl groups[21]. The most likely mechanism is

$$\underset{(3)}{Ar(C{\equiv}C)_nSiEt_3} + OH^- \longrightarrow Ar(C{\equiv}C)_{n-1}C{\equiv}C{:}^- + Et_3SiOH \qquad (1a)$$

$$Ar(C{\equiv}C)_{n-1}C{\equiv}C{:}^- + H_2O \longrightarrow Ar(C{\equiv}C)_nH + OH^- \qquad (1b)$$

the one shown in equation (1) and involves a slow nucleophilic attack on silicon to given the carbanion, which then rapidly abstracts a proton from the solvent. The ease of cleavage increases as n is raised, the increase being particularly marked for the change from $n = 1$ to $n = 2$. For the phenyl derivatives, **3** (Ar = Ph), with $n = 1, 2$

and 3 the relative rates of cleavage are 1 : 240 : 4100. Interestingly, the rates of cleavage of substituted aryl derivatives correlate well with simple Hammett substituent constants, σ [21].

The electronic effect of the butadiynyl group in electrophilic aromatic substitutions has been determined by measuring the rates of acid-catalysed cleavage of the aryl–tin bonds in the aryltrimethylstannanes **4** [22]. The reaction (equation 2) proceeds by way of the benzenonium intermediate **5**, and the relative rates (Table 2) give an indication

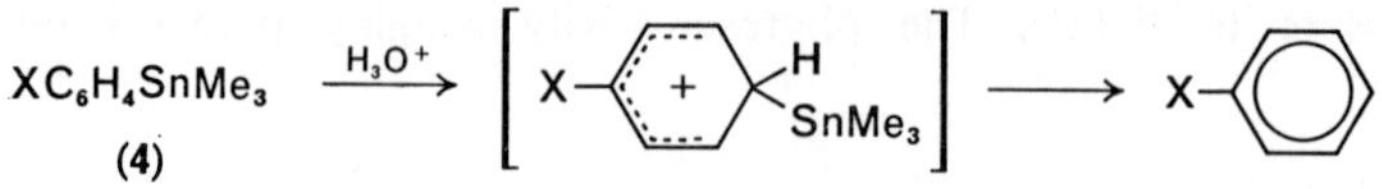

(4) **(5)** (2)

TABLE 2. Relative rates of reaction (2)

X	Relative rate
H	1·00
p-HC≡C	0·425
m-HC≡C	0·288
p-H(C≡C)$_2$	0·263
m-H(C≡C)$_2$	0·207
m-Br	0·195

of the electronic effects of X in the transition state leading to the ion. It is seen that the butadiynyl group deactivates both the *meta* and *para* positions more strongly than does an ethynyl group. Deactivation of the *meta* position is almost as great as that produced by a bromine atom.

$$XC_6H_4CH_2SiMe_3 \xrightarrow[H_2O,\ EtOH]{OH^-} XC_6H_4CH_3 + Me_3SiOH$$

(6)

In the cleavage of the benzyl–silicon bond of **6** by base, negative charge develops on the benzyl carbon in the transition state, and electron withdrawal by X facilitates the reaction. When X is *p*-H(C≡C)$_2$, the rate of cleavage is 3300 times as great as when X = H [22]. This corresponds to a value of 0·72 for the σ^- constant, and indicates that the group can withdraw electrons strongly.

D. Ultraviolet Spectroscopy

Conjugated polyacetylenes exhibit characteristic electronic absorption spectra with the most prominent feature being a very high intensity band with well-defined vibrational fine structure[18, 23]. This band is attributed to the allowed ${}^1\Sigma_g^+ \rightarrow {}^1\Sigma_u^+$ transition[24]. The longest wavelength vibrational peak is always the most intense component of the band, and the intensities of the other components decrease fairly regularly toward shorter wavelengths. The peaks appear at somewhat shorter wavelengths, but with greater intensities than those of the corresponding polyenes. In fact, the intensities of these bands in highly conjugated polyacetylenes rank among the greatest observed for organic compounds—the molar absorptivity in the decayne, *t*-Bu(C≡C)$_{10}$Bu-*t*, for example, reaches 850 000 [13]!

For diacetylene and triacetylene the high-intensity bands are beyond the readily accessible range (Table 4), but with dimethyltriacetylene the bathochromic shift produced by alkyl substitution just brings it into the accessible range (see Table 5). With increasing conjugation the band moves to longer wavelengths and increases in intensity. Unlike polyenes, for which the vibrational spacing (*ca.* 1450 cm^{-1}) is nearly independent of the number of double bonds, the average spacing in polyynes decreases with increasing conjugation from about 2000 cm^{-1} when *n* is 4–6 to about 1700 cm^{-1} when *n* is 12–16 (see Tables 4 and 5).

A second band appears at longer wavelengths with much lower intensity, and is attributed to two overlapping forbidden transitions, ${}^1\Sigma_g^+ \rightarrow {}^1\Sigma_u^-$ and ${}^1\Sigma_g^- \rightarrow {}^1\Delta_u$ [24]. This band also exhibits fine structure, but the pattern is often more complicated than that of the high-intensity band. The components are often labelled from the long wavelength end with the letters A–K [23]. The B peak is often more clearly resolved than the A peak, and the B peak is recorded in Table 3 for purposes of comparison. This band moves to longer wavelengths with increasing conjugation, but, unlike the high-intensity band, does not increase in intensity. Consequently, with highly conjugated polyynes the band may be poorly defined, as in *t*-Bu$(C{\equiv}C)_{12}$Bu-*t*, where it appears as two shoulders on the long-wavelength edge of the high-intensity band[14].

A small bathochromic shift accompanies the change from gas phase to solution (Table 3), but no general trend seems to be followed upon alkyl substitution.

TABLE 3. B peak in the low-intensity band of polyynes, $R(C{\equiv}C)_nR$

n	R	λ_{max} (nm)[b]	Log ε_{max}	Vibrational spacing (cm^{-1})	Solvent	Reference
2	H	231	—	2100	Gas	24
2	H	235	—	2080	P–DMB[a]	24
2	Me	236	2·52	2250	EtOH	25
2	*t*-Bu	239	2·66	2160	MeOH	12
2	Et_3Si	263	2·66	2200	MeOH	2
3	H	276	—	2120	Gas	24
3	H	284	—	2000	P-DMB[a]	11
3	Me	286	2·30	2290	EtOH	26
3	*t*-Bu	283	—	—	MeOH	12
3	Et_3Si	310	2·52	2130	MeOH	2
4	H	322	—	2100	P-DMB[a]	24
4	H	316	—	1960	MeOH	2
4	Me	328	2·25	2240	EtOH	27
4	*t*-Bu	330	2·57	2110	MeOH	12
4	Et_3Si	349	2·28	2050	MeOH	2
5	H	356	—	2050	P-DMB[a]	24
5	Me	348	2·32	2110	EtOH	15
5	*t*-Bu	364	2·43	2190	MeOH	28
6	*t*-Bu	395	2·28	2170	MeOH	12
8	*t*-Bu	437	2·43	1850	Hexane	13
10	*t*-Bu	471	2·34	1800	Hexane	13

[a] P–DMB = mixture of pentane and 2,2-dimethylbutane.

[b] Because of the complexity of this region in many spectra, the choice of the B peak is often open to question. Comparisons between compounds should be made with caution.

Lowering the temperature to -150 °C produces a bathochromic shift, increased intensity of the peaks and better resolution[24].

In butadiyne, a third band system appears in the far ultraviolet (125–145 nm) with intensity nearly equal to that of the ultra-high-intensity band, and is attributed to the first $^1\Sigma_g^+ \rightarrow {}^1\Pi_u$ transition[24]. Also, with highly conjugated polyynes a third, fine-structured band may appear on the short λ tail of the ultra-high-intensity band. In the dodecayne, *t*-Bu(C≡C)$_{12}$Bu-*t*, the absorption occurs at 250–300 nm [14].

I. Unsubstituted polyynes, H(C≡C)$_n$H

The spectra of unsubstituted polyynes (7) have been measured for the homologues with n = 2–10 and n = 12. Gas-phase spectra have been recorded for 7(2)–7(4) from

H(C≡C)$_n$H

(7)(n)

the normal ultraviolet range down to 110 nm [24]. Complete spectra have been recorded for solutions of 7(2)–7(5), but for $n \geqslant 6$ stability considerations do not permit the preparation of solutions of high-enough concentration to permit accurate measurements of the low-intensity bands, and only the high-intensity bands have been reported for these compounds[2].

TABLE 4. Highest intensity peak for unsubstituted polyynes, H(C≡C)$_n$H

n	λ_{max} (nm)	Log ε_{max}	Average vibrational spacing (cm^{-1})	Solvent	Reference
2	165	—	—	Gas	24
3	183	—	—	Gas	24
4	207	—	2060	Gas	24
	226	—	2110	Pentane	24
	226	5·25	2170	MeOH	2
5	251	—	2160	MeOH	2
6	274	5·47	1970	MeOH	2
	275	5·51	1990	Hexane	2
7	295	—	2100	MeOH	2
8	315	5·48	1870	MeOH	2
	316	5·54	1870	Hexane	2
9	332	—	1840	MeOH	2
10	348	—	1810	MeOH	2
12	375	—	1730	MeOH	2

The locations of the highest intensity peaks for these polyynes, along with available intensity data and average spacings of the vibrational peaks are given in Table 4. The bathochromic shift and the decrease in average vibrational spacing that accompany increased conjugation can be seen in the table. Data for octatetrayne show a large red shift accompanying the change from gas-phase to solution spectra, and data for several of the higher polyynes reveal a small bathochromic shift when the solvent is changed from methanol to hexane[2].

A Lewis–Calvin plot of n versus λ^2 for 7(4)–7(12) in MeOH gives an excellent straight line with slope $12{\cdot}2 \times 10^3$ nm^2/triple bond[2].

2. Dialkyl polyacetylenes

Data for the ultra-high-intensity band for disubstituted polyynes $R(C{\equiv}C)_nR$ are collected in Table 5; comparison with the values for unsubstituted polyynes reveals that substitution produces a bathochromic shift and an increase in intensity. The red shift increases in the order Me < *t*-Bu < Et_3Si; the difference between the dimethyl and di-*t*-butyl derivatives remains fairly constant, but that between the di-*t*-butyl and bis(triethylsilyl) derivatives decreases steadily with increasing values of *n*. The greater shift for silylated derivatives has been attributed to more effective conjugative interactions in the excited state[29].

TABLE 5. Highest intensity peak for disubstituted polyynes, $R(C{\equiv}C)_nR$

n	R	λ_{max} (nm)	Log ε_{max}	Vibrational spacing (cm^{-1})	Solvent	Reference
3	Me	207	5·13	—	EtOH	26
	t-Bu	213	5·15	—	MeOH	12
	Et_3Si	230	5·04	—	MeOH	2
4	Me	234	5·45	2020	EtOH	27
	t-Bu	240	5·54	2110	MeOH	12
	Et_3Si	256	5·29	2060	MeOH	2
5	Me	261	5·55	2030	EtOH	15
	t-Bu	265	5·65	2030	MeOH	28
	Et_3Si	278	—	1990	MeOH	2
6	Me	284	5·65	2010	EtOH	15
	t-Bu	289	5·70	1970	MeOH	28
	Et_3Si	298	5·50	2020	MeOH	2
7	*t*-Bu	311	5·72	1880	Ether	12
	Et_3Si	317	—	1930	MeOH	2
8	*t*-Bu	330	5·85	1860	Hexane	13
	Et_3Si	335	5·52	1850	MeOH	2
10	*t*-Bu	363	5·93	1730	Hexane	13
	Et_3Si	365	—	1800	MeOH	2
12	*t*-Bu	387	—	1690	Hexane	14
	Et_3Si	388	—	1690	MeOH	2
16	Et_3Si	426	—	1650	MeOH	2

For the three dimethyl derivatives reported (n = 4, 5, 6) the average vibrational spacing remains fairly constant, but for the other two series the spacing decreases with increasing *n*, from *ca.* 2000 cm^{-1} (n = 4–6) to 1650 cm^{-1} (n = 16). A distinct solvent dependence is found for the bis(triethylsilyl) derivatives, bathochromic shifts of 1–2 nm and concomitant increases in intensity being observed for all maxima when the solvent is changed from methanol to hexane[2]. Dramatic changes in intensity may occur with change of solvent. For example, the molar absorptivity of the most intense band in the spectrum of *t*-Bu$(C{\equiv}C)_5$Bu-*t* has the value 465 000 in cyclohexane, but this decreases to 233 000 in a 50 : 50 mixture of cyclohexane–CS_2 [30]. A bathochromic shift from 268 nm to 276 nm also occurs with this change of solvent. A model has been proposed which rationalizes the effects of non-polar solvents[30].

Lewis–Calvin plots of λ^2 *versus* *n* are linear for the dimethyl and di-*t*-butyl derivatives[14]. There is some indication of deviation from linearity in the latter series for n = 12, but this may be a solvent effect. For the bis(triethylsilyl) series the plot is linear through n = 8, but a definite downward curvature appears for higher members[2].

3. Diaryl polyacetylenes

The general pattern of bands observed for aliphatic polyynes persists for diaryl polyacetylenes, i.e. a group of bands with medium intensity at long wavelengths and a group of higher intensity bands at shorter wavelengths. The long-wavelength bands increase in intensity from $\varepsilon = \sim 200$ to $\varepsilon = \sim 2000$–50 000, and retain their distinct vibrational fine structure. Thus with $Ph(C{\equiv}C)_4Ph$, four peaks appear in the 300–400 nm range with an average vibrational spacing of 2090 cm^{-1} and $\log \varepsilon = \sim 4{\cdot}4$ [16].

The shorter wavelength bands have significantly lower intensities than do their aliphatic counterparts, but with very long conjugated chains the two series approach each other. For example, the highest intensity peak (λ = 386 nm) for $Ph(C{\equiv}C)_{10}Ph$ has $\log \varepsilon = 5{\cdot}20$ [14]. However, the high-intensity bands of diarylpolyynes do not exhibit the distinct fine structure observed with the aliphatic derivatives[16].

TABLE 6. Longest wavelength peaks (λ_L) for diaryl polyacetylenes, $Ar(C{\equiv}C)_nAr$

n	Ar[a]	λ_L (nm)	Log ε	Solvent	Reference
2	Ph	327	4·44	EtOH	16
	Mes	341	4·56	Hexane	14
	1-Nap	375	4·54	THF	35
	1-An	430	4·46	THF	36
	9-An	470	4·50	THF	37
3	Ph	358	4·31	EtOH	16
	1-Nap	397	4·60	THF	35
	1-An	440	4·62	THF	36
	9-An	479	4·66	THF	37
4	Ph	397	4·33	EtOH	16
	1-Nap	422	4·49	THF	35
	1-An	456	4·61	THF	36
	9-An	491	4·65	THF	37
6	Ph	460	3·94	EtOH	16
	Mes	469	4·02	MeOH	14
	1-Nap	479	4·27	THF	35
	1-An	494	4·50	THF	36
	9-An	523	4·65	THF	37
8	Ph	509	4·45	EtOAc	16
	Mes	522	3·67	$CHCl_3$	14
10	Ph	549	3·23	$CHCl_3$	14

[a] Mes = mesityl = 2,4,6-trimethylphenyl; 1-Nap = 1-naphthyl; 1-An = 1-anthryl; 9-An = 9-anthryl.

A third area of high-intensity absorption appears at 250–300 nm for $Ph(C{\equiv}C)_{10}Ph$, and presumably corresponds to the bands observed in this region for $t\text{-}Bu(C{\equiv}C)_{10}Bu\text{-}t$ [14].

The location and intensity of the longest wavelength peak (λ_L) in the spectra of selected diarylpolyynes are presented in Table 6. It can be seen that the value of λ_L depends on the nature of the aryl ring, and also on the point of attachment. Akiyama,

Nakagawa and Nakasuji have synthesized twelve series of diarylpolyynes, $Ar(C{\equiv}C)_nAr$, with $n = 1$–6 and Ar ranging from monocyclic to tetracyclic aryl groups, and have found that the longest wavelength peaks fail to give linear Lewis–Calvin plots of λ^2 *versus* n [31–34]. Instead, linear relationships are obtained by plotting λ-*versus* n^x, where x depends on the nature of Ar, and ranges in value from 1 to 2. Thus, for Ar = 1-naphthyl, $x = 1{\cdot}5$, for Ar = 2-naphthyl, $x = 1{\cdot}3$ and for Ar = 1- or 9-anthryl, $x = 2{\cdot}0$. A rationalization of the linear relationships on the basis of HMO calculations has been given[31]. Whereas earlier workers[16] reported that for $Ph(C{\equiv}C)_nPh$, plots of λ^2 *versus* n are reasonably linear for lower values of n, with a slight downward curvature becoming evident when $n = 6$ or 8, these workers report a *linear* relationship between λ and n [31]. However, if the recently determined value of λ_L for $n = 10$[14] is included in the plot, better agreement is obtained for the λ_L^2 *versus* n plot.

4. Polyenepolyynes

The spectra of conjugated polyenepolyynes are often very complex and difficult to analyse. When the acetylenic part is the major chromophore, e.g. in an enepolyyne, the spectrum resembles that of a pure polyyne, but when the ethylenic portion constitutes the major chromophore, the spectrum is closer to that of a regular polyene[18, 23]. Much of the information about polyenepolyynes can be found in works dealing with naturally occurring polyacetylenes[1].

E. Vibrational Spectroscopy

In principle the number of triple-bond stretching vibrations for a conjugated polyyne is expected to be the same as the number of triple bonds[18]. For symmetrical diynes, one mode should be i.r.-active and one Raman-active; for symmetrical triynes, one should be i.r.-active and two Raman-active, etc. Because of the low force constant of the single bonds which separate the acetylene units compared to that of a triple bond, the splittings are small, and only with some of the lower polyynes is the anticipated number of absorption bands observed (see Table 7). The intensity of the C≡C bands in the i.r. increases with the number of triple bonds, and with $Ph(C{\equiv}C)_6Ph$, for example, it is the strongest band in the spectrum[16].

F. Nuclear Magnetic Resonance Spectroscopy

1. Proton magnetic resonance

The signal for the acetylenic proton in conjugated polyynes is shifted slightly downfield from the position for related monoalkynes. As seen in Table 8, the shift amounts to 0·05 p.p.m. on going from acetylene to butadiyne, and an additional shift of 0·08 p.p.m. occurs on going to hexatriyne. However, no change in position occurs on the attachment of the next triple bond, and the signals for triacetylene and tetraacetylene appear at the same position.

The protons of methyl groups attached to a triple bond reach resonance at about δ 2, and show a slight downfield shift in conjugated polyynes. For example, the shift amounts to 0·09 p.p.m. going from propyne to 1,3-pentadiyne.

The chemical shift of acetylenic protons in conjugated polyynes shows marked solvent effects. For example, the signal for 1,3-pentadiyne appears at δ 1·75, 2·80 and 3·50 in CCl_4, acetone and DMF, respectively[40].

Unusually strong long-range coupling occurs in polyacetylenes, and has been observed for protons separated by nine chemical bonds[41]. CNDO and INDO

TABLE 7. Triple-bond stretching frequencies for conjugated polyacetylenes, $R(C{\equiv}C)_nR'$

n	R	R′	$\nu_{C\equiv C}$ (cm^{-1}) [a]	Reference
2	H	H	2184 vs (R), 2020 m	38
	t-Bu	H	2320 w, 2230 m, 2060 w	14
	Et_3Si	H	2190, 2140	2
	Mes	H	2215 m, 2190 w, 2070 w	14
	Me	Me	2210	15
	Ph	Ph	2220	16
	Et_3Si	Et_3Si	2070	2
3	H	H	2201 vs (R), 2125 w, 2019 vs (R)	10
	Me	Me	2220	15
	Ph	Ph	2200	16
	Mes	Et_3Si	2170 m, 2125 s, 2075 m	14
	Et_3Si	Et_3Si	2170, 2160	2
4	Me	Me	2236	15
	Ph	Ph	2205	16
	Mes	Mes	2193 w, 2190 s, 2075 w	14
	Mes	Et_3Si	2200 m, 2135 s, 2065 m	14
	Et_3Si	Et_3Si	2180, 2045	2
5	Me	Me	2220	15
6	Me	Me	2206	15
	Ph	Ph	2180 s, 2166 s	16
	Mes	Mes	2175 m, 2100 m, 2065 w	14
8	Mes	Mes	2180 w, 2100 m	14
10	Ph	Ph	2185 w, 2070 m, 2020 w	14

[a] R = Raman-active; all other bands i.r.-active.

TABLE 8. Proton chemical shifts in polyacetylenes

Compound[a]	δ (p.p.m.)	Solvent	Reference
$H{-}C{\equiv}C{-}H$	2·01	$CHCl_3$ [b]	39
$H{-}(C{\equiv}C)_2{-}H$	2·06	$CHCl_3$ [b]	11
$H{-}(C{\equiv}C)_3{-}H$	2·14	$CHCl_3$ [b]	11
$H{-}(C{\equiv}C)_4{-}H$	2·14	$CHCl_3$ [b]	11
$CH_3C{\equiv}C{-}H$	1·88	$CHCl_3$ [b]	39
$CH_3(C{\equiv}C)_2{-}H$	1·97	$CHCl_3$ [b]	39
	1·75	CCl_4 [c]	40
$CH_3(C{\equiv}C)_3{-}H$	1·87	CCl_4 [c]	40
$HC{\equiv}C{-}CH_3$	1·88	$CHCl_3$ [b]	39
$H(C{\equiv}C)_2{-}CH_3$	1·97	$CHCl_3$ [b]	39
t-Bu$(C{\equiv}C)_2{-}H$	1·83	CCl_4	14
$Ph(C{\equiv}C)_2{-}H$	2·30	CCl_4 [c]	39
$Mes(C{\equiv}C)_2{-}H$ [d]	2·03	CCl_4	14
$Et_3Si(C{\equiv}C)_2{-}H$	2·00	CCl_4	2

[a] The protons under consideration are the ones attached by the extended bond.
[b] At −50 °C. [c] At infinite dilution. [d] Mes = mesityl.

calculations for conjugated diynes and triynes give satisfactory agreement with experimental values and indicate that the coupling is transmitted through the π system[42].

2. Carbon-13 magnetic resonance

Values of carbon-13 chemical shifts have been reported for a number of polyacetylenes and representative examples are summarized in Table 9[43–46]. The carbon in the α position to a triple bond is shielded, at least partially, as a result of the diamagnetic anisotropy of the triple bond, and in simple alkynes the signal for the α carbon is found some 10–14 p.p.m. upfield from the position in the corresponding alkane. In conjugated diynes the signals of the interior *sp*-hybridized carbons are about 12–14 p.p.m. upfield from their position in the corresponding monoalkyne, and thus it is evident that the shielding of the α position is roughly independent of the hybridization of the α carbon[46]. For example, the shift, 65·5 p.p.m., of C-4 in 3,5-octadiyne (**8**) may be compared with the value of 80 p.p.m. for C-4 in 3-octyne[43]. The values given in Table 9 for the higher polyynes, **9**, **10** and **11**, show that the shielding of the inner carbons increases with successive triple bonds but appears to approach a limiting value. Nevertheless, the chemical shifts of all non-equivalent acetylenic carbons in the pentayne **11** are different!

$$\overset{1}{C}H_3\overset{2}{C}H_2\overset{3}{C}\equiv\overset{4}{C}-C\equiv CCH_2CH_3$$

(8)

$$\overset{1}{C}H_3\overset{2}{C}\equiv\overset{3}{C}-\overset{4}{C}\equiv C-C\equiv C-CH_3$$

(9)

$$(\overset{1}{C}H_3)_3\overset{2}{C}-\overset{3}{C}\equiv\overset{4}{C}-\overset{5}{C}\equiv\overset{6}{C}-C\equiv C-C\equiv C-C(CH_3)_3$$

(10)

$$(\overset{1}{C}H_3)_3\overset{2}{C}-\overset{3}{C}\equiv\overset{4}{C}-\overset{5}{C}\equiv\overset{6}{C}-\overset{7}{C}\equiv C-C\equiv C-C\equiv C-C(CH_3)_3$$

(11)

TABLE 9. Carbon-13 chemical shifts[a] in polyacetylenes

	Carbon number[b]						
Compound	1	2	3	4	5	6	7
(**8**)[c]	13·8	13·2	78·6	65·5			
(**9**)[d]	4·4	74·8	65·0	60·0			
(**10**)[c]	30·3	28·7	88·6	64·7	62·2	61·9	
(**11**)[c]	30·3	28·3	88·5	64·6	62·3	62·1	61·8

[a] Expressed in p.p.m. from tetramethylsilane.
[b] Carbons 3–7 (2–7 in **9**) are acetylenic carbons.
[c] Reference 45.
[d] Reference 43.

The low-field position for the C-5 and C-7 signals in **12** signifies that the triple bonds are able to transmit the electron-withdrawing effect of the ester function[45].

$$n\text{-}C_3H_7-\underset{90}{C}\equiv\overset{65{\cdot}3}{C}-\underset{86{\cdot}7}{C}\equiv\overset{71}{C}-CH=CH-CO_2CH_3$$

(12)

Coupling constants for ^{13}C–^{1}H have been measured, and are found to be very useful in making assignments[44, 45]. As seen in Scheme 1, the coupling of alkyl and acetylenic hydrogens with the carbons of a polyyne chain decreases regularly with increasing number of intervening bonds. Olefinic protons on the other hand show an irregular pattern as indicated[45].

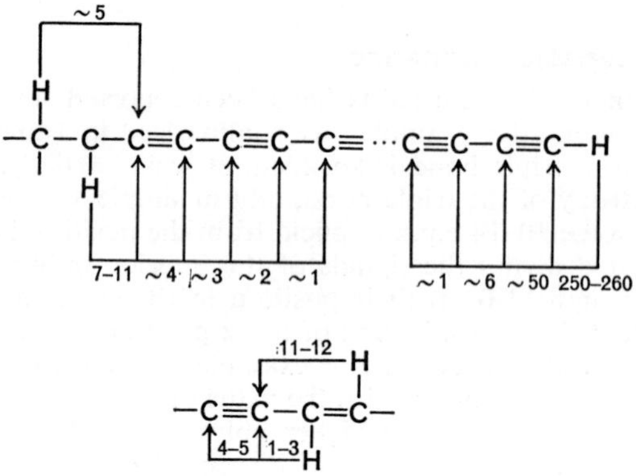

SCHEME 1. Coupling of ^{1}H with acetylenic ^{13}C (values in Hz).

G. Mass Spectrometry

The principal mass spectrometric studies of polyacetylenes have been carried out with naturally occurring compounds, and the technique has proved to be a powerful one for structure determination in this area[1]. A significant molecular ion peak is generally observed. Fragmentation by cleavage of the polyyne chain between triple bonds is an unfavourable process[47]. There is a great tendency to form highly unsaturated hydrocarbons by loss of hydrogen, and loss of C_2H_2 is also a predominant type of fragmentation[48].

H. Photoelectron Spectroscopy

Ultraviolet photoelectron spectra for some of the lower polyynes have been measured, and the π ionization potentials obtained from the spectra are summarized in Table 10[49, 50]. The first ionization potential ($I_{v,1}$), which corresponds to removal of

TABLE 10. Vertical ionization potentials of polyacetylenes from photoelectron spectroscopy

Compound	$I_{v,1}$ (eV)	$I_{v,2}$ (eV)	$I_{v,3}$ (eV)	Reference
HC≡CH	11·40	—	—	49
$H(C{\equiv}C)_2H$	10·17	12·62	—	49
$H(C{\equiv}C)_3H$	9·50	11·55	12·89	50
$CH_3C{\equiv}CH$	10·37	—	—	49
$CH_3(C{\equiv}C)_2H$	9·51	12·01	—	50
$CH_3C{\equiv}CCH_3$	9·59	—	—	50
$CH_3(C{\equiv}C)_2CH_3$	8·91	11·46	—	50
$CH_3(C{\equiv}C)_3CH_3$	8·60	10·63	12·10	49

an electron from the highest occupied π molecular orbital, decreases with increasing conjugation, as can be seen by comparing the values for the series acetylene, butadiyne, hexatriyne or 2-butyne, 2,4-hexadiyne, 2,4,6-octatriyne. Replacement of the terminal hydrogens by methyl groups causes a reduction in ionization potential as a result of inductive and/or hyperconjugative interactions.

Photoelectron spectra have been measured for other, more complex polyynes[51, 52]. The values obtained for a large number of polyynes can be correlated in a simple fashion in terms of a model based on linear combination of bonding orbitals[50, 51, 53].

III. SYNTHESIS

A. Oxidative Coupling

1. Copper derivatives

The oxidative coupling of acetylenes, reported originally by Glaser and subsequently modified by others, has played a major role in the development of polyacetylene chemistry. The reaction has been reviewed comprehensively, and these reviews should be consulted for details and lists of earlier references[3, 54–56].

Glaser's original report described the air oxidation of copper phenylacetylide to give diphenylbutadiyne[57]. Subsequently it was found that the same reaction could be accomplished more conveniently by simply bubbling oxygen or air through a solution containing the alkyne, CuCl and NH_4Cl (equation 3). In the Hay modification,

$$2RC{\equiv}CH+\tfrac{1}{2}O_2 \xrightarrow[NH_4Cl]{CuCl} R(C{\equiv}C)_2R+H_2O \qquad (3)$$

the complex of copper(I) chloride with *N,N,N′,N′*-tetramethylethylenediamine (TMEDA) is used as catalyst with superior results overall[58].

Copper(II) is the actual oxidant in these reactions, the oxygen simply serving to regenerate this ion from the Cu(I) state, as summarized by equations (4) and (5). In

$$2RC{\equiv}CH+2Cu(II) \longrightarrow R(C{\equiv}C)_2R+2H^++2Cu(I) \qquad (4)$$

$$2Cu(I)+\tfrac{1}{2}O_2+H_2O \longrightarrow 2Cu(II)+2OH^- \qquad (5)$$

the Eglinton modification advantage is taken of this fact, and coupling is accomplished without need for air or oxygen by using excess copper(II) acetate in pyridine–methanol[55, 56].

Because copper(II) is such a mild oxidant, coupling can be accomplished satisfactorily with terminal alkynes which contain almost any other type of functional group. The reaction is not limited to monoalkynes, and has been widely used for converting diynes to tetraynes, triynes to hexaynes, etc. Many diarylpolyynes have been synthesized by sequences utilizing oxidative coupling at some stage, and typical examples are illustrated in Scheme 2[31].

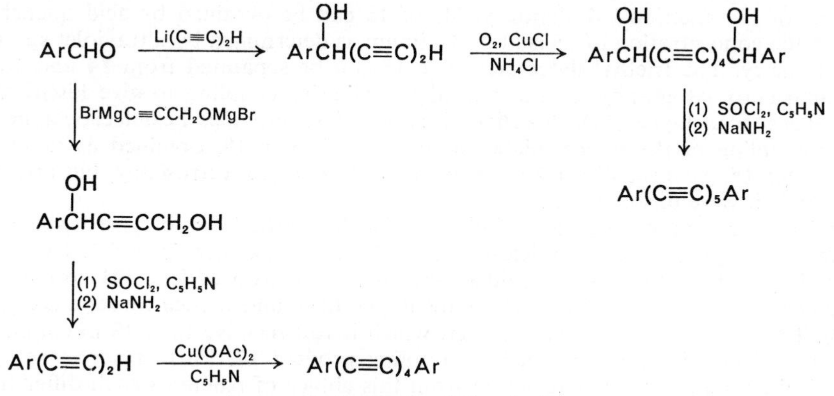

SCHEME 2. Synthesis of diarylpolyynes.

With few exceptions, oxidative coupling cannot be directly used for the synthesis of terminal polyacetylenes, because the products are more reactive toward further coupling than their precursors and uncontrolled chain growth occurs. The use of the triethylsilyl group as a protective group, however, has provided an elegant solution to this problem, and has permitted the synthesis of long-chain polyynes not accessible by other routes[2, 14, 59]. This group is stable under the conditions of the Hay coupling, and can be readily removed after the coupling is accomplished by treatment with dilute alkali at room temperature. The synthesis of 1,3,5,7-octatetrayne (**16**) shown in Scheme 3 illustrates a typical sequence[2]. Butadiynyl(triethyl)silane (**13**),

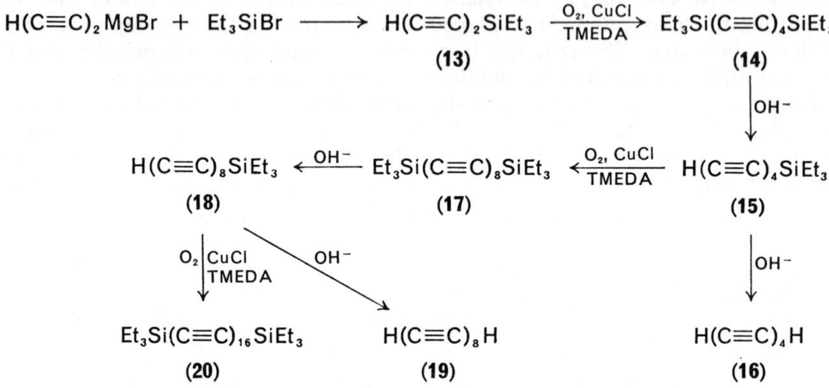

SCHEME 3

readily obtained from $H(C{\equiv}C)_2MgBr$ and Et_3SiBr as shown, undergoes coupling to give bis(triethylsilyl)octatetrayne (**14**) in 80% yield. Removal of the blocking groups is accomplished by brief treatment with very dilute aqueous methanolic alkali, and octatetrayne (**16**) is readily separated from the other cleavage products, Et_3SiOH and $(Et_3Si)_2O$, by column chromatography. The cleavage is quantitative, and the tetrayne **16** is obtained from the chromatographic separation as a solution in petroleum ether.

An added bonus to this synthetic scheme arises from the fact that the rate of cleavage of the bis-silyl derivative **14** is twice that of the monosilyl derivative **15**, and consequently substantial concentrations of the latter are present at intermediate stages of the reaction. Moderate yields of **15** can be obtained by acid quenching when the concentration of **15** is at a maximum as determined by ultraviolet spectrophotometry. The triethylsilyloctatetrayne **15** can be separated from **14** and **16** by column chromatography, and subjected to oxidative coupling to give bis(triethylsilyl)hexadecaoctayne (**17**). Alkaline cleavage of **17** provides hexadecaoctayne **19**. Hay coupling of the intermediate monosilyl derivative **18**, obtained as described above for **15**, gave the dimer **20** as shown by ultraviolet spectroscopy, but attempts to purify it were unsuccessful.

Mixed couplings can be successfully used for the synthesis of polyynes with an odd number of triple bonds if a judicious choice of reaction partners is made[2]. Thus, Hay coupling of **15** with a twelve-fold excess of triethylsilylacetylene (**21**) provides a practical route to bis(triethylsilyl)decapentayne (**23**), and thence to pentaacetylene (**24**). The use of the large excess of **21**, which is less reactive than **15** in couplings, serves to minimize the symmetrical coupling product **17**. Furthermore, the symmetrical coupling products resulting from this choice of reactants each differ from the desired product by three -yne units, and this facilitates the chromatographic separation of pure **24**. Unsubstituted polyynes, $H(C{\equiv}C)_nH$, with $n = 4$–10, and 12, have been synthesized through the use of these procedures, and their ultraviolet spectra have been recorded. They are not stable at room temperature when free of solvent, and all work was performed with dilute solutions.

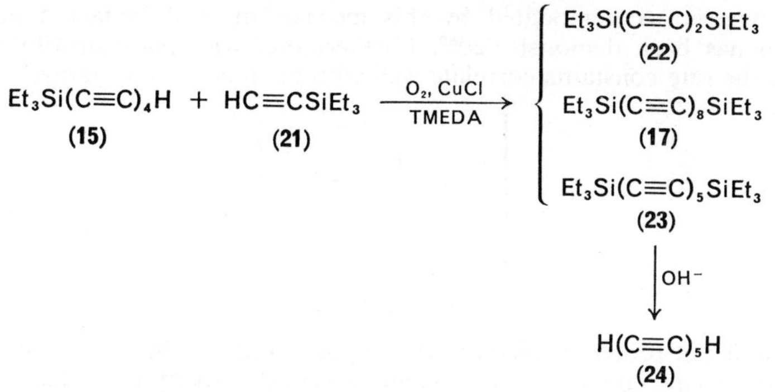

Use of the triethylsilyl derivatives, $Et_3Si(C{\equiv}C)_nH$ ($n = 1, 2, 4$), as one component in mixed oxidative couplings permits extension of terminal polyyne chains by up to four -yne units in a single step, as illustrated by the synthesis of the dodecayne **28** [14]. The mixture of products obtained by Hay coupling of excess *t*-butyldiacetylene (**25**) with the tetrayne **26** was treated with base and then chromatographed to give the hexayne **27**. Only traces of **28** were obtained from **27** by Hay coupling, but somewhat better results were obtained with the Eglinton technique and **28** was obtained as a red-brown crystalline solid which decomposed at *ca.* 50 °C.

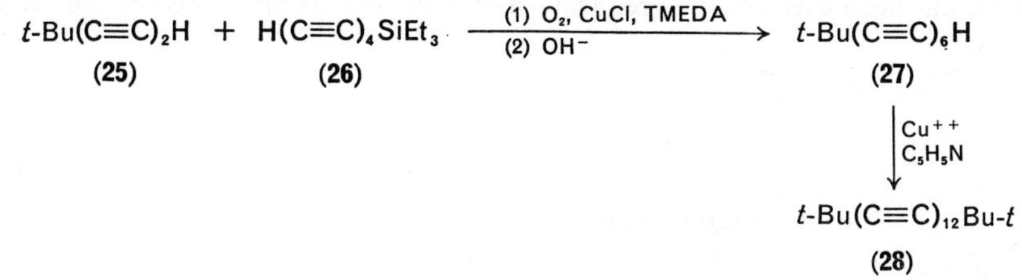

Several studies of the kinetics and effects of structure on reactivity lend support to a mechanism of oxidative coupling of the type first proposed by Bohlmann and coworkers[58, 60–63]. The rate is second order with respect to Cu(II) and alkyne, and varies inversely with $[H^+]^2$. This is interpreted in terms of rapid steps involving displacement of a solvent molecule or other ligand from the coordination sphere of Cu(II) by an alkyne molecule, followed by acid dissociation of the coordinated alkyne to give an acetylide complex. In the rate-determining step, copper(II) is reduced and simultaneously the alkynyl groups are coupled. These steps are summarized in equations (6), (7) and (8), where L represents a ligand—solvent, for

$$RC{\equiv}CH + Cu(II){\cdot}L \underset{}{\overset{K}{\rightleftharpoons}} Cu(II){\cdot}RC{\equiv}CH + L \qquad (6)$$

$$Cu(II){\cdot}RC{\equiv}CH \overset{K}{\rightleftharpoons} Cu(II){\cdot}RC{\equiv}\bar{C}{:} + H^+ \qquad (7)$$

$$2Cu(II){\cdot}RC{\equiv}\bar{C}{:} \xrightarrow{k} R(C{\equiv}C)_2R + 2Cu(I) \qquad (8)$$

example—initially coordinated to Cu(II). No attempt is made to specify the structure of the complexes, but a dimeric structure (**29**) involving bridging acetylide groups is an attractive possibility[60]. Apparently the concentration of the complex is very low because e.p.r. studies have failed to detect it in solutions in which oxidative coupling was occurring[64]. The increased rate of coupling which accompanies increased acidity of the alkyne is accommodated by this mechanism, and in fact a quantitative correlation has been demonstrated[62]. Furthermore, with *para*-substituted phenylacetylenes the rate constants correlate well with Hammett σ constants.

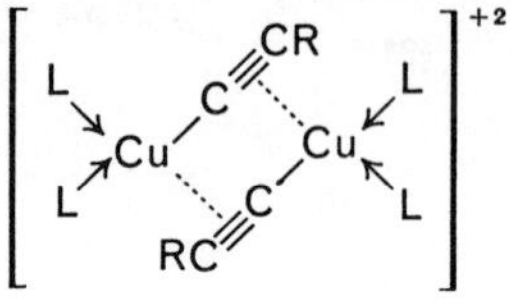

(29)

The coupling of propargyl alcohol with copper(II) acetate in pyridine constitutes a notable exception to the kinetic behaviour described above[65], but it has been shown that the behaviour is peculiar to this particular system[62]. Thus 'normal' kinetic behaviour is found for $Cu(OAc)_2$-pyridine coupling of acetals and ethers of propargyl alcohol, and for propargyl alcohol itself when Cu(II) in aqueous ammonia is used[61].

The kinetics of oxidative polymerization of 1,8-nonadiyne have been studied using oxygen and homogeneous catalysts derived from copper(I) chloride and tertiary amines[66]. A mechanism of the same type as that described above for dimerization was proposed for the polymerization.

Oxidative coupling of compounds having two terminal ethynyl groups per molecule has been used for the preparation of polymers, as illustrated by equation (9)

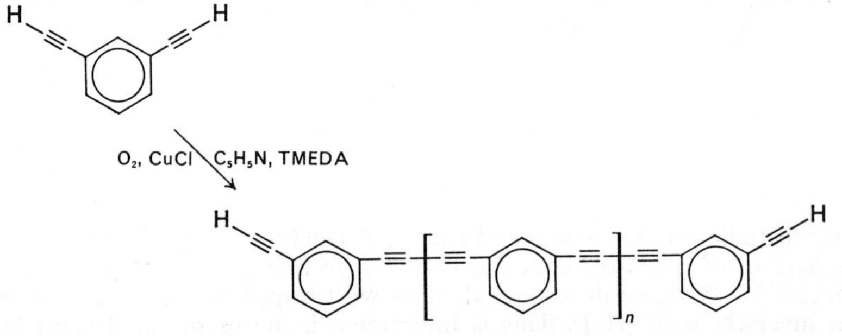

(9)

for the case of *m*-diethylnylbenzene[67]. The polymer, which contains 70 or more monomer units, is nearly colourless and can be formed into a tough transparent film. The corresponding *ortho* and *para* isomers have also been polymerized and copolymerized[67–70].

2. Other organometallic derivatives

The oxidative coupling of other organometallic derivatives of alkynes has been reported. For example, when metal carbonyl derivatives **30**, formed by mixing

$$\underset{\textbf{(30)}}{\mathrm{Li[PhC{\equiv}CM(CO)}_n]} \xrightarrow[\mathrm{MeOH}]{\mathrm{Br_2}} \mathrm{Ph(C{\equiv}C)_2Ph}$$

lithium phenylacetylide with the metal carbonyl, are treated with bromine or iodine, diphenylbutadiyne is obtained in good yield[71]. Oxidation of complex aluminium acetylides **31** with copper(II) bromide provides the corresponding diynes[72].

$$\underset{\textbf{(31)}}{LiAl(C{\equiv}CR)_4} \xrightarrow{CuBr_2} R(C{\equiv}C)_2R$$

Among the many methods that have been reported, the oxidation of dialkyldialkynylborates with iodine is particularly interesting because of its potential in synthetic work[73, 74]. Treatment of bromodicyclohexylborane (**32**) with an alkynyllithium gives the borate **33**, which without being isolated, is cooled to −78 °C and treated with iodine to give the symmetrical diyne **34** in good yield[73]. Because of the low migratory aptitude of the cyclohexyl group, only minor amounts of the monoalkyne **35** are formed.

$$\underset{\textbf{(32)}}{(C_6H_{11})_2BBr} \xrightarrow{RC{\equiv}CLi} \underset{\textbf{(33)}}{Li^+[(C_6H_{11})_2B(C{\equiv}CR)_2]^-} \xrightarrow{I_2} \underset{\textbf{(34)}}{R(C{\equiv}C)_2R} + \underset{\textbf{(35)}}{C_6H_{11}C{\equiv}CR}$$

It is possible to prepare unsymmetrical conjugated diynes by modifying this procedure[74]. Treatment of 1-alkynyldisiamylborane (**36**) with lithium alkyne gives the corresponding borate complex **37**, and when this reacts with iodine at −78 °C, the diyne **38** is obtained in good yield. For example, 3,5-dodecadiyne (**38**; R = Et,

$$\underset{\textbf{(36)}}{Sia_2BC{\equiv}CR} \xrightarrow{R'C{\equiv}CLi} \underset{\textbf{(37)}}{Li[Sia_2B(C{\equiv}CR)(C{\equiv}CR')]} \xrightarrow{I_2} \underset{\textbf{(38)}}{R(C{\equiv}C)_2R'}$$

$$Sia = (CH_3)_2CHCH(CH_3)^-$$

R′ = *n*-Hex) is obtained in 95% yield, and 1-phenyl-1,3-decadiyne (**38**; R = Ph, R′ = *n*-Hex) is obtained in 79% yield. The method holds promise as a useful alternative to the Cadiot–Chodkiewicz synthesis, especially for purely aliphatic diynes, which are obtained in low yields by this route.

B. Cadiot–Chodkiewicz Coupling

The coupling of a terminal alkyne with a 1-bromoalkyne in the presence of a copper(I) salt and an amine base (B), referred to as the Cadiot–Chodkiewicz coupling[75], is of particular synthetic importance because of the facile route it provides to unsymmetrical polyacetylenes with an even or odd number of triple bonds (equation 10). The reaction has been reviewed and these reviews should be

$$RC{\equiv}CH + BrC{\equiv}CR' + B \xrightarrow{Cu^+} R(C{\equiv}C)_2R' + BH^+Br^- \qquad (10)$$

consulted for details, discussions of mechanism, and complete lists of earlier references[54, 56, 75].

The reaction is carried out by slowly adding the 1-bromoalkyne to a solution containing the terminal alkyne, amine, copper(I) chloride and hydroxylamine hydrochloride. The amine, usually ethylamine, is used in excess, e.g. 1·8 moles/mole of alkyne, and catalytic quantities (1–5 mol %) of copper(I) chloride are used. One of the side-reactions is the self-coupling of the bromoalkyne induced by Cu(I) which in turn is oxidized to Cu(II) (equation 11). The hydroxylamine salt serves to reduce the copper back to the cuprous state.

$$2RC{\equiv}CBr + 2Cu^+ \longrightarrow R(C{\equiv}C)_2R + 2Cu^{2+} + 2Br^- \qquad (11)$$

The reaction has found limited use for the direct synthesis of terminal polyacetylenes, as illustrated by the formation of phenylhexatriyne (70%) from butadiyne and bromophenylacetylene (equation 12)[76]. The concomitant formation of diphenyloctatetrayne (30%) in this reaction illustrates the major drawback to this route, i.e. further coupling of the initial product.

$$PhC{\equiv}CBr + H(C{\equiv}C)_2H \xrightarrow[EtNH_2]{CuCl} Ph(C{\equiv}C)_3H \longrightarrow Ph(C{\equiv}C)_4Ph \quad (12)$$

The most general route to terminal polyynes involves the use of a protecting group which can be readily removed after the coupling and, as in oxidative couplings, the triethylsilyl group is admirably suited to this purpose[14, 59, 77–79]. Thus, 1-phenyl-4-triethylsilyl-1,3-butadiyne (**40**) is obtained in 50% yield from phenylacetylene and bromoethynyltriethylsilane (**39**)[77]. The silyl derivative is converted quantitatively to the free diyne **41** by alkali; repetition of the coupling and cleavage yields 1-phenyl-1,3,5-hexatriyne (**43**). Alternatively, **42** can be acquired directly by coupling

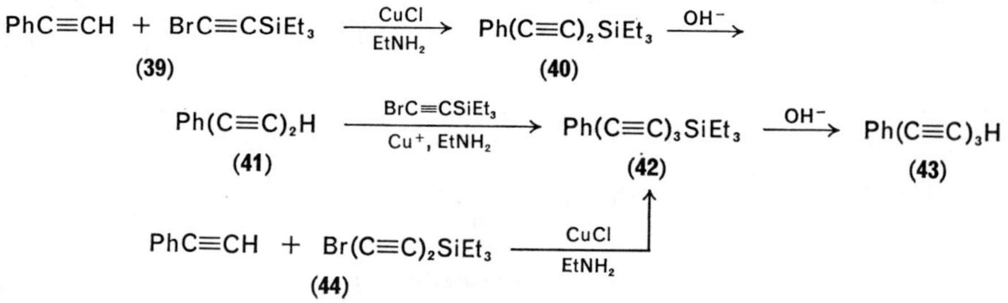

phenylacetylene with 1-bromo-4-triethylsilyl-1,3-butadiyne (**44**), a procedure made more attractive by the recent development of a practical synthesis of **44** [78]. Attempts to couple phenylethynyl bromide with ethynyltriethylsilane were unsuccessful[59].

C. Coupling of Terminal Alkynes with Propargyl, Vinyl and Allenyl Halides

Terminal alkynes couple with propargyl-type halides (**45**) in the presence of copper(I) chloride and ammonia or an amine[80]. Two types of coupling products have been observed, allenynes **46** and 1,4-diynes **47**. When R^3 is hydrogen, the

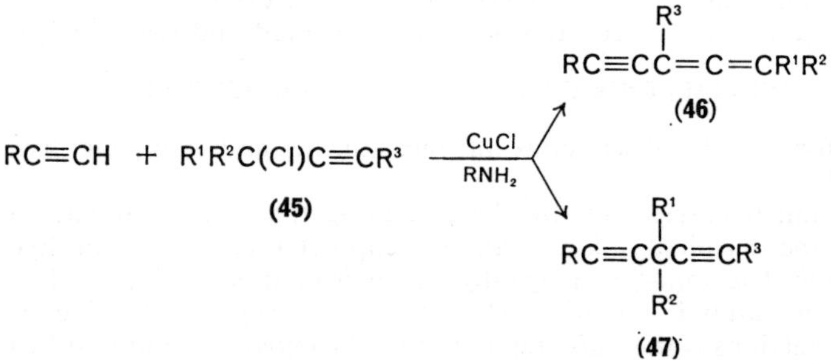

principal product is the allenyne **46**, but if R^3 is an alkyl group the diyne **47** predominates. Thus the coupling of 2-methyl-3-butyn-2-ol (**48**) with 3-chloro-3-methyl-1-butyne in the presence of *t*-butylamine gives **49** in 70% yield. Under the same conditions the product obtained in 60% yield from the coupling of **48** with 1-chloro-2-butyne is almost entirely (95%) the diyne **50** [80].

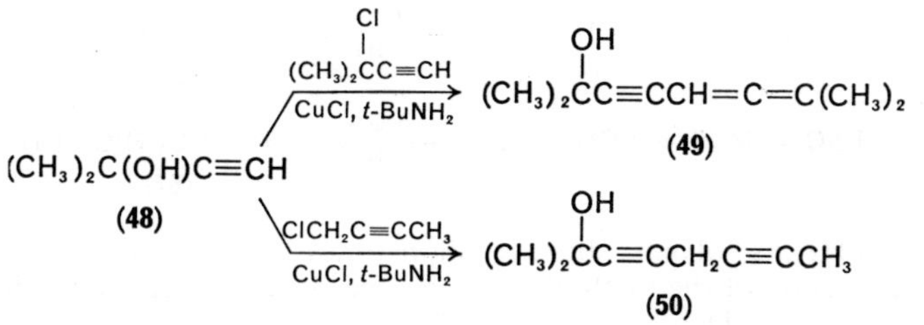

A limited number of enediynes have been prepared by the reaction of copper acetylides with diiodoethylene in pyridine or DMF[81, 82]. For example, *trans*-1,6-diphenyl-3-hexene-1,5-diyne (**52**) is obtained in 90% yield when copper phenylacetylide (**51**) and *trans*-1,2-diiodoethylene are warmed in pyridine. It was reported

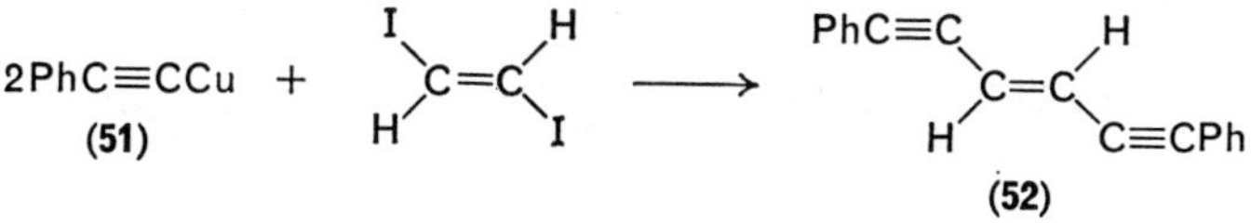

that the tetraethynyl derivative, $(PhC{\equiv}C)_2C{=}C(C{\equiv}CPh)_2$, is obtained from **51** and tetraiodoethylene[81], but apparently this is not correct[83, 94].

Allenediynes are obtained from the coupling of allenic bromides with terminal diynes[79]. Thus from the bromide **53** and butadiynyl(trimethyl)silane (**13**) in the presence of copper(I) bromide and tri-*n*-butylamine, the silylated derivative **54** was obtained in 70% yield. Removal of the silyl group by treatment with dilute methanolic alkali for 10 s afforded the free allenediyne **55** in 84% yield.

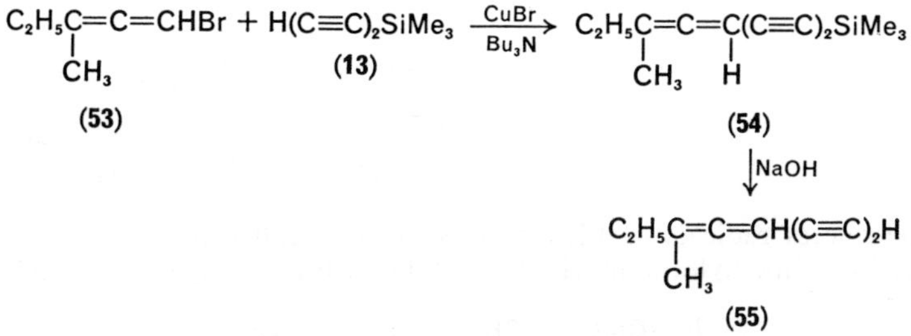

D. Couplings Involving Grignard Reagents

The coupling of 1-alkynyl Grignard reagents with propargyl halides, promoted by copper chloride, provides the most general route to 1,4-diynes, often referred to as 'skipped diynes'[84]. Because of the great tendency of 1,4-diynes to rearrange in the presence of base, synthetic methods involving strongly basic reactants such as $RC{\equiv}CNa$ or basic conditions for work-up are not satisfactory. For this reason attempts by earlier workers to synthesize the parent member of the series, 1,4-pentadiyne (**56**), led to 1,3-pentadiyne instead, and the first satisfactory synthesis of pure 1,4-pentadiyne was not reported until 1969[85]. Optimum conditions, as developed for the synthesis of 1-phenyl-1,4-pentadiyne (**57**), include the use of CuCl

$$HC{\equiv}CMgBr + BrCH_2C{\equiv}CH \xrightarrow[\text{THF}]{\text{CuCl}} \underset{\textbf{(56)}}{HC{\equiv}CCH_2C{\equiv}CH}$$

$$PhC{\equiv}CMgBr + BrCH_2C{\equiv}CH \xrightarrow[\text{THF}]{\text{CuCl}} \underset{\textbf{(57)}}{PhC{\equiv}CCH_2C{\equiv}CH}$$

as promoter, THF as solvent, short reaction times and neutral conditions during work-up[86]. Extension of the method to the synthesis of skipped triynes is illustrated by the synthesis of 5,8,11-hexadecatriyne (**58**)[87].

$$2BuC{\equiv}CMgBr + BrCH_2C{\equiv}CCH_2Br \xrightarrow{\text{CuCl}} \underset{\textbf{(58)}}{BuC{\equiv}CCH_2C{\equiv}CCH_2C{\equiv}CBu}$$

A related reaction (13), which leads to products in which the skipped carbon is quaternary as in **59**, involves tertiary propargylic chlorides[88].

$$PhC{\equiv}CMgBr + CH_3C{\equiv}CCCl(CH_3)_2 \xrightarrow[\text{HgCl}_2]{\text{CuCl}} \underset{\textbf{(59)}}{PhC{\equiv}CC(CH_3)_2C{\equiv}CCH_3} \qquad (13)$$

Tertiary acetylenic chlorides undergo self-coupling when they are treated with methylmagnesium bromide and cobalt(II) chloride, i.e. radical-generating conditions[89]. Under these conditions, 4-chloro-4-methyl-2-pentyne (**60**) gave mainly the diyne **62** along with smaller amounts of another hydrocarbon tentatively identified

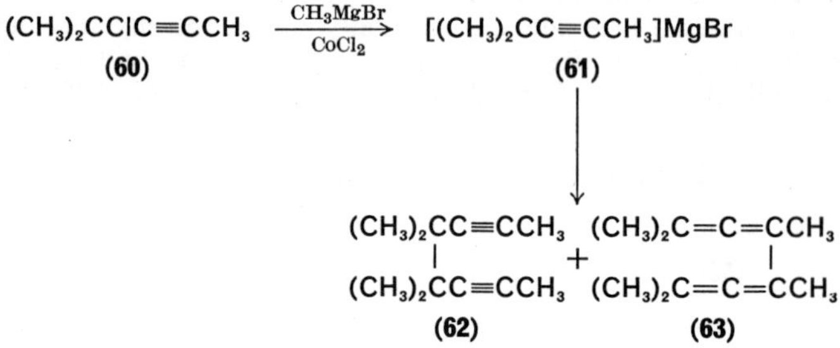

as **63**. The role of radicals in this type of process is called into question by studies of the reaction of methyllithium with **64**[89, 90]. The self-coupling product **65** is obtained

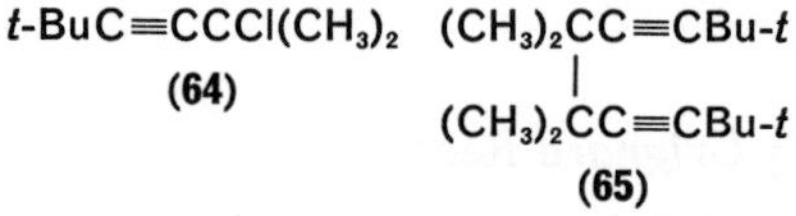

in *ca.* 15% yield, along with other cross-coupling and elimination products. Failure to detect ethane as a product, absence of CIDNP, absence of products containing an allene grouping and lack of dependence of product composition on the order of mixing the reactants are taken as evidence against a radical mechanism[90]. It is suggested that the reaction may involve preliminary halogen–metal exchange followed by attack of the resulting carbanion on a second molecule of halide. The same type of coupling product (**67**) is formed when 6-chloro-6-methyl-2,4-heptadiyne (**66**) is treated with methyl- or ethylmagnesium bromide (prepared from sublimed

$$(CH_3)_2CCl(C{\equiv}C)_2CH_3 \quad \textbf{(66)} \qquad \begin{matrix}(CH_3)_2C(C{\equiv}C)_2CH_3\\ |\\ (CH_3)_2C(C{\equiv}C)_2CH_3\end{matrix} \quad \textbf{(67)}$$

Mg) *without* added cobalt(II) chloride[91]. Ethane is formed when the methyl Grignard is used, and approximately equal amounts of ethylene and ethane are formed with the ethyl Grignard reagent. A functional exchange mechanism was proposed which involves radicals complexed with magnesium of the Grignard reagent or magnesium halide.

Compounds containing four alkynyl groups attached to a single atom (C, Si, Ge) have been obtained by the coupling of alkynyl Grignard reagents with the appropriate tetrahalide[92].

$$4CH_3C{\equiv}CMgBr + MX_4 \longrightarrow M(C{\equiv}CCH_3)_4$$

$$M = C, Si, Ge$$

E. Other Couplings and Dimerizations

Tetraethynylethanes **69** are obtained in fair-to-good yields when the bromodiynes **68** are treated with potassium iodide in acetone[93]. The reaction also occurs, but in lower yield, when magnesium is used as the reducing agent. The tetraethynylethanes

$$\underset{\textbf{(68)}}{2(RC{\equiv}C)_2CHBr} \xrightarrow{KI} \underset{\textbf{(69)}}{(RC{\equiv}C)_2CHCH(C{\equiv}CR)_2} + I_2$$

$$R = Me, Ph, t\text{-}Bu, Et_3Si$$

69 can be converted to the corresponding ethylenes **70** by oxidation of the lithium derivative with *t*-butyl hypochlorite[94]. These compounds provide good examples of cross-conjugated systems with a planar arrangement of the π-electron skeleton.

When 3-bromo-1,5-diphenyl-1,4-pentadiyne (**68**, R = Ph) is treated with potassium *t*-butoxide, a complex mixture of products, **70**, **71** and **72** (R = Ph) is

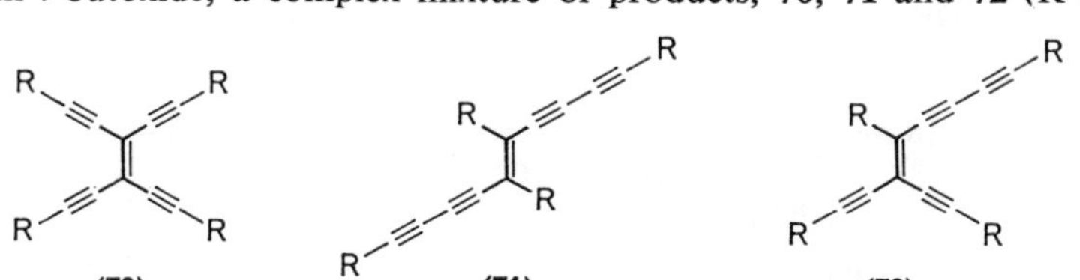

(70) **(71)** **(72)**

formed[83]. These products can be rationalized in terms of the dimerization of the carbene intermediate **73** (R = Ph)[94]. Similar mixtures are obtained by dimerization of carbenes generated by the pyrolysis of the tosylhydrazone derivatives **74**

$$[R{-}\ddot{C}{-}C{\equiv}C{-}C{\equiv}CR \longleftrightarrow RC{\equiv}C{-}\ddot{C}{-}C{\equiv}CR \longleftrightarrow RC{\equiv}C{-}C{\equiv}C{-}\ddot{C}R]$$

(73)

$$\begin{matrix}N{-}N(Li)Ts\\ \|\\ RC{\equiv}C{-}C{-}C{\equiv}CR\end{matrix}$$

(74) R = *t*-Bu, Me_3Si

(R = *t*-Bu, Me_3Si)[95]. When the pyrolysis is carried out in the presence of olefins, the addition is non-stereospecific, signifying the presence of the triplet carbene (**75**).

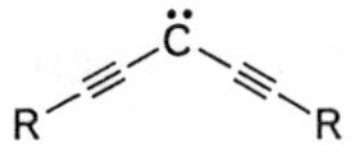

(**75**)

F. Elimination

The synthesis of acetylenes by elimination reactions has been reviewed recently[96], and only a survey of the more important methods that can be applied to the synthesis of polyacetylenes will be presented here.

The development of the dehydrohalogenation of 1,4-dichloro-2-butyne (**76**) as an efficient synthesis of butadiyne paved the way for the synthesis of a wide range of polyacetylene compounds in the 1950s[3, 97]. The reaction can be accomplished by heating the dichloride **76** with alkali, and the butadiyne, which is obtained in yields as high as 98%, is condensed in a cold trap. Although the diyne can be stored at low temperatures, safe practice calls for its use soon after it is prepared. Alternatively, the dehydrohalogenation of **76** can be accomplished with sodium amide in liquid ammonia, and in this case the mono- or disodium salt of butadiyne is obtained, depending on the proportion of base used. These salts may be alkylated directly, condensed with carbonyl compounds, etc., as illustrated in Scheme 4[97–99].

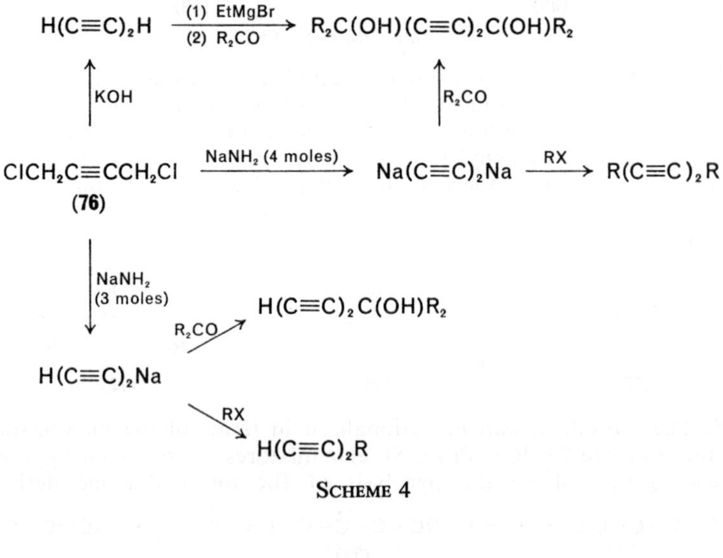

SCHEME 4

The synthetic sequence used for butadiyne has been adapted to the preparation of higher polyynes, $H(C{\equiv}C)_nH$ (**79**, $n = 3, 4, 5$)[15, 26, 100]. The diol **77**, obtained by condensation of the appropriate polyyne with formaldehyde, or preferably by oxidative coupling when $n = 5$, is converted to the corresponding dichloride **78** with thionyl chloride and pyridine. Low-temperature dehydrohalogenation with sodium amide gives the polyyne **79** with one additional triple bond. Yields decrease rapidly with increasing number of triple bonds, and drop to approximately 1% with decapentayne, $H(C{\equiv}C)_5H$.

$$\underset{(\mathbf{77})}{HOCH_2(C{\equiv}C)_{n-1}CH_2OH} \xrightarrow[C_5H_5N]{SOCl_2} \underset{(\mathbf{78})}{ClCH_2(C{\equiv}C)_{n-1}CH_2Cl} \xrightarrow{NaNH_2} \underset{(\mathbf{79})}{H(C{\equiv}C_nH}$$

As with butadiyne, the sodium salts of the polyynes **79** which are present in the reaction mixture after dehydrohalogenation can be alkylated or condensed with carbonyl compounds. The resulting derivatives are more stable and can usually be obtained in somewhat higher yields than the parent polyynes, but the improvement is only slight in the pentaacetylene case, where, for example, the dimethyl derivative, $CH_3(C{\equiv}C)_5CH_3$, is obtained in 3% yield[15].

By the use of di-secondary glycols it is possible to obtain disubstituted polyynes with greater numbers of conjugated triple bonds. The classical work of Bohlmann on the synthesis of di-*t*-butylpolyynes involved extensive use of this approach[12], as illustrated in Scheme 5 for the heptayne **81**. In this case the much milder base sodium

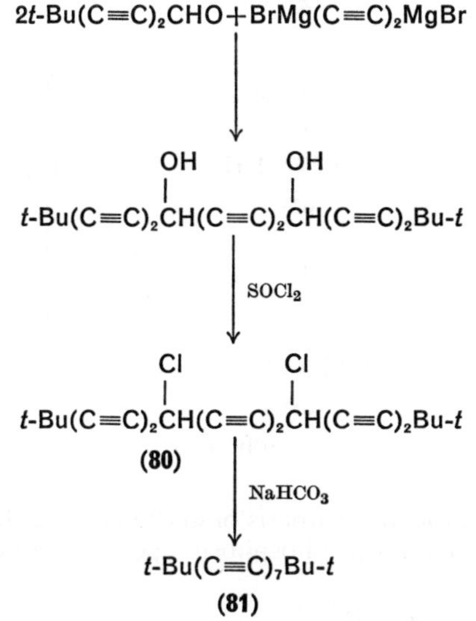

SCHEME 5

bicarbonate was capable of effecting dehydrohalogenation of the dichloride **80**. The heptayne **81** was obtained as a yellow crystalline solid which decomposed slowly above 150 °C. Through combinations of dehydrohalogenation and oxidative coupling Jones and coworkers were able to extend the synthesis to the decayne **84** as summarized in Scheme 6[13]. Dehydrohalogenation of the dichloride **82** was accomplished by chromatography over alkaline alumina, and oxidative coupling of the resulting pentayne **83** yielded the decayne **84**. Interestingly, treatment of the dichloride **82** itself with copper(II) acetate and pyridine gave the decayne **84** directly.

Sequences involving dehydrohalogenation and oxidative coupling have been used extensively in the synthesis of diarylpolyynes, $Ar(C{\equiv}C)_nAr$[16, 31, 101].

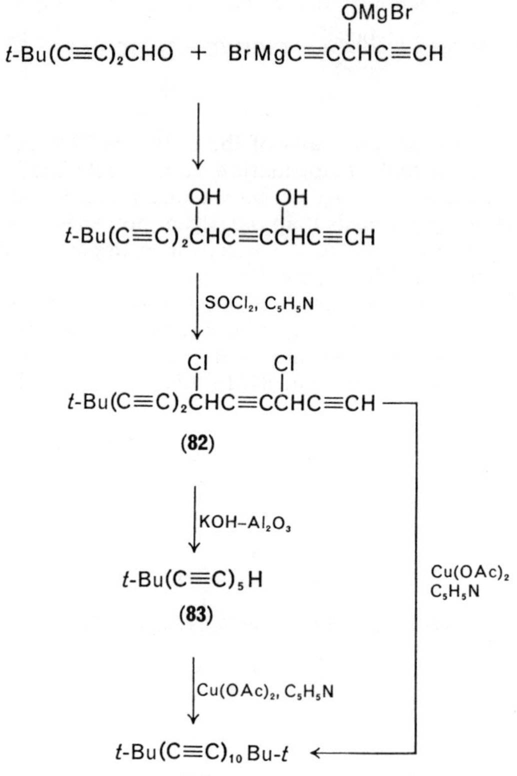

(84)

SCHEME 6

A route that is useful for the synthesis of arylbutadiynes involves treatment of the acetoacetyl derivative **85** with phosphorus pentachloride followed by sodium amide[14, 102, 103].

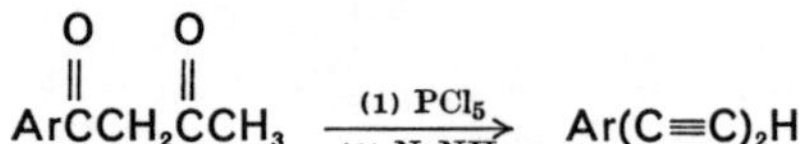

(85)

G. Pyrolysis of Hydrocarbons

Diacetylene is formed as a significant by-product in the commercial synthesis of acetylene by the pyrolysis of methane and other hydrocarbons[3]. Smaller amounts of triacetylene are also formed[104]. Procedures have been devised for removing and recovering the diacetylene[3, 105].

Vinylacetylene and diacetylene are primary products of the pyrolysis of acetylene itself; in the range 700–1200 K, vinylacetylene is the initial product, while diacetylene is the primary molecular product in the 1600–2400 K range[106, 107]. Many other hydrocarbons are formed as secondary products, e.g. benzene, methane, ethylene and butadiene. A free-radical chain mechanism which accounts for the low- and high-temperature behaviour has been proposed[108]. A bimolecular disproportionation of two acetylene molecules giving a vinyl **86** and an ethynyl radical **87** is proposed for the initiation step. It has also been proposed that the initiation step in the high-temperature process involves unimolecular dissociation of acetylene to $^{\bullet}C_2H$ and H [109]. Addition of the ethynyl radical **87** to acetylene giving **88**, followed by loss of a hydrogen atom, furnishes butadiyne (**89**). The hydrogen atom also participates in the chain propagation as indicated in equation (17). At lower temperatures,

$$2C_2H_2 \longrightarrow \underset{(\mathbf{86})}{C_2H_3^{\bullet}} + \underset{(\mathbf{87})}{C_2H^{\bullet}} \tag{14}$$

$$\mathbf{87} + C_2H_2 \longrightarrow \underset{(\mathbf{88})}{C_4H_3^{\bullet}} \tag{15}$$

$$C_4H_3^{\bullet} \longrightarrow H^{\bullet} + \underset{(\mathbf{89})}{C_4H_2} \tag{16}$$

$$H^{\bullet} + C_2H_2 \longrightarrow H_2 + \mathbf{87} \tag{17}$$

addition of the vinyl radical (**86**) to acetylene leads ultimately to vinylacetylene, but in the higher temperature range **86** undergoes dissociation to acetylene and a hydrogen atom instead.

A procedure has been described for the continuous synthesis of diacetylene by passing acetylene through an electrical discharge[110]. Yields as high as 23% and acetylene through-puts of several grams per minute can be realized. Acetylene is ordinarily converted into polymeric material by glow discharges, but when the reaction tube is filled with specially treated glass rings, only volatile compounds are formed[111]. Vinylacetylene and butadiyne are the principal gaseous products, and a mechanism has been proposed for their formation which involves cyclobutadiene as an intermediate, as shown in equation (18).

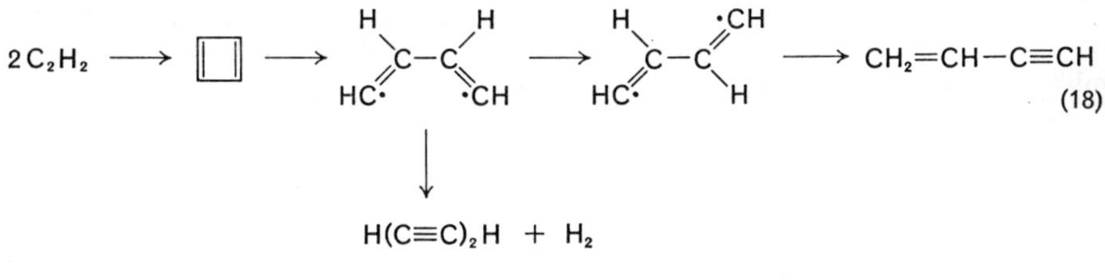

(18)

IV. REACTIONS

A. Addition of Nucleophilic Reagents

Nucleophilic attack occurs more readily on conjugated polyynes than on simple alkynes as evidenced not only by greater rates of reaction, but also by the addition of nucleophilic reagents that fail to react with simple alkynes. Thus cyanide ion, alkyllithium reagents, malonic ester and lithium aluminium hydride add readily to conjugated tetraynes and pentaynes[112].

I. Oxygen nucleophiles

The base-catalysed addition of alcohols to butadiyne occurs under much milder conditions than are required for similar additions to acetylene. The use of a dilute solution of KOH in the alcohol at 60–120 °C is a common procedure. The products formed first are 1-alkoxy-1-buten-3-ynes (**90**), but a second molecule may add, especially at higher temperatures, to give an acetal of 2-butynal **91**, or in some cases mixtures of **91**, **92** and **93** [3, 113].

$$HC{\equiv}CCH{=}CHOR \quad CH_3C{\equiv}CCH(OR)_2 \quad CH_2{=}C{=}CHCH(OR)_2$$

(**90**) (**91**) (**92**)

$$HC{\equiv}CCH_2CH(OR)_2$$

(**93**)

The stereoselectivity of addition depends on the nature of the alcohol and the conditions, including the nature of the solvent[113]. Thus, in the presence of 2% KOH in excess alcohol as solvent, methanol or ethanol adds exclusively *anti* to butadiyne giving **94** (R = Me, Et). A sample of the *syn* adduct **95** (R = Et) failed to isomerize when it was heated at 150 °C with dilute alkali, thus ruling out the possibility of *syn* addition followed by isomerization.

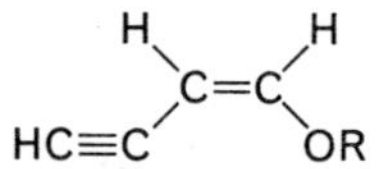

(**94**)

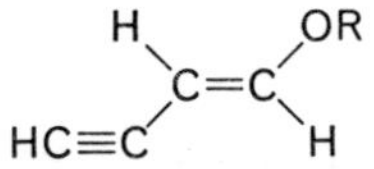

(**95**)

With *n*-propyl alcohol, some *syn* addition occurs and the products **94** and **95** (R = Pr) are formed in the ratio 87 : 13. This same ratio prevails approximately for the addition of the isomeric butyl alcohols, but substantially larger proportions of *syn* addition products are formed with *n*-pentyl and *n*-hexyl alcohols. For example, with *n*-hexyl alcohol the ratio of **94** to **95** (R = *n*-Hex) is 53 : 47[113].

Even with methanol, some *syn* addition occurs when the reaction is carried out in dioxane[114]. Thus *anti* addition appears to be favoured in strongly protic solvents (MeOH, EtOH), but *syn* addition becomes significant as the solvent becomes less hydroxylic (higher alcohols), or aprotic (dioxane). We shall see this same trend appear when the addition of thiols is considered.

1,2- and 1,3-Glycols add to butadiyne in the presence of KOH giving mixtures of isomeric cyclic acetals[115]. The addition of ethylene glycol, for example, gives a mixture of **96**, **97** and **98** in the ratio 46 : 32 : 22.

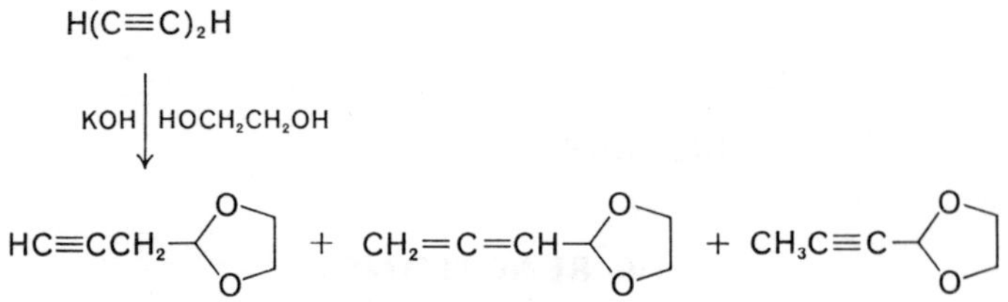

(**96**) (**97**) (**98**)

The reaction of phenol with butadiyne gives amorphous condensation products[116], but aminophenols add in the presence of KOH in DMSO–dioxane giving *cis*- and *trans*-aminophenoxybutenynes (**99**)[117].

$$HC{\equiv}CCH{=}CHOC_6H_4NH_2$$

(**99**)

β-Diethylaminoethanol adds to butadiyne at room temperature even in the absence of alkali giving **100**, free of any *syn* addition product as judged from the infrared spectrum[118].

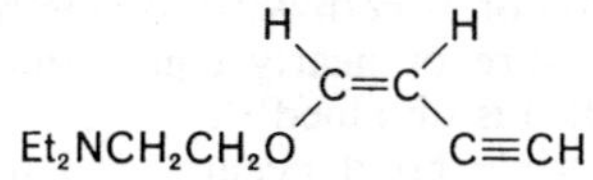

(100)

1.3-Pentadiyne shows reactivity comparable to that of butadiyne, but exclusive *anti* addition of alcohols occurs giving **101** irrespective of the alcohol or solvent[119].

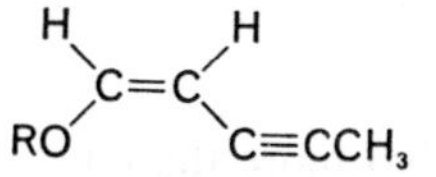

(101)

When both hydrogens of butadiyne are replaced by methyl groups the reactivity toward nucleophilic addition drops substantially and more drastic conditions are required. Addition of methanol to 2,4-hexadiyne (**102**, $n = 2$) requires the use of concentrated alkali at temperatures above 100 °C to give **103** (R = Me)[120]; addition

$CH_3(C{\equiv}C)_nCH_3$ **(102)** $CH_3C{\equiv}CCH{=}C(OR)CH_3$ **(103)**

of ethanol occurs on prolonged boiling with concentrated ethanolic KOH giving **103** (R = Et)[121]. In the case of ethanol, at least, only one stereoisomer is obtained, and although the n.m.r. spectrum did not permit an unequivocal assignment, it seems most likely that it is the one formed by *anti* addition.

As the number of conjugated triple bonds is increased, an increase in reactivity toward nucleophiles is observed[120]. The reactivity of 2,4,6-octatriyne (**102**, $n = 3$) toward methanol addition, which gives **104**, is somewhat greater than that of 2,4-hexadiyne, but a very large increase is noted for 2,4,6,8-decatetrayne (**102**, $n = 4$). The product **105** from the addition to (**102**, $n = 4$) is a mixture of *cis–trans* isomers,

$CH_3(C{\equiv}C)_2CH{=}C(OCH_3)CH_3$ **(104)** $CH_3(C{\equiv}C)_3CH{=}C(OCH_3)CH_3$ **(105)**

but whether this is a result of non-stereoselective addition or isomerization of the initial product was not established[120].

2. Sulphur nucleophiles

Thiols add to butadiyne in the presence of base under mild conditions to give 1-alkylthio-1-buten-3-ynes **106**. A second mole of thiol can be added, but the reaction is usually slower and it is possible to obtain the monoadduct in good yield.

$HC{\equiv}CCH{=}CHSR$ **(106)**

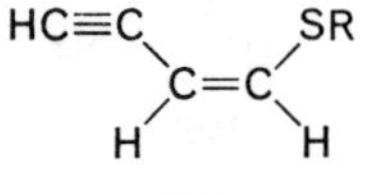

(107)

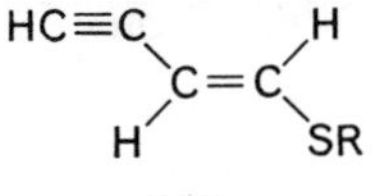

(108)

Both the rate of addition and stereoselectivity are strongly affected by the solvent. The rate is highest in DMF and lowest in methanol[3], but high stereoselectivity is found only when protic solvents are used. In alcoholic solutions *anti* addition occurs giving the *cis* isomer in at least 98% yield[122]. Exclusive *anti* addition of ethanethiol occurs in methanol giving **107** (R = C_2H_5), but amounts of the *syn* addition product **108** (R = C_2H_5) ranging from 15% to 35% arise when DMF, acetone, THF or dioxan is used[3]. The addition of 2-methyl-2-propanethiol in THF is almost totally nonstereoselective, and a mixture of nearly equal amounts of the *cis* and *trans* isomers **107** and **108** (R = *t*-Bu) is obtained[122].

Addition to the terminal triple bond occurs with monosubstituted butadiynes **109**, giving the *anti* addition product **110** when R and R′ are alkyl groups and when

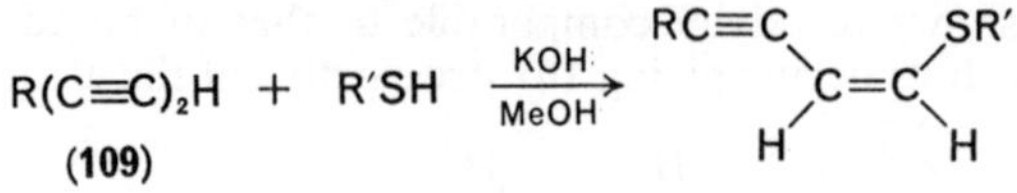

(109) **(110)**

methanol is used as solvent[123]. No more than 4–5% of the *trans* isomer is formed. The same high degree of stereoselectivity is found for the addition of 1-butane- and 1-hexanethiols to 1-phenyl-1,3-butadiyne (**109**, R = Ph) with KOH in methanol at 70 °C [124]. The addition of methanethiol to 5-phenyl-1,3-pentadiyne (**109**, R = $PhCH_2$) in the presence of sodium methanethiolate, using excess methanethiol as solvent, is exclusively *anti* and gives **110** (R = $PhCH_2$, R′ = CH_3)[125].

The addition of methanethiol to 2,4,6,8-decatetrayne (**111**) is not stereoselective, and a mixture of *cis–trans* isomers **112** is obtained[4]. Addition of a second mole, which occurs at the other terminal triple bond and gives **113**, is also non-stereoselective.

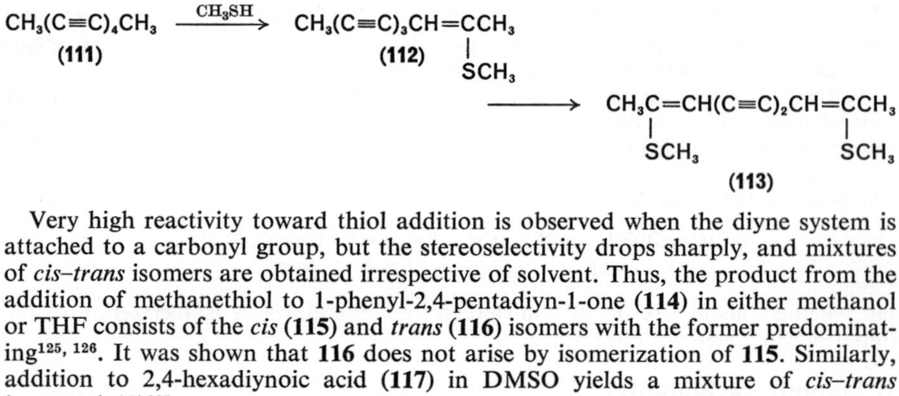

(111) **(112)** **(113)**

Very high reactivity toward thiol addition is observed when the diyne system is attached to a carbonyl group, but the stereoselectivity drops sharply, and mixtures of *cis–trans* isomers are obtained irrespective of solvent. Thus, the product from the addition of methanethiol to 1-phenyl-2,4-pentadiyn-1-one (**114**) in either methanol or THF consists of the *cis* (**115**) and *trans* (**116**) isomers with the former predominating[125, 126]. It was shown that **116** does not arise by isomerization of **115**. Similarly, addition to 2,4-hexadiynoic acid (**117**) in DMSO yields a mixture of *cis–trans* isomers (**118**)[127].

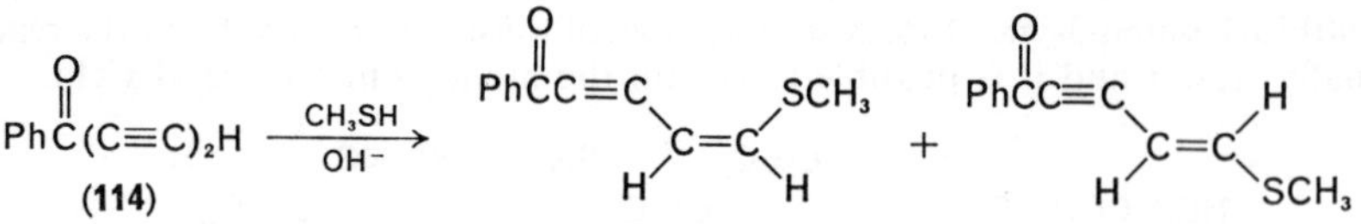

(114) **(115)** **(116)**

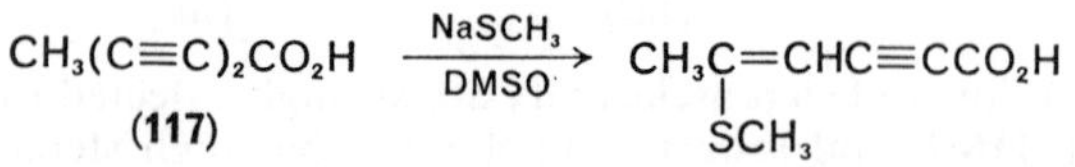

(117) **(118)**

Addition of thiols to disubstituted butadiynes occurs readily, and based on cases where product configuration has been ascertained, appears to follow the usual rule of *anti* addition. Thus, addition of toluene-ω-thiol to **119** gives the *Z* mono adduct **120** and *Z,Z* diadduct **121**[128]. Although additions to unsymmetrically disubstituted

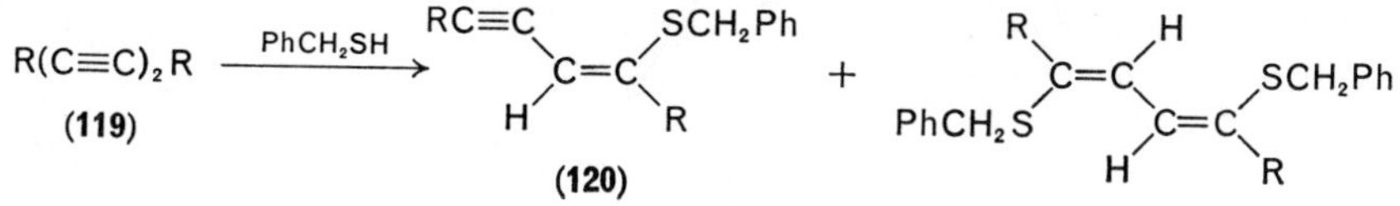

(119) (120) (121)

butadiynes have been reported to be regioselective, it is difficult to rationalize the orientations in some cases. Addition of ethanethiol to **122** (R = *i*-Pr, *t*-Bu) gives adducts (**123**) in which the ethylthio group is adjacent to the isopropyl or *t*-butyl

$$\underset{(122)}{CH_3(C{\equiv}C)_2R + EtSH} \xrightarrow[\text{MeOH, 120°C}]{\text{KOH}} \underset{(123)}{CH_3C{\equiv}CCH{=}C(SEt)R}$$

group[129], and it has been suggested that this orientation results from the greater inductive electron release by these groups. In the addition of methanethiol to the triyne derivative **124**, however, the methylthio group becomes attached next to the electron-withdrawing hydroxymethyl group giving **125**[130], whereas complete reversal

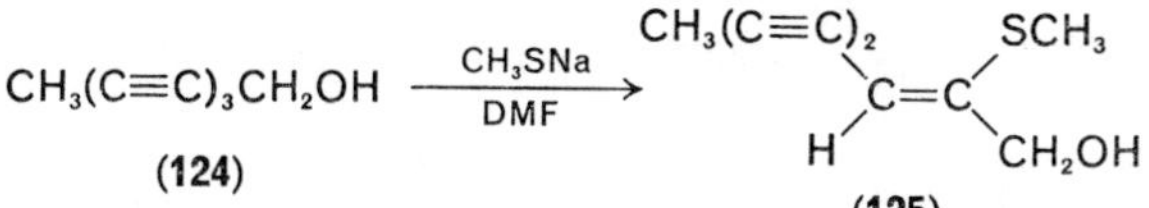

(124) (125)

of orientation occurs in thiol additions to **126**, and the alkylthio group becomes attached to one of the internal acetylenic carbons[124, 131]. Addition of 1-butanethiol gives **127**, while 1-hexanethiol gives **128**.

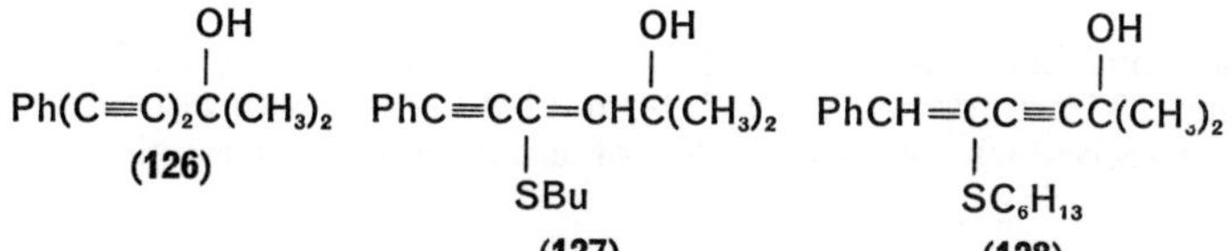

(126) (127) (128)

3. Nitrogen nucleophiles

Primary and secondary aliphatic amines react readily with butadiyne under mild conditions and without the necessity of added catalyst. A wide variety of secondary amines has been used including, for example, dimethyl-, diethyl-, di-*n*-butyl- and diallylamine, as well as the heterocyclic derivatives pyrrolidine, piperidine and morpholine. The product in each case is the corresponding 1-(*N,N*-dialkylamino)-1-buten-3-yne **129**[132–135].

$$H(C{\equiv}C)_2H + RR'NH \longrightarrow \underset{(129)}{RR'NCH{=}CHC{\equiv}CH}$$

With primary amines, addition of a second mole is very rapid and usually only the diadduct **131**, a mixture of enamine–imine tautomers, is isolable[136, 137]. Only in the case of *t*-butylamine has the monoadduct **130** (R = *t*-Bu) been reported[138]. Aromatic amines fail to react[118].

$$H(C{\equiv}C)_2H + RNH_2 \longrightarrow RNHCH{=}CHC{\equiv}CH$$

(130)

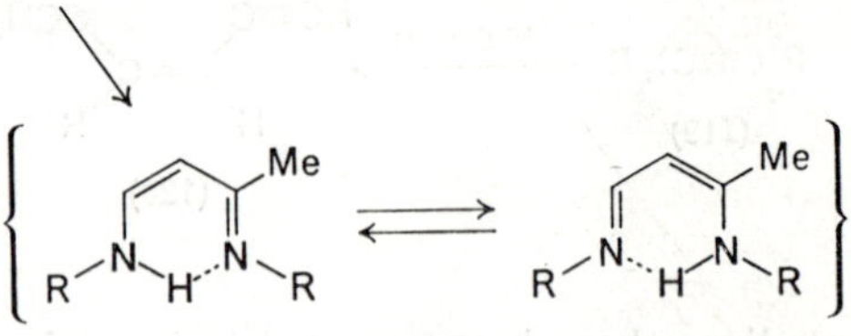

(131)

With 1,3-pentadiyne and primary amines it is possible to obtain either the mono- or diadduct. Thus, refluxing a solution of the diyne in excess *n*-propylamine produces *cis*-1-propylamino-1-penten-3-yne (**132**, R = Pr), whereas heating at 90–100 °C in THF yields the diadduct **134** (R = Pr) presumably by prototropic rearrangement of the initially formed diadduct **133**[139]. Analogous behaviour is found for *n*-butylamine.

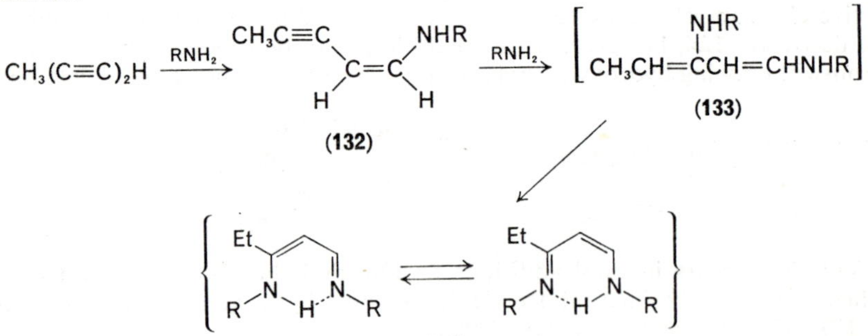

(132) **(133)** **(134)**

The triple bond in aminoalkenynes undergoes hydration very rapidly, and aminovinyl ketones **135** are formed when 1,3-butadiyne reacts with aqueous solutions of primary and secondary amines. Yields of aminobutenones ranging from 40% to

$$H(C{\equiv}C)_2H \xrightarrow[H_2O]{RR'NH} [HC{\equiv}CCH{=}CHNRR'] \longrightarrow CH_3\overset{\overset{\displaystyle O}{\|}}{C}CH{=}CHNRR'$$

(135)

60% are obtained upon addition of diethylamine, di-*n*-propylamine or *n*-butylamine to butadiyne in aqueous DMF[140]. When the reaction of 1,3-pentadiyne with aqueous dimethylamine or diethylamine is carried out at room temperature, mixtures of the monoadduct **136** and the hydration product **137** are obtained. The latter becomes the sole product when the reaction time is extended or the temperature is raised, signifying that **137** is formed by hydration of **136**[141].

$$CH_3(C{\equiv}C)_2H \xrightarrow[H_2O]{R_2NH} CH_3C{\equiv}CCH{=}CHNR_2 + CH_3CH_2\overset{\overset{\displaystyle O}{\|}}{C}CH{=}CHNR_2$$

(136) **(137)**

The reaction of aqueous diethylamine with **138** at room temperature gave only the monoadduct **139**, and it was necessary to raise the temperature to achieve hydration. In this case, as often happens, the ketone **140** suffered partial elimination, and the

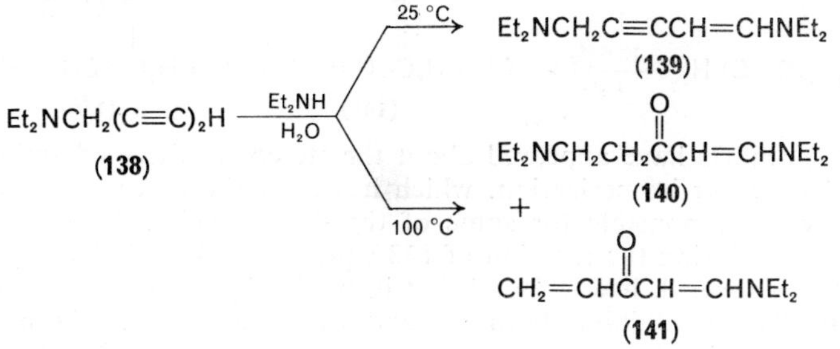

divinyl ketone **141** was also formed[142]. In the reaction of aqueous dimethylamine with the dimethylamino analogue **142** hydration occurred even at room temperature, and **143** and **144** were formed.

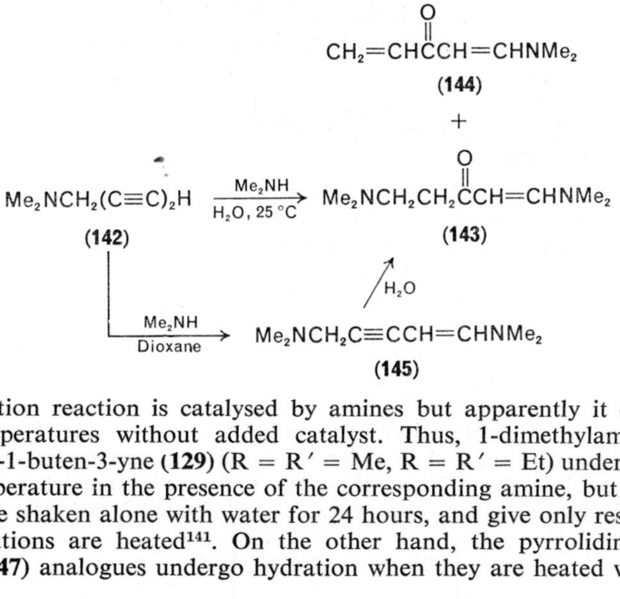

The hydration reaction is catalysed by amines but apparently it can occur at elevated temperatures without added catalyst. Thus, 1-dimethylamino- and 1-diethylamino-1-buten-3-yne (**129**) (R = R′ = Me, R = R′ = Et) undergo hydration at room temperature in the presence of the corresponding amine, but fail to do so when they are shaken alone with water for 24 hours, and give only resins when the aqueous solutions are heated[141]. On the other hand, the pyrrolidino (**146**) and piperidino (**147**) analogues undergo hydration when they are heated with water[133].

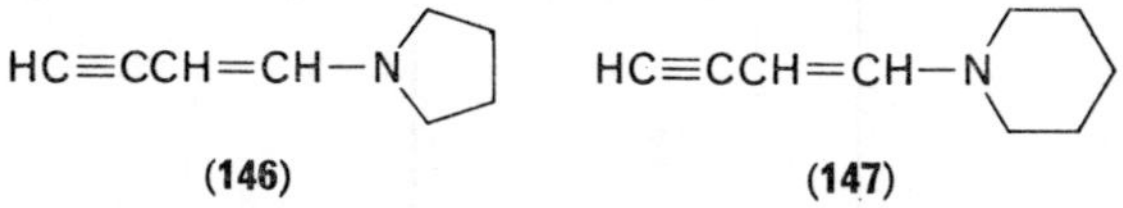

Hydration of **145** occurs at room temperature, but this molecule of course contains a tertiary amine function which can catalyse the addition. A related example of the effectiveness of amines in catalysing the addition of hydroxylic derivatives is the previously cited addition of β-dimethylaminoethanol to butadiyne which occurs in the absence of added alkali.

Besides the expected aminovinyl ketone **148**, the reaction of 1,3-pentadiyne with aqueous solutions of ethylamine also furnishes an equal amount of the isomeric adduct **149** [139]. Formation of **149** apparently involves initial addition of amine to the internal triple bond and subsequent hydration of the terminal alkyne linkage, i.e. the reverse order from that observed in non-aqueous media.

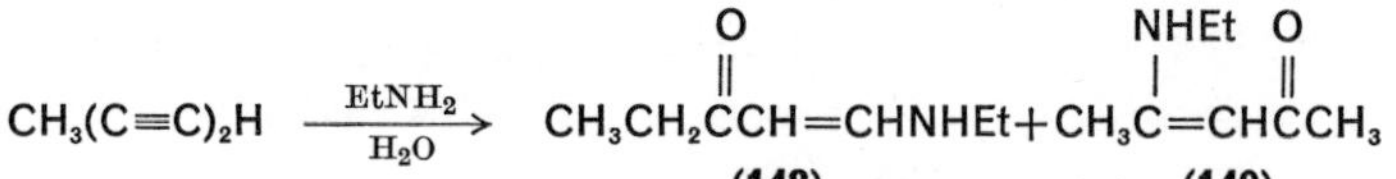

(148) (149)

Conflicting reports have appeared about the stereochemistry of amine addition, but thermal *cis–trans* isomerization, which has been shown to occur with certain adducts, may be responsible for some of the discrepancies. Thus, although *anti* addition prevails during the reaction of **138** with aqueous diethylamine and the *cis* isomer of **139** is formed, it was found that isomerization occurs during distillation, with the distillate containing both *cis* and *trans* isomers[142]. Thermal *cis–trans* isomerization has also been observed with the monoadducts of secondary amines and 1,3-pentadiyne[141].

On the other hand, a mixture of *cis–trans* isomers **151** is formed during the reaction of **150** with aqueous dimethylamine at room temperature, and it was shown that the individual stereoisomers are not interconverted by heating[143]. Only the *trans* isomer of **151** is obtained when the addition is carried out in dioxane at 120 °C.

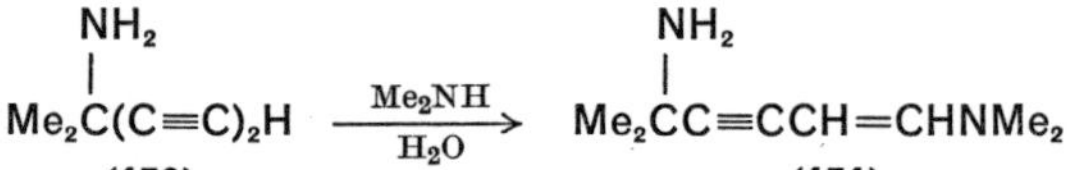

(150) (151)

In some cases products of high stereochemical purity are formed. The monoadduct **129** (R = R′ = Et) from diethylamine and butadiyne is the *cis* isomer containing only 1–5% of the *trans* isomer[135], and the diadducts of butadiyne with primary amines possess *cis* geometry[137, 139]. For other studies in which *cis–trans* mixtures were obtained, it is not possible to ascertain from the reports whether or not both stereoisomers were present prior to distillation.

A 3 : 2 mixture of *cis–trans* isomers is obtained from the addition of secondary amines to butadiyne in dioxane[144]. The ratio remains constant during the course of the reaction signifying that the isomers are formed in this ratio. This, coupled with the second-order kinetics observed and large negative values for the activation entropy ($\Delta S^{\ddagger} \simeq -50$ e.u.), led to the postulation of a mechanism involving rate-determining attack by the amine on the diyne, followed by stereochemical equilibration of the dipolar ion and proton transfer, as illustrated in Scheme 7.

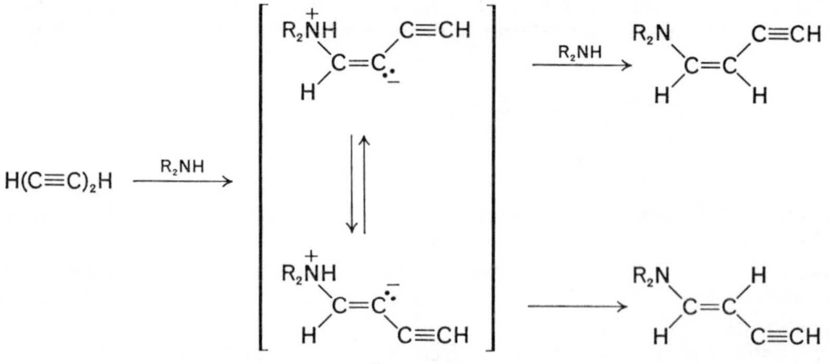

SCHEME 7

Relatively little work has been done on reactions of amines with disubstituted butadiynes. Aqueous dimethylamine and 2,4-hexadiyne react when heated to give the aminovinyl ketone **150** [145], but 2,4-octadiyne fails to react with aqueous diethylamine[141]. Low solubility of the hydrocarbon in the aqueous phase may be responsible for lack of reaction in the latter case.

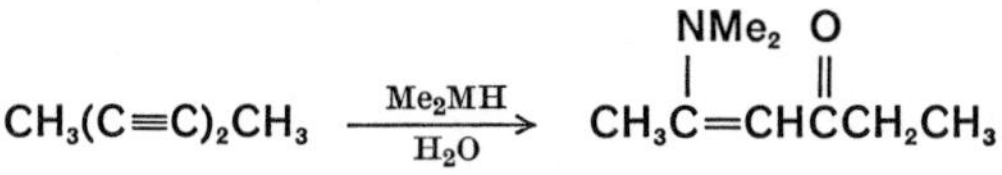

(150)

Addition of amines to carbonyl-activated diynes occurs with great ease, and often the initial adducts react further to give cyclic products. *Syn* addition occurs in the reaction of piperidine with 1-phenyl-2,4-pentadiyne-1-one **(151)** in either ethanol or ether as solvent giving **152** [146].

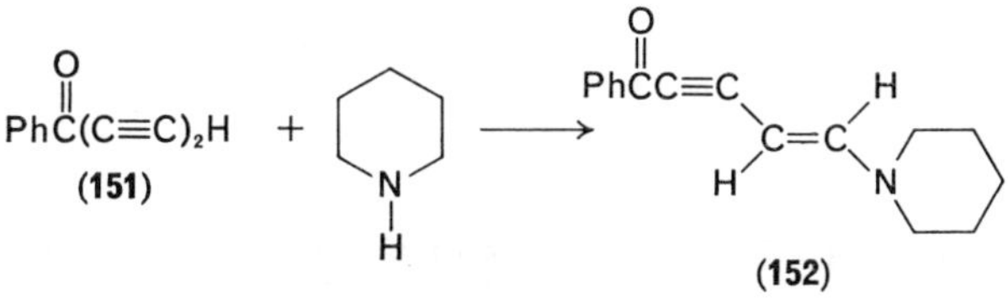

(151) **(152)**

4. Formation of heterocyclics

Polyynes have served as starting materials for the synthesis of a wide variety of heterocyclic ring systems. The reactions used involve addition to triple bonds, and any of the common mechanistic pathways may be followed, i.e. nucleophilic, electrophilic or free radical attack as well as concerted cycloadditions. Although the evidence does not permit unequivocal classification of many of the reactions into one of these categories, the ones considered here are those which most likely involve nucleophilic attack at some stage. In a formal sense the reactions amount to successive additions of a divalent nucleophile to two triple bonds; the first involves intermolecular and the second intramolecular attack, as illustrated in equation (19) for the addition of H_2S to a diyne.

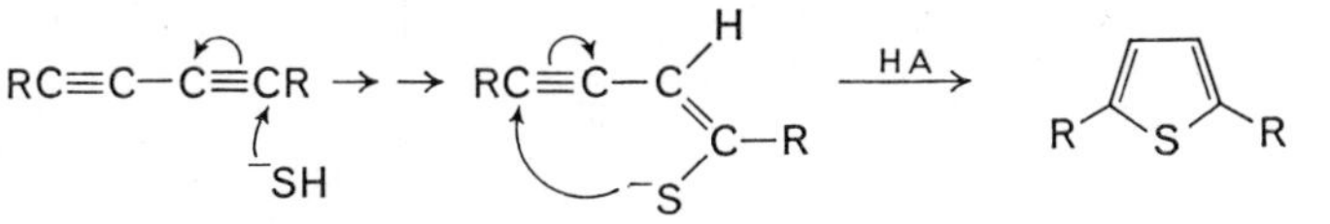

(19)

Thiophenes are readily obtained by addition of hydrogen sulphide to butadiyne, its mono- and disubstituted derivatives, as well as to substituted triynes and tetraynes[147, 148]. The reactions (equation 20) are carried out in weakly alkaline solution, and provide the corresponding thiophenes in yields of 50–85%. Typical examples are shown in Table 11. In the case of 1-phenyl-1,3,5-heptatriyne **(153)** addition to the alkyl-substituted triple bond prevails giving **154**.

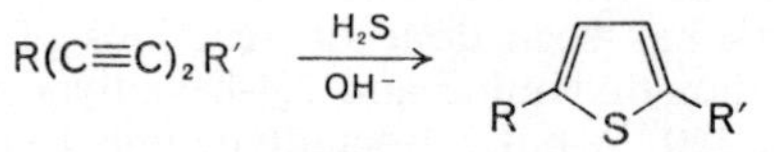

(20)

TABLE 11. Yields for equation (20).

R	R′	Yield (%)
H	H	20
CH_3	CH_3	70
Ph	C≡CPh	83
Ph	$(C{\equiv}C)_2Ph$	75

Addition of hydrogen selenide[149] and hydrogen telluride[150] to diynes has been used for the synthesis of selenophenes and tellurophenes.

$Ph(C{\equiv}C)_3CH_3$

(153)

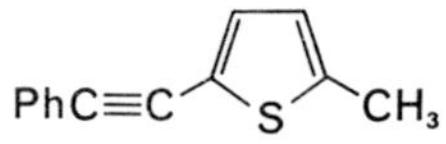

(154)

Furan derivatives are formed when aqueous solutions of amines react with diacetylenic alcohols and glycols. Thus when **155** is heated with aqueous dimethylamine, the aminofuran **158** is formed[145]. The initial steps involve amine addition and hydration giving **156** which suffers dealdolization to give the ketone **157**. In the case of tertiary glycols such as **159**, a similar sequence of steps followed by hydrolysis of the intermediate enamine produces the furanone **160** [151].

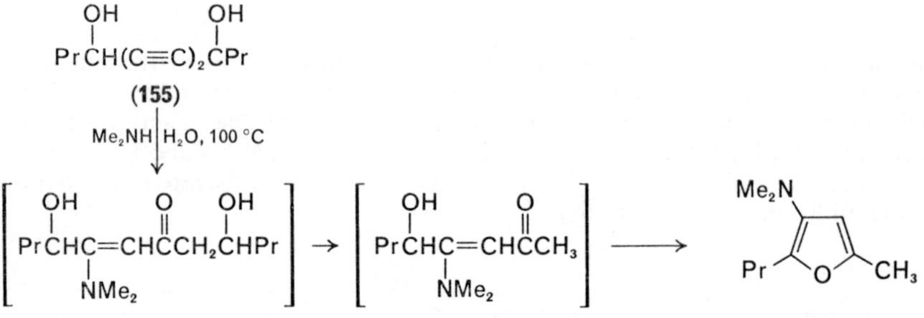

(156) **(157)** **(158)**

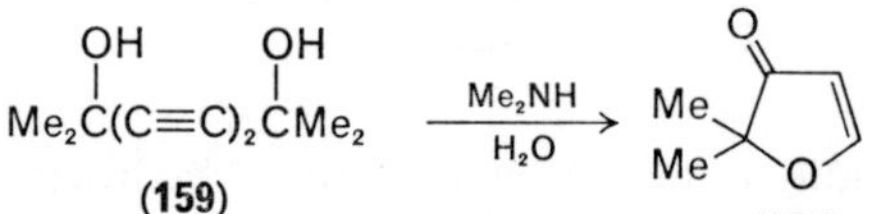

(159) **(160)**

Dioxolanes **162** are formed by the reaction of diacetylenic alcohols with aldehydes and ketones in the presence of base[152]. The initial step involves formation of the hemiacetal **161** which then cyclizes by intramolecular addition of OH to the adjacent triple bond.

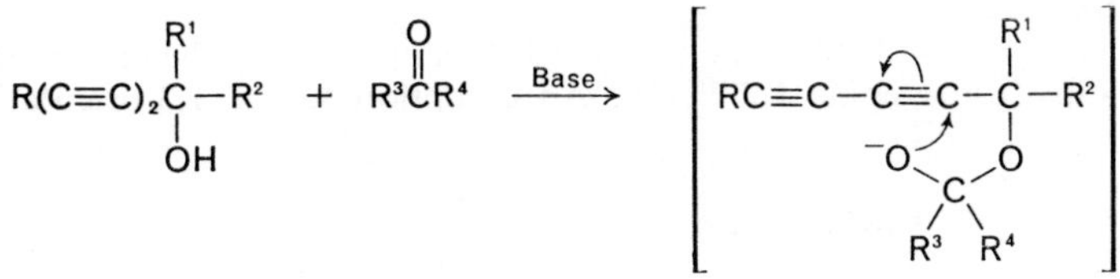

(**161**)

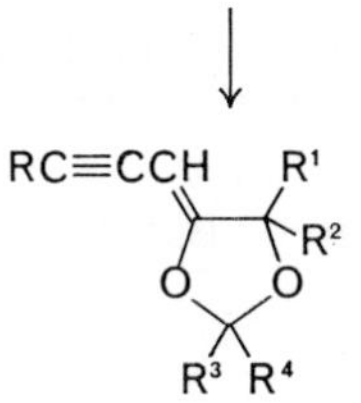

(**162**)

The reaction of butadiyne or its mono- or disubstituted derivatives with ammonia or primary amines in the presence of copper(I) chloride at elevated temperatures gives good yields of pyrrole or 1,2-, 2,5-, or 1,2,5-substituted pyrroles **163** [153].

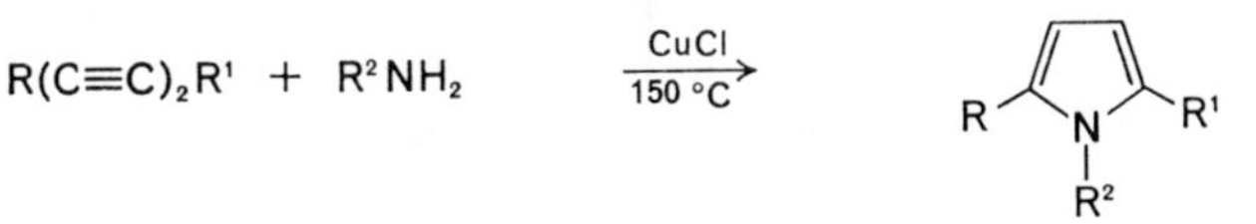

(**163**)

Hydrazine and substituted hydrazines add to conjugated diynes to give pyrazoles, as illustrated by the formation of 3(5)-methylpyrazole (**164**) from butadiyne and

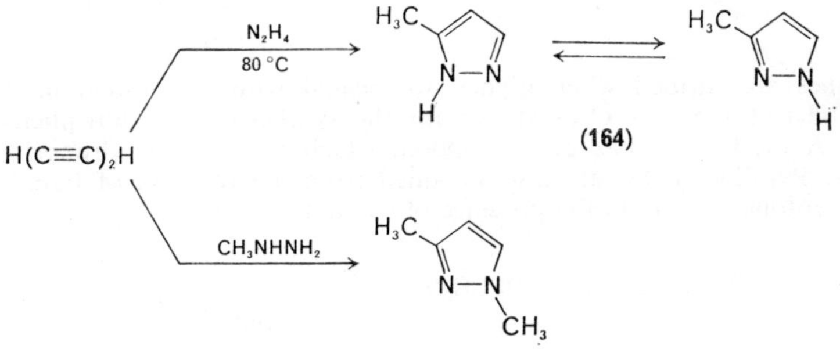

(**164**)

(**165**)

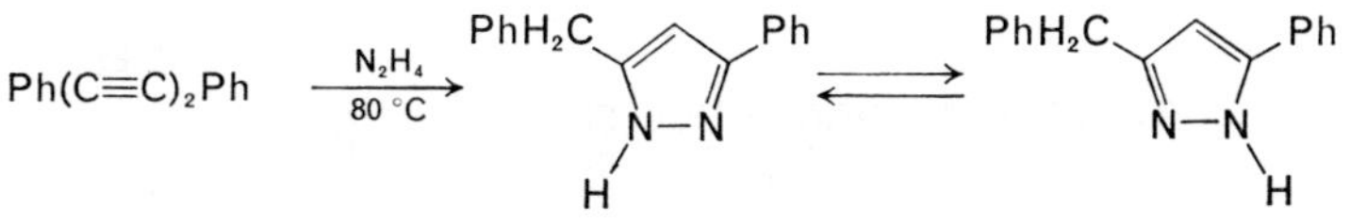

(**166**)

hydrazine and of **166** from diphenylbutadiyne and hydrazine[137, 154, 155]. 1,3-Dimethylpyrazole (**165**) is formed by the reaction of butadiyne and methylhydrazine at room temperature[156].

Ethylenediamine adds to butadiyne in a 1,3 manner giving the tautomeric mixture of methyldihydrodiazepins **167** [155, 157]. An exception to the 1,3-orientation rule is

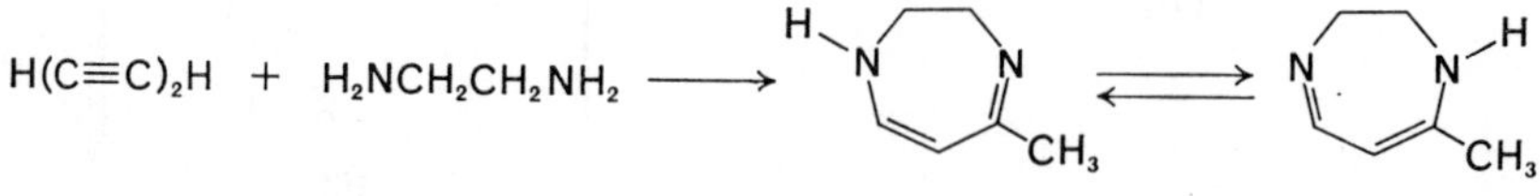

(167)

found in the addition of the diamine **168** to butadiyne. Mixtures of the monoadduct, itself an equilibrating mixture of enyne **169** and triene **170**, and the cyclic diadduct **171** are formed[158].

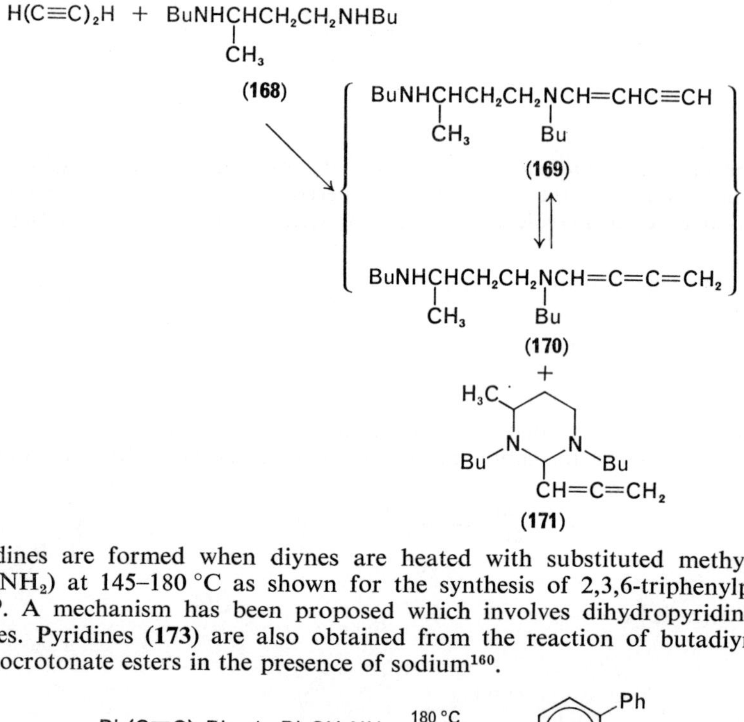

(171)

Pyridines are formed when diynes are heated with substituted methylamines (RCH_2NH_2) at 145–180 °C as shown for the synthesis of 2,3,6-triphenylpyridine (**172**)[159]. A mechanism has been proposed which involves dihydropyridine intermediates. Pyridines (**173**) are also obtained from the reaction of butadiyne with β-aminocrotonate esters in the presence of sodium[160].

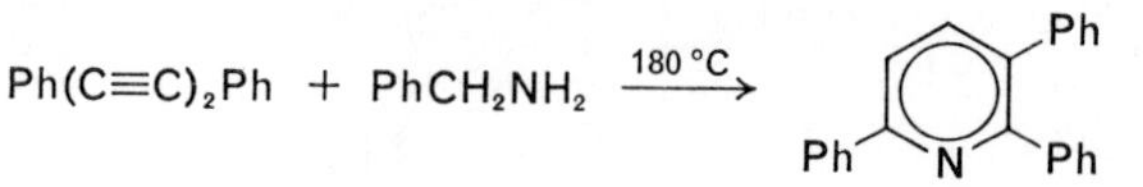

(172)

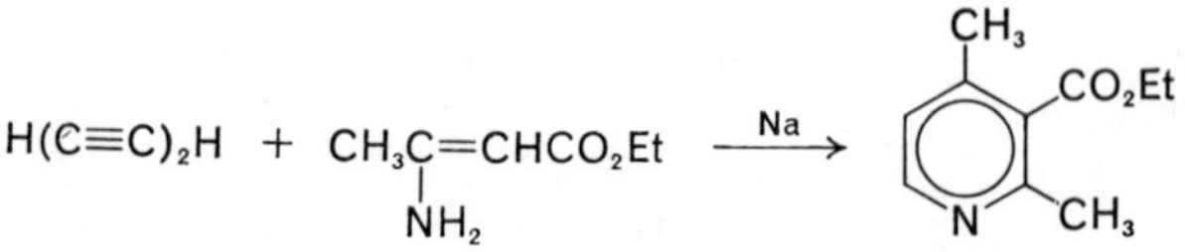

(173)

1,3-Diazines are formed by the addition of guanidine and its derivatives to diynes in the presence of base, as illustrated by the preparation of **174** from 1,3-hexadiyne and guanidine[161].

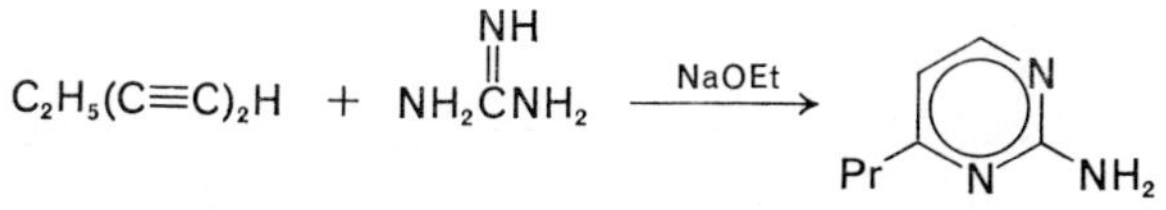

(174)

B. *Addition of Electrophilic Reagents*

I. Halogens

Butadiyne reacts with halogens (Cl_2, Br_2, I_2) even under mild conditions, but unless precautions are taken the initially formed dihalide reacts further and polyhalogenated derivatives become the major products. By using a ten-fold excess of butadiyne at −50 °C it is possible to obtain the dihalo derivatives **175** in 83–95% yield[162]. Under the same conditions iodine chloride gives mainly the diadduct **176**, along with a small amount of the monoadduct **177**. It is also possible to obtain the dichloro derivative **175** (X = Cl) in 20% yield by using copper(II) chloride as the chlorinating agent[163].

$$\underset{\textbf{(175)}\ (X = Cl, Br, I)}{CHX{=}CXC{\equiv}CH} \qquad \underset{\textbf{(176)}}{CHI{=}CClCCl{=}CHI} \qquad \underset{\textbf{(177)}}{CHI{=}CClC{\equiv}CH}$$

With monosubstituted diynes, addition occurs predominantly at the terminal triple bond, although the selectivity depends on the nature of the halogen and the substituent[164]. The dichloro derivatives **178a** and **178b** were obtained in a 3 : 1 ratio from the reaction of chlorine with excess 1,3-pentadiyne at −40 °C. In spite of the fact that a two-fold excess of pentadiyne was used, tetra- and higher polychlorides

$$CH_3(C{\equiv}C)_2H + Cl_2 \xrightarrow{-40\ ^\circ C} \underset{\textbf{(178a)}}{CH_3C{\equiv}CCCl{=}CHCl} + \underset{\textbf{(178b)}}{CH_3CCl{=}CCl{-}C{\equiv}CH}$$

were the major products, and the dichlorides were obtained in only 30% yield. Addition of iodine occurs exclusively at the terminal triple bond giving **179**, whereas both **180a** and **180b** are formed by addition of iodine chloride. The orientation of addition in **180b** is interesting. The ratio of **180a** to **180b**, which was

$$\underset{\textbf{(179)}}{CH_3C{\equiv}CCI{=}CHI} \qquad \underset{\textbf{(180a)}}{CH_3C{\equiv}CCCl{=}CHI} \qquad \underset{\textbf{(180b)}}{CH_3CCl{=}CI{-}C{\equiv}CH}$$

4 : 1 when a two-fold excess of pentadiyne was used, approached 1 : 1 when a very large excess of hydrocarbon was used, signifying that **180b** is consumed faster than **180a** in forming the tetrahalide. This same factor may have affected the ratio of **178a** to **178b** in the chlorination study. Both iodine and iodine chloride add exclusively to the terminal triple bond in 5,5-dimethyl-1,3-hexadiyne (**181**) giving **182** and **183** respectively[164].

$$\underset{\textbf{(181)}}{t\text{-}Bu(C{\equiv}C)_2H} \qquad \underset{\textbf{(182)}}{t\text{-}BuC{\equiv}CCI{=}CHI} \qquad \underset{\textbf{(183)}}{t\text{-}BuC{\equiv}CCCl{=}CHI}$$

Bromination of 2,4-hexadiyne (**184**) at −50 °C yields the dibromide **185** [129], and chlorination of a variety of diynes with copper(II) chloride furnishes the *trans* dichlorides **186** in good yield[165]. Chlorination with copper(II) chloride has also

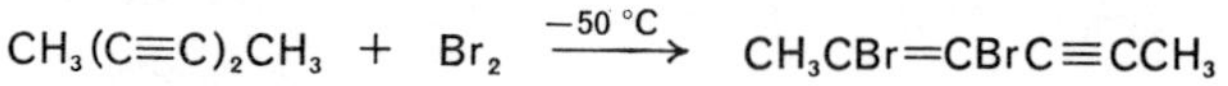

(**184**) (**185**)

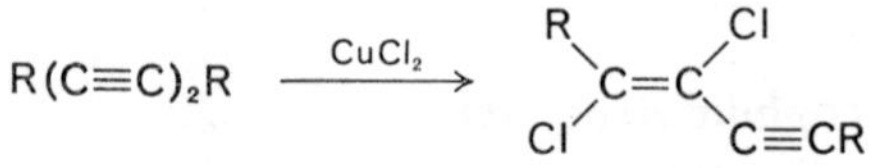

(**186**)

been used for preparing dichloro and tetrachloro adducts of 2,4-hexadiyne-1,6-diol and its simple ethers[166]. The tetrabromide (**187**) obtained from 2,4-hexadiyne-1,6-diol is of interest because it can be obtained in optically active forms as a consequence of the high barrier to rotation about the central carbon–carbon bond[167, 168].

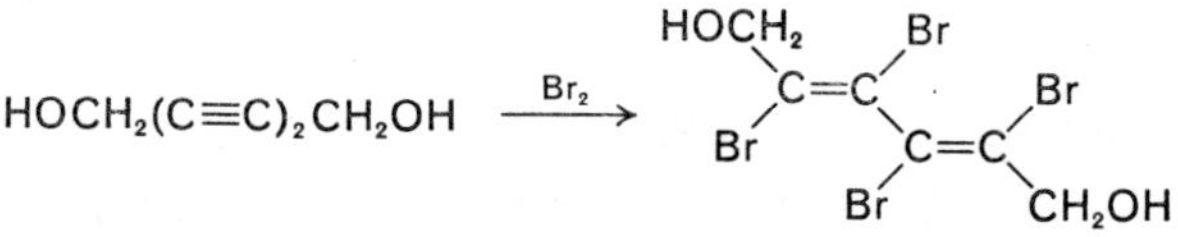

(**187**)

1,4-Addition occurs in the low-temperature bromination of diarylbutadiynes. Thus, diphenylbutadiyne yields the triene **188** of unspecified stereochemistry[169], and di-*p*-tolylbutadiyne provides a mixture of *cis*(**189a**) and *trans*(**189b**) isomers[170].

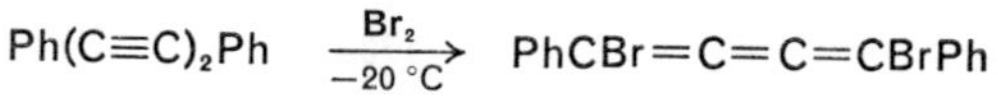

(**188**)

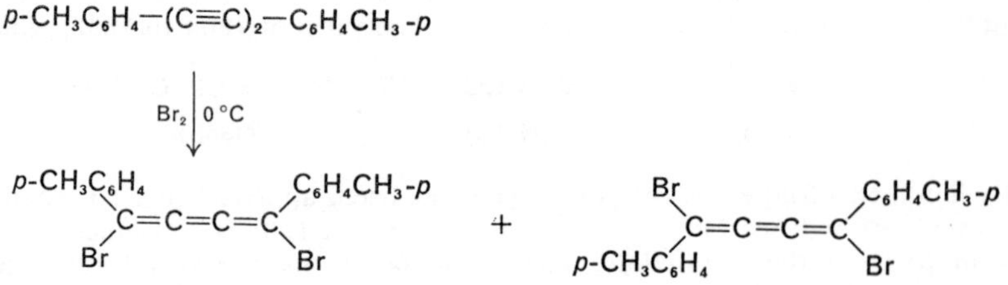

(**189a**) (**189b**)

Cyclization occurs during the reaction of 1,2-bis(phenylethynyl)benzene (**190**) with a wide variety of reagents[171]; for example, reaction with bromine furnishes **192**. Presumably, the intermediate vinyl cation **191** is attacked by Br^- from the least hindered direction as indicated in equation (21).

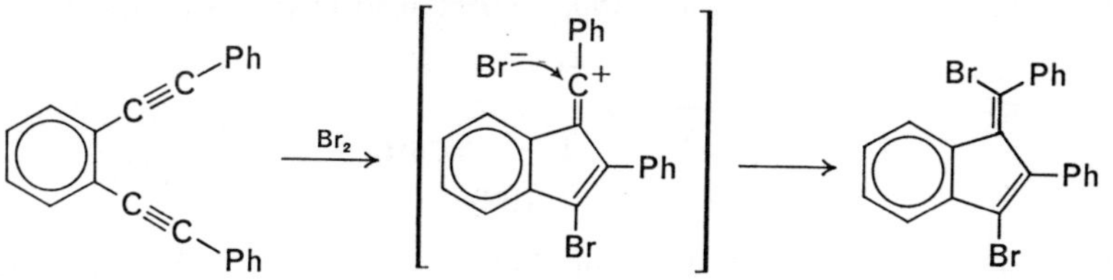

(21)

(190) (191) (192)

The rate of bromine addition to conjugated polyynes parallels that of nucleophilic additions, i.e. the rate increases with increasing number of triple bonds[172]. This is opposite to the trend found for typical electrophilic additions, and it has been postulated that bromination involves an initial nycleophilic attack by bromine (equation 22). Absence of chlorine incorporation in the product obtained by bromination of 2,4,6-octatriyne in the presence of benzyltriethylammonium chloride has been cited as evidence for this hypothesis[172].

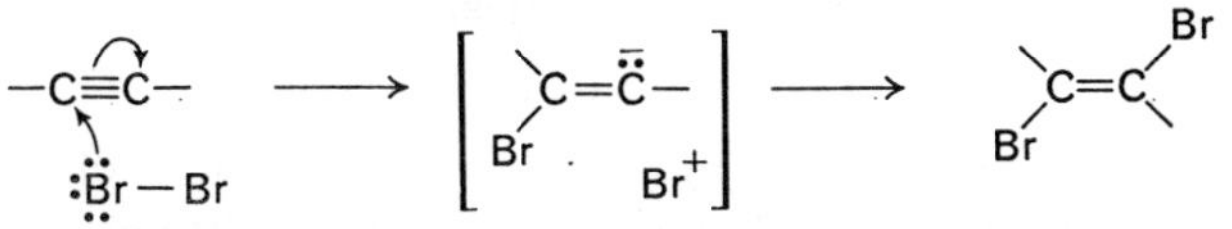

(22)

The orientation of addition of iodine chloride to butadiyne[162], however, corresponds to an electrophilic attack. A study of the kinetics of bromination of various mono- and dialkyl-substituted butadiynes has been interpreted in terms of an electrophilic mechanism[173]. The observed rates increased with increasing inductive electron-releasing power of the substituents. However, the effects are small, and under the conditions used (two-fold excess of hydrocarbon) it seems likely that significant polybromination occurred[164]. Marked catalysis by bromide ion was observed[173] and interpreted in terms of electrophilic attack by bromine on a complex of acetylene with Br^-:

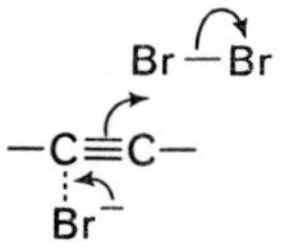

2. Hydrogen halides

Addition of hydrogen halides to conjugated diynes occurs very slowly, and unlike the behaviour in halogen addition, the rate decreases with increasing number of triple bonds[172]. 2,4,6,8-Decatetrayne fails to react with HBr even under fairly drastic conditions. Because of the low rate of the initial addition, polyhalogenated products often predominate over monoadducts.

Addition of HBr or HI to butadiyne in ether or pentane at -40 °C in the presence of hydroquinone occurs with an *anti*-Markownikoff orientation giving *cis*-1-halo-1-butene-3-yne (**193**)[174]. With 1,3-pentadiyne, addition of HI occurs at the terminal

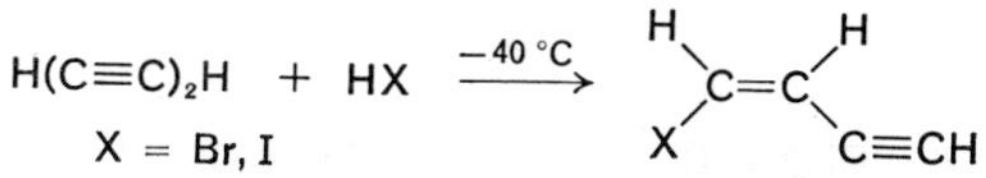

(193)

triple bond exclusively, although in this case products corresponding to both orientations of addition, **194** and **195**, are obtained in the ratio 87 : 13.

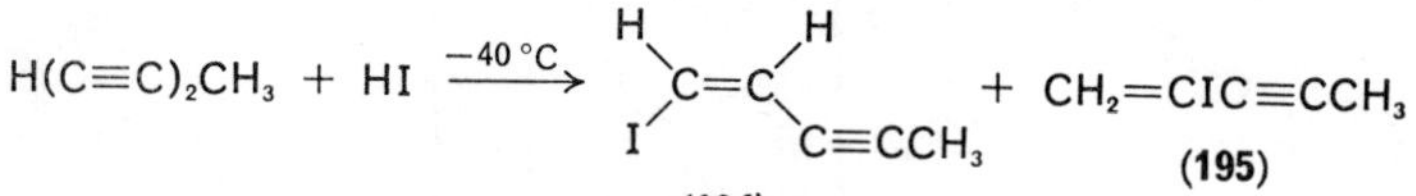

(194) (195)

By using concentrated aqueous solutions of hydrohalic acids in the presence of mercury(I) or copper(I) halides, the orientation of addition becomes predominantly or exclusively Markownikoff[174, 175]. Thus **196** and **197** are formed in a 63 : 37 ratio from butadiyne, and the Markownikoff-type products **198** are formed exclusively from 1,3-pentadiyne.

$$H(C{\equiv}C)_2H \xrightarrow[CuBr]{HBr(aq.)} \underset{(196)}{CH_2{=}CBrC{\equiv}CH} + \underset{(197)}{BrCH{=}CHC{\equiv}CH}$$

$$H(C{\equiv}C)_2CH_3 \xrightarrow[CuX]{HX(aq.)} \underset{(198)}{CH_2{=}CXC{\equiv}CCH_3}$$

$$X = Cl, Br, I$$

Interpretation of the results for hydrogen halide addition is difficult. By analogy with alkene reactions there has been a tendency to refer to the anti-Markownikoff additions as being free-radical reactions. However, in view of the unusually strong electron-withdrawing effects of the ethynyl and butadiynyl groups[22], the situation may be more complex with the acetylenic derivatives. It seems likely that the reversal of orientation brought about by copper(I) and mercury(I) salts involves interaction of the metal ions with alkyne linkages.

3. Water

Acid-catalysed hydration of conjugated diynes is slower than that of simple alkynes, but still occurs under relatively mild conditions in the presence of sulphuric acid or especially sulphuric acid and mercury(II) sulphate. Thus 2,3-butanedione (**200**) is obtained readily from butadiyne, without isolation of the intermediate monohydration product, 3-butyn-2-one (**199**).

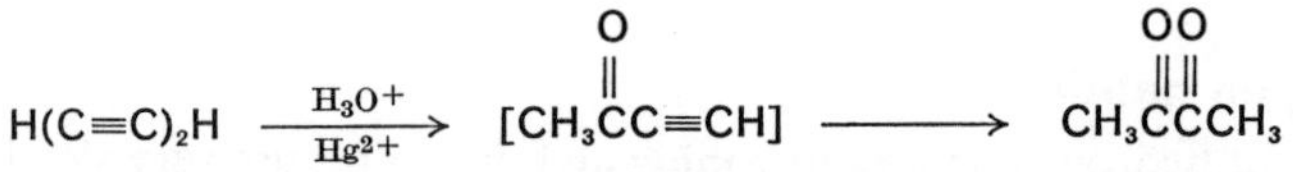

(199) (200)

Conjugated triynes react more slowly than the analogous diynes, but 4,6-octadiyn-3-one (**202**) can be obtained from 2,4,6-octatriyne (**201**) by treatment with cold, concentrated sulphuric acid followed by water[120]. Under the same conditions, only decomposition products and no simple hydration products are obtained from 2,4,6,8-decatetrayne (**203**). Interestingly, the severely hindered pentayne **204** can be heated with concentrated sulphuric acid for extended periods without suffering change[120].

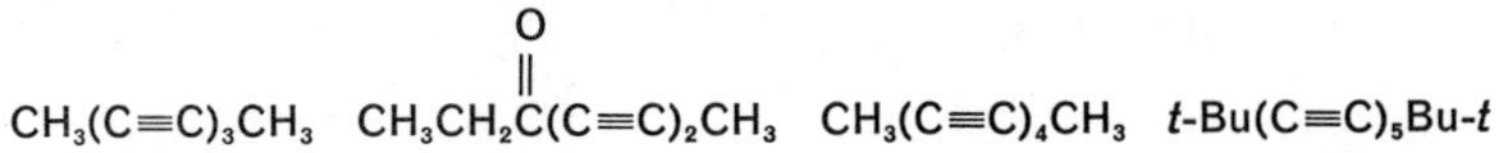

(201) (202) (203) (204)

Hydroboration followed by oxidation provides an indirect route for hydration of diynes and constitutes an efficient route to acetylenic ketones as demonstrated by the synthesis of **206** in 75% yield from diyne **205** [176].

$$\underset{\textbf{(205)}}{Bu(C\equiv C)_2Bu} \xrightarrow[(2)\ H_2O_2,\ OH^-]{(1)\ Sia_2BH} \underset{\textbf{(206)}}{BuCH_2\overset{\overset{\displaystyle O}{\|}}{C}C\equiv CBu}$$

C. Reduction

Partial catalytic hydrogenation of polyynes to give polyenes with *cis* configuration has been accomplished with a variety of substrates. Palladium catalysts, particularly the Lindlar catalyst[177], are generally the most satisfactory. Side-reactions include over-hydrogenation, isomerization of *cis* to *trans* isomers and double-bond migration.

Hydrogenation of skipped polyynes presents a severe test of the selectivity of the reaction, and here the outcome seems to depend on the number of triple bonds[178]. Thus, reduction of 10,13-nonadecadiynoic acid (**207**) or its methyl ester using Lindlar's catalyst and a small amount of quinoline gives the *cis,cis* dienoic acid having a purity of 98–99%, the only impurities being 1% of the *cis* monoenoic acids and less than 0·3% of conjugated dienes. With 8,11,14-eicosatriynoic acid (**208**) the product contains 3–5% of over-hydrogenated products while 1–3% of the double bonds have the *trans* configuration; with 5,8,11,14-eicosatetraynoic acid (**209**) these amounts increase to 8–10% and 2–4% respectively.

Other workers have reported that substantial amounts (*ca.* 10%) of double-bond isomers are formed even with the diynoic acid **210** [179].

$$\underset{\textbf{(207)}}{C_5H_{11}(C\equiv CCH_2)_2(CH_2)_7CO_2H} \qquad \underset{\textbf{(208)}}{C_5H_{11}(C\equiv CCH_2)_3(CH_2)_5CO_2H}$$

$$\underset{\textbf{(209)}}{C_5H_{11}(C\equiv CCH_2)_4(CH_2)_2CO_2H} \qquad \underset{\textbf{(210)}}{C_6H_{13}(C\equiv CCH_2)_2(CH_2)_5CO_2H}$$

Hydrogen adds exclusively to the terminal triple bond of monosubstituted conjugated diynes but the initially formed enynes react rapidly with hydrogen, and even from the beginning, products of over-hydrogenation appear[180, 181]. After the absorption of one mole of hydrogen by 1,3-pentadiyne in the presence of $Pd(CaCO_3)$, the product contains 67% 1-penten-3-yne. No products have been detected which correspond to the initial addition of hydrogen to the internal triple bond.

The non-silylated triple bond in derivatives such as **211** is reduced preferentially[182], and because of the ease with which the alkynyl–silicon bond is cleaved, a route is made available for the selective reduction of the internal triple bond of polyynes which also contain terminal alkyne linkages. Desilylation can be accomplished with dilute base[183], silver nitrate[184] or fluoride salts[185]. An example which illustrates the sequence is the conversion of **211** to **212** in 53% yield[186].

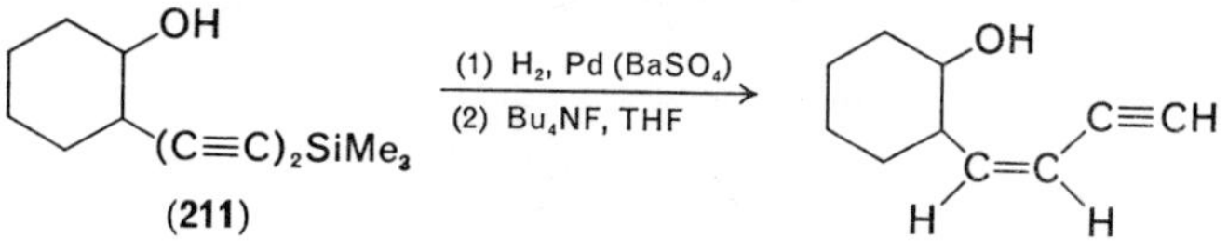

(211) (212)

Conjugated diynes can be reduced to *cis* enynes by the well-known hydroboration–protonolysis sequence[176]. *cis*-5-Dodecen-7-yne (**215**) is obtained from the diyne **213** in 76% yield as shown in equation (23). The orientation of addition of the disiamylborane was established by using CH_3CO_2D in the second step. Addition of a second mole of disiamylborane to **214** is very slow, but reduction to the *cis,cis* diene can be accomplished by using dicyclohexylborane instead [176].

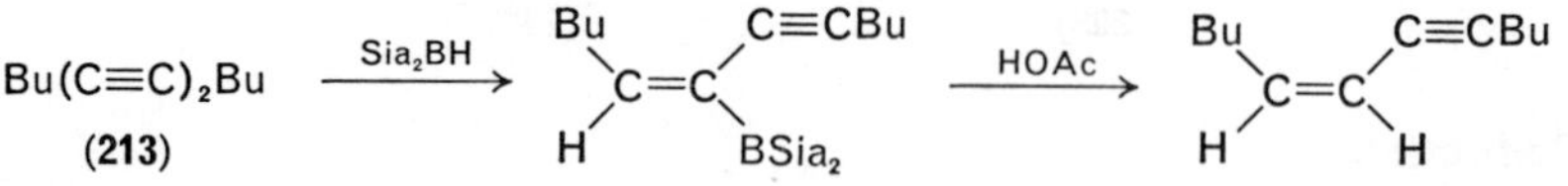

(213) (214) (215) (23)

Simple alkynes are not reduced by lithium aluminium hydride at a significant rate, but polyynes with four or more conjugated triple bonds react readily under mild conditions[112]. Thus, the tetrayne **216** is reduced at room temperature within a few

t-Bu(C≡C)$_4$Bu-*t* (**216**) *t*-Bu(C≡C)$_3$CH=CHBu-*t* (**217**)

minutes and gives the enetriyne **217** in good yield, while the pentayne **218** can be reduced to either **219** or **220** by varying the amount of hydride.

t-Bu(C≡C)$_5$Bu-*t* (**218**) *t*-Bu(C≡C)$_4$CH=CHBu-*t* (**219**) *t*-BuCH=CH(C≡C)$_3$CH=CHBu-*t* (**220**)

D. Cycloaddition

Conjugated diynes can function as dienophiles in Diels–Alder reactions, and adducts have been obtained in which one or both of the triple bonds participates. Most of the studies have involved the use of substituted cyclopentadienones, e.g. tetraphenylcyclopentadienone ('tetracyclone'), which form adducts that undergo decarbonylation to give aromatic hydrocarbons. Butadiyne itself reacts with tetracyclone to give hexaphenylquaterphenyl (**221**) while diphenylbutadiyne gives the mono(**222**)- and di(**223**)-adduct [187, 188]. Bis(4-biphenylyl)butadiyne gives the mono-adduct **224** [189].

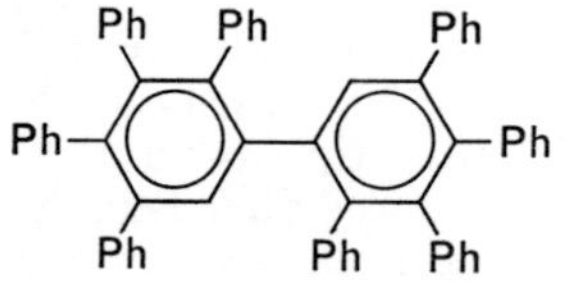

(221)

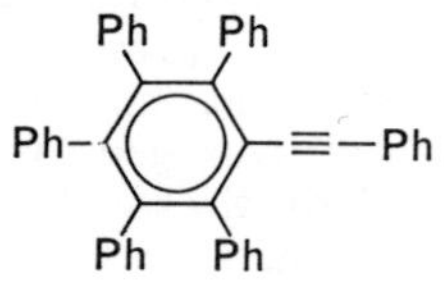

(222)

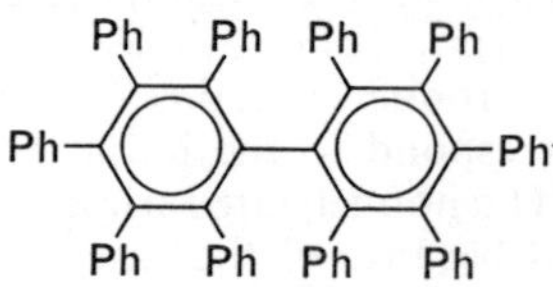

(223)

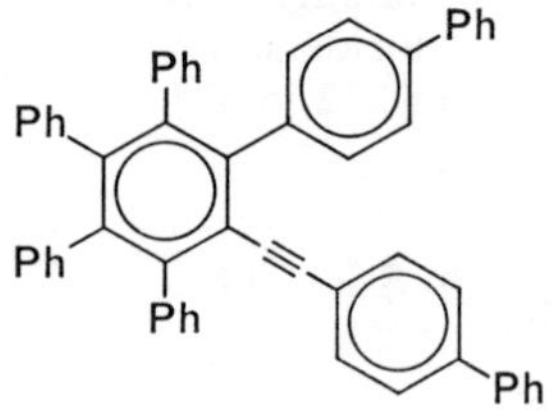

(224)

Diynes can also participate in 1,3-dipolar cycloadditions. The monoadduct **225** is obtained when equimolar proportions of diazomethane and butadiyne are mixed and allowed to stand, while the diadduct **226** is obtained when a 2 : 1 ratio is used[190]. Addition of diazomethane to 1,3-pentadiyne occurs principally at the unsubstituted triple bond to give **227** [191]. The greater reactivity of the double bond

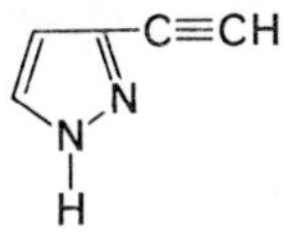

(225)

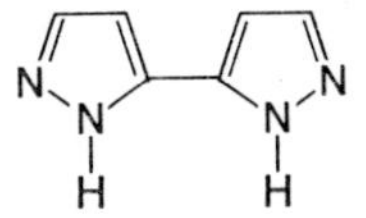

(226)

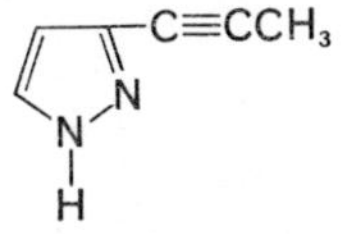

(227)

over the triple bonds in 1-hexene-3,5-diyne is demonstrated by the formation of **228** when the hydrocarbon reacts with a limited amount of benzonitrile oxide[192]; the terminal triple bond will participate in cycloaddition, however, as evidenced by the formation of **229** when the reactants are mixed in 1 : 1 ratio.

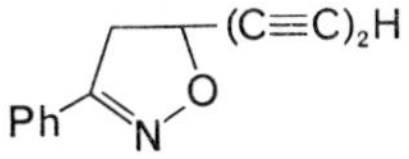

(228)

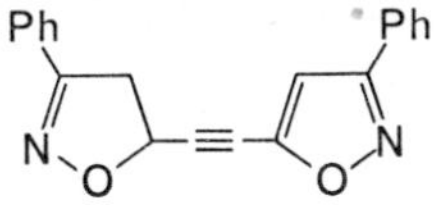

(229)

E. Dimerization of Polyacetylenic Aldehydes

Polyacetylene aldehydes undergo a curious dimerization with loss of carbon monoxide[193–195]. The reaction occurs spontaneously when concentrated solutions of the aldehydes **230** are allowed to stand at room temperature giving both *Z* and *E* isomers of the dimeric aldehydes **231**. *p*-Substituted 5-phenyl-2,4-pentadiynals **(232)**

$$2R(C{\equiv}C)_2CHO \longrightarrow RC{\equiv}CCH{=}C(CHO)(C{\equiv}C)_2R + CO$$

(230) → (231)

$R = Me, Pr, CH_3C{\equiv}C, Ph, PhC{\equiv}C$

are more stable than the unsubstituted derivative and can be isolated as crystalline solids. Thus, whereas dimerization of **232** (X = H) is complete in a few minutes at room temperature, the corresponding reaction of the substituted derivatives occurs slowly in boiling benzene or toluene.

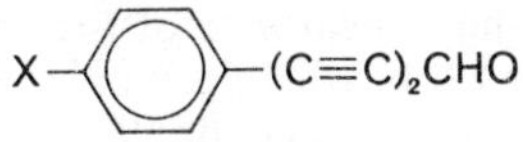

(232)

$X = H, CH_3, Cl, Br, NO_2$

F. Solid-state Polymerization

Certain diacetylene derivatives undergo a remarkable polymerization reaction in the solid state under the influence of heat, ultraviolet light, X-rays or γ radiation[196, 197]. The reaction involves 1,4-addition of the conjugated triple bonds and produces a polymer in which the back-bone is a planar, polyconjugated system **234**. The monomer molecules are aligned in the crystal in a ladder-like manner with the linear diyne system forming the rungs. The polymerization reaction involves tilting of successive molecules, and may proceed with almost no change in lattice parameters.

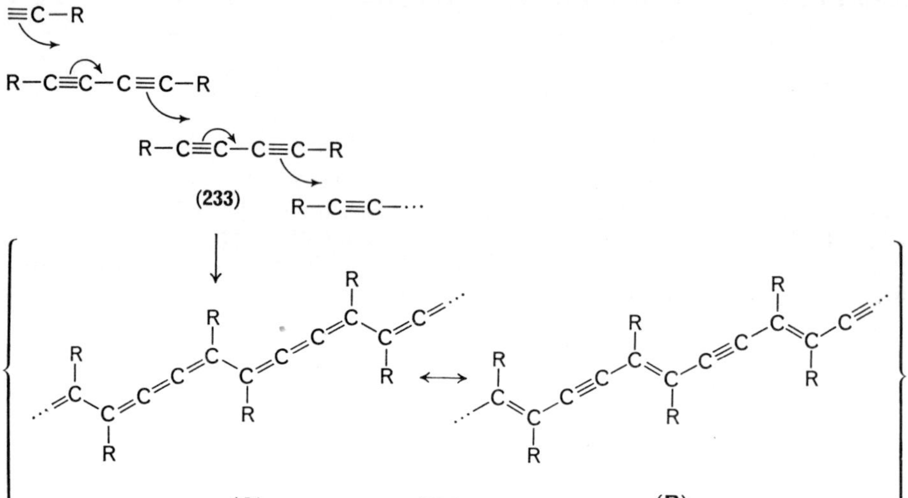

(233)

(A) (234) (B)

(**a**) R = $PhNHCO_2CH_2$
(**b**) R = $TsOCH_2$

A single crystal of monomer becomes a nearly defect-free single crystal of the polymer[198]. The two most commonly used monomers are the phenylurethane and tosylate derivatives of 2,4-hexadiyne-1,6-diol, **233a** and **233b**, but the reaction has been accomplished with a variety of symmetrical and unsymmetrical diacetylene derivatives[199].

X-Ray analysis confirms the structure shown for the polymer, and the bond distances found for the chain indicate that **B** is the major contributing structure[200, 201]. Intense bands for C=C and C≡C in the Raman spectra also indicate that **B** is the major contributor, but the relatively low frequencies for these vibrations as well as the linear correlation found between the two frequencies for various polymers suggest that **A** makes a significant contribution[199]. Both frequencies are found to increase with decreasing phase perfection.

Dramatic changes occur in the appearance of the crystals during polymerization. When crystals of the monomer **233b**, which are pale yellow, are heated to 50 °C the colour changes to red, then to black, and finally a golden metallic lustre appears which is characteristic of the polymer. Absorption spectra recorded during the polymerization show that very long polymer chains are present even at low conversions, and it appears that each initiation step leads almost instantaneously to a long polymer chain[202]. A gradual bathochromic shift during the polymerization, which was originally believed to be caused by an increase in chain length as the polymerization progressed, has been shown to result from contraction of the lattice during polymerization[203]. The absence of an e.s.r. signal during or after the polymerization of a single crystal of **233b** suggests that the lengths of the polymer chains are comparable to the length of the crystal[204]. Weak paramagnetism has been observed in polycrystalline samples of **234b**, and has been interpreted in terms of the departure of polymer chains from an equilibrium conformation[205].

The ribbon-like fibrous polymers formed from **233b** exhibit properties similar to those of metallic and ceramic whiskers[206]. The polymers are semiconductors and exhibit photoconductivity[207]. SCF calculations indicate that polydiacetylenes have a nearly free-electron-like valence band and are best described as wide band-gap semiconductors[208].

A low-temperature splitting of some of the lines in the resonant Raman spectrum of **234b** has been reported recently[209]. Dramatic changes in the frequency and intensity of the C=C and C≡C bands of another polydiacetylene **234**, R = $EtNHCO_2(CH_2)_4$, have been noted with changes in temperature as the polymer passes through a phase transition[210].

Patents have been issued for the use of polydiacetylene derivatives as coloured photoresist films and electrophotographic print-out materials[211–213].

G. Electrophilic Substitution

Cleavage of the alkynyl–metal bond in organometallic derivatives initiated by such agents as H^+, Ag^+, RCO^+ and halogens may be considered formally as electrophilic substitutions.

Cleavage of the alkynyl–metal bond by aqueous acids, as represented by equation (24), occurs under mild conditions when M is Ge, Sn or Pb. The kinetics

$$R(C{\equiv}C)_nMR'_3 + H_2O \xrightarrow{H^+} R(C{\equiv}C)_nH + R'_3MOH \quad (24)$$

have been studied for the cleavage of **235** in aqueous methanolic perchloric acid where $n = 2$ or 3 and Ar is a phenyl or monosubstituted phenyl ring[214]. The mechanism proposed involves rate-determining protonation of the terminal acetylenic carbon followed by rapid nucleophilic attack of solvent on germanium.

$$\underset{\textbf{(235)}}{Ar(C{\equiv}C)_{n-1}{-}C{\equiv}CGeEt_3} + H_3O^+ \longrightarrow Ar(C{\equiv}C)_{n-1}{-}\overset{+}{C}{=}CHGeEt_3 + H_2O$$

$$Ar(C{\equiv}C)_{n-1}{-}\overset{+}{C}{=}CHGeEt_3 + H_2O \longrightarrow Ar(C{\equiv}C)_nH + Et_3Ge\overset{+}{O}H_2$$

The relative rates of cleavage of $Ph(C{\equiv}C)_nGeEt_3$ were found to be 3100 ($n = 1$), 13 ($n = 2$) and 1 ($n = 3$). The decrease in rate as n is changed from 1 to 3, again illustrative of decreasing susceptibility to electrophilic attack with increasing number of conjugated triple bonds, is attributed to decreasing effectiveness of the phenyl group in stabilizing the carbonium ion as well as to the electron-withdrawing inductive effect of the additional alkynyl groups.

Acid cleavage of the tin and lead analogues **236** also occurs readily, kinetic studies showing the rate of reaction of the lead derivative to be approximately three times that of the tin derivative[92].

$$Ph_3M(C{\equiv}C)_2Ph$$

(236) M = Sn, Pb

The alkynyl–silicon bond is not cleaved readily by protonic acids, but cleavage can be accomplished by Ag^+ under mild conditions[184]. For example, treatment of **237** with aqueous ethanolic silver nitrate followed by liberation of the free alkyne from the silver salt by aqueous KCN gave **238** in 80% yield.

$$\underset{\textbf{(237)}}{PrC{\equiv}CCH_2C{\equiv}CSiMe_3} \longrightarrow \underset{\textbf{(238)}}{PrC{\equiv}CCH_2C{\equiv}CH}$$

Friedel–Crafts acylation of the bis(trimethylsilyl)polyynes **239** results in cleavage of one of the trimethylsilyl groups giving **240** in good yields[215]. The trimethylsilyl group in **240** is cleaved by treatment with very mild base, e.g. aqueous borax, and the method constitutes a convenient synthesis of the acyl polyynes **241**. Electron withdrawal by the acyl group in **240** prevents cleavage of the second trimethylsilyl group when excess acid chloride is used.

$$\underset{\textbf{(239)}\ n\,=\,2,\,4}{Me_3Si(C{\equiv}C)_nSiMe_3} \xrightarrow[AlCl_3]{RCOCl} \underset{\textbf{(240)}}{RC(=O)(C{\equiv}C)_nSiMe_3} \xrightarrow{OH^-} \underset{\textbf{(241)}}{RC(=O)(C{\equiv}C)_nH}$$

Butadiynyllithium (**242**) undergoes electrophilic attack by halogens at low temperatures giving monohalobutadiynes **243** [216].

$$H(C{\equiv}C)_2H \xrightarrow{BuLi} \underset{\textbf{(242)}}{H(C{\equiv}C)_2Li} \xrightarrow{X_2} \underset{\textbf{(243)}\ X\,=\,Cl,\,Br,\,I}{H(C{\equiv}C)_2X}$$

H. The Diyne Reaction

Diynes in which the distance between the internal acetylenic carbons, (b) and (c) in **244**, is not greater than *ca.* 3·4 Å react with certain transition metal compounds, particularly tris(triphenylphosphine)rhodium(I) chloride, to give complexes of type

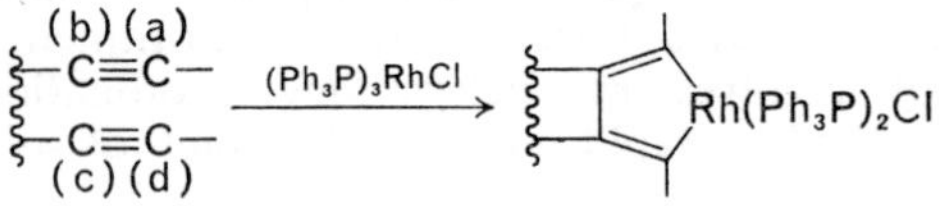

(244) (245)

245. The nature of the skeleton joining the alkyne functions seems to be unimportant and may include sp^3- or sp^2-hybridized carbon atoms as well as heteroatoms. The complexes **245** react with alkynes, halogens, carbon monoxide, isonitriles, etc., to give a variety of carbocyclic and heterocyclic ring systems. These reactions, referred to as the 'diyne reaction', have been studied extensively by Müller and coworkers, and the studies have been reviewed recently[217]. Unless indicated otherwise, the examples cited below are taken from the review article. The generality of the reactions and the wide variety of complex products made available thereby make them of great value to the synthetic organic chemist.

The complexes are prepared simply by heating the diyne with tris(triphenylphosphine)rhodium(I) chloride, abbreviated RhL_3Cl, in an inert solvent. For example, the complex **246** is obtained in 98% yield by heating the reactants in xylene for 30 min.

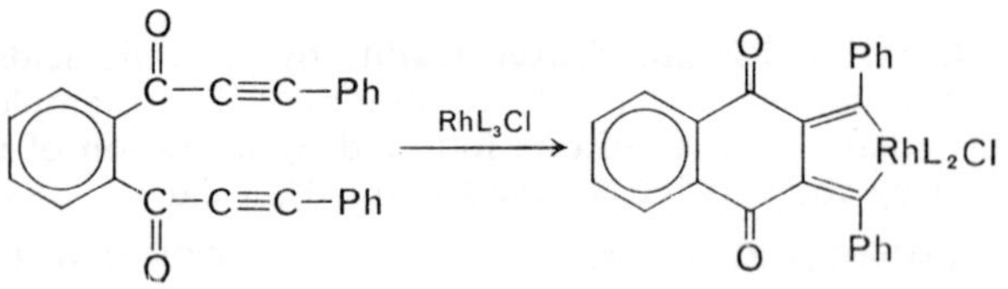

(246)

A new aromatic ring is created in one step when the complexes react with alkynes, as illustrated by the conversion of **246** to the anthraquinone **247** in 68% yield upon reaction with 2-butyne. The reaction is very general for alkynes and fails only when the alkyne is severely hindered sterically or when a heteroatom is present which coordinates strongly with transition metals, e.g. groups containing phosphorus(III). Cycloalkynes, including strained cyloalkynes generated *in situ* from selenadiazoles, react with **246**, but benzyne fails to react[218]. Reaction with oxygen, sulphur, selenium or tellurium gives the corresponding heterocycles **248** [219], while reaction with carbon

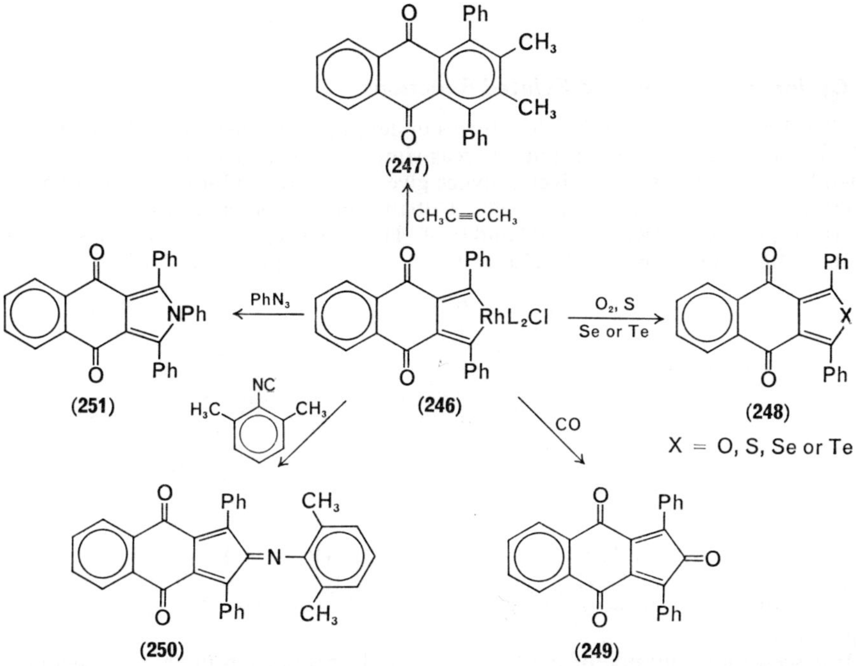

monoxide, carried out by bubbling the gas through a benzene solution of **246** at 60 °C, gives the trione **249** in 60% yield. 2,6-Dimethylphenyl isocyanide reacts with **246** to give the imine **250**, and phenyl azide gives the nitrogen heterocycle **251**, perhaps by way of phenylnitrene as an intermediate. Additional extensions of the reaction have appeared recently[220–222].

In the presence of dicarbonylbis(triphenylphosphine)nickel(0), the diyne **252** reacts with monoalkynes to give quinones **253** [223], and in the presence of nickel tetracarbonyl, **252** dimerizes to give **254** in 80% yield along with a small amount of

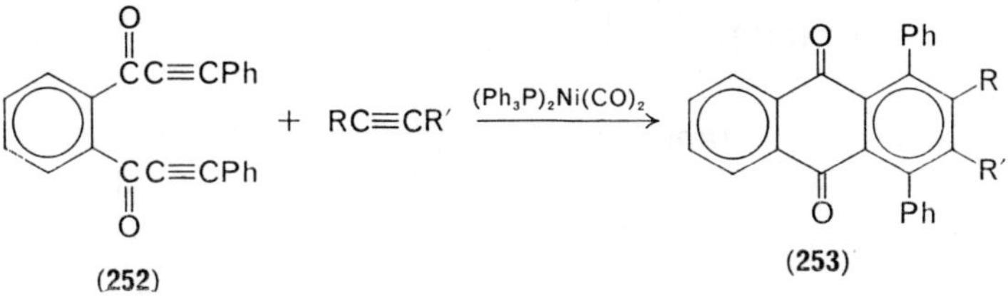

255, while 1,2,3,4-tetraphenylanthraquinone **253** (R = R′ = Ph) is obtained from the reaction of **252** with excess diphenylacetylene in the presence of this catalyst[224].

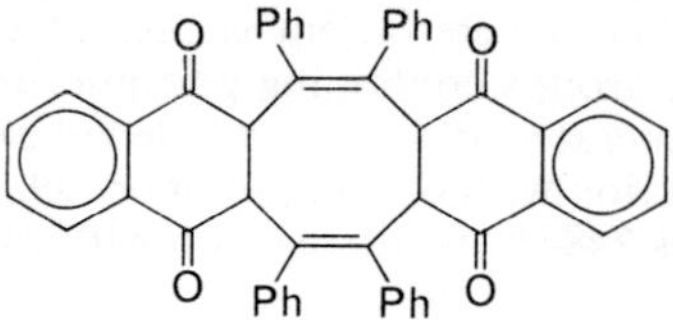

(254)

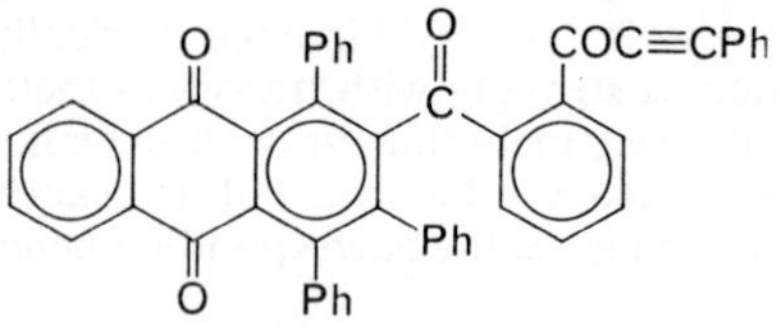

(255)

I. Cyclotrimerization and Related Reactions

Like simple alkynes, conjugated diynes undergo cyclotrimerization in the presence of transition metal catalysts, but whereas simple alkynes give symmetrically substituted benzenes almost exclusively, diynes give symmetrical (**256**) and unsymmetrical (**257**) isomers in many cases[225, 226]. Cobalt complexes such as $[Co(CO)_4]_2Hg$, $Co_2(CO)_8$, $Co_2(CO)_6[R(C{\equiv}C)_2R]$ and $(\eta^5\text{-}C_5H_5)Co(CO)_2$, and the rhodium complex, $(\eta^5\text{-}C_5H_5)Rh(CO)_2$, give both **256** and **257**. For example, with 2,4-hexadiyne and

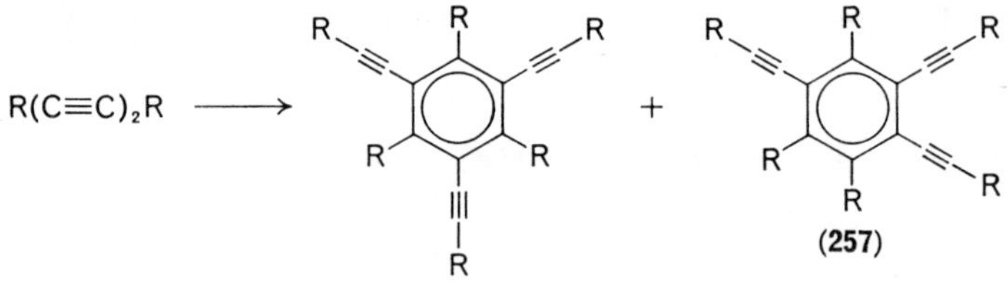

(257)

(256)

$(\eta^5\text{-}C_5H_5)Co(CO)_2$ in a 25 : 1 ratio at 120 °C, **256** (R = Me) and **257** (R = Me) are obtained in 21% and 47% yields, respectively[226]. Interestingly, the catalyst $Ni(CO)_2(PPh_3)_2$ produces solely the unsymmetrical isomers **257** (R = Me, Ph) in high yield[227].

It is seen that only one of the two alkyne linkages in each molecule takes part in the cyclization, and when R is larger than hydrogen, the remaining triple bond in the alkynyl substituents of **256** and **257** fails to participate in further reactions. The triethylnylbenzenes **256** (R = H) and **257** (R = H), however, react in the presence of the trimerization catalysts to give polymers, and consequently attempted cyclotrimerization of butadiyne itself gives a polymeric product[225]. It is possible to obtain the simple triethylnylbenzenes indirectly from the bis(trimethylsilyl) derivatives[225].

Dimers (**259**) and trimers (**260**) have been obtained from non-conjugated diynes in the presence of transition metal complexes[228–230]. With η^5-cyclopentadienylcobalt

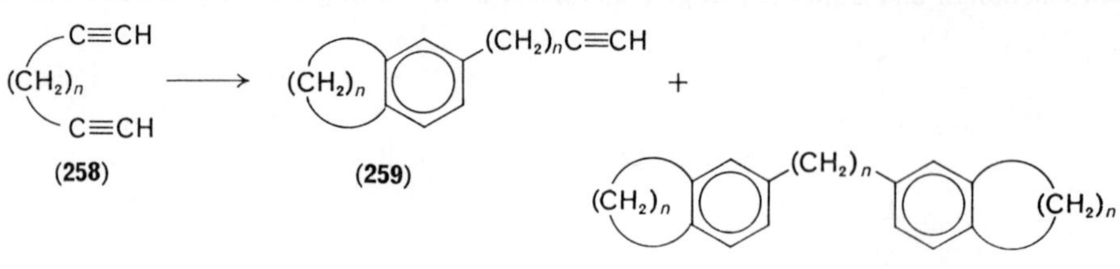

(258) **(259)**

(260)

dicarbonyl catalysts, $CpCo(CO)_2$, moderate yields of the trimer **260** are obtained[230], whereas with $[Co(CO)_4]_2Hg$ a large amount of polymer and only a small yield of dimer are obtained[228].

Triynes **261** in which the alkyne linkages are properly spaced can undergo intramolecular 'cyclotrimerization' to aromatics **262** and **263** in the presence of Ziegler-type catalysts[231].

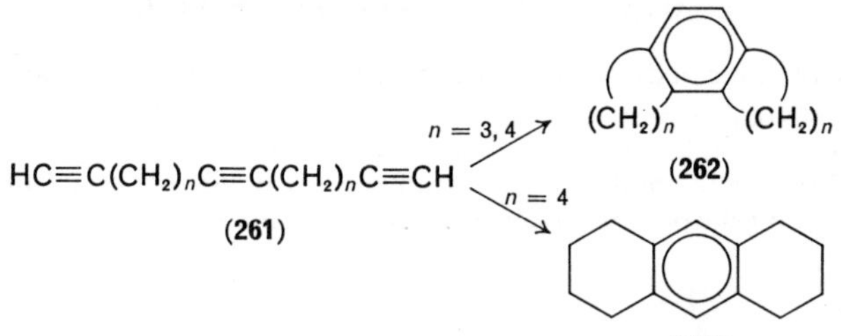

The cooligomerization of diynes and simple alkynes provides a convenient route to bicyclic derivatives, many of which would be very difficult to obtain by other routes[229, 230, 232, 233]. In the presence of $CpCo(CO)_2$, 1,5-hexadiyne reacts with simple alkynes to give benzocyclobutene derivatives **264** [230, 233]. The formation of the bis(trimethylsilyl) derivative **264** ($R = R' = SiMe_3$), the most highly strained

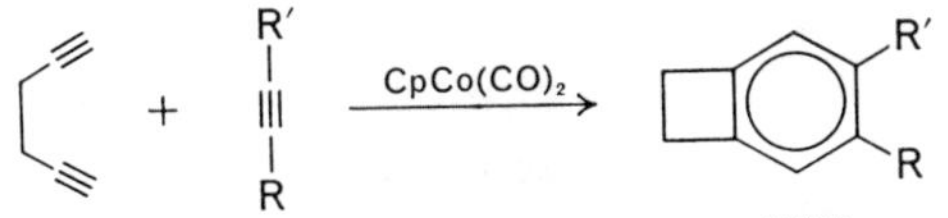

benzocyclobutene synthesized to date, is of special significance because of the ease with which the trimethylsilyl groups can be replaced by reaction with electrophilic reagents.

Indan (**265**, $n = 3$) and tetralin (**265**, $n = 4$) derivatives are obtained from 1,6-heptadiyne or 1,7-octadiyne with simple alkynes in the presence of $CpCo(CO)_2$, and although the yields are not high (14–50%) the products are easily obtained pure by column chromatography[232]. The synthetic applications of these cyclizations have been reviewed recently[233b].

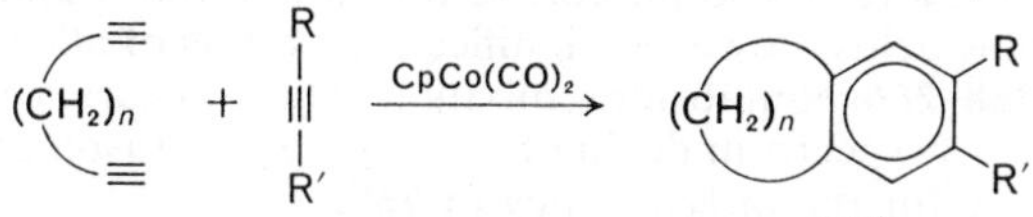

J. Prototropic and Related Rearrangements

The prototropic rearrangements of conjugated diynes closely parallel those observed for simple alkynes, the principal differences being in the rates of reaction and the greater complexity of products made possible by the additional unsaturation. Migration of a triple bond from the terminal position toward the centre of the chain is widely observed with simple alkynes, and similar behaviour has been reported for diynes.

Conjugated diynes with a terminal triple bond give products in which the triple bonds remain conjugated as illustrated by the isomerization of 1,3-hexadiyne to

2,4-hexadiyne and of 1,3-octadiyne to 2,4-octadiyne (equations 25 and 26)[234]. The isomeric diynes are obtained in 90–95% yield under conditions ranging from 0·1N ethanolic KOH at 100 °C for 2 h to 2N ethanolic KOH at room temperature for 1 week.

$$C_2H_5(C{\equiv}C)_2H \xrightarrow{\text{KOH}} CH_3(C{\equiv}C)_2CH_3 \qquad (25)$$

$$Bu(C{\equiv}C)_2H \xrightarrow{\text{KOH}} Pr(C{\equiv}C)_2CH_3 \qquad (26)$$

Four types of intermediates may be postulated for these rearrangements, *viz.* a tetraene, two dienynes and a skipped diacetylene, as shown in Scheme 8[235]. The available evidence indicates that the interconversions occur by a carbanionic mechanism. Substantial build-up of intermediates is not observed in most of these reactions, and the relative importance of the routes has not been established.

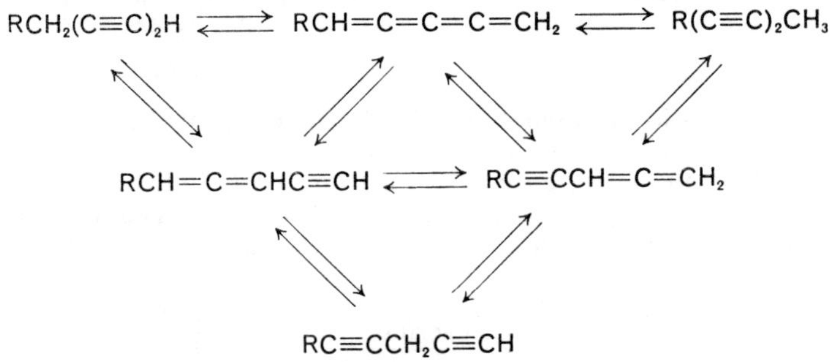

SCHEME 8

Bushby and Whitham carried out a thorough study of the isomerization of 2,4-heptadiynoic acid (**266**) to 3,5-heptadiynoic acid (**268**), and from kinetic and spectroscopic data were able to show that the principal pathway for the isomerization involves the intermediate tetraenoic acid **267**[236].

$$C_2H_5(C{\equiv}C)_2CO_2H \quad CH_3CH{=}C{=}C{=}C{=}CHCO_2H \quad CH_3(C{\equiv}C)_2CH_2CO_2H$$

(266) **(267)** **(268)**

Equilibration of the heptadiynoic acids gives an equilibrium mixture containing 96% **268** and 4% **266**, a ratio comparable to that found for 3- and 2-pentynoic acids. Unlike the pentynoic acids, however, significant amounts of allenic isomers were not detected in the **268**–**266** equilibrium mixture. The appearance of a shoulder at 290–295 nm in the u.v. spectrum during the early stages of isomerization of **266** may be taken as evidence for the intermediacy of **267**.

Attachment of two alkynyl groups to a CH_2 group confers an unusually high acidity on the protons, and as a consequence 1,4-diynes undergo base-catalysed rearrangement under very mild conditions. Thus, 1-phenyl-1,4-pentadiyne (**269**) gives 1-phenyl-1,3-pentadiyne (**271**) in good yield upon treatment with ethanolic

$$PhC{\equiv}CCH_2C{\equiv}CH \longrightarrow PhC{\equiv}CCH{=}C{=}CH_2 \longrightarrow Ph(C{\equiv}C)_2CH_3$$

(269) **(270)** **(271)**

alkali at room temperature[86]. The reaction occurs in two stages, the first being approximately 3 times as fast as the second, and it is possible to isolate **270** by quenching the reaction when the concentration of **270** reaches a maximum.

An interesting variation of this rearrangement appears when the diynol **272** is treated with base[237]. Ketone **274** is obtained, apparently by way of the intermediate allenol **273**.

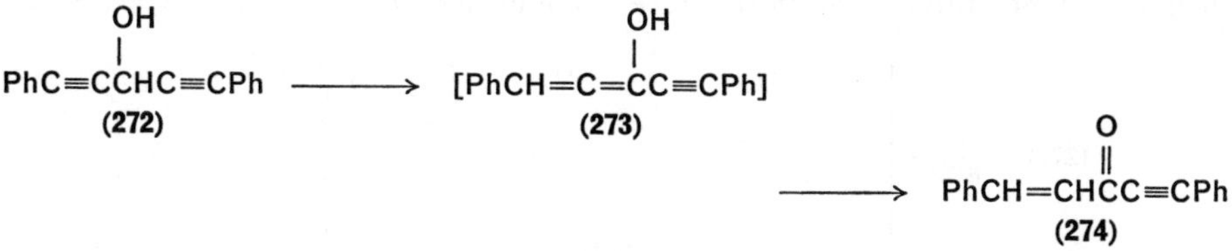

1,5-Diynes require somewhat more vigorous conditions for isomerization than 1,4-diynes. Thus, 1,5-hexadiyne is stable to sodium ethoxide in ethanol or 0·03M potassium *t*-butoxide in *t*-butyl alcohol at room temperature, but isomerization is effected at higher temperatures or by the use of a higher concentration of potassium *t*-butoxide[238]. The reaction is complex, but many of the intricacies have been unravelled, and the relationships are depicted in Scheme 9[121, 238].

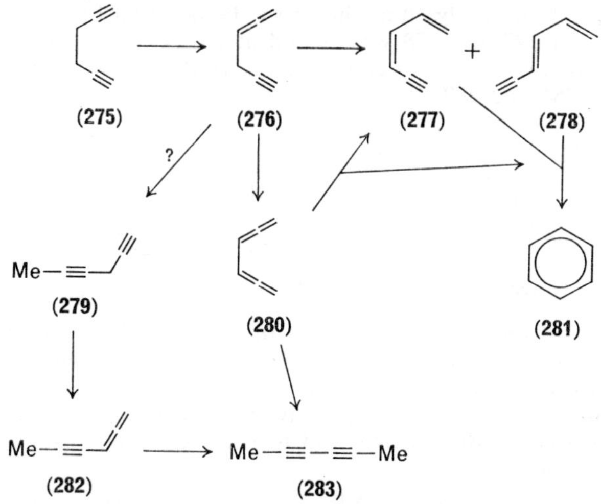

SCHEME 9. Base-catalysed isomerization of 1,5-hexadiyne.

cis- and *trans*-1,3-Hexadien-5-yne (**277** and **278**) are formed as major products, along with small amounts of 2,4-hexadiyne (**283**), when **275** is heated with potassium *t*-butoxide in *t*-butyl alcohol or ethanolic sodium ethoxide. In addition a small amount of 1,2,4,5-hexatetraene (**280**) was present in the reaction mixture involving sodium ethoxide. 1,2-Hexadien-5-yne (**276**) the product expected from the first isomerization step was not detected, but this is understandable because it was demonstrated in separate experiments that **276** isomerizes much faster than the starting diyne. When **276** is treated with *t*-BuOK/*t*-BuOH at room temperature, reaction is immediate, producing a mixture of the dienynes **277** and **278**; on the other hand, when NaOEt is used, **280** is the exclusive product[238].

1,2,4,5-Hexatetraene (**280**) is stable to ethanolic NaOEt at room temperature, but reacts at 65 °C to give **277**, **278** and **283** in the ratio 21 : 22 : 31. Thus **280** occupies a key position in the scheme relating the various C_6H_6 isomers, but it is argued that the rearrangement of **276** may lead directly to **277** and **278** and does not have to proceed by way of **280** [238]. The same hybrid anion **284** can be formed from both **276** and **280**. Furthermore, if the major pathway for conversion of **275** to **277** and **278** involved **280**, then one would anticipate that 2,4-hexadiyne (**283**) would be formed in major amounts, rather than in the trace amounts actually observed.

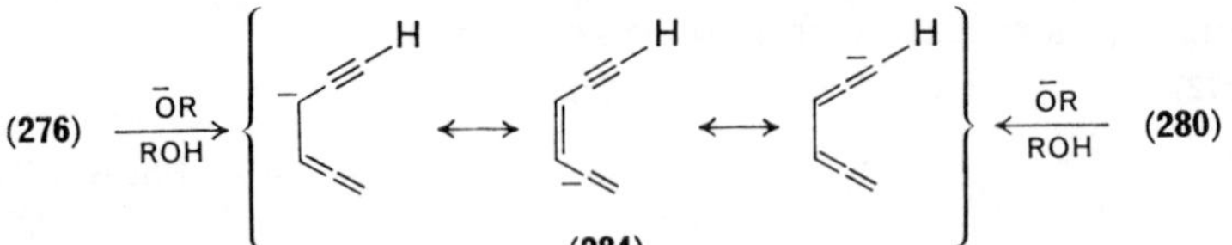

(284)

Another possible route for the formation of **283**, by way of **279** and **280**, was investigated. When **279** was treated with *t*-BuOK/*t*-BuOH it reacted immediately and gave **283** as the sole product, but **282** was obtained along with **283** under milder conditions (0·02M NaOEt/EtOH).

Benzene was obtained in low yield as the sole non-polymeric product when **275**, **276** or a mixture of **277** and **278** was heated at 165 °C with *t*-BuOK in diglyme. The cyclization is base-catalysed as shown by the fact that benzene was not formed when a mixture of **277** and **278** was heated at 165 °C in diglyme alone.

Tetraethynylethane derivatives **285** isomerize in the presence of *t*-BuOK/*t*-BuOH at 40 °C to give **286** and **287** [239].

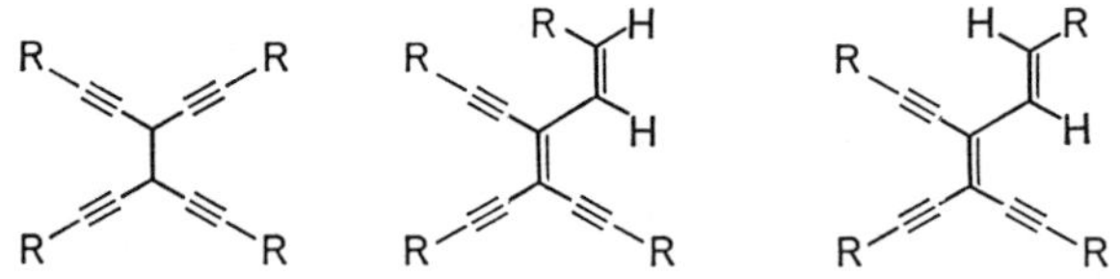

(285) (286) (287)

When 1,6-heptadiyne (**288**) is heated with *t*-BuOK/*t*-BuOH, toluene (**294**) and *trans*-1,3-heptadien-5-yne (**292**) are the major products[121]. Hopf discovered that smooth isomerization to **292** and **294** occurs only when freshly sublimed *t*-BuOK is used, and through the use of aged, less active base he was able to isolate and identify the intermediates **289**, **290** and **293**, and to show that the *cis*-dien-yne **293** is the precursor of toluene[240].

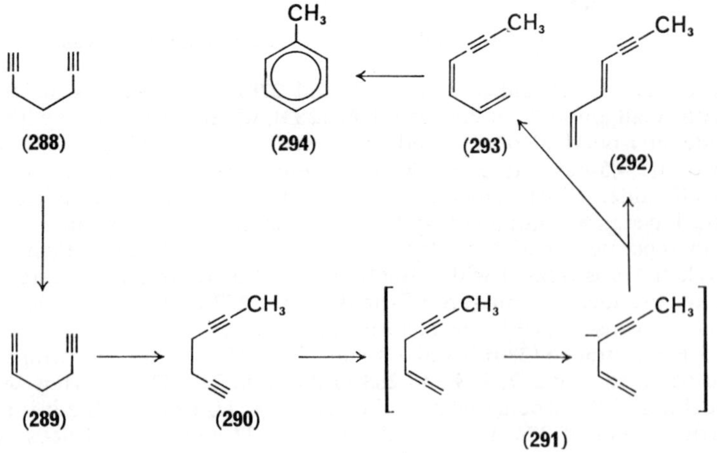

Bispropargyl ethers, sulphides and amines undergo an interesting variety of base-catalysed cycloaddition and dimerization reactions. The primary products obtained from **295** by treatment with *t*-BuOK in THF for a short period of time have been shown to be **296** [241]. If the reactions are carried out under more vigorous conditions or for longer periods of time, further prototropic rearrangement to the naphthalene derivatives **297** occurs. The findings are explained by an initial rearrangement to the

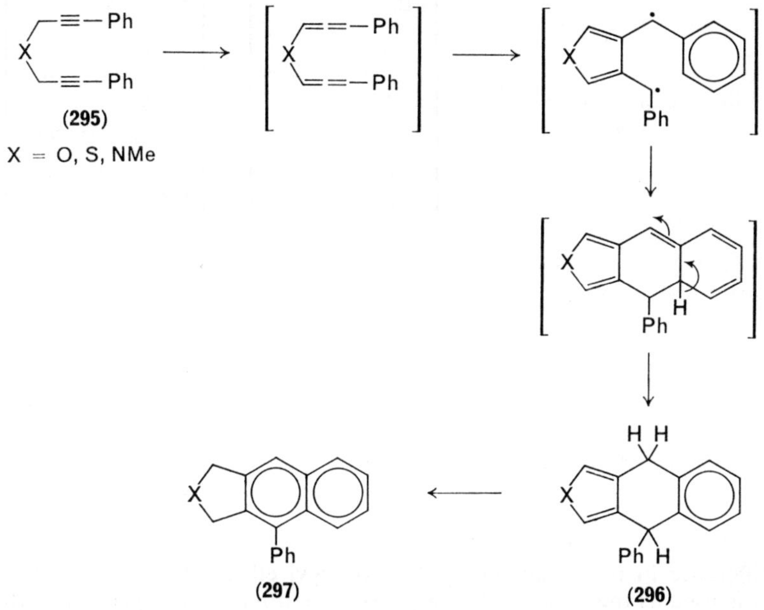

bisallene, which then cyclizes by a diradical pathway, or perhaps by a concerted [2+2+2] cycloaddition. Subsequent prototropic rearrangement gives the products **296**, as indicated. Support for this mechanism is provided by studies of the unsubstituted and *t*-butyl-substituted derivatives **298** [241]. With these derivatives

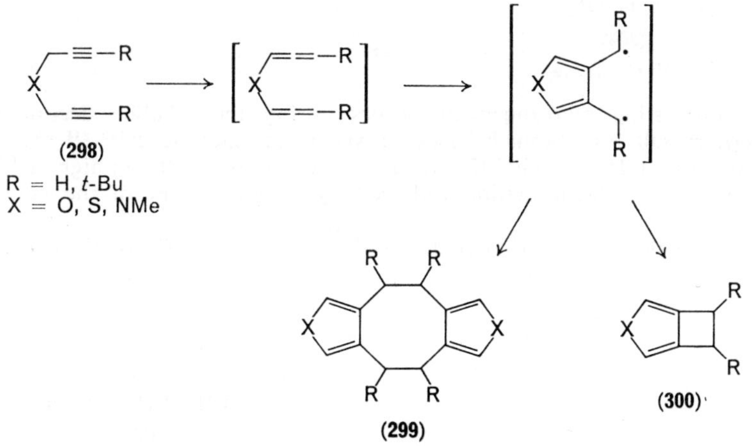

cyclization involving the benzene ring is not possible and products resulting from dimerization (**299**) or ring closure (**300**) of the diradical are obtained instead.

The product of reaction of the bromodiyne **301** with sodium sulphide is the thienocyclobutene **303**, and it has been proposed that this product is formed from the initial thioether **302** by the same type of sequence as that proposed for the formation of **300** [242].

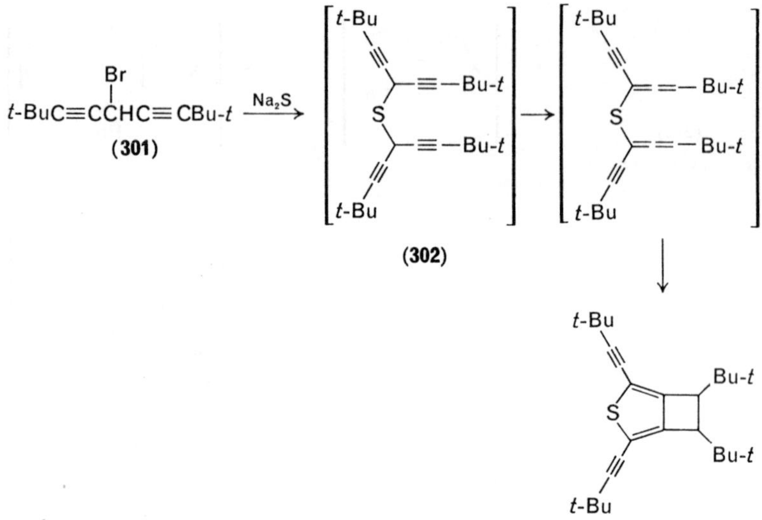

(**301**) (**302**) (**303**)

A difference in the behaviour of bispropargyl ethers **304** appears when butyllithium is used as the base[243]. Here a [2, 3] sigmatropic rearrangement of the intermediate carbanion **305** occurs giving the alcohols **306**.

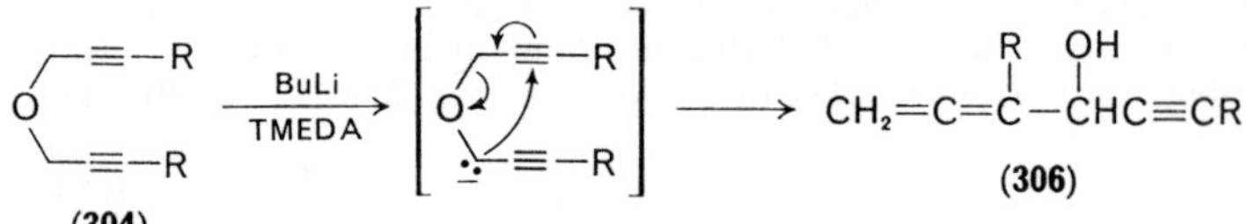

(**304**) (**305**) (**306**)

R = H, Me, $SiMe_3$

An interesting rearrangement occurs during the alkaline decomposition of quaternary salts of Mannich bases derived from diacetylenes[192, 244, 245]. When it is heated with KOH, the salt **307** gives 1-hexene-3,5-diyne (**309**) in high yield, presumably by initial 1,6-elimination followed by rearrangement of the pentaene **308**.

$$CH_3(C{\equiv}C)_2CH_2\overset{+}{N}(Me)Et_2I^- \xrightarrow{KOH} [CH_2{=}C{=}C{=}C{=}C{=}CH_2]$$

(**307**) (**308**)

$$\downarrow$$

$$CH_2{=}CH(C{\equiv}C)_2H$$

(**309**)

When the alkadiynyl chain contains more than seven carbons, mixtures of ene-diynes are obtained.

In the case of aryl-substituted derivatives **310**, the use of KOH or *t*-BuOK as base produces **311** whereas $NaNH_2$ gives **312** [246].

$ArCH_2(C{\equiv}C)_2CH_2\overset{+}{N}(Me)Et_2I^-$ **(310)** $\quad Ar(C{\equiv}C)_2CH{=}CH_2$ **(311)** $\quad ArCH{=}CH(C{\equiv}C)_2H$ **(312)**

K. Thermal Rearrangement

1,5-Hexadiyne rearranges at elevated temperatures (220–400 °C) to give 3,4-bismethylenecyclobutene (**315**) in nearly quantitative yields[247, 248]. The reaction involves a slow [3,3] sigmatropic rearrangement to the bisallene **314**, followed by

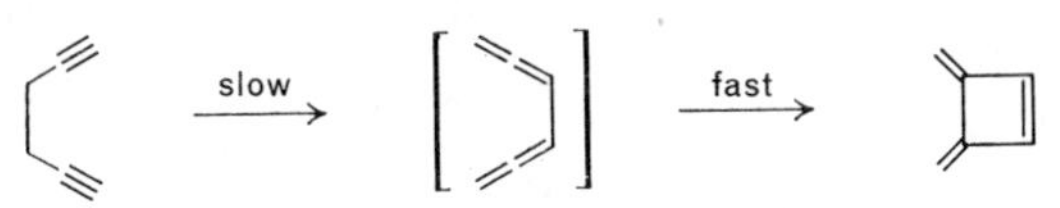

(313) **(314)** **(315)**

rapid cyclization. 1,2,4,5-Hexatetraene (**314**) has been synthesized and shown to rearrange to **315** at a rate much faster than the rearrangement of **313** [249]. The rearrangement of **313** is first order with rate constant given by

$$k\ (s^{-1}) = 2{\cdot}59 \times 10^{11} \exp(-34\ 400/RT).$$

The *A* factor is unusually small for a reaction in which a single internal rotation is frozen out in the transition state, but this has been accounted for in terms of the unusually large moment of inertia for rotation about the central carbon–carbon bond[250].

When **313** is subjected to very high temperatures (400–600 °C) additional products, principally benzene and fulvene, begin to appear[251]. These have been shown to be secondary products which arise as a result of the reversibility of the cyclization step leading to **315**, and the different modes of cyclization which are accessible to **314** at the elevated temperatures[252].

Substituted bismethylenecyclobutenes **317** are obtained by rearrangement of derivatives **316** in which the acetylenic hydrogens of 1,5-hexadiyne are replaced by

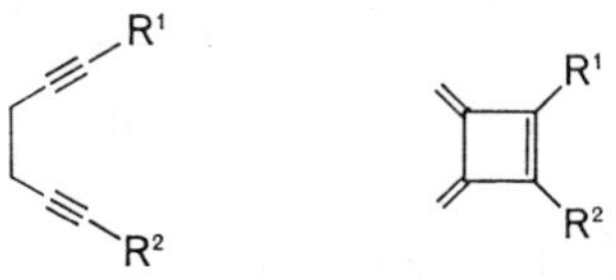

(316) **(317)**

simple alkyl groups[248, 253–257], CF_3 [258], Me_3Si [259], Br [260], CO_2Me and CN [250, 258, 259]. The reactivity is decreased substantially by alkyl substitution as illustrated by the relative rates 137 : 37 : 1 for 1,5-hexadiyne, 1,5-heptadiyne and 2,6-octadiyne[258]. The rearrangement proceeds normally when one hydrogen is replaced by *t*-butyl or trimethylsilyl, but fails with the bis(trimethylsilyl) derivative **316** ($R^1 = R^2 = Me_3Si$) [259].

Studies with molecules carrying substituents on the interior carbons have demonstrated that conrotation occurs preferentially in the cyclization step, in agreement with orbital symmetry considerations[248, 258, 261]. Thus, (*Z,E*)-3,4-bisethylidenecyclobutene (**319**, R = Me) is obtained from *meso*-3,4-dimethyl-1,5-hexadiyne (**318**, R = Me), while the *E,E* (**321**, R = Me) and *Z,Z* isomers (**322**, R = Me) are obtained from the racemic diyne **320** (R = CH_3). At lower temperatures (220 °C) **321** and **322** are formed

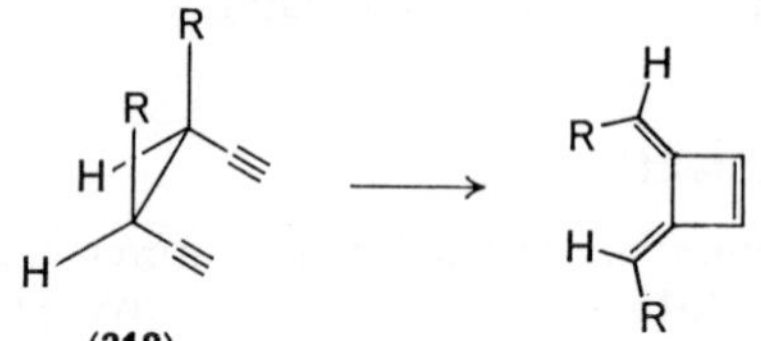

(318) (319)

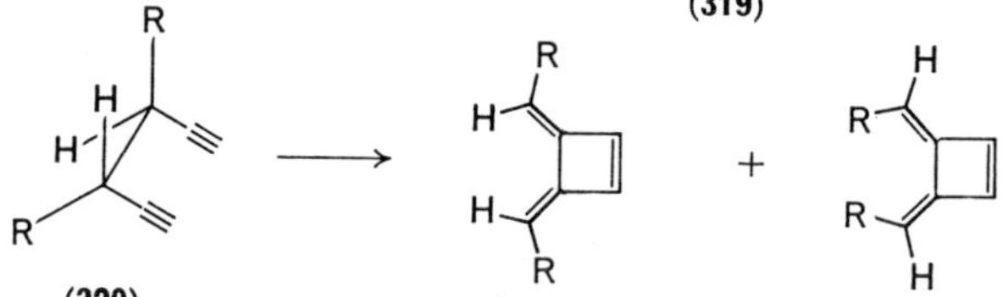

(320) (321) (322)

in nearly equal amounts (55 : 45) under kinetic control. At higher temperatures, however, thermodynamic control prevails and **321** becomes the predominant product[258]. The trimethylsilyl (TMS) ether of the *meso* glycol **318** (R = OTMS) rearranges at 375 °C giving a product consisting of 99% **319** (R = OTMS) and 1% **322** (R = OTMS), while the product from the racemic diyne **320** (R = OTMS) at 310 °C consists of 49% **321** (R = OTMS), 51% **322** (R = OTMS) and 1% **319** (R = OTMS). The stereospecificity of these reactions decreases significantly as the temperature is raised, but one of the most interesting features is the preference for the *Z,Z* isomer **322** (R = OTMS) over the *E,E* isomer **321** (R = OTMS) under conditions where thermodynamic control is expected to prevail[261].

Rearrangement of 1,5-hexadiyn-3-ol gives the aldehyde **323** and phenol[262]. With the diol **324**, however, the intermediate enol **325** tautomerizes instead of cyclizing and the acyclic diketone **326** is obtained.

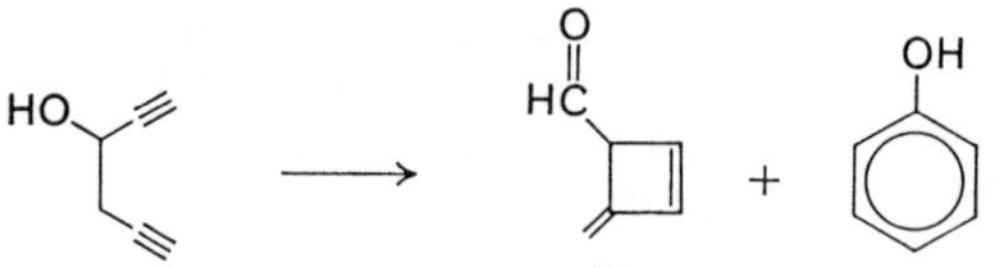

(323)

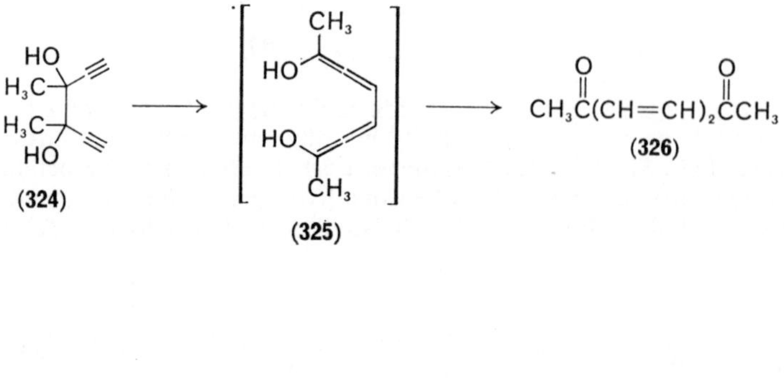

(324) (325) (326)

Rearrangement also occurs with molecules in which the ethynyl groups are attached to three- and four-membered rings. Both *cis*- and *trans*-1,2-diethynylcyclopropane (**327**, X = CH_2) give **328** (X = CH_2), the *trans* isomer requiring a significantly higher temperature than the *cis*[263, 264]. The product molecules are formed in a highly excited vibrational state, and when the reaction is carried out at low pressures, **329** and **330** are also formed by rearrangement of **328** in the excited state.

(**327**) (**328**) (**329**) (**330**)

X = CH_2, O, S

Rearrangement of *cis*- and *trans*-1,2-diethylnyloxirane (**327**, X = O) occurs under mild conditions giving **328** (X = O)[265], but with the sulphur analogue **327** (X = S) only the *cis* isomer rearranges to **328** (X = S)[266]; the *trans* isomer undergoes desulphurization on heating and gives 3-hexene-1,5-diyne.

1,2-Dihydropentalene (**335**) (95%) and a small amount (2·5%) of bicyclo[4.2.0]-octa-1,5,7-triene (**334**) are formed upon pyrolysis of *cis*-1,2-diethylnylcyclobutane (**333**)[267]. The fragmentation product, 1-buten-3-yne (**332**) is the major component (52%) of the pyrolysate from *trans*-1,2-diethylnylcyclobutane (**331**), but the rearrangement products **335** and **334** are also formed to the extent of 42% and 4·5%, respectively. These products are accounted for as illustrated in Scheme 10[267].

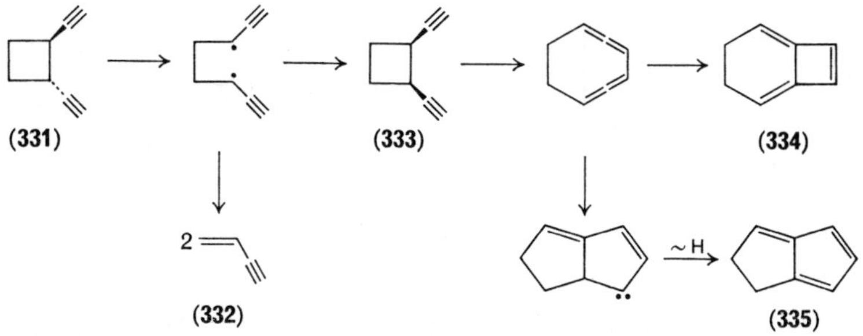

SCHEME 10

The *cis*-dialkynylcyclobutene **336** undergoes the customary conrotatory ring opening at 80 °C giving **337** [268], but in the case of the *trans* isomer **338**, the initial product (**339**) recyclizes, giving the benzocyclobutadiene derivative **340** [269]. The formation of **340** from **338** is complete within a few minutes at 110 °C, whereas

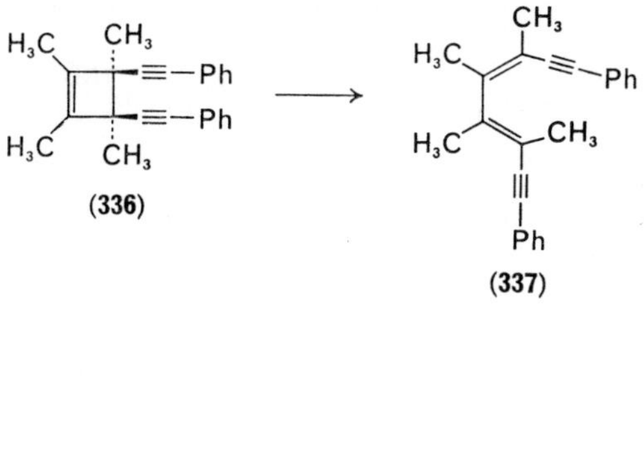

its dimerization, which yields **341**, is relatively slow at this temperature. Consequently, it is possible to isolate and characterize **340** [269].

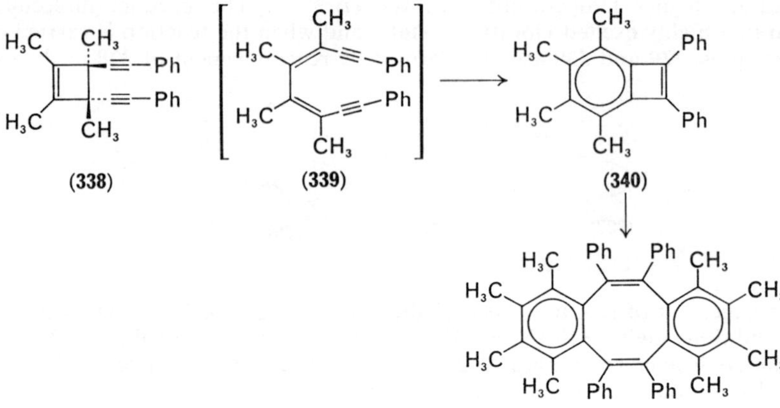

(**338**) (**339**) (**340**) (**341**)

Comparable behaviour has been observed with **343**, the unsubstituted analogue of **339**. When the bis(trimethylsilyl) derivative **342** is hydrolysed with dilute base, the product obtained is the benzocyclobutadiene dimer **344** presumably formed by the pathway indicated[270].

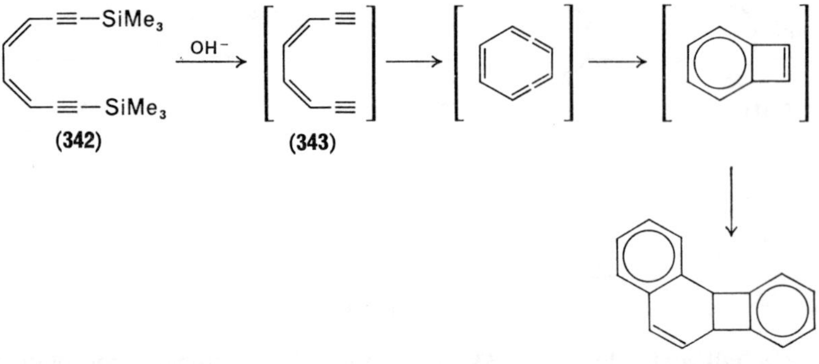

(**342**) (**343**) (**344**)

The tricyclic derivative containing two cyclobutadiene rings **346** is obtained as a blue, high-melting solid in nearly quantitative yield by heating **345** in boiling xylene[271].

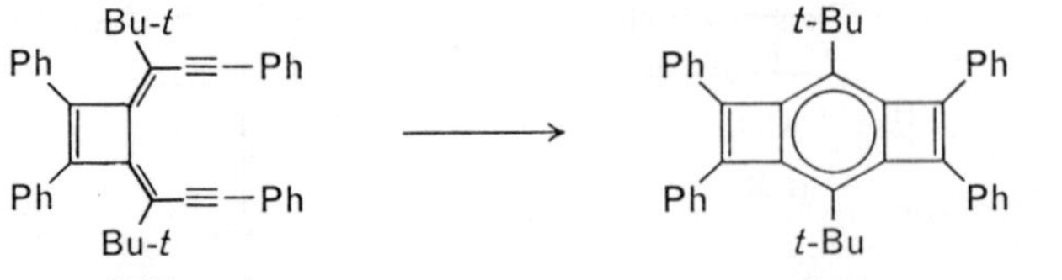

(**345**) (**346**)

The rearrangement that occurs when thioethers of type **347** are heated in the presence of secondary amines provides an interesting analogy to 1,5-diynes[272]. The products are **349**, **350** and **351**, and it is postulated that these arise by the reaction of the amine with the thioketene **348**, which, in turn, is formed by a [3, 3] sigmatropic rearrangement of **347** [272].

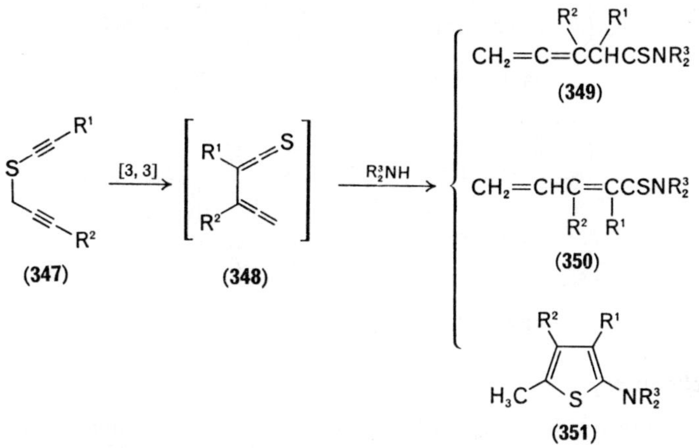

Rapid equilibration between **352** and **354** occurs at 200 °C, presumably via the benzenediyl (*p*-benzyne) intermediate **353** [273]. The intermediate **353** has a sufficiently

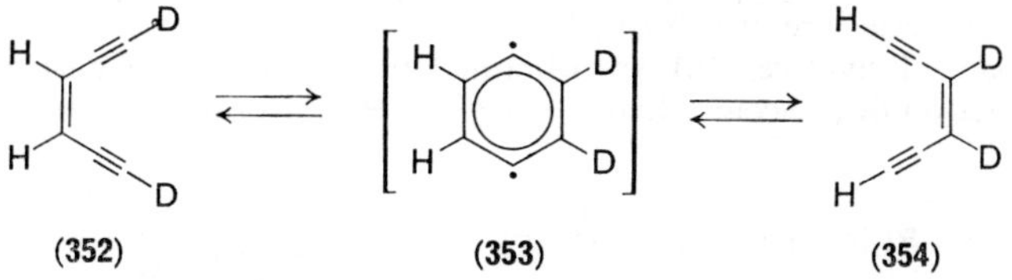

long life-time to permit reaction with a variety of trapping reagents. The benzenoid analogue **355** of **352** does not change when it is heated in boiling benzene or DMF[274].

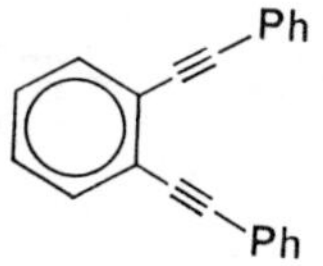

(355)

1,8-Bis(phenylethynyl)naphthalene **(356)**, with parallel triple bonds, undergoes [2+2+2] cycloaddition with intramolecular hydrogen migration to give **357** [274–276], and similar behaviour is found for the analogue with crossed triple bonds **(358)**, which furnishes **359**. These same products are obtained by photochemical rearrangements, although small amounts of azulenic isomers and dimers are also formed from **356**.

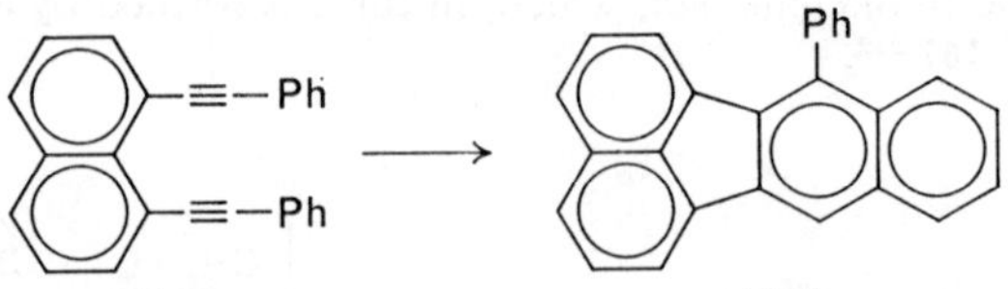

(356) **(357)**

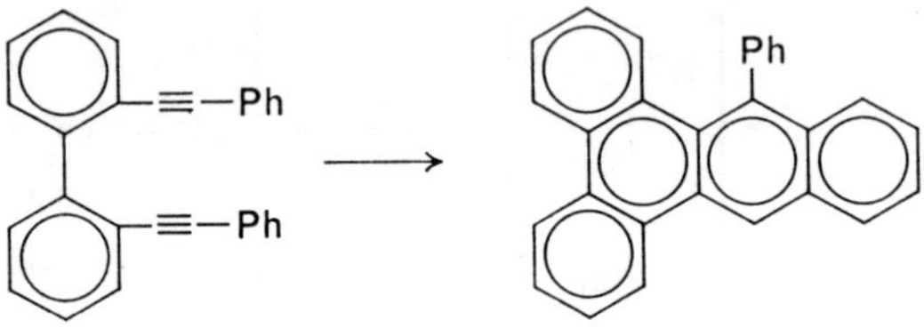

(358) **(359)**

L. Photochemistry

When they are irradiated in hydrogen–donor solvents, conjugated diynes act as hydrogen acceptors and are reduced to conjugated enynes[277]. Complex mixtures are obtained as a result of the fact that free radicals are formed from both the diyne and a solvent moelcule in the initial hydrogen abstraction step. Thus irradiation of 5,7-dodecadiyne in pentane gives the reduction product, *cis*- and *trans*-5-dodecen-7-yne, along with branched-chain decanes formed by dimerization of pentyl radicals, and addition products, $C_{17}H_{30}$, of a molecule of pentane to the diyne. Large amounts of polymeric material are also formed.

Cyclopropane derivatives **361** are obtained by irradiation of derivatives **360** [278]. While the reaction does occur when $n = 1$ (R = Ph), it proceeds much more readily

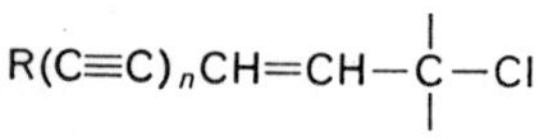

(360)

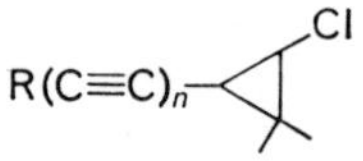

(361)

with molecules containing two or three alkyne linkages. A mixture of stereoisomers **363** and **364** is obtained from **362**, and, in general, low stereoselectivity is observed.

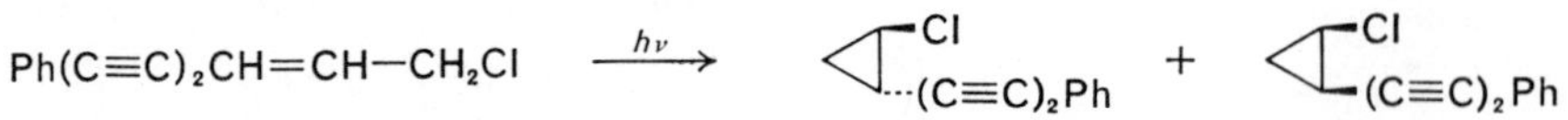

(362) **(363)** **(364)**

Ultraviolet irradiation of dialkynylbenzenes produces polymers along with small amounts of dimers. The dimer fraction consists of azulenes and, in some cases, naphthalenes. The product distribution in the dimer fraction depends on the relative orientation of the alkynyl groups as well as the substituents on the alkynyl groups and on the ring. *o*-Diethynylbenzene (**365**) produces azulenes **366** and **367** [279], while *p*-diethynylbenzene yields **368** and **369** [280]. These products can be accounted for in terms of the mechanism originally proposed for the photodimerization of diphenylacetylene[281]. The cyclobutadiene derivative **370**, formed by a $({}_{\pi}2_s + {}_{\pi}2_s)$ head-to-head

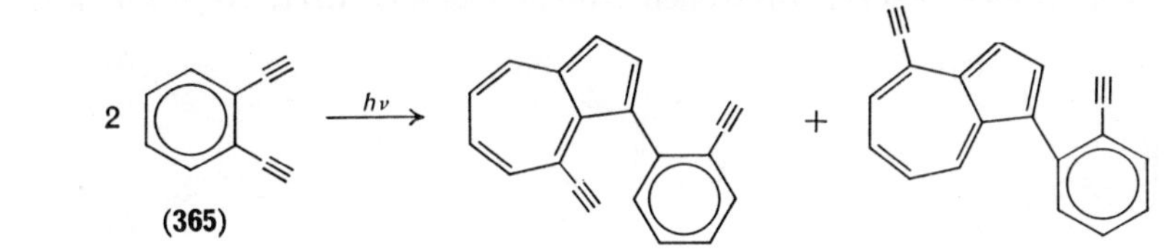

(365) (366) (367)

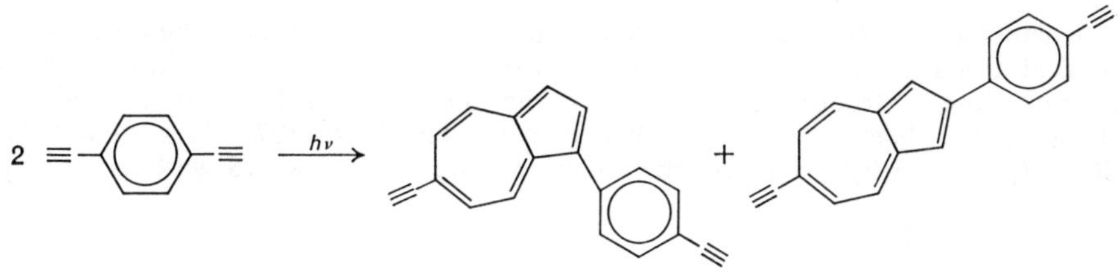

(368) (369)

cycloaddition, rearranges to the bicyclobutane derivative **371**. Cleavage of bonds a and c in **371** provides **366**, **367** and **368**. Formation of **369** can be accounted for in terms of an initial head-to-tail dimerization followed by a sequence of steps similar to those above[280].

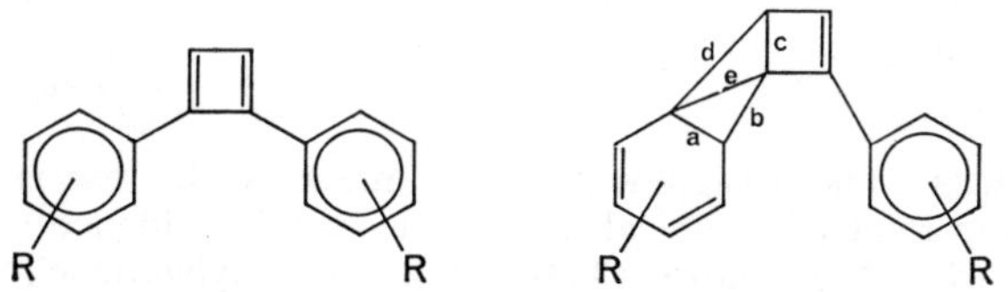

(370) (371)

Irradiation of *p*-bis(phenylethynyl)benzene (**372**) gives one azulene derivative **373** and a naphthalene derivative **374**. Formation of **374** can be accounted for in terms of a bicyclobutane intermediate similar to **371**. Cleavage of bonds corresponding to c and e in **371** and subsequent hydrogen migration would lead to **374** [280].

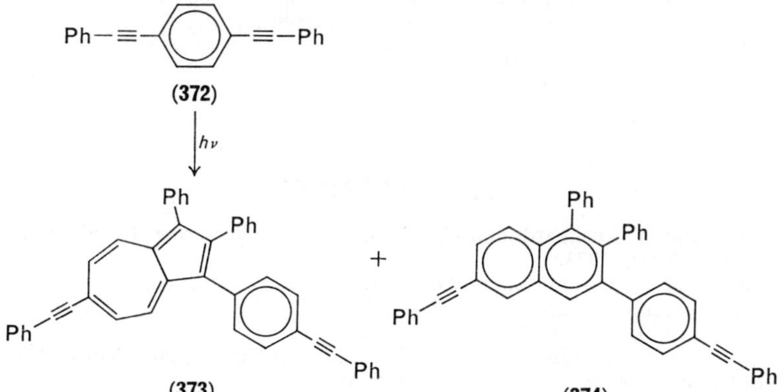

(372) (373) (374)

Behaviour similar to that of **365** is observed with *o*-dipropynylbenzene **(375)** and 1,2-bis(phenylethynyl)cyclohexene **(376)**, but *o*-bis(arylethynyl)benzenes **(377)** show different behaviour[274, 282, 283]. *o*-Bis(phenylethynyl)benzene **(377**, Ar = Ph**)** gives a

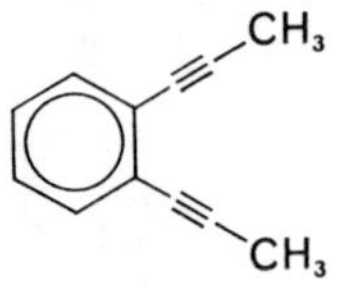

(375)

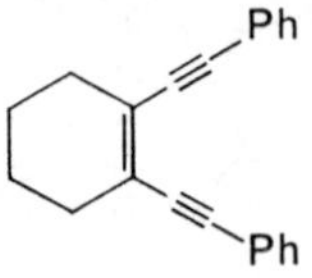

(376)

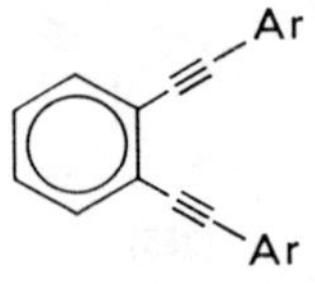

(377)

green azulenic type dimer called 'verdene', but unlike the dimers described previously, chemical and spectroscopic evidence points to the absence of alkyne linkages, and the 'ring-closed' structure **378** (Ar = Ph) has been assigned to verdene. Similar behaviour has been noted for other diaryl derivatives (**377**: Ar = *p*-NCC_6H_4, *p*-BrC_6H_4 and 2,6-$F_2C_6H_3$), and the corresponding structures **378** were assigned to

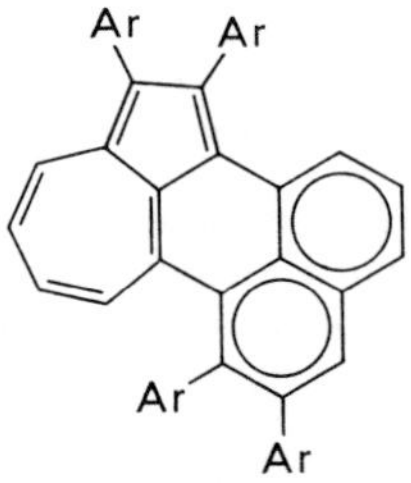

(378)

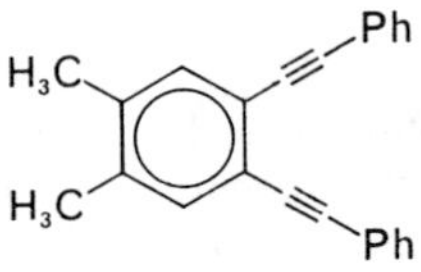

(379)

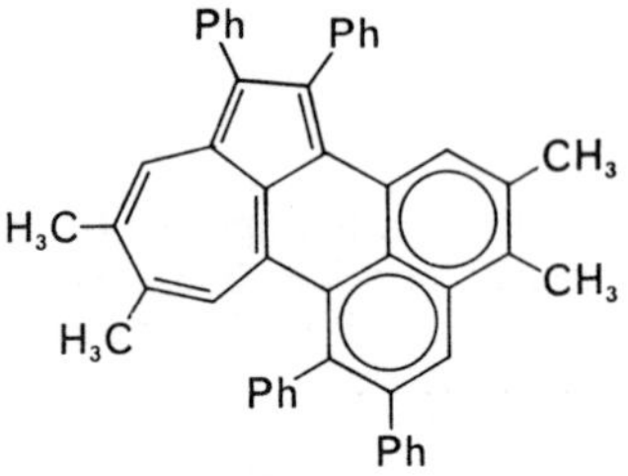

(380)

the photoproducts. The assignments were made with reservations, however, because the same kind of chemical and spectroscopic evidence points toward an analogous 'ring-closed' structure **(380)** for 'tetramethylverdene', the photodimer of **379**, but X-ray analysis shows the structure to be **381** [284].

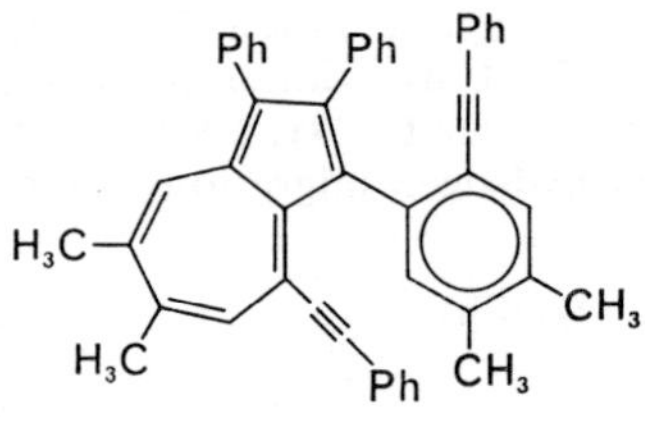

(381)

V. REFERENCES

1. F. Bohlmann, T. Burkhardt and C. Zdero, *Naturally Occurring Acetylenes*, Academic Press, New York, 1973.
2. R. Eastmond, T. R. Johnson and D. R. M. Walton, *Tetrahedron*, **28**, 4601 (1972).
3. M. F. Shostakovskii and A. V. Bogdanova, *The Chemistry of Diacetylenes*, John Wiley and Sons, New York, 1974.
4. H. G. Viehe (Ed.), *Chemistry of Acetylenes*, Marcel Dekker, New York, 1969.
5. F. Bohlmann, *Angew. Chem.*, **65**, 385 (1953).
6. F. Bohlmann and H. J. Mannhardt, *Fortschr. Chem. Org. Naturstoffe*, **14**, 1 (1957).

7. R. J. Tedeschi and A. E. Brown, *J. Org. Chem.*, **29**, 2051 (1964).
8. P. E. Driedger and H. V. Isaacson, *Chem. Eng. News*, **50** (Mar. 20), 51 (1972).
9. M. Ford and D. R. M. Walton, *Chem. Eng. News*, **50** (July 24), 67 (1972).
10. E. Bjarnov, D. H. Christensen, O. F. Nielsen, E. Augdahl, E. Kloster-Jensen and A. Rogstad, *Spectrochim. Acta, Part A*, **30**, 1255 (1974).
11. E. Kloster-Jensen, *Angew. Chem. Int. Ed. Eng.*, **11**, 438 (1972).
12. F. Bohlmann, *Chem. Ber.*, **86**, 657 (1953).
13. E. R. H. Jones, H. H. Lee and M. C. Whiting, *J. Chem. Soc.*, 3483 (1960).
14. T. R. Johnson and D. R. M. Walton, *Tetrahedron*, **28**, 5221 (1972).
15. C. L. Cook, E. R. H. Jones and M. C. Whiting, *J. Chem. Soc.*, 2883 (1952).
16. J. B. Armitage, N. Entwistle, E. R. H. Jones and M. C. Whiting, *J. Chem. Soc.*, 147 (1954).
17. F. Sondheimer, R. M. McQuilkin and P. J. Garratt, *J. Amer. Chem. Soc.*, **92**, 6682 (1970).
18. J. Dale in *Chemistry of Acetylenes* (Ed. H. G. Viehe), Marcel Dekker, New York, 1969, pp. 3–96.
19. B. F. Coles, P. B. Hitchcock and D. R. M. Walton, *J. Chem. Soc. Dalton Trans.*, 442 (1975).
20. G. H. Mansfield and M. C. Whiting, *J. Chem. Soc.*, 4761 (1956).
21. C. Eaborn, R. Eastmond and D. R. M. Walton, *J. Chem. Soc. (B)*, 127 (1971).
22. C. Eaborn, A. R. Thompson and D. R. M. Walton, *J. Chem. Soc. (B)*, 357 (1970).
23. E. S. Stern and C. J. Timmons, *Introduction to Electronic Absorption Spectroscopy in Organic Chemistry*, 3rd ed., St. Martin's Press, New York, 1971, pp. 88–92.
24. E. Kloster-Jensen, H. J. Haink and H. Christen, *Helv. Chim. Acta*, **57**, 1731 (1974).
25. J. B. Armitage, C. L. Cook, N. Entwistle, E. R. H. Jones and M. C. Whiting, *J. Chem. Soc.*, 1998 (1952).
26. J. B. Armitage, C. L. Cook, E. R. H. Jones and M. C. Whiting, *J. Chem. Soc.*, 2010 (1952).
27. J. B. Armitage, E. R. H. Jones and M. C. Whiting, *J. Chem. Soc.*, 2014 (1952).
28. F. Bohlmann, *Chem. Ber.*, **86**, 63 (1953).
29. H. Bock and H. Seidl, *J. Chem. Soc. (B)*, 1158 (1968).
30. H. Christen and E. Kloster-Jensen, *Helv. Chim. Acta*, **56**, 1752 (1973).
31. M. Nakagawa, S. Akiyama, K. Nakasuji and K. Nishimoto, *Tetrahedron*, **27**, 5401 (1971), and references cited therein.
32. S. Akiyama and M. Nakagawa, *Bull. Chem. Soc. Japan.*, **45**, 259 (1972).
33. K. Nakasuji, S. Akiyama and M. Nakagawa, *Bull. Chem. Soc. Japan.*, **45**, 883 (1972).
34. K. Nakasuji, S. Akiyama and M. Nakagawa, *Bull. Chem. Soc. Japan.*, **45**, 875 (1972).
35. K. Nakasuji, S. Akiyama, K. Akashi and M. Nakagawa, *Bull. Chem. Soc. Japan.*, **43**, 3567 (1970).
36. S. Akiyama and M. Nakagawa, *Bull. Chem. Soc. Japan.*, **40**, 340 (1967).
37. S. Akiyama and M. Nakagawa, *Bull. Chem. Soc. Japan.*, **43**, 3561 (1970).
38. A. V. Jones, *Proc. Roy. Soc. London, Ser. A*, **211**, 285 (1952).
39. E. Kloster-Jensen and R. Tabacchi, *Tetrahedron Letters*, 4023 (1972).
40. P. Jouve and M. P. Simonnin, *C. R. Hebd. Séances Acad. Sci. Ser. C*, **257**, 121 (1963).
41. E. I. Snyder and J. D. Roberts, *J. Amer. Chem. Soc.*, **84**, 1582 (1962).
42. I. R. Peat and W. F. Reynolds, *Can. J. Chem.*, **51**, 2968 (1973).
43. D. E. Dorman, M. Jautelat and J. D. Roberts, *J. Org. Chem.*, **38**, 1026 (1973).
44. R. Zeisberg and F. Bohlmann, *Chem. Ber.*, **108**, 1040 (1975).
45. R. Zeisberg and F. Bohlmann, *Chem. Ber.*, **107**, 3800 (1974).
46. M. T. Hearn, J. *Mag. Reson.*, **19**, 401 (1975).
47. R. T. Aplin and S. Safe, *Chem. Comm.*, 140 (1967).
48. F. Bohlmann, C. Zdero, H. Bethke and D. Schumann, *Chem. Ber.*, **101**, 1553 (1968).
49. C. Baker and D. W. Turner, *Proc. Roy. Soc. London, Ser. A*, **308**, 19 (1968).
50. F. Brogli, E. Heilbronner, V. Hornung and E. Kloster-Jensen, *Helv. Chim. Acta*, **56**, 2171 (1973).
51. F. Brogli, E. Heilbronner, J. Wirz, E. Kloster-Jensen, R. G. Bergman, K. P. Vollhardt and A. J. Ashe, *Helv. Chim. Acta*, **58**, 2620 (1975).
52. P. Bischof, R. Gleiter, H. Hopf and F. T. Lenich, *J. Amer. Chem. Soc.*, **97**, 5467 (1975).

53. E. Heilbronner, V. Hornung, J. P. Maier and E. Kloster-Jensen, *J. Amer. Chem. Soc.*, **96**, 4252 (1974).
54. P. Cadiot and W. Chodkiewicz in *Chemistry of Acetylenes* (Ed. H. G. Viehe), Marcel Dekker, New York, 1969, pp. 597–647.
55. W. G. Nigh in *Oxidation in Organic Chemistry*, Part B (Ed. W. S. Trahanovsky), Academic Press, New York, 1973, pp. 11–31.
56. G. Eglinton and W. McRae, *Adv. Org. Chem.*, **4**, 225 (1963).
57. C. Glaser, *Ber.*, **2**, 422 (1869); *Ann.*, **137**, 154 (1870).
58. A. S. Hay, *J. Org. Chem.*, **27**, 3320 (1962).
59. R. Eastmond and D. R. M. Walton, *Chem. Comm.*, 204 (1968).
60. F. Bohlmann, H. Schönowsky, E. Inhoffen and G. Grau, *Chem. Ber.*, **97**, 794 (1964).
61. L. G. Fedenok, V. M. Berdnikov and M. S. Shvartsberg, *Zh. Org. Khim*, **11**, 2492 (1975).
62. L. G. Fedenok, V. M. Berdnikov and M. S. Shvartsberg, *Zh. Org. Khim.*, **10**, 922 (1974).
63. L. G. Fedenok, V. M. Berdnikov and M. S. Shvartsberg, *Zh. Org. Khim.*, **9**, 1781 (1973).
64. E. G. Derouane, J. N. Braham and R. Hubin, *Chem. Phys. Letters.*, **25**, 243 (1974).
65. A. A. Clifford and W. A. Waters, *J. Chem. Soc.*, 3056 (1963).
66. H. J. Kevelam, K. P. de Jong, H. C. Meinders and G. Challa, *Makromol. Chem.*, **176**, 1369 (1975).
67. A. S. Hay, *German Patent* 1,158,272 (1963); *Chem Abstr.*, **60**, 7953 (1964).
68. D. M. White, *German Patent*, 2,207,354 (1972); *Chem Abstr.*, **78**, 17487 (1973).
69. D. M. White, *Polym. Prepr. Amer. Chem. Soc. Div. Polym. Chem.*, **12**, 155 (1971).
70. D. M. White, *U.S. Patent*, 3,821,153 (1974); *Chem Abstr.*, **81**, 153721 (1974).
71. I. Rhee, M. Ryang and S. Tsutsumi, *Tetrahedron Letters.*, 4593 (1969).
72. V. V. Gavrilenko, L. L. Ivanov and L. I. Zakharkin, *Zh. Obshch. Khim.*, **37**, 550 (1967).
73. A. Pelter, K. Smith and M. Tabata, *J. Chem. Soc. Chem. Commun.*, 857 (1975).
74. J. A. Sinclair and H. C. Brown, *J. Org. Chem.*, **41**, 1078 (1976).
75. W. Chodkiewicz, *Ann. Chim.* (*Paris*) [13], **2**, 819 (1957).
76. F. Bohlmann, P. Herbst and H. Gleinig, *Chem. Ber.*, **94**, 948 (1961).
77. R. Eastmond and D. R. M. Walton, *Tetrahedron*, **28**, 4591 (1972).
78. B. N. Ghose and D. R. M. Walton, *Synthesis*, 890 (1974).
79. P. D. Landor, S. R. Landor and P. Leighton, *J. Chem. Soc. Perkin Trans.*, 1, 1628 (1975).
80. A. Sevin, W. Chodkiewicz and P. Cadiot, *Bull. Soc. Chim. Fr.*, 913 (1974).
81. J. Burdon, P. L. Coe, C. R. Marsh and J. C. Tatlow, *J. Chem. Soc. Perkin Trans.*, *1*, 639 (1972).
82. L. Yu. Ukhin, A. M. Sladkov and V. I. Gorshkov, *Zh. Org. Khim.*, **4**, 25 (1968).
83. Y. Hori, K. Noda, S. Kobayashi and H. Taniguchi, *Tetrahedron Letters*, 3563 (1969).
84. W. J. Gensler and A. P. Mahadevan, *J. Amer. Chem. Soc.*, **77**, 3076 (1955).
85. D. A. Ben-Efraim and F. Sondheimer, *Tetrahedron*, **25**, 2823 (1969).
86. H. Taniguchi, I. M. Mathai and S. I. Miller, *Org. Syn.*, **50**, 97 (1970).
87. W. J. Gensler and A. P. Mahadevan, *J. Amer. Chem. Soc.*, **78**, 167 (1956).
88. G. D. Gracheva and A. I. Zahkarova, *Zh. Org. Khim.*, **2**, 965 (1966).
89. T. L. Jacobs and P. Prempree, *J. Amer. Chem. Soc.*, **89**, 6177 (1967).
90. R. S. Macomber, *J. Org. Chem.*, **38**, 816 (1973).
91. L. Skattebøl, *Tetrahedron*, **21**, 1357 (1965).
92. J. C. Masson, M. Le Quan and P. Cadiot, *Bull. Soc. Chim. Fr.*, 1085 (1968).
93. H. Hauptmann, *Tetrahedron Letters*, 3587 (1974).
94. H. Hauptmann, *Angew. Chem. Int. Ed. Eng.*, **14**, 498 (1975).
95. H. Hauptmann, *Tetrahedron*, **33**, 1293 (1976).
96. G. Köbrich and P. Buck in *Chemistry of Acetylenes* (Ed. H. G. Viehe), Marcel Dekker, New York, 1969, pp. 99—168.
97. J. B. Armitage, E. R. H. Jones and M. C. Whiting, *J. Chem. Soc.*, 44 (1951).
98. J. B. Armitage, E. R. H. Jones and M. C. Whiting, *J. Chem. Soc.*, 1993 (1952).
99. F. Bohlmann, *Chem. Ber.*, **84**, 545 (1951).
100. F. Bohlmann, *Chem. Ber.*, **84**, 785 (1951).
101. H. Schlubach and V. Franzen, *Justus Liebigs Ann. Chem.*, **572**, 116 (1951).

102. V. N. Andrievskii, I. L. Kotlyarevskii and M. S. Schvartsberg, *Izv. Akad. Nauk SSSR, Ser. Khim.*, 882 (1966).
103. M. S. Shvartsberg, S. F. Vasilevskii, V. G. Kostrovskii and I. L. Kotlyarevskii, *Khim. Geterosikl. Soedin.*, **5**, 1055 (1969).
104. W. Hunsmann, *Chem. Ber.*, **83**, 213 (1950).
105. D. G. Walker, D. A. Keyworth and J. R. Sudduth, *U.S. Patent*, 3,870,482 (1974); *Chem. Abstr.*, **82**, 155274 (1975).
106. I. D. Gay, G. B. Kistiakowsky, J. V. Michael and H. Niki, *J. Chem. Phys.*, **43**, 1720 (1965).
107. C. F. Cullis and N. H. Franklin, *Proc. Roy. Soc. London, Ser. A*, **280**, 139 (1964).
108. M. H. Back, *Can. J. Chem.*, **49**, 2199 (1971).
109. B. N. Al'tschuler, *Kinet. Katal.*, **15**, 835 (1974).
110. R. Grziwa, G. Rosskamp and R. N. Schindler, *Z. Naturforsch., Part B*, **29**, 442 (1974).
111. G. Rosskamp and H. Suhr, *Justus Liebigs Ann. Chem.*, 1478 (1975).
112. F. Bohlmann, E. Inhoffen and J. Politt, *Justus Liebigs Ann. Chem.*, **604**, 207 (1957).
113. Yu. A. Zaichenko, L. N. Reshetova and I. A. Maretina, *Zh. Org. Khim.*, **7**, 2461 (1971).
114. N. A. Pogorzhel'skaya, I. D. Dobreva, A. N. Belyaeva, I. A. Maretina and A. A. Petrov, *Zh. Org. Khim.*, **4**, 1143 (1968).
115. A. N. Volkov, A. N. Khudyakova, V. V. Keiko and G. N. Mazneva, *Zh. Org. Khim.*, **10**, 953 (1974).
116. M. F. Shostakovskii, A. V. Bogdanova and G. I. Plotnikova, *Dokl. Akad. Nauk SSSR*, **124**, 107 (1959).
117. A. N. Volkov, L. V. Sokolyanskaya, E. O. Tsetlina and R. Ya. Mushii, *Zh. Org. Khim.*, **9**, 64 (1973).
118. M. F. Shostakovskii, I. A. Chekulaeva, L. V. Kondrat'eva and B. V. Lopatin, *Izv. Akad. Nauk SSSR, Ser. Khim.*, 2217 (1962).
119. B. P. Gusev and V. F. Kucherov, *Izv. Akad. Nauk SSSR, Ser. Khim.*, 208 (1974).
120. F. Bohlmann and H. Viehe, *Chem. Ber.*, **88**, 1017 (1955).
121. D. A. Ben-Efraim and F. Sondheimer, *Tetrahedron*, **25**, 2837 (1969).
122. E. N. Prilezhaeva, G. S. Vasil'ev and V. N. Petrov, *Izv. Akad. Nauk SSSR, Ser. Khim.*, 2217 (1967).
123. Yu. I. Porfir'eva, L. B. Sokolov and A. A. Petrov, *Zh. Obshch. Khim.*, **34**, 1881 (1964).
124. A. N. Volkov, Yu. M. Skvortsov, I. I. Danda and M. F. Shostakovskii, *Zh. Org. Khim.*, **6**, 897 (1970).
125. F. Bohlmann, H. Bornowski and D. Kramer, *Chem. Ber.*, **96**, 584 (1963).
126. I. L. Mikhelashvili and E. N. Prilezhaeva, *Zh. Org. Khim.*, **10**, 2524 (1974).
127. F. Bohlmann, W. von Kap-herr, C. Rybak and J. Repplinger, *Chem. Ber.*, **98**, 1736 (1965).
128. W. Schroth, F. Billig and G. Reinhold, *Angew. Chem. Int. Ed. Eng.*, **6**, 698 (1967).
129. Yu. I. Porfir'eva, E. S. Turbanova and A. A. Petrov, *Zh. Obshch. Khim.*, **34**, 3966 (1964).
130. F. Bohlmann and H. Hummel, *Chem. Ber.*, **101**, 2506 (1968).
131. M. F. Shostakovskii, A. N. Volkov, A. N. Khudyakova, *Khim. Atsetilena, Tr. Vses. Konf.*, 94 (1968); *Chem Abstr.*, **79**, 18257 (1973).
132. A. A. Petrov and I. A. Maretina, *Zh. Obshch. Khim.*, **29**, 2458 (1959).
133. I. A. Maretina, M. A. Kirillova, F. S. Khamzin and A. A. Petrov, *Zh. Org. Khim.*, **4**, 1138 (1968).
134. M. F. Shostakovskii, I. A. Chekulaeva and L. V. Kondrat'eva, *Izv. Akad. Nauk SSSR, Ser. Khim.*, 1690 (1959).
135. M. F. Shostakovskii, I. A. Chekulaeva, L. V. Kondrat'eva and B. V. Lopatin, *Dokl. Akad. Nauk SSSR*, **135**, 101 (1960).
136. A. A. Petrov and I. A. Maretina, *Zh. Obshch. Khim.*, **30**, 696 (1960).
137. W. Schroth, J. Peschel and Z. Zschunke, *Z. Chem.*, **9**, 110 (1969).
138. W. Schroth, J. Peschel and Z. Zschunke, *Z. Chem.*, **9**, 108 (1969).
139. B. P. Gusev and V. F. Kucherov, *Izv. Akad. Nauk SSSR, Ser. Khim.*, 204 (1974).
140. A. N. Volkov, L. V. Sokolyanskaya and R. Ya. Mushii, *Zh. Org. Khim.*, **9**, 1599 (1973).
141. E. A. El'perina, B. P. Gusev and V. F. Kucherov, *Izv. Akad. Nauk SSSR, Ser. Khim.*, 1803 (1966).

142. B. P. Gusev, E. A. El'perina and V. F. Kucherov, *Izv. Akad. Nauk SSSR*, 2345 (1967).
143. B. P. Gusev, E. A. El'perina and V. F. Kucherov, *Khim. Geterotsikl. Soedin.*, **7,** 1530 (1971).
144. A. A. Ul'yanov and I. A. Maretina, *Zh. Org. Khim.*, **7,** 2052 (1971).
145. B. P. Gusev and V. F. Kucherov, *Izv. Akad. Nauk SSSR, Ser. Khim.*, 206 (1974).
146. R. Mestres, *J. Chem. Soc., Perkin I,* 805 (1972).
147. K. E. Schulte, J. Reisch and L. Hörner, *Chem. Ber.*, **95,** 1943 (1962).
148. K. E. Schulte, J. Reisch, W. Herrmann and G. Bohn, *Arch. Pharm.*, **296,** 456 (1963).
149. R. F. Curtis, S. N. Hasnain and J. A. Taylor, *Chem. Comm.*, 365 (1968).
150. F. Fringuelli and A. Taticchi, *Ann. Chim. (Rome)*, **62,** 777 (1972).
151. B. P. Gusev, I. I. Nazarova, and V. F. Kucherov, *Khim. Atsetilena*, 88 (1968); *Chem. Abstr.*, **71,** 61082 (1969).
152. H. Meister, *Chem. Ber.*, **98,** 2862 (1965).
153. K. E. Schulte, J. Reisch and H. Walker, *Chem. Ber.*, **98,** 98 (1965).
154. E. G. Darbinyan, Yu. B. Mitardzhyan and S. G. Matsoyan, *Arm. Khim. Zh.*, **23,** 640 (1970); *Chem. Abstr.*, **74,** 76362 (1971).
155. W. W. Paudler and A. G. Zeiler, *J. Org. Chem.*, **34,** 999 (1969).
156. E. G. Darbinyan, Yu. B. Mitardzhyan, G. A. Eliazyan and S. G. Matsoyan, *Arm. Khim. Zh.*, **24,** 743 (1971); *Chem Abstr.*, **76,** 14424 (1971).
157. W. Schroth, J. Peschel and A. Zschunke, *Z. Chem.*, **9,** 143 (1969).
158. I. B. Bystrova, I. A. Chekulaeva, A. V. Ignatenko and V. A. Ponomarenko, *Izv. Akad. Nauk SSSR.*, 1163 (1973).
159. A. J. Chalk, *Tetrahedron*, **30,** 1387 (1974).
160. Yu. A. Zaichenko, I. A. Maretina and A. A. Petrov, *Zh. Org. Khim.*, **8,** 1342 (1972).
161. M. A. Kirillova, I. A. Maretina and A. A. Petrov, *Zh. Org. Khim.*, **7,** 14 (1971).
162. L. A. Dorofeeva, Yu. I. Porfir'eva and A. A. Petrov, *Zh, Org. Khim.*, **9,** 449 (1973).
163. G. M. Mkryan, N. A. Papazyan, E. S. Voskanyan, D. G. Rafaelyan and G. A. Chukhaddzhyan, *Armyansk. Khim. Zh.*, **19,** 192 (1966); *Chem. Abstr.*, **65,** 10479 (1966).
164. L. A. Dorofeeva, Yu I. Porfir'eva and A. A. Petrov, *Zh. Org. Khim.*, **8,** 2002 (1972).
165. D. G. Rafaelyan, A. P. Adamyan, N. A. Papazyan and G. M. Mkryan, *Arm. Khim. Zh.*, **28,** 585 (1975); *Chem. Abstr.*, **84,** 4383 (1976).
166. G. M. Mkryan, D. G. Rafaelyan and N. A. Papazyan, *Arm. Khim. Zh.*, **27,** 748 (1974); *Chem. Abstr.*, **82,** 86037 (1975).
167. M. Rösner and G. Köbrich, *Angew. Chem. Int. Ed. Eng.*, **13,** 741 (1974).
168. M. Rösner and G. Köbrich, *Angew. Chem. Int. Ed. Eng.*, **14,** 708 (1975).
169. H. H. Schlubach and E. W. Trautschold, *Justus Liebigs Ann. Chem.*, **594,** 67 (1955).
170. H. Fischer and H. Fischer, *Tetrahedron Letters*, 435 (1969).
171. H. W. Whitlock, Jr. and P. E. Sandvick, *J. Amer. Chem. Soc.*, **88,** 4525 (1966).
172. F. Bohlmann, H. Sinn, J. Politt and E. Inhoffen, *Chem. Ber.*, **89,** 1281 (1956).
173. G. D. Mel'nikov, Yu. I. Porfir'eva and A. A. Petrov, *Zh. Org. Khim.*, **8,** 1557 (1972).
174. Yu. I. Porfir'eva, L. A. Vasil'eva, E. S. Turbanova and A. A. Petrov, *Zh. Org. Khim.*, **5,** 591 (1961).
175. A. A. Petrov, Yu. I. Porfir'eva and E. S. Turbanova, *Zh. Org. Khim.*, **3,** 1339 (1967).
176. G. Zweifel and N. L. Poston, *J. Amer. Chem. Soc.*, **92,** 4068 (1970).
177. H. Lindlar and R. Dubuis, *Org. Syn.*, **46,** 89 (1966).
178. A. Steenhoek, B. H. Van Wijngaarden and H. J. J. Pabon, *Rec. Trav. Chim.*, **90,** 961 (1971).
179. J. Jacob and G. Grimmer, *Tetrahedron Letters*, 1633 (1969).
180. S. P. Mel'nikova, L. N. Vorob'ev and A. A. Petrov, *Zh. Org. Khim.*, **8,** 252 (1972).
181. S. P. Mel'nikova, L. N. Vorob'ev and A. A. Petrov, *Zh. Org. Khim.*, **7,** 1327 (1971).
182. L. B. Sokolov, L. N. Vorob'ev, Yu. I. Porfir'eva and A. A. Petrov, *Zh. Org. Khim.*, **1,** 1544 (1965).
183. C. Eaborn and D. R. M. Walton, *J. Organomet. Chem.*, **4,** 217 (1965).
184. H. M. Schmidt and J. F. Arens, *Rec. Trav. Chim.*, **86,** 1138 (1967).
185. E. J. Corey and R. A. Ruden, *Tetrahedron Letters*, 1495 (1973).
186. A. B. Holmes, R. A. Raphael and N. K. Wellard, *Tetrahedron Letters*, 1539 (1976).
187. W. Dilthey, W. Schommer, W. Höschen and H. Dierichs, *Ber.*, **68,** 1159 (1935).
188. M. A. Ogliaruso and E. I. Becker, *J. Org. Chem.*, **30,** 3354 (1965).

189. W. Ried and V. B. Saxena, *Justus Liebigs Ann. Chem.*, **739,** 159 (1970).
190. R. Kuhn and K. Henkel, *Justus Liebigs Ann. Chem.*, **549,** 279 (1941).
191. L. B. Sokolov, Yu. I. Porfir'eva and A. A. Petrov, *Zh. Org. Khim.*, **1,** 610 (1965).
192. K. Schulze, *Z. Chem.*, **15,** 216 (1975).
193. A. Gorgues and E. Levas, *C. R. Hebd. Séances Acad. Sci., Ser. C*, **266,** 41 (1968).
194. A. Gorgues and E. Levas, *C. R. Hebd. Séances Acad. Sci., Ser. C*, **268,** 1905 (1969).
195. R. K. Bentley, U. Graf, E. R. H. Jones, R. A. M. Voss, V. Thaller and R. A. Vere Hodge, *J. Chem. Soc. (C)*, 683 (1969).
196. G. Wegner, *Z. Naturforsch. B*, **24,** 824 (1969).
197. G. Wegner, *Advan. Chem. Ser.*, No. 129, 255 (1973).
198. J. Kaiser, G. Wegner and E. W. Fischer, *Isr. J. Chem.*, **10,** 157 (1972).
199. R. H. Baughman, J. D. Witt and K. C. Yee, *J. Chem. Phys.*, **60,** 4755 (1974).
200. E. Hädicke, E. C. Mez, C. H. Krauch, G. Wegner and J. Kaiser, *Angew. Chem. Int. Ed. Eng.*, **10,** 266 (1971).
201. D. Kobelt and E. F. Paulus, *Acta Cryst., Sect. B*, **30,** 232 (1974).
202. D. Bloor, L. Koski, G. C. Stevens, F. H. Preston and D. J. Ando, *J. Mater. Sci.*, **10,** 1678 (1975).
203. D. Bloor, F. H. Preston and D. J. Ando, *Chem. Phys. Letters*, **38,** 33 (1976).
204. D. Bloor, D. J. Ando, F. H. Preston and G. C. Stevens, *Chem. Phys. Letters*, **24,** 407 (1974).
205. G. C. Stevens and D. Bloor, *J. Polym. Sci., Polym. Phys. Ed.*, **13,** 2411 (1975).
206. R. H. Baughman, H. Gleiter and N. Sendfeld, *J. Polym. Sci., Polym. Phys. Ed.*, **13,** 1871 (1975).
207. W. Schermann and G. Wegner, *Makromol. Chem.*, **175,** 667 (1974).
208. D. S. Bourdeaux, *Chem. Phys. Letters*, **38,** 341 (1976).
209. D. N. Batchelder and D. Bloor, *Chem. Phys. Letters*, **38,** 37 (1976).
210. G. J. Exarhos, W. M. Risen, Jr. and R. H. Baughman, *J. Amer. Chem. Soc.*, **98,** 481 (1976).
211. E. Fischer and G. Wegner, *German Patent*, 1,940,690 (1971); *Chem. Abstr.*, **74,** 118421 (1971).
212. E. Fischer and G. Wegner, *German Patent*, 1,940,692 (1971); *Chem. Abstr.*, **74,** 127782 (1971).
213. A. R. Guevara and P. M. Borsenberger, *German Patent*, 2,253,931 (1973); *Chem. Abstr.*, **79,** 47834 (1973).
214. C. Eaborn, R. Eastmond and D. R. M. Walton, *J. Chem. Soc. (B)*, 752 (1970).
215. D. R. M. Walton and F. Waugh, *J. Organomet. Chem.*, **37,** 45 (1972).
216. E. Kloster-Jensen, *Tetrahedron*, **22,** 965 (1966).
217. E. Müller, *Synthesis*, 761 (1974).
218. E. Müller and G. Odenigbo, *Justus Liebigs Ann. Chem.*, 1435 (1975).
219. E. Müller, E. Luppold, and W. Winter, *Synthesis*, 265 (1975).
220. E. Müller, E. Luppold and W. Winter, *Chem. Ber.*, **108,** 237 (1975).
221. E. Müller and W. Winter, *Justus Liebigs Ann. Chem.*, 605 (1975).
222. E. Müller and W. Winter, *Justus Liebigs Ann. Chem.*, 1876 (1975).
223. E. Müller, A. Scheller, W. Winter, F. Wagner and H. Meier, *Chem. Ztg.*, **99,** 155 (1975).
224. F. Wagner and H. Meier, *Tetrahedron*, **30,** 773 (1974).
225. W. Hübel and R. Merenyi, *Chem. Ber.*, **96,** 930 (1963).
226. R. S. Dickson and L. J. Michel, *Aust. J. Chem.*, **28,** 285 (1975).
227. A. J. Chalk and R. A. Jerussi, *Tetrahedron Letters*, 61 (1972).
228. A. J. Hubert and J. Dale, *J. Chem. Soc.*, 3160 (1965).
229. V. Yu. Vasil'ev and V. O. Reikhsfel'd, *Tr. Khim. Met. Inst., Akad. Nauk Kaz. SSR*, **18,** 29 (1972); *Chem. Abstr.*, **79,** 18461 (1973).
230. K. P. C. Vollhardt and R. G. Bergman, *J. Amer. Chem. Soc.*, **96,** 4996 (1974).
231. A. J. Hubert, *J. Chem. Soc. (C)*, 1984 (1967).
232. R. L. Hillard III and K. P. C. Vollhardt, *Angew. Chem. Int. Ed. Eng.*, **14,** 712 (1975).
233. (a) W. G. L. Aalbersberg, A. J. Barkovich, R. L. Funk, R. L. Hillard III and K. P. C. Vollhardt, *J. Amer. Chem. Soc.*, **97,** 5600 (1975). (b) K. P. C. Vollhardt, *Acc. Chem. Res.*, **10,** 1 (1977).
234. B. P. Gusev, E. A. El'perina and V. F. Kucherov, *Izv. Akad. Nauk SSSR, Ser. Khim.*, 1659 (1965).

235. R. J. Bushby, *Quart. Rev.*, **24,** 585 (1970).
236. R. J. Bushby and G. H. Whitham, *J. Chem. Soc.* (*B*), 563 (1970).
237. K. G. Migliorese, Y. Tanaka and S. I. Miller, *J. Org. Chem.*, **39,** 739 (1974).
238. H. Hopf, *Chem. Ber.*, **104,** 3087 (1971).
239. H. Hauptmann, *Tetrahedron Letters*, 1931 (1975).
240. H. Hopf, *Tetrahedron Letters*, 1107 (1970).
241. P. J. Garratt and S. B. Neoh, *J. Amer. Chem. Soc.*, **97,** 3255 (1975).
242. H. Hauptmann, *Tetrahedron Letters*, 3589 (1974).
243. M. Huche and P. Cresson, *Tetrahedron Letters*, 367 (1975).
244. K. Schulze, U. Pfüller, H. Tichek and M. Mühlstadt, *Tetrahedron Letters*, 2999 (1968).
245. K. Schulze, U. Pfüller and M. Mühlstädt, *Tetrahedron*, **31,** 1455 (1975).
246. K. Schulze, A. Hiller and M. Mühlstädt, *J. Prakt. Chem.*, **318,** 381 (1976).
247. W. D. Huntsman and H. J. Wristers, *J. Amer. Chem. Soc.*, **85,** 3308 (1963).
248. W. D. Huntsman and H. J. Wristers, *J. Amer. Chem. Soc.*, **89,** 342 (1967).
249. H. Hopf, *Angew. Chem. Int. Ed. Eng.*, **9,** 732 (1970).
250. W. D. Huntsman, *Intra-Science Chem. Rept.*, **6,** 151 (1972).
251. J. E. Kent and A. J. Jones, *Aust. J. Chem.*, **23,** 1059 (1970).
252. T. J. Henry and R. G. Bergman, *J. Amer. Chem. Soc.*, **94,** 5103 (1972).
253. H. A. Brune and H. P. Wolff, *Tetrahedron*, **24,** 4861 (1968).
254. H. A. Brune, H. P. Wolff and H. Hüther, *Tetrahedron*, **25,** 1089 (1969).
255. H. A. Brune, H. P. Wolff and H. Hüther, *Tetrahedron*, **27,** 3949 (1971).
256. H. A. Brune, H. P. Wolff and H. Hüther, *Z. Naturforsch.*, (*B*), **26,** 765 (1971).
257. B. A. W. Coller, M. L. Heffernan and A. J. Jones, *Aust. J. Chem.*, **21,** 1807 (1968).
258. W. D. Huntsman and D. G. Onderak, unpublished results; D. G. Onderak, Ph.D Dissertation, Ohio University, 1972.
259. W. D. Huntsman and N. D. Conrad, unpublished results; N. D. Conrad, M.S. Thesis, Ohio University, 1975.
260. R. Criegee and R. Huber, *Chem. Ber.*, **103,** 1855 (1970).
261. N. Manisse and J. Chuche, *Tetrahedron Letters*, 3095 (1975).
262. A. Viola and J. McMillan, Abstracts, 158th National Meeting of the American Chemical Society, Houston, Texas, Feb. 1970, ORGN 050.
263. M. B. D'Amore, R. G. Bergman, M. Kent and E. Hedeya, *J. Chem. Soc., Chem. Commun.*, 49 (1972).
264. R. G. Bergman, *Acc. Chem. Res.*, **6,** 25 (1973).
265. K. P. C. Vollhardt and R. G. Bergman, *J. Amer. Chem. Soc.*, **94,** 8950 (1972).
266. K. P. C. Vollhardt and R. G. Bergman, *J. Amer. Chem. Soc.*, **95,** 7538 (1973).
267. L. Eisenhuth and H. Hopf, *Chem. Ber.*, **108,** 2635 (1975).
268. E. Müller and A. Huth, *Tetrahedron Letters*, 1031 (1972).
269. H. Straub, *Angew. Chem. Int. Ed. Eng.*, **13,** 405 (1974).
270. G. H. Mitchell and F. Sondheimer, *J. Amer. Chem. Soc.*, **91,** 7520 (1969).
271. F. Toda and M. Ohi, *J. Chem. Soc., Chem. Commun.*, 506 (1975).
272. J. Meijer, P. Vermeer, H. J. T. Bos and L. Brandsma, *Rec. Trav. Chim.*, **93,** 26 (1974).
273. R. R. Jones and R. G. Bergman, *J. Amer. Chem. Soc.*, **94,** 660 (1972).
274. E. Müller, J. Heiss, M. Sauerbier, D. Streichfuss and R. Thomas, *Tetrahedron Letters*, 1195 (1968).
275. B. Bossenbroek and H. Shechter, *J. Amer. Chem. Soc.*, **89,** 7111 (1967).
276. H. A. Staab and J. Ipaktschi, *Chem. Ber.*, **104,** 1170 (1971).
277. D. A. Ben-Efraim, *Tetrahedron Letters*, 957 (1967).
278. F. Bohlmann, W. Skuballa, C. Zdero, T. Kühle and P. Steirl, *Justus Liebigs Ann. Chem.*, **745,** 176 (1971).
279. E. Müller, M. Sauerbier and G. Zountsas, *Tetrahedron Letters*, 3006 (1969).
280. G. Clauss and W. Ried, *Chem. Ber.*, **108,** 528 (1975).
281. G. Büchi, C. W. Perry and E. W. Robb, *J. Org. Chem.*, **27,** 4106 (1962).
282. E. Müller, J. Heiss, M. Sauerbier and G. Ege, *Justus Liebigs Ann. Chem.*, **723,** 61 (1969).
283. E. Müller, M. Sauerbier, D. Streichfuss, R. Thomas, W. Winter, G. Zountsas and J. Heiss, *Justus Liebigs Ann. Chem.*, **750,** 63 (1971).
284. N. Brodherr, P. Narayanan, K. Zechmeister and W. Hoppe, *Justus Liebigs Ann. Chem.*, **750,** 53 (1971).

CHAPTER **14**

Natural acetylenes

SIR EWART R. H. JONES and VIKTOR THALLER
The Dyson Perrins Laboratory, Oxford University, Oxford, England

I. INTRODUCTION

Although the compound known today as dehydromatricaria ester (**1**) was obtained crystalline from essential oils as long ago as 1826 [1], the first correct structure assignment for a natural acetylene, tariric acid (octadec-6-ynoic acid), a component of the seed fat of the Simarubaceae *Picramnia tariri* DC., was published in 1902 [2]. The elucidation of the structure of lachnophyllum ester (**2**) from the Compositae *Lachnophyllum gossipinum* Bge. appeared in 1935 [3] and was followed from 1941 onwards by those of several acetylenes from Compositae species headed by matricaria

$$CH_3(C{\equiv}C)_3{-}CH\overset{t}{=}CH{-}CO_2CH_3 \qquad \textbf{(1)}$$

$$CH_3(CH_2)_2(C{\equiv}C)_2{-}CH\overset{c}{=}CH{-}CO_2CH_3 \qquad \textbf{(2)}$$

$$CH_3CH\overset{c}{=}CH{-}(C{\equiv}C)_2{-}CH\overset{c}{=}CH{-}CO_2CH_3 \qquad \textbf{(3)}$$

t = *trans*, c = *cis*

ester (**3**)[4] to bring the number of known natural acetylenes by 1950 to *ca*. 10. From then onwards their numbers increased rapidly, mainly by the contributions from the schools of Sörensen, Bohlmann and Jones, so that by 1976 over 700 were known. (The early history of natural acetylene research has been described in some detail by Sörensen[1a] and Bu'Lock[5].)

Acetylenes are widespread in Nature. Although the majority occur in 15 out of the 600 or so higher plant families and in Basidiomycete fungi, they also occur in algae, microbial cultures and even in an animal. The most comprehensive treatment of natural acetylene chemistry is to be found in a book by Bohlmann, Burkhardt and

Zdero[6]. Recently discovered natural acetylenes appear in the *Chemical Society Specialist Reports on Aliphatic Chemistry* for the years 1972 and 1973 [7] and will presumably figure in future editions. The pre-1966 era of natural acetylene chemistry is covered comprehensively by two reviews in German[8] and the pre-1964 period in a review in English[5]. A number of shorter articles have also appeared, one recently covering exclusively the non-polyacetylenic acetylenes[9]. A table of fungal polyacetylenes is also available[10].

II. CLASSIFICATION OF NATURAL ACETYLENES

The majority of natural acetylenes known today are polyacetylenes. This name encompasses what now appears to be a biogenetically uniform group of secondary metabolites, usually not strictly poly-ynes[11] (e.g. the esters **2** and **3**), which originate from oleic acid and are found in the roots and the aerial parts of plants and in fungi. In what follows, aspects of their chemistry are discussed prefaced by a survey of natural acetylenes of different origin.

From the seed fats of a few tree families (e.g. Simarubaceae, Santalaceae, Olacaceae) were isolated several C_{18} and C_{17} fatty acids with differently-situated triple bonds, of which tariric acid and the acid **4** are examples[5]. These acids originate

$$CH_3(CH_2)_3CH{=}CH{-}(C{\equiv}C)_2(CH_2)_7CO_2H \qquad \mathbf{(4)}$$

by a pathway which seems to differ from that leading to the polyacetylenes. Of a different order of significance is crepenynic acid (**5**), the major fatty acid in the seed

$$CH_3(CH_2)_4C{\equiv}CCH_2CH^{c}{=}CH(CH_2)_7CO_2H \qquad \mathbf{(5)}$$

fat of *Crepis foetida* (Compositae)[12]. This acid is of crucial importance in polyacetylene biogenesis; its presence was proved in some fungal mycelia and is assumed in tissues of polyacetylene-producing organisms.

Almost certainly derived from fatty acids are the seed constituents of some Lauraceae like the C_{17} ethynyl triol **6** from the avocado pear[9] and the laurencin

$$HC{\equiv}C(CH_2)_{11}CH(OH)CH_2CH(OH)CH_2OH \qquad \mathbf{(6)}$$

(**7**)-like bromine-containing oxygen heterocycles from red seaweeds[9]. The recently increased activity in the analysis of constituents from marine organisms is likely further to enlarge the number of laurencin-like compounds[9a].

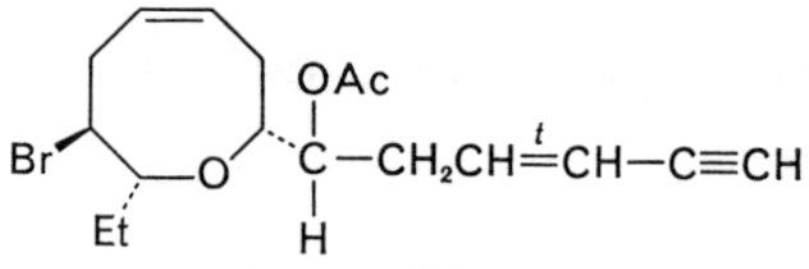

(**7**)

Carbon–carbon triple bonds are cropping up in ever-increasing numbers in established classes of natural products like the carotenoids, terpenes, amino acids and, most recently, alkaloids: **8**, **10** and **11** are typical representatives[9].

The end groups of acetylenic carotenoids like alloxanthin (**8**), found in algae and marine organisms, are structurally related to the end groups of fucoxanthin (**9**), the most abundant natural carotenoid[13]. The allene and acetylene bonds are known to be biogenetically linked in polyacetylenes and the same seems likely to apply to

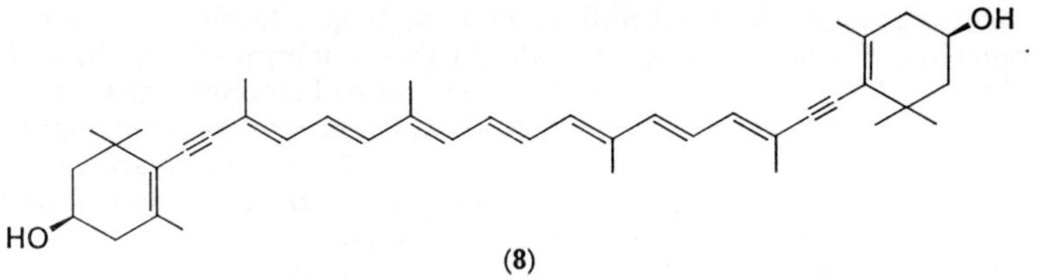

(8)

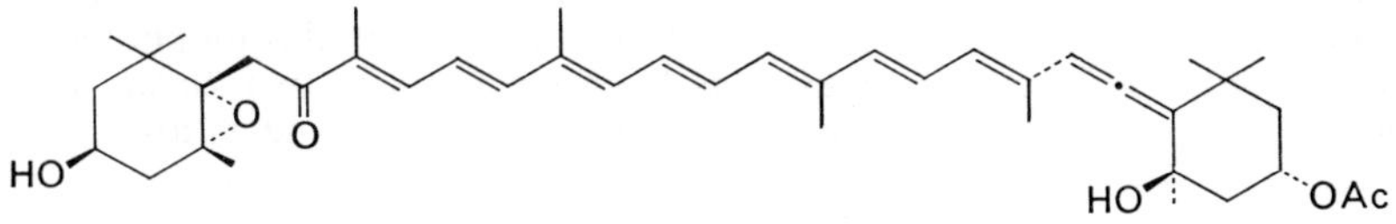

(9)

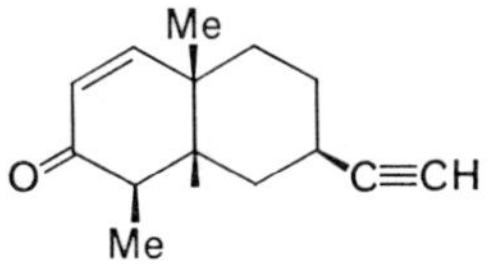

(10)

$HC{\equiv}CCH(OH)CH_2CH(NH_2)CO_2H$

(11)

carotenoids. Acetylenic carotenoids might thus be even more widespread than we are aware of today.

Histrionicotoxin, one of the spiropiperidine alkaloids from the skin of the Colombian frog *Dendrobates histrionicus*[14], containing two *cis*-but-3-en-1-yne groups, is the first acetylenic alkaloid and also the first acetylene of animal origin. (The acid **12**, present in the secretion of the soldier beetle, *Chauliognatus lecontei*[15],

$$CH_3CH^{c}{=}CH{-}(C{\equiv}C)_2(CH_2)_2CO_2H \qquad \textbf{(12)}$$

and acetylenic carotenoids in the mussel *Mytilus edulis*[16] almost certainly originate from the plant diet of these organisms.)

Cellocidin (**13**) and siccaine (**14**) are representatives of non-polyacetylenic microbial metabolites. The growth conditions for these acetylene-producing microorganisms

$H_2NOCC{\equiv}CCONH_2$

(13)

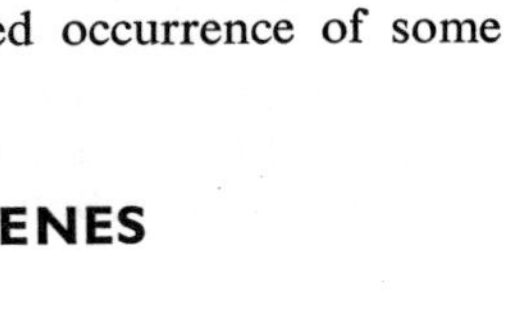

(14)

are often difficult to reproduce and the reported occurrence of some microbial acetylenes is not easy to confirm[9, 17].

III. POLYACETYLENES

A. Distribution and Detection

About 85% of the known polyacetylenes were isolated from the roots and aerial parts of higher plants (up to a few grams per kilogram dry tissue). They are common in two major families, the Umbelliferae and the largest family of flowering plants, the Compositae, in which they occur in all 13 tribes, but especially in the *Heliantheae*, *Anthemideae* and *Cynereae*. They are fairly widespread amongst the Campanulaceae and Araliaceae and have been found sporadically in several other plant families. Basidiomycete species produce the remaining 15% of known polyacetylenes, mostly in cultures (up to 60 mg per litre). Mycelia are a poor source though polyacetylenes have been isolated[18] from sporophores of wild fungi.

Polyacetylenes are readily detected in crude tissue or culture extracts when chromophores with the unique and intense ultraviolet absorption are present (from $-[C{\equiv}C]_2-CH{=}CH-$ or $-[C{\equiv}C]_2-CO-$ to longer chromophores; *cf.* the spectra recorded on crude ether extracts from the roots of two *Dahlia* hybrids in Figure 1—the recognizable maxima due to the main components are indicated in

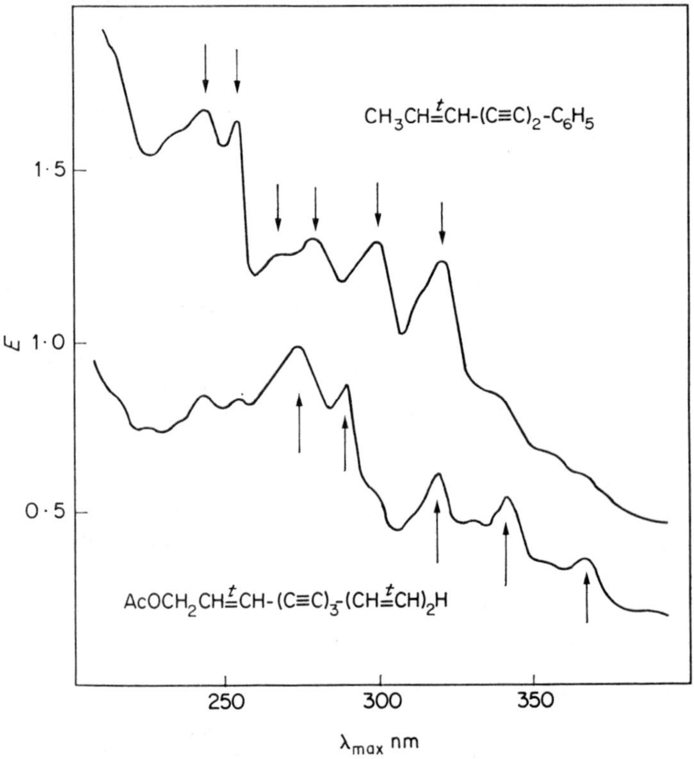

FIGURE 1. Ultraviolet absorption of crude ether extracts from roots of two *Dahlia* hybrids.

each case). The ultraviolet absorptions of compounds with atypical chromophores like carlina oxide (**15**) (λ_{max} 250 nm, ε 18 000)[8] are much more difficult to recognize whilst the convenient detection of minor constituents with just one, two or even

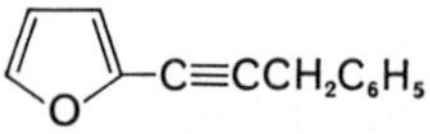

(15)

three triple bonds as chromophores has been hitherto impossible; it might be accomplished in the future with the help of laser Raman spectroscopy.

B. Polyacetylene Structures

Polyacetylenes comprise a wide range of combinations of differing chain lengths (C_6–C_{18}), degrees of unsaturation [$(CH{=}CH)_a{-}(C{\equiv}C)_b{-}(CH{=}CH)_c$, for example, $a = 1, b = 2$ and $c = 2$ and $a = 0, b = 5$ and $c = 1$], and with a considerable number of functional groups and cyclic systems in varying relationship to the chromophore.

The number of possible combinations is enormous; it accounts for the large number of polyacetylenes found, and suggests that even larger numbers are likely to be encountered. Some combinations are favoured, some are more frequent in higher plants or exclusive to them, others are more prone to occur in fungi. The crepenynate pathway (see below) from oleate envisages the same initial stages for both plant and fungal polyacetylenes. The much larger number of polyacetylenes found in higher plants than in fungi must therefore reflect the higher organization of the former and the opportunity for secondary transformations of common precursors in the different tissues and species. Fungal polyacetylenes are demonstrably the products of the stress conditions under which the cultures are made to grow. The smaller number of variants are most likely due partly to the more primitive organisms involved and their similar response to dietary deficiencies and partly to their excretion into the culture medium where oxidation and chain shortening appear to be favoured. Some structural differences between polyacetylenes from higher plants and fungi are depicted in Table 1. Formulae **1, 2, 3, 12** and **16–41** illustrate

TABLE 1. Some structural differences between plant and fungal polyacetylenes

	Plants	Fungi
Chain lengths	C_{9-18}	C_{6-14}
(Major groups)	(C_{13}, C_{14}, C_{17})	(C_{8-11})
Allenes	Rare	Frequent
O-Heterocycles (incl. epoxides)	Many	Rare
Benzenoid compounds	Many	None
S-Compounds	Many	Few
Acetates	Many	None
Free CO_2H	Rare	Many

some of the structural features and permutations found in natural polyacetylenes (**16–30** are plant polyacetylenes and **31–41** fungal polyacetylenes). The structural formulae are drawn so as to indicate actual or probable biogenetic relationship to oleic acid $CH_3(CH_2)_7CH^c{=}CH(CH_2)_7CO_2H$.

$$H_2C{=}CH{-}CO{-}(C{\equiv}C)_2CH_2CH\overset{c}{=}CH(CH_2)_6CH_3 \qquad \textbf{(16)}$$

$$HOH_2C(CH_2)_2(C{\equiv}C)_2{-}(CH\overset{t}{=}CH)_3CH(OH)(CH_2)_2CH_3 \qquad \textbf{(17)}$$

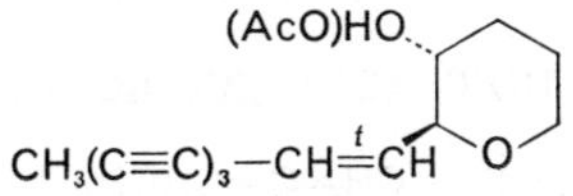

(18)

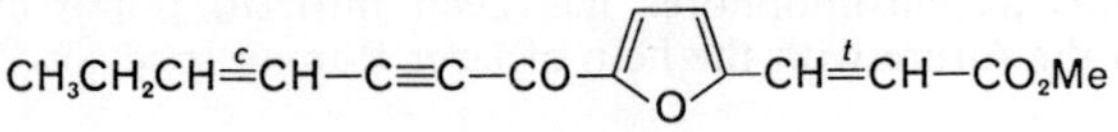
(19)

$$CH_3(C{\equiv}C)_5{-}CH{=}CH_2 \qquad (20)$$

$$CH_3(C{\equiv}C)_4{-}CH{=}CHCH(Cl)CH_2OAc \qquad (21)$$

$$CH_3CH^t{=}CH{-}(C{\equiv}C)_3{-}CH^t{=}CHCH(OH)CH_2OH \qquad (22)$$

$$CH_3(C{\equiv}C)_3{-}C_6H_5 \qquad (23)$$

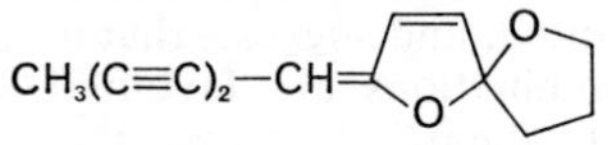
(24)

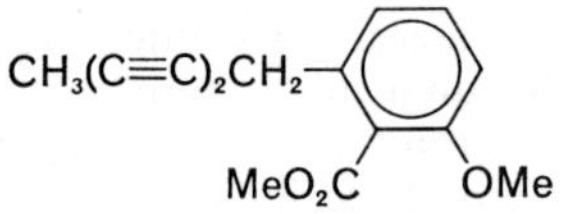
(25)

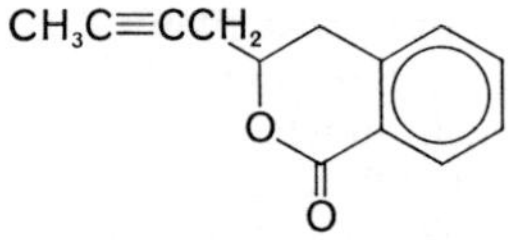
(26)

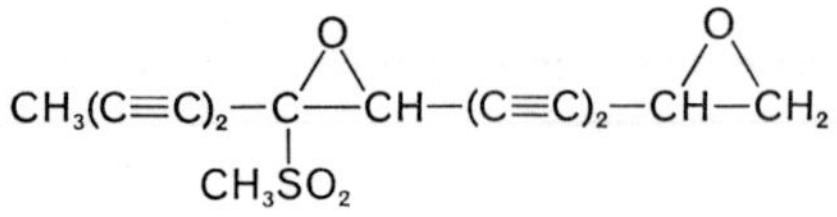
(27)

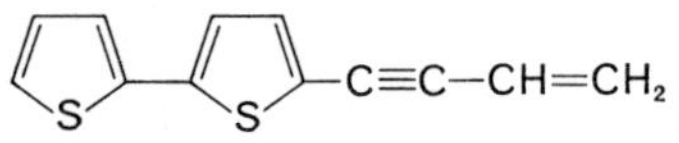
(28)

$$CH_3(C{\equiv}C)_2{-}CO{-}C_6H_5 \qquad (29)$$

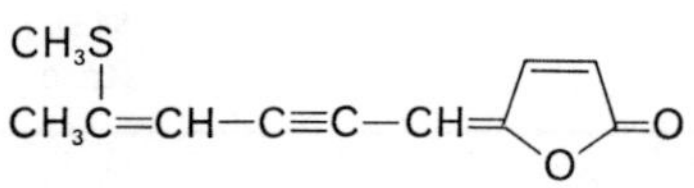
(30)

$$HO_2C{-}(C{\equiv}C)_3CH_2CH^c{=}CH(CH_2)_3CO_2H \qquad (31)$$

$$CH_3(C{\equiv}C)_4CH(OH)CH(OH)CH(OH)CH_2OH \quad (2S, 3S, 4S) \qquad (32)$$

$$HC{\equiv}C{-}C{\equiv}C{-}CH{=}C{=}CH{-}CH^c{=}CH{-}CH^t{=}CHCH_2CO_2H \qquad (33)$$

$$HC{\equiv}CCH_2(C{\equiv}C)_2{-}CH^c{=}CHCH_2CO_2H \qquad (34)$$

$$CH_3(C{\equiv}C)_3{-}CH^t{=}CHCH_2OH \qquad (35)$$

$$HO_2C{-}(C{\equiv}C)_3(CH_2)_2CO_2H \qquad (36)$$

$$H(C{\equiv}C)_2{-}CH{=}C{=}CHCH_2CH_2OH \qquad (37)$$

$$CH_3CH_2CH(OH)(C{\equiv}C)_2CH(OH)CH_2OH \quad (2R, 7S) \qquad \textbf{(38)}$$

$$HOCH_2(C{\equiv}C)_3{-}CONH_2 \qquad \textbf{(39)}$$

$$NC{-}(C{\equiv}C)_2{-}CH^t{=}CH{-}CO_2H \qquad \textbf{(40)}$$

$$H(C{\equiv}C)_3H \qquad \textbf{(41)}$$

A large number of polyacetylenes have retained obviously straight-chain structures and are readily recognized by their typical ultraviolet absorptions (e.g. **1, 2, 3, 20, 22, 31, 32**, etc.).

In a considerable number of polyacetylenes the straight chains have been converted into both homo- and heterocyclic structures. In some, like the benzenoid **23**, the tetrahydropyran **18**, the spiroketal **24** and frutescin (**25**), the polyacetylene structure is still clearly recognizable though not always detectable in the ultraviolet spectrum (e.g. frutescin, **25**[19] and the spiroketal **24**[20]). It is more difficult to recognize the relationship of compounds like carlina oxide (**15**)[21], capillarin (**26**)[22], or the dithienyl **28**[23], in which most of the original chromophore participates in the ring formation, with the straight-chain polyacetylenes. Some exotic structures like a dithio[24] or thiethanone[25] ring and epoxysulphone[26] (cf. **27**) containing polyacetylenes have also been encountered. No nitrogen heterocyclic examples have so far been discovered.

C. *Structural Elucidation and Synthesis*

Spectra are today crucial in structure elucidation on account of the small amounts of material usually available and the poor stability of many polyacetylenes in the condensed phase. The ultraviolet spectra, notable for the typical sharp fine structures associated with many poly-yne and poly-yn-ene chromophores (*cf.* the chromophore allocation in Figure 1; tables for typical poly-yn-ene maxima exist[6, 8]), often determine the unsaturated part of the molecules. A few unequivocal chemical reactions like the manganese dioxide and periodate oxidations for allylic/propargylic alcohols and acetylenic/ethylenic α,β-glycols, respectively, along with the associated bathochromic shifts of the ultraviolet maxima, provide additional information about the groups attached to the chromophore. For the non-acetylenic parts of the molecules ^1H n.m.r. has been of the greatest help. In the infrared even the usually weak —C≡C— stretching band becomes very strong indeed in yn-ones of the falcarinone (**16**)[27] and wyerone (**19**)[28] types and can even serve for their detection. The use of ^{13}C n.m.r. has not yet been used in the structure determination of natural polyacetylenes[28a]. Mass spectra often enable the determination of the molecular formulae of polyacetylenes (they are notoriously awkward to combust) and also give some typical fragmentation patterns. By contrast one cannot but admire the discovery of the structure of lachnophyllum ester (**3**)[3] on the basis of the correct interpretation of its molecular refractivity and the information obtained from a few incisive chemical reactions; albeit with amounts of material a thousand times greater than those used today in polyacetylene research.

Many reactions familiar to acetylene and polyene chemistry have been used in the synthesis of natural polyacetylenes. The longer poly-yn-ene chains are usually unstable and the tendency is to form them as late as possible in the fabrication of the molecules. Generally, terminal fragments are prepared first by taking advantage of such simple acetylenes and diacetylenes as are commercially available or relatively easily synthesized[29]. Two reactions are then predominantly used to join these fragments: the Cadiot–Chodkiewicz coupling[30] which permits the asymmetric linking of terminal acetylenes and the Wittig reaction[31] which permits the introduction of a double bond into the conjugated system. The application of these two reactions is illustrated in the synthesis of the chiral fungal metabolite **42**[32] in Scheme 1.

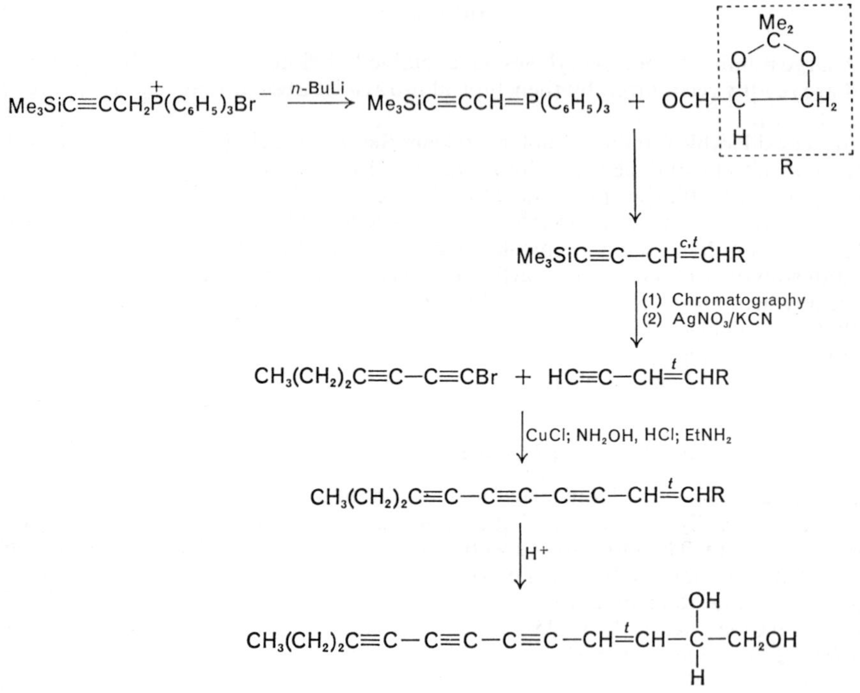

(42)

SCHEME 1. Synthesis of the chiral fungal metabolite **42** from *Fistulina pallida* cultures.

D. Biogenesis

The currently accepted hypothesis for the biogenesis of polyacetylenes, first proposed[33] and experimentally supported[34] by Bu'Lock, involves primarily the desaturation of the distal half of the oleate chain (C-10–C-18) via the α-en-δ-yne system of crepenynic acid **(5)** (Scheme 2). The other types of transformations adumbrated include chain shortening, usually by the classical α- or β-oxidations of fatty acids at the proximal end, rearrangement and/or oxidation of the skipped system, extension of the chromophore, ω-oxidation at C-18 and chain shortening at the distal end by deformylation or decarboxylation, functionalization, cyclization (e.g. thiophen from —C≡C—C≡C—, furan from —C≡C—CH=CHCHOH—), etc., in sequences which may be peculiar to both metabolite and organism and which lead to the great variety of structures encountered in Nature and illustrated by the examples in this chapter. As has already been stressed above, a greater number of these transformations is encountered in plants than in fungi.

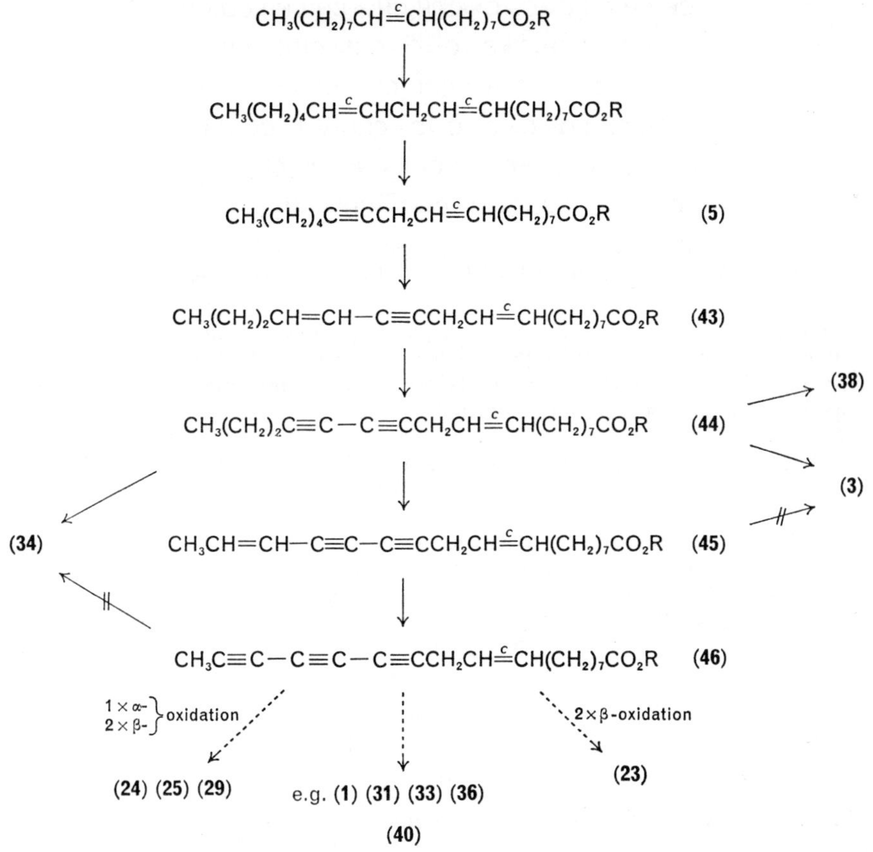

SCHEME 2. Proposed pathways for some plant and fungal polyacetylenes on the basis of biosynthetic experiments.

Biosynthetic experiments with likely precursors in plants and fungal cultures are the main source of our knowledge of the pathways involved. In fungal cultures biosynthetic experiments are easier and give higher incorporations than those with plants, but with fungal cultures, as with other microorganisms[35], a variety of alternative sequences may be available. Thus the fungus *Clitocybe rhizophora* incorporates all the precursors listed in Scheme 3 with similar efficiencies (1·5–4%)[36] and the experiments carried out so far give little information about the stage (C_{18} or C_{10}) at which, and the means by which, oxygen is introduced at C-7 of the triol **38**. On the other hand, some rather unique metabolites like diatretyne 2 (**40**)[37] (C-9–C-16 of oleate) and drosophilin C (**34**)[38] (C-8–C-18) appear to be formed by very specific processes, at least in the ultimate stages of their biogenesis. For example, the C_{18} diyne skipped-ene ester **44** is incorporated into drosophilin C but the corresponding 17,18-dehydro analogue is not[38]. Desaturation at C-17 and C-18 and chain shortening might be linked processes as was demonstrated in the case of matricaria ester **(3)** (see below).

$$CH_3CH_2CH_2C{\equiv}C{-}C{\equiv}CCH_2CH^c{=}CH(CH_2)_7CO_2CH_3$$

$$CH_3CH_2CH_2C{\equiv}C{-}C{\equiv}C{-}CH^t{=}CHCO_2CH_3$$

$$CH_3CH_2CO{-}C{\equiv}C{-}C{\equiv}CCH_2CH^c{=}CH(CH_2)_7CO_2CH_3$$

$$CH_3CH_2CO{-}C{\equiv}C{-}C{\equiv}C{-}CH^t{=}CH{-}CO_2CH_3$$

$$CH_3CH_2CH(OH)C{\equiv}C{-}C{\equiv}CCH_2CH^c{=}CH(CH_2)_7CO_2CH_3$$

$$CH_3CH_2CH(OH)C{\equiv}C{-}C{\equiv}C{-}CH^t{=}CH{-}CO_2CH_3$$

SCHEME 3. Possible C_{18} and C_{10} precursors incorporated (1·5–4 %) by cultures of the fungus *Clitocybe rhizophora* into $CH_3CH_2CH(OH)(C{\equiv}C)_2CH(OH)CH_2OH$ (**38**).

In most biosynthetic experiments with plants great specificity has been observed throughout the pathway and sequences leading to individual metabolites could be envisaged with a high degree of certainty. Chain shortening of the C_{18} triyne skipped-ene **46** was thus found to occur by either

(*i*) one α- and two β-oxidations (in that order) to a C_{13} triyne skipped-ene intermediate which is converted in *Chrysanthemum frutescens* L. into benzenoid acetylenes[39, 40] (e.g. **25** and **29**) and in *C. flosculosum* L. into the spiroketal **24** [39], or

(*ii*) two β-oxidations to a C_{14} triyne skipped-ene intermediate which in *Coreposis lanceolata* L. yields phenylheptatriyne (**23**)[39, 41]. Pathways leading to the benzenes have been proposed, and that for **23** probably involves the hydroxy keto-ester **47** [41].

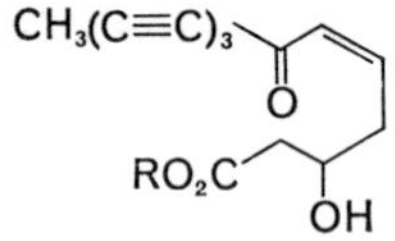

(47)

Another example of the specificity encountered in plants is found in the difference between the pathways leading to $CH_3C{\equiv}CR$ and $CH_3CH{=}CHR$ polyacetylenes. Unlike the case of dehydromatricaria ester (**1**) for which **46** and the C_{18} triynene keto ester **48** are precursors[42], the C_{18} enediyne skipped-ene ester **45** (Scheme 2) is

$$CH_3(C{\equiv}C)_3{-}CH{=}CH{-}CO(CH_2)_7CO_2R$$

(48)

not incorporated into matricaria ester (**3**). Since the esters **44** and **49** are incorporated, the 8-ene formation of matricaria ester appears to occur *after* rearrangement and oxidation and is linked to the chain-shortening process.[42]

$$CH_3(CH_2)_2(C{\equiv}C)_2{-}CH{=}CH{-}CO(CH_2)_7CO_2R$$

(49)

No real evidence exists concerning the *in vivo* formation of the carbon–carbon triple bond but dehydrogenation *via cis* double bonds was favoured speculatively[34] and appears probable on account of the similar incorporations observed for linoleate and crepenynate (**5**)[37, 43] and the better incorporations of 14-*cis*- than 14-*trans*-dehydrocrepenynate (**43**) into several fungal metabolites (e.g. **35**, **36**, **40**)[44]. Biosynthetic experiments with leaf homogenates of *Chrysanthemum flosculosum*[45] indicate that the enzymes required for the desaturation of oleic acid are located within the chloroplasts whilst the final oxidation of the C_{13} triyne skipped-ene intermediate and the cyclization to the spiroketal **24** were effected by extracellular enzymes.

For the biosynthetic experiments mentioned a considerable number of potential precursors specifically labelled with ^{14}C and ^{3}H have been synthesized, e.g. References 39–41, 46.

E. Physiological Properties

No obvious physiological role can be allocated to polyacetylenes in the organisms which produce them. They have been detected very early in the life of plants, e.g. ene-tetrayn-ene polyacetylenes in *Dahlia* seedlings[47] and the ketone **19** in broad bean (*Vicia faba* L.) seedlings[28]. The ketone **19** may act as a systemic fungicide in the seedlings and later in the grown plant; attack by pathogenic fungi causes a several hundred-fold increase of its concentration in the leaves[48]. Similarly, a twenty-fold increase in the concentration of the diol **22** occurs in safflower (*Carthamus tinctorium* L.) on infection with pathogenic fungi[49]. Acetylenic ketones in general appear to be fungicidal; another and an outstanding example is capillin **(29)** (first isolated from the Compositae *Artemisia capillaris* Thunb.[50]) which is active against fungi of the skin, *cf.* Reference 51. Nematocidal activity is shown by the dithienyl **28** from *Tagetes* species, *cf.* Reference 51.

Several polyacetylenes are known to be generally toxic. Thus the plant extract used by the natives of the Lower Amazon Basin as fish poison on their arrow heads contains the tetrahydropyran alcohol **18** as active principle[52] (it also occurs in dahlias[53]), whilst the high toxicity of the water hemlock (*Cicuta virosa* L.) is due to cicutoxin **(17)**[54].

The first fungal polyacetylenes like mycomycin **(33)**[55] and agrocybin **(39)**[56] were detected and isolated on account of their antibiotic properties; however, their comparative instability precluded any practical application.

An extensive review article on synthetic and natural acetylenes as potential drugs is available[51].

F. Taxonomical Implications

The relatively easy detection of small amounts of polyacetylenes makes them an obvious choice for chemical–taxonomical investigations, and Sörensen[57] and Bohlmann[6] have both discussed in detail possible implications of the distribution of polyacetylenes. The occurrence of related polyacetylenes restricted to individual plant families or tribes can be illustrated by a few of the C_{14} polyacetylenes identified so far only in species of the Campanulaceae family[58] (Scheme 4).

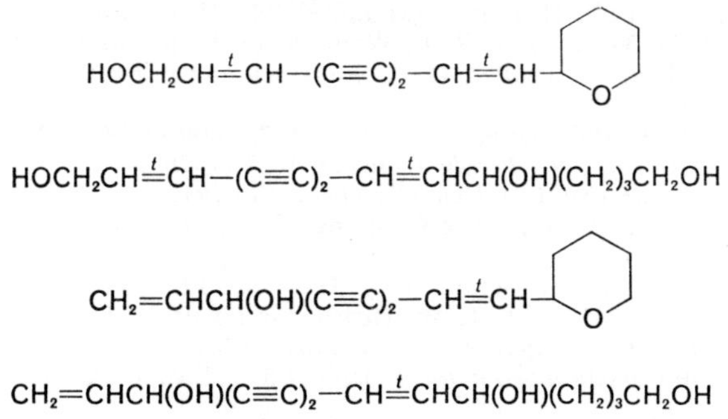

SCHEME 4. Polyacetylenes from Campanulaceae species.

IV. REFERENCES

1a. N. A. Sörensen, *Proc. Chem. Soc.*, 98 (1961). 1b. *Cf.* Reference 6, p. 1.
2. A. Arnaud, *Bull. Soc. Chim. Fr.*, [3], **27**, 484, 489 (1902).
3. V. V. Vil'yams, V. S. Smirnov and V. P. Gol'mov, *Zh. Obshch. Khim.*, **5**, 1195 (1935).
4. N. A. Sörensen and J. Stene, *Ann. Chem.*, **549**, 80 (1941).
5. J. D. Bu'Lock in *Progr. Org. Chem.*, **6**, 86 (1964).
6. F. Bohlmann, T. Burkhardt and C. Zdero, *Naturally Occurring Acetylenes*, Academic Press, London, 1973.
7. *Aliphatic Chemistry*, Vol. 2 (Ed. W. Parker), The Chemical Society, London, 1974, p. 213; Vol. 3 (Ed. A. McKillop), 1975, p. 256.
8. F. Bohlmann, H. Bornowski and C. Arndt in *Fortschr. Chem. Forsch.*, **4**, 138 (1962); F. Bohlmann in *Fortschr. Chem. Forsch.*, **6**, 65 (1966).
9. R. A. M. Ross, *Phytochemistry*, **11**, 221 (1972).
9a. D. J. Faulkner, *Tetrahedron*, **33**, 1421 (1977).
10. E. R. H. Jones and V. Thaller in *Microbial Products*, *Handbook of Microbiology*, Vol. 3, (Ed. A. I. Laskin and H. A. Lechevalier), Chemical Rubber Company, Cleveland, 1973, p. 63.
11. *Cf.* E. R. H. Jones, *Proc. Chem. Soc.*, 199 (1960).
12. K. L. Mikolajczak, C. R. Smith, jun., M. O. Bagby and I. A. Wolff, *J. Org. Chem.*, **29**, 318 (1964).
13. B. C. L. Weedon, *Rev. Pure Appl. Chem.*, **20**, 51 (1970).
14. T. Tokuyama, K. Uenoyama, G. Brown, J. W. Daly and B. Witkop, *Helv. Chim. Acta*, **57**, 2597 (1974); J. W. Daly, B. Witkop, T. Tokuyama, T. Nishikawa and I. L. Karle, *Helv. Chim. Acta*, **60**, 1128 (1977).
15. J. Meinwald, Y. C. Meinwald, A. M. Chalmers and T. Eisner, *Science*, **160**, 890 (1968).
16. A. Khare, G. P. Moss and B. C. L. Weedon, *Tetrahedron Letters*, 3921 (1973).
17. E. R. H. Jones, J. W. Keeping, M. G. Pellatt and V. Thaller, *J. Chem. Soc., Perkin Trans. I*, 148 (1973).
18. I. W. Farrell, J. W. Keeping, M. G. Pellatt and V. Thaller, *J. Chem. Soc., Perkin Trans. I*, 2642 (1973).
19. F. Bohlmann and K.-M. Kleine, *Chem. Ber.*, **95**, 602 (1962).
20. F. Bohlmann, P. Herbst, C. Arndt, H. Schönowsky and H. Gleinig, *Chem. Ber.*, **94**, 3193 (1961).
21. F. Bohlmann and W. v. Kap-herr, *Chem. Ber.*, **99**, 148 (1966).
22. F. Bohlmann and K.-M. Kleine, *Chem. Ber.*, **95**, 39 (1962); S. Noguchi and N. Sugiyama, *J. Chem. Soc. Japan*, **81**, 654 (1960) [*Chem. Abstr.*, **55**, 8398 (1961)].
23. J. H. Uhlenbroek and J. D. Bijloo, *Rec. Trav. Chim.*, **78**, 382 (1959); *cf.* also Reference 8.
24. F. Bohlmann and K.-M. Kleine, *Chem. Ber.*, **98**, 3081 (1965).
25. F. Bohlmann and W. Skuballa, *Chem. Ber.*, **106**, 497 (1973).
26. F. Bohlmann, K.-M. Rode and C. Zdero, *Chem. Ber.*, **100**, 537 (1967).
27. F. Bohlmann, C. Arndt, H. Bornowski and K.-M. Kleine, *Chem. Ber.*, **94**, 958 (1961).
28. C. H. Fawcett, D. M. Spencer, R. L. Wain, A. G. Fallis, Sir Ewart R. H. Jones, M. Le Quan, C. B. Page, V. Thaller, D. C. Shubrook and P. M. Whitham, *J. Chem. Soc. (C)*, 2455 (1968).
28a. R. Zeisberg and F. Bohlmann, *Chem. Ber.*, **107**, 3800 (1974); M. T. W. Hearn and J. L. Turner, *J. Chem. Soc., Perkin Trans. II*, 1027 (1976).
29. L. Brandsma, *Preparative Acetylenic Chemistry*, Elsevier, Amsterdam, 1971.
30. G. Eglinton and W. McCrae, 'The Coupling of Acetylenic Compounds', *Advan. Org. Chem.*, **4**, 225 (1963).
31. A. Maercker, 'The Wittig Reaction', *Org. Reactions*, **14**, 270 (1965).
32. M. Ahmed, G. C. Barley, M. T. W. Hearn, Sir Ewart R. H. Jones, V. Thaller and J. A. Yates, *J. Chem. Soc., Perkin Trans. I*, 1981 (1974).
33. J. D. Bu'Lock in *Comparative Phytochemistry* (Ed. T. Swain), Academic Press, London, 1966, p. 79.
34. J. D. Bu'Lock and G. N. Smith, *J. Chem. Soc. (C)*, 332 (1967).

35. For example, M. Fryberg, A. C. Oehlschlager and A. M. Unrau, *J. Amer. Chem. Soc.*, **95**, 5747 (1973).
36. C. A. Higham, Sir Ewart R. H. Jones and V. Thaller, unpublished work.
37. Sir Ewart R. H. Jones, V. Thaller and J. L. Turner, *J. Chem. Soc., Perkin Trans. I*, 424 (1975).
38. M. Ahmed, Sir Ewart R. H. Jones and V. Thaller, *J. Chem. Research*, 1977 (S) 125; (M) 1579.
39. F. Bohlmann, R. Jente and R. Reinecke, *Chem. Ber.*, **102**, 3283 (1969).
40. F. Bohlmann, R. Jente, W. Lucas, J. Laser and H. Schulz, *Chem. Ber.*, **100**, 3183 (1967).
41. F. Bohlmann, H. Bonnet and R. Jente, *Chem. Ber.*, **101**, 855 (1968).
42. F. Bohlmann and T. Burkhardt, *Chem. Ber.*, **105**, 521 (1972).
43. G. C. Barley, A. C. Day, U. Graf, Sir Ewart R. H. Jones, I. O'Neill, R. Tachikawa, V. Thaller and R. A. Vere Hodge, *J. Chem. Soc. (C)*, 3308 (1971), and unpublished work.
44. I. W. Farrell, C. A. Higham, Sir Ewart R. H. Jones and V. Thaller, unpublished work.
45. F. Bohlmann and H. Schulz, *Tetrahedron Letters*, 4795 (1968).
46. G. C. Barley, Sir Ewart R. H. Jones, V. Thaller and R. A. Vere Hodge, *J. Chem. Soc., Perkin Trans. I*, 151 (1973); A. G. Fallis, M. T. W. Hearn, Sir Ewart R. H. Jones, V. Thaller and J. L. Turner, *J. Chem. Soc., Perkin Trans. I*, 743 (1973).
47. C. T. Bedford, D. Bhattacharjee, J. R. F. Fairbrother, Sir Ewart R. H. Jones, S. Safe and V. Thaller, *J. Chem. Soc., Perkin Trans. I*, 735 (1976).
48. C. H. Fawcett, R. D. Firn and D. M. Spencer, *Physiological Plant Pathology*, **1**, 163 (1971).
49. E. H. Allen and C. A. Thomas, *Phytochemistry*, **10**, 1579 (1971).
50. K. Imai, *J. Pharm. Soc. (Japan)*, **76**, 405 (1956).
51. K. E. Schulte and G. Rücker in *Progress in Drug Research*, Vol. 14 (Ed. E. Jucker), Birkhäuser Verlag, Basel and Stuttgart, 1970, p. 387.
52. S. C. Cascon, W. B. Mors, B. M. Tursch, R. T. Aplin and L. J. Durham, *J. Amer. Chem. Soc.*, **87**, 5237 (1965).
53. C. Chin, M. C. Cutler, Sir Ewart R. H. Jones, J. Lee, S. Safe and V. Thaller, *J. Chem. Soc. (C)*, 314 (1970).
54. E. F. L. J. Anet, B. Lythgoe, M. H. Silk and S. Trippett, *J. Chem. Soc.*, 309 (1953).
55. E. A. Johnson and K. L. Burdon, *J. Bacteriol.*, **54**, 281 (1947); W. D. Celmer and I. A. Solomons, *J. Amer. Chem. Soc.*, 74, 1870 (1952).
56. F. Kavanagh, A. Hervey and W. J. Robbins, *Proc. Nat. Acad. Sci. U.S.*, **36**, 102 (1950); J. D. Bu'Lock, E. R. H. Jones, G. H. Mansfield, J. W. Thompson, and M. C. Whiting, *Chem. Ind.*, 990 (1954).
57. N. A. Sörensen in *Chemical Plant Taxonomy* (Ed. T. Swain), Academic Press, London, 1963, p. 219.
58. Sh. O. Badanyan, R. K. Bentley, J. K. Jenkins, Sir Ewart R. H. Jones and V. Thaller, *J. Chem. Soc., Perkin Trans. I*, 145 (1973).

CHAPTER **15**

Cyclic acetylenes

M. Nakagawa
Osaka University, Toyonaka, Osaka 560, Japan

I. INTRODUCTION

Essentially the same methods of synthesis of open-chain acetylenes had been adopted for the preparation of cyclic acetylenes, e.g. cyclopentadecyne (**1**) and cycloheptadecyne (**2**) were obtained by the reaction of potassium hydroxide with the corresponding 1-bromo-cycloalkenes[1].

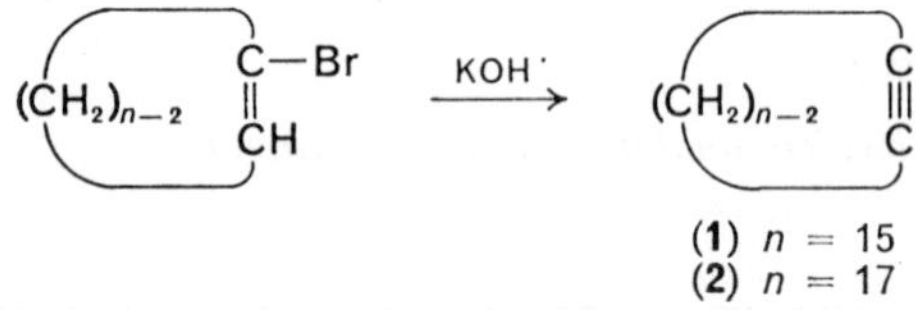

Similarly cyclododecyne (**4**) was synthesized from 1,2-dichlorocyclododecane (**3**)[2]. Civetone (**7**), a natural perfume of animal origin, was prepared by partial reduction followed by acid hydrolysis of the ketal of the cycloheptadecyne derivative **6** obtained by dehydrobromination of the dibromo ketal **5**[3].

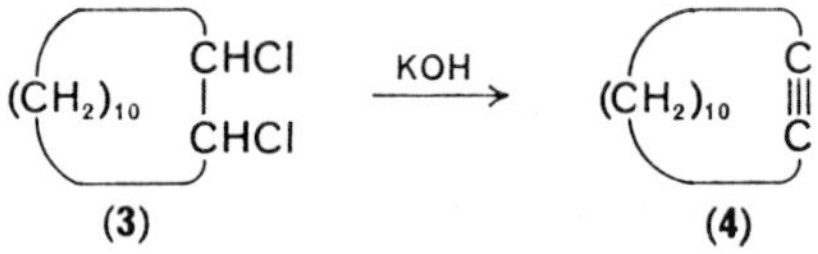

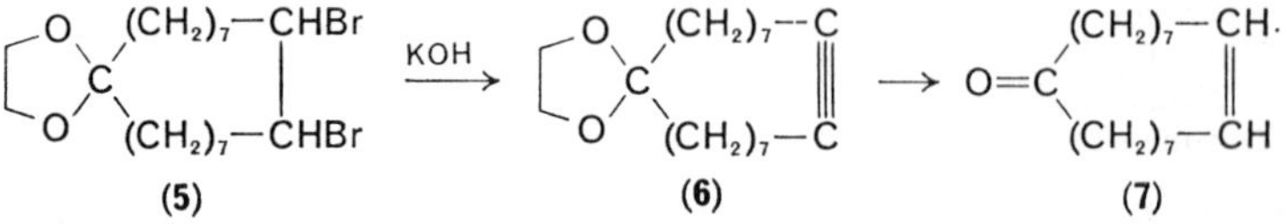

However, cyclic acetylenes had been regarded as a special group of compounds accessible only with difficulty until recent developments of acetylene chemistry, when a wide variety of cyclic compounds were prepared and their properties extensively studied.

The four carbon atoms ($-CH_2-C{\equiv}C-CH_2-$) in a disubstituted acetylene are linear, owing to the *sp* hybridization of the acetylenic carbon atoms. Consequently, incorporation of the linkage in a small or a medium-sized ring system may cause a remarkable ring strain. Hence comparison of the physical and chemical properties of strained cyclic acetylenes with those of open-chain analogues is an interesting problem. The straight and rigid acetylenic bond may exert a prominent restriction on the conformation of cyclic acetylenes, and in some of these the triple bond seems to be held rigidly in a spatial position proximate to other groups in the cyclic system, thus offering model substances for the study of transannular interactions. Finally, the studies on fully conjugated cyclic polyenpolyynes (dehydroannulenes) are interesting, especially in connection with their aromaticity or antiaromaticity.

This chapter is concerned with the above-mentioned three problems of general interest, because transannular phenomena are the subject of a separate chapter.

II. SHORT-LIFE CYCLIC ACETYLENES

A. Ring Strain and Reactivity

Inspection of molecular models reveals that cyclononyne should be slightly strained, while all smaller cycloalkynes should be strongly strained. In fact up to the present time, cycloheptyne is the smallest cycloalkyne to have been isolated[4, 81]. Since the reactivity of medium-sized cycloalkynes increases with decreasing ring size (see Section III), one may expect the formation of smaller cycloalkynes as highly reactive reaction intermediates.

B. Evidence for their Intermediacy

1,2-Dibromocyclohexene (**8**) yields **11** and **12** on treatment with magnesium[5]. This result can be best explained by assuming the formation of cyclohexyne (**9**) and dimerization thereof to give cyclobutadiene (**10**). Addition of **9** to **10** and dimerization of **10** should give **11** and **12**, respectively. However, the formation of **11** and **12** cannot

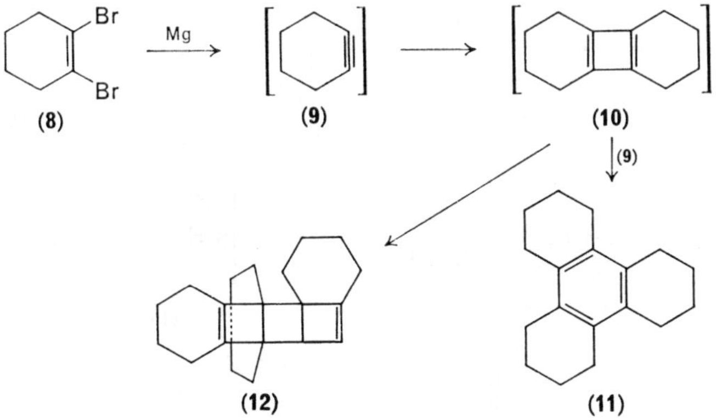

be regarded as an unequivocal proof for the intervention of (**9**). Several alternative pathways leading to the same products have been suggested by Krebs[6]. Thus, for example, **10** could be formed by stepwise elimination of bromine from two molecules of **8**.

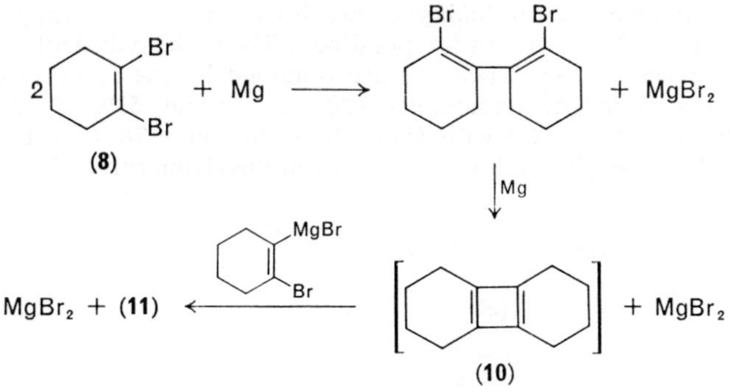

Trapping reactions and isotopic labelling have been used to exclude the above and related mechanisms and to confirm the occurrence of cycloalkynes in a similar manner as in benzyne research.

The reaction of 1,2-dibromocycloalkenes **(13)** with magnesium in the presence of 1,3-diphenylbenzo[*c*]furan **(16)** affords a cycloalkyne adduct **(17)**[7]. Also, the mercuric oxide oxidation of bishydrazone, which has been used extensively in the preparation of open-chain[8, 9] and medium[4c] or large cyclic acetylenes[10–17], is adapted for small ring bishydrazones **(18)**[18]. Treatment of **18** with mercuric oxide in the presence of **16** or of phenyl azide results in the adducts, **17** and **19**, respectively. It seems

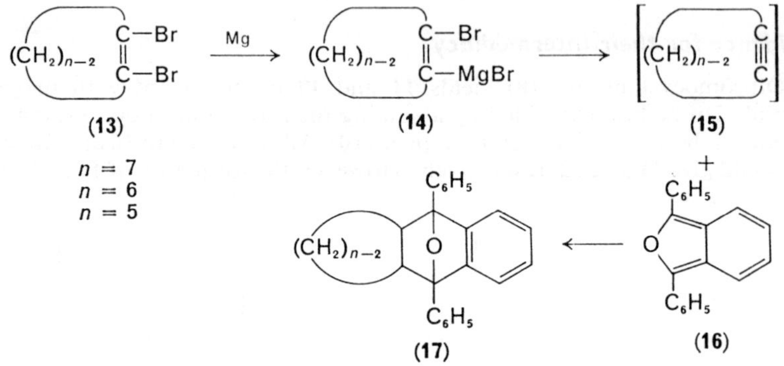

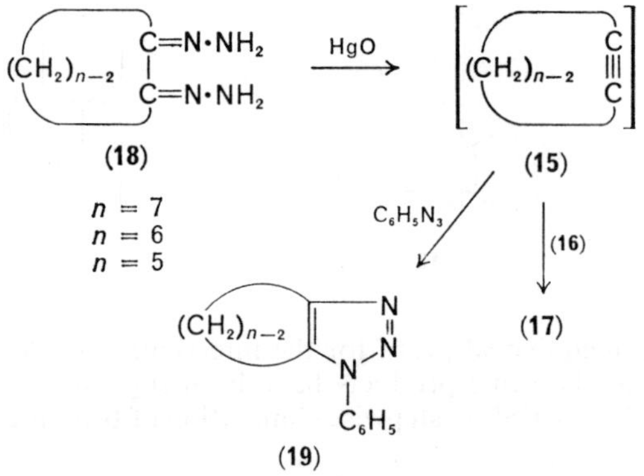

reasonable to assume that cycloalkynes are involved in these trapping reactions, even though alternative pathways are possible[19]. The diphenylbenzo[*c*]furan adducts **(17)** of cyclohexyne and cyclopentyne are obtained by the reaction of the corresponding bromomethylenecycloalkanes **(20,** $n = 6$ and 5)[20]. Oxidation of the aminotriazole derivatives **(21)** with lead tetraacetate at −76 °C in the presence of tetracyclone **(22)** also gives adducts **23**[21]. Lithiotosylaminotriazoles **(24)** gives the

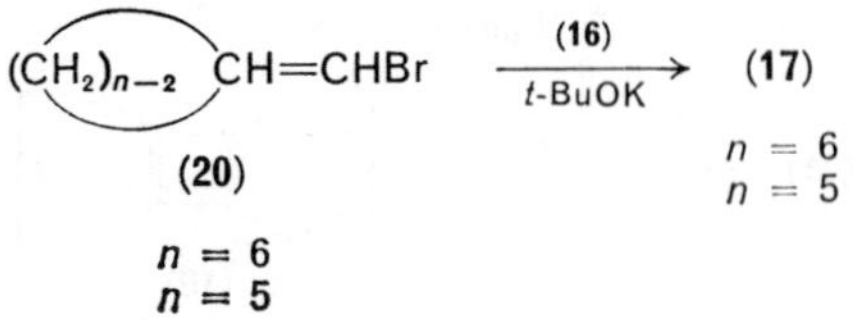

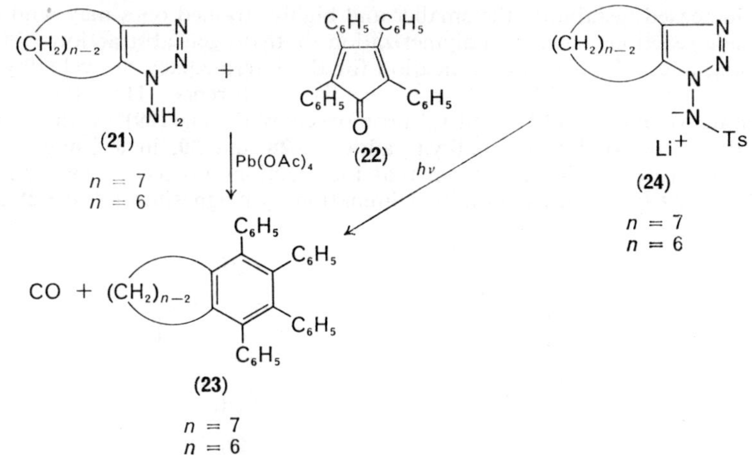

same adducts (**23**) by photolysis in the presence of tetracyclone[22]. The yields of the adducts of the above-mentioned trapping reactions are summarized in Table 1.

TABLE 1. Yields of cycloalkyne adducts. I: Grignard reaction; II: oxidation of bishydrazone; III: base elimination of bromomethylene derivative; IV: oxidation of aminotriazole; V: photolysis of tosylaminotriazole anion

	Diphenylbenzofuran adduct (%)			Tetracyclone adduct (%)	
Cycloalkyne	I[a]	II[b]	III[c]	IV[d]	V[e]
Cycloheptyne	64	26	—	93	56
Cyclohexyne	50·5	7	35	88	54
Cyclopentyne	2·1	0·5	12	—	—

[a] Reference 7.
[b] Reference 18.
[c] Reference 20.
[d] Reference 21.
[e] Reference 22.

Decrease of the yields of adducts in various trapping reactions with decreasing ring size offers further support for the intermediate occurrence of cycloalkynes, since if the trapping reagent were to add to the intermediate precursor of the cycloalkyne, for instance **25** or **26**, such a significant difference in product yield would not be expected.

The observed trend can be accounted for in two different ways[7]; the smaller cycloalkynes may be formed in lower yields owing to the more rapid decomposition of their precursors, or the yields of all cycloalkynes may be about the same, but, owing to their increased instability, the smaller and highly strained ones may tend to take part in side-reactions, such as polymerization or hydrogen abstraction.

Trapping is usually a strong indication for the intermediacy of cycloalkyne, but adducts may be formed without their intermediate occurrence. Thus, the reaction of 1,2-dibromocyclobutene **(27)**[23] and 1,2-bromoacenaphthylene **(29)**[24] with magnesium yielded the corresponding cycloalkyne adducts, **28** and **30**, in 8% and 4% yield, respectively. However, it was shown that the reactions proceeded via addition of **16** to **27** and **29** followed by bromine elimination by magnesium to give **28** and **30**.

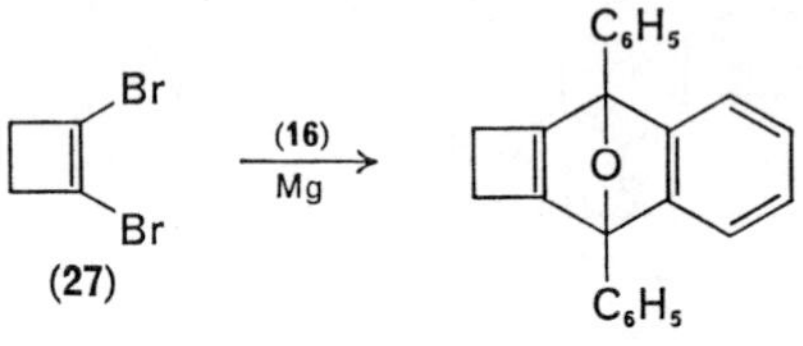

(27) (28)

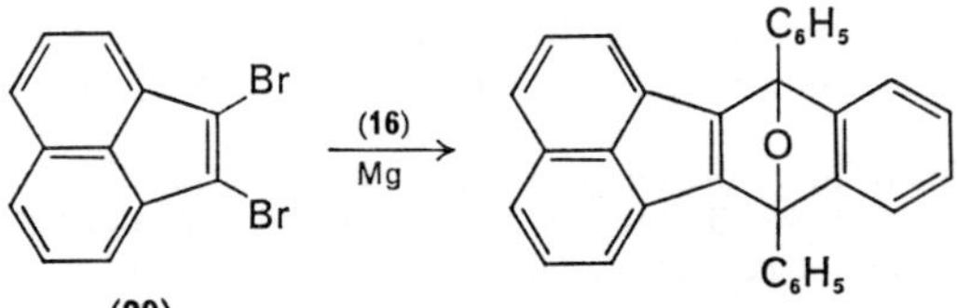

(29) (30)

The reaction of phenyllithium with 1-chlorocyclohexene **(31)**[25–27] and 1-chlorocyclopentene **(34)**[27, 28] gives 1-phenylcyclohexene **(33)** and 1-phenylcyclopentene **(36)**, respectively. Roberts and his coworkers studied these two reactions with ^{14}C-labelled

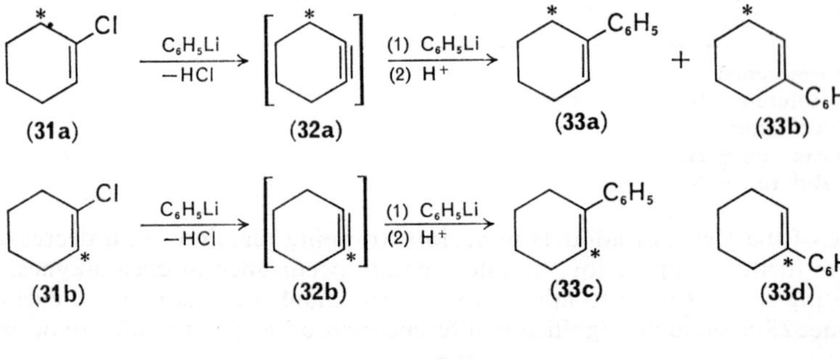

(31a) (32a) (33a) (33b)

(31b) (32b) (33c) (33d)

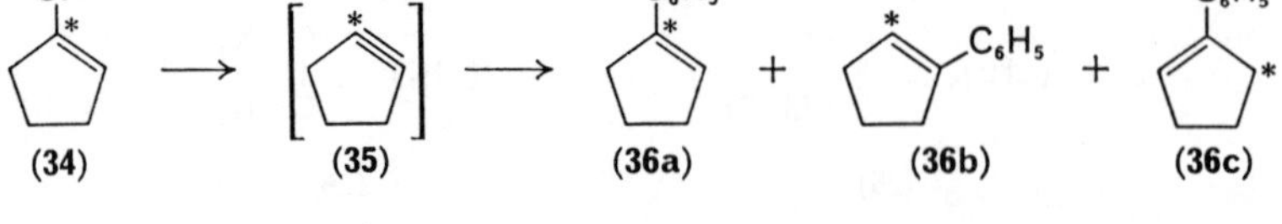

(34) (35) (36a) (36b) (36c)

1-chlorocycloalkenes[26–28]. A mixture (1 : 1) of **31a** and **31b** labelled with ^{14}C in the positions denoted by an asterisk was treated with phenyllithium in ether at 150 °C. Benzoic acid obtained by oxidative degradation of the resulting 1-phenylcyclohexene showed 23% of the specific activity of the starting material. Similarly, 1-chlorocyclopentene-1-^{14}C (**34**) afforded rearranged product (**36c**, 14·9%) together with the expected **36a** (48·9%) and **36b** (36·2%) showing that partial equilibration took place under the reaction conditions. A direct nucleophilic substitution of chlorine by phenyllithium is excluded by these results. The observed ^{14}C distribution can be explained by intervention of cyclohexyne (**32**) (the calculated specific activity of benzoic acid is 25%) and cyclopentyne (**35**). However, intermediacy of cyclic allenes such as 1,2-cyclohexadiene (**38**) and 1,2-cycloheptadiene cannot be excluded, because the formation of **38** [29] and **40** [30] from the corresponding 1-bromocycloalkenes (**37** and **39**) on treatment with potassium *t*-butoxide has been confirmed by trapping reactions. The observed ^{14}C distribution in the 1-phenylcyclopentene (**36**) can be explained by intermediate occurrence of 1,2-cyclopentadiene provided that half of the phenyllithium adds to the middle and the other half to the ends of the allenic

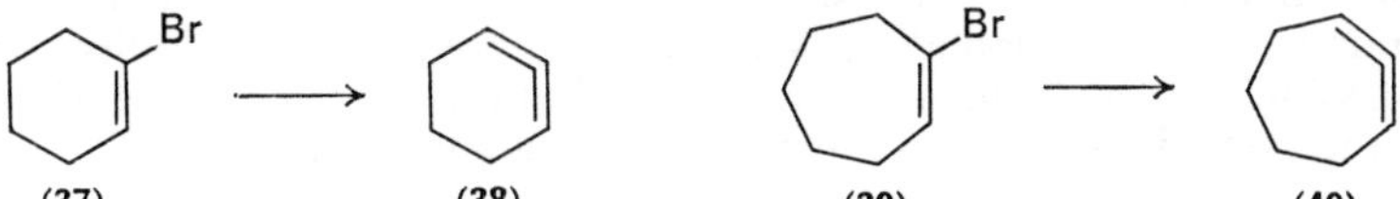

(**37**) (**38**) (**39**) (**40**)

(**41**) (**42**)

linkage, and the lithio-3-phenylcyclopentene-2-^{14}C (**41**) formed rearranges to 1-lithio-2-phenylcyclopentene-1-^{14}C (**42**). However, it was found that 2-chloro-3-methylcyclohexene (**43**) and 2-chloro-3-methylcyclopentene (**44**) gave the corresponding phenylcycloalkenes on treatment with phenyllithium, whereas the isomeric chlorides (**45** and **46**), possessing no olefinic hydrogens, yielded no phenylcycloalkenes under the same reaction conditions[31]. Furthermore, the deuterium content in 1-phenylcyclopentene (**36**) obtained by the reaction of 1-chlorocyclopentene-2,5,5-d_3 (**47**) with phenyllithium was found to be 1·84 (methylene) and 0·14 (olefinic) by n.m.r. spectroscopy, while no change of deuterium content was observed in the unreacted **47** [32]. These results clearly indicate that the main pathway for the formation of 1-phenylcyclopentene does not contain a cycloallene intermediate.

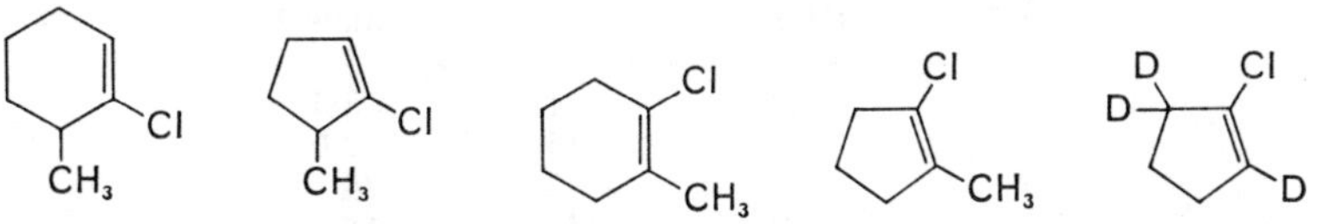

(**43**) (**44**) (**45**) (**46**) (**47**)

Another alternative mechanism to cycloalkyne formation in trapping reactions is the addition of the trapping reagent to the cycloalkyne precursor followed by an elimination.

Owing to the rapid decomposition of the intermediate precursor of the unstable cycloalkyne, such as **26** or **27**, kinetic investigations to confirm the intermediacy of cyclic acetylenes could not be performed. However, 1-lithio-2-bromocyclopentene (**48**) was found to be fairly stable at room temperature. The kinetic measurements indicate that **48** loses lithium bromide in a first-order reaction ($k = 2 \times 10^{-5}$ s^{-1} at 20 °C in ether), and the Arrhenius energy of activation for this reaction was estimated to be 30 kcal/mole [33, 34]. When lithium chloride was added to a decomposing solution of **48**, 1-chlorocyclopentene **(34)** was isolated indicating the reaction is reversible[34]. The fact that an addition of 1,3-diphenylbenzo[*c*]furan **(16)** does not increase the rate of lithium bromide elimination from **(48)** proves the generation of cyclopentyne **(35)**.

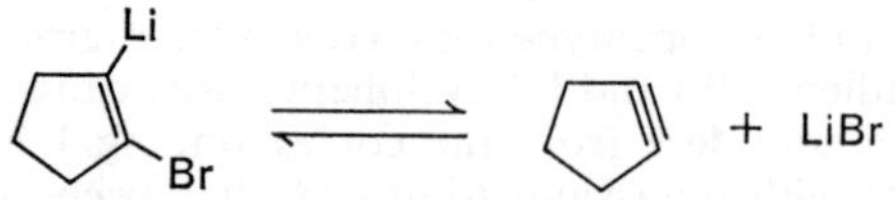

(48) **(35)**

The fact that the rate of dehydrobromination by potassium *t*-butoxide from **49** [35], **50** [36] and **51** [36] is not affected by the addition of trapping agent indicates the intervention of the corresponding cycloalkyne, although the possibility of addition of 1-potassio-2-bromoalkene to the trapping agent cannot be excluded, if the formation of the potassio derivative is the rate-determining step.

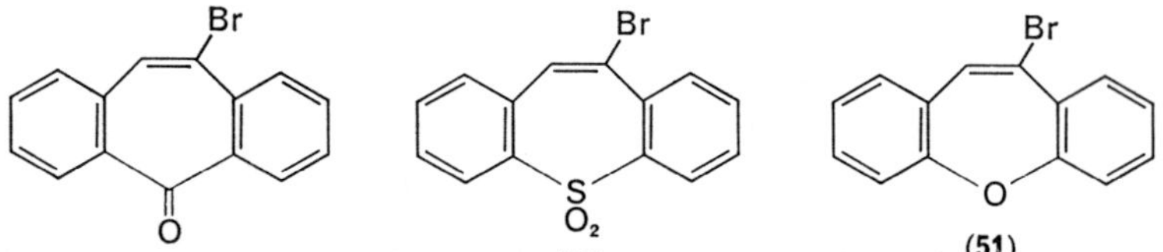

(49) **(50)** **(51)**

The relative stabilities of cycloheptyne and cyclohexyne generated from 1-amino-4,5-cycloalkeno-1,2,3-triazoles **(21)** and lead tetraacetate[21] were investigated. After the nitrogen evolution, which occurred instantaneously on addition of lead tetraacetate, had ceased, the trapping reagent was added after different time intervals to the reaction mixture, and yields of adduct **(23)** were determined. The results

TABLE 2. The yields of adducts **23**

n	Temperature (°C)	Time (min)	Yields of adducts **23**
7	−25	1	3·4
7	−25	15	0·6
7	−76	2	74
7	−76	10	68
7	−76	60	49
7	−76	180	35
6	−76	2	0·6
6	−110	3	3·8
6	−110	15	2·1

summarized in Table 2 clearly show that the stability of cycloalkyne decreases markedly with decreasing ring size. The half-life of cycloheptyne was estimated to be about 1 h at −75 °C, while that of cyclohexyne was only a few seconds at −110 °C.

Competitive addition of phenyllithium and lithium piperidide to cyclooctyne, cycloheptyne, cyclohexyne and cyclopentyne has been studied in order to gain information on the relative stabilities of cyclic acetylenes[37].

All attempts to find evidence for the intervention of cyclobutyne have failed up to the present date[23, 28, 38] and cyclopentyne is the smallest cycloalkyne whose intermediacy has been firmly proved.

C. *Reactions*

1. Polar addition

Nucleophilic additions to the acetylenic bond in various short-life cyclic acetylenes including optically active derivatives[44] have been extensively studied[20, 39–44]. The formation of 1,2-di-bromocycloheptene was also observed in the bromine oxidation of the aminotriazole derivative **21** ($n = 7$)[21].

2. Cycloaddition

As previously mentioned, short-life cycloalkynes add to reactive dienes such as 1,3-diphenylbenzo[*c*]furan (**16**) and tetracyclone to give Diels–Alder-type adducts, and the reaction has frequently been used to establish the intermediacy of short-life cyclic acetylenes. The addition reactions have found synthetic applications[35, 45, 46, 48]. The addition of 1-diethylaminobutadiene to cycloalkynes provides an interesting synthetic route for benzo annelation[47]. Intervention of dehydrobullvalene was also confirmed by the formation of Diels–Alder-type adducts[49, 50]. At present only azides are used as 1,3 dipoles in the addition reaction with short-life cycloalkynes[18, 35, 51].

3. Isomerization and oligomerization

Isomerization of stable cycloalkynes to the isomeric cyclic allenes under basic conditions indicates that the equilibrium shifts towards allene with decreasing ring sizes[53]. The observed trend suggests that the short-life cycloallenes should be more stable than the corresponding cycloalkynes. The formation of short-life cycloallenes has been proved by the structures of the dimers and the trapping products[29, 30, 53, 54].

Short-life cycloalkynes appear to form cyclobutadienes by dimerization in the absence of trapping reagent. Although the resulting cyclobutadienes have neither been isolated nor trapped, the structures of oligomeric products may be conveniently interpreted by assuming the dimerization of the short-life cycloalkyne to a cyclobutadiene in the first step[55, 56]. Addition of short-life cycloalkynes to the butadienes[57, 58] give rise to the trimers[5, 55, 59–65]. Dimerization of the cyclobutadiene to form the tetramer of the original cycloalkyne has also been observed[55, 56]. The formation of hexamers of cycloheptyne has been reported[21]. Thermal decomposition of cycloheptenocyclopropenone at 250 °C gives only a trimer[64] suggesting that only at low temperatures can sufficient cyclobutadiene accumulate to allow formation of hexamer.

III. MEDIUM-RING ACETYLENES

Medium-ring acetylenes (8- to 11-membered) with one triple bond are isolable compounds. Eight- and ten-membered cyclic acetylenes containing more than one unsaturated bond have been prepared and isolated. Isolable substituted cycloheptyne derivatives will also be discussed in this section.

A. Physical Properties

The enthalpies of hydrogenation of some medium-ring acetylenes to the corresponding cycloalkanes are summarized in Table 3 [66, 67]. It has been suggested that the result for cyclooctyne (69·1 kcal/mole) may be somewhat in error. Cyclooctyne,

TABLE 3. Enthalpies of hydrogenation of cycloalkynes at 25 °C

Cycloalkyne	ΔH (kcal/mole)
Cyclooctyne	69·1
Cyclononyne	61·9
Cyclodecyne	56·5
Cycloundecyne	57·2
Cyclododecyne	61·7
4-Octyne	62·8
1,8-Cyclotetradecadiyne	125·4

in spite of careful purification, took up only about 90% of the theoretical amount of hydrogen. It is suspected that the cyclooctyne was partially polymerized in the calorimeter[68]. Considerable strain in cyclooctyne is reflected in the high ΔH value. The ΔH value for cyclononyne is about the same as for 4-octyne or cyclododecyne, but is 4–6 kcal/mole higher than the value for cyclodecyne and cycloundecyne. Since considerable Pitzer strain and transannular hydrogen interaction have been shown to be present in medium-ring cycloalkanes by measurements of heats of combustion[69–72] and X-ray analyses[73], the low ΔH values for cyclodecyne and cycloundecyne can probably be ascribed to the increase of Pitzer strain and hydrogen interaction in the hydrogenation products, cyclodecane and cycloundecane.

Medium-ring cycloalkynes are in equilibrium with the corresponding cyclic allenes in various basic media. The composition of the equilibrium mixture at 100 °C in *t*-butanol using potassium *t*-butoxide as a base is shown in Table 4 [52].

TABLE 4. The equilibrium composition of cycloalkyne–cycloallene mixtures

Cycloalkyne	% Cycloalkyne in the mixture
Cycloundecyne	74
Cyclodecyne	35
Cyclononyne	7

The data show that the allene becomes more stable than the acetylene as the ring size decreases. This fact seems to be attributable to the Baeyer strain, since in an allene linkage only three carbons must be in a straight alignment compared to four in an acetylene.

The medium-ring acetylenes show typical C≡C stretching vibration[10–15, 74] at *ca.* 2210 cm^{-1}. The fairly strong band in cyclooctyne in the same region suggests that an angle strain at the triple bond would render the C≡C stretching vibration more unsymmetric. This fact indicates that, in spite of an appreciable strain, the essential character of acetylenic bond is preserved in cyclooctyne.

The n.m.r. spectrum of 4,4,7,7-tetramethylcyclooctyne (**52**) at room temperature shows only three signals corresponding to the α-methylene groups, the methyl groups and the regular methylene groups. The expected splitting of the three signals can be observed at −70 °C. From the coalescence temperature (−56 °C) of the two methyl signals, $\Delta F^{\ddagger} = 11{\cdot}8 \pm 0{\cdot}5$ kcal/mole was calculated for the averaging process. The n.m.r. spectral behaviour of **52** indicates that the bond angles at the acetylenic bond are considerably bent, rendering flexibility to the eight-membered ring[75].

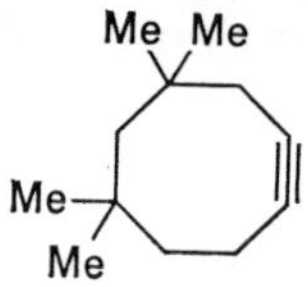

(52)

B. *Isolable Seven-membered Acetylenes*

1. 3,3,6,6-Tetramethyl-1-thiacycloheptyne

As stated in the preceding section, the intermediacy of cycloheptyne has been proved unequivocally by trapping reactions and kinetic studies. However, the isolation of the parent cycloheptyne[74, 75] could not be achieved, presumably owing to rapid addition reactions to the highly reactive acetylenic bond even in diluted solutions[7, 18, 21, 22, 32, 35, 37, 45, 46, 49c, 64, 77].

Krebs and coworkers have synthesized 3,3,6,6-tetramethyl-1-thiacycloheptyne (**55**)[78]. The thiacycloheptanedione (**53**) derived from the corresponding acyloin was converted into the bishydrazone (**54**). Oxidation of **54** with mercury(II) oxide gave **55** in 5·5% yield together with the *cis*-olefin (**56**, 6·9%). In the absence of oxygen

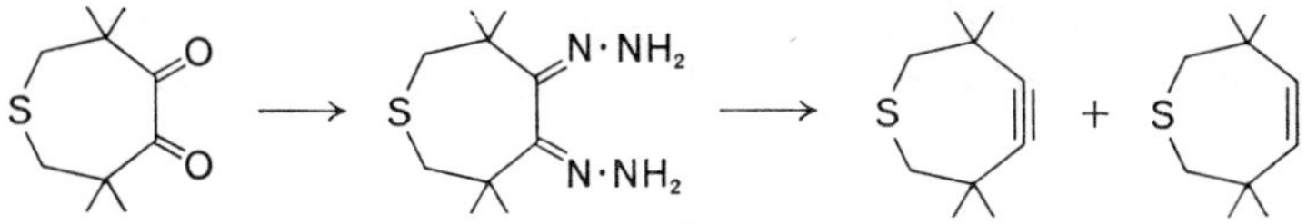

(53) (54) (55) (56)

the cycloheptyne (**55**) could be kept at −80 °C without decomposition for several days. Also no di- or oligomerization reaction of **55** could be observed even at 140 °C. Diphenylbenzo[*c*]furan (**16**) gave **57** by reaction with **55**. However, tetracyclone yielded no adduct. Isocyanides react with **55** to give cyclopropene derivatives (**58**)[79]. It is to be noted that the n.m.r. spectrum indicates the high conformational

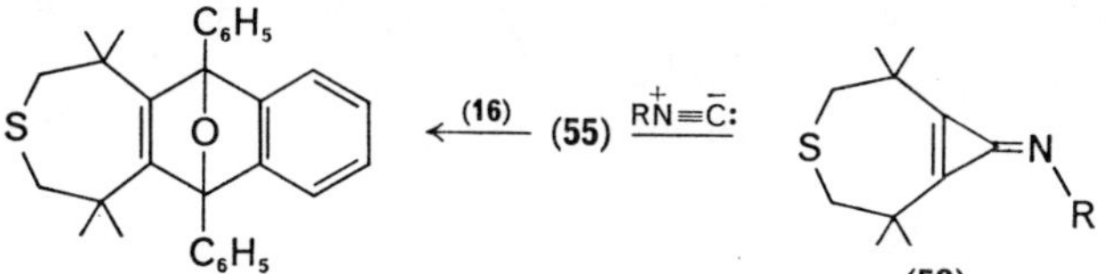

(57) (58)

(**a**) R = *p*-$O_2NC_6H_4$
(**b**) R = C_6H_5
(**c**) R = cyclohexyl

mobility of **55** at room temperature in spite of the presence of a considerable ring strain[78]. The fact that the seven-membered cyclic acetylene (**55**) can be isolated is attributable to a suppression of intermolecular reactions by the steric hindrance of the *gem*-dimethyl groups at the positions adjacent to the strained and reactive acetylenic bond. Interestingly, a stable cyclobutadiene derivative (**59**) has been obtained[80] by treatment of **55** with $[(C_6H_5CN)_2PdCl_2]$ in THF. The reaction of the complex **59** with ethylene-bis(diphenylphosphane), $(C_6H_5)_2PCH_2CH_2P(C_6H_5)_2$, yielded the cyclobutadiene derivative (**60**) as yellowish crystals. The cyclobutadiene **60**, which can be regarded as a tetra-*t*-butyl-cyclobutadiene derivative, was found to be sensitive to oxygen, and a furan derivative (**61**) was isolated from the oxygenated products.

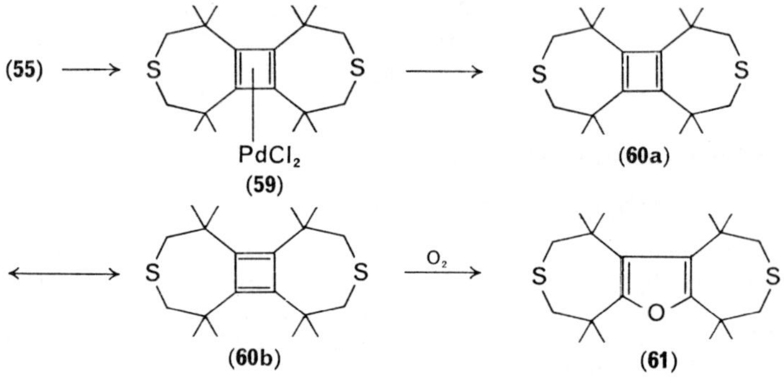

2. 3,3,7,7-Tetramethylcycloheptyne

The synthesis of 3,3,7,7-tetramethylcycloheptyne (**64**), a carbocyclic analogue of **55**, has also been performed[81]. The bishydrazone (**63**) derived from tetramethylcycloheptane-1,2-dione (**62**)[83] was oxidized with lead tetraacetate and **64** was obtained together with **65**, **66**, **67** and **68**. The last compound (**68**) is a dimer of **64** with

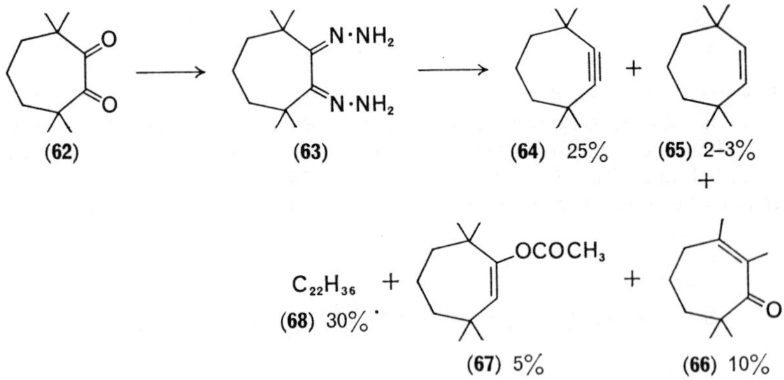

unidentified structure. It was found that **64** has higher reactivity than the thia analogue (**55**). The cycloheptyne **64** reacts with phenyl azide, diphenylbenzofuran (**16**) and *p*-nitro-phenylisocyanide to give the corresponding 1 : 1 adducts, (**69**), (**70**) and

(71), respectively. The dimer **72** was obtained within 1 h on standing from neat **64**. Treatment of **64** with carbon disulphide yielded the 2 : 2 adduct **(73)**. The rate of dimerization of **64** was found to be 10^7–10^8 times slower than that of cycloheptyne. From this result, the enthalpy of activation of **64** for dimerization was estimated to be 9–10 kcal/mol higher than that of cycloheptyne[81].

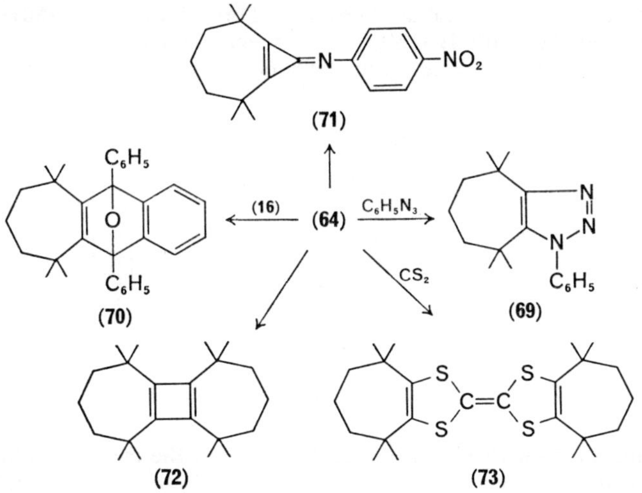

C. Eight-membered Acetylenes

1. Cyclooctyne

Cyclooctyne **(75)** can be prepared according to the usual methods of synthesis of acetylenic compounds. For instance, oxidation of the bishydrazone of cyclooctane-1,2-dione **(74)**[10–16, 74] or debromination of 1,2-dibromocyclooctene **(76)** with magnesium[76] or the photolysis of the anion of 1-tosylamino-1,2,3-triazole derivative **(77)**[23] gave cyclooctyne **(75)**.

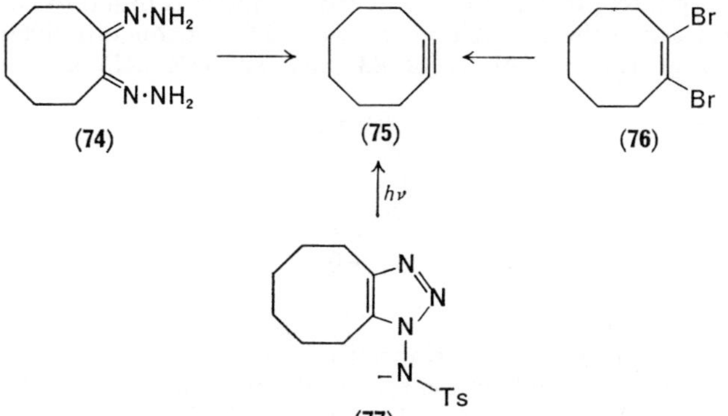

(77)

Cyclooctyne, being considerably strained, shows a high reactivity to various reagents such as bromine[76] and butyllithium[7, 84], and its Diels–Alder reactions occur under milder conditions than with unstrained acetylenes[7, 18, 84].

On treatment with aqueous silver nitrate it forms a stable crystalline complex **(78)**, from which **75** can be recovered almost quantitatively by addition of ammonia[76]. Anhydrous nickel(II) bromide converts cyclooctyne into the trimer **(79)** with a quantitative yield[86]. However, when the reaction was carried out in the presence of a trace of water, the dimeric cyclobutadiene–nickel bromide complex **(80)** was obtained in 9·4% yield, together with **79** (85%)[86]. The spectroscopic properties of **80** showed close similarity with those of the tetramethylcyclobutadiene–nickel chloride complex[87].

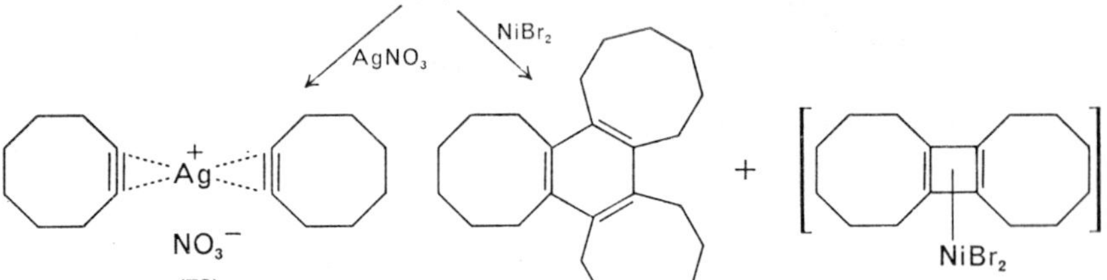

(78) **(79)** **(80)**

The reaction of **75** with nickel tetracarbonyl gives the cyclopentadienone–nickel complex **(81)**, thermal decomposition of which yields the cyclopentadienone **82** [86].

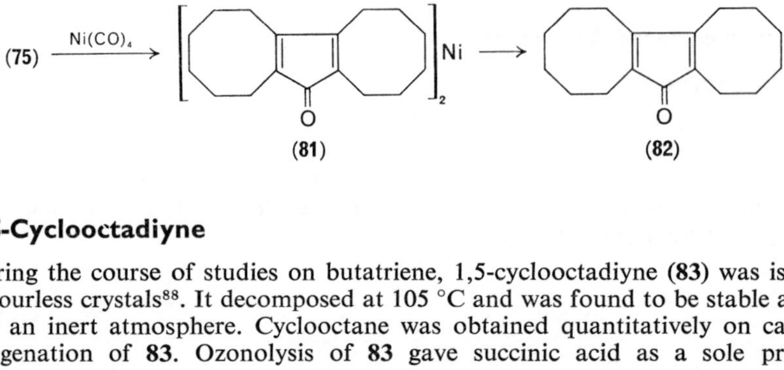

(81) **(82)**

2. 1,5-Cyclooctadiyne

During the course of studies on butatriene, 1,5-cyclooctadiyne **(83)** was isolated as colourless crystals[88]. It decomposed at 105 °C and was found to be stable at 0 °C under an inert atmosphere. Cyclooctane was obtained quantitatively on catalytic hydrogenation of **83**. Ozonolysis of **83** gave succinic acid as a sole product.

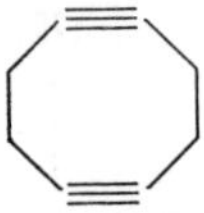

(83)

The ^{1}H- and ^{13}C-n.m.r. spectra of **83** exhibit signals at δ 2·62 ($CDCl_3$), and δ 20·2 (t) (J = 140 Hz) and 85·8 (s) p.p.m., respectively. The photoelectron spectrum of **83** shows three overlapped bands in the region of 9·0 to 10·0 eV in addition to a band at 10·1 eV. The splitting of the π band has been ascribed to splitting of the degenerate π molecular orbitals by ring strain[89]. The preliminary results of X-ray crystal structure analysis of **83** show that the molecule has a centre of symmetry and holds a high planarity.

3. Benzo derivatives

It was anticipated that 5,6-didehydro-11,12-dihydrodibenzo[*a.e*]cyclooctene (**85**), having a large ring strain, should be less stable than cyclooctyne. However, **85** obtained from the dibromo compound **84** on treatment with potassium *t*-butoxide and *n*-methylpiperidine was found to be stable[90]. **85a** could be kept without

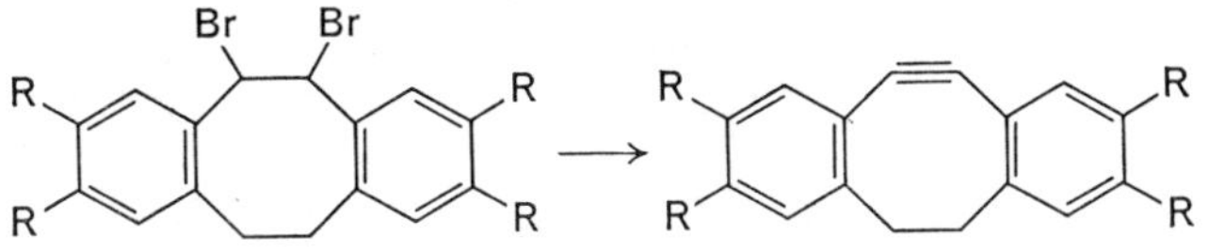

(84) **(85)**

(**a**) R = H; (**b**) R = OCH_3

decomposition for a few days at room temperature, and **85b** showed no change for three years. In spite of its unusual stability, **85b** reacts on heating with azides, cyclopentadiene and piperidine to give the corresponding adducts, **86**, **87** and **88**,

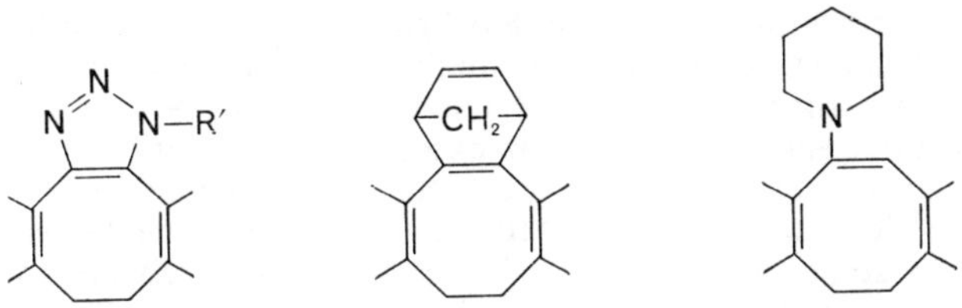

(86) **(87)** **(88)**

R′ = C_6H_5, CH_3, $COOC_2H_5$, $CH_2COOC_2H_5$

respectively. The tetramethoxy derivative (**85b**) gives the known dibenzocyclooctadiene (**89**)[91] and cyclooctatriene (**90**)[91] derivatives on catalytic hydrogenation[90]. The bathochromic shift of the electronic spectrum of **85b**, which is similar to that of diphenylacetylene, was ascribed to the bending of the triple bond[90].

(89) **(90)**

Recently **85a** has been prepared by a different route[92]. Dibenzocyclooctadienone (**91**) is converted into the semicarbazone (**92**), which gives the selenadiazole derivative (**93**) on treatment with selenium dioxide. Thermolysis of **93** yields **85a** as crystals stable at room temperature. The ^{1}H-n.m.r. spectrum of **85a** indicates the conformational stability of the eight-membered ring, in contrast to the conformationally mobile 4,4,7,7-tetramethylcyclooctyne (**95**) [coalescence temperature: −56 °C (60 MHz), −30 °C (220 MHz)][75]. The selenadiazole **93** gives the adduct **94** on heating with tetracyclone[92].

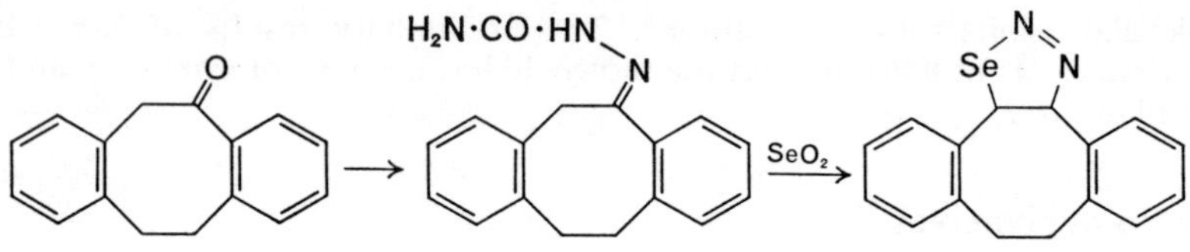

(91) (92) (93)

(93)

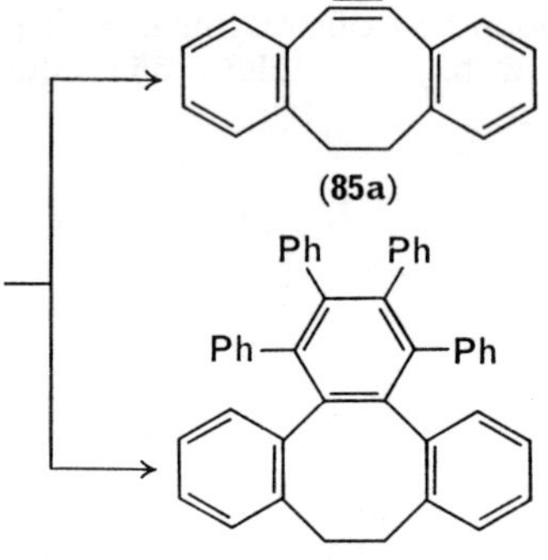

(85a)

(94)

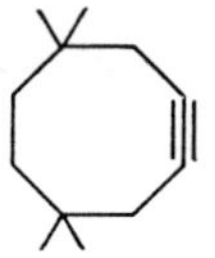

(95)

The isolation of 1,5-cyclooctadiyne **(83)**[88] gave a strong impulse to the studies on eight-membered acetylenes containing more than one unsaturated bond in the cyclic system. Dehydrobromination of the tetrabromide **97**, prepared from *sym*-dibenzocyclooctatetraene **(96)** with an excess of potassium *t*-butoxide at room temperature, gives *sym*-dibenzo-1,5-cyclooctadiene-3,7-diyne **(98)** as pale yellow crystals, which decomposes at *ca.* 110 °C. The diacetylene **98** is comparatively stable, although some decomposition is observed after two days on standing at room temperature exposed to light and air. The structure of **98** was confirmed by

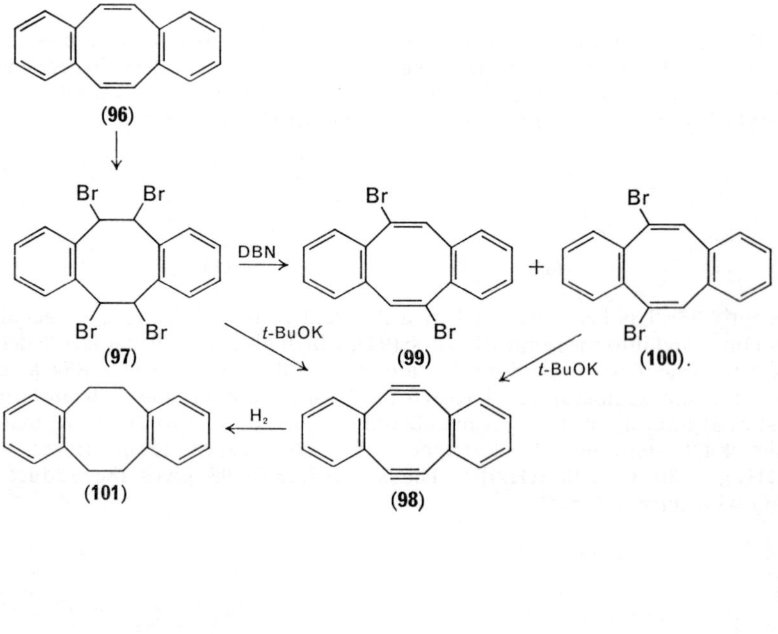

the i.r., n.m.r. and mass spectral data, and by smooth hydrogenation to dibenzocyclooctadiene (**101**)[93]. Alternatively, treatment of **97** with DBN leads to a mixture of dibromides, **99** and **100**. Further dehydrobromination of the mixture with potassium *t*-butoxide yields **98** [93]. X-ray structure analysis revealed that, in the crystalline state, this has a substantially planar eight-membered ring[94]. Bromination of **96** with one mole of bromine gives the known **102** [95]. Treatment of **102** with potassium *t*-butoxide gives *sym*-dibenzo-1,3,5-cyclooctatrien-7-yne (**103**) as golden yellow crystals, which decompose at *ca.* 85 °C [93]. Treatment with 1,5-diazabicyclo-[3.4.0]non-5-ene gives **104**, which, on further dehydrobromination with potassium *t*-butoxide, yields **103** [93]. The monoacetylene **103** is very unstable, and the crystals decompose after a few minutes on standing at room temperature. Catalytic partial and full hydrogenation of **103** give **96** and **101**, respectively[93].

Treatment of (**103**) in THF-d_8 with potassium mirror at -20 °C gives a deep green solution of the dipotassium salt of the dianion **105** [93].

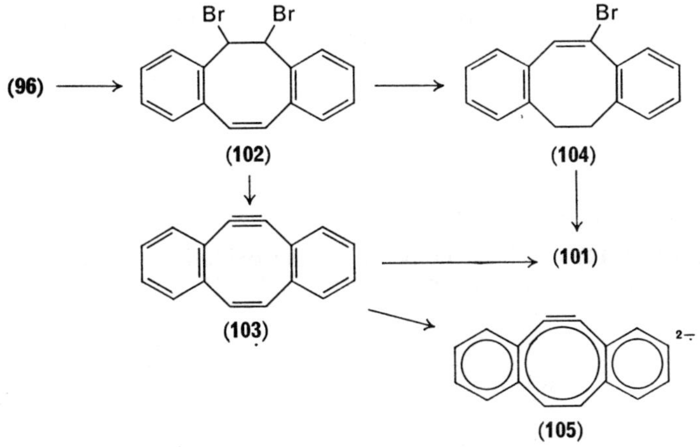

5,10-Dibromobenzocyclooctatetraene (**106**) was obtained from the photobromination product of biphenylene[96]. A solution of **106** in THF was treated with potassium *t*-butoxide for 30 s at room temperature to give the diacetylene **107** as a yellow liquid, which decomposed in a few minutes at 0 °C. The diacetylene **107** exhibits an electronic spectrum closely related to that of the dibenzo analogue (**98**). The bis adducts, **108** and **109** [97], are obtained by the reactions with diphenylbenzofuran (**16**) and tetracyclone, respectively[96]. Because the eight-membered ring in **98** has been proved to hold an almost planar structure[94], **103** and **107** also presumably contain a planar conjugated eight-membered ring. In fact, **98, 103** and **107** show complex electronic spectra, indicating the presence of highly conjugated systems, as compared with the simple spectrum of the non-planar **96**. The high-field shift of the signals of both aromatic and olefinic protons of **98** and **103** and the olefinic signal of **107** in their ^{1}H-n.m.r. spectra seem to reflect the effect of induction of paramagnetic ring current in the planar 8π electron systems*. On the contrary, the olefinic and a part of the benzenoid proton signals of **105**, which is a 10π electron system, were observed at a lower field than those of **96** despite the presence of two negative charges. The dianion **105** clearly sustains a diamagnetic ring current, and the ^{1}H-n.m.r. spectrum closely resembles that of the corresponding dianion of **96** [98].

* For ^{1}H-n.m.r. spectra of [4n]- and [4n+2]π electron systems, see Section V.

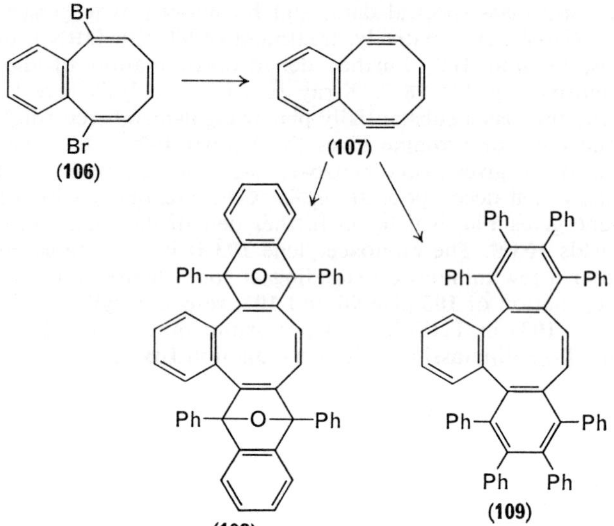

(106) (107) (108) (109)

D. Ten-membered Acetylenes

Cyclodecyne (**111**) has been prepared by the oxidation of the bis-hydrazone of 1,2-cyclodecanedione (**110**) with mercury(II) oxide[10–16, 74].

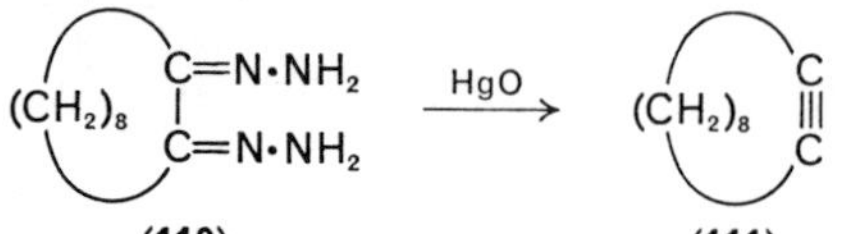

(110) (111)

Open-chain acetylenes yield pure *trans* alkenes on reduction with alkali metal in liquid ammonia. However, reduction of cyclodecyne (**111**) with sodium in liquid ammonia led to a mixture of *cis*- (**112**, $>90\%$) and *trans*-cyclodecene (**113**, $>4\%$)[99, 100]. The formation of the *cis* isomer (**112**) was attributed to the reduction of 1,2-cyclodecadiene (**114**) formed by a rapid isomerization of the starting cycloalkyne, while the formation of the *trans* isomer (**113**) was ascribed to the direct reduction of **111**.

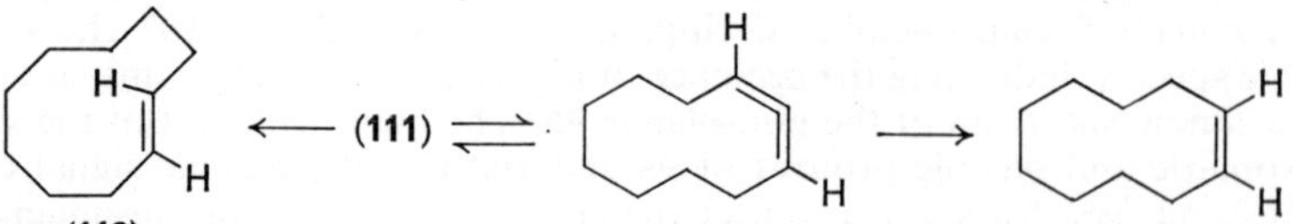

(113) (114) (112)

It is to be noted that cyclododecyne, under the same conditions, gave a mixture of *cis* and *trans* isomers with the latter being the major component[99], while cyclononyne[11] yielded only *trans* isomer. Since (see Table 4) the cycloalkyne–cycloallene equilibria under basic conditions shift to the allene side with decreasing ring size, these results indicate that the isomerization of cycloalkyne and the reduction of cycloallene should be fast compared to the reduction of the cycloalkyne. Larger cycloalkynes behave like an open-chain acetylene, e.g. cyclotetradecyne yields only *trans*-cyclotetradecene[99].

Various ten-membered acetylenes containing more than one unsaturation have been prepared. The reaction of bis(chloromethyl) ether with acetylenedimagnesium bromide gave 1,6-dioxa-3,8-cyclodecadiyne (**115**) in *ca.* 2% yield together with trioxacyclopentadecatriyne (**116**, *ca.* 05%)[101, 102]. A chair conformation (**117**) has been proposed for **115** on the basis of the X-ray diffraction data[102].

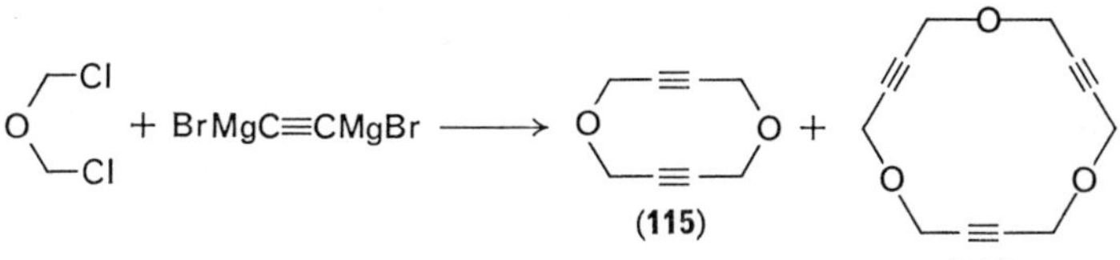

(**115**) (**116**)

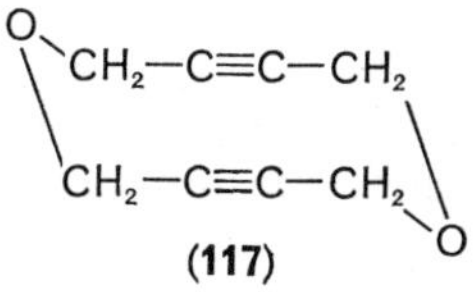

(**117**)

Oxidative coupling of 1,5,9-decatriyne (**118**)[103] with copper(II) acetate in pyridine[104] gave a mixture of **121a** and **121b** in addition to the cyclic dimer **120**. The former compounds (**121a** and **121b**) arose from the addition of acetic acid to the unstable cyclic monomer, 1,3,7-cyclodecatriyne (**119**)[105]. Hydrogenation of the enol acetate mixture over platinum catalyst gave cyclodecanol and bicyclic hydrocarbons formed by a transannular cyclization[105].

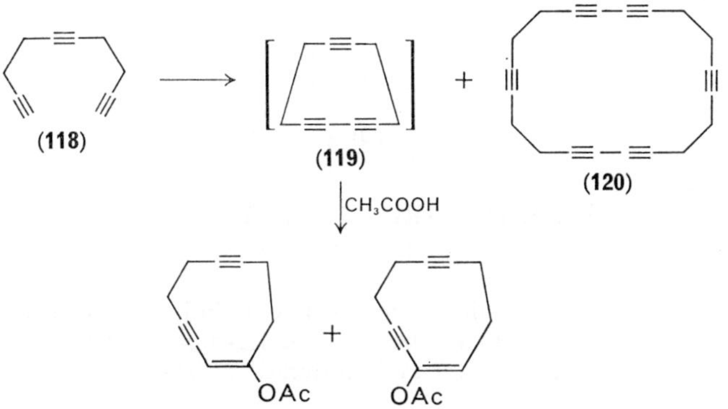

(**118**) (**119**) (**120**) (**121a**) (**121b**)

All-*cis*-1,6-dichloro-1,3,6,8-cyclodecatetraene (**123**) and all-*cis*-1,6,6-trichloro-1,3,8-cyclodecatriene (**124**) were obtained by the reaction of *cis,cis*-3,8-cyclodeca-diene-1,6-dione (**122**) with phosphorus pentachloride[106]. Treatment of **123** with lithium diisopropylamide led to all-*cis*-1-chloro-1,3,8-cyclodecatrien-6-yne (**125**) and naphthalene (**126**). The same products (**125** and **126**) were obtained by a similar treatment of **124** [107]. The ten-membered cyclic acetylene **125** is an unstable colourless liquid, being completely decomposed after standing for 16 h in the neat state or in a solution[107].

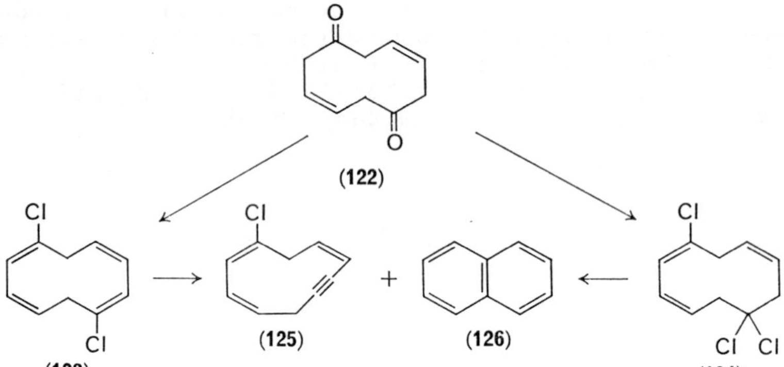

(122) (123) (125) (126) (124)

When a mixture of 9,10-epoxy-1-decalone (**127**) and *p*-toluenesulphonylhydrazine in ethanol was kept at 50 °C for 2 h, 5-cyclodecyn-1-one (**128**) was obtained in a yield of 50% [108]. This fragmentation reaction has been used for the preparation of 4-cyclopentadecyn-1-one[109].

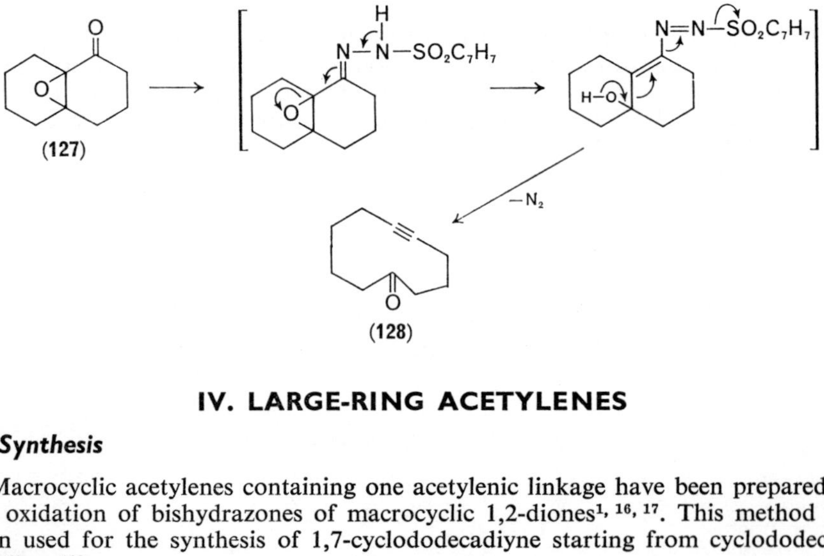

(127) (128)

IV. LARGE-RING ACETYLENES

A. Synthesis

Macrocyclic acetylenes containing one acetylenic linkage have been prepared by the oxidation of bishydrazones of macrocyclic 1,2-diones[1, 16, 17]. This method has been used for the synthesis of 1,7-cyclododecadiyne starting from cyclododecyn-1,2-dione[110].

In 1961, Wotiz[111] and Dale[112] independently reported the new syntheses of non-conjugated cyclic polyynes. When α,ω-dibromides are treated with a mixture of sodium acetylide and disodium acetylide in liquid ammonia, cyclic diynes (**129**) are obtained together with the linear polyynes (**130**). Sodium acetylide, being a chain terminator, prevents the formation of large amounts of linear polyynes. 1,8-Cyclotetradecadiyne (**129**, n = 5) and the 22-membered dioxadiyne (**131**) were obtained by this method.

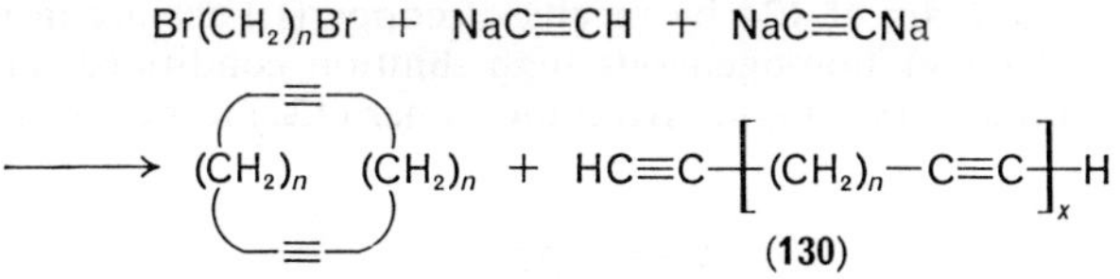

(130)

(129)

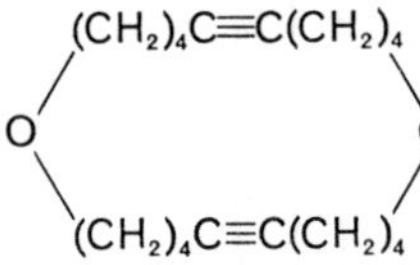

(131)

The reaction of α,ω-dibromoalkanes with mono- or disodioalkadiynes was also used to give the same products[111]. Dale[112] used essentially the same two-step reaction. The generalized reaction can be expressed as follows:

$$2Na(C{\equiv}C)_xH + Br(CH_2)_mBr \longrightarrow H(C{\equiv}C)_x(CH_2)_m(C{\equiv}C)_xH$$

$$Na(C{\equiv}C)_x(CH_2)_m(C{\equiv}C)_xNa + Br(CH_2)_nBr \longrightarrow$$

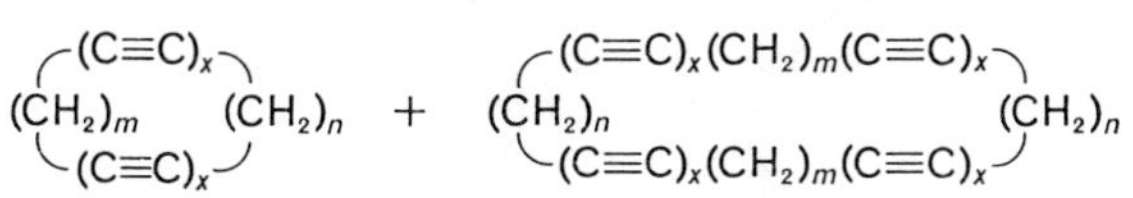

(132) (133)

Cyclic non-conjugated diynes and tetraynes containing 11–26 carbon atoms and cyclic tetraynes having two conjugated diyne units in the range of 16–26 carbon atoms have been prepared by this reaction sequence[112].

Formation of diphenyldiacetylene by the oxidation of the copper(I) salt of phenylacetylene was observed by Glaser over a century ago[113]. The modern refinement of the reaction involves shaking an ethynyl compound with an aqueous

$$2C_6H_5C{\equiv}CCu \xrightarrow{[O]} C_6H_5C{\equiv}CC{\equiv}CC_6H_5$$

copper(I) chloride–ammonium chloride solution in an atmosphere of oxygen (the so-called Glaser reaction). More recently oxidation of an ethynyl compound with copper(II) acetate in pyridine has been proposed[10]. When the Glaser coupling reaction is applied to α,ω-diethynyl compounds (**134**, n = 3, 4 or 5), cyclic dimers (**135**, n = 3, 4 or 5) are formed in addition to linear oligomers (**136**)[114–118].

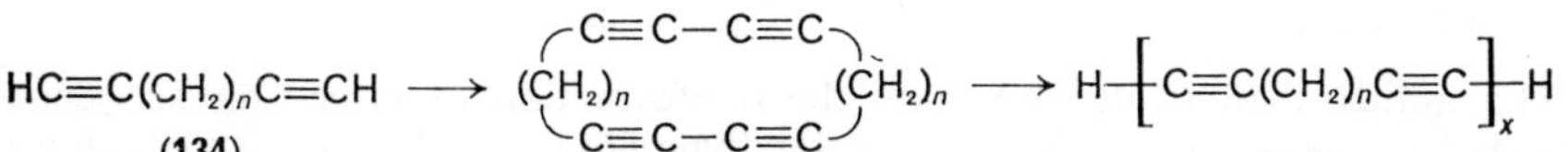

(134) (135) (136)

The oxidative coupling of **134** by means of copper(II) acetate in pyridine, which can be performed under homogeneous high dilution conditions, produced a wide variety of cyclic oligomers, ranging from monomer **(137)** to hexamer **(141)**[104, 119, 120].

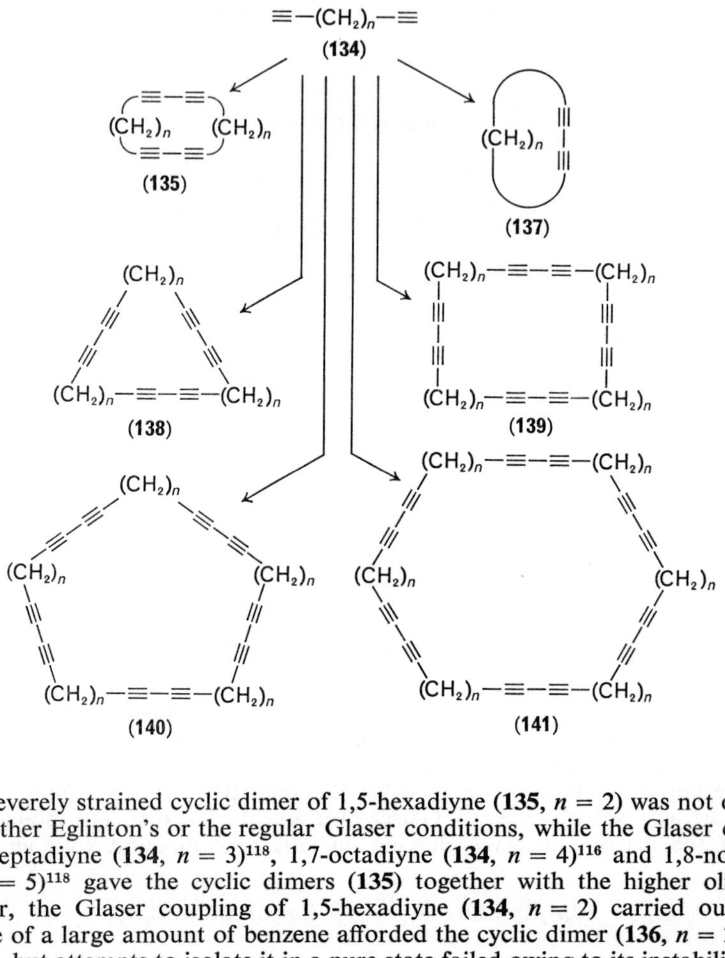

The severely strained cyclic dimer of 1,5-hexadiyne (**135**, $n = 2$) was not obtained under either Eglinton's or the regular Glaser conditions, while the Glaser coupling of 1,6-heptadiyne (**134**, $n = 3$)[118], 1,7-octadiyne (**134**, $n = 4$)[116] and 1,8-nonadiyne (**134**, $n = 5$)[118] gave the cyclic dimers **(135)** together with the higher oligomers. However, the Glaser coupling of 1,5-hexadiyne (**134**, $n = 2$) carried out in the presence of a large amount of benzene afforded the cyclic dimer (**136**, $n = 2$) in the solution, but attempts to isolate it in a pure state failed owing to its instability[121, 122].

A wide variety of macrocyclic polyacetylenes has been prepared by the oxidative coupling and converted into dehydroannulenes, which will be discussed in Section V.

B. Properties

1. Conformation

The influence of triple bonds on the conformation of large-ring acetylenes is reflected in the melting points. Cycloalkadiyne (**142**, $x = 1$) and cycloalkatetrayne (**142**, $x = 2$) show a significant alternation of melting points (Figure 1)[73, 112]. Since

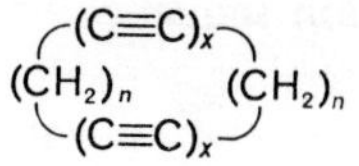

(**142**)

14-, 18-, 22- and 26-membered rings can exist in strainless conformations containing only staggered bonds, while this is not possible for 12-, 16-, 20- and 24-membered rings, the melting point alternation can be reasonably ascribed to the difference of

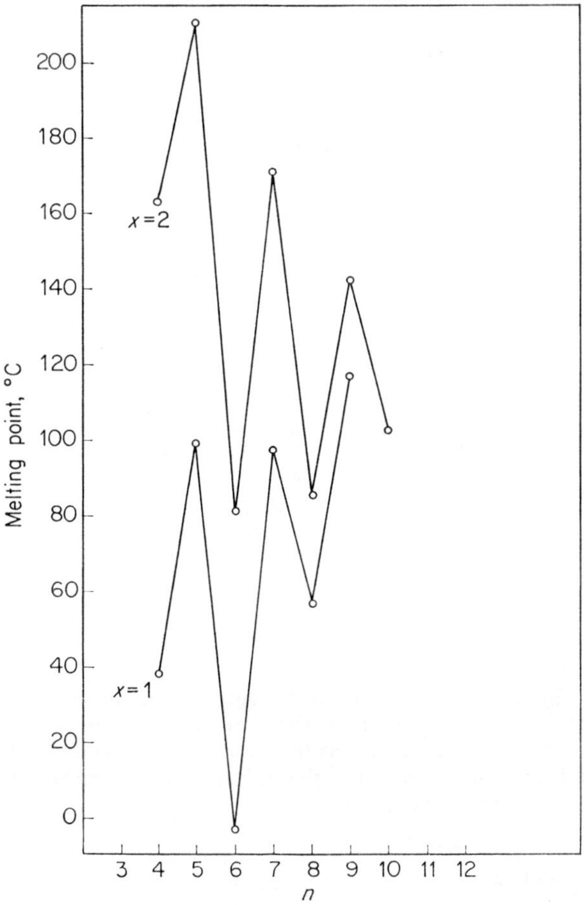

FIGURE 1. Alternation of melting points of the cycloalkadiynes (**142**, $x = 1$) and the cycloalkatetraynes (**142**, $x = 2$)[112]. Reproduced by permission of the Chemical Society.

conformation. An analogous alternation of melting points has been observed in a series of *o,p*-bridged cyclic tolans (**144**) which were synthesized by the Fritsch–Buttenberg–Wiechell rearrangement of the cyclic intermediate (**143**) prepared according to the following reaction sequence:[123]

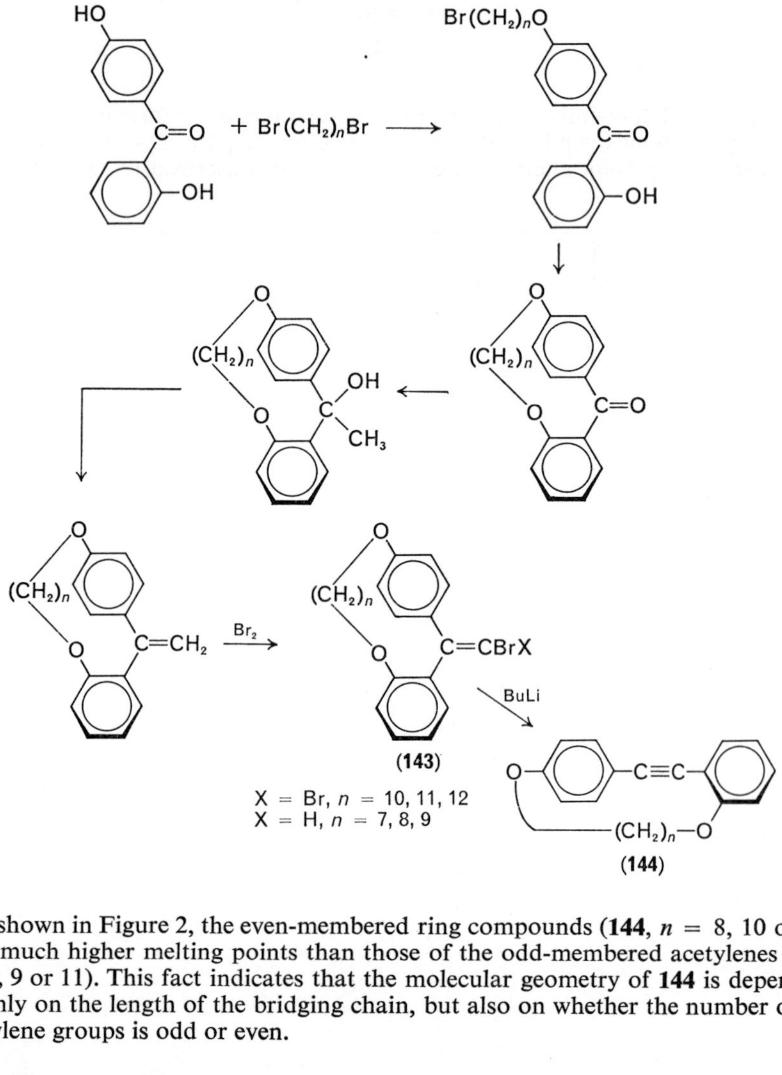

(143)

X = Br, n = 10, 11, 12
X = H, n = 7, 8, 9

(144)

As shown in Figure 2, the even-membered ring compounds (**144**, n = 8, 10 or 12) show much higher melting points than those of the odd-membered acetylenes (**144**, n = 7, 9 or 11). This fact indicates that the molecular geometry of **144** is dependent not only on the length of the bridging chain, but also on whether the number of the methylene groups is odd or even.

2. Strained systems

Because a diacetylenic linkage has a linear alignment of six carbon atoms (C—C≡C—C≡C—C), incorporation of the linkage in a monocyclic system may, at least in some cases, cause a distortion or a bending of the linear alignment, which should result in a change of physical properties.

Misumi and his coworkers have prepared 1,3-cyclotetradecadiyne (**145**) and 1,3-cyclotridecadiyne (**146**) by Eglinton's oxidative coupling under high dilution conditions of the corresponding α,ω-diynes[124]. 1,3-Tridecadiyne (**146**), the smallest

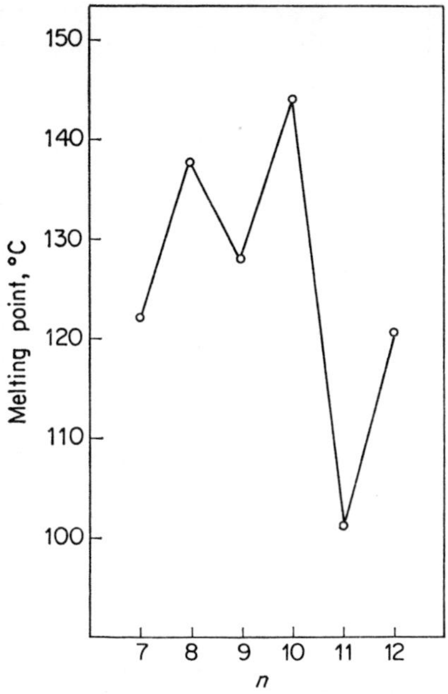

FIGURE 2. Alternation of melting points of *o,p'*-bridged cyclic tolans (**144**)[123]. Reproduced by permission of the Chemical Society of Japan.

monocyclic conjugated diacetylene known so far, was found to be an unstable compound in contrast to the stable **145**. Examination of molecular models reveals that **145** is strain-free, whereas **146** is a highly strained molecule. As illustrated in Figure 3, the strainless **145** exhibited an electronic spectrum with a distinct vibrational

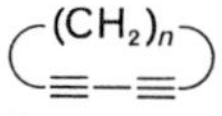

(**145**) $n = 10$
(**146**) $n = 9$

fine structure, while the strained **146** gave a structureless absorption curve without significant difference between the position of absorption maxima. The observed anomaly of the spectrum of **146** can be reasonably attributed to the restricted vibration and the distortion of the diacetylenic linkage in **146**.

The effects of twist or bending of diphenyldiacetylene chromophore systems on the electronic spectra have been studied in detail. *o,o'*-Bridged polymethylene ether derivatives of diphenyldiacetylene (**147**)[125] and *p,p'*-bridged analogues (**148**)[126] have been prepared by Eglinton's oxidative coupling of the corresponding α,ω-diethynyl compounds under high dilution conditions.

p,p'-Bridged cyclic tolans (**149**)[127] and *o,p'*-bridged cyclic tolans (**144**)[128] were synthesized by the Fritsch–Buttenberg–Wiechell rearrangement of the corresponding 1,1-bis(hydroxyphenyl)haloethylene polymethylene ether derivatives.

The electronic spectra of these cyclic acetylenes show characteristic features of diphenyldiacetylene and diphenylacetylene chromophores. However, they also

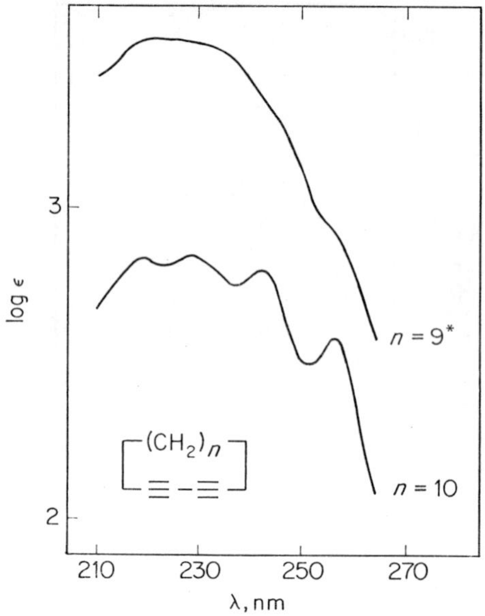

FIGURE 3. Electronic spectra of the cycloalkadiynes **145** and **146** in cyclohexane[124]. * Owing to the instability of **146**, the spectrum was measured with a solution of unknown concentration. Reproduced by permission of the Chemical Society of Japan.

exhibit an interesting change with variation in the length of the bridging chain.[129] The locations of the longest wavelength absorption maxima (λ_{max}) and the absorption intensities (ε) of **147, 148** and **149** are summarized in Table 5 together with those of the open-chain reference compounds. Features of the molecular geometry as revealed by examination of the molecular models are also given.

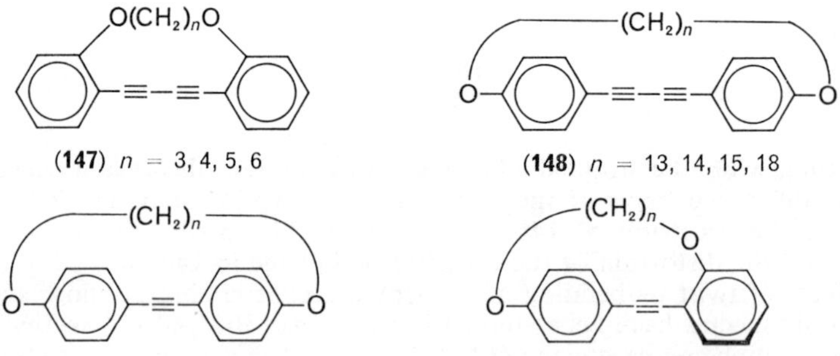

(**147**) *n* = 3, 4, 5, 6

(**148**) *n* = 13, 14, 15, 18

(**149**) *n* = 11, 12, 13, 14, 18

(**144**) *n* = 7, 8, 9, 10, 11, 12

The hyperchromism of the longest wavelength absorption bands observed in the higher homologues of **147, 148** and **149** as compared with the ε values of the respective open-chain analogues can be reasonably ascribed to the enhanced coplanarity of the two phenyl groups as a result of the ring formation. The fact that the maximum ε values are attained in **147** (n = 5), **148** (n = 15) and **149** (n = 13) is consistent with the above-mentioned argument, because these molecules are strain-free and hold rigid and planar structures due to the presence of a bridging chain of adequate length. However, marked decrease of the absorption intensities is observed with the decrease of chain length. The significant hypochromism of the longest wavelength bands in the lower homologues of **147**, **148** and **149** should, therefore, be attributed to the increase of the ring strain.

TABLE 5. Molecular geometry and electronic spectral data of cyclic diphenyldiacetylenes and cyclic tolans

Compound	Molecular geometry	λ_{max} (nm)	$\varepsilon \times 10^{-2}$
147, $n = 3$	Highly strained	337	159
147, $n = 4$	Strained, rigid	353	199
147, $n = 5$	Rigid, planar, strain-free	353	365
147, $n = 6$	Slightly flexible, strain-free	349	336
o,o'-Dimethoxydiphenyldiacetylene		349	286
148, $n = 13$	Highly strained, rigid, non-planar	345·5	367
148, $n = 14$	Strained, rigid, non-planar	344	425
148, $n = 15$	Strain-free, rigid, planar	343·5	514
148, $n = 18$	strain-free, planar	342·5	465
p,p'-Di-*n*-butoxydiphenyldiacetylene		340·5	407
149, $n = 11$	Highly strained, rigid, non-planar	318	334
149, $n = 12$	Highly strained, rigid	317	369
149, $n = 13$	Strain-free, rigid, planar	317	443
149, $n = 14$	Strain-free, rigid, planar	316	426
149, $n = 18$	Strain-free, flexible	315	355
p,p'-Di-*n*-butoxytolan		313	319

In the case of **144**, the decrease of chain length should increase the twisting of the two phenyl groups, i.e. the molecular model of **144** ($n = 7$) shows that the interplanar angle of the two phenyl groups should be almost rectangular. Therefore, the hypochromism observed in **144** should be regarded as the superposition of the effect of ring strain and the effect of twisting of the two phenyl groups.

With regard to the effect of ring strain on the location of λ_{max}, the above-mentioned four series of cyclic acetylenes, **147**, **148**, **149** and **144**, show striking contrast. In the case of **149** and **148**, the increase of ring strain results in bathochromic shifts of the longest wavelength bands (λ_{max}). On the contrary, decrease in ring size exerts hypsochromic effects on the λ_{max} of **147** and of **144**. The hypsochromism observed in **144** cannot be ascribed only to the increase of ring strain, because the twisting of the two phenyl groups should also result in the increase of the transition energy of the longest wavelength bands of **144**. However, the difference in the spectral behaviour between the *o,o'* series (**147**) and the *p,p'* series (**148** and **149**) seems to be attributable to the difference in the mode of ring strain in these two series of cyclic acetylenes. The strained molecules of **147** are held in planar conformation regardless of the magnitude of the ring strain as illustrated in Figure 4. The diacetylenic linkage in **147** ($n = 3$) seems to be forced to bend significantly due to the short bridging chain, but the two phenyl nuclei should still be held in a coplanar position. Consequently, the *p* orbitals which contribute to the conjugation of the entire chromophore system are also held in a parallel position independent of the magnitude of the ring strain.

In the *p,p′* series (**148** and **149**), as shown in Figure 4, the two benzene nuclei should deviate from the coplanar position according to the increase in ring strain. Therefore, the *p* orbitals in the strained members of **148** and **149**, which contribute to the conjugation of the entire chromophore system, should deviate from the parallel position.

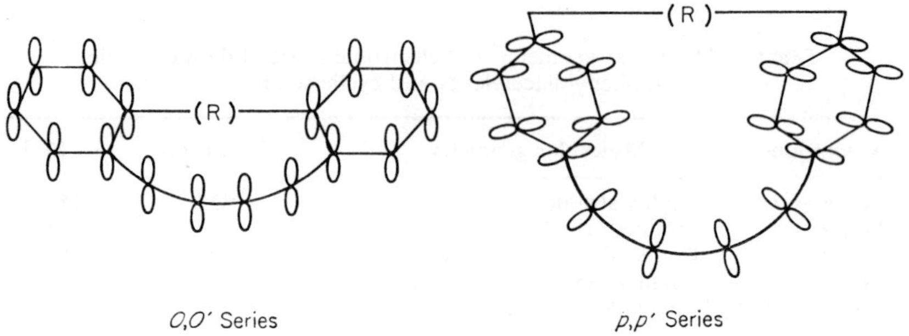

FIGURE 4. Schematic illustration of the *p* orbitals in the strained *o,o′*-bridged and *p,p′*-bridged diphenyldiacetylenes (**147** and **148**) (R = $-O(CH_2)_nO-$)[129]. Reproduced by permission of the Chemical Society of Japan.

On the basis of this consideration, it was assumed that the increase in energy of the ground state of the strained molecule of the *o,o′* series (**147**) should be smaller than that of the *p,p′* series (**148** and **149**), whereas the increase in energy of its excited state should be larger than or almost the same as that of the *p,p′* series (**148** and **149**). Consequently, as illustrated in Figure 5, the transition energy ($\Delta E'$)

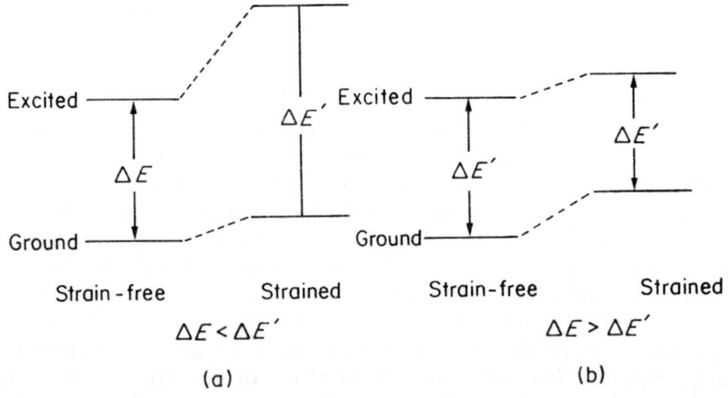

FIGURE 5. Transition energies of (a) *o,o′*-bridged and (b) *p,p′*-bridged diphenyldiacetylenes[129]. Reproduced by permission of the Chemical Society of Japan.

of the strained *o,o′*-bridged molecule (**147**) becomes larger than that of the strain-free molecule (ΔE). Conversely, the transition energy ($\Delta E'$) of the strained *p,p′*-bridged molecule (**148** and **149**) becomes smaller than that of the strainless molecule (ΔE), i.e. the increase of ring strain in the *p,p′* series and in the *o,o′* series should result in bathochromic and hypsochromic shifts, respectively[129].

V. DEHYDROANNULENES

A. Dehydroannulenes

1. Synthesis

Oxidative coupling of 1,5-hexadiyne (**150**) with copper(II) acetate in pyridine yielded a mixture of the cyclic trimer **151**, the tetramer **152**, the pentamer **153** and the hexamer **154** in addition to other compounds[119, 120, 130]. The cyclic trimer **151**

$HC{\equiv}C{\cdot}CH_2{\cdot}CH_2{\cdot}C{\equiv}CH$

(**150**)

↓

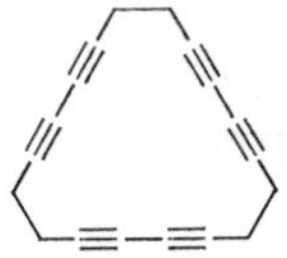

(**151**)

\+

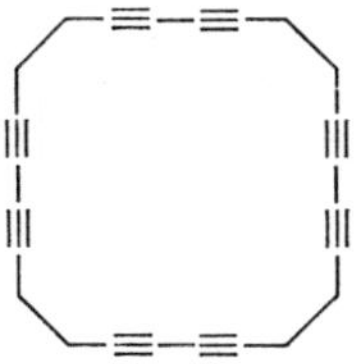

(**152**)

\+

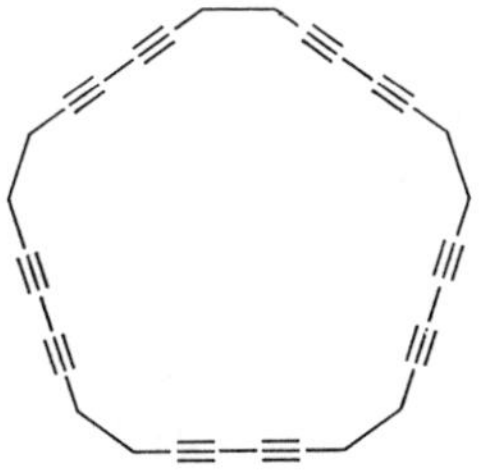

(**153**)

\+

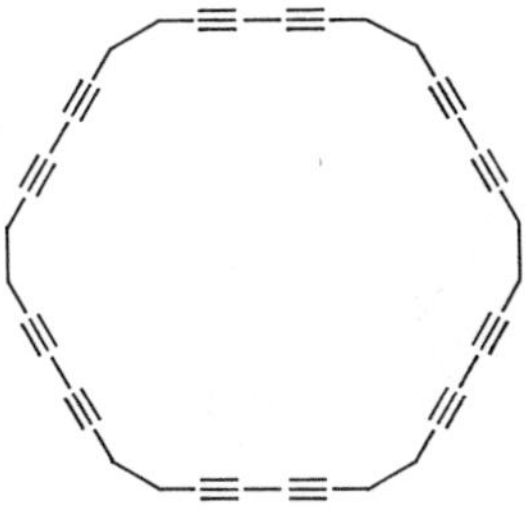

(**154**)

gave fully conjugated 18-membered hexaentriyne, trisdehydro[18]annulene, isomer I (**155**) in 50% yield on treatment with potassium *t*-butoxide in *t*-butanol at 90 °C for 30 min[131, 132]. A more careful examination of the rearranged products revealed that they contained isomer II, a geometrical isomer of **155** and the tetrakisdehydro-[18]annulene, a dehydrogenated product of **155**[133]. Partial hydrogenation of **155** yielded the [18]annulene (**156**) in *ca.* 30% yield[134, 135]. The success of the synthesis

(**151**) ⟶

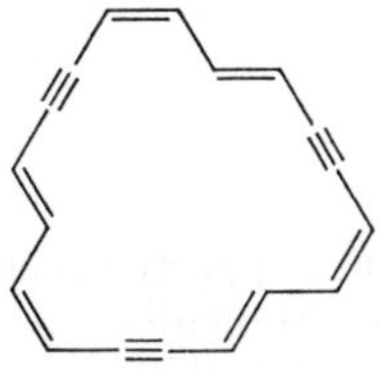

(**155**)

⟶

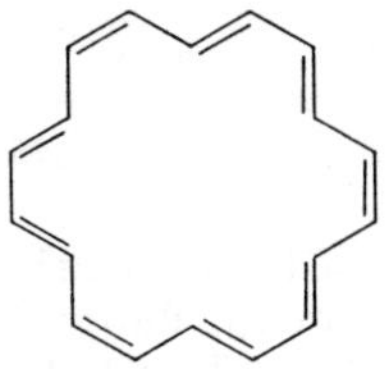

(**156**)

of the fully conjugated macrocyclic systems, **155** and **156**, reported by Sondheimer in 1959[134], was an important milestone in the history of non-benzenoid aromatic chemistry.

The prototropic rearrangement of the cyclic tetramer **152**, the pentamer **153** and the hexamer **154** on treatment with potassium *t*-butoxide gave the tetrakis-dehydro[24]- **(157)**[135], the pentakisdehydro[30]- **(159)**[135] and the hexakisdehydro[36]- **(161)** annulenes[130]. Catalytic half-reduction of the triple bonds in **157** and **159** gave [24]annulene **(158)** and [30]annulene **(160)**, respectively[135], although their configurations were not firmly established.

(152) →

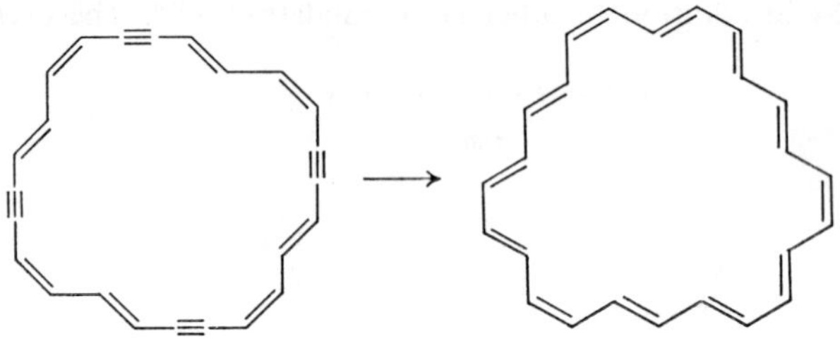

(157) (158)

(153) →

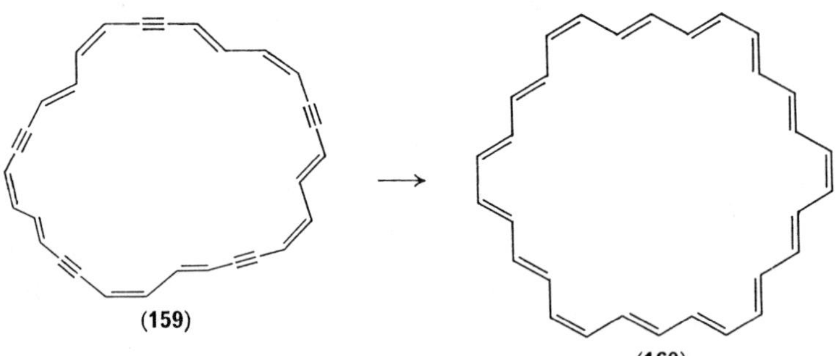

(159) (160)

(154) →

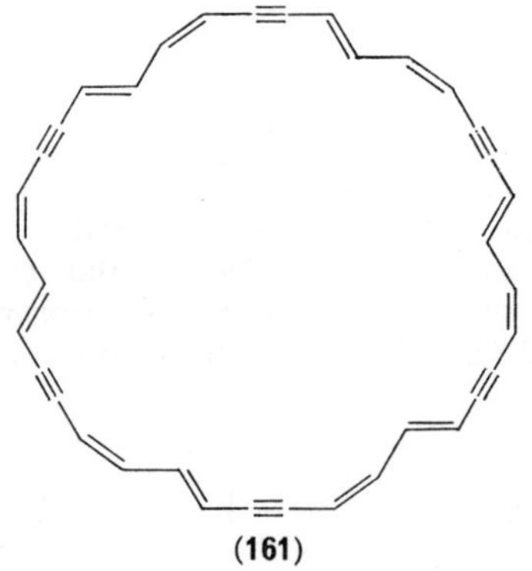

(161)

Treatment of 1,5,9-decatriyne **(162)** with copper(II) acetate in pyridine yielded the 20-membered cyclic dimer **163** and the 30-membered cyclic trimer **166**[136]. The bisdehydro[20]annulene **164** and the trisdehydro[30]annulene **167** were obtained by the prototropic rearrangement from **163** and **166**, respectively. Half-reduction of the dehydroannulenes **(164** and **167)** yielded [20]- **(165)** and [30]- **(160)** annulenes, respectively[136].

Combining the oxidative coupling reaction of α,ω-diethynyl compounds to give cyclic poly-ynes with the base-induced prototropic rearrangement of the cyclic polyenpoly-ynes, Sondheimer and his coworkers have synthesized a wide variety of dehydroannulenes.

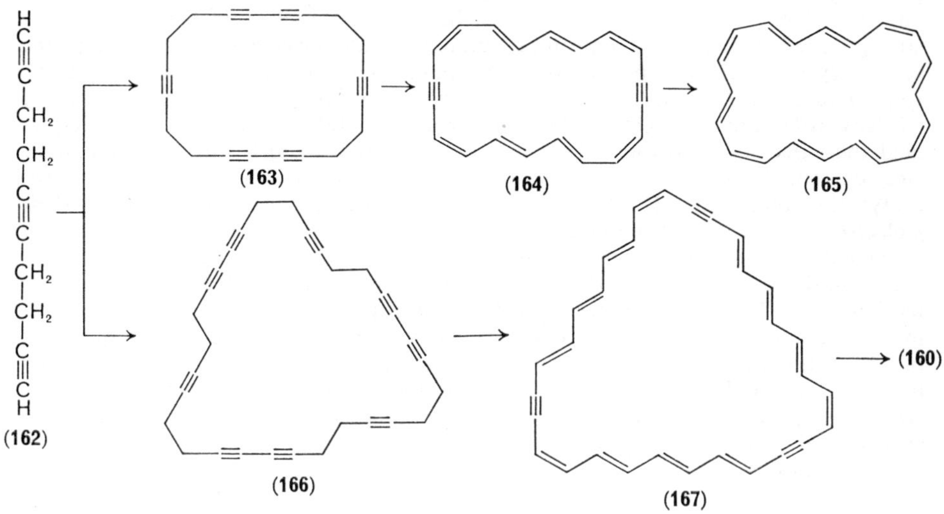

2. Aromaticity and antiaromaticity

Although no generally accepted definition of aromaticity exists a compound is considered to be aromatic if it possesses a large (negative) resonance energy which is apparently the consequence of π-electron delocalization in the ground state of the molecule. Therefore, determination of heats of hydrogenation or combustion is the most direct criterion for aromaticity. Other criteria are diminished bond alternation in the conjugated cyclic system and the presence of a magnetically induced diamagnetic ring current*. Consequently, X-ray structure analysis and n.m.r. spectroscopy are also important tools for testing the aromaticity of a compound. The Hückel molecular orbital method (HMO) predicts aromaticity for the annulenes with $(4n+2)\pi$ electrons in the ring, provided that the molecules hold planar or near-planar configuration (Hückel's rule)[137]. The measurements of ^{1}H-n.m.r. spectra performed on [18]annulene **(156)** and trisdehydro[18]annulene **(155)** gave dramatic results.[140, 141]. As shown in Table 6, the intra-annular (inner) protons of [18]annulene

TABLE 6. 60 MHz ^{1}H-n.m.r. parameters of [18]annulene **(156)** and trisdehydro[18]annulene **(155)** (τ value)

Compound	Inner protons	Outer protons	Solvent	Temperature (°C)
[18]Annulene	12·20	0·97	THF-d_8	0
Trisdehydro[18]annulene, Isomer I **(155)**	8·26	2·98, 2·44 2·44	$CDCl_3$	60

* The terms 'diatropic' and 'paratropic' have been proposed by Professor F. Sondheimer for the molecules showing the effect of dia- or paramagnetic ring current in the n.m.r. spectrum, and molecules showing no ring current effect are named 'atropic'; *cf. Acc. Chem. Res.*, **5**, 81 (1972).

(**156**) exhibit signals at an extremely high field, whereas the outer protons resonate at a much lower field than the usual vinyl protons. This result clearly shows that [18]annulene (**156**) (n = 4 in Hückel's rule), sustains a diamagnetic ring current in an applied external magnetic field, i.e. the protons situated inside the ring are strongly shielded, whereas the protons situated outside the ring are deshielded by the secondary magnetic field produced by the diamagnetic ring current in the annulene ring. The same ^{1}H-n.m.r. spectral trend was observed in the spectrum of the trisdehydro[18]annulene (**155**). The inner protons in **155** resonate at a much higher field than the outer protons. This fact indicates that a fairly strong diamagnetic ring current is induced in the conjugated 18-membered ring containing three acetylenic bonds; that is, the *sp*-hybridized acetylenic carbon atoms offer their *p* electrons in the orbitals perpendicular to the molecular plane for the formation of a delocalized π-molecular orbital. As expected, dehydroannulenes have a higher conformational stability than the corresponding annulenes. The ^{1}H-n.m.r. spectra of trisdehydro[18]annulene isomer I (**155**) were found to be almost temperature-independent up to 150 °C, whereas the coalescence temperature of the spectrum of [18]annulene (**156**) was found to be 40 °C, thus indicating a higher conformational mobility of the latter compound[140]. Also, the coalescence temperature of the ^{1}H-n.m.r. spectrum of [14]annulene (**168**) was reported to be −30 °C, while the corresponding tris-[142, 143], bis-[139, 142] and monodehydro[14]annulenes[109, 138, 139, 142] gave non-mobile spectra.

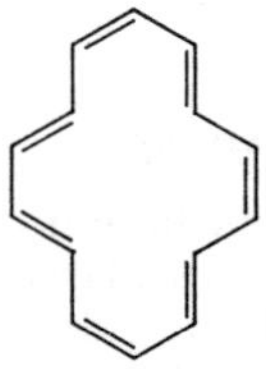

(**168**)

Formation of substituted products by electrophilic reactions is one of the characteristics of benzenoid compounds. It has been shown that a few dehydro-[4*n*+2]annulenes afford substituted annulenes under strictly limited reaction conditions. Monodehydro[14]annulene, on treatment at room temperature with copper(II) nitrate–acetic anhydride, oleum–dioxane and subsequent methylation and acetic anhydride–borontrifluoride etherate, yielded monosubstituted products, **169a**, **169b** and **169c**, respectively. The electrophilic reactions must have resulted in substitution of one of the outer protons, but the exact point of attack has not been determined[144].

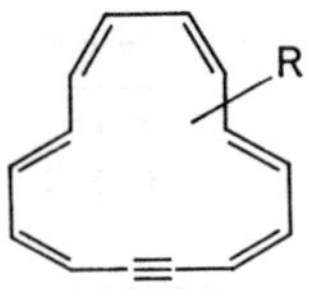

(**169**)

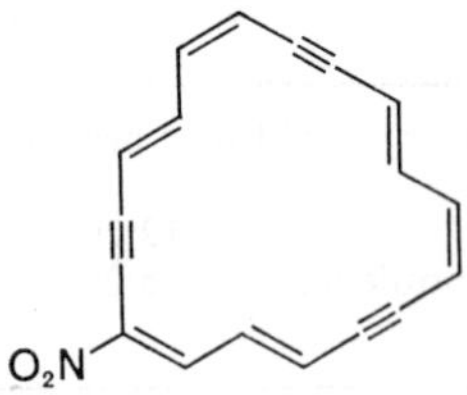

(**170**)

(**a**) R = NO_2
(**b**) R = SO_3CH_3
(**c**) R = $COCH_3$

It has been found that **155** can be converted into 3-nitro-1,3,7-trisdehydro-[18]annulene (**170**) on brief treatment with copper(II) nitrate in acetic anhydride at room temperature.[145] The monoacetyl derivative of **155** was also prepared[145].

Recent quantum-mechanical theory of induced ring currents in fully conjugated monocyclic molecules has shown that paramagnetic ring currents should be induced in planar or near-planar $4n\pi$ systems, in contrast to diamagnetic ring currents in $(4n+2)\pi$ systems[146–148]. Consequently, in a $4n\pi$ system, the ^{1}H-n.m.r. signals of the protons outside the ring should be displaced to high, and those inside to low, field.

Treatment of a solution of cyclododecatetrayne (**132**, $n = 2$) with potassium *t*-butoxide at room temperature gave 1,5-bisdehydro[12]annulene (**171**)[122] and 1,5,9-trisdehydro[12]annulene (**172**)[149], which is produced by dehydrogenation and prototropic rearrangement in a basic medium. The ^{1}H-n.m.r. spectrum of **171** measured at +37 °C exhibits signals at τ 5·73 (H^3, H^6), 5·32 (H^4, H^5), 4·82 (octet, H^2, H^8) and −1·18 (quartet, H^1, H^8)[148]. The lowest field quartet was assigned to the protons attached to the *trans* double bond (H^1, H^8), which rapidly interchanges the outer and inner positions, and the quartet represents an average of the true positions of the inner H^1 and outer H^8 protons. In this case, cooling to −80 °C did not result in the 'non-mobile' structure, only a certain down-field shift (−1·55) and broadening of the quartet being apparent.

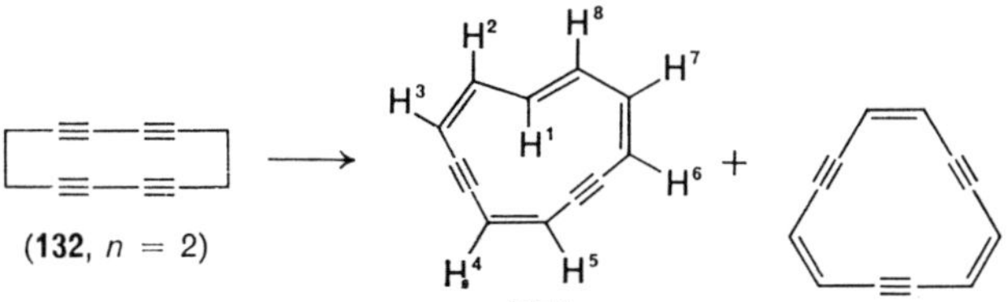

(132, n = 2) (171) (172)

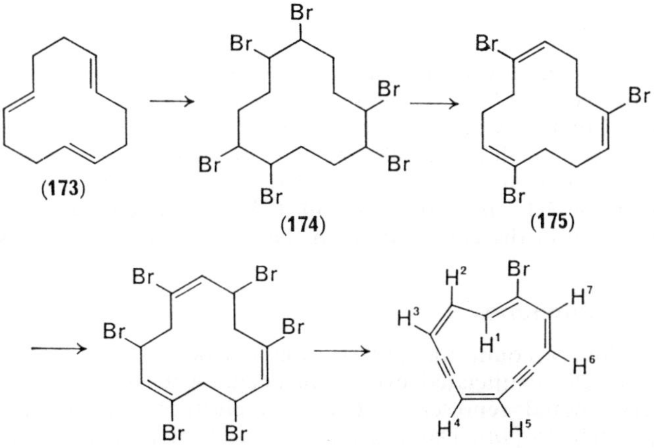

(173) (174) (175)

(176) (177)

Hexabromocyclododecane (**174**) prepared by the bromination of cyclododecatriene (**173**) was treated with a limited amount of sodium ethoxide to give tribromocyclododecatriene (**175**). The hexabromide **176** was prepared on treatment of **175** with NBS. The dehydrobromination of **176** with four equivalents of sodium ethoxide yielded 5-bromo-1,9-bisdehydro[12]annulene (**177**)[150]. Treatment of **176** with an excess of sodium ethoxide gave 1,5,9-trisdehydro[12]annulene (**172**)[151]. Owing to the presence of a bulky bromine atom on the *trans* double bond, conformational interconversion of **177** is not possible. In fact, the ^{1}H-n.m.r. spectrum at room temperature exhibits the inner proton (H^1) signal at τ $-6{\cdot}4$. Since local anisotropy effects of the two triple bonds, if operative, cause an upfield shift[147], the fact that the inner proton (H^1) resonates at such a low field should be attributed to the strong deshielding effect of the paramagnetic ring current induced in the nearly planar 12π electron system.

1,3,7,9,13,15-Hexakisdehydro[18]annulene (**178**), a [$4n+2$]dehydroannulene having a similar geometry to 1,5,9-trisdehydro[12]annulene (**172**), has been synthesized according to the following sequence of reactions[152]:

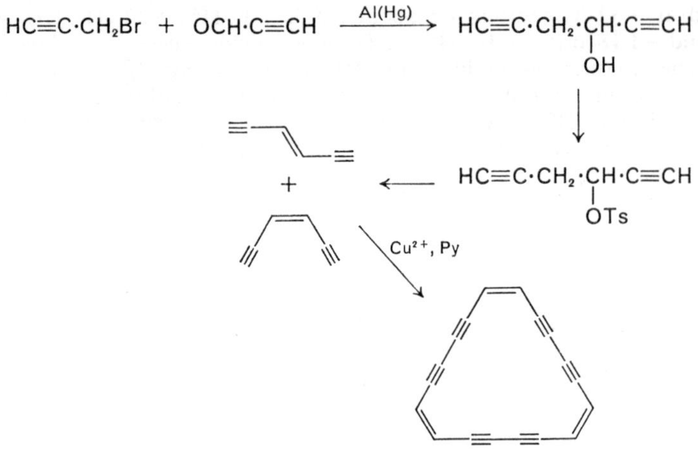

(178)

The ^{1}H-n.m.r. spectrum of **172** shows a singlet at τ 5·55[150], whereas a singlet at τ 2·98 is observed in **178**[158]. Although the possibility cannot be excluded that the high-field signal of **172** is due to local anisotropy, the fact that the protons in **178** exhibit a signal at a much lower field provides strong evidence for the existence of a diamagnetic ring current in **178** and a paramagnetic ring current in **172**. The comparison of the ^{1}H-n.m.r. spectrum of **172** with that of **178** seems to give a conclusive decision in the controversy on the presence of induced ring currents[153].

3. Dehydroannulenones

If we take into account the polarization of the carbonyl group, a dehydroannulenone, a fully conjugated cyclic polyenepolyyne ketone containing ($4n+3$) ring members, should represent a ($4n+2$) π electron system and be aromatic. Conversely, a dehydro[$4n+1$]annulenone should represent a $4n\pi$ electron system and be non-aromatic.

1,2-Diethynylcyclohexene (**180**) prepared from **179** via four steps was converted into the mono-Grignard derivative. The reaction of the latter with ethyl formate led to **181**, which on oxidative coupling and subsequent oxidation of the resulting 13-membered cyclic alcohol with manganese(IV) oxide yielded the tetrakisdehydro-[13]annulenone **182**[154]. If there is an appreciable contribution from the polarized structure **183**, it should sustain a paramagnetic ring current in an external magnetic

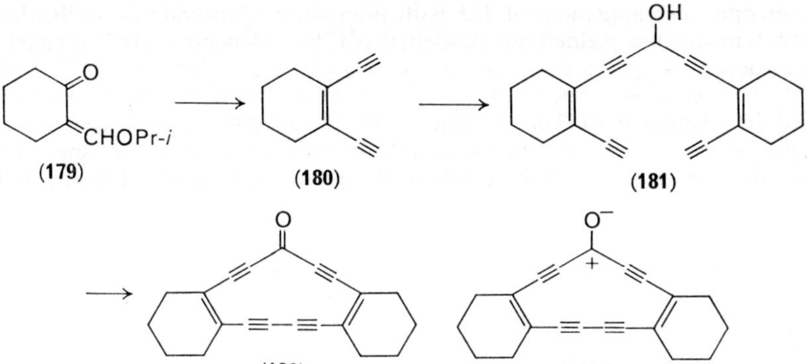

(179) (180) (181) (182) (183)

field. Since **182** contains no protons attached directly to the 13-membered ring, no definite evidence could be obtained from the ^{1}H-n.m.r. spectrum. However, the rather high field signal of allylic methylene protons seemed to be attributable to the shielding effect of paramagnetic ring current in **182** [154].

The attempts to prepare the parent tetrakisdehydro[13]annulenone, via the oxidative coupling of **184**, failed, and it would seem that the instability of the highly

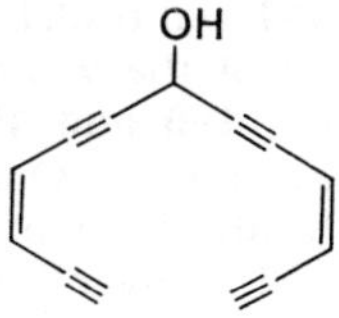

(184)

strained 13-membered cyclic alcohol is to blame[155]. For this reason the non-conjugated 17-membered cyclic ketone **185** has been synthesized according to the route outlined below[155, 156]:

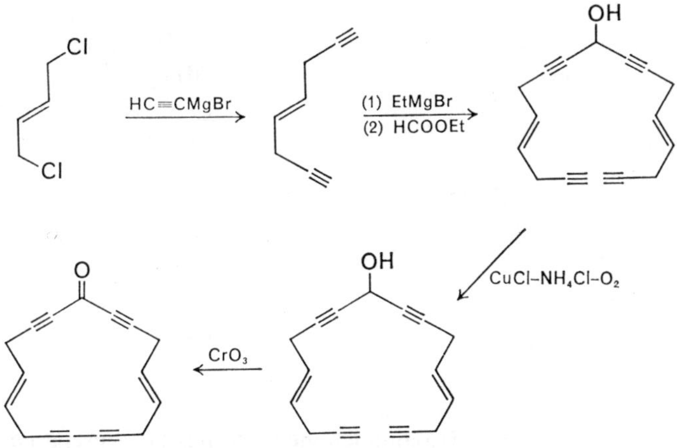

(185)

Prototropic rearrangement of **185** with potassium *t*-butoxide in undistilled THF at a low temperature yielded tetrakisdehydro[17]annulenenone (**187**) formed by the loss of four hydrogen atoms, instead of the expected bisdehydro[17]annulenone (**186**)[156]. The reaction carried out under similar conditions in freshly distilled THF resulted in trisdehydro[17]annulenones, **188** [157] and **189** [155], with the loss of two hydrogen atoms. As shown below in the formulae, the ^{1}H-n.m.r. spectra clearly indicate the paratropicity of these dehydro[17]annulenones, **187, 188** and **189.**

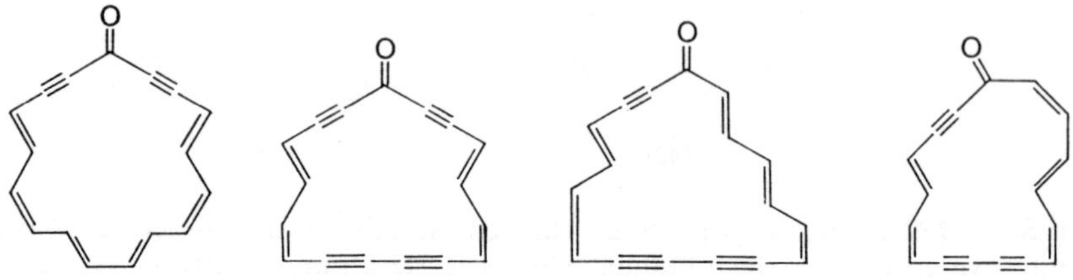

	(186)	(187)	(188)	(189)
N.m.r. spectra.	Inner H: τ =	−0·31	−0·50–−0·10	−0·10–0·42
	Outer H: τ =	3·88–4·87	3·53–4·84	3·75–5·05

The bisdehydro[13]annulenone **191** and bisdehydro[17]annulenone **193** have been prepared by the oxidative coupling of the α,ω-diethynyl compounds obtained by the condensation of the aldehydes, **190** and **192** with acetone, and subsequent oxidation of the resulting cyclic alcohols[158]. Only the 17-membered cyclic ketone **193** was found to be paratropic. The atropicity of **191** seems to be attributable to the poor planarity of the highly strained 13-membered ring.

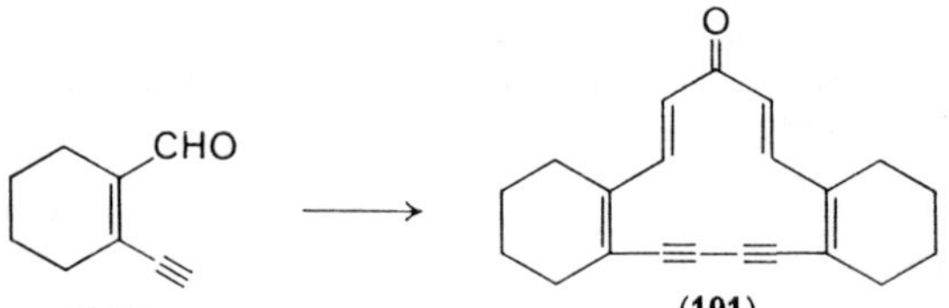

(190) (191)

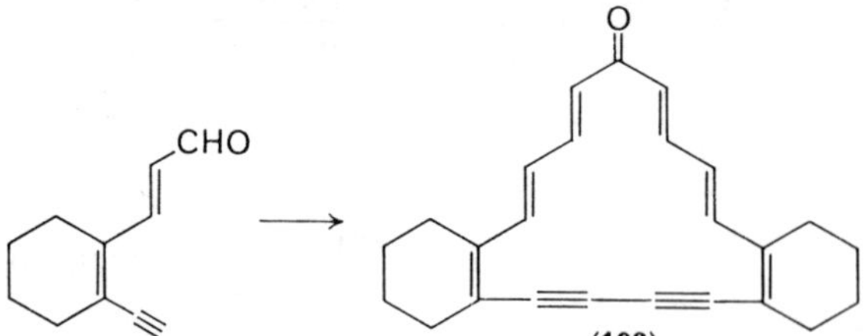

(192) (193)

A diatropic tetrakisdehydro[15]annulenone (**196**) has been prepared by the route outlined as follows:[159]

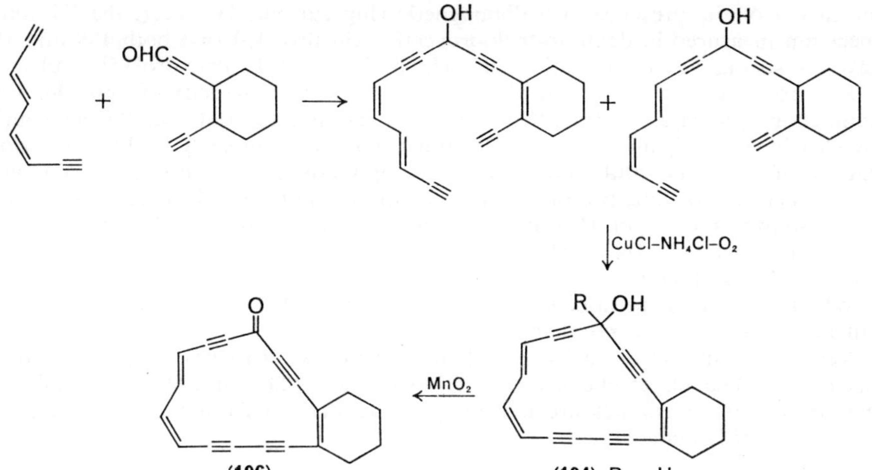

(196)

(194) R = H
(195) R = CH_3

The reaction of methyllithium with **194** yielded **195**, a reference compound for ^{1}H-n.m.r. spectroscopy. The ^{1}H-n.m.r. data indicate that the dehydro[15]annulenone **196** is diatropic and, as expected, the diatropicity is enhanced by protonation of the carbonyl group.

Two geometrical isomers of the bisdehydro[15]annulenone, **198** and **199**, have been synthesized by oxidative coupling of **197**[160]. Neither **198** nor **199** showed any

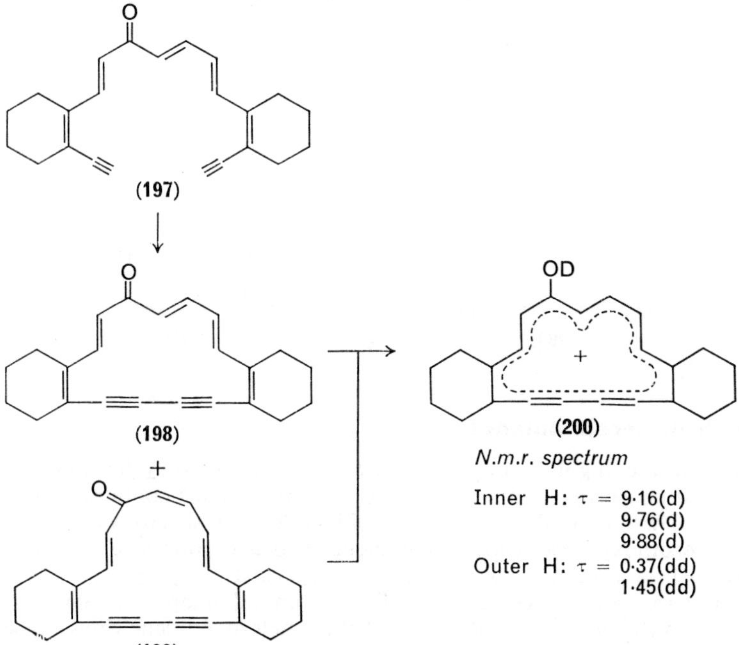

N.m.r. spectrum

Inner H: τ = 9·16(d)
9·76(d)
9·88(d)

Outer H: τ = 0·37(dd)
1·45(dd)

indication of the presence of a diamagnetic ring current. However, the ^{1}H-n.m.r. spectrum measured in deuterio-trifluoroacetic acid revealed that both **198** and **199** gave the strongly diatropic cation **200**. The finding that the dehydro[15]annulenone **196** is aromatic is in contrast to the apparent lack of aromaticity of lower [$4n+3$]-annulenones, such as tropone[161]. Sondheimer has pointed out that further studies are clearly necessary in order to resolve this apparent discrepancy, and that discrimination of aromatic and non-aromatic compounds by the first-order coupling constants of the olefinic proton n.m.r. resonances is not possible, because in each of the ^{1}H-n.m.r. spectra of **187**, **195** and **196**, H^1 was a doublet (J = 16 Hz), H^2 a double doublet (J = 16, 12 Hz), H^3 a double doublet (J = 12, 10 Hz) and H^4 a doublet (J = 10 Hz)[155].

Whether an annulenedione shows some quinonoid character is a very interesting, but not fully explored, problem.

Recently, some tetrakisdehydro[18]annulenediones (**201–204**) have been synthesized[162, 163]. The electrochemical reduction of **201**, **203** and **204** clearly indicates that these annulenediones are indeed quinones derived from the aromatic tetrakisdehydro[18]annulene[164].

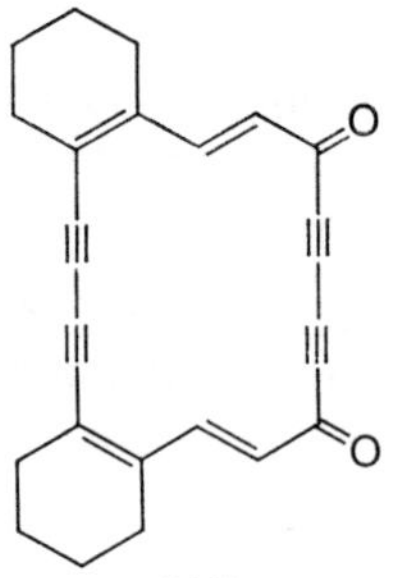

(201)

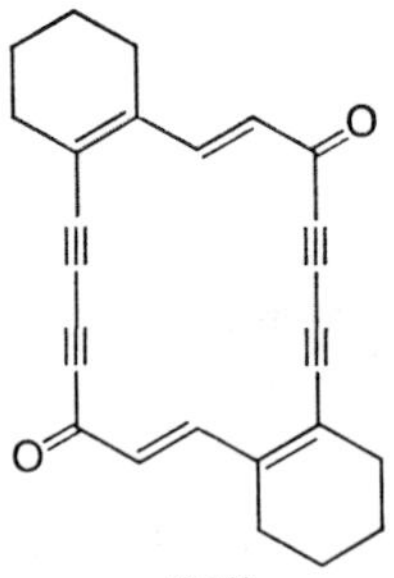

(202)

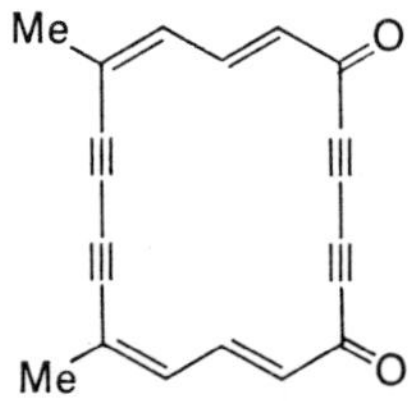

(203)

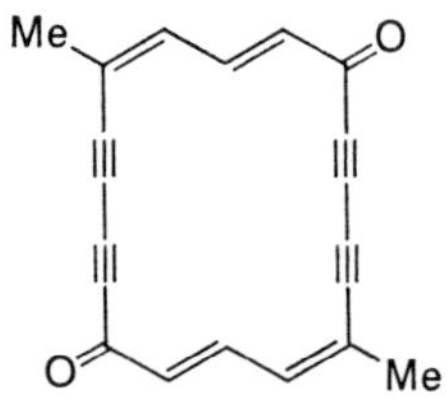

(204)

4. Dehydroheteroannulenes

In view of the highly aromatic nature of some heterocyclic compounds, the synthesis and properties of macrocyclic conjugated systems containing one or more hetero atoms, heteroannulenes, are of considerable interest. However, the ^{1}H-n.m.r. spectra of *N*-ethoxycarbonylaza[13]annulene[165], oxa[17]annulene[166] and *N*-ethoxycarbonylaza[17]annulene[167], which can be regarded as [14]π, [18]π and [18]π electron systems, respectively, revealed little, if any, diamagnetic ring current. This may be due to the non-planar structure of these heteroannulene rings caused by the steric interaction of the inner protons, and the diminished difference in energy between the localized and delocalized forms owing to the presence of relatively strongly electron-withdrawing groups[168].

Considering this situation Beeby and Sondheimer have carried out the first synthesis of a bisdehydroaza[17]annulene having a methylene bridging chain[169].

The di-*trans* isomer **206** was isolated from the products of the reaction of **190** with the bisphosphonium salt **205** in the presence of lithium ethoxide. The reaction of **207** (obtained by the oxidative coupling of **206**, with dimethyl sulphate followed by reduction with sodium hydrosulphite) yielded the cyclic compound, containing the dihydropyridine nucleus, **208a**. Lithium aluminium hydride reduction of **207** yielded

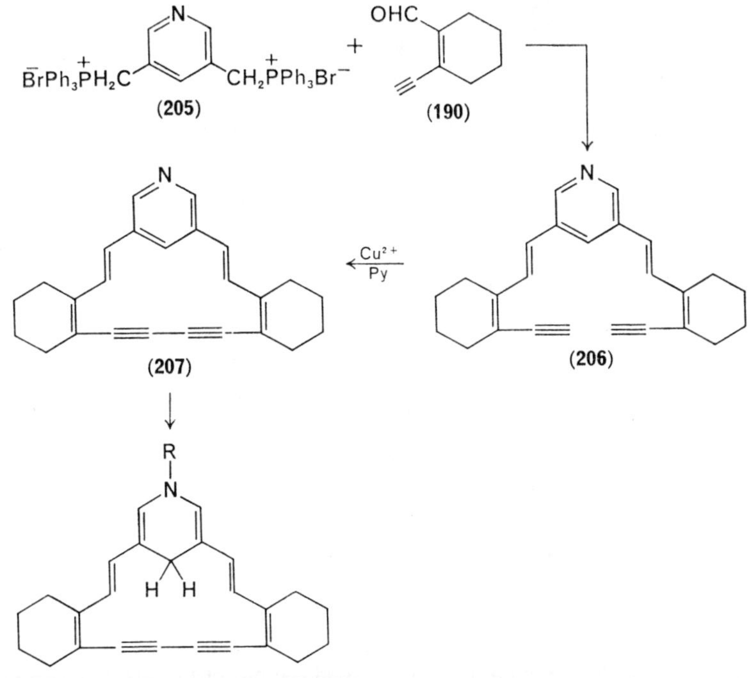

(208) (a) R = Me, (c) R = COOEt
(b) R = H, (d) R = COMe

208b. Treatment of **208b** with ethyl chlorocarbonate or acetyl chloride afforded **208c** or **208d**, respectively. The reaction of **207** with an excess of alkyllithium (RLi) gave a 1,4-addition product (**209**) which yielded **210a–d** [170] or **211a–c** [171] on treatment with methyl iodide or sodium bicarbonate solution, respectively. The dihydropyridine derivatives **211a–c** were readily converted into the salts **212a–c** by the reaction with potassium mirror[171].

As expected, the bisdehydroaza[17]annulenes **208a–d** proved to be diatropic. The magnitude of ring current was found to decrease with increasing electronegativity of the N substituent (**a**>**b**>**c**>**d**)[169]. In accordance with the above observation, a much stronger diatropicity was found for **212**. The ^{1}H-n.m.r. spectral behaviour of alkyl protons disposed within the cavity of the π electron cloud in **210** and **212** [172] clearly showed that the shielding effect decreases with the increase of distance of the protons from the plane of the annulene ring[170, 171] (Table 7).

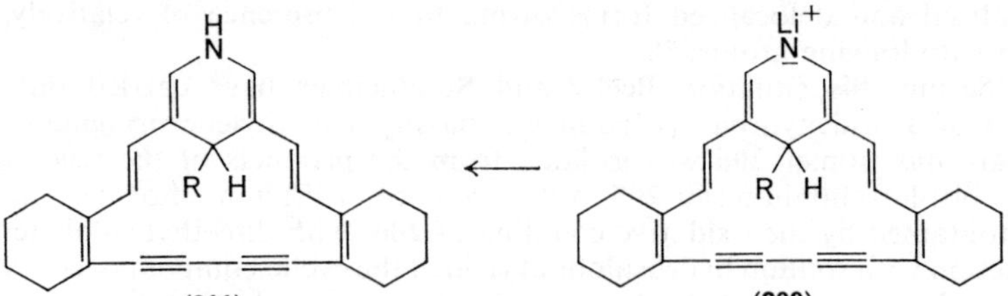

(211) (209)

(a) R = H
(b) R = CH_3
(c) R = C_2H_5

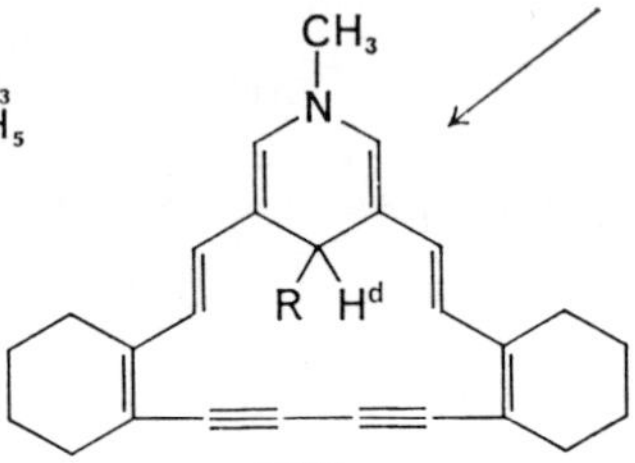

(210)

(a) R = $CH_3{}^{\alpha}$, (b) R = $CH_2{}^{\alpha}\cdot CH_3{}^{\beta}$
(c) R = $CH_2{}^{\alpha}\cdot CH_2{}^{\beta}\cdot CH_3{}^{\gamma}$
(d) R = $CH_2{}^{\alpha}\cdot CH_2{}^{\beta}\cdot CH_2{}^{\gamma}\cdot CH_3{}^{\delta}$

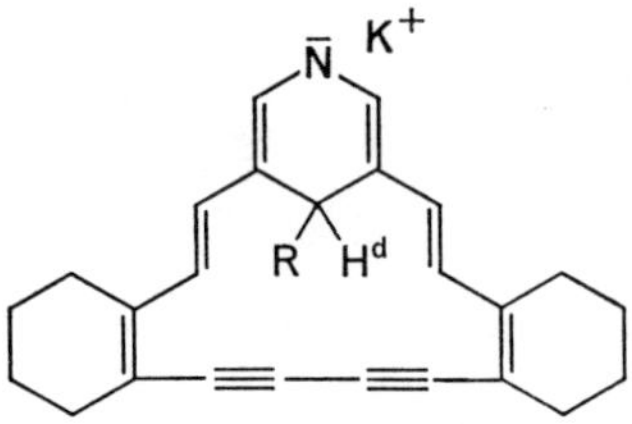

(212) (a) R = H
(b) R = $CH_3{}^{\alpha}$
(c) R = $CH_2{}^{\alpha}$ $CH_3{}^{\beta}$

TABLE 7. ^{1}H-N.m.r. parameters of H^d and the protons of alkyl groups in **210a–d** and **212b–c** (τ values)

Compound	H^d	H^{α}	H^{β}	H^{γ}	H^{δ}
210a	9·49	10·52	—	—	—
210b	9·41	10·14	10·14	—	—
210c	9·40	10·09	9·75	9·75	—
210d	*ca.* 9·4	10·11	9·81	*ca.* 9·4	*ca.* 9·5
212b	13·47	12·41	—	—	—
212c	13·28	11·83	11·24	—	—

Using the vinylogous aldehyde **192** in the place of **190** in the preparation of **208c**, bisdehydroaza[21]annulene **(213)** has been synthesized[173]. The effect of the diamagnetic ring current was clearly observed in the ^{1}H-n.m.r. spectrum of **213** which is a 22π electron system, but the diatropicity was found to be less than that of the 18π system, **208c**.

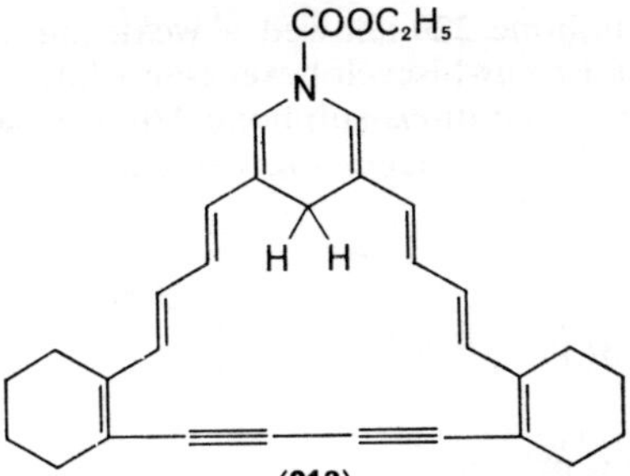

(213)

The methylene-bridged bisdehydroaza[19]annulene **(215)**, a hetero[$4n-1$]annulene with $4n\pi$ electrons, has been synthesized[17]. Comparison of the ^{1}H-n.m.r. spectrum of **215** with that of the open-chain analogue **214** indicates that **215** is a paratropic 20π electron system just as carbocyclic [$4n$]annulenes[174].

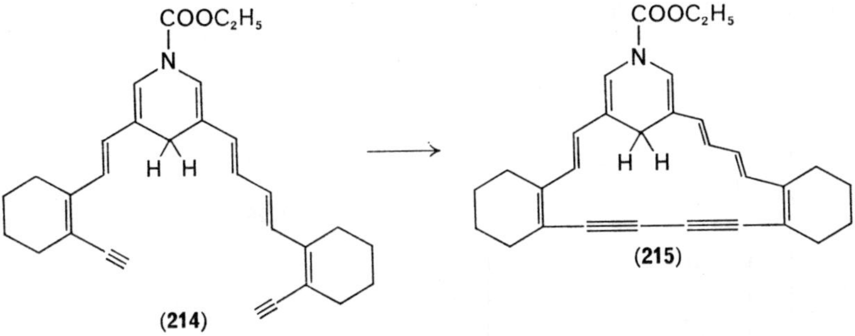

(214) **(215)**

Starting from the aldehyde **192** and the bisphosphonium bromide, $Br^-Ph_3P^+CH_2$-$SCH_2P^+Ph_3Br^-$, bisdehydrothia[17]annulenes **216, 217** and **218** and their oxidation products **219, 220, 221** and **222** have been prepared[175]. A weak diatropicity was

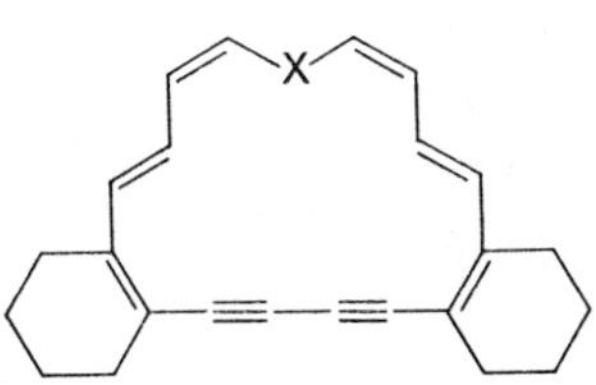

(216) X = S
(219) X = SO
(220) X = SO_2

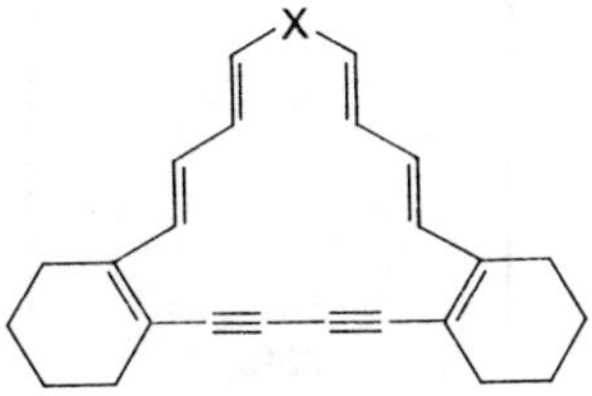

(217) X = S
(221) X = SO_2

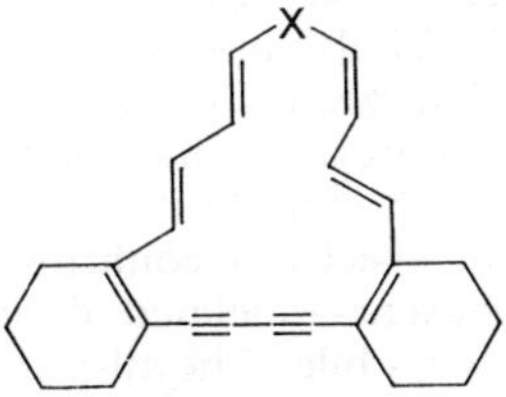

(218) X = S
(222) X = SO_2

found in **217**, but the sulphone **220** showed a weak paratropicity[175]. Since, rather surprisingly, only the all-*trans*-biscyclohexene-annelated bisdehydro[17]annulene **217** was diatropic, and only the di-*cis*-sulphone **220** was paratropic[175], the synthesis of dimethylbisdehydrothia[17]annulenes was carried out according to the following scheme[176]:

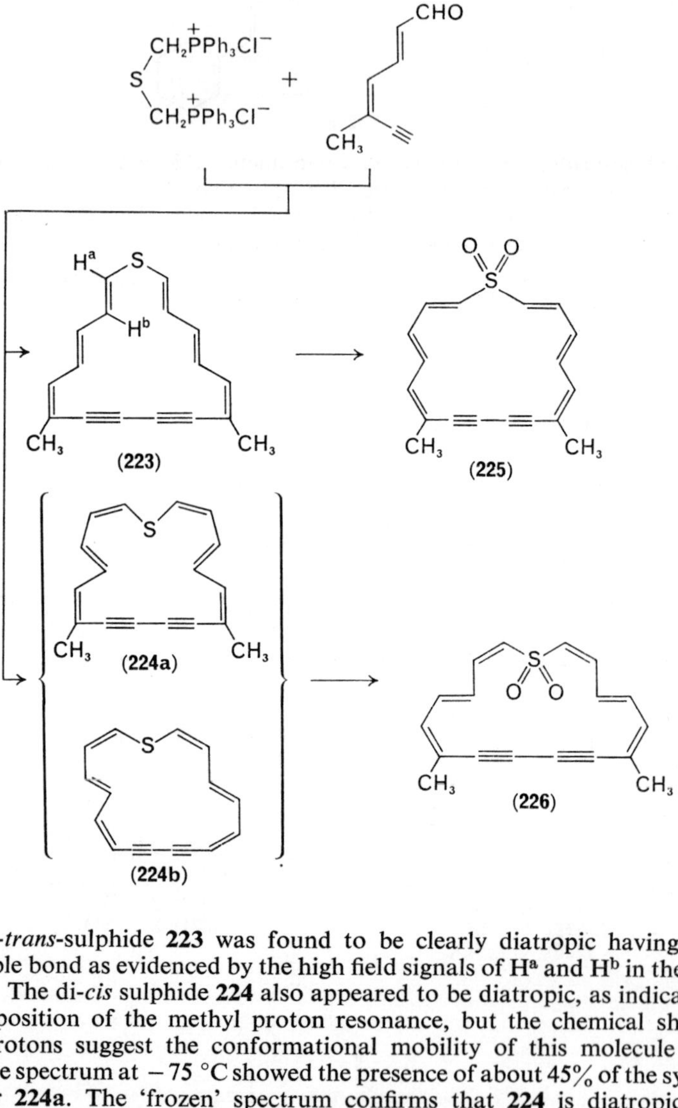

The all-*trans*-sulphide **223** was found to be clearly diatropic having a mobile *trans* double bond as evidenced by the high field signals of H^a and H^b in the ^1H-n.m.r. spectrum. The di-*cis* sulphide **224** also appeared to be diatropic, as indicated by the low-field position of the methyl proton resonance, but the chemical shifts of the olefinic protons suggest the conformational mobility of this molecule (i.e. **224a** and **b**). The spectrum at −75 °C showed the presence of about 45% of the symmetrical conformer **224a**. The 'frozen' spectrum confirms that **224** is diatropic, and undoubtedly also the biscyclohexene-annelated derivative **216** is considered to be diatropic but conformationally mobile. The all-*trans*-sulphone **225** was found to be atropic, whereas the di-*cis* sulphone **226** was clearly paratropic. This fact parallels the observations that the annelated sulphone **221** is atropic and the di-*cis* isomer

220 is paratropic[175], and indicates that the di-*cis* isomers, **220** and **226** can adopt a more planar conformation.

Starting from 3-methyl-2-penten-4-ynal and the bisphosphonium chloride, $Cl^-Ph_3P^+CH_2XCH_2P^+Ph_3Cl^-$, dimethylbisdehydrooxa[13]annulenes **227** and **228**, and dimethylbisdehydrothia[13]annulenes **229** and **230**, the corresponding sulphones, **231** and **232** and the sulphoxide, **233** have been synthesized.[177] Using the resonances of the 'fixed' H^c and the methyl protons as the ring current probe, it was shown that the oxa[13]annulene **227** is at most weakly diatropic, the sulphide **229** is definitely diatropic, and the sulphone **231** is clearly atropic. That the signals of both H^a and H^b in the ^{1}H-n.m.r. spectrum of the diatropic sulphide **229** shifted to higher field as compared to the acyclic model **234** was explained by the contribution of the rotamer **235** and/or **236**, since in a diatropic molecule the shielding effect on an inner proton far exceeds the deshielding effect on an outer proton. The coalescence temperature of H^a and H^b in **229** was found to be about −90 °C. The *cis,trans* isomers, **228**, **230** and **232** were found to be atropic, presumably due to their less planar structure

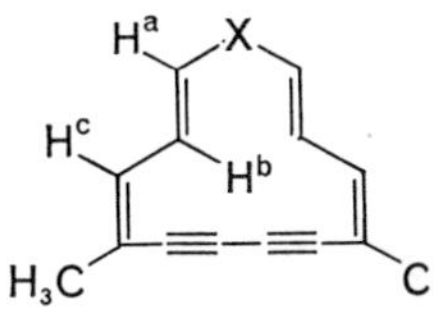

(227) X = O
(229) X = S
(231) X = SO_2
(233) X = SO

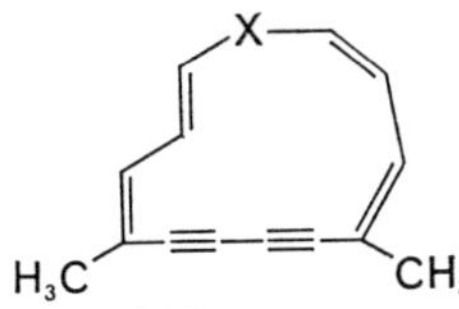

(228) X = O
(230) X = S
(232) X = SO_2

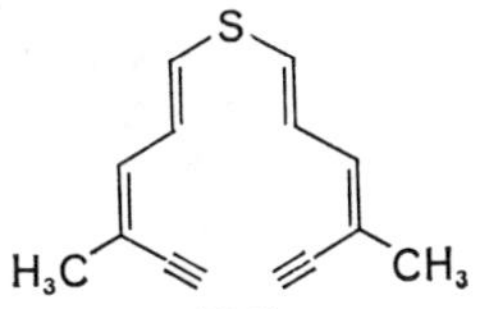

(234)

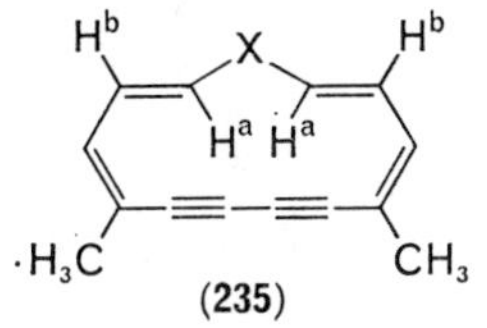

(235)

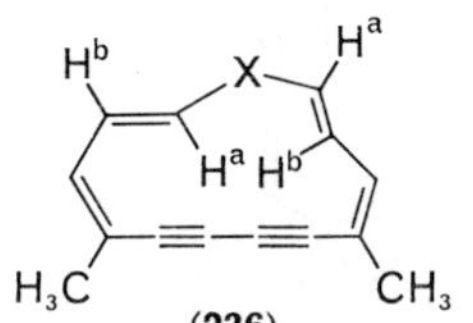

(236)

compared to the di-*trans* analogues, **227**, **229** and **231**. The bisdehydrothia[15]-annulenes **237** and **238** and bisdehydrooxa[15]annulene **239** have been synthesized by the Wittig reaction of the corresponding bisphosphorane ($Ph_3P{=}CHXCH{=}PPh_3$) with a 1 : 1 mixture of the corresponding enyne aldehyde and dienyne aldehyde[178].

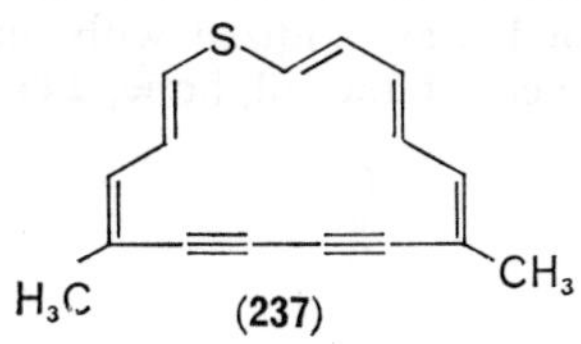

(237)

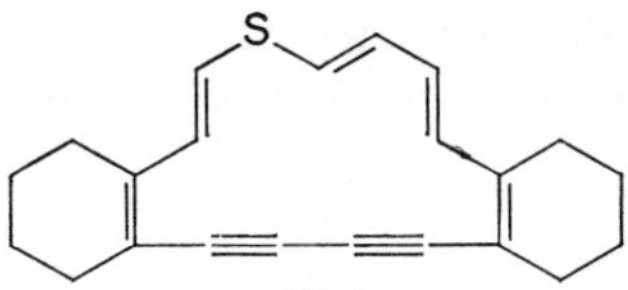

(238)

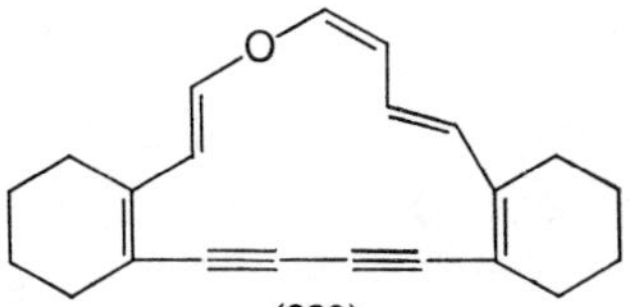

(239)

The ^{1}H-n.m.r. spectra of **237** and **238** do not indicate any conformational mobility, in contrast to bisdehydrothia[13]- (**227**)[177] and [17]annulenes (**223** and **224**)[176]. It was found that the sulphides **237** and **238** are clearly paratropic 16π electron systems. The ^{1}H-n.m.r. spectrum of mono-*cis*-bisdehydrooxa[15]annulene (**239**) shows that it is atropic; molecular models suggest that **239**, unlike all-*trans* analogues (**237** and **238**), cannot be planar.

The methylene-bridged bisdehydrooxa- (**240**) and thia- (**241a,b**) [17]annulenes, which were found to be diatropic, have been prepared according to the following reaction sequence:[179]

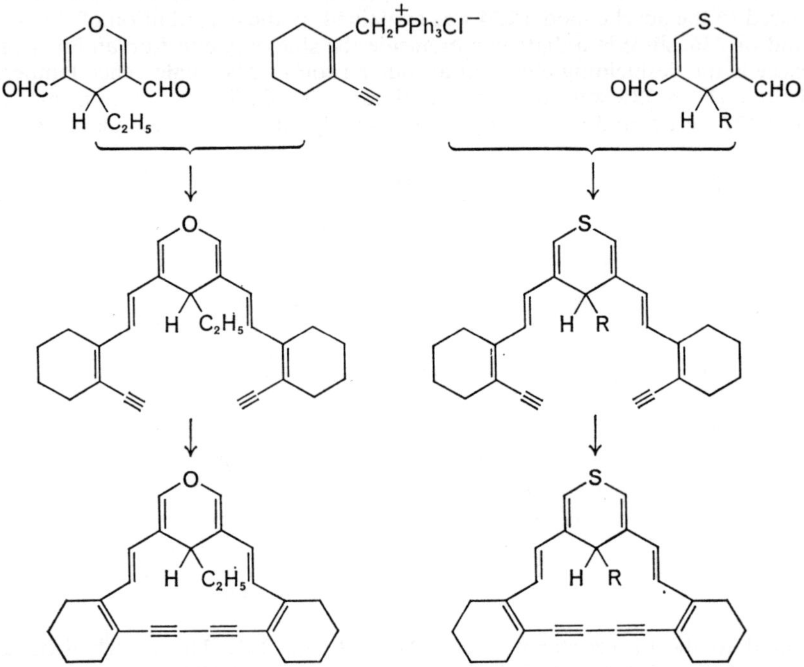

(**240**) (**241**) (**a**) R = H
(**b**) R = C_2H_5

Oxidation of **241a, b** with two molar equivalents of *m*-chloroperbenzoic acid afforded the sulphones **242a** and **b**. When **241a, b** was oxidized with one molar equivalent of the same peracid, *anti* and *syn* isomers of the sulphone, **243** and **244**

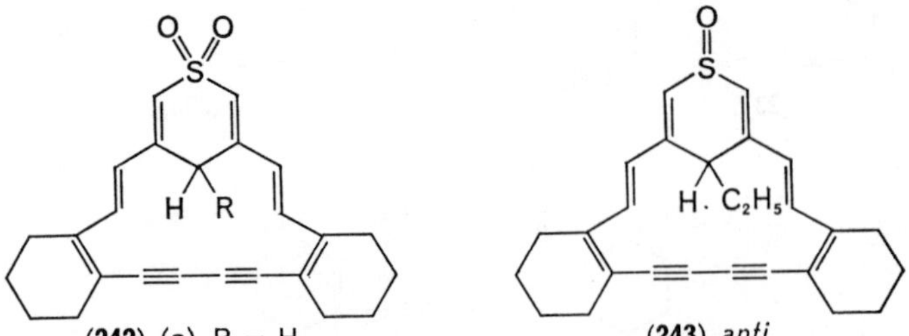

(**242**) (**a**) R = H
(**b**) R = C_2H_5

(**243**) *anti*
(**244**) *syn*

were obtained (Figure 6)[180]. The sulphones **242a, b** were found to be clearly paratropic in contrast to the diatropic sulphides **241a, b**. The stereoisomeric sulphides **243** and **244** show remarkably different ^{1}H-n.m.r. and electronic spectra. The spectrum of

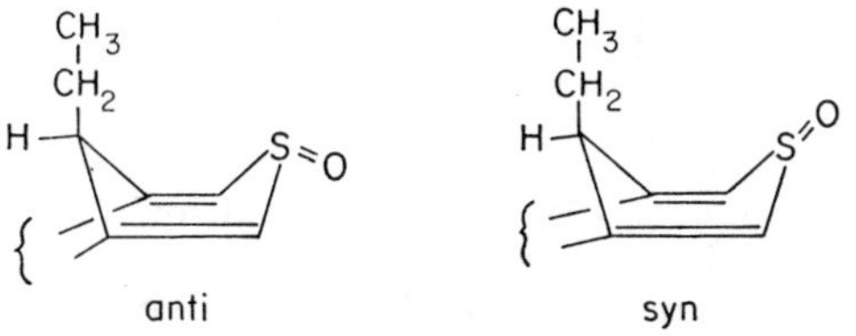

FIGURE 6. *Anti* and *syn* sulphoxides **243** and **244**.

the *syn* isomer **244** is very similar to that of the sulphone **242b** showing paratropicity in the ^{1}H-n.m.r. spectrum. On the other hand, the ^{1}H-n.m.r. spectrum shows that the *anti* isomer **243** is atropic, and also the electronic spectrum is quite different from that of **244**. Considering the favourable disposition of the 2*p* orbital of the axial sulphoxide oxygen in the *syn* isomer **244** for overlap with the π perimeter, the possibility that the *syn* isomer **244** is a paratropic 'Möbius' 18π electron system was suggested [180].

Up to now, no pyridine-type dehydroheteroannulene has been synthesized*.

5. Dehydroannulene anions and cations

The addition of one or two electrons to $[4n+1]$- or $[4n]$annulene should result in an aromatic anion or dianion having $[4n+2]$ π electrons.

The alcohol **245**, obtained by the sodium borohydride reduction of trisdehydro-[17]annulenone **188**, was converted into the methyl ether **246**. Treatment of **246** with

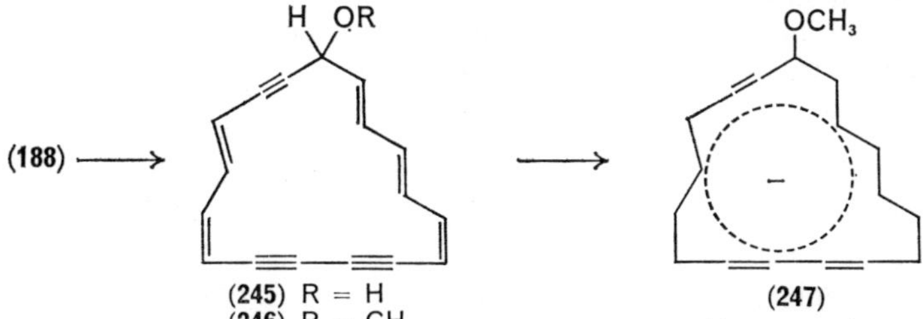

N.m.r. spectrum

Inner H: τ = 18·54–19·09
Outer H: τ = −0·47–2·16

methyllithium in THF-d_8 at a low temperature gave a dark blue solution of the lithium salt of **247** [157]. The anion, being an 18π electron system, was found to be strongly diatropic as shown by the data under the structure. Similarly, the atropic biscyclohexene-annelated bisdehydro[13]annulenone **191** could be converted into the strongly diatropic bisdehydro[13]annulenyl **248** [181]. This result indicates that the anion **248** holds higher planarity than **191**.

* The sole instance of a pyridine-type non-benzenoid heteroaromatic system is *trans*-1,3,15,16-tetramethyl-2-azadihydropyrene, a peripheral 14π electron system, prepared by V. Boekelheide, *J. Am. Chem. Soc.*, 92, 3684 (1970).

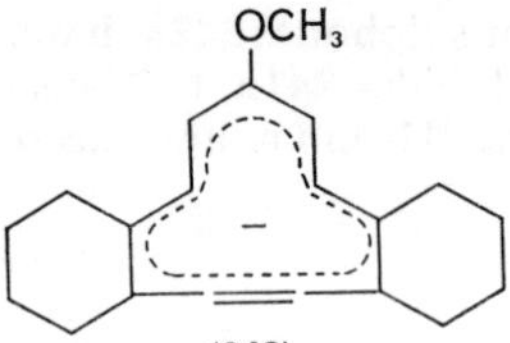

(248)

The bisdehydro[12]annulene **171** [122] and the trisdehydro[12]annulene **172** [149] gave the corresponding diatropic dianions (**249** and **250**) on treatment with potassium[182]. Although, as already mentioned, **171** shows high conformational mobility, the

N.m.r. spectra

(171) 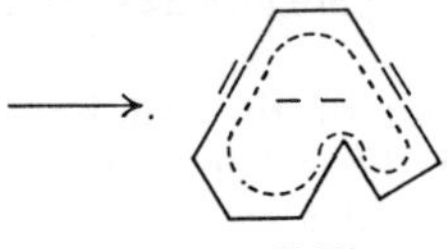

(249)

Inner H	Outer H
$\tau = 16{\cdot}88$	$\tau = 2{\cdot}03–2{\cdot}47$
	$3{\cdot}31–4{\cdot}01$

(172) 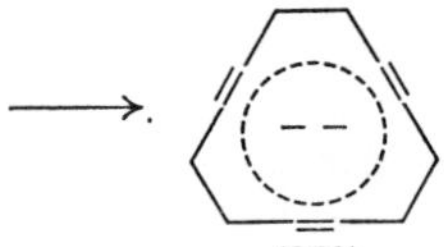

(250)

$\tau = 5{\cdot}58$

anion (**249**) was found to be conformationally stable giving almost temperature-independent ^{1}H-n.m.r. spectra in the range of −35 °C to +30 °C. The higher conformational stability and planarity of the annulenyl anion as compared with the corresponding neutral annulene have been generally observed, *e.g.* [16]annulene having an extremely high conformational mobility[183] gives a planar and conformationally stable dianion[184].

The extremely unstable octakisdehydro[24]annulene **251** yielded the dianion **253** via the radical anion **252** on treatment with potassium mirror[185]. The e.s.r.

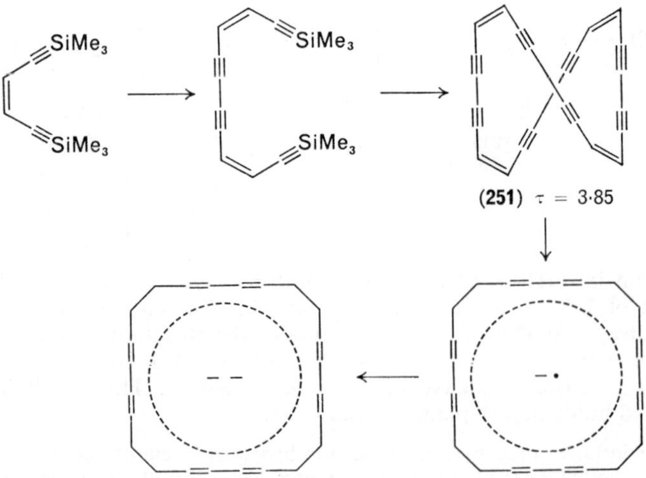

(251) $\tau = 3{\cdot}85$

(253) $\tau = 3{\cdot}88$ (252)

spectrum of the radical anion **252** was found to be consistent with the planar symmetrical structure. The coincident position of the ^{1}H-n.m.r. signals of **251** and **253** is attributed to the balance between the deshielding effect of the diamagnetic ring current and the shielding due to the excess of electron density, as has previously been observed for cyclooctatetraene and its dianion[186].

Cations derived from $[4n+3]$annulenes can be expected to be aromatic $[4n+2]\ \pi$ electron systems like the cyclopropenium ($n = 0$) [187] and tropylium ($n = 1$) cations[188].

Treatment of the alcohols **254** and **256**, obtained by reduction of the bisdehydro-[15]annulenones **198** and **199** [161], with trifluoroacetic acid gave the strongly diatropic bisdehydro[14]annulenium cation (**255**)[189]. The diatropicity of **255** was found to be stronger than that of the protonated species (**200**) of the annulenones **198** and **199**.

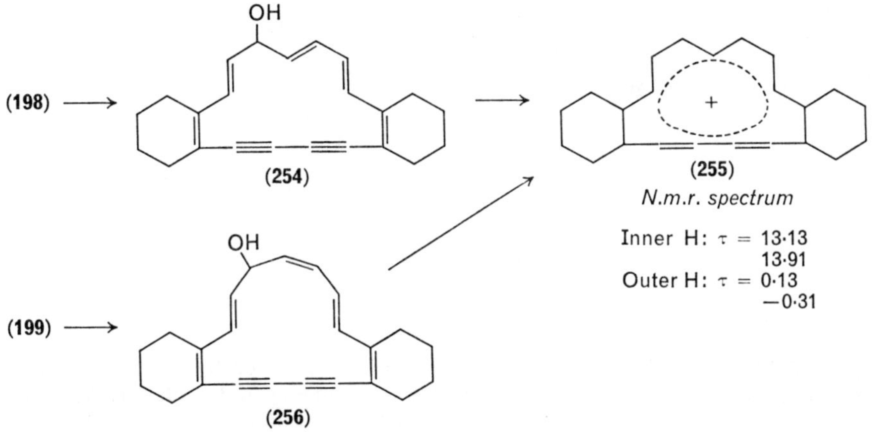

N.m.r. spectrum

Inner H: τ = 13·13
13·91
Outer H: τ = 0·13
−0·31

B. *'Acetylene–cumulene' Dehydroannulenes*

I. Bisdehydro[4*n*+2]annulenes

Oxidative coupling of *trans,trans*-4,10-tetradecadien-1,7,13-triyne (**257**) and subsequent prototropic rearrangement of the coupling products yielded mainly the geometrical isomers of monodehydro[14]annulene (**258** and **259**)[141] and a by-product, 1,8-bisdehydro[14]annulene (**260**)[190, 191]. Formation of **260** indicates that dehydrogenation occurs under the basic conditions of prototropic rearrangement.

$$HC{\equiv}CCH_2CH{=}CHCH_2C{\equiv}CCH_2CH{=}CHCH_2C{\equiv}CH$$

(257)

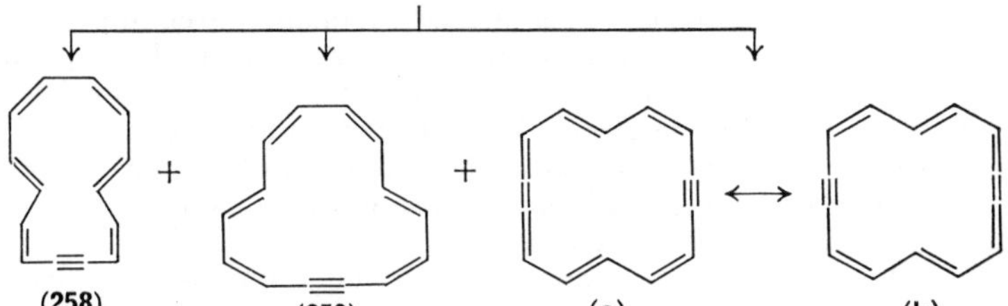

(258) (259) (a) (b)

(260)

The ^{1}H-n.m.r. spectrum of **260** exhibits signals due to the inner protons at τ 15·48 (H^c, t) and to the outer protons at τ 1·46 (H^a, d) and 0·36 (H^b, q) indicating the induction of a strong diamagnetic ring current[139]. The compound **260** was found to be quite stable in the solid state or in solution and gave electrophilic substitution products as summarized in the following scheme[144]:

(260)

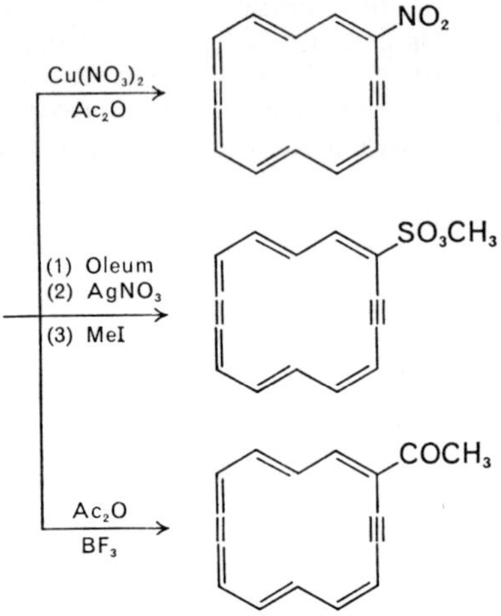

The highly aromatic nature of **260** was further supported by the X-ray structure determination[191]. The molecule has a centre of symmetry and is planar. Apart from the C(*sp*)—C(*sp*) linkage, the C—C bond lengths are about the same as in benzene. The absence of bond alternation indicates a high degree of π-electron delocalization. Consequently, the compound is better represented by a symmetrical formula **260c**, being a resonance hybrid of the Kekulé structures **260a** and **260b**.

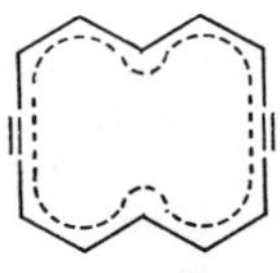

(260c)

Thus, **260**, containing a formal acetylenic and a formal cumulenic linkage in the conjugated system, holds a unique position among dehydroannulenes containing only acetylenic linkages, in which no equivalent resonance structures can be drawn. Symmetrical 'acetylene–cumulene' dehydroannulenes seem to be a good tool for the study of the aromaticity of macrocyclic conjugated systems. An efficient method for their synthesis has been developed by Nakagawa and his coworkers[192].

The *cis*-3-substituted-2-penten-4-ynal (**262**) predominantly formed by anionotropic rearrangement of **261** [193] gave, on aldol condensation with methyl ketones, the diyne ketone **263**. When a solution of the latter in THF was slowly added to a suspension of finely powdered potassium hydroxide in liquid ammonia, a mixture of the diastereomers of the cyclic glycol **264** was obtained. Treatment of **264** in benzene or ether with tin(II) chloride dihydrate in concentrated hydrochloric acid gave the tetrasubstituted 1,8-bisdehydro[14]annulenes **265a–c** [194], which were found to be stable and to form charge-transfer complexes with 2,4,7-trinitrofluorenone (Table 8).

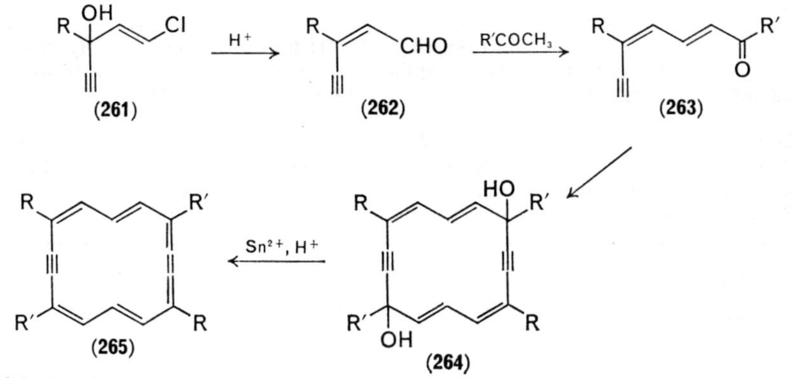

(**a**) R = R' = *t*-Bu
(**b**) R = Ph, R' = *t*-Bu
(**c**) R = R' = Ph

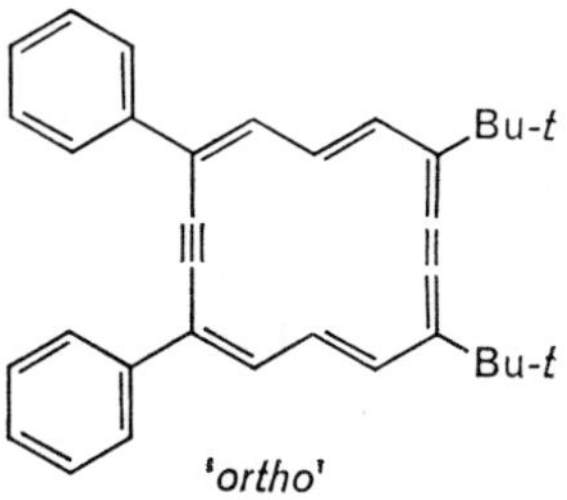

'ortho'

(266)

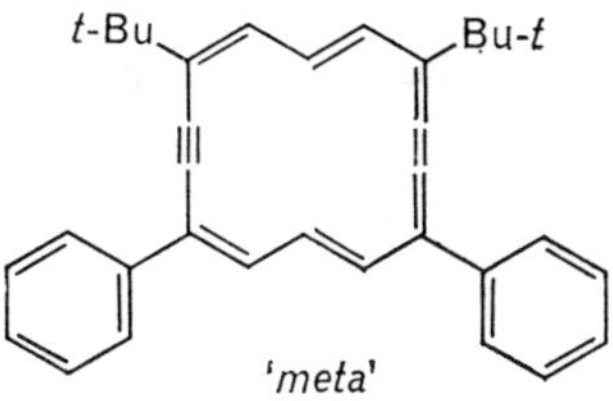

'meta'

(267)

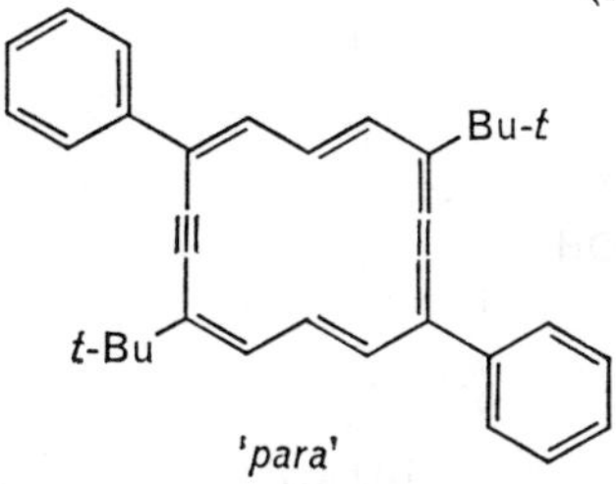

'para'

(265b)

The ^{1}H-n.m.r. spectra summarized in Table 8 indicate that the bisdehydro[14]-annulenes, **265a–c** are strongly diatropic. The electronic spectra of **265a–c** consist of three main absorption bands clearly showing features characteristic of [4*n*+2]-annulenes. Remarkable bathochromic shifts of the electronic spectra were observed with increase in number of phenyl groups (Table 8).

TABLE 8. Properties of tetrasubstituted bisdehydro[14]annulenes **265a–c** and their parent compound **260**

Compound	Colour of crystals	Longest wavelength λ_{max}	^{1}H-n.m.r. τ value		Charge-transfer complex (annulene: trinitrofluorene)
			Outer H	Inner H	
260	Red	586 (2900)[a]	1·46	15·48[b]	—
265a	Red	590 (922)[c]	0·58	14·39[d]	1 : 1
265b	Brown violet	623 (1640)[c]	0·12	13·42[d]	1 : 1
265c	Violet	658 (2520)[c]	0·06	12·56[d]	2 : 1

[a] In isooctane.
[b] In $CDCl_3$.
[c] In THF.
[d] In THF-d_8.

The di-*t*-butyldiphenylbisdehydro[14]annulene **265b** can be regarded as a '*para*' isomer of diphenylbisdehydro[14]annulene corresponding to *p*-terphenyl. The '*ortho*' (**266**) and '*meta*' (**267**) isomers corresponding to *o*- and *m*-terphenyls, respectively, were synthesized by the reaction sequence outlined below[195]:

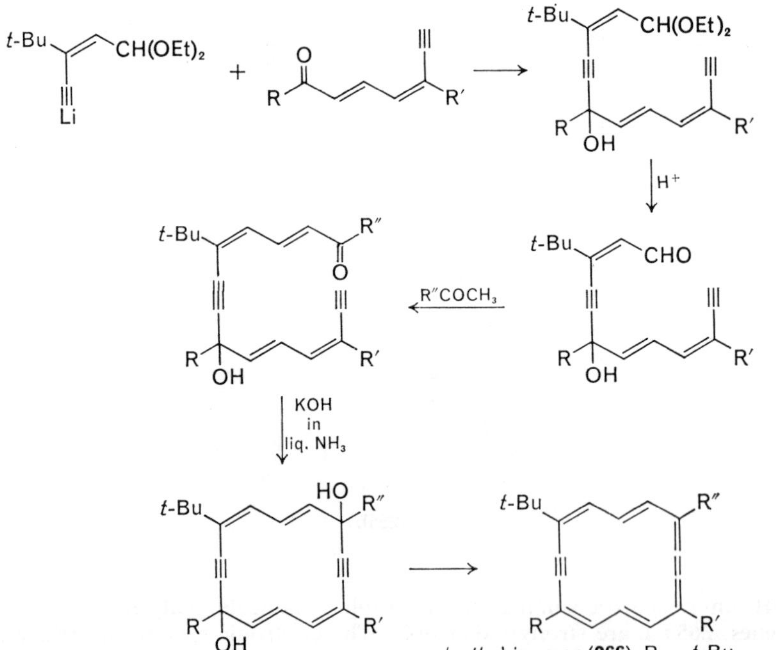

'*ortho*' isomer (**266**) R = *t*-Bu
R′ = R″ = Ph
'*meta*' isomer (**267**) R″ = *t*-Bu
R = R′ = Ph

The positional isomers (**265**, **266** and **672**) showed rather similar electronic and ^{1}H-n.m.r. spectra, and the appreciable differences known for *o*-, *m*- and *p*-terphenyls could not be observed. The dimethoxy derivative of the '*para*' isomer (**268**) was prepared by the reaction sequence outlined below:[195]

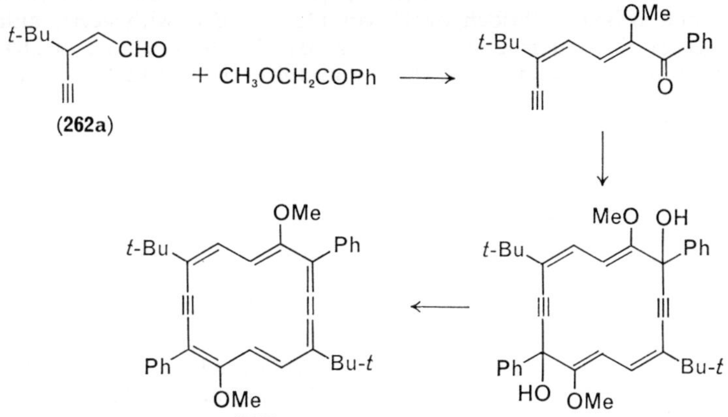

The dimethoxybisdehydro[14]annulene **268** was also found to be strongly diatropic and an intensification of the longest wavelength band (637 nm, ε 4540) was observed as compared with that of **265b** (624 nm, ε 1640), **266** (624 nm, ε 1800) and **267** (624 nm, ε 1270). At first glance, this result seems to suggest that the direction of polarization of the longest wavelength band is perpendicular to the axis which bisects the molecule through the mid-points of C(*sp*)—C(*sp*) linkages. However, measurements of the fluorescence excitation spectra of tetra-*t*-butyl-bisdehydro[14]annulene **265a** and the '*para*' isomer **265b**, the polarized reflection spectrum of a single crystal of tetra-*t*-butylbisdehydro[18]annulene **282** and theoretical calculations have been performed by Tanaka and coworkers[196]. The nature of electronic transitions and the direction of polarization are firmly established on the basis of these investigations. As shown in Figure 7, the longest wavelength

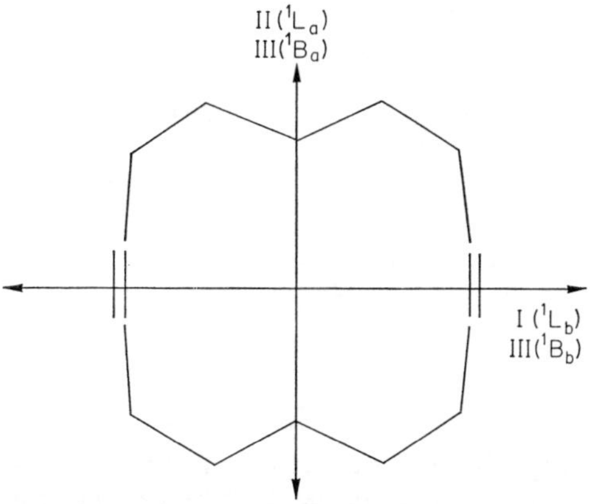

FIGURE 7. Direction of polarization of bisdehydro[14]annulene.

band (I) is ^1L_b species and the direction of polarization is parallel to the axis, the medium wavelength band (II) is ^1L_a species having perpendicular polarization and the short wavelength band (III) consists of parallel polarized ^1B_b species and perpendicular polarized ^1B_a species.

3-Acetyl-7,10,14-tri-*t*-butyl-1,8-bisdehydro[14]annulene (**269**) was obtained on treatment of tetra-*t*-butylbisdehydro[14]annulene (**265a**) with acetic anhydride in the presence of zinc chloride at 140 °C for 30 s [197]. The structure of **269** has been confirmed by a stepwise synthesis[197]. The acetyl derivative (**269**) showed a strong diatropicity.

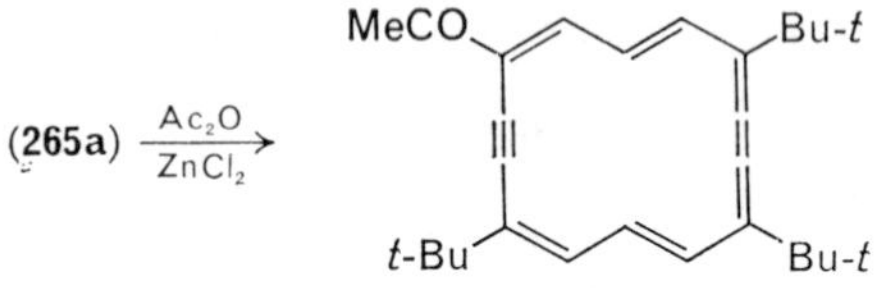

(**269**)

3-Carboxy-7,10,14-tri-*t*-butyl-1,8-bisdehydro[14]annulene (**274**) has been synthesized in order to determine the pK_a' value[197]. The product of the aldol condensation of *t*-butylpentenynal (**262a**) with pyruvic acid was converted into the methyl ester **270**. The ethynyldiene ketone **271** [198] was treated with lithium diethylamide at a low temperature to give the lithio derivative of **271**. The reaction of the lithio derivative with **270** gave **272a**, which was converted into the methoxy methyl ester (**272b**) on treatment with dimethyl sulphate. Treatment of **272b** with potassium hydroxide in liquid ammonia yielded the cyclic glycol **273**. The reduction of **273** with tin(II)

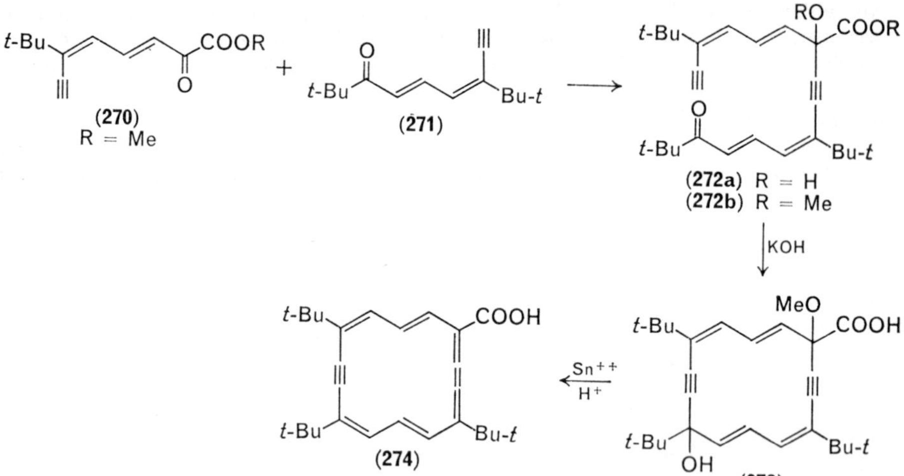

chloride dihydrate and hydrogen chloride gave the strongly diatropic carboxy-bisdehydro[14]annulene **274**. The pK_a' value of **274** was determined to be 5·92 ± 0·11 (at 18 °C) by the spectrophotometric method. The pK_a' values of benzoic and *para*-*t*-butylbenzoic acids were found to be 5·55 ± 0·04 and 5·64 ± 0·07, respectively. Since acetylene and allene carboxylic acids were found to be much stronger acids than the corresponding ethylenic or saturated carboxylic acids (e.g. $CH_3C{\equiv}CCH_2COOH$: 3·60; $CH_2{=}CHC{\equiv}CCH_2COOH$: 3·37; $CH_2{=}C{=}CHCOOH$: 3·69)[199], the fact that the dissociation of the carboxyannulene **274** is much smaller than those of acetylenic and allenic acids and is similar to those of benzoic and substituted benzoic acids can be regarded as a reflection of highly delocalized π-electron system in the 1,8-bisdehydro[14]annulene ring.

The cyclic glycol **273** gave 7,10,14-*t*-butyl-3-methoxy-1,8-bisdehydro[14]annulene (**275**) by decarboxylative aromatization on treatment with hydrogen chloride[200].

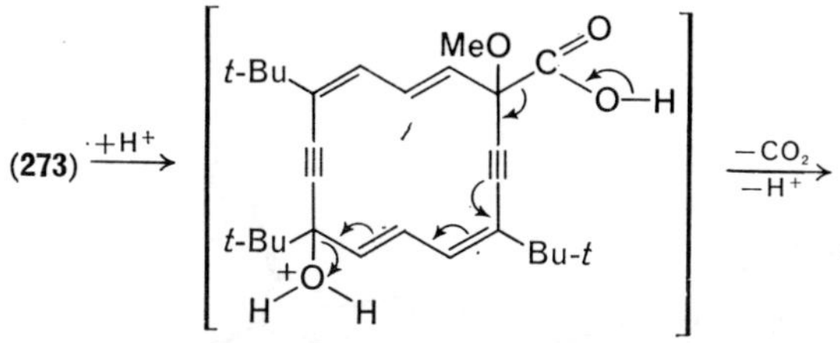

The higher analogue of **265a**, tetra-*t*-butylbisdehydro[18]annulene (**282**), has been synthesized[201]. The reaction of the acetal **276** derived from **262a** with ethyl vinyl ether in the presence of borontrifluoride etherate[202] yielded the ethoxy acetal **278**, which could be converted into the dienyne aldehyde **279** [203]. The trienyne ketone **280** obtained by the aldol condensation of **279** with pinacolone[203] was treated with potassium hydroxide in liquid ammonia. The 18-membered cyclic glycol **281**, which could be separated into two stereoisomers, gave reddish violet crystals of tetra-*t*-butylbisdehydro[18]annulene (**282**) in a high yield on treatment with tin(II) chloride dihydrate in ether saturated with hydrogen chloride. The dehydroannulene **282** was found to be strongly diatropic having a high conformational stability as revealed by the essentially temperature-independent ^{1}H-n.m.r. spectra.

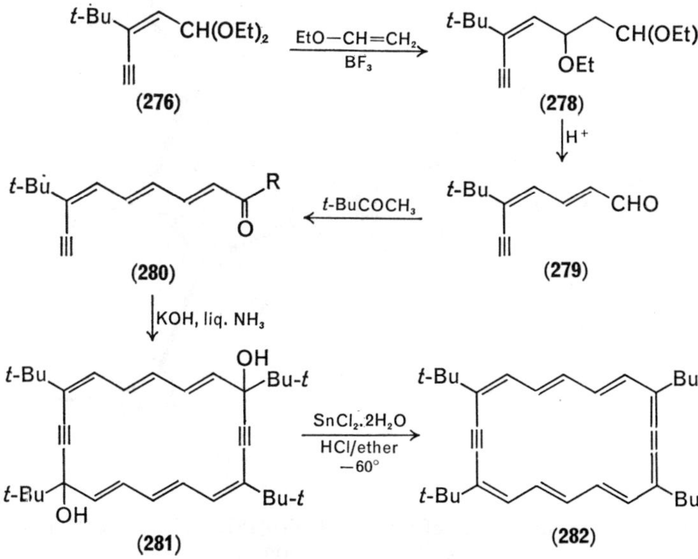

The bond lengths and angles in **282** determined by X-ray structure analysis are shown in Figure 8[204]. The molecule is nearly planar and the polyene part consisting of twelve carbon atoms with six formal double bonds shows almost no bond alternation (average bond length 1·387 Å). The lengths of the formal acetylenic linkage (=C—C≡C—C=) and the formal cumulenic linkage (=C=C=C=C=)

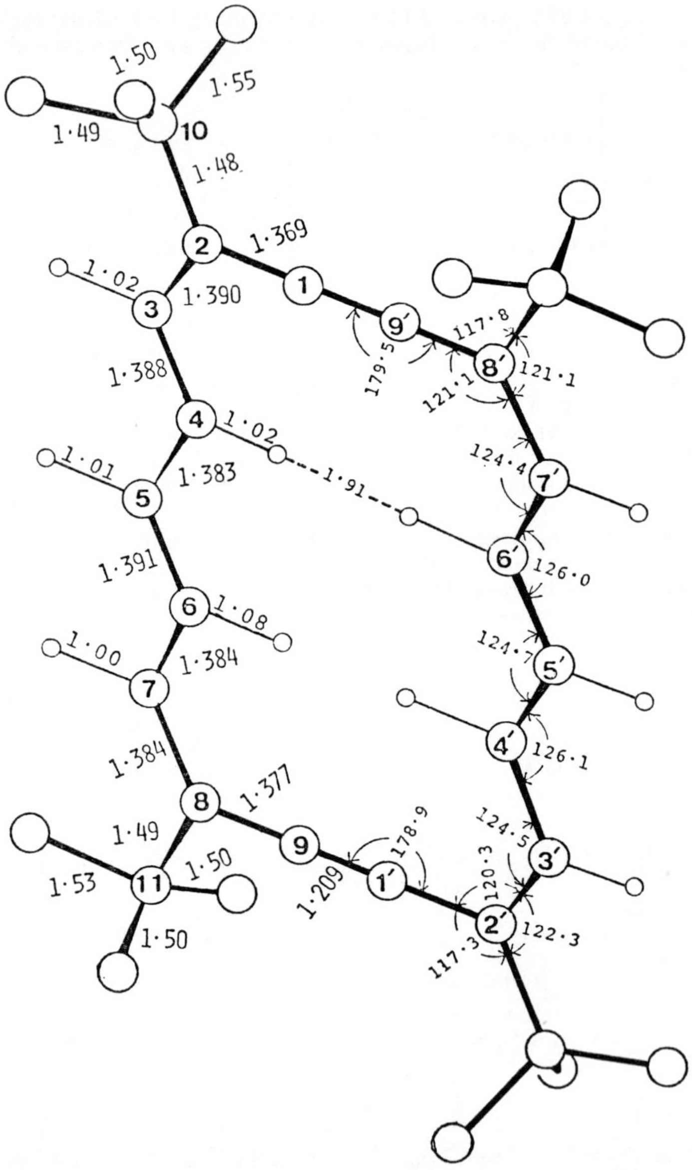

FIGURE 8. Bond lengths and angles of the bisdehydro[18]annulene **282** [204]. Reproduced by permission of Pergamon Press.

are found to be the same. These results indicate that **282** has a highly delocalized 18π electron system and is better represented by the symmetrical formula **283**.

In an analogous reaction sequence, di-*t*-butyldiphenyl- (**284**) and tetraphenyl- (**285**) bisdehydro[18]annulenes[205] have been prepared and proved to be strongly diatropic.

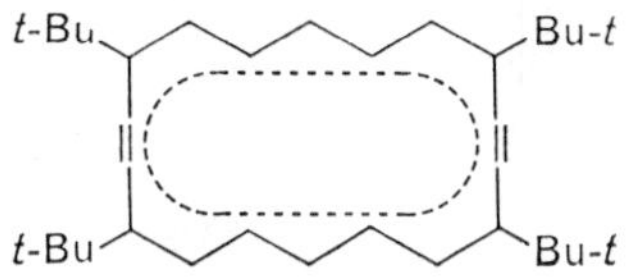

(**283**)

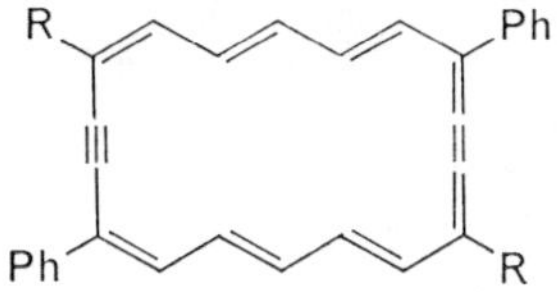

(**284**) R = *t*-Bu
(**285**) R = Ph

The tetraenyne ketone **286** was prepared according to the route outlined below:[206]

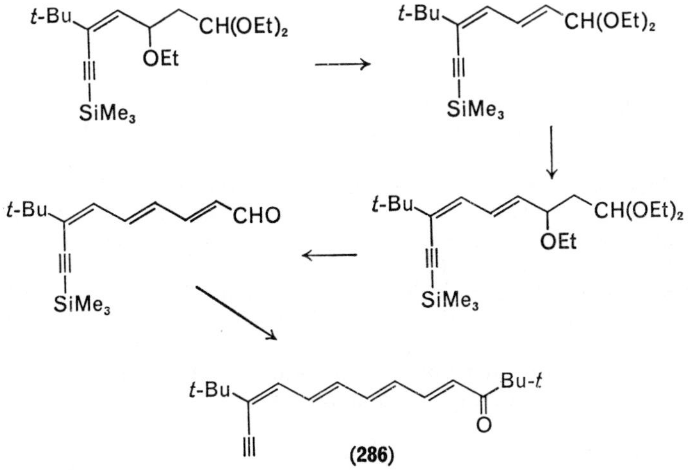

(**286**)

The pentaenyne ketone **287** has been synthesized by a similar reaction[207]. The hexaenyne ketone **288** has been obtained by the condensation of trimethylsilyl derivative of **262a** with crotonaldehyde followed by aldol condensation with pinacolone[208].

Treatment of the polyenyne ketones **286**, **287** and **289** with potassium hydroxide in liquid ammonia, and subsequent reduction of the resulting cyclic glycols with tin(II) chloride and hydrogen chloride, yielded tetra-*t*-butylbisdehydro[22]- (**289**)[206], -[26]- (**290**)[207] and [30]- (**291**)[208] annulenes as highly coloured crystals. Thus, a series of bisdehydro[4*n*+2]annulenes ranging from 14π to 30π electron systems (**265a**, **282**, **289**, **290** and **291**) have been obtained.

The bisdehydro[4*n*+2]annulenes **289**, **290** and **291** were found to be diatropic retaining the conformation shown by the formulae, e.g. the ^{1}H-n.m.r. spectra of the bisdehydro[22]annulene **289** measured at 36 °C and 70 °C showed almost no temperature dependency[206]. This result shows high conformational stability of 'acetylene–cumulene' dehydroannulenes, because the coalescence temperature of the ^{1}H-n.m.r. spectrum of [22]annulene has been reported to be about 20 °C [209].

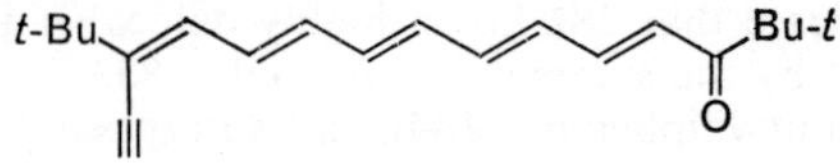

(287)

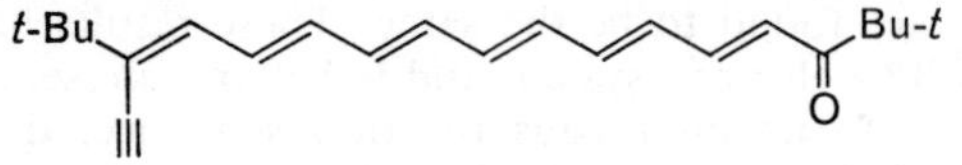

(288)

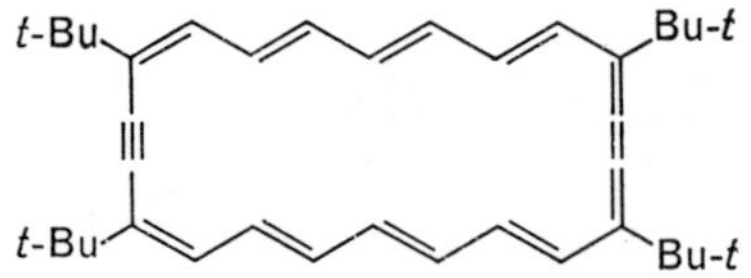

(289)

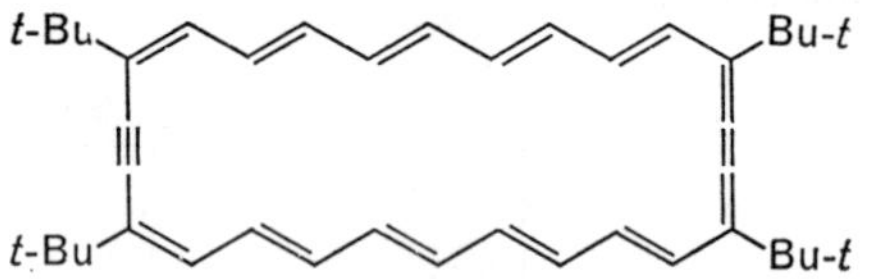

(290)

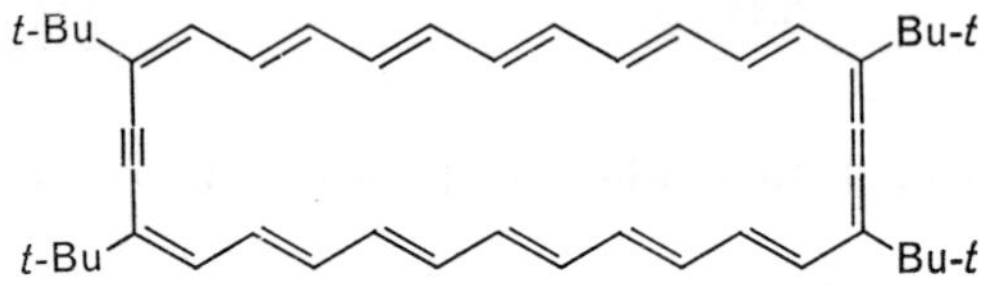

(291)

The results of the X-ray structure analysis of **289** [210] showed an interesting difference from that of **282** [204]; i.e. as shown in Figure 9, the linkage $C^{10}—C^{11}—C^{12}—C^{13}$ has more cumulenic features than the linkage $C^{2}—C^{1}—C^{22}—C^{21}$, which showed enhanced acetylenic features, and also the bonds $C^{10}—C^{9}$ and $C^{13}—C^{14}$ adjacent to the more cumulenic linkage were found to be longer than the bonds $C^{2}—C^{3}$ and $C^{21}—C^{20}$ adjacent to the more acetylenic linkage, as expected from the valence-bond structure. These results indicate an increased bond alternation in the 22π electron system **(289)** compared to the 18π analogue **(282)**.

It has been predicted theoretically that planar $[4n+2]$annulenes will be aromatic up to [22]annulene, but that [26]annulene will no longer be aromatic[211]. The findings that monodehydro[22]annulene[212], [22]annulene[209] and tetra-*t*-butylbisdehydro-[22]annulene **(289)**[206] are diatropic are consistent with the theoretical prediction. Trisdehydro[26]annulene **(292)**[213]* and monodehydro[26]annulene **(293)**[214]* prepared

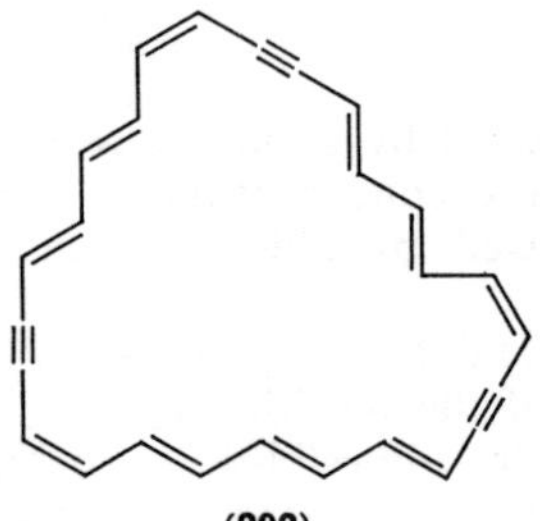

(292)

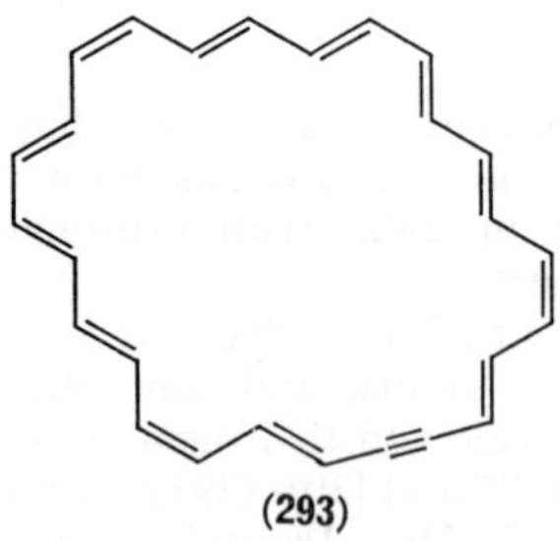

(293)

by Sondheimer were found to be atropic and diatropic, respectively, showing that the atropicity of **292** is due to the perturbation of the three acetylenic linkages, which causes the alternate bond structure to be energetically preferred to the delocalized system. A much more intense diatropicity found in the bisdehydro[26]annulene **290**

* The possible configurations **(292** and **293)** which are consistent with the ratio of inner and outer proton signals are shown.

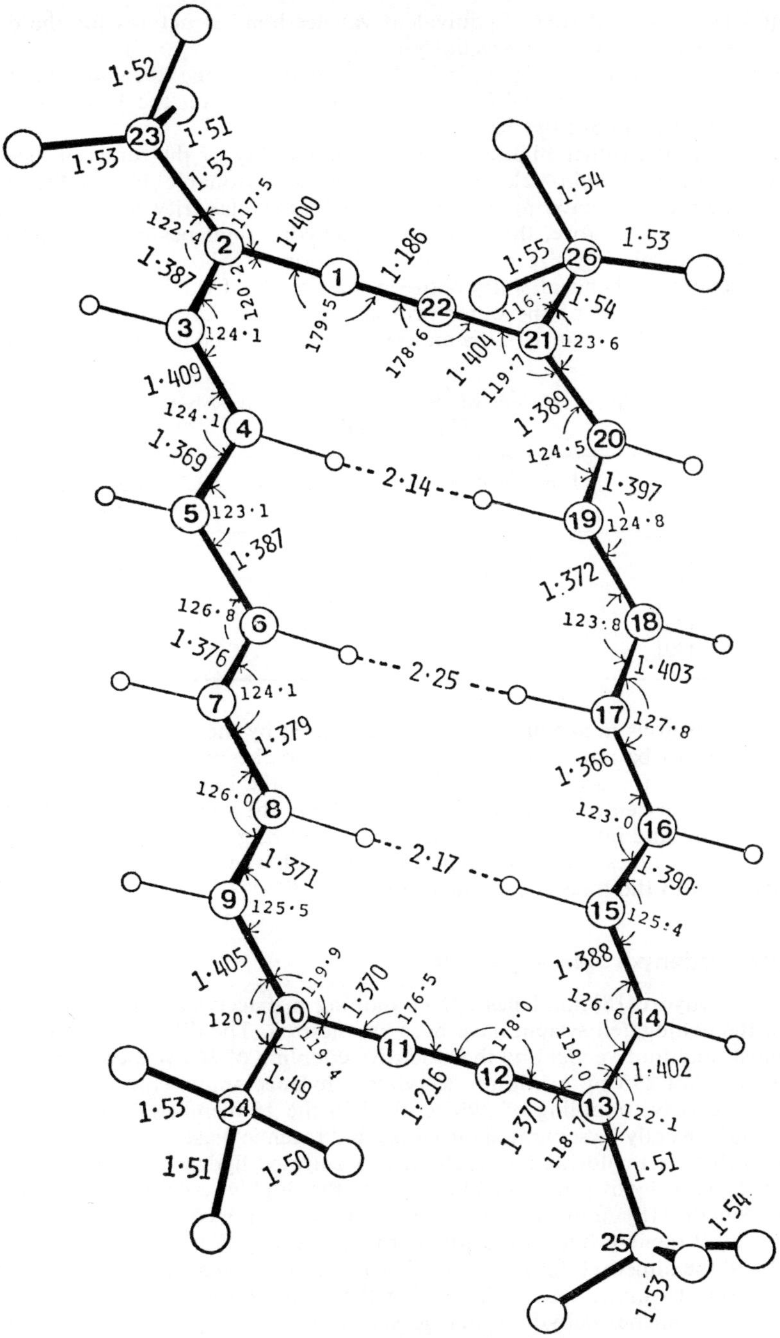

FIGURE 9. Bond lengths and angles of the bisdehydro[22]annulene **289** [210]. Reproduced by permission of Pergamon Press.

indicates an important role of equivalent valence-bond structures for the delocalization of π electrons in macrocyclic systems.

The preparation of the diatropic bisdehydro[30]annulene **291** indicates that the upper limit of aromaticity of [4*n*+2]annulene, at least on the basis of magnetic criteria, should lie above the 30-membered ring.

Considering the rather high conformational stability of the series of 'acetylene–cumulene' bisdehydro[4*n*+2]annulenes, it seems reasonable to assume that the bisdehydroannulenes have approximately the same planarity and essentially the same geometry. Therefore, this series of bisdehydroannulenes makes it possible to study the effect of ring size on the delocalization of a [4*n*+2] π electron system. The differences in chemical shifts between the signals of the inner protons (τ_i) and the lowest field signal of the outer protons (τ_o), which are always located at the position nearest to the centre of the molecule, is summarized in Table 9. The chemical

TABLE 9. The magnitude of chemical shifts of tetra-*t*-butylbisdehydro-[4*n*+2]annulenes

[4*n*+2]	Inner protons τ_i	Outer protons τ_o	Chemical shift $\tau_i-\tau_o$
[14]	14·44	0·68	13·76
[18]	13·42	0·13	13·29
[22]	10·83	0·84	9·99
[26]	8·05	1·77	6·28
[30]	6·5	2·5	4·0

shift of the protons in aromatic compounds caused by the diamagnetic ring current is considered to be approximately proportional to the product of intensity of the ring current (J), the area of the molecule (S) and the inverse cube distance of the protons from the centre of the molecule (R^3) [215]. On the assumption that the distance (R) is constant and independent of the variation of ring size, it has been shown that the ($\tau_i-\tau_o/S$) values decrease monotonously with the increase in ring size[192] in accordance with the theoretical conclusion[146, 147, 211a, c, 216, 217].

2. Tetrakisdehydro[4*n*+2]annulenes

Tetrakisdehydro[18]annulenes **297** containing a diacetylene and a hexapentaene unit in the conjugated system have been synthesized. The diketone **294** obtained in an almost quantitative yield by the oxidative coupling of **263** was converted into the bis-ethynyl diol **295** by a lithium acetylide–ethylenediamine complex[218] in organic solvent. Oxidative coupling of **295** resulted in the 18-membered cyclic glycol **296**, which could usually be separated into *meso* and racemic diastereoisomers. Treatment of **296** with tin(II) chloride dihydrate in concentrated hydrochloric acid or in ether saturated with hydrogen chloride yielded the highly coloured tetrasubstituted tetrakisdehydro[18]annulenes **297**. These, as summarized in Table 10, bearing various substituent groups were found to be strongly diatropic. The tetramethyl-(**297a**) and the dimethyl- (**297b**)diphenyl derivatives showed essentially temperature-independent ^{1}H-n.m.r. spectra indicating the high conformational stability of the 'acetylene–cumulene' dehydroannulene skeleton.

The ^{1}H-n.m.r. spectra of **297a–f** indicate that they have a highly delocalized 18π electron system. Consequently, the formal diacetylene and hexapentaene units incorporated in the aromatic annulene ring should be identical just as the formal double and single bonds in the Kekulé formula of benzene.

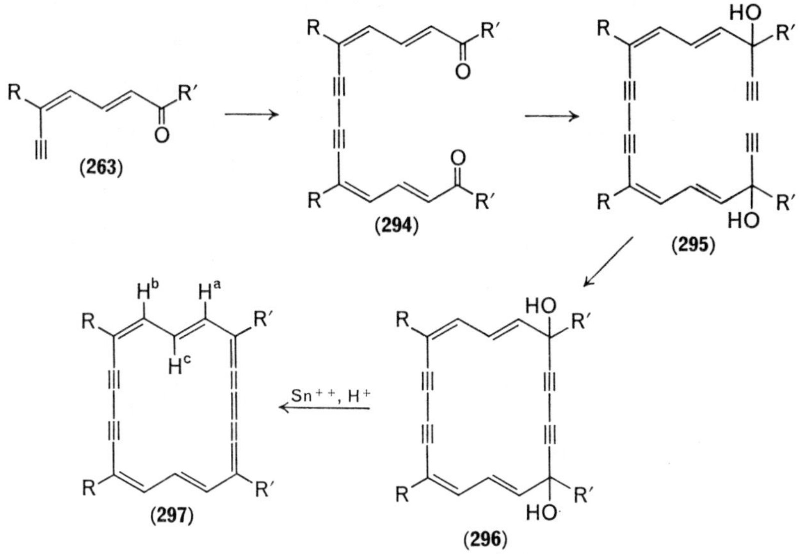

TABLE 10. ¹H-n.m.r. parameters of tetrakisdehydro[18]annulenes (**297**) (τ values)

Compound	R	R′	H^a	H^b	H^c	Me	*t*-Bu	Reference
297a	Me	Me	0·34		15·24	7·42	—	219
297b	Me	Ph	0·12	0·54	14·20	6·56	—	220, 222
297c	Ph	Ph	−0·31		13·19	—	—	221, 222
297d	*p*-MeO—Ph	*p*-MeO—Ph	−0·04		13·00	—	—	223
297e	*t*-Bu	Ph	−0·40	−0·01	13·90	—	7·93	224
297f	*t*-Bu	*t*-Bu	0·02		14·92	—	7·89	220, 225

The isomeric 18-membered cyclic glycols **298** and **299** were converted into di-*t*-butyldiphenylbisdehydro[18]annulenes, **300** and **301**[224]. They showed the same decomposition point (189·0–191·0 °C) and gave superimposable i.r. spectra, as well as identical electronic and ¹H-n.m.r. spectra. Hence, the diacetylene and the hexapentaene linkages are identical as a result of a high degree of π-electron delocalization, although the possibility of a fast bond shift in **300** and **301** cannot be excluded *a priori*. However, the argon laser Raman spectra of **300** and **301** exhibit a single absorption due to the stretching vibration of the C(*sp*)—C(*sp*) bond at 2080 cm^{-1} showing the identity of diacetylenic and cumulenic linkages[224b].

In an analogous reaction sequence, tetra-*t*-butyl- and tetraphenyltetrakisdehydro-[22]annulenes, **302** [226] and **303** [227], have been prepared. These are unstable and their n.m.r. spectra show fairly strong diatropicity.

The temperature-independent ¹H-n.m.r. spectra of **302** indicate the rigid nature of the molecular framework of 'acetylene–cumulene' dehydroannulenes. The ¹H-n.m.r. spectrum of monodehydro[22]annulene[212] exhibits signals of outer protons at τ 1·55–3·75 and those of inner protons at τ 6·55–9·30. A much stronger diatropicity

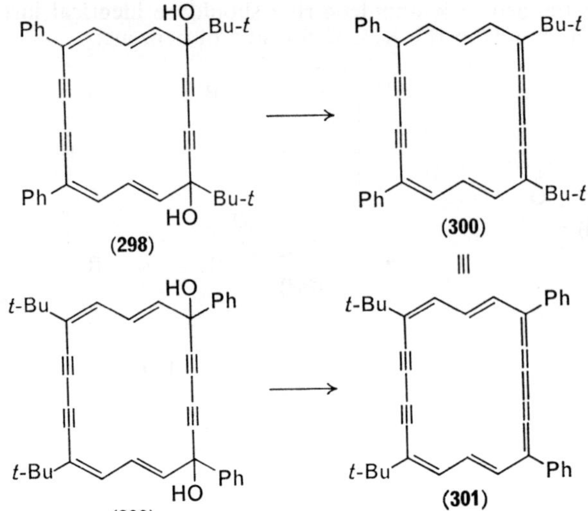

(298) (300) (299) (301)

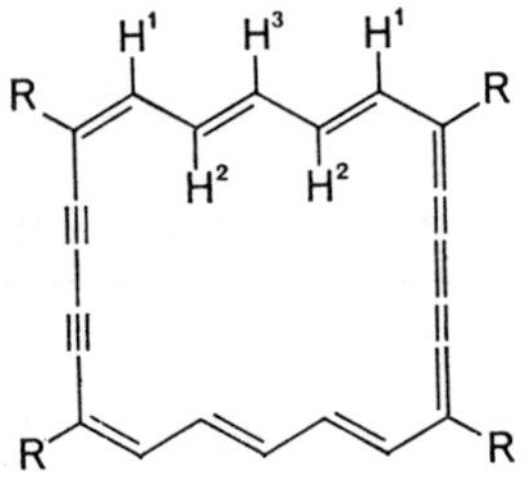

(302), (303)

^{1}*H-N.m.r. spectra* (τ values)

	H¹	H³	H²	*t*-Bu
(**302**) R = *t*-Bu	0·09d (*J* = 13)	−0·4t (*J* = 13)	13·7t (*J* = 13)	7·88s
(**303** R = Ph	−0·72d (*J* = 14)	−0·79t (*J* = 14)	12·80t (*J* = 14)	

found in **302** as compared with that of monodehydro[22]-annulene[212], in spite of the presence of four equivalents of acetylenic bonds, strongly indicates again the important role of the degenerate valence-bond structures for π-electron delocalization. The electronic spectrum of **302** was found to be closely related to that of **289**.

3. Trisdehydro[4*n*]annulenes

'Acetylene–cumulene' dehydro[4*n*]annulenes offer a tool for the study of antiaromaticity in view of the high conformational stability and the strong diatropicity of the [4*n*+2] counterparts.

The reaction of the lithio derivative of the diethyl acetal of the polyenynal **304** with the polyenyne ketone **305** followed by acid hydrolysis gave the hydroxyaldehyde **306**. The ethynyl ketone **307** obtained by the aldol condensation of **306** with pinacolone was ethynylated[218] to give **308**. Oxidative coupling of **308** yielded the cyclic glycols (**309**, $n = 1$, 2 or 3), which in turn yielded the tetra-*t*-butyltris-dehydro[16]- (**310**)[228], -[20]- (**311**)[229], and -[24]- (**312**)[230] annulenes. The high conformational stabilities of **310, 311** and **312** have been revealed by their essentially

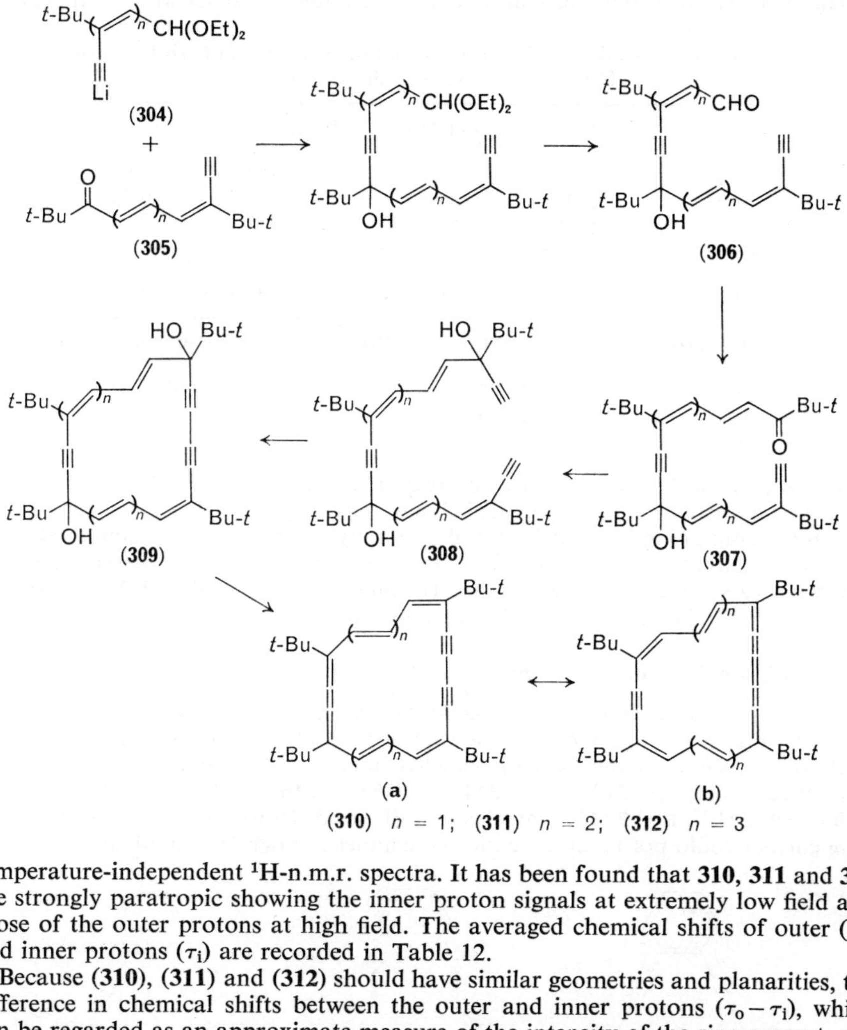

(**310**) $n = 1$; (**311**) $n = 2$; (**312**) $n = 3$

temperature-independent ^{1}H-n.m.r. spectra. It has been found that **310, 311** and **312** are strongly paratropic showing the inner proton signals at extremely low field and those of the outer protons at high field. The averaged chemical shifts of outer (τ_o) and inner protons (τ_i) are recorded in Table 12.

Because (**310**), (**311**) and (**312**) should have similar geometries and planarities, the difference in chemical shifts between the outer and inner protons ($\tau_o - \tau_i$), which can be regarded as an approximate measure of the intensity of the ring current, can be reasonably compared. As shown in Table 11, a marked decrease in ($\tau_o - \tau_i$) values with increase in ring size was observed. It has been predicted theoretically that polyolefinic character of both [$4n+2$]- and [$4n$]annulenes should increase with increase in ring size[211, 217]. The results shown in Table 11 offer experimental verification of the prediction for [$4n$]annulenes.

The ^{13}C chemical shifts of *sp*-hybridized carbon atoms in the paratropic tris-dehydro[16]- (**310**), -[20]- (**311**), and diatropic tetra-*t*-butylbisdehydro[14]- (**265a**) and -[18]- (**282**) annulenes are summarized in Table 12[229]. The bisdehydro[$4n+2$]-annulenes give signals ascribable to *sp*-hybridized carbons at an intermediate field between the region of acetylenic (65–90 p.p.m.)[231, 232] and cumulenic carbons (*ca.* 150 p.p.m.)[232], indicating the presence of a highly delocalized [$4n+2$] electron system, whereas the trisdehydro[$4n$]annulenes (**310** and **311**) exhibit two groups of signals, which are attributable to acetylenic and cumulenic carbon atoms. Appearance

TABLE 11. Chemical shifts of outer and inner protons of trisdehydro-[$4n$]annulenes ($CDCl_3$, 36 °C, τ values)

	[$4n$]	Outer H (τ_o)	Inner H (τ_i)	($\tau_o-\tau_i$)
310	[16]	5·83	−7·10	12·93
311	[20]	5·61	−3·78	9·39
312	[24]	5·30	−1·80	7·10

TABLE 12. ^{13}C chemical shifts of *sp*-hybridized carbons in $CDCl_3$

Trisdehydro[$4n$]annulene, −20 °C			Bisdehydro[$4n+2$]annulene, 36 °C	
[16]-	86·6, 90·5	153·3	[14]-	116·7
[20]-	85·2, 86·5	148·3	[18]-	115·7

of two signals in the acetylenic carbon region seems to be ascribable to the outside and internal carbons in the diacetylenic bond, thus suggesting that the alternate bond structures containing a butatriene and a diacetylene are predominant in the tris-dehydro[$4n$]annulenes (**310**, **311** and presumably **312**). The observed coupling constants ($J = 11$ and 15 Hz) in the ^1H-n.m.r. spectra of **310**, **311** and **312** are consistent with the alternate bond structures.

C. Annelated Dehydroannulenes

The properties of annulenes annelated with a 6π ring system are of considerable interest with respect to the participation of benzenoid π electrons in the macrocyclic π-electron system. In the field of dehydroannulenes, a wide variety of annelated derivatives, such as **313** [233, 234], **314** [234], **315** [235], **316** [235], **317** [235], **318** [236], **319** [236], **320** [237, 238], **321** [239] and **322** [240] have been synthesized. However, the effect of induced ring current could not be observed in these annelated dehydroannulenes.

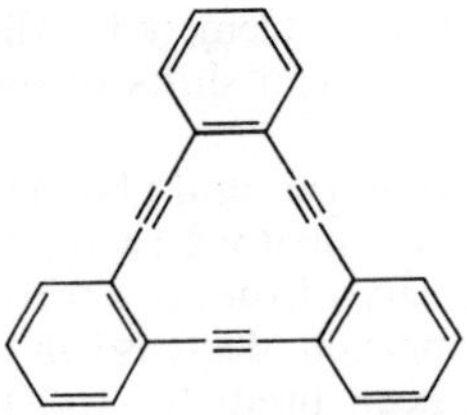

(313)

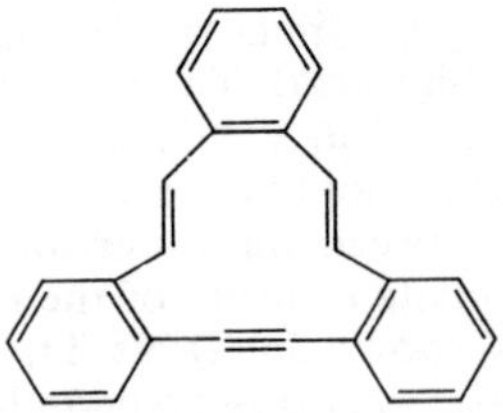

(314)

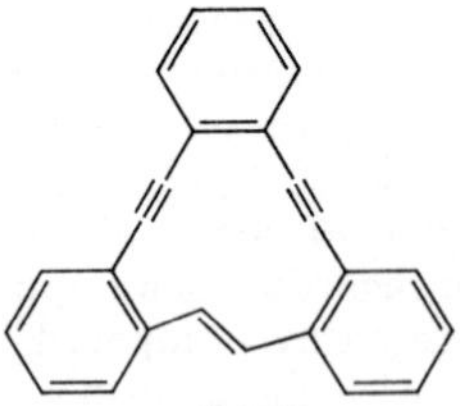

(315)

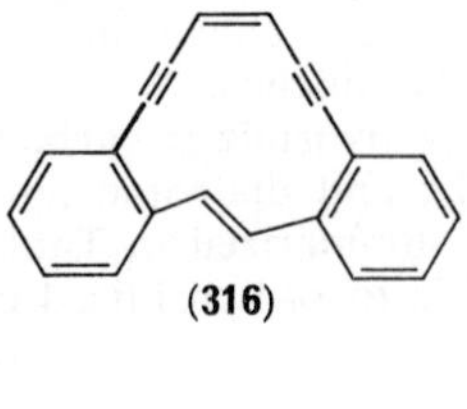

(316)

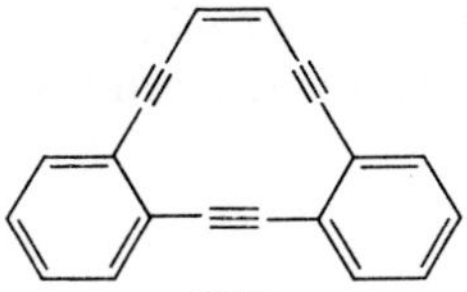

(317)

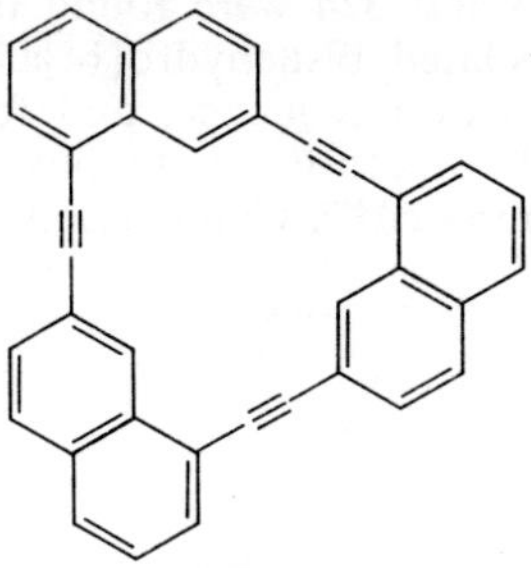

(318)

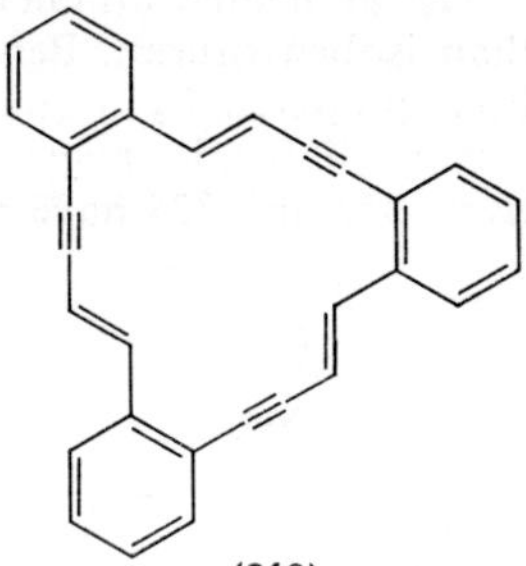

(319)

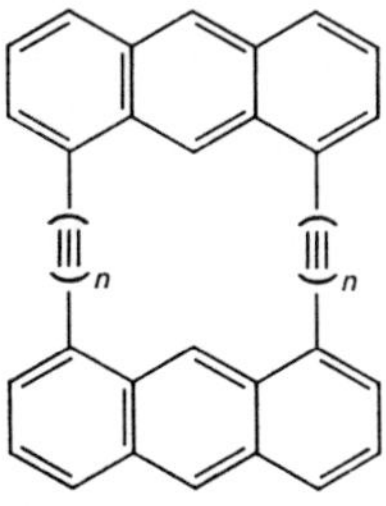

(320) n = 1, 2

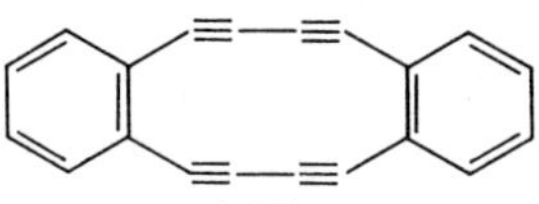

(321)

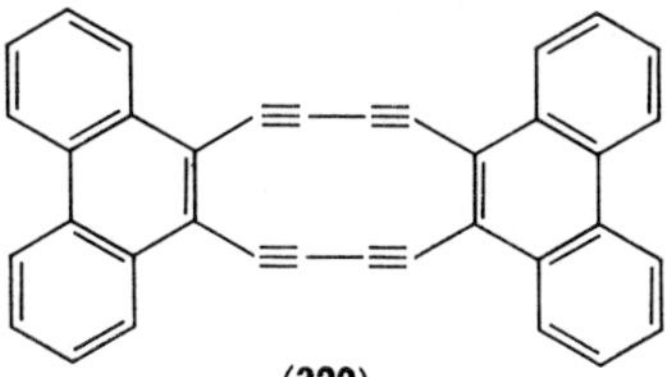

(322)

Bisdehydro[14]annuleno[*c*]furans (**323** and **324**), 14π-electron analogues of isobenzofuran, have been synthesized according to the following scheme[241]:

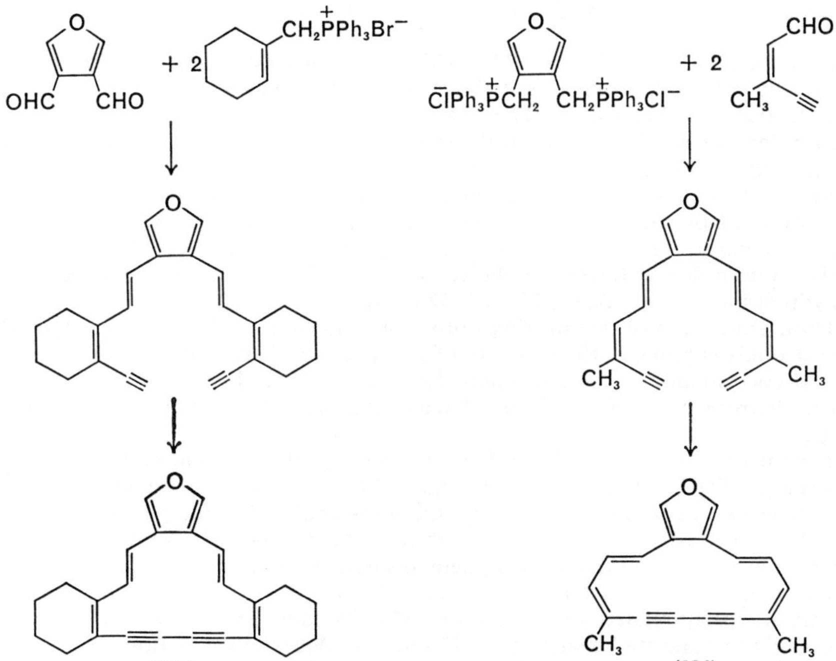

(323) (324)

The annelated annulenes **323** and **324** were found to be more weakly diatropic than isobenzofuran. Benzo-annelated bisdehydro[14]annulenes, **325** and **326** have been prepared by a similar reaction sequence, and found to be moderately and weakly diatropic, respectively[242]. Recently, analogous thiophene and furan derivatives, **327** and **328** have been prepared[243]. Comparison of the ^{1}H-n.m.r. spectra with

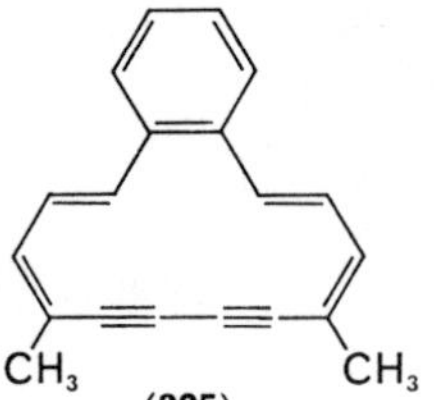

(325)

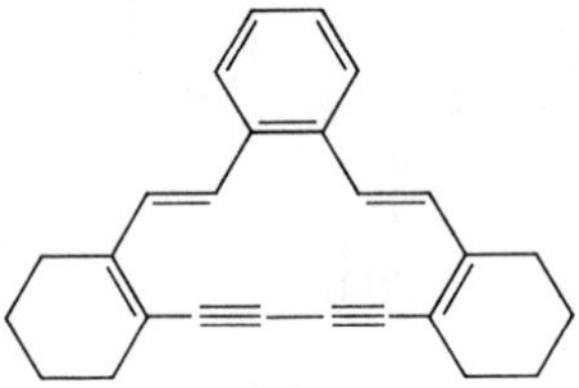

(326)

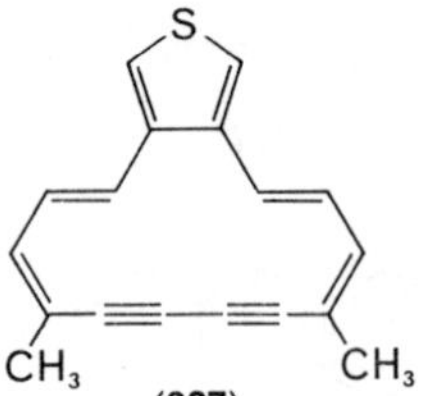

(327)

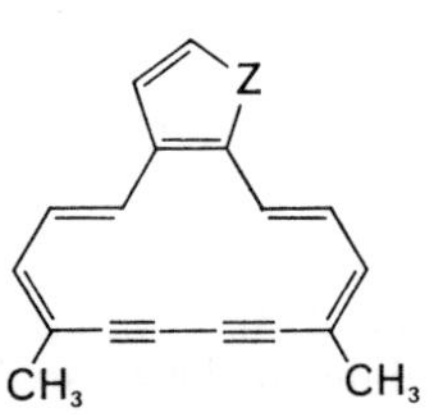

(328) Z = S
(329) Z = O

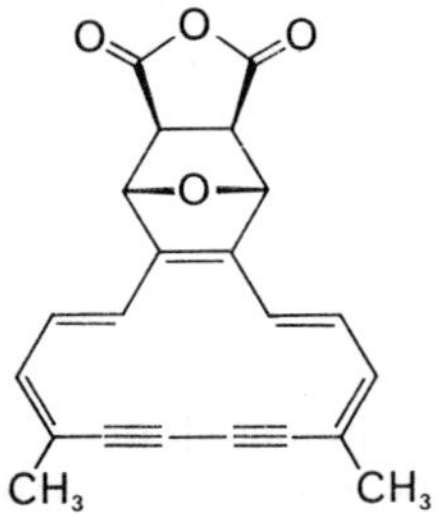

(330)

that of the reference substance **330**, which was obtained by the reaction of maleic anhydride with **323**, revealed that the decrease of diamagnetic ring current is in the order **330 > 329 > 328 > 325 > 327 ≃ 324**.

In order to get information on the effect of annelation of a 6π ring onto macrocyclic 4*n*π systems, annelated monodehydro[12]annulenes, **333–336** have been synthesized using **327** or **328** as a starting material[244]. The reference compound **337** was prepared by the reaction of **334** with maleic anhydride. It was found that the degree of paratropicity decreased in the sequence **337 > 336 > 335 > 333 ≃ 331**. This is the same order as found for the reduction of diatropicity in the annelated bisdehydro[14]annulenes, **324**, **325**, **327**, **328** and **329**.

Thus, annelation of one 6π ring onto a dehydro[4*n*]- or dehydro[4*n* + 2]annulene ring strongly suppresses the tropicity of the macrocyclic ring as compared with that of the corresponding non-annelated dehydroannulene. However, an interestingly strong diatropicity was found in annelated dehydroannulenes fused with two aromatic rings.

Treatment of the cyclic glycol **339**, obtained by dimerization of **338**, with tin(II) chloride in THF saturated with hydrogen chloride gave an extremely air-sensitive deep blue-violet solution[244]. Similarly, the non-annelated bis(dihydronaphtho) (**341**) and the annelated dihydronaphtho-naphtho (**342**) analogues have been obtained as stable compounds like di-*t*-butyldiphenylbisdehydro[14]annulene (**265b**)[245].

The fact that the extremely unstable blue-violet solution gives an electronic spectrum closely related with those of **265a**, **341** and **342** indicates the formation of dinaphtho-annelated bisdehydro[14]annulene (**340**) in the solution. The ^{1}H-n.m.r. spectrum of **340** can be obtained using a solution prepared with THF-d_8 and

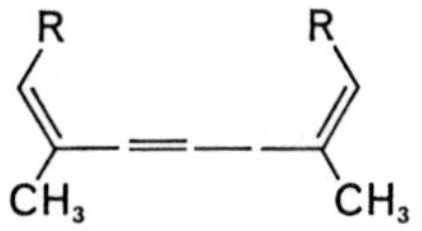

(**331**) R = CHO
(**332**) R = $CH_2P^+Ph_3Br^-$

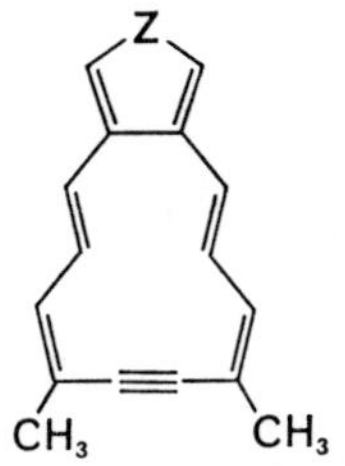

(**333**) R = S
(**334**) R = O

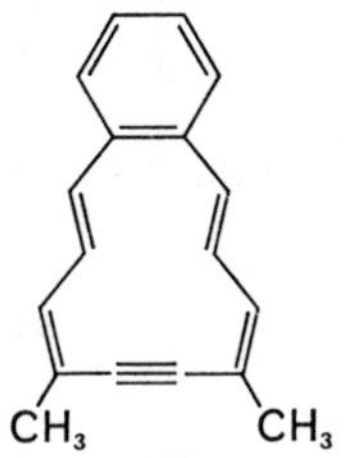

(**335**)

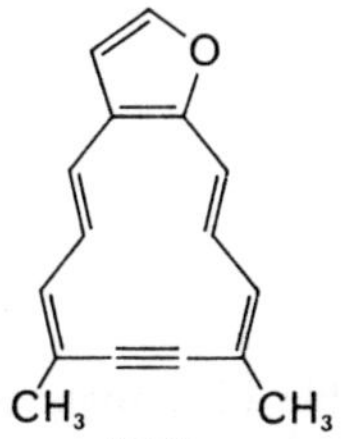

(**336**)

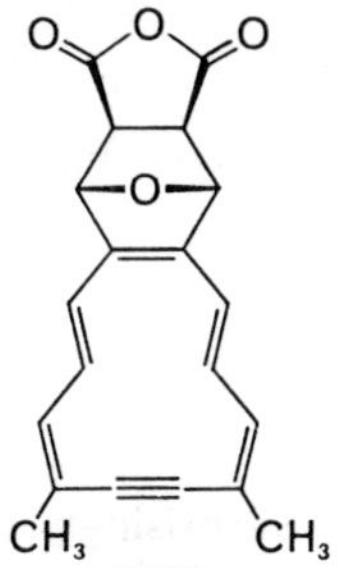

(**337**)

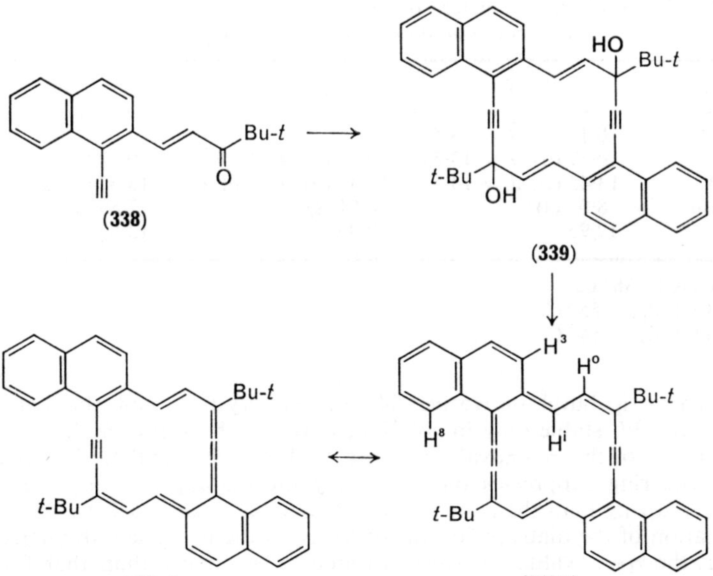

(**338**) (**339**)

(**340a**) (**340b**)

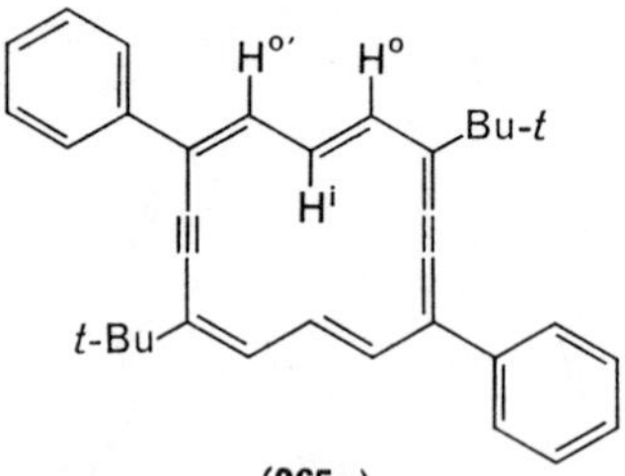

(265a)

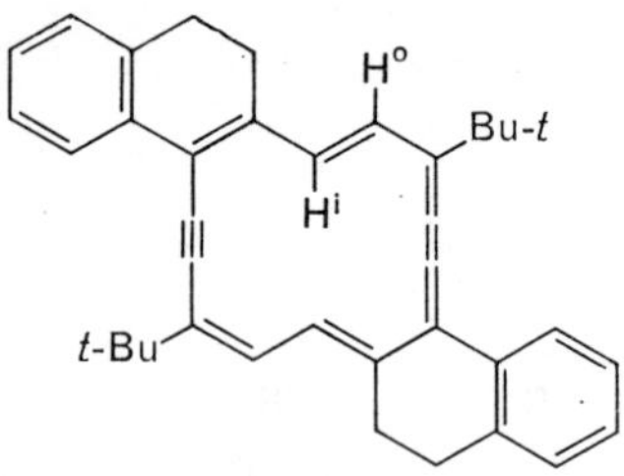

(341)

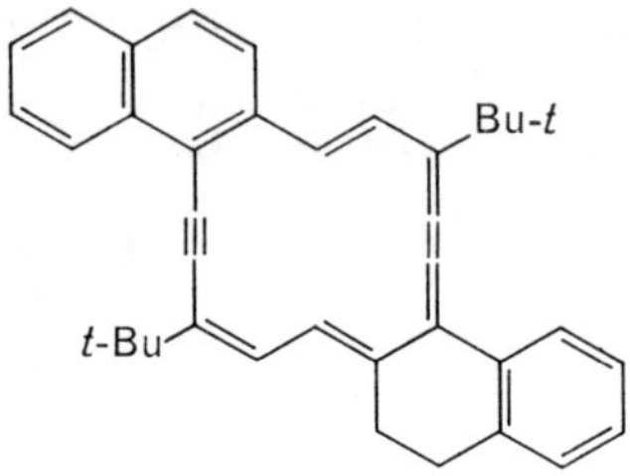

$\tau_i - \tau_o = 11{\cdot}02$

(342)

deuterium chloride. Surprisingly, the unstable bis-annelated dehydroannulene **340** shows a strong diatropicity just as do the stable non-annelated analogues, **265a** and **339** (Table 13). On the other hand, an appreciable suppression of the diatropicity was observed in the mononaphtho derivative **342** as indicated by the n.m.r data.

TABLE 13. ^{1}H-n.m.r. parameters of dinaphthobisdehydro[14]annulene (**340**) and reference compounds **265b** and **341** (τ value)

τ	(**265b**)[a]	(**341**)[b]	(**340**)[c]
H$^{o'}$	0·12 (d), J = 13·5		
H^{o}	0·47 (d), J = 13·5	0·48 (d), J = 14·0	−0·22 (d), J = 15·0
H^{i}	13·42 (t), J = 13·5	13·47 (d), J = 14·0	13·45 (d), J = 15·0
t-Bu	8·02 (s)	8·01 (s)	7·89 (s)
$\tau_i - \tau_o$	12·95	12·99	13·67

[a] $CDCl_3$, 36 °C.
[b] THF-d_8, −55 °C.
[c] THF-d_8, −54 °C.

The benzo-tri-t-butyl derivative **343**[246], red crystals, and the benzonaphtho derivative **344**[247], stable only in solution, have also been prepared.

Comparison of the $\tau_i - \tau_o$-values for **342** and **343** reveals that the annelation of one benzene ring suppresses more strongly the diamagnetic ring current in the 14-membered ring than the annelation of one naphthalene nucleus. An appreciable intensification of the diatropicity was observed in the benzonaphtho derivative **344**. However, the $\tau_i - \tau_o$-value for **344** was found to be smaller than that for the fully symmetrical dinaphtho derivative **340**.

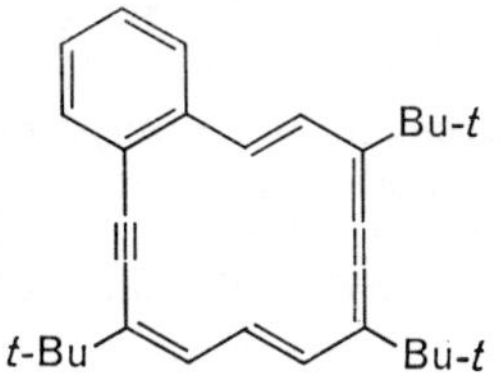

$\tau_i - \tau_o$ = 8·42, −50 °C

(343)

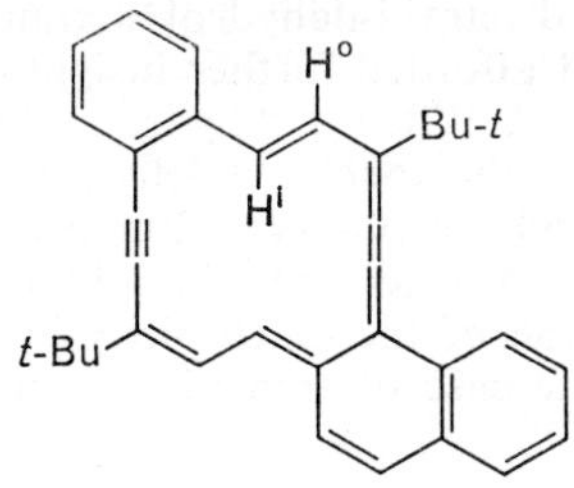

$\tau_i - \tau_o$ = 11·90, −51 °C

(344)

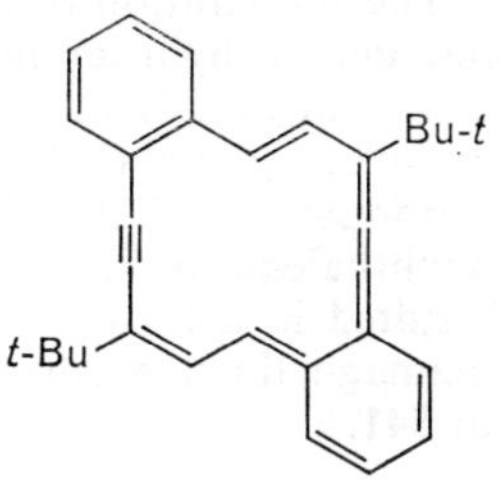

(345)

The electronic spectrum indicates the formation of dibenzo-di-*t*-butylbisdehydro-[14]annulene (**345**), but the ^{1}H-n.m.r. spectrum could not be obtained owing to the extreme instability, although a strong diatropicity was anticipated[247].

The synthesis and properties of the annelated bisdehydro[18]annulenes, **346** [248] and **347** [249], together with the non-annelated bisdehydroannulene **348** [249] have been

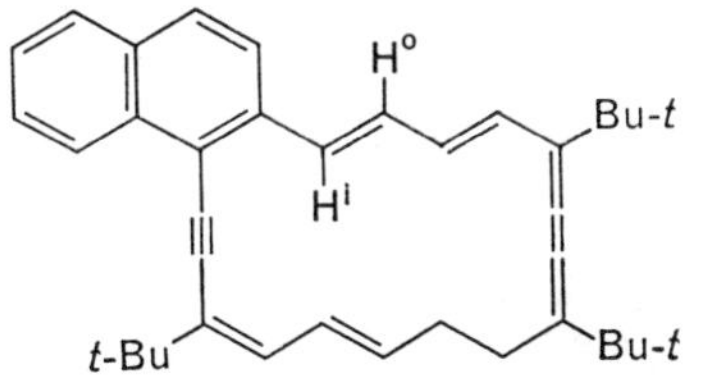

$\tau_i - \tau_o$ = 9·13
t-Bu τ = 8·11

(346)

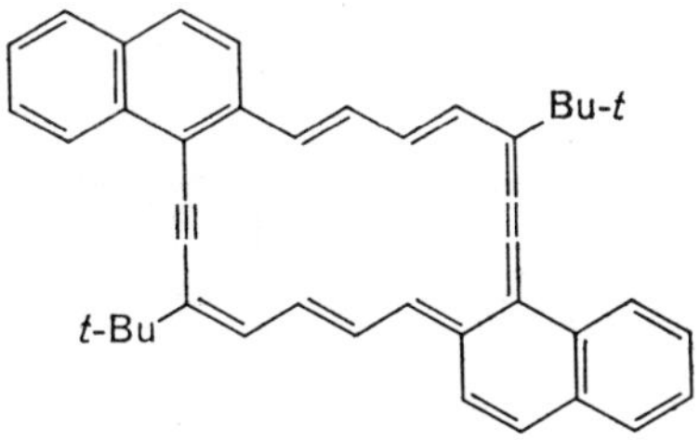

t-Bu τ = 7·89

(347)

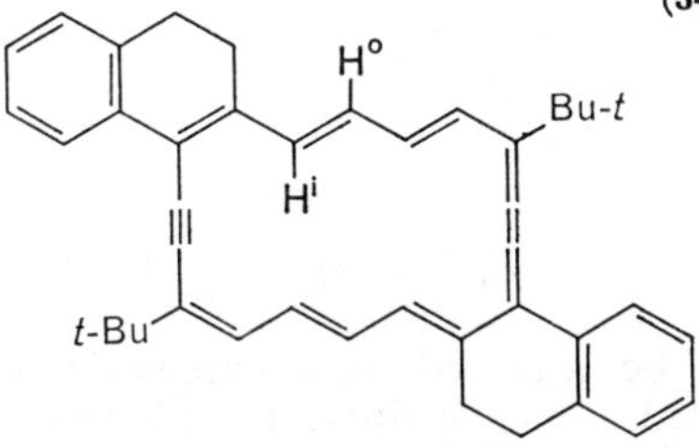

$\tau_i - \tau_o$ = 13·01
t-Bu τ = 7·98

(348)

reported. The mononaphtho derivative **346**, as indicated at the bottom of the formula, was found to be clearly diatropic, although the $\tau_i - \tau_o$-value is smaller than that of the 14π-analogue (**342**). Although the formation of the dinaphtho derivative **347** was confirmed by the electronic spectrum, a clear ^{1}H-n.m.r. spectrum could not be obtained owing to the instability. However, the fact that a singlet ascribable to the *t*-butyl protons could be observed at a fairly low field (τ 7·89, in THF-d_8, at −80 °C) seems to suggest the induction of a strong diamagnetic ring current in **347**.

The preparation of annelated tetrakisdehydro[18]annulenes **349** and **350** and the reference dehydroannulene **351** afforded further insight into the π-electron delocalization in annelated annulenes[250]. When a dihydronaphthalene nucleus in **351** was replaced by naphthalene to form the annelated dehydroannulene **350**, an appreciable suppression of diatropicity in **350** was observed. Further replacement of the dihydronaphthalene by naphthalene to give the dinaphtho-annelated derivative **349** resulted in a further suppression of the diatropicity in contrast to the increase of diamagnetic ring current in the case of transition from **342** to **340** and from **343** to **341**.

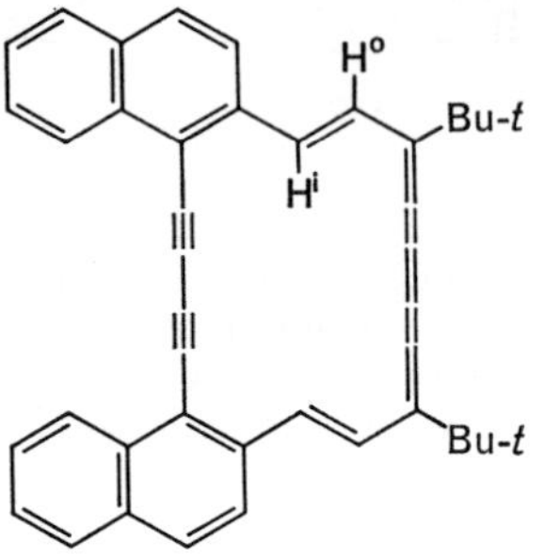

$\tau_i - \tau_o = 7{\cdot}08$

(349)

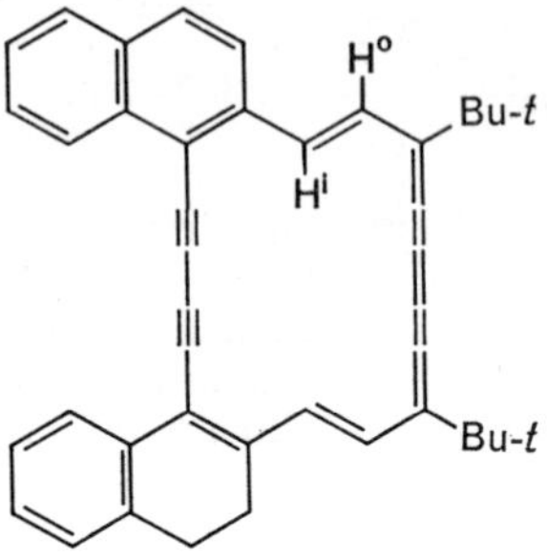

$\tau_i - \tau_o = 10{\cdot}92$

(350)

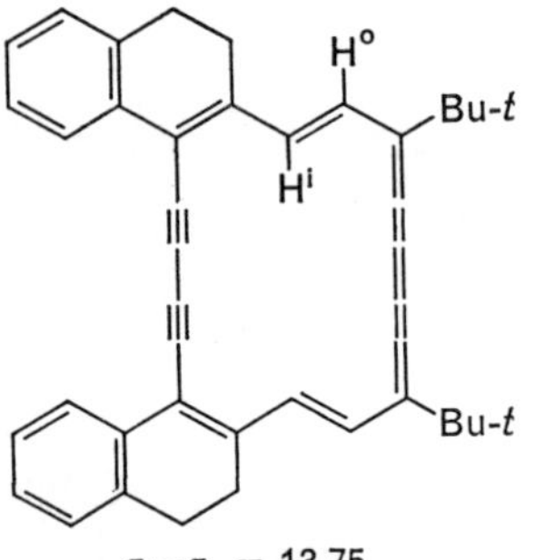

$\tau_i - \tau_o = 13{\cdot}75$

(351)

The $\tau_i - \tau_o$-value can be regarded as an approximate measure of π-electron delocalization in annelated annulene rings. The above-mentioned results obtained on conformationally stable annelated 'acetylene–cumulene' dehydroannulenes seem to give the following conclusions:

(*i*) Annelation of one or more 6π electron rings onto a [4*n*+2]annulene ring decreases progressively the diatropicity of the annulene ring provided that the condition in (*ii*) is not fulfilled.

(*ii*) Annelation with two 6π electron rings at positions which make it possible to write equivalent Kekulé structures does not suppress the diatropicity of the annulene ring.

(*iii*) The degree of suppression of diamagnetic ring current in the annelated annulene ring is proportional to the resonance energy of the 6π electron system which is lost by the participation of two π electrons in the macrocyclic ring (benzenoid to *o*-quinonoid).

(*iv*) The degree of reduction of diamagnetic ring current in the annulene ring is inversely proportional to the magnitude of resonance energy or that of diatropicity of the parent dehydroannulene ring.

Significant suppression of tropicity of bisdehydroannulenones by annelation with one or two benzene nuclei has been reported[251–255].

D. Dehydroannulenoannulenes

In benzenoid chemistry condensed systems such as naphthalene and anthracene have been well known. However, condensed systems of aromatic annulenes corresponding to naphthalene remained unknown until quite recently.

An annulenoannulene (**359**) consisting of two bisdehydro[14]annulenes has been synthesized by Sondheimer and Cresp[256]. Bisdehydro[14]annuleno[*c*]furan (**324**) was treated with lead tetraacetate, and the resulting diacetate **352** was hydrolysed to give the diol (**353**), a potential dialdehyde. The reaction of carbethoxymethylenetriphenylphosphorane with **353** yielded the diester **354**. The diol (**355**), obtained by the reduction of **354**, was oxidized to give the dialdehyde (**356**). The reaction of the Grignard derivative of 3-bromo-1-butyne with **356** gave a diastereoisomeric mixture of the diol **357**. Oxidative coupling of **357** yielded the bicyclic glycol **358** as a mixture of diastereomers. Dehydration of the crude **358** *via* the dimesylate gave the tetrakisdehydro[14]annuleno[14]annulene (**359**) as dark red-brown prisms. The ^{1}H-n.m.r.

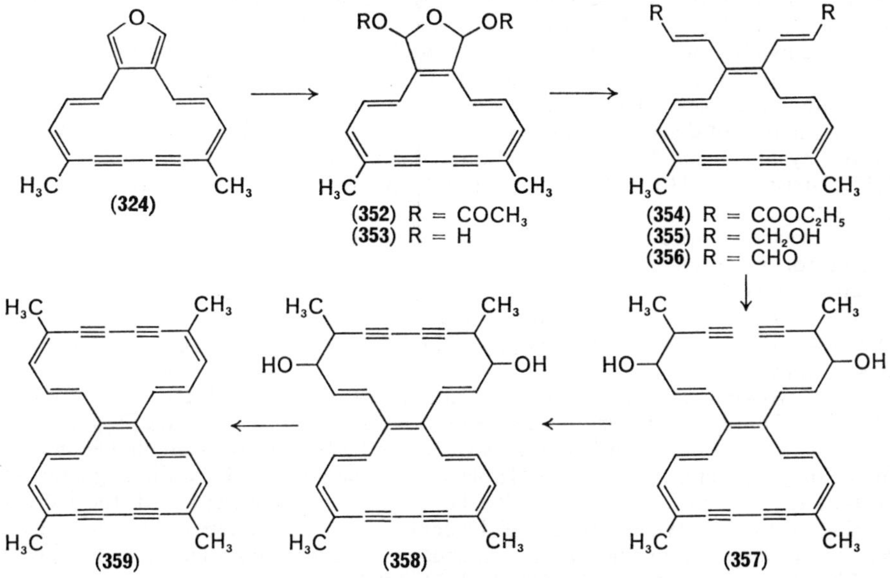

spectrum of the annulenoannulene **359** exhibits signals of H^a, H^b, H^c and methyl protons at τ 6·18, 2·13, 2·69 and 7·52, respectively, indicating that the annulenoannulene is diatropic. The diatropicity of **359** was found to be stronger than those of the annelated analogues **324** and **335** and to be less than those of the bisdehydro-[14]annulene derivatives **354**, **355** and **356**.

At the same time, the synthesis of a condensed system consisting of two 'acetylene–cumulene' tetrakisdehydro[18]annulenes has been reported by Nakagawa and his coworkers[257].

The methyl ketone **360** obtained by the condensation of acetone with 3-*t*-butyl-2-penten-4-ynal (**262a**) was oxidatively coupled to give the diketone **361**. The α,ω-diethynyl compound **362** obtained by the condensation of **361** with **262a** was oxidized by Eglinton's procedure. The 26-membered cyclic diketone **363**, thus obtained, was ethynylated[218] to give a mixture of **364a** and **364b**. Coupling of the high-melting isomer (**364a**) yielded the bicyclic glycol **365** which with tin(II) chloride and ether saturated with hydrogen chloride gave 5,10,18,23-tetra-*t*-butyl-6,8,19,-21,27,29-hexakisdehydro[12.12.4][18]annuleno[18]annulene (**366**) as stable dark-green crystals. The electronic spectrum of **366** consists of three main absorption bands clearly showing features characteristic of [4*n*+2]annulenes. The ^{1}H-n.m.r. spectrum of **366** exhibits inner proton (H^b) signals at τ 12·85 (dd) (J = 13–14), outer proton (H^a and H^c) signals at τ −0·64 (d) (J = 14) and τ −0·06 (J = 13) and the *t*-butyl proton signal at τ 7·81 (s), respectively, thus indicating a strong diatropicity of the annulenoannulene **366**.

An analogous annulenoannulene (**377**) consisting of two bisdehydro[14]annulenes has been synthesized by Akiyama, Iyoda and Nakagawa[258].

The ethynyl ketone **368** was prepared by the aldol condensation of **262a** with **367**. Treatment of **368** with diethyllithium amide followed by the reaction with trimethylsilyl chloride gave **369**. The ethynyl alcohol (**370**) obtained on treatment of **369** with lithium acetylide in THF [259] reaction with diluted sulphuric acid gave the aldehyde **371**. The dimethyl acetal (**372**) derived from **371** was converted into the lithio derivative **373**. The reaction of **373** with **369** followed by an acid treatment yielded the dialdehyde **374**. The reaction of the carbanion derived from diethyl 3,3-dimethyl-2-oxo-butanephosphonate, *t*-BuCOCH$_2$(O)P(OEt)$_2$, with the dialdehyde **374** gave the diketone **375** in a high yield. Treatment of the diketone **375** with potassium hydroxide in liquid ammonia without removal of the protective groups gave a mixture of diastereoisomers of the 22-membered cyclic glycol **376**. Reduction of the cyclic glycol **376** gave the annulenoannulene **377**, 5,8,16,19-tetra-*t*-butyl-6,17,23-trisdehydro[10.10.2][14]annuleno[14]annulene, as stable reddish-purple crystals.

The electronic spectrum of **377** was found to be closely related to that of **265a** except for a bathochromic shift and a hyperchromism in (**376**). The ^{1}H-n.m.r. spectrum reveals that **377** is strongly diatropic showing the inner proton (H^b) signals at τ 12·85 (dd) (J = 13, 14) and those of the outer protons at τ −0·16 (d) (H^a, J = 14) and at τ 0·39 (d) (H^c, J = 13). The signal of the *t*-butyl protons was observed at τ 7·99 (s).

Thus, two kinds of strongly diatropic annulenoannulenes (**366** and **377**) have been obtained. Consequently, whether **366** and **377** are perturbed [26]- and [22]annulenes or annulenoannulenes consisting of two 18π and 14π electron systems becomes an interesting problem. As shown in Table 10, the magnitude of the diamagnetic ring current as estimated approximately by the $\tau_i - \tau_o$-values decreases with the increase of ring size of 'acetylene–cumulene' bisdehydro[4*n*+2]annulenes. The same trend has been observed in the ^{1}H-n.m.r. spectra of tetra-*t*-butyltetrakisdehydro[18]-(**297f**) and tetra-*t*-butyltetrakisdehydro[22]- (**302**) annulene. The observed $\tau_i - \tau_o$-values for **365** ($\Delta\tau$ = 12·91 and 13·49) and for **376** ($\Delta\tau$ = 12·46 and 13·01) seem to be too large, if **366** and **377** are perturbed peripheral [26]π and [22]π electron systems being resonance hybrids of (**365b**↔**365c**) and (**376b**↔**376c**). The increase in $\tau_i - \tau_o$-value of **377** cannot be ascribed to an enhanced planarity caused by bridging between the 1- and 12-positions with an acetylenic linkage, because tetra-*t*-butylbisdehydro-[22]annulene (**289**) shows essentially temperature-independent ^{1}H-n.m.r. spectra, and a highly planar structure of **289** has been shown by the X-ray structure analysis (Figure 9). The electronic spectrum of **377** shows a considerable hypsochromic shift

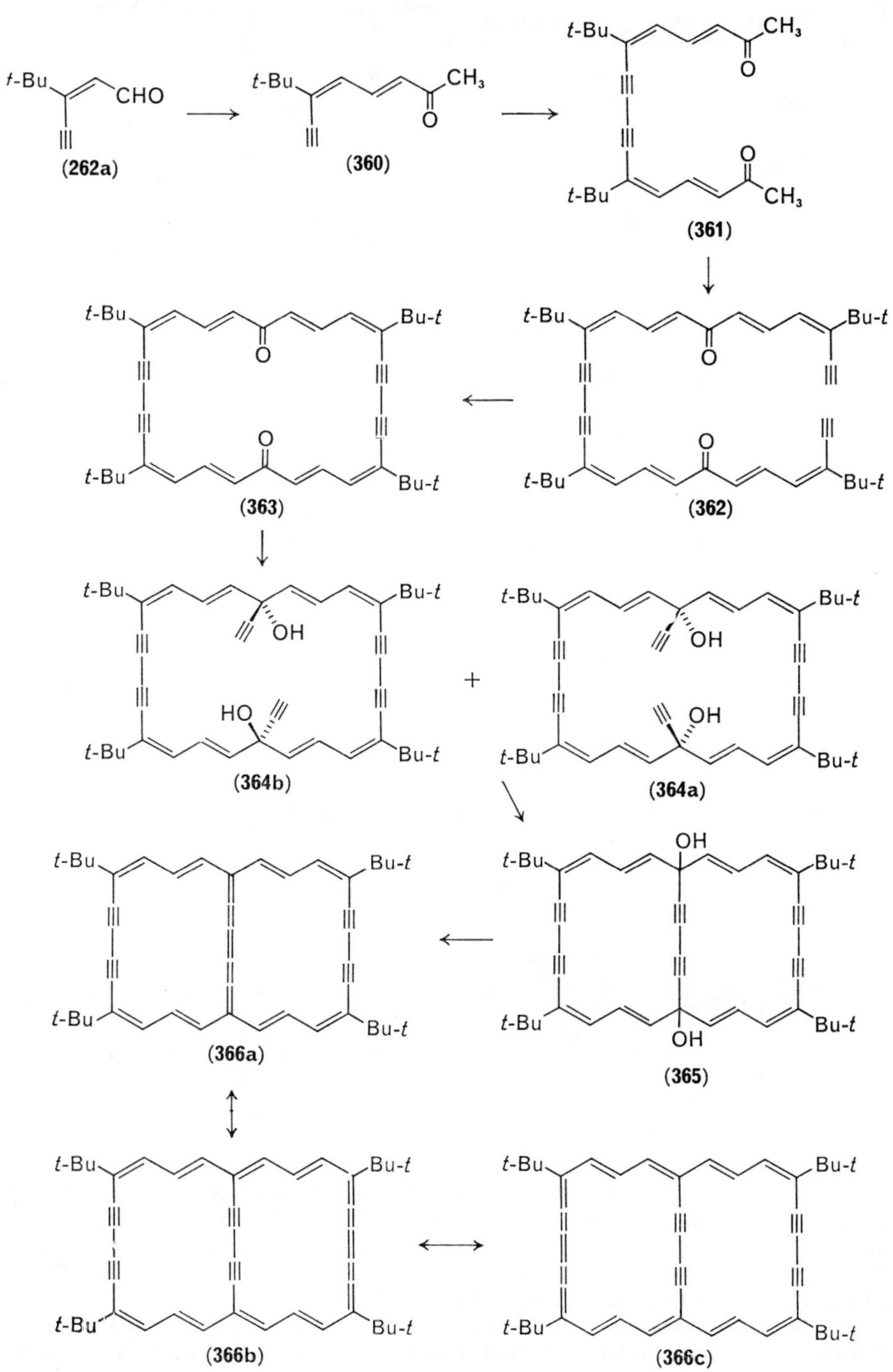
t-Bu
CHO
(262a)
CH3
(360)
(361)
(362)
(363)
(364b)
(364a)
(365)
(366a)
(366b)
(366c)
Bu-t
OH
HO

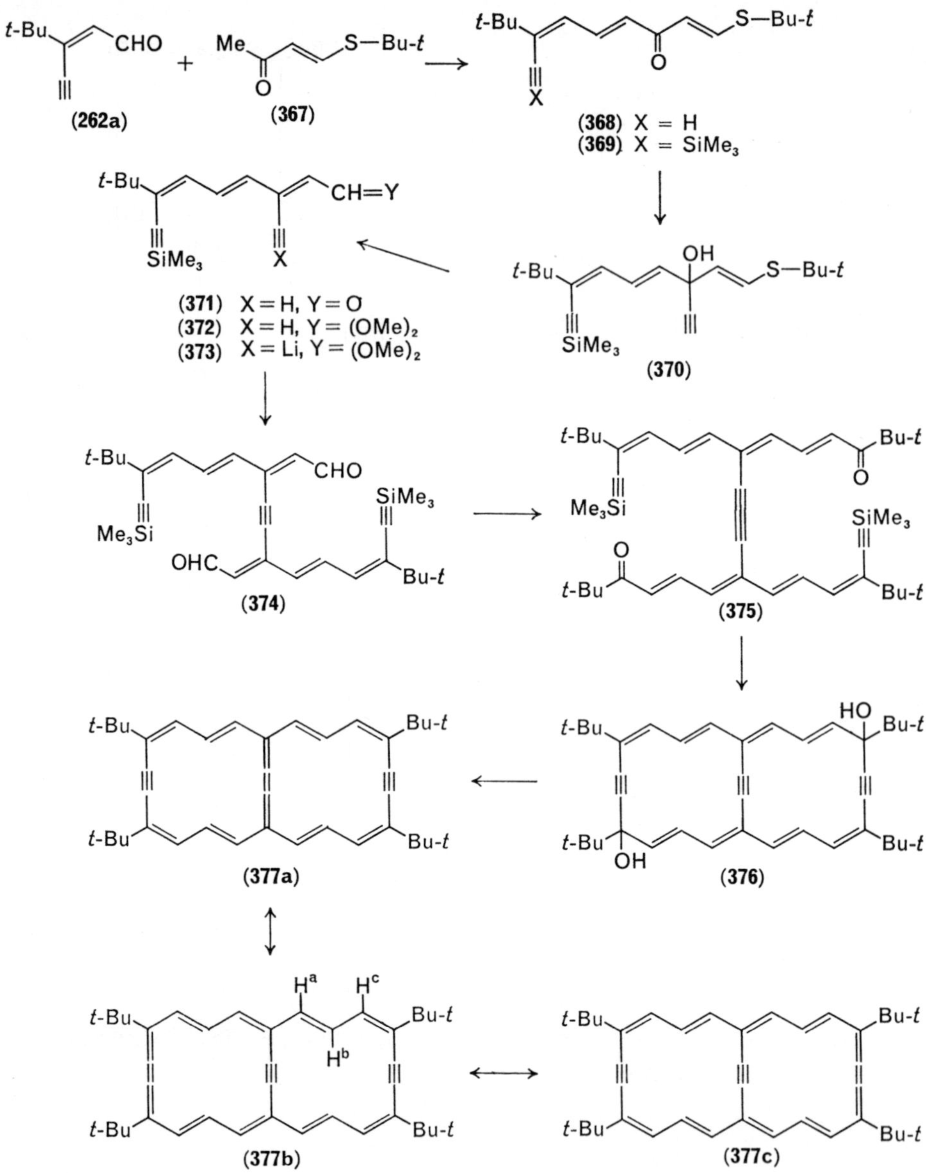

as compared with that of **289**, although the periphery of **377** is the same 22π electron system. The ^{13}C-n.m.r. spectrum of **377** showed signals due to *sp*-hybridized carbon atoms at 102·3 and 119·8 p.p.m. (TMS as an internal standard). The chemical shifts of these signals indicate that all the *sp*-hybridized carbon atoms in **377** have a hybrid character of an acetylene and a cumulene.

The electronic and n.m.r. spectral behaviour of **366** and **377** strongly suggests that these annulenoannulenes are higher analogues of naphthalene being resonance hybrids of valence-bond structures (**366a**↔**366b**↔**366c**) and (**377a**↔**377b**↔**377c**), which may be better represented by the symmetrical formulae, **366d** and **377d**, respectively.

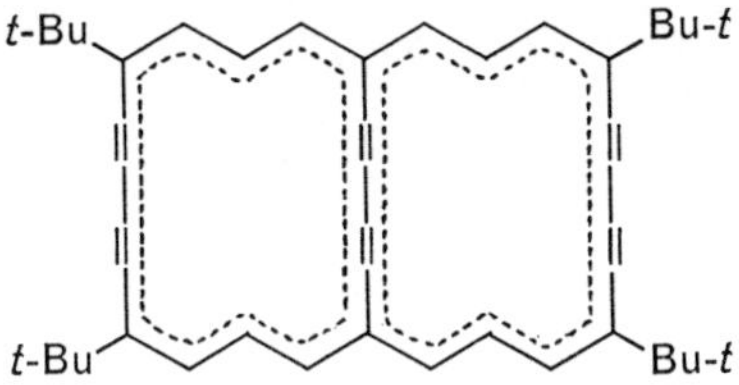

(**366d**)

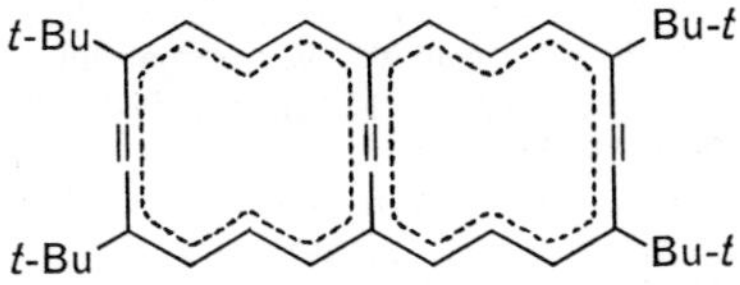

(**377d**)

VI. REFERENCES

1. L. Ruzicka, M. Hübbin and H. A. Boekenoogen, *Helv. Chim. Acta*, **16**, 498 (1933).
2. W. Ziegenbein and W. M. Schneider, *Chem. Ber.*, **98**, 824 (1965).
3. M. Stoll, J. Hulstkamp and A. Rouvé, *Helv. Chim. Acta*, **31**, 543 (1948).

4a. N. A. Domnin, *J. Gen. Chem. USSR*, **8**, 851 (1938).
4b. N. A. Domnin and T. D. Malysheva, *J. Gen. Chem. USSR*, **25**, 325 (1955).
4c. A. T. Blomquist and L. H. Liu, *J. Amer. Chem. Soc.*, **75**, 2153 (1953).

5. G. Wittig and U. Mayer, *Chem. Ber.*, **98**, 471 (1965).
6. A. Krebs in *Chemistry of Acetylenes* (Ed. H. G. Viehe), Marcel Dekker, New York, 1969, Chap. XV, p. 1009.
7. G. Wittig and R. Pohlke, *Chem. Ber.*, **94**, 3276 (1961).
8. T. Curtius, *Ber.*, **22**, 2161 (1889).
9. T. Curtius and K. Theen, *J. prak. Chem.*, **44**, 161 (1891).
10. A. T. Blomquist, R. E. Burge, Jr., L. H. Liu, J. C. Bohrer, A. C. Sucsy and J. Kleis, *J. Amer. Chem. Soc.*, **73**, 5510 (1951).
11. A. T. Blomquist, L. H. Liu and J. C. Bohrer, *J. Amer. Chem. Soc.*, **74**, 3643 (1952).
12. V. Prelog, K. Schenker and W. Künig, *Helv. Chim. Acta*, **36**, 471 (1953).
13. A. T. Blomquist, R. E. Burge, Jr. and A. C. Sucsy, *J. Amer. Chem. Soc.*, **74**, 3636 (1952).
14. V. Prelog, K. Schenker and H. H. Günthard, *Helv. Chim. Acta*, **35**, 1598 (1952).
15. V. Prelog and V. Boarland, *Helv. Chim. Acta*, **38**, 1776 (1955).
16. V. Prelog and M. Speck, *Helv. Chim. Acta*, **38**, 1786 (1955).
17. V. Prelog and S. Polyak, *Helv. Chim. Acta*, **40**, 816 (1957).
18. G. Wittig and A. Krebs, *Chem. Ber.*, **94**, 3260 (1961).
19. A. Krebs in *Chemistry of Acetylenes* (Ed. H. G. Viehe), Marcel Dekker, New York, 1969, Chap. XV, p. 1010–1011.
20. K. L. Erickson and J. Wolinsky, *J. Amer. Chem. Soc.*, **87**, 1142 (1965).
21. G. Wittig and J. Meske-Schüller, *Ann.*, **711**, 65 (1968).
22. F. G. Willey, *Angew. Chem.*, **76**, 144 (1964).
23. G. Wittig and E. R. Wilson, *Chem. Ber.*, **98**, 451 (1965).
24. K. Rasheed, *Tetrahedron*, **22**, 2957 (1966).
25. G. Wittig and G. Harborth, *Ber.*, **77**, 306 (1944).
26. F. Scardiglia and J. D. Roberts, *Tetrahedron*, **1**, 343 (1957).
27. L. K. Montgomery, F. Scardiglia and J. D. Roberts, *J. Amer. Chem. Soc.*, **87**, 1142 (1965).
28. L. K. Montgomery and J. D. Roberts, *J. Amer. Chem. Soc.*, **82**, 4750 (1960).
29. G. Wittig and P. Fritze, *Ann.*, **711**, 82 (1968).
30. G. Wittig and J. Meske-Schüller, *Ann.*, **711**, 76 (1968).
31. L. K. Montgomery and L. E. Applegate, *J. Amer. Chem. Soc.*, **89**, 2952 (1967).
32. L. K. Montgomery, A. O. Clouse, A. M. Crelier and L. E. Applegate, *J. Amer. Chem. Soc.*, **89**, 3453 (1967).

33. G. Wittig, J. Weinlich and E. R. Wilson, *Chem. Ber.*, **98,** 458 (1965).
34. A. Krebs in *Chemistry of Acetylenes* (Ed. H. G. Viehe), Marcel Dekker, New York, 1969, Chap. XV, pp. 1011, 1015.
35. W. Tochtermann, K. Oppenländer and U. Walter, *Chem. Ber.*, **97,** 1318 (1964).
36. W. Tochtermann, K. Oppenländer and M. N. D. Hoang, *Ann.*, **701,** 117 (1967).
37. L. K. Montgomery and L. E. Applegate, *J. Amer. Chem. Soc.*, **89,** 5305 (1967).
38. R. Willstätter and W. V. Schmaedel, *Ber.*, **38,** 1922 (1905).
39. T. Nozoe and Y. Kitahara, *Proc. Japan Acad.*, **30,** 204 (1954); *Chem. Abstr.*, **49,** 13200 (1955).
40. T. Toda, *Bull. Chem. Soc. Japan*, **40,** 588 (1967).
41. T. Yamatani, M. Yasunami and K. Takase, *Tetrahedron Letters*, 1725 (1970).
42. J. Wolinsky, *J. Org. Chem.*, **26,** 704 (1961).
43. W. A. Böll, *Angew. Chem.*, **78,** 755 (1966).
44. P. G. Gassman and J. J. Valcho, *J. Amer. Chem. Soc.*, **97,** 4768 (1975).
45. W. Tochtermann, K. Oppenländer and U. Walter, *Chem. Ber.*, **97,** 1329 (1964).
46. W. Tochtermann, K. Oppenländer and M. N. D. Hoang, *Ann.*, **701,** 117 (1967).
47. A. Krebs and D. Byrd, *Ann.*, **707,** 66 (1967).
48. T. Nakazawa and I. Murata, *Angew. Chem.*, **87,** 742 (1975).
49a. G. Schröder and J. F. M. Oth., *Angew. Chem.*, **79,** 458 (1967).
49b. G. Schröder, H. Röttele, R. Merényi and J. F. M. Oth, *Chem. Ber.*, **100,** 3527 (1967).
49c. J. F. M. Oth, R. Merényi, H. Röttele and G. Schröder, *Chem. Ber.*, **100,** 3538 (1967).
50. G. Schröder, *Angew. Chem.*, **77,** 774 (1965).
51. A. Krebs, *Angew. Chem.*, **77,** 966 (1965).
52. W. R. Moore and H. R. Ward, *J. Amer. Chem. Soc.*, **85,** 86 (1963).
53. W. J. Ball and S. R. Landor, *J. Chem. Soc.*, 2298 (1962).
54. G. Wittig and P. Fritze, *Angew. Chem.*, **78,** 905 (1966).
55. G. Wittig and U. Mayer, *Chem. Ber.*, **96,** 342 (1963).
56. G. Wittig and J. Weinlich, *Chem. Ber.*, **98,** 471 (1965).
57. L. Watts, J. D. Fritzpatrick and R. Pettit, *J. Amer. Chem. Soc.*, **87,** 3253 (1965).
58. L. Watts, J. D. Fritzpatrick and R. Pettit, *J. Amer. Chem. Soc.*, **88,** 623 (1966).
59. A. Favorskii and W. Boshowsky, *Ann.*, **390,** 122 (1912).
60. A. Favorskii, N. A. Domnin and M. F. Chestakowski, *Bull. Soc. Chim. Fr.*, **3,** 1727 (1936).
61. G. Wittig, J. Weinlich and E. R. Wilson, *Chem. Ber.*, **98,** 458 (1965).
62. A. Favorskii, *J. Gen. Chem. USSR*, **6,** 720 (1936); *Chem. Abstr.*, **30,** 6337 (1936).
63. N. A. Domnin, *J. Gen. Chem. USSR*, **9,** 1983 (1939); *Chem. Abstr.*, **34,** 4383 (1940).
64. R. Breslow, L. J. Altman, A. Krebs, E. Mohasci, I. Murata, R. A. Peterson and J. Posner, *J. Amer. Chem. Soc.*, **87,** 1326 (1965).
65. R. J. Böttcher, H. Röttele, G. Schröder and J. F. M. Oth, *Tetrahedron Letters*, 3935 (1968).
66. P. Goebel, J. Sicher, M. Svoboda and R. B. Turner, *Proc. Chem. Soc.*, 237 (1964).
67. J. Sicher, M. Svoboda, J. Závada, R. B. Turner and P. Goebel, *Tetrahedron*, **22,** 659 (1966).
68. A. Krebs in *Chemistry of Acetylenes* (Ed. H. G. Viehe), Marcel Dekker, New York, 1969, Chap. XV, p. 1031.
69. S. Kaarsemaker and J. Coops, *Rec. Trav. Chim.*, **71,** 261 (1952).
70. J. W. Knowlton and F. D. Rossini, *J. Res. Natl. Bur. Standards*, **43,** 113 (1949).
71. R. Spitzer and H. M. Huffman, *J. Amer. Chem. Soc.*, **69,** 211 (1947).
72. J. Coops, H. van Kamp, W. A. Lambregts, B. J. Visser and H. Dekker, *Rec. Trav. Chim.*, **79,** 1226 (1960).
73. For a review, see J. Dale, *Angew. Chem.*, **78,** 1070 (1966).
74. A. T. Blomquist and L. H. Liu, *J. Amer. Chem. Soc.*, **75,** 2153 (1953).
75. A. Krebs, *Tetrahedron Letters*, 4511 (1968).
76. G. Wittig and H. L. Dorsch, *Ann.*, **711,** 46 (1968).
77. W. Tochtermann, G. Schnabel and A. Mannschreck, *Ann.*, **705,** 169 (1967).
78. A. Krebs and H. Kimling, *Tetrahedron Letters*, 761 (1970).
79. A. Krebs and H. Kimling, *Angew. Chem.*, **83,** 401 (1971).
80. H. Kimling and A. Krebs, *Angew. Chem.*, **84,** 952 (1972).

81. A. Krebs and H. Kimling, *Angew. Chem.*, **83**, 540 (1971).
82. N. J. Leonard and P. M. Mader, *J. Amer. Chem. Soc.*, **72**, 5388 (1950).
83. J. Wolinsky and K. Erickson, *Chem. Ind. (London)*., 1953 (1964).
84. V. Franzen and H. J. Joshek, *Ann.*, **703**, 90 (1967).
85. G. Wittig, H. L. Dorsch and J. Meske-Schüller, *Ann.*, **711**, 55 (1968).
86. G. Wittig and P. Fritze, *Ann.*, **712, 79** (1968).
87. R. Criegee and G. Schröder, *Ann.*, **623**, 1 (1959).
88. E. Kloster-Jensen and J. Wirz, *Angew. Chem.*, **85**, 723 (1973).
89. H. Schmidt, A. Schweig and A. Krebs, *Tetrahedron Letters*, 1471 (1974).
90. G. Seiz, L. Pohl and R. Pohlke, *Angew. Chem.*, **81**, 427 (1969).
91. A. R. Battersby and R. Binks, *J. Chem. Soc.*, 2880 (1955).
92. H. Meier and H. Gugel, *Synthesis*, 338 (1976).
93. H. N. C. Wong, P. J. Garratt and F. Sondheimer, *J. Amer. Chem. Soc.*, **96**, 5604 (1974).
94. R. Destro, T. Pilati and M. Simonetta, *J. Amer. Chem. Soc.*, **97**, 658 (1975).
95. M. P. Cava, R. Pohlke, B. W. Erickson, J. C. Rose and G. Fraenkel, *Tetrahedron*, **18**, 1005 (1962); M. Avram, I. G. Dinulescu, D. Dinu, G. Mateescu and C. D. Nenitzescu, *Tetrahedron*, **19**, 309 (1963).
96. H. N. C. Wong and F. Sondheimer, *Angew. Chem.*, **88**, 126 (1976).
97. J. W. Barton, T. A. Chaundri, P. Gaskin and K. E. Whitaker, *J. Chem. Soc., Perkin Trans. I*, 717 (1972).
98. T. J. Katz, M. Yoshida and L. C. Siew, *J. Amer. Chem. Soc.*, **87**, 4516 (1965).
99. M. Svoboda, J. Závada and J. Sicher, *Collect. Czech. Chem. Commun.*, **30**, 413 (1965).
100. J. Sicher, M. Svoboda and J. Závada, *Collect. Czech. Chem. Commun.*, **30**, 421 (1965).
101. R. Lespieau, *Compt. Rend.*, **188**, 502 (1929).
102. F. Sondheimer, Y. Gaoni and J. Bregman, *Tetrahedron Letters*, 25 (1960).
103. F. Sondheimer and Y. Gaoni, *J. Amer. Chem. Soc.*, **84**, 3520 (1962).
104. G. Eglinton and A. R. Galbraith, *Chem. Ind. (London)*., 737 (1956); *J. Chem. Soc.*, 889 (1959).
105. R. Wolovsky and F. Sondheimer, *J. Amer. Chem. Soc.*, **88**, 1525 (1966).
106. C. A. Grob and P. W. Schiess, *Helv. Chim. Acta*, **43**, 1546 (1960).
107. K. Grohmann and F. Sondheimer, *Tetrahedron Letters*, 3121 (1967).
108. M. Tanabe, D. F. Crave and R. L. Dehn, *Tetrahedron Letters*, 3943 (1967).
109. A. Eschenmoser, D. Felix and G. Ohloff, *Helv. Chim. Acta*, **50**, 708 (1967).
110. D. J. Cram and N. L. Allinger, *J. Amer. Chem. Soc.*, 78, 2518 (1956).
111. J. H. Wotiz, R. F. Adams and C. G. Parsons, *J. Amer. Chem. Soc.*, **83**, 373 (1961).
112. J. Dale, A. J. Hubert and G. S. D. King, *J. Chem. Soc.*, 73 (1963).
113. C. Glaser, *Ber.*, **2**, 422 (1869); *Ann.*, **154**, 137 (1870).
114. F. Sondheimer, *Pure Appl. Chem.*, **7**, 363 (1963).
115. F. Sondheimer and Y. Amiel, *J. Amer. Chem. Soc.*, **78**, 4178 (1956).
116. F. Sondheimer and Y. Amiel, *J. Amer. Chem. Soc.*, **79**, 5817 (1957).
117. Y. Amiel, F. Sondheimer and R. Wolovsky, *Proc. Chem. Soc.*, 22 (1957).
118. F. Sondheimer, Y. Amiel and R. Wolovsky, *J. Amer. Chem. Soc.*, **79**, 6263 (1957).
119. F. Sondheimer, Y. Amiel and R. Wolovsky, *J. Amer. Chem. Soc.*, **79**, 4247 (1957).
120. F. Sondheimer, Y. Amiel and R. Wolovsky, *J. Amer. Chem. Soc.*, **81**, 4600 (1959).
121. R. Wolovsky and F. Sondheimer, *J. Amer. Chem. Soc.*, **84**, 2844 (1962).
122. R. Wolovsky and F. Sondheimer, *J. Amer. Chem. Soc.*, **87**, 5720 (1965).
123. M. Kataoka, T. Ando and M. Nakagawa, *Bull. Chem. Soc. Japan*, **44**, 177 (1971).
124. T. Matsuoka, T. Negi, T. Otsubo, Y. Sakata and S. Misumi, *Bull. Chem. Soc. Japan*, **45**, 1825 (1972); T. Matsuoka, T. Negi and S. Misumi, *Synthetic Commun.*, **2**, 87 (1972).
125. F. Toda and M. Nakagawa, *Bull. Chem. Soc. Japan*, **34**, 862 (1961).
126. M. Kataoka, T. Ando, and M. Nakagawa, *Bull. Chem. Soc. Japan*, **44**, 1909 (1971).
127. T. Ando and M. Nakagawa, *Bull. Chem. Soc. Japan*, **44**, 172 (1971).
128. M. Kataoka, T. Ando and M. Nakagawa, *Bull. Chem. Soc. Japan*, **44**, 177 (1971).
129. F. Toda, T. Ando, M. Kataoka and M. Nakagawa, *Bull. Chem. Soc. Japan*, **44**, 1914 (1971).
130. F. Sondheimer and R. Wolovsky, *J. Amer. Chem. Soc.*, **84**, 260 (1962).
131. E. R. H. Jones, B. L. Shaw and M. C. Whiting, *J. Chem. Soc.*, 3212 (1954); B. L. Shaw and M. C. Whiting, *J. Chem. Soc.*, 3217 (1954).

132. F. Sondheimer, D. A. Ben-Efraim and Y. Gaoni, *J. Amer. Chem. Soc.*, **83**, 1682 (1961).
133. R. Wolovsky, *J. Amer. Chem. Soc.*, **87**, 3638 (1965).
134. F. Sondheimer and R. Wolovsky, *Tetrahedron Letters*, 3 (1959).
135. F. Sondheimer, R. Wolovsky and Y. Amiel, *J. Amer. Chem. Soc.*, **84**, 274 (1962).
136. F. Sondheimer and Y. Gaoni, *J. Amer. Chem. Soc.*, **84**, 3520 (1962).
137. E. Hückel, *Z. Physik*, **70**, 204 (1931); *Grundzüge der Theorie ungesättigter und aromatischer Verbindungen*, Verlag Chemie, Berlin, 1938.
138. L. M. Jackman, F. Sondheimer, Y. Amiel, D. A. Ben-Efraim, Y. Gaoni, R. Wolovsky and A. A. Bothner-By, *J. Amer. Chem. Soc.*, **84**, 4302 (1962).
139. F. Sondheimer, *Proc. Roy. Soc.*, **297A**, 173 (1967).
140. Y. Gaoni, A. Melera, F. Sondheimer and R. Wolovsky, *Proc. Chem. Soc.*, 397 (1964).
141. F. Sondheimer and Y. Gaoni, *J. Amer. Chem. Soc.*, **82**, 5765 (1960).
142. F. Sondheimer, I. C. Calder, J. A. Elix, Y. Gaoni, P. J. Garratt, K. Grohmann, G. di Maio, J. Mayer, M. V. Sargent and R. Wolovsky, *Aromaticity, Special Publication No. 21*, Chemical Society, London, 1967, p. 75.
143. J. Mayer and F. Sondheimer, *J. Amer. Chem. Soc.*, **88**, 602 (1966).
144. Y. Gaoni and F. Sondheimer, *J. Amer. Chem. Soc.*, **86**, 521 (1964).
145. I. C. Calder, P. J. Garratt, H. C. Longuet-Higgins, F. Sondheimer and R. Wolovsky, *J. Chem. Soc. (C)*, 1041 (1967).
146. F. Baer, H. Kuhn and W. Regel, *Z. Naturforsch.*, **22a**, 103 (1967).
147. J. A. Pople and K. G. Untch, *J. Amer. Chem. Soc.*, **88**, 4811 (1966).
148. H. C. Longuet-Higgins, Reference 143, p. 109.
149. F. Sondheimer, R. Wolovsky, P. J. Garratt, and I. C. Calder, *J. Amer. Chem. Soc.*, **88**, 2610 (1966).
150. K. G. Untch and D. C. Wysocki, *J. Amer. Chem. Soc.*, **89**, 6386 (1967).
151. K. G. Untch and D. C. Wysocki, *J. Amer. Chem. Soc.*, **88**, 2608 (1966).
152. W. H. Okamura and F. Sondheimer, *J. Amer. Chem. Soc.*, **89**, 5991 (1967).
153. J. I. Musher, *J. Chem. Phys.*, **43**, 4081 (1965); **46**, 1219 (1967); *Advan. Magnetic Resonance*, **2**, 177 (1966).
154. G. M. Pilling and F. Sondheimer, *J. Amer. Chem. Soc.*, **90**, 5610 (1968); **93**, 1977 (1971).
155. F. Sondheimer, *Pure Appl. Chem.*, **28**, 331 (1971).
156. C. W. Brown and F. Sondheimer, *J. Amer. Chem. Soc.*, **91**, 760 (1969).
157. J. Griffiths and F. Sondheimer, *J. Amer. Chem. Soc.*, **91**, 7518 (1969).
158. P. D. Howes, E. LeGoff and F. Sondheimer, *Tetrahedron Letters*, 3691 (1972).
159. G. P. Cottrell, G. H. Mitchell, F. Sondheimer and G. M. Pilling, *J. Amer. Chem. Soc.*, 93, 259 (1971).
160. P. D. Howes, E. LeGoff and F. Sondheimer, *Tetrahedron Letters*, 3695 (1972).
161. D. J. Berteli and T. G. Andrews, *J. Amer. Chem. Soc.*, **91**, 5280 (1969); D. J. Berteli, T. G. Andrews and P. O. Crews, *J. Amer. Chem. Soc.*, **91**, 5286 (1969).
162. K. Yamamoto and F. Sondheimer, *Angew. Chem.*, **85**, 41 (1973).
163. N. Darby, K. Yamamoto and F. Sondheimer, *J. Amer. Chem. Soc.*, **96**, 248 (1974).
164. R. Breslow, D. Murayama, R. Drury and F. Sondheimer, *J. Amer. Chem. Soc.*, **96**, 249 (1974).
165. G. Schröder, G. Flank and J. F. M. Oth, *Angew. Chem.*, **85**, 353 (1973).
166. G. Schröder, G. Plinke and J. F. M. Oth, *Angew. Chem.*, **84**, 472 (1972).
167. G. Schröder, G. Flank and J. F. M. Oth, *Angew. Chem.*, **84**, 474 (1972).
168. F. Sondheimer, *Chimia*, **28**, 163 (1974).
169. P. J. Beeby and F. Sondheimer, *J. Amer. Chem. Soc.*, **94**, 2128 (1972).
170. P. J. Beeby and F. Sondheimer, *Angew. Chem.*, **84**, 896 (1972).
171. P. J. Beeby, J. M. Brown, P. J. Garratt and F. Sondheimer, *Tetrahedron Letters*, 599 (1974).
172. *Cf.* V. Boekelheide and T. A. Hylton, *J. Amer. Chem. Soc.*, **92**, 3669 (1970).
173. P. J. Beeby and F. Sondheimer, *Angew. Chem.*, **85**, 404 (1973).
174. P. J. Beeby and F. Sondheimer, *Angew. Chem.*, **85**, 406 (1973).
175. R. H. McGirk and F. Sondheimer, *Angew. Chem.*, **84**, 897 (1972).
176. R. L. Wife and F. Sondheimer, *Tetrahedron Letters*, 195 (1975).
177. R. L. Wife and F. Sondheimer, *J. Amer. Chem. Soc.*, **97**, 640 (1975).
178. R. L. Wife, P. J. Beeby and F. Sondheimer, *J. Amer. Chem. Soc.*, **97**, 641 (1975).

179. J. M. Brown and F. Sondheimer, *Angew. Chem.*, **86,** 346 (1974).
180. J. M. Brown and F. Sondheimer, *Angew. Chem.*, **86,** 348 (1974).
181. E. LeGoff and F. Sondheimer, *Angew. Chem.*, **84,** 996 (1972).
182. P. J. Garratt, N. E. Rowland and F. Sondheimer, *Tetrahedron*, **27,** 2157 (1971).
183. J. F. M. Oth and J.-M. Gilles, *Tetrahedron Letters*, 6259 (1968).
184. J. F. M. Oth, G. Anthoine and J.-M. Gilles, *Tetrahedron Letters*, 6265 (1968).
185. R. M. McQuilkin, P. J. Garratt and F. Sondheimer, *J. Amer.* Chem. Soc., **92,** 6682 (1970).
186. T. J. Katz, *J. Amer. Chem. Soc.*, **82,** 3784, 3785 (1960).
187. R. Breslow, J. T. Groves and G. Ryan, *J. Amer. Chem. Soc.*, **89,** 5048 (1967); D. G. Farnum, G. Mehta and R. G. Silberman, *J. Amer. Chem. Soc.*, **89,** 5048 (1967).
188. G. Merling, *Ber.*, **24,** 3108 (1891); W. von Doering and L. H. Knox, *J. Amer. Chem. Soc.*, **76,** 3203 (1954); T. Schaefer and W. G. Schneider, *Can. J. Chem.*, **41,** 966 (1963).
189. P. D. Howes and F. Sondheimer, *J. Amer. Chem. Soc.*, **94,** 8261 (1972).
190. F. Sondheimer, Y. Gaoni, L. M. Jackman, N. A. Bailey and R. Mason, *J. Amer. Chem. Soc.*, **84,** 4595 (1962).
191. N. A. Bailey and R. Mason, *Proc. Chem. Soc.*, **A290,** 94 (1966).
192. For a review, see M. Nakagawa, *Pure Appl. Chem.*, **44,** 885 (1975).
193. *Cf.* I. Heilbron, E. R. H. Jones and M. Julia, *J. Chem. Soc.*, 1430 (1949).
194. K. Fukui, T. Nomoto, S. Nakatsuji and M. Nakagawa, *Tetrahedron Letters*, 3157 (1972).
195. T. Nomoto, S. Nakatsuji and M. Nakagawa, *Chemistry Letters*, 839 (1974).
196. J. Tanaka, to be published.
197. T. Satake, S. Nakatsuji, M. Iyoda, S. Akiyama and M. Nakagawa, *Tetrahedron Letters*, 1881 (1976).
198. T. Katakami, S. Tomita, K. Fukui and M. Nakagawa, *Chemistry Letters*, 225 (1972); S. Tomita and M. Nakagawa, *Bull. Chem. Soc. Japan*, **49,** 302 (1976).
199. G. H. Mansfield and M. C. Whiting, *J. Chem. Soc.*, 4761 (1956).
200. T. Satake, S. Nakatsuji, S. Akiyama and M. Nakagawa, *Tetrahedron Letters*, 2153 (1976).
201. M. Iyoda and M. Nakagawa, *Tetrahedron Letters*, 3161 (1972).
202. M. Müller-Cunradi and K. Pieroth, *U.S. Patent*, 2165962 (1936); *Chem. Abstr.*, **33,** 8210 (1939); R. Rüegg, M. Montavon, G. Ryser, G. Saucy, U. Schwieter and O. Isler, *Helv. Chim. Acta*, **42,** 854 (1959) and the references cited therein.
203. M. Iyoda, H. Miyazaki and M. Nakagawa, *J. Chem. Soc., Chem. Commun.*, 431 (1972); *Bull. Chem. Soc. Japan*, **49,** 2306 (1976).
204. C. Kabuto, Y. Kitahara, M. Iyoda and M. Nakagawa, *Tetrahedron Letters*, 2791 (1976).
205. S. Akiyama, T. Nomoto, M. Iyoda and M. Nakagawa, *Bull. Chem. Soc. Japan*, **49,** 2579 (1976).
206. M. Iyoda and M. Nakagawa, *J. Chem. Soc., Chem. Commun.*, 1003 (1972).
207. M. Iyoda and M. Nakagawa, *Tetrahedron Letters*, 4253 (1972).
208. M. Iyoda and M. Nakagawa, *Tetrahedron Letters*, 4743 (1973).
209. R. M. McQuilkin, B. W. Metcalf and F. Sondheimer, *J. Chem. Soc., Chem. Commun.*, 338 (1971).
210. C. Kabuto, Y. Kitahara, M. Iyoda and M. Nakagawa, *Tetrahedron Letters*, 2787 (1976).
211a. M. J. S. Dewar and G. J. Gleicher, *J. Amer. Chem. Soc.*, **87,** 685 (1965).
211b. H. C. Longuet-Higgins and L. Salem, *Proc. Roy. Soc.*, **A251,** 172 (1959).
211c. H. C. Longuet-Higgins and L. Salem, *Proc. Roy. Soc.*, **A257,** 445 (1960).
211d. C. A. Coulson and W. T. Dixon, *Tetrahedron*, **17,** 215 (1962).
212. R. M. McQuilkin and F. Sondheimer, *J. Amer. Chem. Soc.*, **92,** 6341 (1970).
213. C. C. Loznoff and F. Sondheimer, *J. Amer. Chem. Soc.*, **89,** 4247 (1967).
214. B. W. Metcalf and F. Sondheimer, *J. Amer. Chem. Soc.*, **93,** 5271 (1971).
215. J. A. Pople, *J. Chem. Phys.*, **24,** 1111 (1956).
216. T. Nakajima and S. Kohda, *Bull. Chem. Soc. Japan*, **39.** 804 (1966).
217. H. P. Feyges in *Topics in Carbocyclic Chemistry* (Ed. D. Lloyd), Logos Press, London, 1969, p. 269.
218. D. F. Beumel, Jr. and R. F. Harris, *J. Org. Chem.*, **29,** 1872 (1964).

219. J. Ojima, T. Katakami, G. Nakaminami and M. Nakagawa, *Tetrahedron Letters*, 1115 (1968); *Bull. Chem. Soc. Japan*, **49**, 292 (1976).
220. T. Katakami, S. Tomita, K. Fukui and M. Nakagawa, *Chemistry Letters*, 225 (1972).
221. K. Fukui, T. Okamoto and M. Nakagawa, *Tetrahedron Letters*, 3121 (1971).
222. T. Katakami, K. Fukui, T. Okamoto and M. Nakagawa, *Bull. Chem. Soc. Japan*, **49**, 297 (1976).
223. S. Akiyama, M. Fukuoka and M. Nakagawa, Unpublished result.
224a. T. Nomoto, K. Fukui and M. Nakagawa, *Tetrahedron Letters*, 3253 (1972).
224b. T. Nomoto, K. Fukui and M. Nakagawa, *Bull. Chem. Soc. Japan*, **49**, 305 (1976).
225. S. Tomita and M. Nakagawa, *Bull. Chem. Soc. Japan*, **49**, 302 (1976).
226. M. Iyoda, H. Miyazaki and M. Nakagawa, *J. Chem. Soc., Chem. Commun.*, 431 (1972); *Bull. Chem. Soc. Japan*, **49**, 2306 (1976).
227. S. Akiyama, T. Nomoto, M. Iyoda and M. Nakagawa, *Bull. Chem. Soc. Japan*, **49**, 2579 (1976).
228. S. Nakatsuji, M. Morigaki, S. Akiyama and M. Nakagawa, *Tetrahedron Letters*, 1233 (1975).
229. S. Nakatsuji and M. Nakagawa, *Tetrahedron Letters*, 3927 (1975).
230. S. Nakatsuji, S. Akiyama and M. Nakagawa, *Tetrahedron Letters*, 2623 (1976).
231. D. E. Dorman, M. Jautelat and J. D. Roberts, *J. Org. Chem.*, **38**, 1026 (1973); C. Charrier, D. E. Dorman and J. D. Roberts, *J. Org. Chem.*, **38**, 2644 (1973).
232. J. P. C. M. Van Dongen, M. J. A. de Bie and R. Steur, *Tetrahedron Letters*, 1371 (1973).
233. I. D. Campbell, G. Eglinton, W. Henderson and R. A. Raphael, *J. Chem. Soc., Chem. Commun.*, 88 (1966).
234. H. A. Staab and F. Graf, *Tetrahedron Letters*, 751 (1966); *Chem. Ber.*, **103**, 1107 (1970).
235. H. A. Staab and R. Bader, *Chem. Ber.*, **103**, 1157 (1970).
236. K. Endo, Y. Sakata and S. Misumi, *Tetrahedron Letters*, 2557 (1970).
237. S. Akiyama, S. Misumi and M. Nakagawa, *Chem. Ind. (London)*, 346 (1960); *Bull. Chem. Soc. Japan*, **33**, 1293 (1960).
238. S. Akiyama, S. Misumi and M. Nakagawa, *Bull. Chem. Soc. Japan*, **35**, 1829 (1962).
239. O. M. Behr, G. Eglinton, A. R. Galbraith and R. A. Raphael, *J. Chem. Soc.*, 3641 (1960).
240. M. Morimoto, S. Akiyama, S. Misumi and M. Nakagawa, *Bull. Chem. Soc. Japan*, **33**, 1293 (1960).
241. P. J. Beeby, R. Weavers and F. Sondheimer, *Angew. Chem.*, **86**, 163 (1974).
242. R. Weavers and F. Sondheimer, *Angew. Chem.*, **86**, 167 (1974).
243. R. R. Jones, J. M. Brown and F. Sondheimer, *Tetrahedron Letters*, 4183 (1975).
244. M. Iyoda, M. Morigaki and M. Nakagawa, *Tetrahedron Letters*, 817 (1974).
245. M. Iyoda, M. Morigaki and M. Nakagawa, *Tetrahedron Letters*, 3677 (1974).
246. A. Yasuhara, T. Satake, M. Iyoda and M. Nakagawa, *Tetrahedron Letters*, 895 (1975).
247. A. Yasuhara, M. Iyoda, T. Satake and M. Nakagawa, *Tetrahedron Letters*, 3931 (1975).
248. M. Morigaki, M. Iyoda and M. Nakagawa, *Tetrahedron Letters*, 2311 (1975).
249. M. Morigaki, M. Iyoda and M. Nakagawa, *Tetrahedron Letters*, 2315 (1975).
250. M. Iyoda and M. Nakagawa, *Chemistry Letters*, 815 (1975).
251. J. Ojima, Y. Yokoyama and T. Yokoyama, *Chemistry Letters*, 1261 (1974).
252. J. Ojima, Y. Yokoyama and T. Yokoyama, *Chemistry Letters*, 487 (1975).
253. J. Ojima, A. Kimura and T. Yokoyama, *Chemistry Letters*, 207 (1975).
254. J. Ojima, M. Ishiyama, A. Kimura and Y. Yokoyama, *Tetrahedron Letters*, 1909 (1975).
255. R. T. Weavers, R. R. Jones and F. Sondheimer, *Tetrahedron Letters*, 1043 (1975).
256. T. M. Cresp and F. Sondheimer, *J. Amer. Chem. Soc.*, **97**, 4412 (1975).
257. T. Kashitani, S. Akiyama, M. Iyoda and M. Nakagawa, *J. Amer. Chem. Soc.*, **97**, 4424 (1975).
258. S. Akiyama, M. Iyoda and M. Nakagawa, *J. Amer. Chem. Soc.*, **98**, 6410 (1976).
259. M. M. Midland, *J. Org. Chem.*, **40**, 2250 (1975).

CHAPTER **16**

Proximity interactions of acetylenes

S. MISUMI and T. KANEDA
ISIR, Osaka University, Suita, Osaka, Japan

I. INTRODUCTION

Proximity effects between functional groups in the same molecule frequently render the properties of the molecule unusual. A well-known example concerns proximity interactions of acetylenic bonds. Thus, internally hydrogen-bonded *o*-hydroxyphenylacetylene is more volatile (b.p. 75 °C/15 mm) than its methoxy derivative (b.p. 90 °C/15 mm), which is reminiscent of the effect of the hydrogen bond in *o*-nitrophenol.

In this brief chapter it is difficult to refer to all pertinent studies relating to proximity interactions of triple bonds. We shall therefore concentrate particularly on recent studies of physical and chemical properties influenced by intramolecular π–π interactions between triple bonds or between a triple bond and other unsaturated systems.

II. PROXIMITY INTERACTIONS IN SPECTRAL BEHAVIOUR

A large number of cyclic acetylenes[1] have been synthesized to investigate the proximity interactions between two triple bonds or between a triple bond and other unsaturated systems. Such transannular proximity interactions have been observed in some cyclic acetylenes of medium ring size. However, no appreciable evidence of the interaction was detected in the electronic spectra of diacetylenes **1** [2], **2** [2] and **3** [3], where two triple bonds are closely fixed and conjugated to a large chromophore such as naphthalene, even though the triple bonds are within van der Waals' radii of ~2·9 [4, 5] or ~3·0 Å [6].

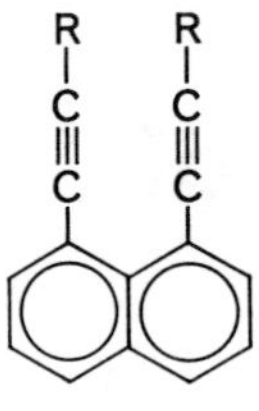

(1)

(a) R = H
(b) R = CH_3
(c) R = C_6H_5

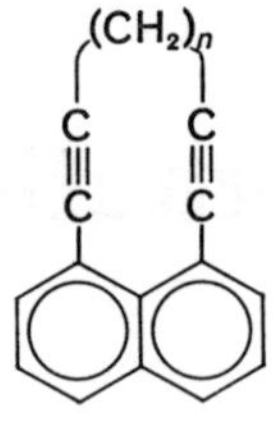

(2)

(n = 3, 4)

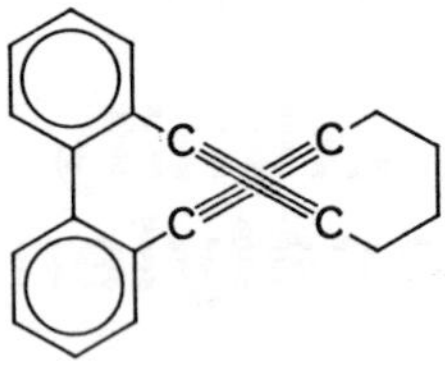

(3)

A. Electronic Absorption Spectra

A number of relatively simple cyclic acetylenes, **4** [7, 8], **5** [9, 10], **6** [11–13], **7** [8] and **8** [14–16], were prepared in order to examine the proximity interaction of triple bonds. However, no precise evidence was observed in their electronic spectra, owing to the considerably high energy requirements of the transition of the monoacetylene chromophore.

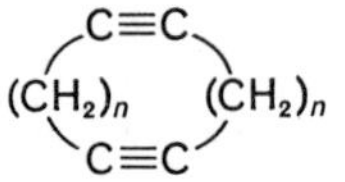

(4)

(a) n = 2[7]
(b) n = 4[8]

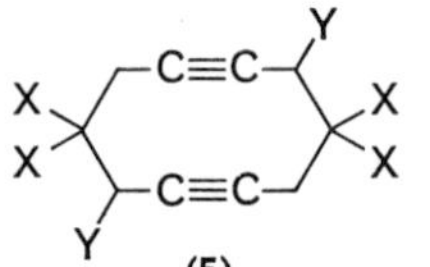

(5)

(a) X = H, Y = OH[9]
(b) X = H, Y = OCH_3[9]
(c) X = $CO_2C_2H_5$, Y = H[10]

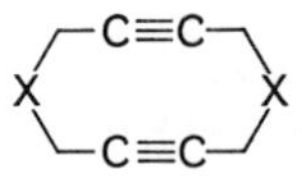

(6)

(a) X = O[11, 12]
(b) X = S, SO, SO_2[13]

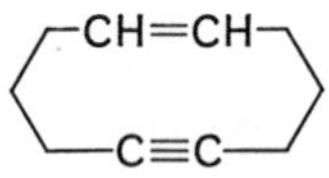

(7)[8]

Dialkyl 1,3-diacetylenes usually absorb in the ultraviolet region, λ_{max} 226, 239 and 253 nm. Of a series of cyclic tetraacetylenes **(8)**, two higher homologues, **8c** and **8d,** show absorption spectra similar to that of acyclic compounds, whereas the trimethylene homologue **(8b)** demonstrates strikingly different features in its absorption spectrum: (*i*) a red shift of 9 nm and (*ii*) appearance of a new band at 246 nm. This is attributed to the marked transannular electronic interaction in **8b,** in

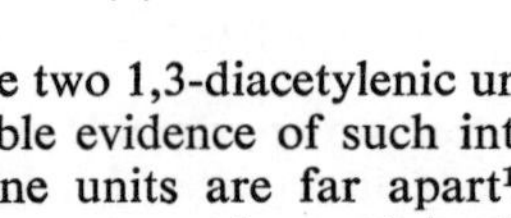

(8)

(a) n = 2[14]
(b) n = 3[15]
(c) n = 4[16]
(d) n = 5[15]

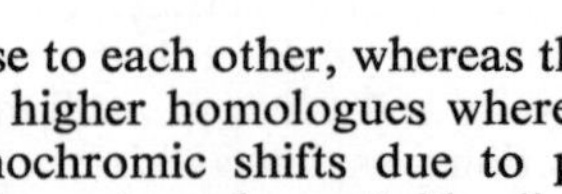

(9)

(a) X = H, Y = $OCOCH_3$
(b) X, Y = CH_2=

which the two 1,3-diacetylenic units are very close to each other, whereas there is no appreciable evidence of such interaction in the higher homologues where the 1,3-diacetylene units are far apart[15]. Similar bathochromic shifts due to proximity interactions were observed in the spectra of the extremely unstable dimethylene homologue (**8a**) and two derivatives of the trimethylene homologue [**9a** [17] (λ_{max} 226, 238, 246 (sh), 263 nm in CH_3OH) and **9b** [18]].

The interaction between a triple bond and an aromatic ring was first studied with paracyclophynes (**10**)[19, 20], but only ambiguous evidence was observed in their spectra[19]. The first positive evidence for such a transannular interaction was shown in the spectra of naphthalenophapolyynes (**11**)[21]. Here all the absorption bands appear at the same positions, but their intensities in the cyclic monomer **11a** are markedly decreased as compared with the intensities per unit chromophore of the cyclic dimer **11b** and trimer **11c**. This hypochromism of **11a** was explained in terms of a dispersion

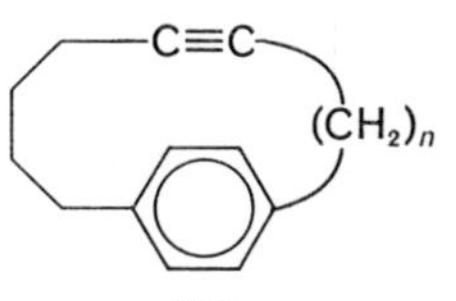

(10)

(**a**) $n = 3$
(**b**) $n = 4$

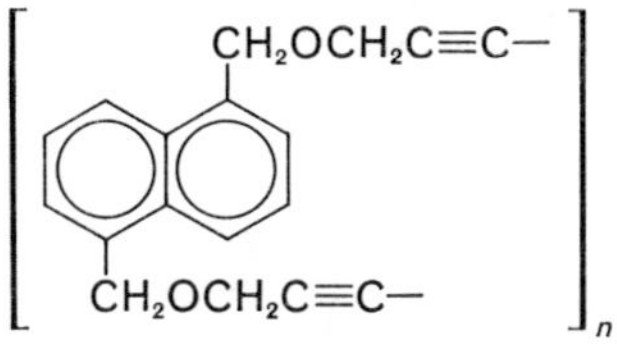

(11)

(**a**) $n = 1$
(**b**) $n = 2$
(**c**) $n = 3$

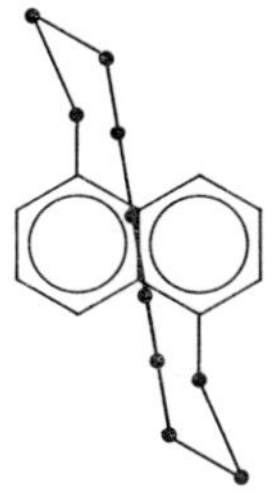

(11a)

force interaction between the closely situated two chromophores as seen in the structure **11a** and found to be in good agreement with the values calculated according to Tinoco and Rhodes[21]. The interaction between diphenylacetylene and *p*-xylene was examined with a few composite cyclic compounds containing the diphenylacetylene moiety[22].

A series of [*n*]paracyclophadiynes or [*m.n*]paracyclophadiynes (**12**) with different numbers of methylenes was prepared for the study of the transannular electronic interaction between a diacetylene unit and a benzene ring[23]. The absorption spectra of **12** show features which are obvious when compared with that of a reference compound **15c** (Figure 1): (*i*) disappearance of vibrational fine structures with decrease of the methylene number, (*ii*) bathochromic shift of the longest wavelength bands and (*iii*) appearance of a new band at 233 nm for **12c** and at 227 nm for **12d**,

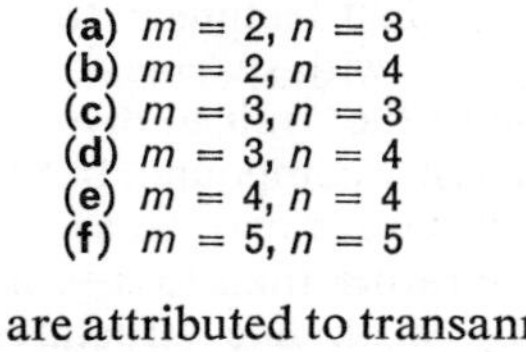

(12)

(**a**) $m = 2, n = 3$
(**b**) $m = 2, n = 4$
(**c**) $m = 3, n = 3$
(**d**) $m = 3, n = 4$
(**e**) $m = 4, n = 4$
(**f**) $m = 5, n = 5$

respectively. These new bands are attributed to transannular π–π interactions between the two chromophores rather than to molecular strain[23a]. Later the new band of **12c** was associated with an intramolecular charge transfer from the benzene ring to the diacetylene group on the basis of theoretical calculations and crystalline spectral measurements[25]. The exact molecular structure of the highly strained **12c** was determined by X-ray crystallographic analysis as shown in Figure 2 [26]. The figure demonstrates a geometry which is favourable for the electronic interaction between the two chromophores as well as for strong bending and close fixing, within van der Waals' radii, of both chromophores. A strainless homologue (**12e**) exhibited a

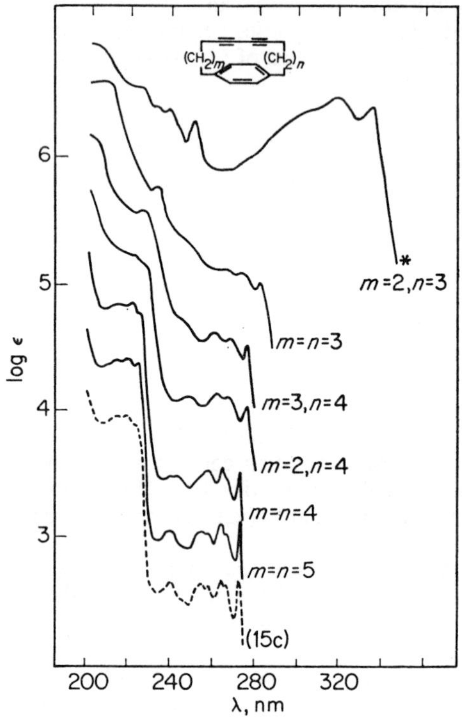

FIGURE 1. Electronic spectra of [*m.n*]paracyclophadiynes (**12**) and **15c**. The intensities of **15c** are reduced to the value per unit chromophore, i.e. ε/2. The curves are displaced upward successively by 0·5 log ε units from the curve immediately below, except the qualitative spectrum (*) of **12a**.

hyperchromic effect similar to the case of **11a** [23a]. In the two highly strained cyclophadiynes, **12a** [68b] and **13** [29], spectral information about the transannular interaction could not be detected because of their instability. Moreover, cyclophadiynes (**14**) and cyclophatetraynes (**15**) show a distinct feature in their electronic spectra (Figure 3). Thus clear, enhanced fine structures with spacings of about 2100 cm^{-1} were observed in the longer wavelength region of the strained diyne **14a** and the strainless tetrayne **15b** compared with that of a reference compound, but **14b** showed a normal spectrum. The appearance of these fine structures may be ascribed mainly to local excitation of the diacetylene chromophore rather than to π–π interaction of the two chromophores[24]. Similar but relatively weak fine structures were also observed in the spectrum of the severely strained cyclophadiyne **16** [24]. On the other hand, two anthracenophadiynes, **17** [27] and **18** [28], showed no remarkable spectral features compared with the corresponding acyclic compounds.

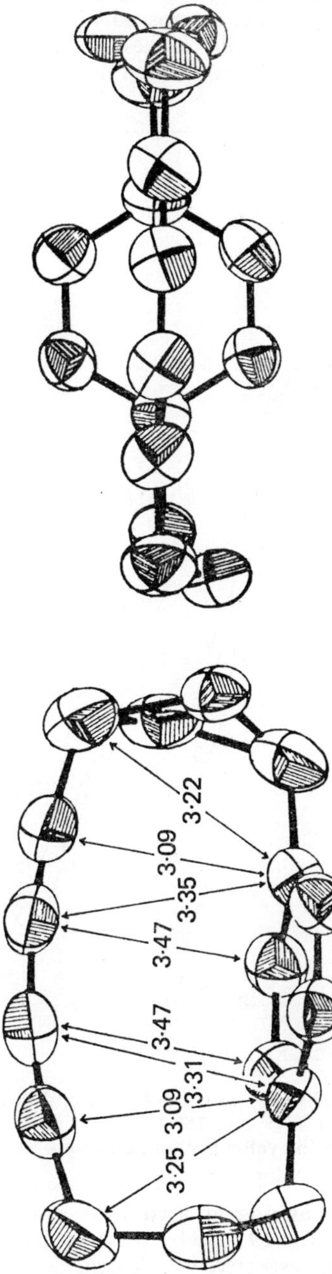

FIGURE 2. Molecular geometries of [3.3]paracyclophadiyne (**12c**). Reproduced, by permission, from Reference 26.

Two diacetylene-bridged triptycenes, **19** and **20**, were prepared in order to study the insertion effect of the diacetylene group on the circular electronic interaction of their parent compound, triptycene. Compared with the corresponding dimethyltriptycenes, **19** shows a red-shift of the longest wavelength maximum by 10 nm, whereas there is no sign of the band-shift in the 1,4-bridged **20**. These results may be

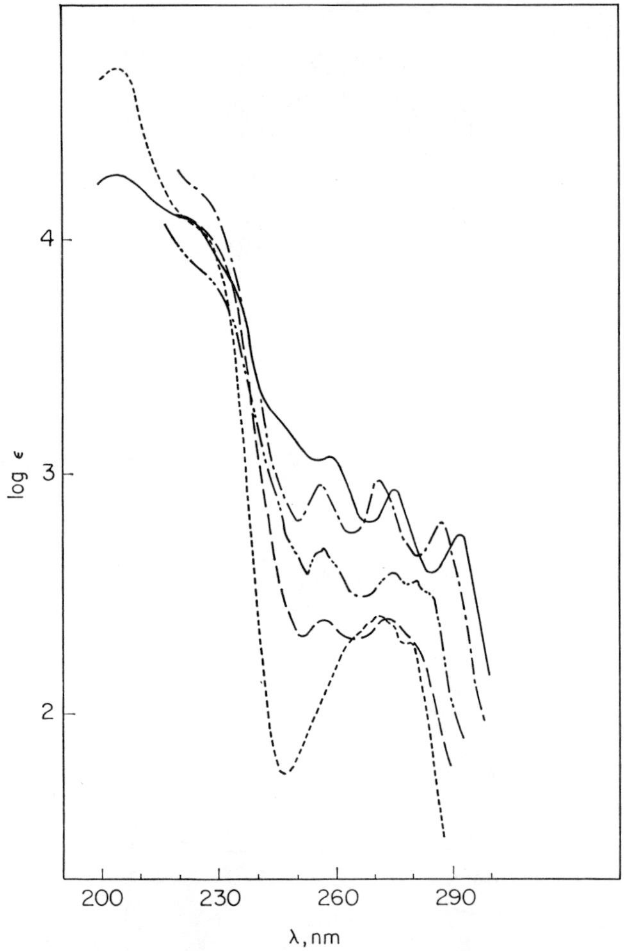

FIGURE 3. Absorption spectra of **14a** (——), **14b** (– – –) and bis(5-pentynyl)durene (· · · · ·) in cyclohexane, **15a** (— · · —) in THF and **15b** (— · — ·) in dioxane. The intensities of **15** are reduced to the value per unit chromophore, i.e. $\varepsilon/2$.

explained by the interaction among at least three π-electronic systems, that is, benzene–diacetylene–benzene for **19** and by the interaction localized in the paracyclophadiyne moiety for **20**, respectively[31]. A consideration of the projections of the molecular models (Figure 4) is suggestive of the difference in the 'short-circuiting effect' caused by the inserted diacetylene group in the two bridged triptycenes.

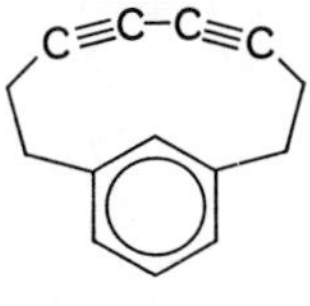

(13)

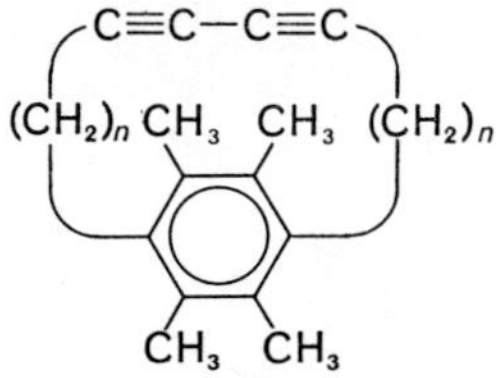

(14)

(**a**) $n = 3$
(**b**) $n = 4$

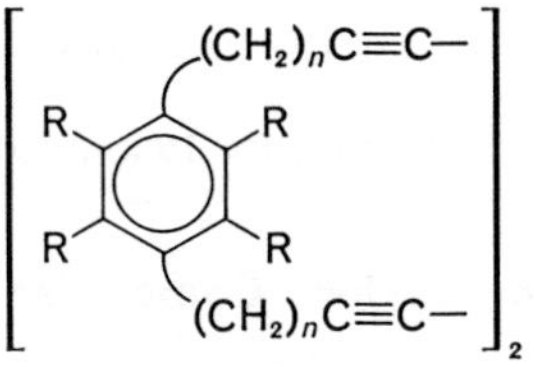

(15)

(**a**) $R = CH_3$, $n = 2$
(**b**) $R = CH_3$, $n = 3$
(**c**) $R = H$, $n = 4$

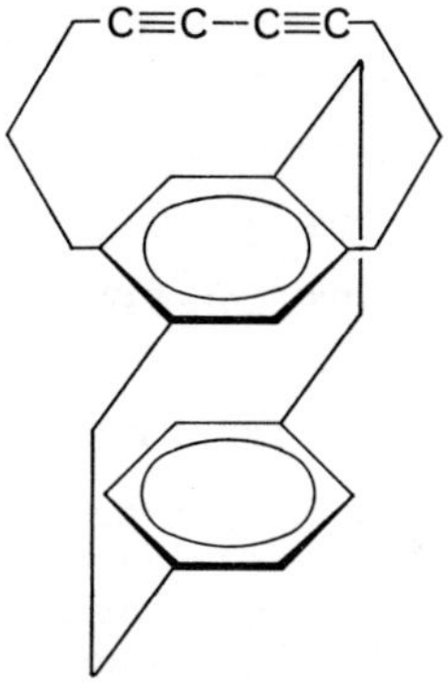

(16)

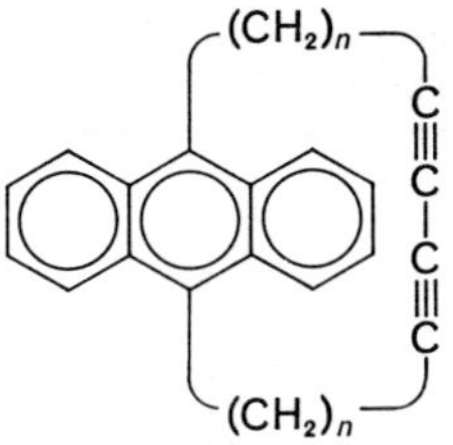

(17)

(**a**) $n = 3$
(**b**) $n = 4$

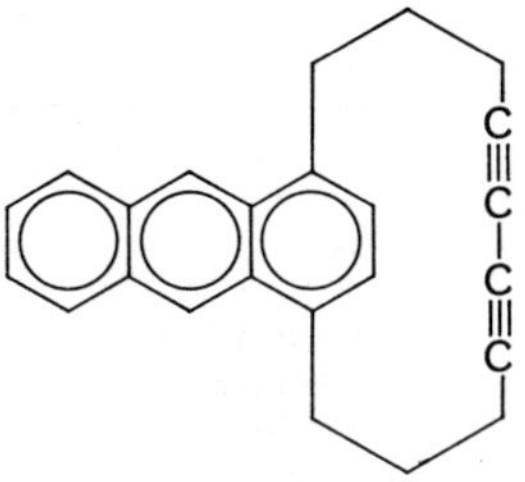

(18)

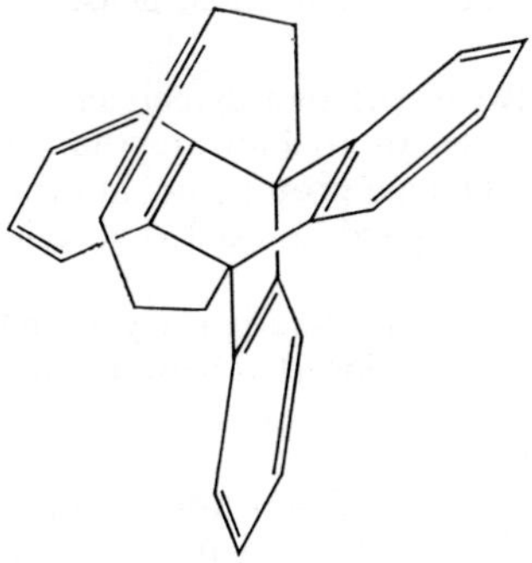

(19)

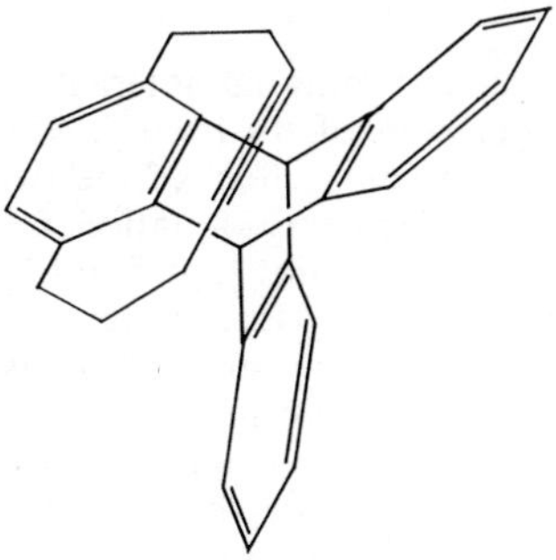

(20)

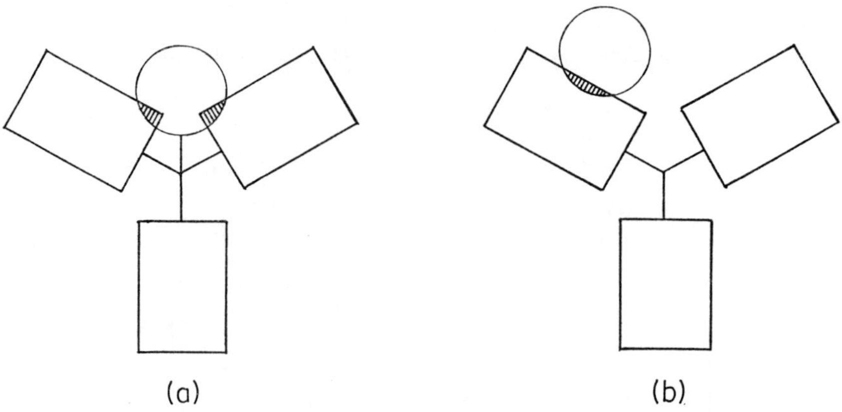

FIGURE 4. Projection of two bridged triptycenes; (a) for **19** and (b) for **20**.

B. Charge Transfer Spectra of Tetracyanoethylene Complexes

The absorption spectra of the charge transfer complexes of 1,3-diacetylenes were recently measured. Table 1 shows maxima of one-to-one tetracyanoethylene (TCNE) complexes of dialkyl-1,3-diacetylene and related compounds[32b]. The cyclic tetrayne (**8b**)–TCNE complex shows a normal absorption spectrum regardless of the closed, parallel conformation of the two diacetylene groups in **8b**. This is in striking contrast

TABLE 1. Absorption maxima of tetracyanoethylene complexes with 1,3-diacetylenes and alkynes in methylene chloride

Acetylene	λ_{max} (nm)
(**8**) (n = 3)	428
(**8**) (n = 4, 9, 10)	423
1,3-Cyclotridecadiyne	431
1,3-Cyclotetradecadiyne	430
5,7-Dodecadiyne	424 (K = 7·0, ε = 1100)[a]
5-Decyne	370 (K = 3·9, ε = 500)[a]
3-Octyne	367

[a] K and ε were determined by Benesi–Hildebrand plots.

to the marked transannular interaction between the two diacetylene groups in the electronic spectrum of **8b** itself[15]. This fact seems to indicate that in the TCNE complex the proximity effect of the two diacetylene groups, or, in other words, the transannular π-electronic donation from the non-complexed diacetylene group to the complexed one, is not significant[32b].

The one-to-one TCNE complexes of a series of [$m.n$]paracyclophadiynes (**12**), where m and n are the numbers of methylene groups inserted between the benzene and diacetylene groups, were investigated for the π–π transannular interaction between the diacetylene and an aromatic ring (Table 2). Each of these complexes shows an absorption maximum in the narrow region of 425–431 nm, except for the **12c** ($m = n = 3$) complex which exhibits a distinct additional maximum at 555 nm [32]. Curve analyses of these spectra indicated that the spectrum of the **11d** complex also

TABLE 2. Absorption maxima of tetracyanoethylene complexes with [*m.n*]paracyclophadiynes (**12**) in methylene chloride

[*m.n*]Paracyclophadiyne	λ_{max} (nm)^{a}	
12 ($m = n = 3$)	427 (425)	555 (555)
12 ($m = 2, n = 4$)	430 (402)	(470)
12 ($m = 3, n = 4$)	431 (415)	(510)
12 ($m = n = 4$)	429 (410)	(480)
12 ($m = n = 5$)	426 (410)	(480)

^{a} λ_{max} in parentheses is the absorption maximum resolved by a curve resolver.

contained an additional maximum near 510 nm and those of the **12b**, **12e** and **12f** complexes at 470–480 nm. Compared with the theoretically well-established spectrum of the *p*-xylene–TCNE complex[30], the longer wavelength maxima of the **12c**– and **12d**–TCNE complexes are probably associated with a parallel structure (**21**) where transannular π-electron transfer from the diacetylene group toward the complexing benzene ring may be favoured. On the other hand, such a transannular electron transfer is improbable in the case of a perpendicular structure (**22**) and a diacetylene-site complex (**23**) because of unfavourable π-orbital overlap[32b]. A crystalline 2 : 1 TCNE complex of **11e** was isolated[32b] and found to be a molecular structure (**24**) where the TCNE molecule was sandwiched between two benzene rings[33].

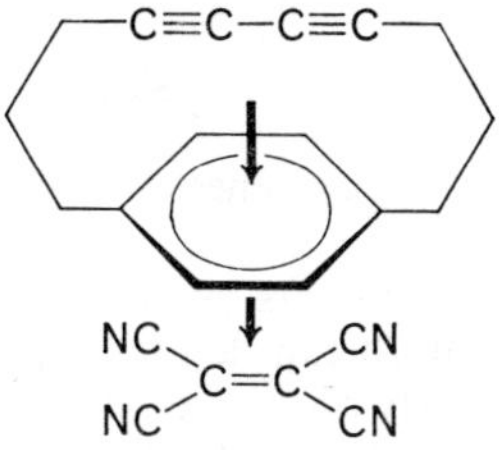

(**21**)

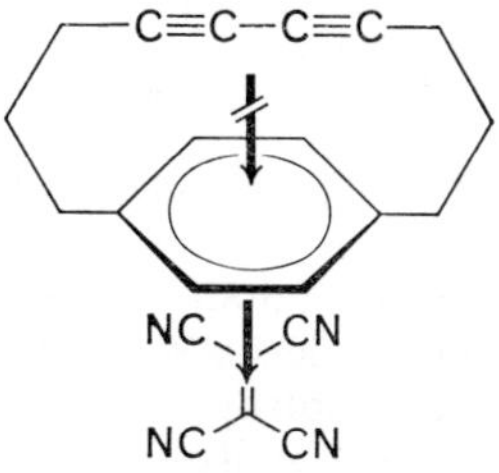

(**22**)

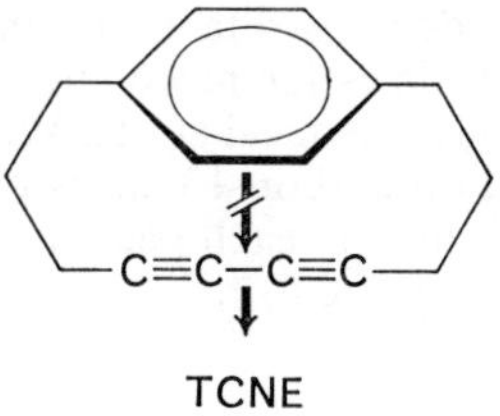

(**23**)

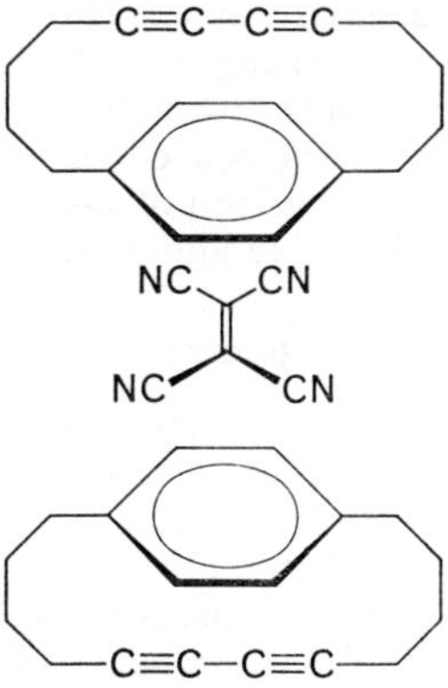

(**24**)

C. Carbon-13 Nuclear Magnetic Resonance Spectra

Carbon-13 resonance spectroscopy is an excellent tool for obtaining direct information about acetylenic *sp*-carbon participation in proximity interactions.

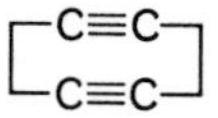

(4a)

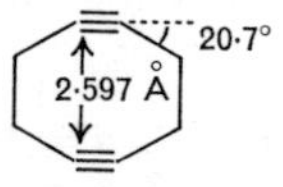

(25)

The *sp*-carbons of **4a** are deshielded by 14·5 p.p.m. from the chemical shift (81·3 p.p.m.)[34] of those of cyclotridecyne. This remarkable effect has been ascribed to partial olefinic character of the triple bonds caused by large molecular deformation[7]. In fact the *cis*-olefinic configuration has recently been confirmed by X-ray structure analysis of **25** [7]. As to the chemical shifts of acetylenic *sp*-carbons in cyclic tetraacetylenes, both of the inner and outer *sp*-carbons of **8b** are deshielded by *ca.* 3 p.p.m. relative to the corresponding carbons of acyclic diacetylenes (Figure 5).

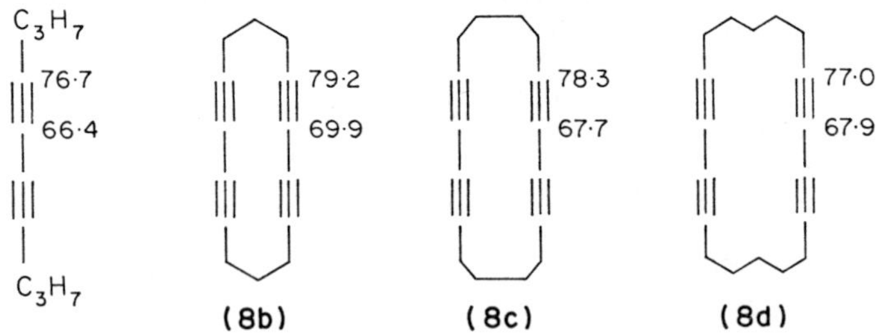

FIGURE 5. Carbon-13 chemical shifts of cyclic tetraacetylenes (**8**), δ (TMS) [34].

For larger rings, **8c** and **8d**, the differences in chemical shifts of *sp*-carbons are about 1 p.p.m. The larger deshielding of the *sp*-carbons of **8b** can be explained in terms of the diamagnetic anisotropy effect of the transannularly positioned diacetylene group, which was estimated by Roberts and coworkers[34], in addition to the ring strain effect.

Similarly, the deshielding shifts of *sp*-carbons are found with paracyclophadiyne (**12c**) and anthracenophadiyne (**17a**) in spite of the strong shielding effect due to the transannular aromatic ring. These facts clearly demonstrate that rehybridization of the *sp*-carbons arising from their molecular strain is far more effective than the anisotropy effect of the aromatic rings on carbon-13 chemical shifts. Weak but clear ring current effects of benzenoid aromatic nuclei on carbon-13 shifts of *sp*-carbons were found by comparison of *sp*-carbon chemical shifts in each pair of rigid, **12c** and **17a**, and strainless, **12e** and **17b**, cyclophadiynes (Figure 6)[35].

D. Photoelectron Spectra

The photoelectron spectra of cyclic diacetylenes, **4a** and **6a** and a reference compound, cyclooctyne were measured in order to study the proximity interaction of acetylenic bonds[36]. The observed values of the lowest vertical ionization potentials were in the order **6a** > **4a** > cyclooctyne. The photoelectron bands were assigned by semiempirical calculations, MINDO/2 and SPINDO, using X-ray crystallographical data, and their relative sequence and positions were explained in terms of through-bond and through-space interactions between π and σ orbitals.

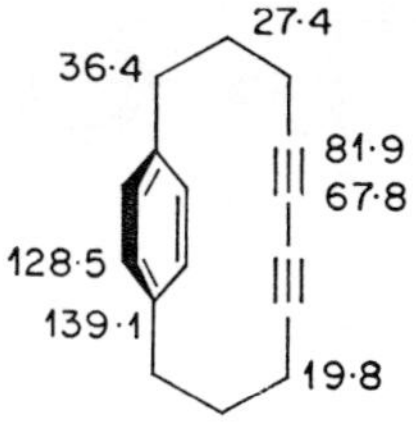

(12c)

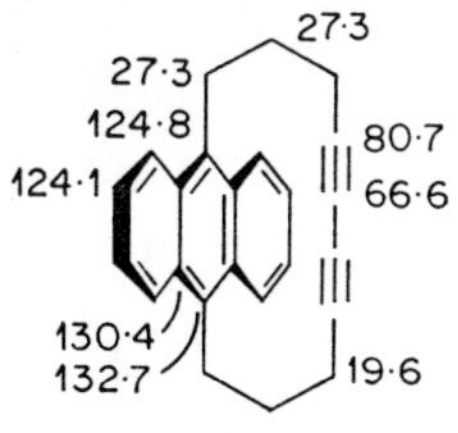

(17a)

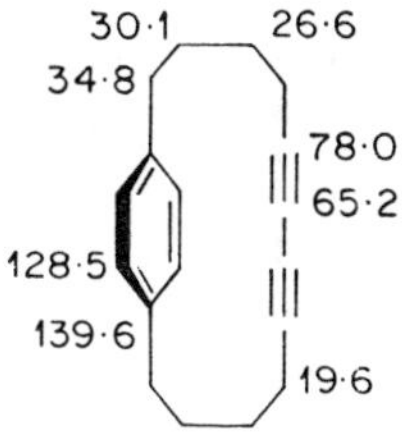

(12e)

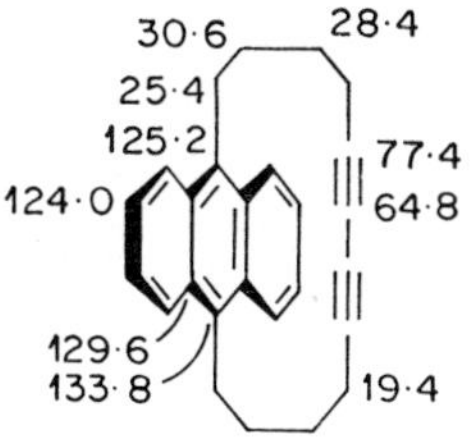

(17b)

FIGURE 6. Carbon-13 chemical shifts of cyclophadiynes, δ (TMS).

III. PROXIMITY INTERACTIONS IN CHEMICAL REACTIONS

Many ring systems have been prepared by cycloaddition of acetylenic compounds, following concerted or multi-step intra- or intermolecular reaction mechanisms. In particular, the placing in close proximity of triple bonds or a triple bond and another unsaturated system, such that intramolecular cycloaddition might lead to four–seven-membered rings, would seem of interest. This section deals with transannular carbon–carbon bond formation of triple bonds in acyclic and cyclic systems.

A. Triple Bond–Triple Bond Interactions

1. Acyclic acetylenes

Since the appearance of the theoretical consideration[37] that the formation of tetrahedrane from two acetylenes would be a photochemically allowed process, a number of papers dealing with attempted syntheses of tetrahedranes have been published. It was reported that 2,2′-bis(phenylethynyl)biphenyl (**26**) was transformed thermally and photochemically into an isomeric hydrocarbon, for which a tetrahedrane (**27**) or a cyclobutadiene (**28**) structure would be possible[38]. Actually, the structure of the hydrocarbon was confirmed to be 9-phenyl-1,2 : 3,4-dibenzoanthracene (**29**) derived from transannular bond formation[39].

Irradiation of the closely fixed diacetylene (**30**), which seemed to provide a greater possibility of formation of a tetrahedrane (**31**), gave unchanged **30** in almost quantitative yield[40], whereas 2,2′-diethynylbiphenyl (**32**) afforded the dibromide **34** as the result of an unexpected intramolecular carbon–carbon bond formation when hydrogen bromide in acetic acid was added[41].

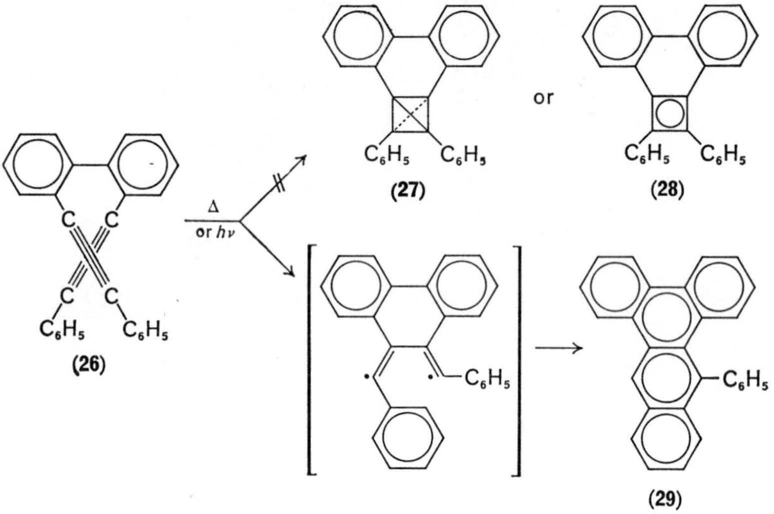

(26) (27) (28) (29)

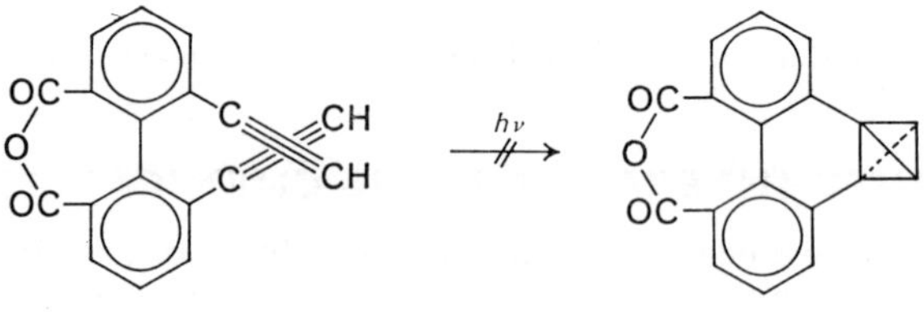

(30) (31)

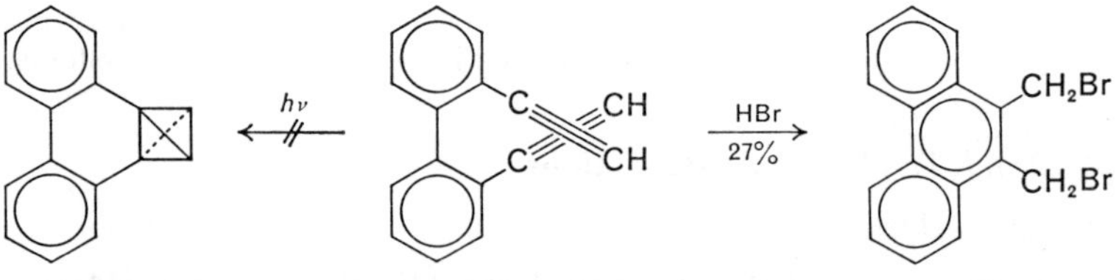

(33) (32) (34)

An azulene (**36**) having two intramolecularly crossing ethynyl groups was obtained together with its isomer (**37**) on irradiation of *o*-diethynylbenzene (**35**) and no further photochemical reaction was observed[42]. On the other hand, *o*-bisphenylethynylbenzene (**38**) gave an azulenophenalene system, verdene (**39**), on irradiation[43]. The mechanism of the formation of **39** was considered as a result of further cycloaddition of two triple bonds crossed in an intermediate azulene derivative (**40**). This photochemical reaction is reminiscent of the Büchi reaction[44] in which diphenylacetylene yields an azulene system.

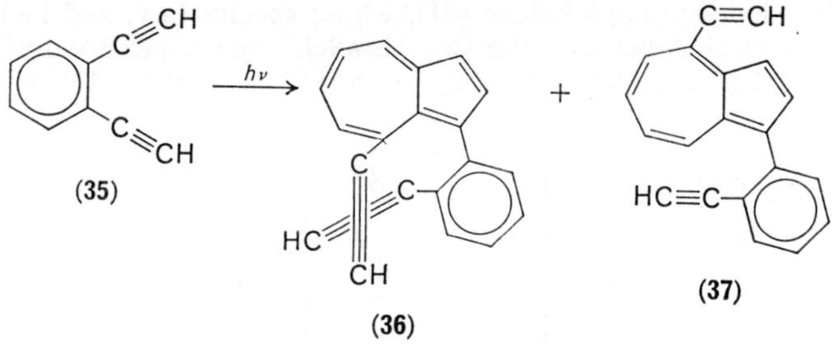
C≡CH
C≡CH
hν
(35)
HC≡
CH
(36)
+
C≡CH
HC≡C
(37)

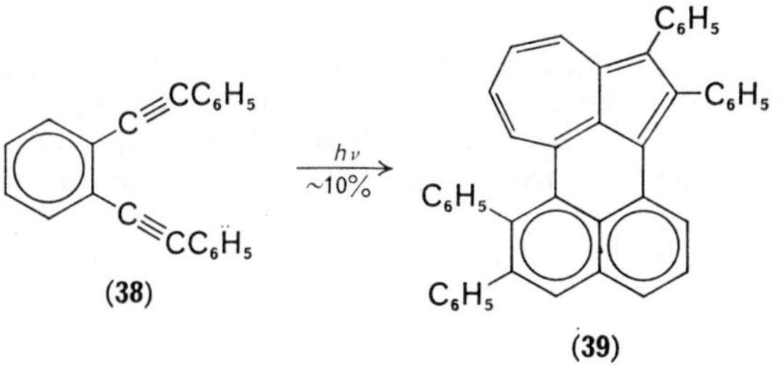
C≡CC6H5
C≡CC6H5
(38)
hν
~10%
C6H5
C6H5
C6H5
C6H5
(39)

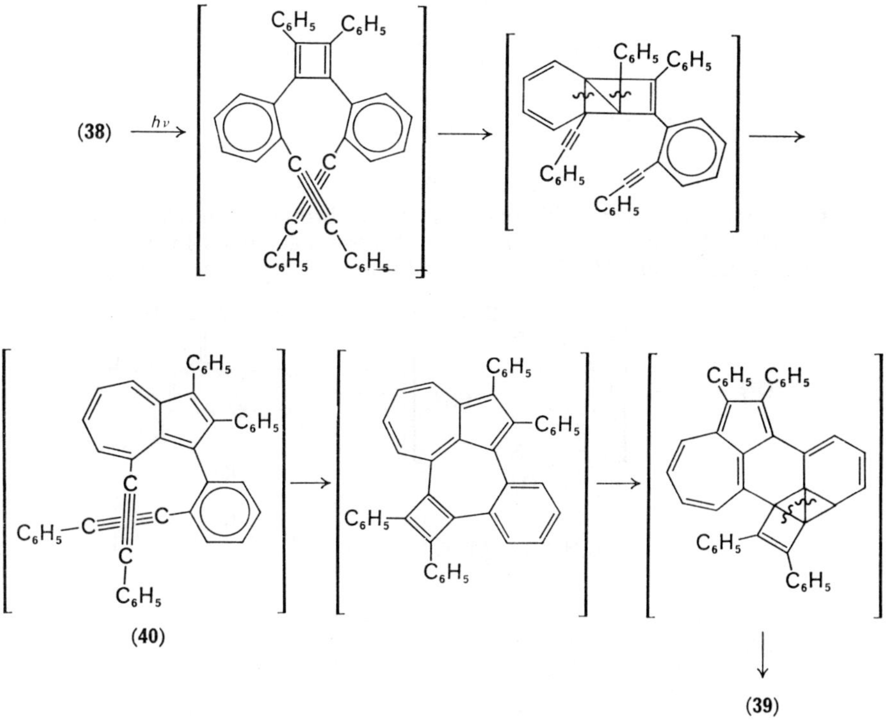
(38)
hν
C6H5
C6H5
C6H5
(40)
(39)

1,8-Bis(phenylethynyl)naphthalene **(41)**, whose spectra (u.v. and i.r.) showed no appreciable interaction between the two parallel, close triple bonds[45b], gave both thermally and photochemically 7-phenylbenzo[*k*]fluoranthene **(42)** in good yields,

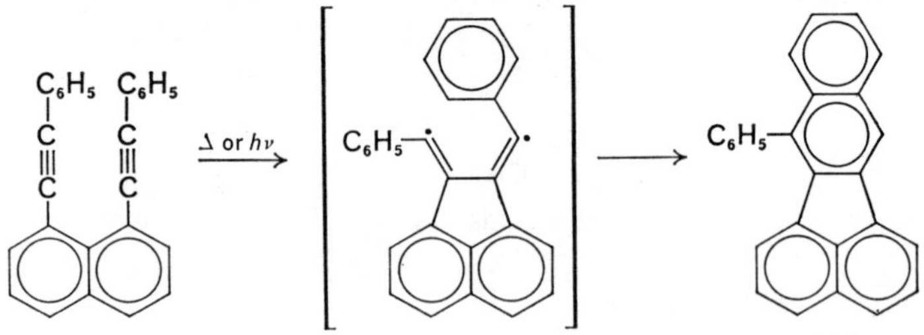

(41) **(42)**

i.e. 80% and 87%, on irradiation in 'skellysolve B' and cyclohexane, respectively; quantitative yields were obtained under reflux with pyridine or acetic anhydride[45, 46]. In the case of 1,8-bis(2′,4′,6′-trimethylphenylethynyl)naphthalene, in which the *ortho* methyl groups hindered the isomerization to the fluoranthene system, a deep green azulenic compound **(43)** was obtained in 8·5% yield under similar irradiative conditions[45b]. The mechanism by which **41** is converted to **42** is not clear and some attempts to isolate intermediates have as yet been unsuccessful.

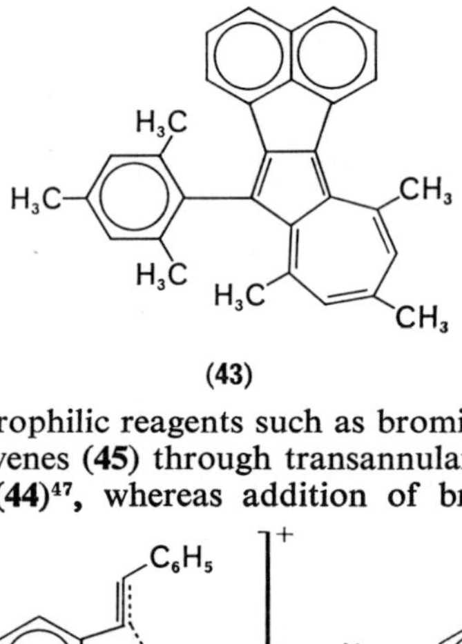

(43)

On treatment with electrophilic reagents such as bromine and hydrogen bromide, **38** gave diphenylbenzofulvenes **(45)** through transannular bond formation between proximate triple bonds **(44)**[47], whereas addition of bromine to **41** yielded the

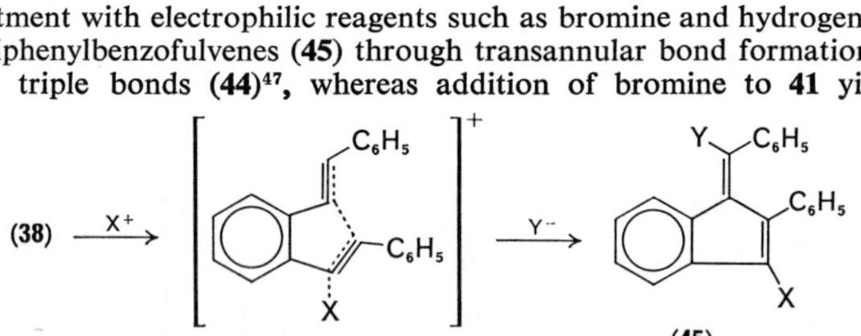

(44) **(45)**

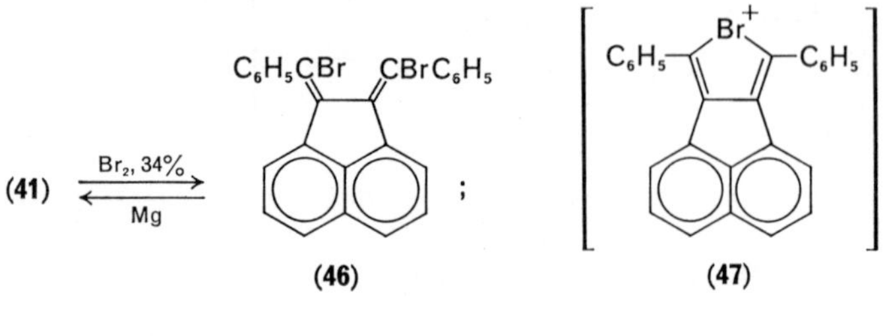

(46) **(47)**

acenaphthene derivative (**46**), possibly by a 1,4-transannular process involving **47**[46b]; **46** is conversely debrominated by magnesium to the original compound **41**. Similar transannular bond formation involving triple bonds was found in the bromination of the triacetylenic compound **48** to yield **49** which afforded a diketone (**50**) on hydrolysis[48d].

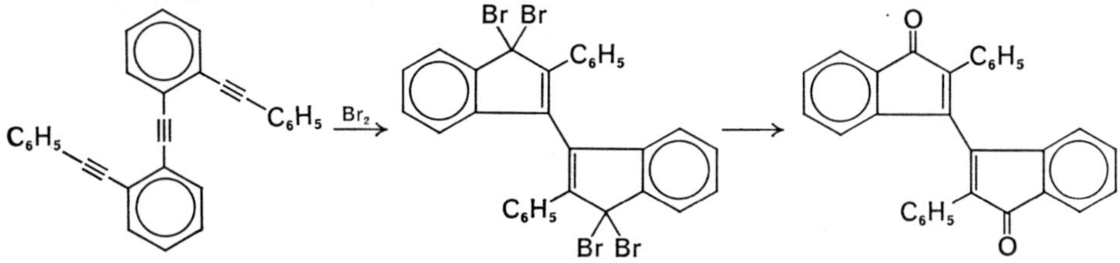

(**48**) (**49**) (**50**)

The diethynyl compounds **26**, **38** and **41** showed a different type of ring closure on thermal decomposition of their tetrachloroplatinate complexes to give the 3-phenylbenzofulvene derivatives **51**, **52** and **53**, respectively[49, 50]. **53** was also quantitatively obtained by treatment of **41** with mercuric acetate in acetic acid–sulphuric acid[45].

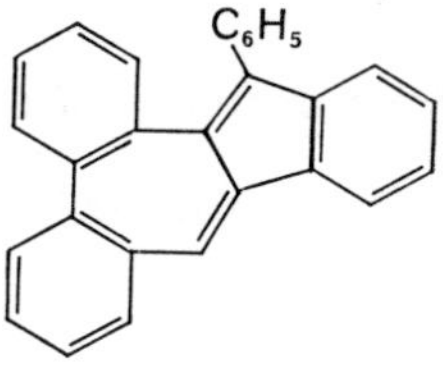

(**51**)

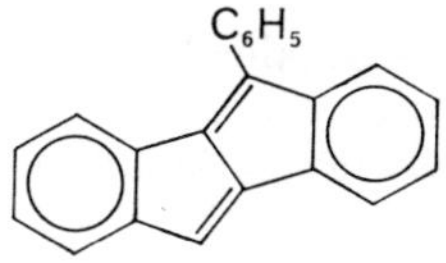

(**52**)

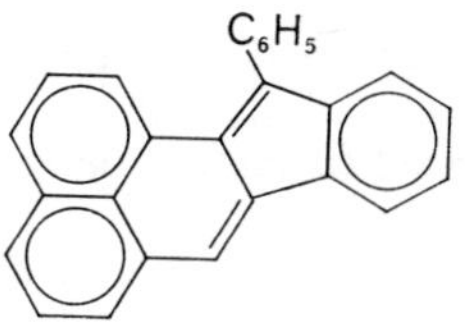

(**53**)

A number of diethynyl compounds (**54**), where two triple bonds are conjugated with a carbonyl group and/or an aromatic ring, were transformed to polycyclic compounds (**56**) via transition metal complexes (**55**), especially those of rhodium[51].

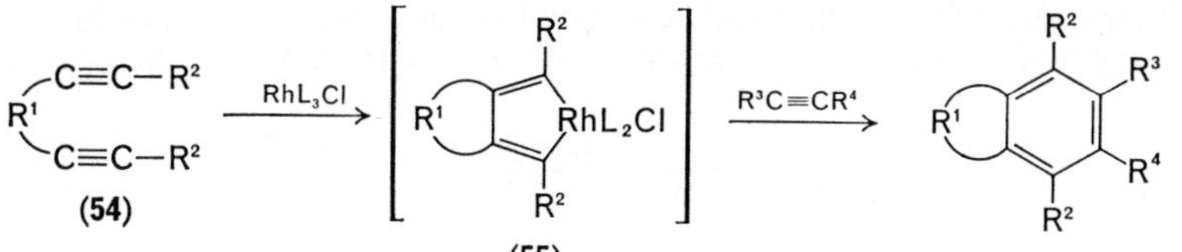

(**55**) (**56**)

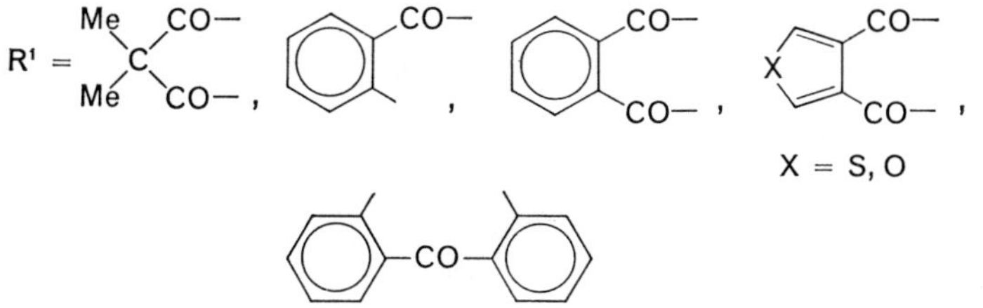

X = S, O

2. Cyclic acetylenes

Base-catalysed prototropic rearrangements of cyclic polyacetylenes have often been used to synthesize annulene and dehydroannulene systems. In such reactions, transannular carbon–carbon bond formation between proximate triple bonds occasionally takes place to give polycyclic compounds. Treatment with a base of cyclic polyacetylenes **57** and **59** gave benzenoid aromatic compounds **58** and **61**, with the desired dehydro[16] and [18]annulenes, respectively[52].

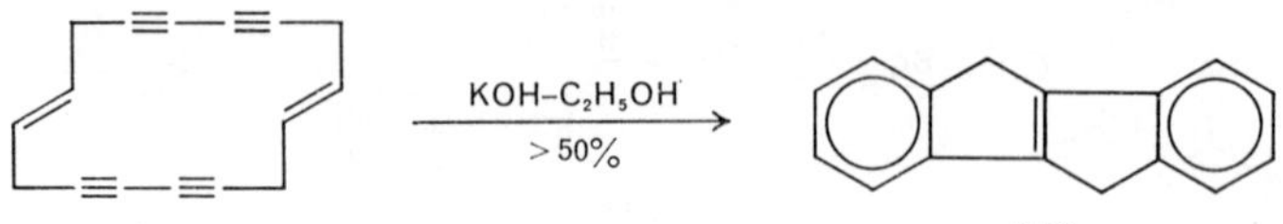

(57) **(58)**

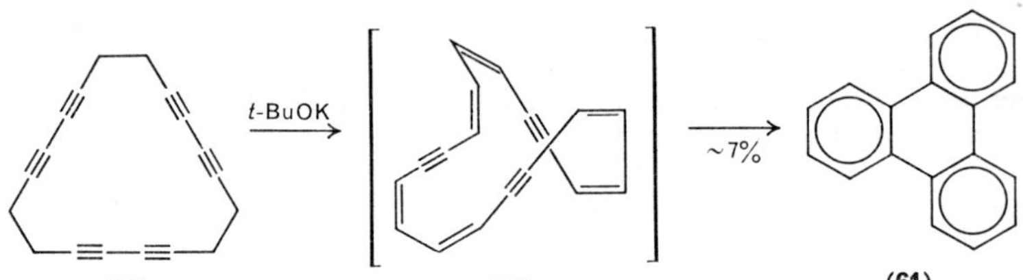

(59) **(60)** **(61)**

Similarly, biphenylene **(63)** was obtained in *ca.* 7% yield, from treatment of the thermally unstable tetrayne **8a** with potassium *t*-butoxide, probably by transannular bond formation between the triple bonds closely placed in the intermediate **62** [14, 53].

(8a) **(62)** **(63)**

Refluxing of the cyclic tetrayne disulphonate **64** with 7% potassium hydroxide in aqueous methanol gave a mixture of three benzazulenes **66**, **67** and **68**, by cyclization of the intermediate **65**, which could be isolated under milder conditions by extraction with strong acids from the organic solvent in a similar manner to azulene[54].

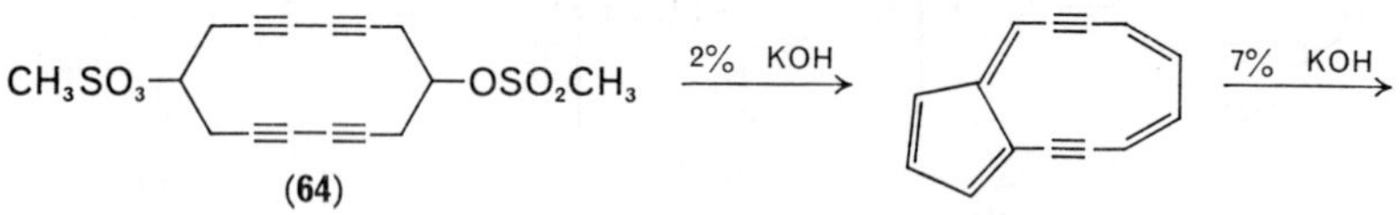

(64)

(65) 15–20%

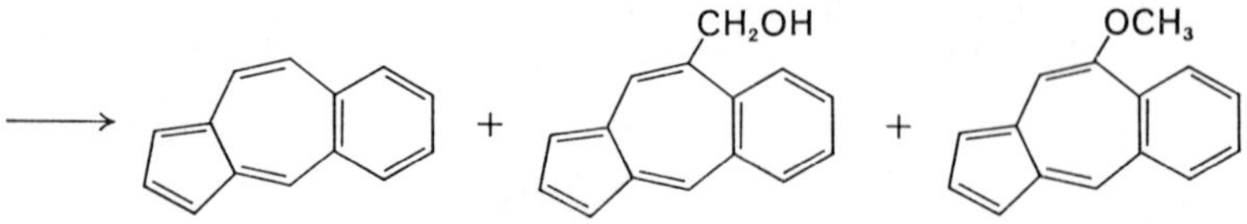

(66) ~34% **(67)** ~9% **(68)** ~5%

It is well known that the transannular carbon–carbon bond formation is often induced in catalytic hydrogenation of severely strained cyclic acetylenes where triple bonds are closely placed to each other. Hydrogenation of the highly strained tetrayne (**69**) with 10% Pd–C gave the polycyclic compounds, **70** and **71**, in addition to the expected hexadecahydro compound (50% yield)[55]. On the other hand, Birch reduction of **69** gave another perhydro product (**72**) in good yield[55b]. From a similarly constructed tetrayne (**73**) a polycyclic compound (**74**) was obtained by catalytic hydrogenation over platinum[56].

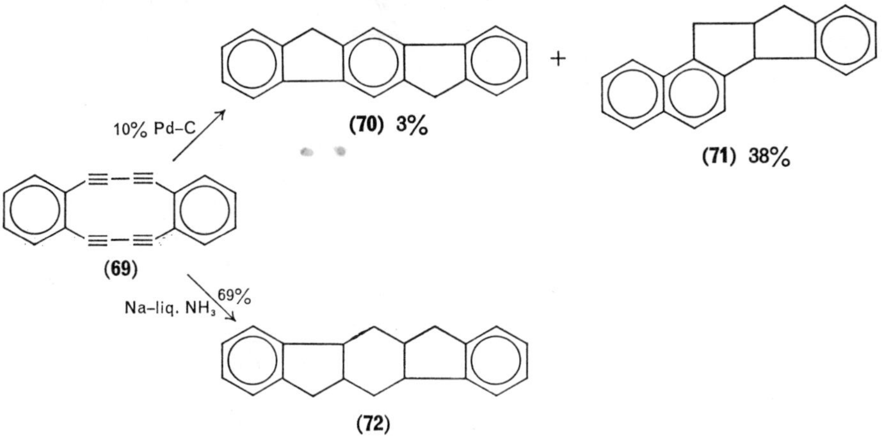

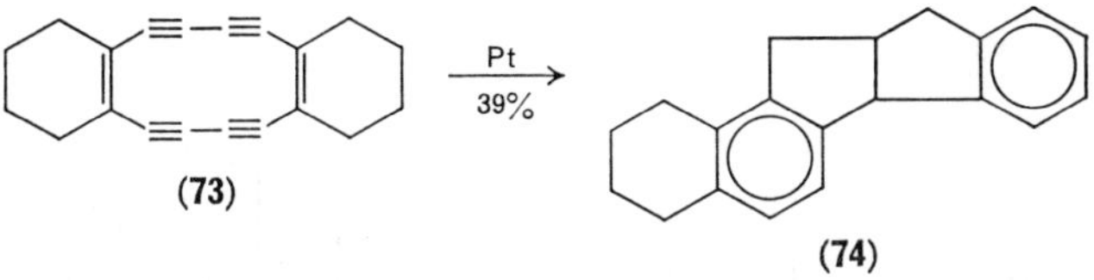

Catalytic hydrogenation of the diacetylene **65** gave a benzazulene (**66**) when interrupted after a few minutes of hydrogenation, while longer treatment resulted in formation of a mixture of **75, 76** and **77** [54].

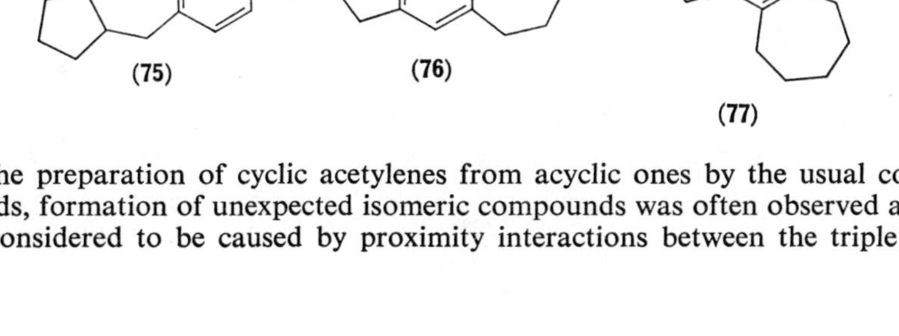

In the preparation of cyclic acetylenes from acyclic ones by the usual coupling methods, formation of unexpected isomeric compounds was often observed and has been considered to be caused by proximity interactions between the triple bonds

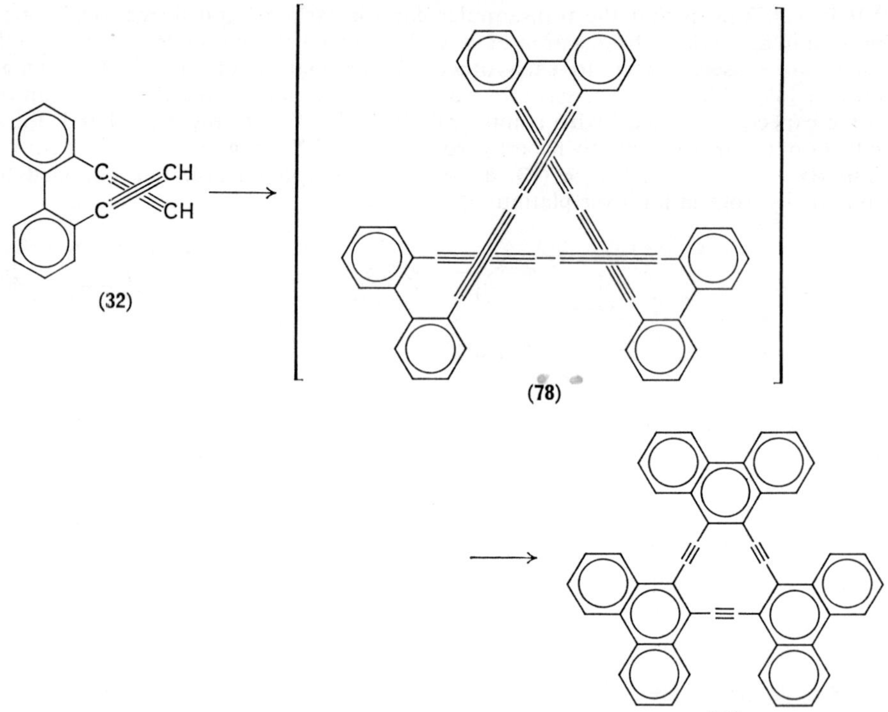
C
C
CH
CH
(32)
(78)
(79)

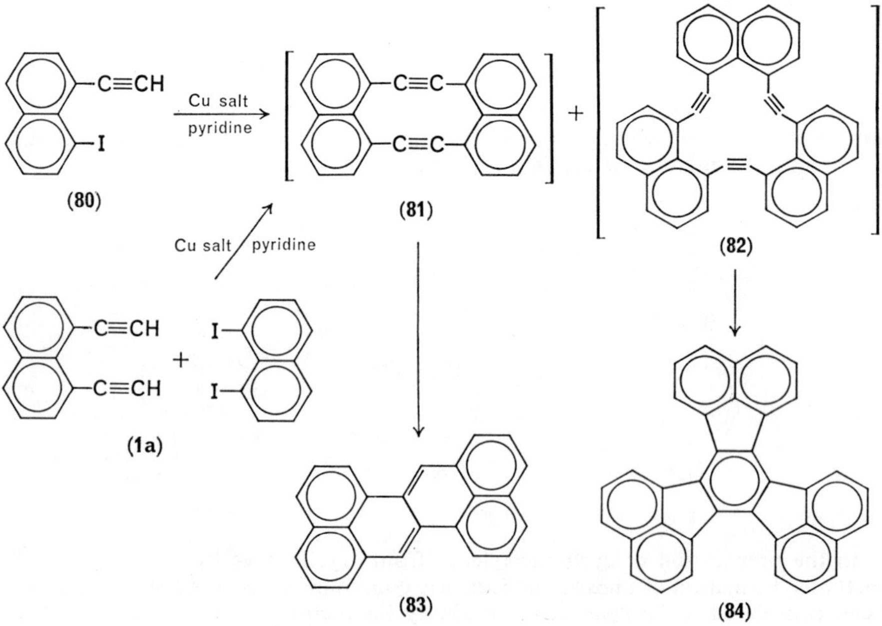
C≡CH
I
Cu salt
pyridine
C≡C
C≡C
(80)
(81)
(82)
Cu salt
pyridine
C≡CH
C≡CH
I
I
(1a)
(83)
(84)

closely placed to each other. The Eglinton oxidative coupling[41a] of diethynylbiphenyl (**32**) gave triphenanthro-tridehydro[12]annulene (**79**) in place of expected triply-crossed hexayne (**78**)[41b].

Coupling of 8-iodo-1-naphthylacetylene (**80**) according to the Castro method[57a] afforded two polycyclic compounds, zethrene (**83**) and decacyclene (**84**), which would be derived from intramolecular cyclization of the intermediates, **81** and **82**, respectively[57]. All other attempts to prepare **81**, e.g. the Castro coupling of **1a** with 1,8-diiodonaphthalene, led to **83** and its derivatives because of the extreme proximity of the two parallel triple bonds, within van der Waals' radii[58].

The oxidative coupling of the dipropargylamine **85** according to the Hay method[59] resulted in a one-step synthesis of the 1,5,9-tridehydro[12]annulene derivative **87** [60]. Since normal oxidative coupling of 1,6-heptadiyne yields its cyclic trimer[61], the formation of the annulene is very interesting and is explained by valence isomerization of triple bonds brought close together due to the geminal methyl groups in **86**.

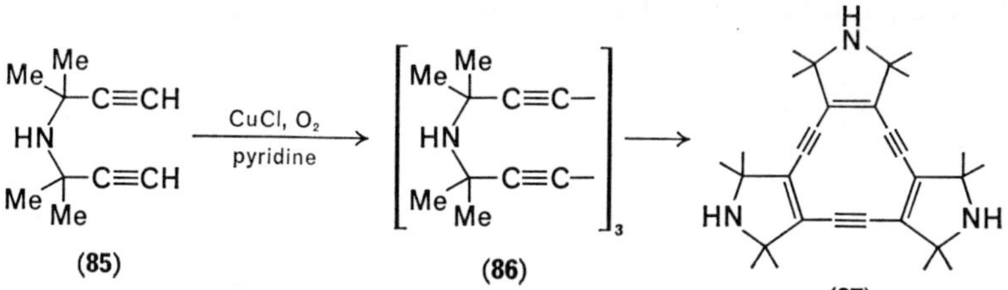

(85) **(86)** **(87)**

The cyclic diacetylene 5,6,11,12-tetradehydrotetrabenzo[a,c,g,i]cyclododecahexaene **89** is particularly interesting for its two triple bonds are closely fixed and cross each other, and the four *sp*-carbons occupy the apices of an expected tetrahedrane (**90**). In view of this, **89** was synthesized from bis(2-bromophenyl)acetylene (**88**) by

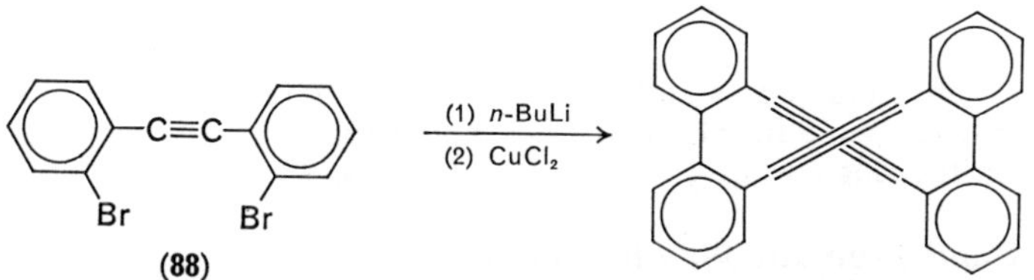

(88) **(89)**

metalation with *n*-butyllithium followed by treatment with anhydrous cupric chloride[62]. Nevertheless, according to the intense Raman absorption band of $\nu_{C\equiv C}$ at 2220 cm^{-1}, the compound obtained shows the character of triple bonds and not of a tetrahedrane. However, there is an interesting problem regarding the

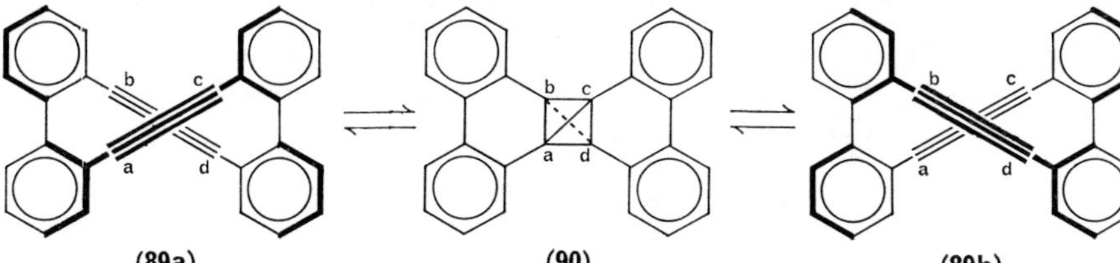

(89a) **(90)** **(89b)**

chirality of the molecule. The possibility exists of a novel rearrangement between **89a** and **89b** through an intermediate tetrahedrane (**90**). Although such a racemization has not been observed so far, a transannular carbon–carbon bond formation was recently found in the addition of bromine and hydrobromic acid to **89** [63].

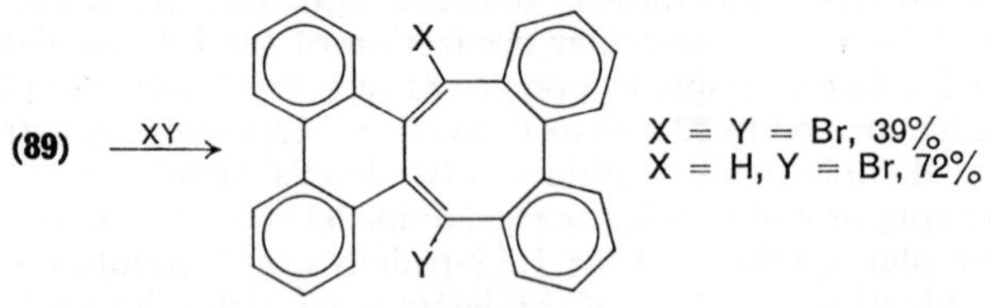

(**91**)

An attractive transformation of cyclic diacetylenes (**4**) to cyclobutadienes (**92**) was attempted considering the proximity interaction, but it has been unsuccessful so far. However, transition metal complexes such as the π-cyclopentadienyl cobalt

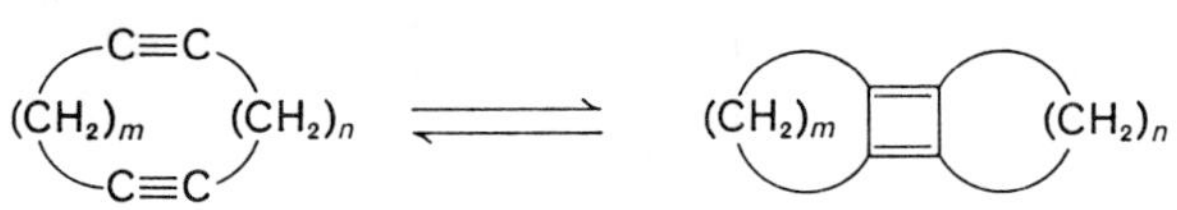

(**4**) (**92**)

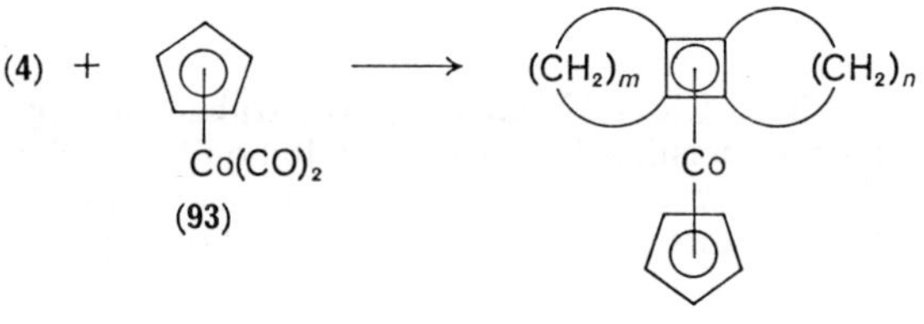

(**93**) (**94**)

complex **94** were obtained as stable derivatives by refluxing **4** with an equimolar amount of π-cyclopentadienyl cobalt dicarbonyl (**93**)[64–67]. The yields are strongly dependent on the ring size of 4: 86% for $m = n = 4$, 2% for $m = n = 5$.

B. Triple Bond–Aromatic Ring Interactions

In the course of a study concerning charge transfer complexes of [*n*]paracyclophadiyne or [*m.n*]paracyclophadiyne (**12**)[23] with tetracyanoethylene (TCNE), it was

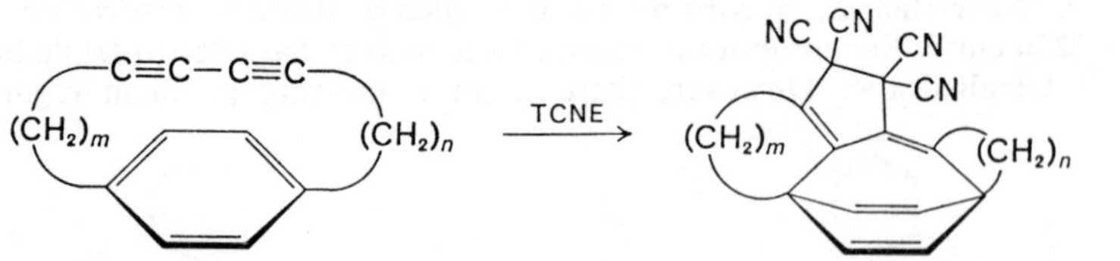

(**12**) (**95**)

$m \leqslant n$		Yield (%)
2	3	—
2	4	76
3	3	73
3	4	34
4	4	0

found that a novel multicycloaddition took place thermally among the three isolated, unsaturated systems. The structure (**95**) of the products was clearly confirmed by spectral analyses and the formation was proposed to be a 1,3-mode cycloaddition (Figure 7)[68]. No such cycloaddition reaction was observed for **12** with $m = n = 4$ or

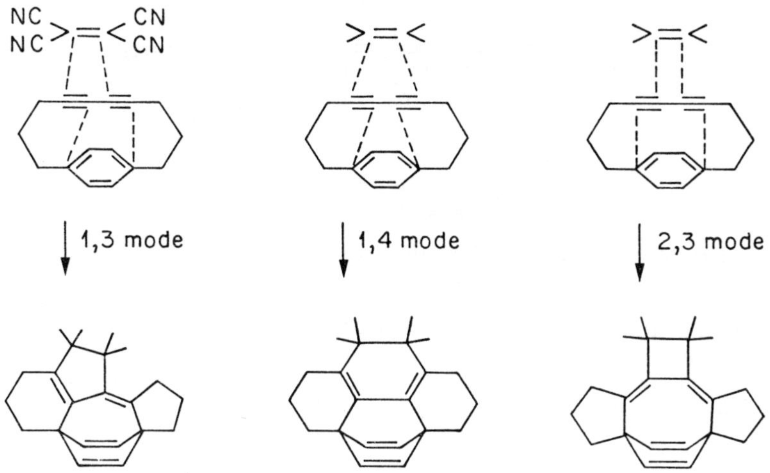

FIGURE 7. Cycloaddition modes of TCNE to [3.3]paracyclophadiyne.

more methylene numbers, while the highly strained cyclophadiyne (**16**) reacted smoothly with TCNE even at room temperature to give **96** [68]. We first thought that

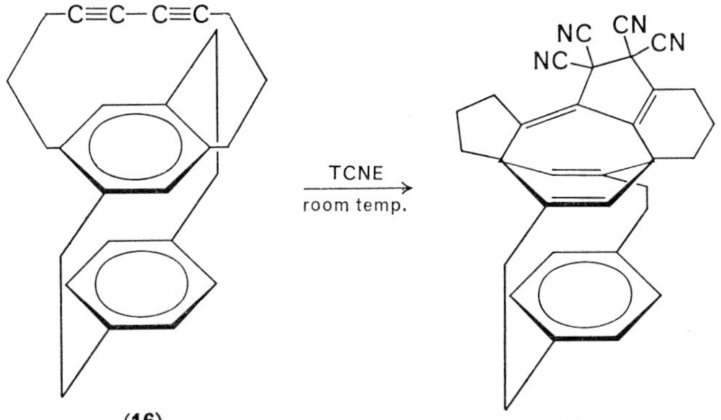

(**16**) (**96**)

they were intriguing examples of a thermally allowed [π4s + π2a + π2s + π2a] pericyclic process, provided that these multicycloadditions proceeded in a concerted manner. However, the mechanism of these reactions has recently been explained by molecular orbital theory, that is, in terms of three-system interaction among the HOMO of the benzene ring, the HOMO of the diyne group and the LUMO of TCNE as shown in Figure 8 [71], in which highly electron-deficient TCNE allows the HOMO–HOMO interaction between the diyne group and the benzene ring to contribute to stabilization and bond formation between them.

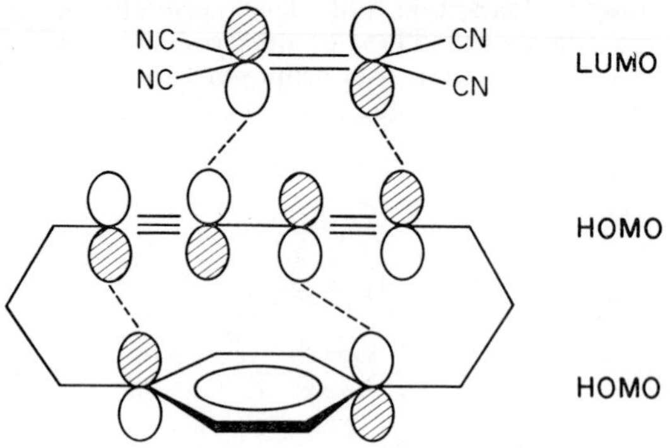

FIGURE 8. The three-system interaction orbital set for multicycloaddition of [3.3]paracyclophadiyne with TCNE.

In the case of the 1,4-anthracenophadiyne **18**, TCNE was exclusively added to the 9 and 10 positions of the anthracene ring to give quantitative yield of an adduct **(97)**[28]. When the adduct **97** was refluxed in tetrachloroethane, an isomeric one-to-one adduct **(98)** was obtained in 38% yield, probably via a retro Diels–Alder process

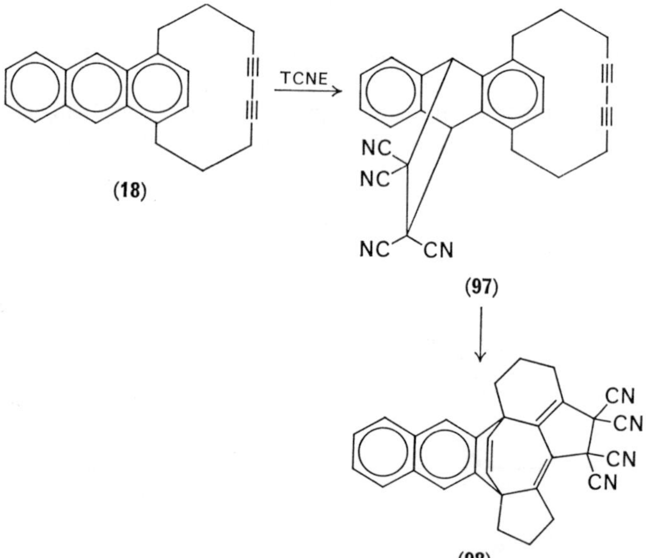

(18) **(97)** **(98)**

followed by the above-mentioned pericyclic reaction. Another anthracenophadiyne **(17a)** also gave a pericyclic adduct with TCNE. This compound gave a novel photochemical reaction on irradiation in benzene[69], yielding a unique photodimer, the [4]radiallene derivative **(100)**, in quantitative yield. Moreover, irradiation of **17a** in a large excess of furan or cyclopentadiene afforded quantitative yields of the multi-cycloadducts **(102)**[69]. The mechanism in the photochemical dimerization and cycloadditions of **17a** is an interesting problem. Extensive studies of these reactions[70] recently demonstrated the extremely strained butatriene **99** to be the most likely intermediate in these photo-induced reactions on the basis of the following results: (*i*) all the reactions are independent of the concentration of the starting material (**17a**), (*ii*) the intermediate (**99**) is yielded on repeated irradiation of **17a** in glycerol–ethanol (1 : 1) matrix at low temperature and reverts quantitatively to the starting

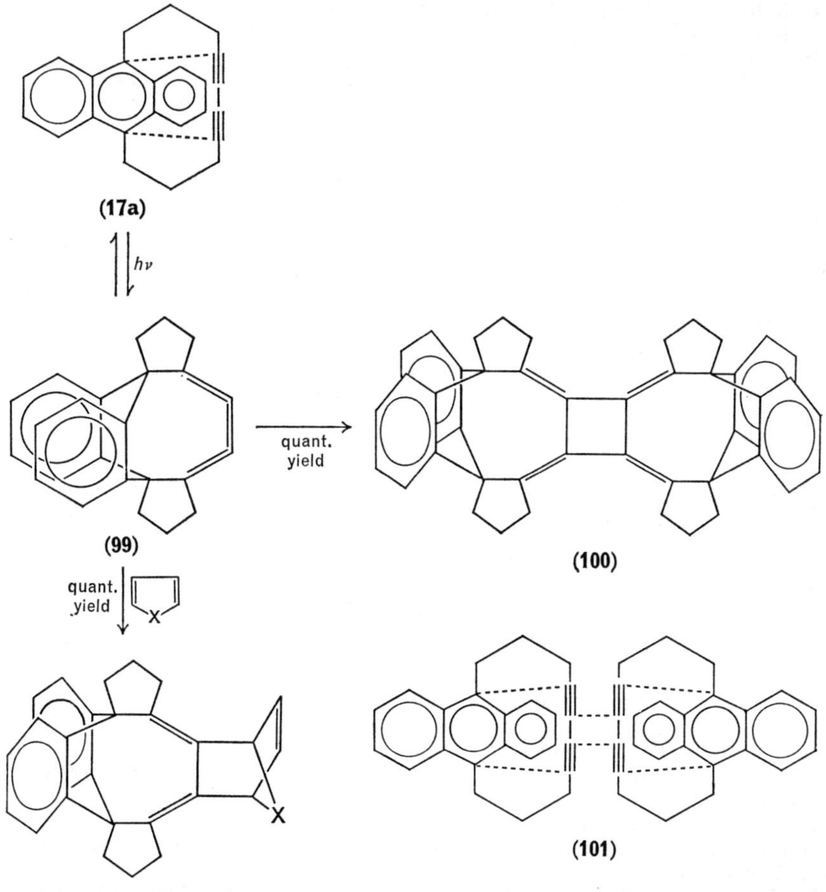

(**102**) (X = CH_2, O)

material by irradiation with shorter wavelength light and (*iii*) the intermediate (**99**) in the same matrix gave its dimer (**100**) on warming up to room temperature without irradiation.

Consequently, it is concluded that the photochemical dimerization proceeds by a two-step mechanism involving [π4s + π4s] and [π2s + π2s] processes and not by the one-step process of [π4s + π2a + π2a + π4s + π2a + π2a] as shown in **101**.

IV. REFERENCES

1. For reviews of cyclic acetylenes: A. Krebs in *Chemistry of Acetylenes* (Ed. H. G. Viehe), Marcel Dekker, New York, 1969, p. 987; M. Nakagawa, this volume, Ch. 15.
2. A. Nissen and H. A. Staab, *Chem. Ber.*, **104,** 1191 (1971).
3. H. A. Staab, H. Mack and A. Nissen, *Chem. Ber.*, **105,** 2310 (1972).

4. A. E. Jungk and G. M. J. Schmidt, *Chem. Ber.*, **104,** 3272 (1971).
5. A. E. Jungk, *Chem. Ber.*, **105,** 1595 (1972).
6. H. Irngartinger, *Chem. Ber.*, **105,** 1184 (1972).
7. E. Kloster-Jensen and J. Wirz, *Helv. Chim. Acta*, **58,** 162 (1975).
8. D. J. Cram and N. L. Allinger, *J. Amer. Chem. Soc.*, **78,** 2518 (1956).
9. C. B. Reese and A. Shaw, *Chem. Comm.*, 331 (1972).
10. A. Etournaud and H. Wyler, *Helv. Chim. Acta*, **56,** 625 (1973).
11. R. Lespieau, *Compt. Rend.*, **188,** 502 (1929).
12. F. Sondheimer, Y. Gaoni and J. Bregman, *Tetrahedron Letters*, 25 (1960).
13. G. Eglinton, I. A. Lardy, R. A. Raphael and G. A. Sim, *J. Chem. Soc.*, 1154 (1964).
14. R. Wolovsky and F. Sondheimer, *J. Amer. Chem. Soc.*, **87,** 5720 (1965).
15. F. Sondheimer, Y. Amiel and R. Wolovsky, *J. Amer. Chem. Soc.*, **79,** 6263 (1957).
16. F. Sondheimer and Y. Amiel, *J. Amer. Chem. Soc.*, **78,** 4178 (1956); **79,** 5817 (1957).
17. F. Sondheimer and Y. Gaoni, *J. Amer. Chem. Soc.*, **81,** 6301 (1959).
18. B. E. Looker and F. Sondheimer, *Tetrahedron*, **27,** 2567 (1971).
19. D. J. Cram and M. F. Antar, *J. Amer. Chem. Soc.*, **80,** 3109 (1958).
20. D. J. Cram and M. Cordon, *J. Amer. Chem. Soc.*, **77,** 4090 (1955).
21a. I. Tinoco, Jr., *J. Amer. Chem. Soc.*, **82,** 4785 (1960); W. Rhodes, *J. Amer. Chem. Soc.*, **83,** 3609 (1961).
21b. T. Ando and M. Nakagawa, *Bull. Chem. Soc. Japan*, **40,** 363 (1967).
22. F. Toda and M. Nakagawa, *Bull. Chem. Soc. Japan*, **33,** 230 (1960); **34,** 862, 874 (1961).
23a. T. Matsuoka, T. Negi, T. Otsubo, Y. Sakata and S. Misumi, *Bull. Chem. Soc. Japan*, **45,** 1825 (1972).
23b. T. Matsuoka, T. Negi and S. Misumi, *Synth. Comm.*, **2,** 87 (1972).
24. T. Kaneda, T. Ogawa and S. Misumi, unpublished results.
25. T. Takabe, M. Tanaka and J. Tanaka, *Bull. Chem. Soc. Japan*, **47,** 1917 (1974).
26. T. Aono, K. Sakabe, N. Sakabe, C. Katayama and J. Tanaka, *Acta Cryst.*, **B31,** 2389 (1975).
27. T. Inoue, T. Kaneda and S. Misumi, unpublished results.
28. Y. Yasufuku, T. Kaneda and S. Misumi, unpublished results.
29. A. J. Hubert and J. Dale, *J. Chem. Soc.*, 86 (1963).
30. D. D. Holder and C. C. Thompson, *Chem. Comm.*, 277 (1972).
31. Y. Yasufuku, T. Kaneda and S. Misumi, unpublished work.
32a. T. Matsuoka, Y. Sakata and S. Misumi, *Tetrahedron Letters*, 2549 (1970).
32b. T. Kaneda and S. Misumi, *Bull. Chem. Soc. Japan*, **50** (1977) in press.
33. K. Harata, T. Aono, K. Sakabe, N. Sakabe and J. Tanaka, *Acta Cryst.*, **A28,** S14 (1972).
34. C. Charrier, D. E. Dorman and J. D. Roberts, *J. Org. Chem.*, **38,** 2644 (1973). Chemical shifts were converted to the ones relative to δ (TMS) using a relation δ (TMS) = 193·7 − δ (CS_2) (p.p.m.).
35. T. Kaneda, T. Inoue, Y. Yasufuku and S. Misumi, *Tetrahedron Letters*, 1543 (1975).
36. G. Bieri, E. Heilbronner, E. Kloster-Jensen, A. Schmelzer and J. Wirz, *Helv. Chim. Acta*, **57,** 1265 (1974).
37. R. Hoffmann and R. B. Woodward, *J. Amer. Chem. Soc.*, **87** 2046 (1965); *Acc. Chem. Res.*, **1,** 17 (1968).
38. S. A. Kandil and R. E. Dessy, *J. Amer. Chem. Soc.*, **88,** 3027 (1966).
39. E. H. White and A. A. F. Sieber, *Tetrahedron Letters*, 2713 (1967).
40. M. S. Newman and M. W. Logue, *J. Org. Chem.*, **36,** 1398 (1971).
41a. G. Eglinton and A. R. Galbraith, *Chem. Ind.* (*London*), 737 (1956); *J. Chem. Soc.*, 889 (1959); for a review, see G. Eglinton and W. McCrae, *Advances in Organic Chemistry*, Vol. 4, Interscience, New York, 1963, p. 225.
41b. H. A. Staab, H. Mack and E. Wehinger, *Tetrahedron Letters*, 1465 (1968).
42. E. Müller, M. Sauerbier and G. Zountsas, *Tetrahedron Letters*, 3003 (1969).
43. E. Müller, J. Heiss and M. Sauerbier, *Ann. Chem.*, **723,** 61 (1969); E. Müller, W. Winter, G. Zountsas and J. Heiss, *Ann. Chem.*, **735,** 99 (1970).
44. G. Büchi, C. W. Perry and E. W. Robb, *J. Org. Chem.*, **27,** 4106 (1962).
45a. J. Ipaktschi and H. A. Staab, *Tetrahedron Letters*, 4403 (1967).
45b. H. A. Staab and J. Ipaktschi, *Chem. Ber.*, **104,** 1170 (1971).
46a. B. Bossenbroek and H. Shechter, *J. Amer. Chem. Soc.*, **89,** 7111 (1967).

46b. B. Bossenbroek, D. C. Sanders, H. M. Curry and H. Shechter, *J. Amer. Chem. Soc.*, **91,** 371 (1969).
47. H. W. Whitlock, Jr. and P. E. Sandvick, *J. Amer. Chem. Soc.*, **88,** 4525 (1966); H. W. Whitlock, Jr., P. E. Sandvick, L. E. Overman and P. B. Reichardt, *J. Org. Chem.*, **34,** 879 (1969).
48a. E. Müller, H. G. Fritz, K. Munk, H. Straub and H. Geisel, *Tetrahedron Letters*, 5167 (1969).
48b. E. Müller and G. Zountsas, *Tetrahedron Letters*, 4531 (1970).
48c. T. J. Barton, A. J. Nelson and J. Clardy, *J. Org. Chem.*, **36,** 3995 (1971).
48d. B. J. Whitlock and H. W. Whitlock, Jr., *J. Org. Chem.*, **37,** 3559 (1972).
48e. H. A. Staab and B. Draeger, *Chem. Ber.*, **105,** 2320 (1972).
49. E. Müller, J. Heiss, M. Sauerbier, D. Streichfuss and R. Thomas, *Tetrahedron Letters* 1195 (1968).
50. E. Müller, K. Munk, P. Ziemek and M. Sauerbier, *Ann. Chem.*, **713,** 40 (1968).
51. E. Müller, *Synthesis*, 761 (1974) and references citied therein; cf. E. Müller, E. Luppold and W. Winter, *Synthesis*, 265 (1975); F. Wagner and H. Meier, *Synthesis*, 324 (1975); H. Straub and J. Hambrecht, *Synthesis*, 425 (1975).
52. Y. Amiel and F. Sondheimer, *Chem. Ind* (*London*), 1162 (1960); F. Sondheimer and Y. Gaoni, *J. Amer. Chem. Soc.*, **83,** 4863 (1961); F. Sondheimer and R. Wolovsky, *J. Amer. Chem. Soc.*, **84,** 260 (1962).
53. F. Sondheimer, R. Wolovsky, P. J. Garratt and I. C. Calder, *J. Amer. Chem. Soc.*, **88,** 2610 (1966).
54. J. Mayer and F. Sondheimer, *J. Amer. Chem. Soc.*, **88,** 602, 603 (1966).
55a. O. M. Behr, G. Eglinton and R. A. Raphael, *Chem. Ind.* (*London*), 699 (1959).
55b. O. M. Behr, G. Eglinton, A. R. Galbraith and R. A. Raphael, *J. Chem. Soc.*, 3614 (1960).
56. G. M. Pilling and F. Sondheimer, *J. Amer. Chem. Soc.*, **93,** 1970 (1971).
57a. R. D. Stephens and C. E. Castro, *J. Org. Chem.*, **28,** 2163, 3313 (1963); J. Burdon, P. L. Coe, C. R. Marsh and J. C. Tatlow, *Chem. Comm.*, 1259 (1967); K. Endo, Y. Sakata and S. Misumi, *Bull. Chem. Soc. Japan*, **44,** 2465 (1971); for a review, see P. Cadiot and W. Chodokiewicz in *Chemistry of Acetylenes* (Ed. H. G. Viehe), Marcel Dekker, New York, 1969, p. 597.
57b. H. A. Staab, A. Nissen and J. Ipaktschi, *Angew. Chem.*, **80,** 241 (1968); *Angew. Chem. Int. Ed.*, **7,** 226 (1968); H. A. Staab, J. Ipaktshi and A. Nissen, *Chem. Ber.*, **104,** 1182 (1971)
58. R. H. Mitchell and F. Sondheimer, *Tetrahedron*, **26,** 2141 (1970).
59. A. S. Hay, *J. Org. Chem.*, **25,** 1275 (1960).
60. J. H. van Roosmalen, E. Jones and H. J. Kevelam, *Tetrahedron Letters*, 1865 (1972).
61. F. Sondheimer, Y. Amiel and R. Wolovsky, *J. Amer. Chem. Soc.*, **81,** 4600 (1959).
62. H. A. Staab and E. Wehinger, *Angew. Chem.*, **80,** 240 (1968); *Angew. Chem. Int. Ed.*, **7,** 225 (1968).
63. H. A. Staab, E. Wehinger and W. Thorwart, *Chem. Ber.*, **105,** 2290 (1972).
64. R. B. King and A. Efraty, *J. Amer. Chem. Soc.*, **92,** 6071 (1970); **94,** 3021 (1972).
65. R. B. King and C. W. Eavenson, *J. Organometal. Chem.*, **16,** 75 (1969); R. B. King and I. Haiduc, *J. Amer. Chem. Soc.*, **94,** 4044 (1972); R. B. King, I. Haiduc and C. W. Eavenson, *J. Amer. Chem. Soc.*, **95,** 2508 (1973).
66. H. B. Chin and R. Bau, *J. Amer. Chem. Soc.*, **95,** 5068 (1973).
67. R. E. Davis, B. L. Barnett, R. G. Amiet, W. Merk, J. S. McKennis and R. Pettit, *J. Amer. Chem. Soc.*, **96,** 7108 (1974).
68a. T. Kaneda, T. Ogawa and S. Misumi, *Tetrahedron Letters*, 3373 (1973).
68b. T. Kaneda, T. Ogawa and S. Misumi, unpublished results.
69. T. Inoue, T. Kaneda and S. Misumi, *Tetrahedron Letters*, 2969 (1974).
70. T. Hayashi, M. Mataga, T. Inoue, T. Kaneda, M. Irie and S. Misumi, *J. Amer. Chem. Soc.*, **99,** 523 (1977).
71. S. Inagaki, H. Fujimoto and K. Fukui, *J. Amer. Chem. Soc.*, **98,** 4693 (1976).

CHAPTER **17**

The electrochemistry of the carbon–carbon triple bond

J. H. P. UTLEY
Queen Mary College, London, England
and
R. LINES
Chemical Centre, University of Lund, Lund, Sweden

I. GENERAL INTRODUCTION

The isolated carbon–carbon triple bond is not easily reduced cathodically or oxidized anodically. Appropriate substitution can lower the potentials required for addition or removal of an electron but even so electron transfer usually occurs at potentials at which further electrochemical oxidation or reduction of products is likely.

A. Ease of Reduction

For reduction, relevant data from polarographic and cyclic voltammetric experiments are summarized in Tables 1 and 2, respectively. For the results in Table 1 the variety of solvents and reference electrodes used makes comparisons difficult. It is clear, however, that even with the activation of a phenyl substituent (entries 6, 7, 9–14) reduction occurs at very cathodic potentials. In this context it is worth noting that in aprotic solvents at *ca.* -3 V (*vs.* S.C.E.) it becomes difficult to distinguish between direct electron transfer to the alkyne and the production of the cathode of solvated electrons. Under the latter conditions the indirect electroreductions[8] show many of the characteristics of dissolving metal reductions (see Section II.B). Even at extreme cathodic potentials it is not clear that an electron is added to the triple bond; the e.s.r. spectra of the radical anions of dimesitylacetylene and (2,4,6,2′,4′,6′-hexa-*t*-butyldiphenyl)acetylene have been interpreted in terms of equal distribution of the odd electron in the aromatic rings[4].

TABLE 1. Polarographic data for reduction of alkynes

Compound	$-E_{\frac{1}{2}}$ (V)	Reference electrode	Solvent–electrolyte	Reference
(1) $HC{\equiv}CCO_2CH_3$	1·35	S.C.E.	H_2O–KCl	1
(2) $HC{\equiv}CCO_2H$	1·45	S.C.E.	H_2O–KCl–HCl (10^{-2} M)	1
(3) n-$C_3H_7C{\equiv}CCOCH_3$	1·99	S.C.E.	DMF–Bu_4NBF_4 (0·5M)	2
(4) n-$C_3H_7C{\equiv}CCO_2CH_3$	2·26	S.C.E.	DMF–Bu_4NBF_4 (0·5M)	2
(5) n-$C_3H_7C{\equiv}CCO_2Li$	2·31	S.C.E.	DMF–Bu_4NBF_4 (0·5M)	2
(6) $PhC{\equiv}CH$	1·97	Hg pool	DMF–Bu_4NI (0·16M)	3
(7) $PhC{\equiv}CPh$	1·69	Hg pool	DMF–Bu_4NI (0·16M)	3
	2·48	S.C.E.	1,2-Dimethoxy-ethane–Bu_4NClO_4 (0·1M)	4
(8) *trans*-$PhCH{=}CHPh$	1·64	Hg pool	DMF–Bu_4NI (0·16M)	3
(9) 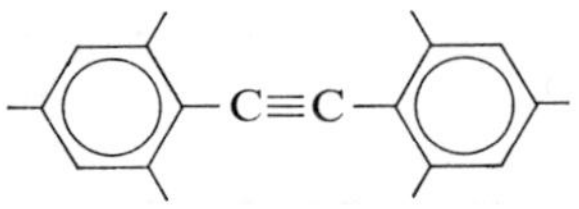	2·70	S.C.E.	1,2-Dimethoxy-ethane–Bu_4NClO_4 (0·1M)	4
(10) 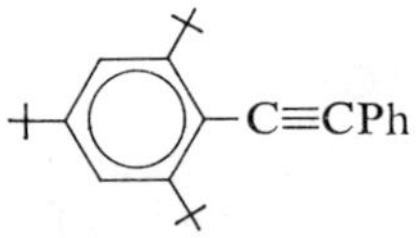	2·75	S.C.E.	1,2-Dimethoxy-ethane–Bu_4NClO_4 (0·1M)	4
(11) 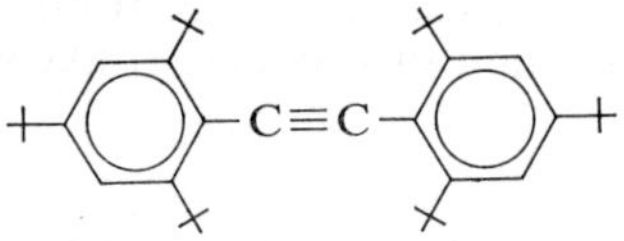	2·93	S.C.E.	1,2-Dimethoxy-ethane–Bu_4NClO_4 (0·1M)	4
(12) $PhC{\equiv}C(CH_2)_3CH_3$	2·65	S.C.E.	DMF–Bu_4NClO_4 (0·05M)	5
	1·88	Cd/$CdCl_2$	DMF–Bu_4NClO_4 (0·1M)	6
(13) $PhC{\equiv}C(CH_2)_4Cl$	1·77	Cd/$CdCl_2$	DMF–Bu_4NClO_4 (0·1M)	6
(14) $PhC{\equiv}C(CH_2)_4Br$	(2·35), 2·60, 2·80	S.C.E.	DMF–Bu_4NClO_4 (0·05M)	5

TABLE 2. Cyclic voltammetry of acetylenes[7]

Compound	Solvent[a]	Cathode[b]	E (V)[c]	$(E-E/2)$ (mV)	ν ($V\ s^{-1}$)	Comments
(1) $MeO_2CC{\equiv}CCO_2Me$	DMF/0·1M TBAI	Pt bead	−1·02	*ca.* 150	0·4	Irreversible up to 150 $V\ s^{-1}$
		Vit. C	−0·83	*ca.* 70		
(2) $MeO_2CC{\equiv}CCO_2Me$	PC/0·1M TBAI	Hg drop	−0·8	100	0·2	Irreversible
		Vit. C	−0·9	130		
(3) $MeO_2CC{\equiv}CCO_2Me$	CH_2Cl_2/0·1M TBAP	Hg drop	−1·2	*ca.* 60	0·2	Irreversible
		Vit. C	−1·26	*ca.* 60		
(4) $MeO_2CC{\equiv}CCO_2Me$	CH_3CN/0·1M TBAP	Pt bead	> −1·4	—	0·4	—
		Vit. C				
(5) $MeO_2CC{\equiv}CCO_2Me$	MeOH/0·1M TBAI	Hg drop	−0·92	100		
			−1·06		0·2	Irreversible
		Vit. C	−1·14	*ca.* 40		
(6) $PhC{\equiv}CCO_2Me$	DMF/0·2M TBAP	Vit. C	−1·9	140	0·3	Irreversible
(7) $PhC{\equiv}CCO_2Me$	CH_3CN/0·1M TBAOAc	Hg drop	−1·4	140	0·2	Irreversible; green coloration at electrode
		Pt bead	−1·5	170		
(8) *trans*-$PhCH{=}CHCO_2Me$	CH_3CN/0·1M TBAOAc	Hg drop	−1·32	—	0·2	Irreversible
		Pt bead	−1·37			
(9) $PhC{\equiv}CPh$	CH_3CN/0·1M TBAOAc	Hg drop	−1·8	130	0·2	Irreversible
		Pt bead	−1·8	*ca.* 200		

[a] PC = propylene carbonate, TBAI = *n*-Bu_4NI, TBAP = *n*-Bu_4NClO_4, TBAOAc = *n*-$Bu_4NOAc.HOAc$.
[b] Vit. C = vitreous carbon.
[c] *vs.* Ag wire.

Conjugation between the triple bond and the carbonyl function lowers the reduction potential considerably whereas alkyl substitution makes reduction more difficult (entries 1–5). A comparison between the half-wave potentials for reduction of PhC≡CPh (1·69 V, *vs.* Hg pool) and *trans*-PhCH=CHPh (1·65 V) substantiates the fact that, at least for this case, a likely product of reduction is more vulnerable to electroreduction than the starting material. In practice electrolyses in protic media aimed at producing alkene from alkyne usually proceed to give alkane.

For reduction of $PhC{\equiv}C(CH_2)_4X$ (X = Cl, Br; entries 13, 14) the relatively low half-wave potentials relate, for the bromide, to cathodic cleavage of the carbon–bromine bond, but for the chloride it is likely that the radical anion of the alkyne is produced, which allows nucleophilic intramolecular displacement of chloride (see Section II.A).

Cyclic voltammetric results are available[7] for a few compounds only (Table 2), but the general conclusions based on polarography are confirmed. In each case the reductions are of activated alkynes and are irreversible, i.e. a fast chemical reaction follows electron transfer. It is also significant that, again, an acetylene ($PhC{\equiv}CCO_2Me$, entry 7) is less easily reduced than the corresponding alkene (*trans*-$PhCH{=}CHCO_2Me$, entry 8), which would be the first-formed product of cathodic hydrogenation.

B. Mechanism of Electroreduction

The mechanism of cathodic hydrogenation, which requires a proton donor, is most probably that given in Scheme 1; the reduction potentials are in the order $E_2 < E_1 \sim E_3 < E_4$. This mechanism is analogous with that well established for the cathodic hydrogenation of carbonyls and polycyclic hydrocarbons[9], and activated

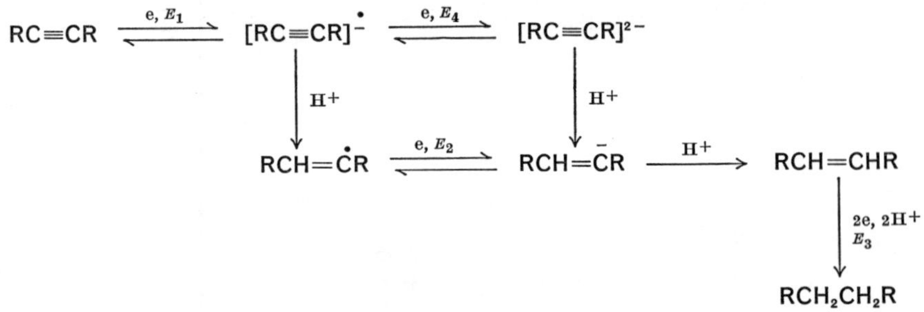

SCHEME 1

alkenes[10]. For hindered alkynes (Table 1, entries 9, 10) reversible one-electron reduction to the radical anions has been observed[4] and in those cases the radical anions are stable at room temperature. Dianions are formed at considerably greater cathodic potentials than are the radical anions (e.g. −1·96 V (Hg pool), *cf.* −1·69 V for PhC≡CPh[3]). Reduction of diphenylacetylene in DMF, at high current density and therefore probably at the second wave, gives in the presence of carbon dioxide products that have been rationalized in terms of trapping of the dianion[3] (Section II.A, Scheme 4).

An alternative mode of reduction involves hydrogenation by cathodically generated hydrogen with the metal electrode surface acting as a hydrogenation catalyst. These reactions have been well reviewed[11] but are discussed briefly in Section II.B.

C. Ease of Oxidation

For the electrochemical oxidation and reduction of alkynes and alkenes an analogy may be drawn with their relative reactivities towards electrophilic and nucleophilic attack. Alkynes are the more easily attacked by nucleophiles and are slightly easier to reduce. Alkynes are, however, much less prone to electrophilic attack than alkenes and are correspondingly more difficult to oxidize electrochemically.

Electron transfer to an anode involves the removal of an electron from the highest occupied molecular orbital and, in the absence of solvent, the ease of this process is reflected in the first ionization potential (I.P.). Electrochemical oxidation must perforce involve a solvent but despite this complication there is a remarkably linear empirical relationship[12] between gas-phase ionization potentials and oxidation half-wave potentials ($E_{\frac{1}{2}}$) referred to the Ag/Ag^+ electrode in acetonitrile. For a considerable number and range of organic compounds the best linear plot of $E_{\frac{1}{2}}$ *vs.* I.P. obeys the equation, $E_{\frac{1}{2}} = 0{\cdot}92(\text{I.P.}) - 6{\cdot}20$. Using this equation and experimental or calculated I.P. values culled from the literature, $E_{\frac{1}{2}}$ values for a number of alkenes and alkynes have been calculated and displayed in Table 3. The calculated $E_{\frac{1}{2}}$ values

TABLE 3. Ionization potentials (I.P.) and oxidation half-wave potentials ($E_{\frac{1}{2}}$)

Compound	I.P. (eV) Calc.[13]	I.P. (eV) Exp.[14]	$E_{\frac{1}{2}}$ (V) (Calc.[a]; *vs.* Ag/Ag^+, MeCN)[12]
t-BuC≡CBu-*t*	7·98	—	1·14
Cyclohexene	—	8·95	2·03 (1·98)
$CH_3CH{=}CHCH_3$	—	9·13	2·20 (2·21)
$CH_3C{\equiv}CCH_3$	9·28	—	2·34
$CH_3CH_2CH{=}CH_2$	—	9·58	2·61 (2·78)
$CH_3CH_2C{\equiv}CH$	10·14	10·18	3·17
$CH_3C{\equiv}CH$	10·35	10·36	3·33
$CH_2{=}CH_2$	—	10·51	3·47 (2·90)
$HC{\equiv}CH$	11·41	11·41	4·30

[a] Values calculated according to $E_{\frac{1}{2}} = 0{\cdot}92(\text{I.P.}) - 6{\cdot}20$; figures in parentheses are values quoted in Reference 12.

cannot be relied upon and clearly the equation used does not hold well for the higher values. However, the pattern is clear; oxidation of alkynes is difficult and probably occurs at potentials beyond those at which solvents and electrolytes usually oxidize (*ca.* 3·0 V *vs.* Ag/Ag^+). Substitution by alkyl groups is expected to lower the oxidation potentials considerably but this has not yet been tested experimentally.

There are few reported examples of preparatively significant electrochemical oxidations of acetylenes and it is doubtful whether any of them involve initial electron transfer from the triple bond. The few significant examples are discussed in Section III.

II. CATHODIC REDUCTION

Electrochemical methods often provide clean and efficient alternatives to conventional synthetic procedures. One such area which has been explored is the cathodic reduction of acetylenes and this section attempts to summarize the results of experiments using various electrolysis conditions.

There are two methods of electrochemically reducing acetylenes, namely, direct charge transfer to the triple bond from the cathode and the electrolytic generation of an intermediate which attacks the acetylene. The first method (direct reduction) has the advantage that mechanistic studies using, for example, cyclic voltammetry and coulometry can be carried out, while the second method (indirect reduction) appears to offer more scope for product control and has been more extensively investigated.

A. Direct Reduction

Direct reduction of acetylenes requires the use of aprotic solvents or cathodes of high hydrogen overvoltage, for example, Hg or Pb. Isolated triple bonds, like isolated double bonds, reduce beyond the accessible potential range of the more common electrolytic solvents*. Electrolytic reduction is therefore confined to activated acetylenic compounds, that is, those containing double bonds or electron-withdrawing substituents conjugated with the triple bond. House and coworkers[2] have formulated empirical rules for estimating the reduction potentials of α,β-unsaturated carbonyl compounds including α,β-acetylenic carbonyl compounds. The general feature of direct reductions is the complete saturation of the triple bond and representative examples are given in Table 4.

The reduction mechanism of diphenylacetylene has been variously interpreted. Laitinen and Wawzonek[16], using aqueous dioxane, proposed protonation *via* the dianion **1** following a slow electron transfer to the anion radical (Scheme 2).

$$\mathrm{PhC{\equiv}CPh + e \rightleftharpoons Ph\dot{C}{=}\bar{C}Ph}$$

$$\mathrm{Ph\dot{C}{=}\bar{C}Ph + e \longrightarrow Ph\bar{C}{=}\bar{C}Ph} \quad \mathbf{(1)}$$

$$\mathrm{Ph\bar{C}{=}\bar{C}Ph + 2H_2O \longrightarrow PhCH{=}CHPh + 2OH^-}$$

$$\mathrm{PhCH{=}CHPh \xrightarrow{2e,\ 2H_2O} PhCH_2CH_2Ph}$$

SCHEME 2

Wawzonek and Wearing[3] in a later study using DMF proposed a similar mechanism but with the formation of the anion radical involving slow electron transfer. Surprisingly, the polarographic data gave no evidence for slow electron transfer and, indeed, reversible behaviour of this compound has recently been observed[17] in THF using cyclic voltammetry. Sioda and coworkers[18] observed two waves for diphenyl acetylene in DMF; the first corresponded to the transfer of three electrons, the second to one electron. The mechanism was formulated as involving the protonation of anion radicals in an ECECE (*E*lectron transfer, *C*hemical reaction, *E*lectron transfer, etc.) process (Scheme 3) and this seems to be quite reasonable in view of the electron affinity of the radical relative to that of the acetylene.

The electrolysis of diphenyl acetylene[3] in the presence of CO_2 gave both diphenylmaleic anhydride and diphenylfumaric acid and these products were cited as evidence for dianion formation. The results, however, can easily fit an anion radical mechanism (Scheme 4).

* Benzene has recently found use for voltammetry[15] and because of its large electroactive range and inertness may provide a medium for mechanistic studies, although the high resistance of the solvent precludes preparative experiments.

TABLE 4. Direct reduction of acetylenes

Compound	Electrode/solvent	Products (%)	Reference
PhC≡CPh	Hg/DMF and 75% dioxane	$PhCH_2CH_2Ph$ (68), $PhCH_2$—CHPh—CHPh—CH_2Ph (6·1)	3
	Hg/DMF/CO_2	Diphenylmaleic anhydride (4), diphenylfumaric acid (8), *meso*-diphenylsuccinic acid	3
CH≡CC(Me)=CHMe	Hg/75% dioxane	EtC(Me)=CHMe	29
PhC≡CBu-*n*	Hg/DMF	$PhCH_2CH_2$Bu-*n*, *cis*/*trans* PhCH=CHBu-*n*, *trans*-$PhCH_2$CH=CHPr-*n*	6
HO_2CC≡CCO_2H	Hg/HCl, KCl	*rac*-HO_2CCH(Me)CH(Me)CO_2H	19
EtO_2CC≡CCO_2Et	Hg/HCl, KCl	$EtO_2CCH_2CH_2CO_2Et$	19
PhC(O)C≡CH	Hg/25% EtOH	PhC(O)CH=CH_2	20
Me_3SiC≡CC(O)Ph	Hg/20% EtOH	Me_3SiCH=CHCH(OH)Ph	22
PhC≡C$(CH_2)_4$Br	Hg/DMF	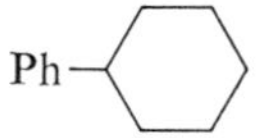 (23), 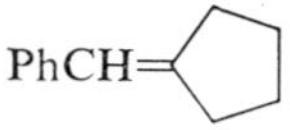 (20), 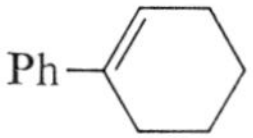 (11), 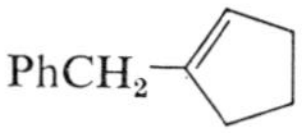 (7), PhC≡CBu-*n* (17), PhCH=CHBu-*n* (13)	5

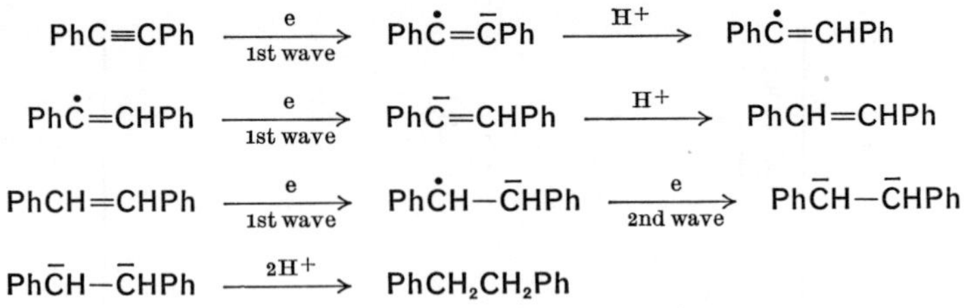

SCHEME 3

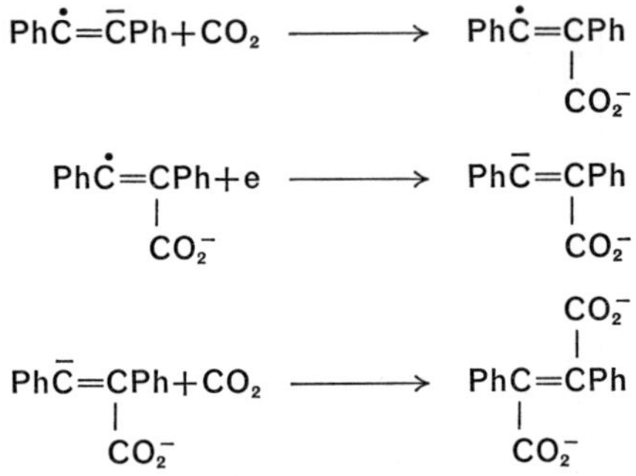

SCHEME 4

Moore and Peters[6] have observed an interesting rearrangement reaction during the electrolysis of 1-phenyl-1-hexyne in DMF at a Hg cathode. When the acetylene was reduced at concentrations of less than 2mM, 1-phenylhexane was formed by the usual 4e process; however, at higher concentrations (2–10mM) the acetylene underwent an isomerization to 1-phenyl-1,2-hexadiene (**2**). This allene, being readily reduced, was converted initially into a mixture of alkenes and finally into 1-phenylhexane (Scheme 5).

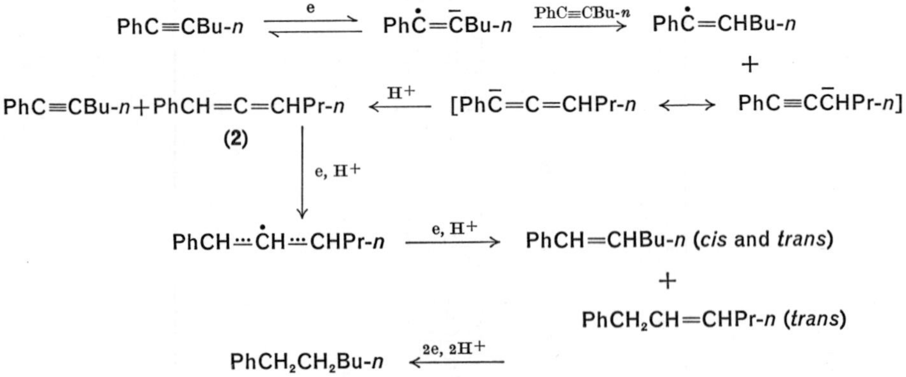

SCHEME 5

The reduction of acetylene dicarboxylic acid[19] (or its monoethyl ester) leads to an interesting hydrodimerization reaction (Scheme 6).

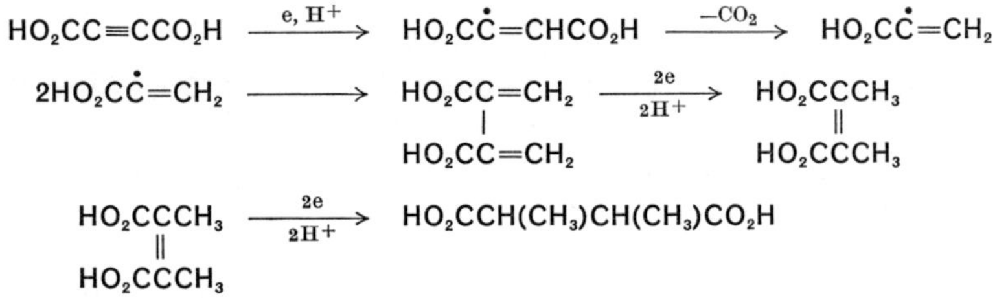

SCHEME 6

α,β-Acetylenic carbonyl compounds undergo several cathodic reactions the nature of which depend on the pH of the electrolyte. In an early study of the reduction of phenylbenzoylacetylene, Prévost and coworkers[20] suggested the following dimerization processes (Scheme 7). The potential difference ($E_2 - E_1$) for the formation of saturated *vs.* unsaturated ketones was *ca.* 0·4 V over the pH range 1·3–8·6.

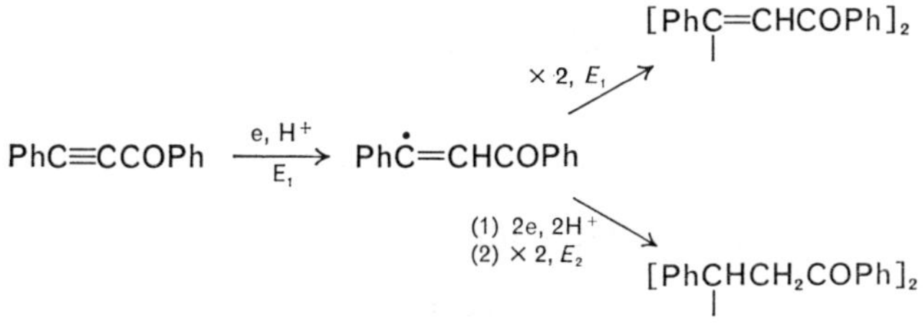

SCHEME 7

This system was reinvestigated for the pH range 2–12 by Degrand and coworkers[21]. They formulated a more complex reduction mechanism, the ultimate products depending on the electrode potential and solution pH. The first step in the reduction of phenylbenzoylacetylene or its 3,4-dimethoxy derivative probably involves the formation of acrylophenones which were further reduced either to the electroinactive acetylenic alcohols or to chalcones which underwent further reduction to the saturated alcohols. Scheme 8 summarizes the pathways believed to be involved.

It is noteworthy that the products from the above reaction are modified[22] if a silicon atom is incorporated in the position α to the triple bond (Scheme 9). The reaction appears to have general applicability, as α,β-unsaturated alcohols were obtained[22] when the phenyl group was replaced by Me, $CH{=}CH_2$ or $C{\equiv}CCH{=}CH_2$.

The electroreductive cyclization of some acetylenic halides in DMF has been reported[5] by Moore and Peters. 6-Bromo-1-phenyl-1-hexyne (**3**) gave three polarographic waves at −2·35 V, −2·60 V and −2·80 V (*vs.* S.C.E.). The first wave was correlated with C–Br fission (*n*-hexyl bromide was reduced at −2·29 V) while the two remaining waves corresponded to triple bond reduction (1-phenyl-1-hexyne gave waves at −2·65 V and −2·88 V)*. The electrolysis reaction mixture contained both five- and six-membered carbocycles as well as straight-chain reduction products

* In another, similar, investigation only one reduction wave was reported[6] for 1-phenyl-1-hexyne in DMF solution.

(Table 4). The formation of the six-membered ring compounds is unique to the electrochemical method, since chemical reduction of the acetylenic halide by, for example, butyllithium yields solely benzylidene cyclopentane (**4**). It was suggested[5]

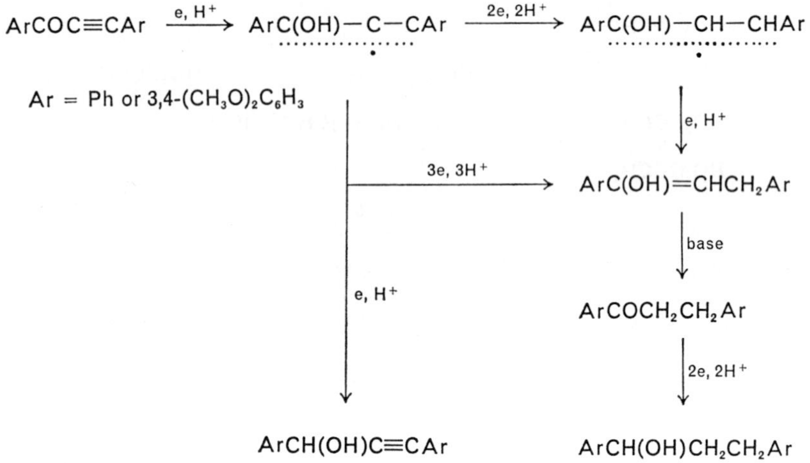

SCHEME 8

SCHEME 9

that the electrode played an important role in the cyclization process either by the formation of organomercury intermediates or by the creation, for the adsorbed radical, of a different environment from that of the homogeneous cyclization process. Scheme 10 was proposed for the formation of the five-membered carbocycles.

First wave (−2·40 V); C–Br *cleavage*

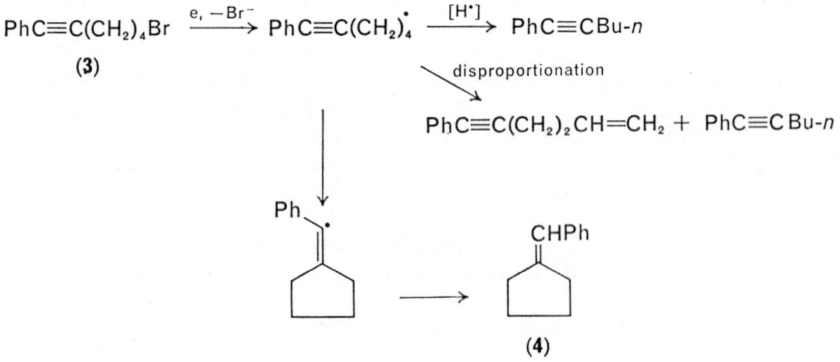

Second and third waves (−2·60 V, −2·80 V); *triple bond reduction*

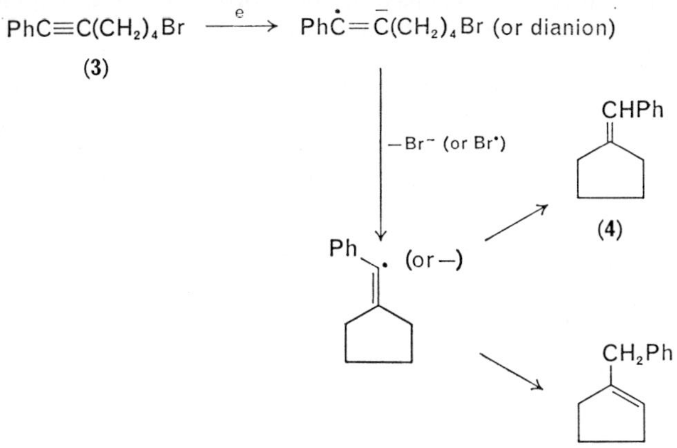

SCHEME 10

B. Indirect Reduction

As described previously, indirect reductions involve the generation of a reagent which reacts with the acetylene. The most common reagent is hydrogen although considerable interest has been shown in dissolving metal (solvated electron) reductions. Table 5 presents some examples of indirect reductions. Reductions

TABLE 5. Indirect reduction of acetylenes

Compound	Electrode/solvent	Products (%)	Reference
PhC≡CH	Spongy Ni/10% H_2SO_4	PhEt, PhCH=CH_2	30
PhC≡CPh	Spongy Ni/10% H_2SO_4	*cis*-PhCH=CHPh (80)	30
n-PrC≡CPr-*n*	Spongy Ni/EtOH, H_2SO_4	*cis*-4-octene (80)	30
	Pt/LiCl, $MeNH_2$	*trans*-4-octene (47) *cis*-4-octene (1)	26
Me_2(CH≡C)COH	Ni/alkaline soln.	Me_2(CH_2=CH)COH (80–90)	25
Et_2C(OH)C≡CH	Cu/Ag alloy/NaOH, EtOH	Et_2C(OH)CH=CH_2 (80)	31
Me_3SiC≡CPh	Pt/LiCl, $MeNH_2$	PhC≡CH (47), PhEt (38)	27
CH≡CH	Cr(II)Cl_2/HCl	CH_2=CH_2 (90)	28

using aqueous acid or alkaline solutions involve hydrogen production at low overvoltage cathodes (Pt, Ni or Co) and acetylenic compounds are reduced in a manner analogous to catalytic hydrogenations. Thus isolated triple bonds can be reduced to give the *cis* alkene, for non-terminal acetylenes (*cf.* direct reduction). An interesting device has been described by Lee and Cashmore[23] for carrying out highly selective and stereospecific hydrogenations. The substrate, e.g. but-2-yn-1,4-diol, is circulated inside a Au/Pd alloy tube whilst hydrogen is cathodically evolved on the outside of the tube. Catalytic reduction takes place following diffusion of hydrogen into the alloy. In the example cited, the *cis* alkene is obtained exclusively. Cathodic hydrogenations usually result, however, in mixtures of both the alkene and alkane and although advantage can be taken of the different rates of hydrogenation of the acetylenic and olefinic bonds on the Pt group metals[24], Ag or Cu cathodes are the metals of choice for selective reduction to alkenes. The reduction pathway on these metals (or Cu/Ag alloys) may involve formation of organometallic intermediates, for example, Lebedeva[25] observed the formation of an organocopper compound during the reduction of dimethylethynyl methanol at a copper cathode. The mechanism of reduction has not been carefully investigated, however, and these compounds may only be side-products; it is worth remembering that with reduction at mercury cathodes small quantities of organomercury compounds are often found.

Solvated electron reductions are performed by electrolysing a solution of LiCl and the acetylene in a basic solvent, for example, $MeNH_2$ or HMPA. In the absence of substrate a deep blue colour develops around the cathode which apparently contains Li^+ bound to a negatively charged solvent complex. This blue complex reacts almost as rapidly as it is formed when a reducible substrate is present. Reduction under these conditions does not differ in principle from that which occurs with alkali metals in liquid ammonia[8]. Benkeser and Tincher[26] found the following results for the indirect reduction of alkyl and aryl acetylenes in a $MeNH_2$/LiCl electrolyte:

Dialkyl acetylenes ⟶ *trans* olefins.

Non-conjugated aromatic acetylenes ⟶ aromatic olefins (*trans* for non-terminal).

Conjugated aromatic acetylenes ⟶ alkylbenzenes.

The stereochemical outcome is consistent with the reductions proceeding in bulk solution, remote from the electrode.

The same workers[27] have also reported the cathodic reduction of an acetylenic silicon compound. (Phenylethynyl)trimethylsilane underwent C–Si cleavage when reduced at a Pt cathode in a $MeNH_2$/LiCl electrolyte (Scheme 11). It is interesting

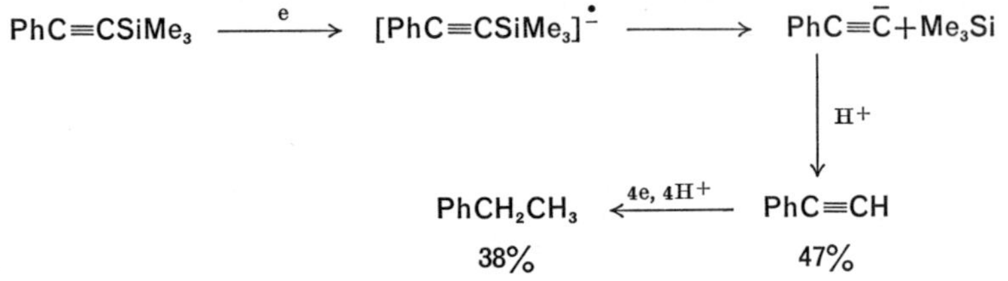

SCHEME 11

to note that the initial reaction did not involve the reduction of the acetylenic bond; however, if isopropyl alcohol was added to the electrolyte, C–Si cleavage was suppressed and complete reduction of the triple bond took place (Scheme 12). It seems likely that under these conditions rapid protonation of the anion radical was responsible for the change in mechanism.

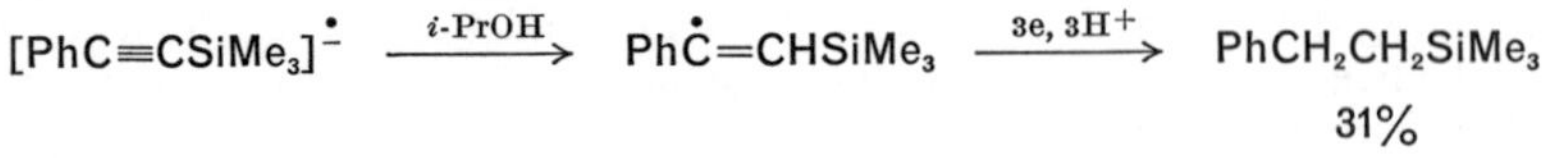

SCHEME 12

Chromium(II) chloride[28] has been suggested as an effective and selective reducing agent for acetylenic bonds and one which may be produced *in situ* at a cathode. Thus acetylene is reduced to ethylene with 90% current efficiency (Scheme 13).

Since the Cr(II) compound can easily be regenerated by electrochemical reduction, the method has been explored as a possibly useful industrial reaction. A major difficulty with the process, however, which is a factor against commercial exploitation, was found to be the slow rate of reaction between the Cr(II) chloride and the acetylene. As a laboratory method it is of interest provided that acid-sensitive substrates, for example acetylenic alcohols[32], are avoided.

$$2CrCl_2 + C_2H_2 + 2HCl \longrightarrow 2CrCl_3 + C_2H_4$$

SCHEME 13

III. ANODIC OXIDATION

The anodic oxidation of acetylene at a gold anode and in aqueous solution has been studied in great detail[33, 34]. In the presence of H_2SO_4, Na_2SO_4 or NaOH, acetylene was partially oxidized at 353 K to give polymers and carbon dioxide. For well-behaved electrochemical reactions the empirical relationship between electrode potential (E) and current density (i) is of the form $E = a + b \log i$ where a and b are constants characteristic of a given reaction. In the case of acetylene oxidation in aqueous solution a discontinuity was observed[34] in the plots of E *vs.* $\log i$ (Tafel curve) which indicates a change of mechanism as anode potential increases. Under the conditions for the partial oxidation of acetylene to carbon dioxide the predominant electrode reaction was oxygen evolution following $OH^- \rightarrow OH^\bullet + e$; it must be concluded therefore that decomposition of the acetylene is *via* radical attack and not *via* electron transfer from the triple bond.

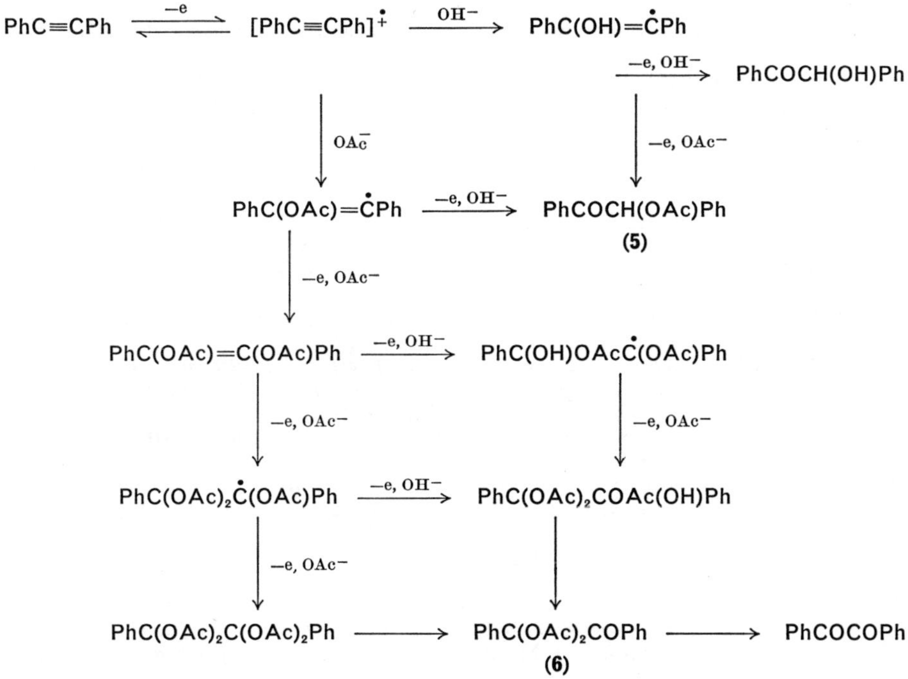

SCHEME 14

The only other alkyne the anodic oxidation of which has been studied in detail is diphenylacetylene. Again it is difficult to be certain about the mechanism involved; the phenyl group and the triple bond may be electroactive but, because of the conjugation between them, it is probably more rigorous to consider electron transfer from the molecule as a whole. The relevant oxidation potential is apparently 2 V (*vs.* S.C.E.), the potential employed for anodic cyanation[35].

Benzoin acetate (**5**) and the keto diacetate (**6**) are the major products (according to gas chromatographic analysis) of oxidation[36] of diphenylacetylene at a carbon anode and in acetic acid containing sodium acetate. The keto-diacetate is readily converted into benzil. Under these conditions acetate ion would be oxidized at $\leqslant 2{\cdot}2$ V (*vs.* S.C.E.) and, because diphenylacetylene is preferentially oxidized, it is likely that the aromatic compound is discharged, as for cyanation, at *ca.* 2·0 V. From the figures given in Table 3 it seems unlikely that localized oxidation of the triple bond is taking place even though it is the triple bond which is acetoxylated. The probable mechanism of formation of the products is indicated in Scheme 14.

In contrast, anodic cyanation of diphenylacetylene occurs exclusively at the aromatic ring and the triple bond remains intact[35]; 4-cyanodiphenylacetylene is formed in 60% yield. The reaction was run at 2 V (*vs.* S.C.E.) for 3·9 F mol^{-1} using a platinum anode in methanol containing sodium cyanide. Anodic cyanation results in preferred attack at the aromatic nucleus in other systems; toluene, mesitylene and hexamethylbenzene give nuclear substitution and little side-chain cyanation under similar conditions[37] in contrast with the corresponding acetoxylation or methoxylation reactions.

IV. REFERENCES

1. V. P. Gul'tyai, S. G. Mairanovskii, N. K. Lisitsina and M. V. Mavrov, *Izv. Akad. Nauk USSR, Ser. Khim.*, **7**, 1655 (1968).
2. H. O. House, L. E. Huber and M. J. Umen, *J. Amer. Chem. Soc.*, **94**, 8471 (1972).
3. S. Wawzonek and D. Wearring, *J. Amer. Chem. Soc.*, **81**, 2067 (1959).
4. H. E. Zimmerman and J. R. Dodd, *J. Amer. Chem. Soc.*, **92**, 6507 (1970).
5. W. M. Moore and D. G. Peters, *Tetrahedron Letters*, 453 (1972).
6. W. M. Moore and D. G. Peters, *J. Amer. Chem. Soc.*, **97**, 139 (1975).
7. R. Lines and J. H. P. Utley, unpublished results.
8. O. J. Jacobus and J. F. Eastham, *J. Amer. Chem. Soc.*, **87**, 5799 (1965); M. M. Baizer (Ed.), *Organic Electrochemistry*, Marcel Dekker, New York, 1973, Chap. XXV.
9. M. M. Baizer (Ed.), *Organic Electrochemistry*, Marcel Dekker, New York, 1973, Chaps. V and VIII.
10. M. M. Baizer (Ed.), *Organic Electrochemistry*, Marcel Dekker, New York, 1973, Chap. XIX; V. G. Puglisi and A. J. Bard, *J. Electrochem. Soc.*, **119**, 829 (1972); E. Lamy, L. Nadjo and J. M. Saveant, *J. Electroanal. Chem. Interfacial Electrochem.*, **42,** 189 (1973); L. A. Avaca and J. H. P. Utley, *J. Chem. Soc., Perkin Trans. II*, 161 (1975).
11. A. P. Tomilov, *Russian Chemical Reviews*, **31,** 569 (1962).
12. L. L. Miller, G. D. Nordblom and E. A. Mayeda, *J. Org. Chem.*, **37,** 916 (1972).
13. L. S. Levitt, *Chem. Ind.* (*London*), 637 (1973).
14. R. W. Kiser, in *Introduction to Mass Spectrometry and its Applications*, Prentice-Hall, Englewood Cliffs, N.J., 1965, p. 308; T. Nakayama and K. Watanabe, *J. Chem. Phys.*, **40,** 558 (1964).
15. R. Lines and V. D. Parker, in the press.
16. H. A. Laitinen and S. Wawzonek, *J. Amer. Chem. Soc.*, **64,** 1765 (1942).
17. S. Wawzonek and D. Wearring, *J. Amer. Chem. Soc.*, **81,** 2067 (1959).
18. R. E. Sioda, D. O. Cowan and W. S. Koski, *J. Amer. Chem. Soc.*, **89,** 230 (1967).
19. I. Rosenthal, J. R. Hayes, A. J. Martin and P. J. Elving, *J. Amer. Chem. Soc.*, **80,** 3050 (1958).
20. C. Prévost, P. Souchay and M. Chauvelier, *Bull. Soc. Chim. Fr.*, 714 (1951).

21. C. Degrand, B. Grosjean and P. L. Compagnon, *Bull. Soc. Chim. Fr.*, 281 (1973).
22. N. M. Deriglazov, B. V. Prokop'ev, V. B. Pukhnarevich and N. V. Komarov, *Khim. Atsetilena, Tr. Vses. Konf., 3rd*, 1968, 276 (1972); *Chem. Abstr.*, **79**, 26532f (1973).
23. J. B. Lee and P. Cashmore, *Chem. Ind. (London)*, **42,** 1758 (1966).
24. J. Kato and M. Sakuma, *J. Electrochem. Soc. Japan*, **25,** 126 (1957).
25. A. I. Lebedeva, *Zh. Obshch. Khim.*, **18,** 1161 (1948).
26. R. A. Benkeser and C. A. Tincher, *J. Org. Chem.*, **33,** 2727 (1968).
27. R. A. Benkeser and C. A. Tincher, *J. Organometallic Chem.*, **13,** 139 (1968).
28. I. L. Kotlyarevskii, *Trudy Vostochno-Sibirskogo Filiala Akad. Nauk SSSR, No. 4*, 67 (1956).
29. A. A. Petrov and V. P. Petrov, *Zh. Obshch. Khim.*, **29,** 3987 (1959).
30. K. N. Campbell and E. E. Young, *J. Amer. Chem. Soc.*, **65,** 965 (1943).
31. A. I. Lebedeva and G. A. Mishnina, *Zh. Obshch. Khim.*, **21,** 1124 (1951).
32. A. E. Favorskii and A. I. Lebedeva, *Bull. Soc. Chim. Fr.*, **6,** 1347 (1939).
33. J. W. Johnson, H. Wroblowa and J. O'M. Bockris, *J. Electrochem. Soc.*, **111,** 863 (1964).
34. J. W. Johnson, J. L. Reed and W. J. Jones, *J. Electrochem. Soc.*, **114,** 572 (1967).
35. K. Yoshida and T. Fueno, *J. Chem. Soc., Chem. Comm.*, 711 (1970).
36. W. J. Koehl, *J. Org. Chem.*, **32,** 614 (1967).
37. L. Eberson and S. Nilsson, *Discuss. Faraday Soc.*, **45,** 242 (1968).

CHAPTER **18**

The preparation of acetylenes and their protection

DAVID A. BEN-EFRAIM
The Weizmann Institute of Science, Rehovot, Israel

I. INTRODUCTION

Due to the comparatively large measure of unsaturation inherent in the carbon–carbon triple bonds, compounds containing them are extremely valuable synthetic intermediates, which can readily be converted to other products. There is therefore great significance in the development of methods of formation and introduction of this multiple bond into organic structures. These methods have been extensively reviewed in the past three decades in books and reviews[1]. The chemistry and synthesis of special groups of acetylenic compounds have also been amply reviewed in recent years[2]. Brandsma's book[3] is a practical handbook detailing numerous laboratory procedures which lead to acetylenes and to compounds related to them. It also describes the handling of alkali metal amides in liquid ammonia, and should be on the shelf of every chemist interested in acetylenes.

The classical method of formation of the triple bond is by elimination of a stable entity from a more saturated structure. In the past the major elimination route has involved the removal of hydrogen halide molecules, i.e. dehydrohalogenation. In recent years organic chemists have imaginatively and ingeniously constructed more complex structures from which other stable moieties could be eliminated, mainly thermally, to furnish acetylenes. These newer methods should still be investigated as to the extent of their applicability and should encourage the planning of even more sophisticated elimination procedures, operating under mild conditions in the presence of sensitive groups contained in the substrates. They should also suppress the concurrent formation of allenes and conjugated dienes, which accompany many elimination reactions.

Besides the generation of the triple bond by elimination, the ethynyl and alkynyl groups can be introduced into existing substrates by substitution, mainly by nucleophilic substitution. A final section in this chapter treats the protection of the C—H bond of terminal acetylenes and of the triple bond itself. Work published in the past decade is stressed. The patent literature has not been consulted but numerous patent references will be found in References 1g, h and 4. The coupling of acetylenic compounds (Glaser coupling and Cadiot–Chodkiewicz coupling) as well as the preparation of acetylenes by prototropic rearrangement are not covered as they may well be treated elsewhere in this volume. The coverage in this chapter extends until approximately the middle of the year 1976.

II. ACETYLENES BY ELIMINATION REACTIONS

Elimination can formally start from saturated compounds (equations 1 and 2) or from unsaturated substrates (equations 3 and 4). As pointed out, dehydrohalogenation has been until now the major route of elimination. However, other moieties

$$\text{—CAB—CXY—} \longrightarrow \text{—C}\equiv\text{C—} \tag{1}$$

$$\rangle\text{CA—CBXY} \longrightarrow \text{—C}\equiv\text{C—} \tag{2}$$

$$\text{—CA=CB—} \longrightarrow \text{—C}\equiv\text{C—} \tag{3}$$

$$\rangle\text{C=CAB} \longrightarrow \text{—C}\equiv\text{C—} \tag{4}$$

have been eliminated, particularly in recent years. These include acids, bases, halogens, alcohols and mercaptans, CO, CO_2, SO_2, Ph_3PO, and above all molecular nitrogen.

A. Dehydrohalogenations

Hydrogen halide eliminations take place from 1,1- and 1,2-dihaloalkanes (equations 5 and 6), as well as from vinyl halides (equations 7 and 8). Dehydrohalogenation

$$\beta\text{-Elimination: —CHXCHX—} \xrightarrow{-2\text{HX}} \text{—C}\equiv\text{C—} \tag{5}$$

$$\beta\text{-Elimination: —CH}_2\text{CX}_2\text{—} \xrightarrow{-2\text{HX}} \text{—C}\equiv\text{C—} \tag{6}$$

$$\beta\text{-Elimination: —CH=CX—} \xrightarrow{-\text{HX}} \text{—C}\equiv\text{C—} \tag{7}$$

$$\alpha\text{-Elimination: } \rangle\text{C=CHX—} \xrightarrow{-\text{HX}} \text{—C}\equiv\text{C—} \tag{8}$$

according to equation (8) involves a rearrangement and is treated separately in Section II.A.9. 1,*n*-Eliminations are also observed and some of them are shown at appropriate locations. To effect the elimination the substrate is treated with a base. A variety of weaker and stronger bases have been used for this purpose and the material reviewed is to a large extent classified according to the bases used. The most popular bases have been until now sodium amide in liquid ammonia and potassium hydroxide and alkoxides in alcoholic solvents. In recent years, however, other bases have been introduced as dehydrohalogenating agents, such as quaternary ammonium hydroxides (Section II.A.6) and fluoride ions (Section II.A.7). The right combination of base, solvent and temperature is of utmost significance in selective, partial or total elimination of hydrogen halide, as is convincingly illustrated in Scheme 1 [5]. This sequence of reactions demonstrates several generalizations: (*i*) that the elimination of a molecule of hydrogen halide from a 1,2-dihaloalkane is faster than from a vinyl halide, the latter demanding a stronger base (*t*-BuOK in

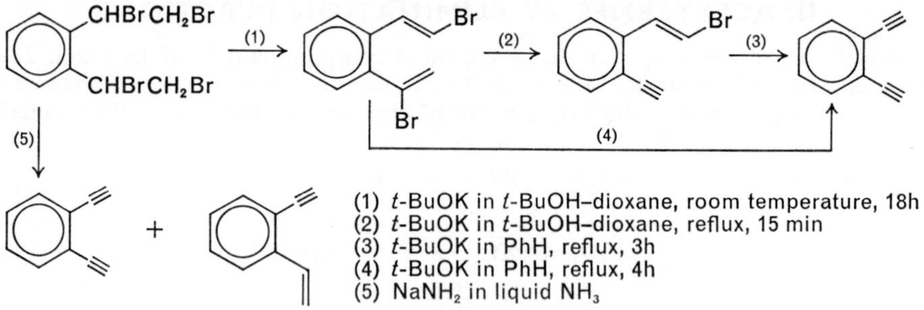

SCHEME 1

benzene *vs* *t*-BuOK in *t*-BuOK–dioxane); (*ii*) that *trans* elimination from vinyl halides is much more favoured than *cis* elimination; and (*iii*) that sodium amide in liquid ammonia is a much more efficient eliminating agent, operating under milder conditions than oxygen bases. Further examples which emphasize the stepwise dehydrohalogenation of 1,2-dihaloalkanes by sequential utilization of weaker and stronger bases involve the preparation of phenylpropargyl aldehyde (equation 9)[6],

$$\text{PhCH}{=}\text{CHCHO} \xrightarrow{\text{Br}_2} \text{PhCHBrCHBrCHO} \xrightarrow{\text{K}_2\text{CO}_3,\ \text{AcOH}} \text{PhCH}{=}\text{CBrCHO}$$

$$\xrightarrow{\text{HC(OEt)}_3} \text{PhCH}{=}\text{CBrCH(OEt)}_2 \xrightarrow[\text{reflux}]{\text{KOH, abs. EtOH}} \text{PhC}{\equiv}\text{CCH(OEt)}_2$$

$$\xrightarrow{\text{dil. H}_2\text{SO}_4} \text{PhC}{\equiv}\text{CCHO} \qquad (9)$$

1,8-diethynylnaphthalene (Scheme 2)[7], and *sym*-dibenzo-1,5-cyclooctadiene-3,7-diyne and *sym*-dibenzo-1,3,5-cyclooctatrien-7-yne (Scheme 3)[8]. If the hydrogen of

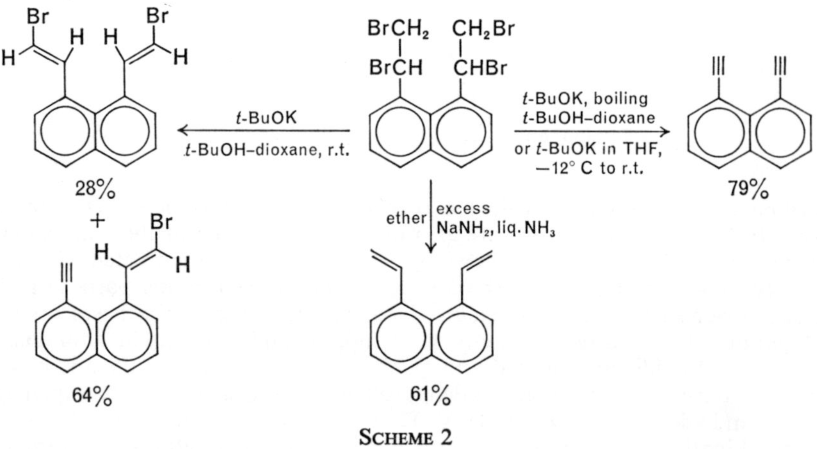

SCHEME 2

the hydrogen halide to be eliminated is made more acidic by electronegative substituents, elimination will take place even with an aqueous base (equations 10 [9] and 11 [10]).

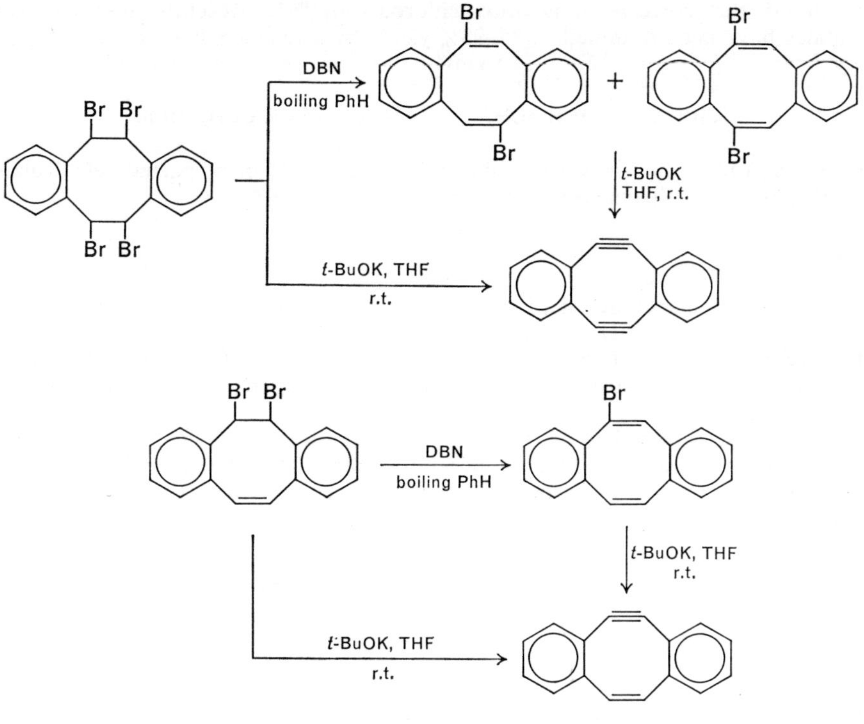

DBN = 1,5-diazobicyclo[4.3.0]non-5-ene

SCHEME 3

$$o\text{-NCC}_6\text{H}_4\text{CHBrCHBrCO}_2\text{H} \xrightarrow{10\%\ \text{aq. KOH}} o\text{-NCC}_6\text{H}_4\text{C}{\equiv}\text{CCO}_2\text{H} \tag{10}$$

$$\text{R}^1\text{R}^2\overset{+}{\text{N}}(\text{CH}_2\text{CH}{=}\text{CClMe})_2\ \text{X}^- \xrightarrow{\text{aq. NaOH}} \text{R}^1\text{R}^2\overset{+}{\text{N}}(\text{CH}_2\text{C}{\equiv}\text{CMe})_2\ \text{X}^- \tag{11}$$

Elimination of hydrogen halide from trifluoromethyldihaloalkanes by potassium hydroxide to yield acetylenes is found to proceed faster with the bromo and iodo compounds than with the chloro compounds. Hydrogen fluoride is not eliminated at all[11].

The mechanism and stereochemsitry of dehydrohalogenations from vinyl halides have been extensively reviewed[1k, 12]. It should however be pointed out that the effects of particular bases and solvents and of temperature cannot always be predicted with confidence as to rate of reaction and product distribution (acetylene, allene, diene)[13]. Therefore a variety of combinations of base, solvent and reaction conditions should be tried in order to obtain satisfactory results.

I. Starting materials

vic-Dihaloalkanes are generally obtained by bromination of the appropriate olefin[14a], and *gem*-dihaloalkanes are obtained by chlorination of aldehydes and ketones with phosphorus pentachloride, which occasionally also yields a mixture containing the corresponding monochloroalkene[14b, c]. Recently terminal *gem*-dihalides have been obtained in 95–98% yields by a reaction between a halide and vinyl bromide (equation 12)[14d]. The vinyl halides are generally obtained by a Wittig

$$R^1R^2R^3CCl + BrCH{=}CH_2 \xrightarrow[-20\text{ to }-10\,^\circ C]{AlCl_3} R^1R^2R^3CCH_2CHClBr \qquad (12)$$

reaction of a phosphorane with an aldehyde or ketone[14e], or by partial dehydrohalogenation of dihaloalkanes (see, for example, Schemes 1–3).

2. Oxygen bases

The bases used include alkali metal carbonates (Na_2CO_3, K_2CO_3), alkali metal hydroxides (NaOH, KOH), and alkali metal alkoxides, such as EtONa and *t*-BuOK. They are generally used in excess and in high concentration. Of these KOH has been in the past one of the most popular, and it was used either in a refluxing solution or in the molten state.

A drawback of these bases is that in the preparation of terminal aliphatic acetylenes they may cause, at the high temperatures used (100–200 °C), partial prototropic rearrangement to the 2-alkynes via the intermediate allenes, as illustrated in equation (13)[15]. Hence, these dehydrohalogenation reagents are preferably used to

$$Me_2CHCH_2CHCl_2 \xrightarrow[170\,^\circ C]{KOH,\ HOCH_2CH_2OH} \underset{40\%}{Me_2CHC{\equiv}CH} + \underset{29\%}{Me_2C{=}C{=}CH_2} \qquad (13)$$

$$Me_2C{=}C{=}CH_2 \xrightarrow[170\,^\circ C]{NaNH_2,\ isooctane} Me_2CHC{\equiv}CH$$

prepare arylacetylenes. The thermodynamic equilibrium generally favours the more stable disubstituted acetylene over the terminal acetylene[16] and since the elimination is generally carried out in solution under true equilibrium conditions and at higher temperatures, the isomerization is facilitated. The shift of the equilibrium in the opposite direction, i.e. towards the 1-alkyne, in eliminations by sodium amide in liquid ammonia, will be discussed in Section II.A.3.

The triple bond, being more electrophilic than the double bond, is more susceptible to attack by strong nucleophiles. Hence, it might in the presence of a base in an alcohol add a molecule of the latter to yield a vinyl ether and thus its own yield would be decreased, as illustrated in equation (14), where only the vinyl ether

$$\underset{(1)}{PhCHBrCHBrCO_2Et} \xrightarrow{EtONa,\ EtOH} PhC(OEt){=}CHCO_2Et \qquad (14)$$

is formed as major product[17]. However, in the case of elimination from the corresponding acid, the alkyne is obtained in 77–81% yield (equation 15) because the

$$PhCHBrCHBrCO_2H \xrightarrow[reflux]{KOH,\ MeOH} PhC{\equiv}CCO_2H \qquad (15)$$

carboxylate ion formed diminishes the electrophilicity of the triple bond[18]. One way to overcome this drawback is by the use of a base without solvent (e.g. equation 16)[19].

$$PhCH{=}CHBr \xrightarrow[200\,^\circ C]{KOH} \underset{67\%}{PhC{\equiv}CH} \qquad (16)$$

A more ingenious route involves the dehydrobromination of ester **1** in a benzene solution containing a catalytic amount of ethanol and using sodium hydride as base. The ethoxide ion, being a stronger base in benzene than in ethanol, deprotonates the substrate and is converted to ethanol. As a result much less ethoxide is available for addition to the triple bond, thus affording the ethyl phenylpropiolate and the ethyl β-ethoxycinnamate in *ca.* 56% and 27% yield, respectively[20].

Dehydrohalogenation by oxygen bases from vinyl halides proceeds readily by *trans* elimination via an *E*2-type of mechanism. *cis*-Elimination is sluggish or does not occur at all. Thus *trans* elimination from bromovinyl ethers **2** (R = alkyl) furnishes the acetylenes **3** in a fast reaction in about 90% yield, whereas elimination from **4** (R = alkyl) is very sluggish and yields a mixture of the acetylene **3** and allene **5** (equations 17 and 18)[21]. Similar observations were shown in Schemes 1 and

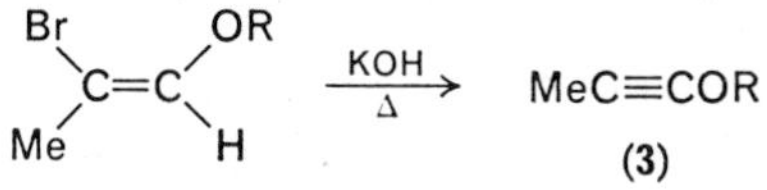

(2) **(3)** (17)

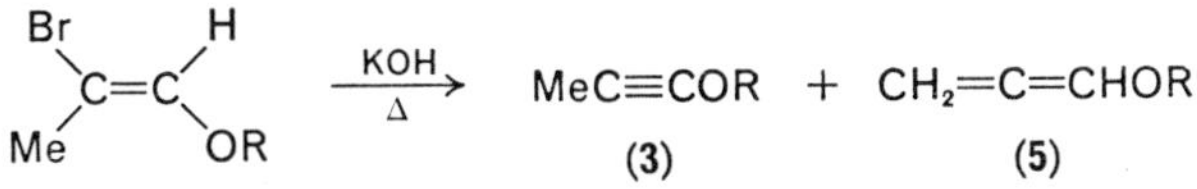

(4) **(3)** **(5)** (18)

2. The *cis* isomer of the vinyl bromide **6** yields the corresponding acetylene **7** by a fast *trans* elimination, whereas the *trans* isomer does not react at all (equation 19)[22].

$$\underset{\textbf{(6)}}{\mathrm{CHBr\overset{c}{=}CHCH_2NHBu}\text{-}n} \xrightarrow[77\,^\circ\mathrm{C}]{\text{NaOH, aq. dioxane}} \underset{\textbf{(7)}}{\mathrm{HC{\equiv}CCH_2NHBu}\text{-}n} \quad (19)$$

By contrast, elimination from both isomers using sodium amide in liquid ammonia proceeds smoothly to yield the acetylene **7** [22]. In the case of the β-bromovinyl ketones **8** both *trans* and *cis* elimination proceed in good yields from the respective isomers, but *trans* elimination is faster by one order of magnitude (equation 20)[23].

$$\underset{\textbf{(8)}\ \ (cis \text{ and } trans)}{i\text{-}\mathrm{PrCBr{=}CHCOMe}} \xrightarrow[\text{r.t.}]{\mathrm{K_2CO_3}\text{, aq. MeOH}} i\text{-}\mathrm{PrC{\equiv}CCOMe} \quad (20)$$

The kinetics of dehydrohalogenation from the configurational isomers of vinyl halides have been determined for numerous reactions. Thus *cis*-*p*-nitro-β-bromostyrene in the presence of ethanolic NaOH is converted quantitatively by *trans* elimination to *p*-nitrophenylacetylene within a few minutes (equation 21), whereas the *trans* isomer hardly reacts at all in that short time. However, the latter affords 1,1-diethoxy-2-*p*-nitrophenylethane in high yield when kept under the above conditions for 20 days (equation 22). The mechanism of the latter reaction could not

(21)

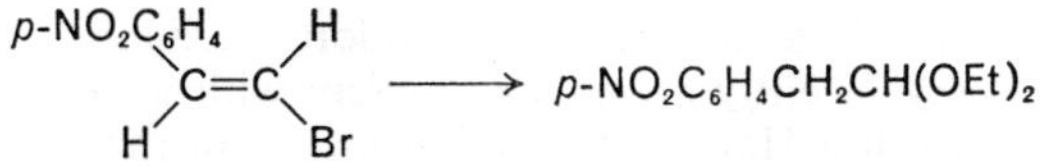

(22)

be elucidated but it was found that the formation of the acetylene from the *cis*-haloolefin (by *trans* elimination) was 2300 times as rapid as the formation of the acetal from the *trans* haloolefin[24]. Schemes 4 [25] and 5 [13b] illustrate that *trans* elimination is faster than *cis* or α-elimination, but that the relative rates of reaction are solvent- and base-dependent. Additional data are found in Banthorpe's book[26].

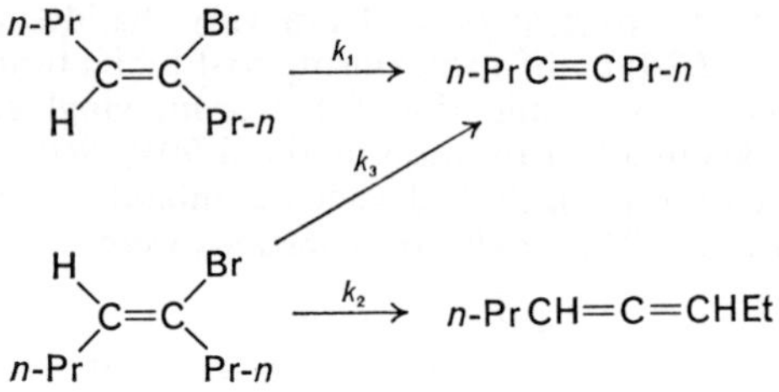

$k_1/k_2 \sim 44$; $k_2/k_3 \sim 1500$; $k_1/k_3 \sim 7 \times 10^4$

Base = MeONa–MeOH–DMSO

SCHEME 4

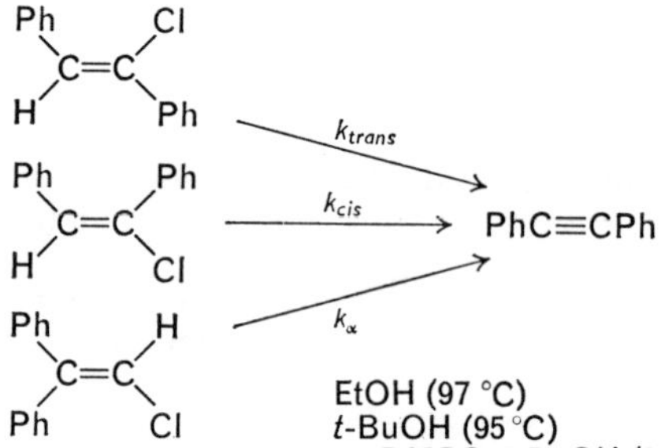

Base = *t*-BuOK

	k_{trans}	k_{cis}	k_α
EtOH (97 °C)	2×10^4	100	1
t-BuOH (95 °C)	3000	600	1
5M DMSO–*t*-BuOH (25 °C)	10^8	6×10^5	1
DMSO (19 °C)	250	40	1

SCHEME 5

Numerous papers of recent years indicate that the mechanisms of dehydrohalogenation of vinyl halides are mainly of the *E*2-type, but that other mechanisms may also compete. Dehydrochlorination of vinyl chlorides with MeONa or EtONa in the corresponding alcohols is shown to proceed by an *E*2 mechanism[27]. In the case of the *cis*- and *trans*-β-chloro-4-nitrostyrenes elimination with methanolic MeONa competes favourably with substitution only in the *cis* isomer. The primary isotope effect, k_H/k_D is about 1·6–2·2 and there is no H, D exchange, which points to an *E*2-like mechanism with a large carbanionic character[28]. Similar isotope effect studies have shown that elimination from *cis*-β-halostyrenes in a series of bases (MeO^- in MeOH to t-BuO^- in *t*-BuOH) involves a variable *E*2 mechanism[29]. On the other hand, studies based on H, D exchange, isotope effects, order of reaction and relative reaction rates in elimination from *cis*- and *trans*-chlorostyrenes with alkali metal alkoxides have been taken as evidence for three competing mechanisms, namely, *E*2, *E*1cB, and *E*1cB–HBA [30]. Bordwell has concluded that in the elimination of HBr from *cis*-1,2-dichloroethylene a reversible anion mechanism, $(E1cB)_R$, is involved (equation 23)[31]. Very recently an *E*2 mechanism with an *E*1cB-like transition state has been inferred from a Hammett correlation, i.e. in the case of elimination from

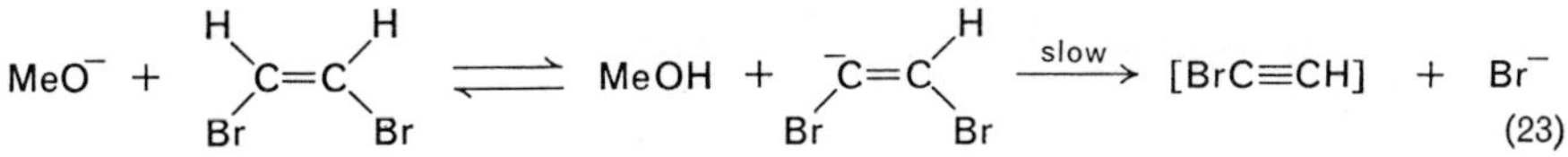

(23)

para-substituted methyl β-chlorocinnamates by MeONa and EtONa in their respective alcohols[32].

a. *Alkali metal carbonates*. Generally these relatively weak bases can only induce dehydrohalogenation from 1,2-dihaloalkanes to the corresponding vinyl halides, as illustrated in equation (9)[6] (Section II.A) or in equation (24)[33]. By contrast, elimination from β-bromo α,β-unsaturated ketones with K_2CO_3 in aqueous MeOH proceeds smoothly to yield α,β-acetylenic ketones (see equation 20)[23].

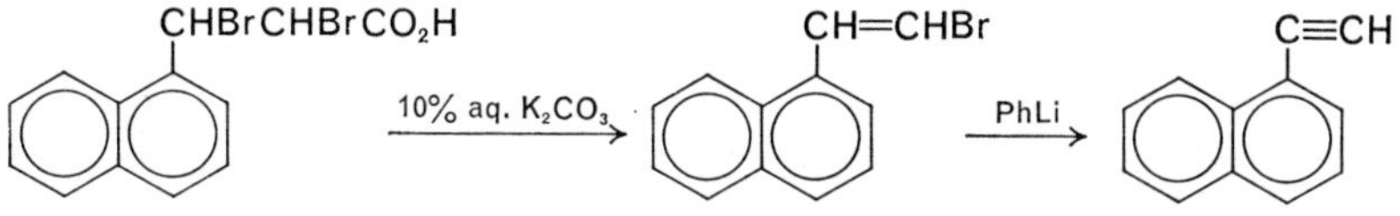

(24)

b. *Alkali metal hydroxides*. Use of these bases without solvent has the advantage of preventing alcohol addition to the formed acetylene (Section II.A.2). In fact many types of functionalized acetylenes have been obtained by distilling them out of the mixtures of their halogenated precursors with KOH pellets at temperatures between 150 and 200 °C. In this manner were obtained arylacetylenes (equation 16)[19], 1-alkynyl ethers (equation 25)[21a, 34], 1-alkynyl thioethers (equation 26)[35], 1-nitroacetylenes (equation 27)[36] and highly fluorinated alkylacetylenes (equation 28)[37].

$$\text{MeCHBrCHBrOEt} \xrightarrow[\text{reflux}]{\text{PhNEt}_2,\ \text{PhH}} \text{MeCBr{=}CHOEt} \xrightarrow[150\text{–}180\ ^\circ\text{C}]{\text{powdered KOH}} \text{MeC}{\equiv}\text{COEt} \qquad (25)$$

$$\text{EtSCH{=}CHBr} \xrightarrow[150\ ^\circ\text{C}]{\text{KOH}} \text{EtSC}{\equiv}\text{CH} \qquad (26)$$

$$t\text{-BuC}{\equiv}\text{CH} \xrightarrow{\text{N}_2\text{O}_4+\text{I}_2,\ \text{ether}} \underset{\textit{cis}\ \text{and}\ \textit{trans}}{t\text{-BuCI{=}CHNO}_2} \xrightarrow[100\ ^\circ\text{C},\ 0.1\ \text{mm}]{\text{KOH}} \underset{94\%}{t\text{-BuC}{\equiv}\text{CNO}_2} \qquad (27)$$

$$\text{R}_f\text{CH{=}CHPh} \xrightarrow{\text{Br}_2} \text{R}_f\text{CHBrCHBrPh} \xrightarrow[\text{distillation}]{\text{powdered KOH}} \underset{60\text{–}90\%}{\text{R}_f\text{C}{\equiv}\text{CPh}} \qquad (28)$$

Many more dehydrohalogenations are, however, being carried out in aqueous or alcoholic solution under reflux. Under these conditions the reaction is particularly suitable for the preparation of acetylenes which will not prototropically isomerize, such as arylacetylenes. The following examples illustrate the large number of structural types which can be converted to acetylenes while surviving the strong reaction conditions (equations 9, 29–35). A facile conversion of aryl ketones and

$$\text{MeCCl{=}CHCH}_2\text{OH} \xrightarrow[\text{reflux}]{\text{aq. NaOH}} \underset{40\%}{\text{MeC}{\equiv}\text{CCH}_2\text{OH}} \qquad (\text{Ref. 38})\ (29)$$

$$\text{PhCH}_2\text{COMe} \xrightarrow{\text{PCl}_5} \text{PhCH{=}CClMe} \xrightarrow[\text{reflux}]{\text{NaOH, abs. EtOH}} \underset{\substack{74\% \\ \text{overall yield}}}{\text{PhC}{\equiv}\text{CMe}}$$

(Ref. 39) (30)

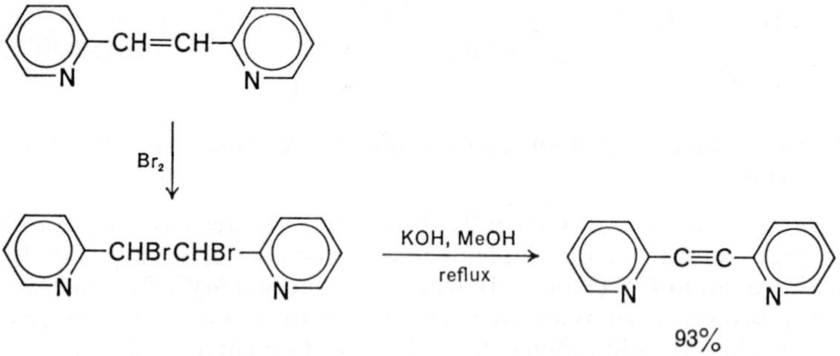

(Ref. 40) (31)

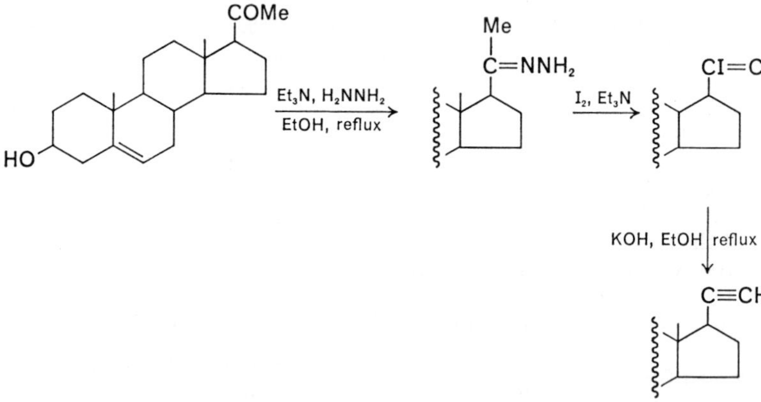

(Ref. 41) (32)

$$\mathrm{Me(CH_2)_7CH{=}CH(CH_2)_7CO_2Me \xrightarrow{Br_2} Me(CH_2)_7CHBrCHBr(CH_2)_7CO_2Me}$$

$$\xrightarrow[\text{(2) HCl}]{\text{(1) KOH–}n\text{-C}_5\text{H}_{11}\text{OH, 150 °C}} \mathrm{Me(CH_2)_7C{\equiv}C(CH_2)_7CO_2H} \quad \text{(Ref. 42)} \quad (33)$$

32–42%

$$\mathrm{ClCH_2CH{=}CClMe} \xrightarrow[n\text{-butyl Cellosolve, 165–170 °C}]{\text{KOH, HOCH}_2\text{CH}_2\text{OH}} \mathrm{HC{\equiv}CCH{=}CH_2} \quad \text{(Ref. 43)} \quad (34)$$

43–48%

$$\mathrm{ClCH_2C{\equiv}CCH_2Cl} \xrightarrow[\text{reflux}]{\text{aq. KOH, dioxane}} \mathrm{HC{\equiv}CC{\equiv}CH} \quad \text{(Ref. 44)} \quad (35)$$

60%

α,β-unsaturated aryl ketones to arylacetylenes which involves the use of the Vilsmeyer complex is shown in equation (36)[45].

$$\mathrm{ArCOMe + [Me_2NCHCl]^+PO_2Cl_2^- \longrightarrow [ArCCl{=}CHCH{=}NMe_2]^+PO_2Cl_2^-}$$

$$\xrightarrow{\text{H}_2\text{O}} \mathrm{ArCCl{=}CHCHO} \xrightarrow[\text{90 °C}]{\text{5\% aq. NaOH, dioxane}} \mathrm{ArC{\equiv}CH} \quad (36)$$

64–98%

In spite of the examples cited above, it has been reported that occasionally only one mole of hydrogen halide is eliminated under the above reaction conditions (equation 37)[46].

$$\mathrm{Me(CH_2)_7CHBrCH_2Br} \xrightarrow[\text{reflux}]{\text{KOH, EtOH}} \mathrm{Me(CH_2)_7CBr{=}CH_2 + Me(CH_2)_7CH{=}CHBr} \quad (37)$$

cis and *trans*

It is interesting to compare yields of an acetylene obtained from the same precursor under different reaction conditions (Scheme 6)[47]. It is found that dehydrohalogenation in solution is more efficient and that alkoxides are superior to

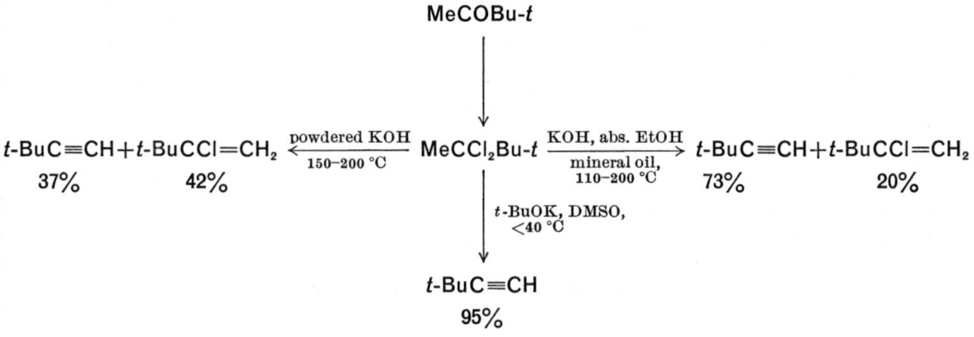

SCHEME 6

hydroxides as bases. Improvements in yields can be achieved by using the alkali metal hydroxides in dipolar aprotic solvents such as DMSO, in which they are much stronger bases. They can then readily dehydrohalogenate *gem*- and *vic*-dihaloalkanes at 130–160 °C to furnish the acetylenes in high yield. Dichloroalkanes react somewhat more slowly than the dibromides, and the temperature may be lowered when alkoxides replace KOH or NaOH. *t*-Butylacetylene is thus obtained in 91% yield (equation 38)[14d].

$$t\text{-BuCH}_2\text{CHClBr} \xrightarrow[130\text{–}160\ ^\circ\text{C}]{\text{KOH, DMSO}} \underset{91\%}{t\text{-BuC}\equiv\text{CH}} \qquad (38)$$

c. *Alkali metal alkoxides*. These bases have proved to be efficient in dehydrohalogenating both hindered acyclic and strained cyclic systems as already illustrated in Schemes 1–3 and 6. As indicated in Section II.A.2.b this efficiency is enhanced by using the alkoxides in aprotic dipolar solvents. Further examples are given in equations (39)–(41). The utilization of *t*-BuOK in the synthesis of *t*-butoxyacetylenes

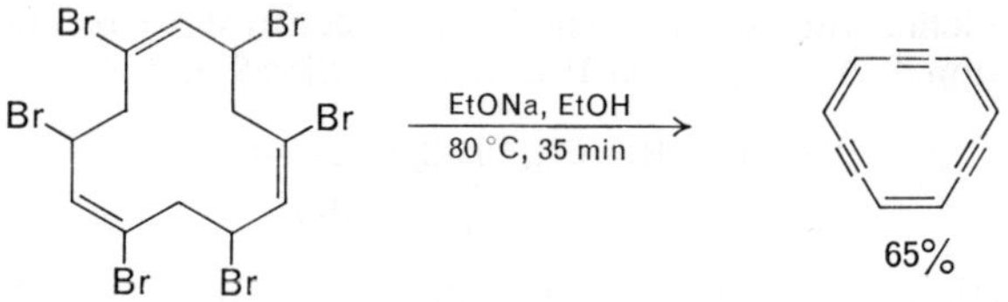

(Ref. 48) (39)

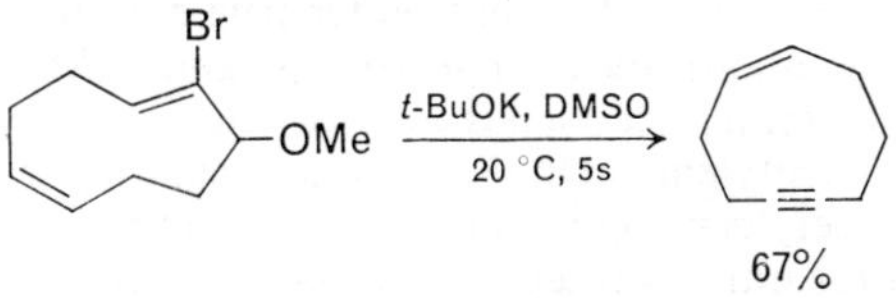

(Ref. 49) (40)

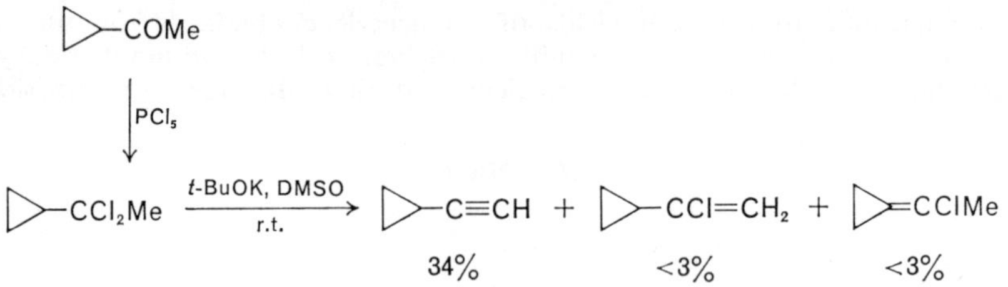

(Ref. 14c)

(41)

in which the free base serves as a dehydrohalogenating agent as well as a nucleophile is exemplified in equation (42)[50].

$$\text{PhCH=CClF} \xrightarrow{\text{2 eq. } t\text{-BuOK}} \text{PhC}\equiv\text{COBu-}t \quad (42)$$

ca. 50%

3. Alkali metal amides

Of the alkali metal amides sodium amide is more frequently used than any other amide. In the past it was used in mineral oil at elevated temperatures (equation 43)[51]

$$n\text{-}C_7H_{15}CCl_2Me \xrightarrow[\text{170 °C}]{\text{NaNH}_2\text{, mineral oil}} n\text{-}C_7H_{15}C\equiv CH \quad (43)$$

but at present it is most widely used in liquid ammonia at its boiling point (−33 °C) or below. It is a powerful elimination agent in that solvent operating under mild conditions. Sodium amide in liquid ammonia may also be used in an autoclave at room temperature under high pressure. It is desirable to prepare the sodium amide directly from sodium in the liquid ammonia[3] and to use three equivalents of it in the preparation of terminal acetylenes from dihaloalkanes, in order to precipitate the sodium acetylide and thus avoid possible prototropic isomerization (equation 44).

$$RCHXCH_2X + 3NaNH_2 \longrightarrow RC\equiv CNa + 2NaX + 3NH_3 \quad (44)$$

The acetylide so formed can be directly alkylated to an internal acetylene without work-up. A further advantage of dehydrohalogenation of vinyl halides with $NaNH_2$ in liquid ammonia or in a dipolar aprotic solvent (e.g. DMSO) is that both the *cis* and *trans* haloolefins furnish the acetylene, in contrast to reaction with oxygen bases[22] (see Section II.A.2, equation 19). Bromoolefins **9** and *cis*-**10** are converted to

n-$C_8H_{17}CBr=CH_2$ **(9)** n-$C_8H_{17}CH=CHBr$ **(10)**

1-decyne in 45 min on treatment with sodium amide in DMSO at 65–70 °C, whereas *trans*-**10** requires 9 h [46].

It was pointed out in Section II.A.2 that under true equilibrium conditions, as in dehydrohalogenation by alkali metal hydroxides and alkoxides in aqueous or alcoholic solution, the terminal acetylenes formed may prototropically isomerize to the more stable internal acetylenes. On dehydrohalogenation with $NaNH_2$ in liquid ammonia the terminal acetylenes which are formed are precipitated as their sodium salts and thus the equilibrium is shifted and no isomerization takes place. Likewise no isomerization was observed when internal acetylenes were prepared by dehydrohalogenation with $NaNH_2$ in liquid ammonia, although occasionally one compound or another was reported to yield isomerized products[52]. It has recently been shown that on dehydrohalogenation with $NaNH_2$ in DMSO the terminal acetylene formed isomerizes only very slowly to the internal acetylene. Under these conditions the sodium acetylides are soluble and they can under the existing equilibrium conditions isomerize to the more stable internal acetylenes. The following example (equation 45)

$$RCH_2CHBrCH_2Br \xrightarrow[65\text{–}70\ ^\circ C,\ 9h]{NaNH_2,\ DMSO} RCH_2C{\equiv}CH \quad 72\text{–}94\%$$

$$RCH_2CHBrCH_2Br \xrightarrow[65\text{–}70\ ^\circ C,\ 30h]{NaNH_2,\ DMSO} RC{\equiv}CMe \quad \text{high yields} \qquad (45)$$

$R = n\text{-}C_8H_{17},\ n\text{-}C_{11}H_{23}$ to $C_{14}H_{29}$

illustrates how either a terminal or internal acetylene can be obtained preparatively pure and in high yield from the same precursor by utilizing shorter or longer reaction times[46]. This method works well for long-chain alkynes which are either inaccessible or accessible in low yields only on dehydrohalogenation with $NaNH_2$ in liquid ammonia at high temperature in an autoclave, conditions which promote isomerization. Another example where $NaNH_2$ causes isomerization only at high temperature is given in equation (46)[53]. As is shown in equations (47)–(53), many important

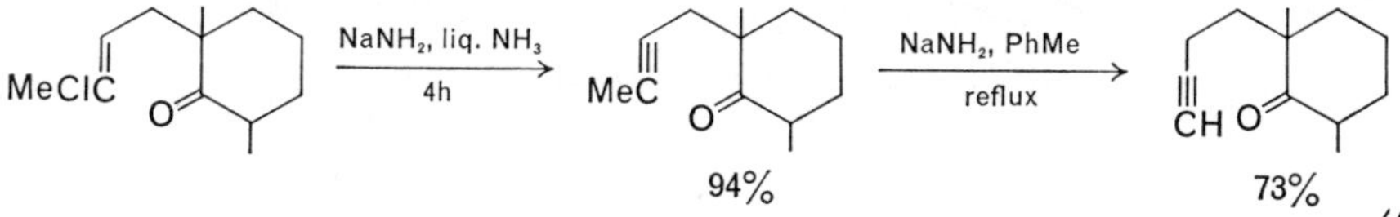

(46)

$$ClCH_2CH{=}CHCH_2Cl \xrightarrow{NaNH_2,\ liq.\ NH_3} HC{\equiv}CCH{=}CH_2 \quad 84\% \qquad \text{(Ref. 54)} \quad (47)$$

$$ClCH_2C{\equiv}CCH_2Cl \xrightarrow{NaNH_2,\ liq.\ NH_3} HC{\equiv}CC{\equiv}CH \quad 78\text{–}92\% \qquad \text{(Refs. 44 and 55)} \quad (48)$$

$$ClCH_2(C{\equiv}C)_2CH_2Cl \xrightarrow{NaNH_2,\ liq.\ NH_3} H(C{\equiv}C)_3H \qquad \text{(Ref. 56)} \quad (49)$$

$$RCHCl(C{\equiv}C)_nCHClR \xrightarrow{NaNH_2,\ liq.\ NH_3} R(C{\equiv}C)_{n+1}R \qquad \text{(Ref. 57)} \quad (50)$$

R = 9-anthryl, $n = 1, 2, 4$

$$CH_2{=}CHCH_2CH_2CH{=}CH_2 \xrightarrow{2\ eq.\ Br_2} CH_2BrCHBrCH_2CH_2CHBrCH_2Br \xrightarrow{NaNH_2,\ liq.\ NH_3} HC{\equiv}CCH_2CH_2C{\equiv}CH \quad 56\% \qquad \text{(Ref. 58)} \quad (51)$$

$$CH_2{=}CHCH_2CH_2CH{=}CH_2 \xrightarrow{1\ eq.\ Br_2} CH_2BrCHBrCH_2CH_2CH{=}CH_2 \xrightarrow{NaNH_2,\ liq.\ NH_3} \underset{65\%}{HC{\equiv}CCH_2CH_2CH{=}CH_2} + \underset{8\%}{HC{\equiv}CCH{=}CHEt} \qquad \text{(Ref. 59)} \quad (52)$$

$$CH_2{=}CHCH_2CH_2CH{=}CHCH_2CH_2CH{=}CH_2 \xrightarrow[\text{(2). } NaNH_2\text{, liq. } NH_3]{\text{(1). 3 eq. } Br_2} \underset{25\%}{HC{\equiv}CCH_2CH_2C{\equiv}CCH_2CH_2C{\equiv}CH} \quad \text{(Ref. 60)} \quad (53)$$

acetylenic intermediates may be smoothly prepared in high yield and practically without isomerization by dehydrohalogenation with $NaNH_2$ in liquid ammonia. In the case of vinylacetylene and diacetylene yields are much higher than on dehydrohalogenation with KOH (Section II.A.2.b).

A variety of functionalized acetylenes have been obtained by dehydrohalogenation with $NaNH_2$ in liquid ammonia with the functions remaining intact, as illustrated in several examples (equations 54–56). Stearolic acid[61] and 2-butyn-1-ol[62] were obtained

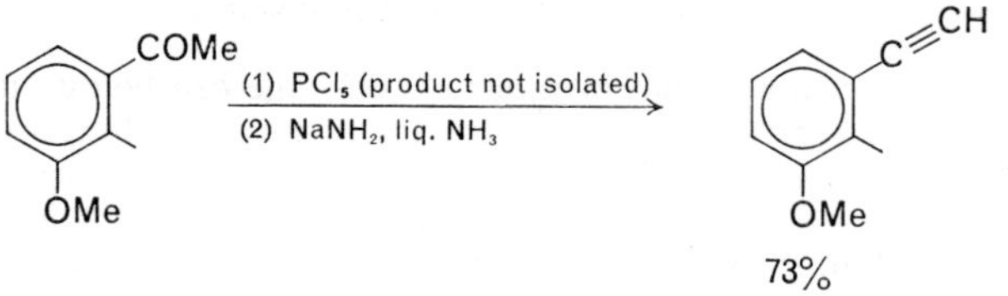

(Ref. 63) (54)

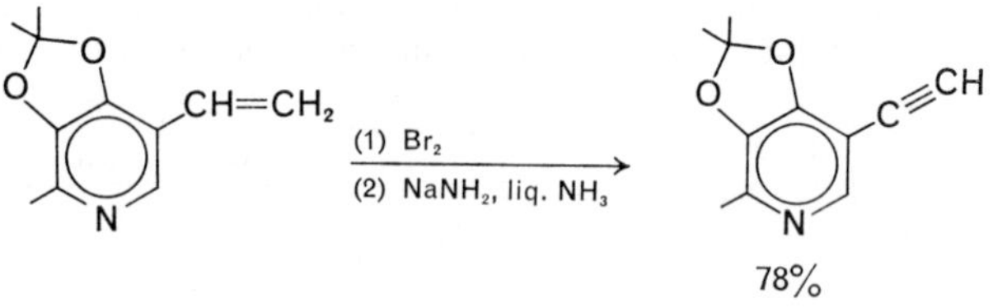

(Ref. 64) (55)

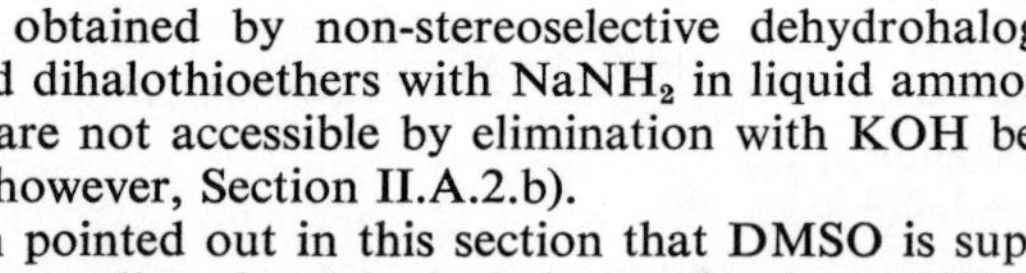

(Ref. 65) (56)

in 52–62% and 75–85% yields, respectively, as compared with substantially lower yields on dehydrohalogenation with KOH (see Section II.A.2.b). 1-Alkynyl ethers and thioethers were obtained by non-stereoselective dehydrohalogenation from 2-halovinyl ethers and dihalothioethers with $NaNH_2$ in liquid ammonia[34, 35]. Many of these compounds are not accessible by elimination with KOH because of their thermal lability (see, however, Section II.A.2.b).

It has already been pointed out in this section that DMSO is superior to liquid ammonia as a reaction medium for dehydrohalogenation by $NaNH_2$. Another such polar aprotic solvent is HMPT. It furnishes good yields of acetylenes at room temperature (equations 57–58).

Lithium amide in liquid ammonia has been as successful as sodium amide in inducing dehydrohalogenation of dihaloalkyl ethers and halovinyl ethers to 1-alkynyl ethers[68]. Lithium dialkylamides have also found use in the preparation of aryl- and alkylacetylenes in high yields[69], and of protected acetylenic sugars[70]. They have also been utilized for concurrent elimination and substitution in the synthesis of ynamines[71], obtained in 30–40% overall yield from the corresponding aldehydes

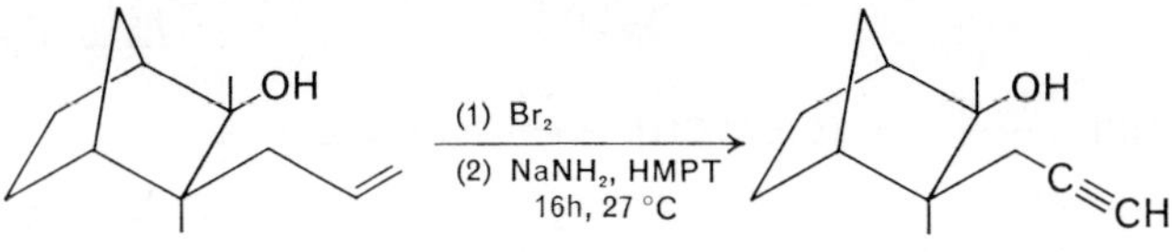

48% (Ref. 66) (57)

$$Et_2NCH_2CBr{=}CH_2 \xrightarrow[\text{r.t.}]{NaNH_2,\ HMPT} Et_2NCH_2C{\equiv}CH$$

73% (Ref. 67) (58)

via fluoroolefins (equation 59)[72]. Mixtures of ynamines and ketene *N,N*-acetals (1,1-bisdialkylaminoalkenes) which can be separated by distillation were obtained

$$ClCF_2CO_2Na + Ph_3P \longrightarrow Ph_3P{=}CF_2 \xrightarrow{RCHO} RCH{=}CF_2 \xrightarrow[\text{ether}]{LiNEt_2} RC{\equiv}CNEt_2 \quad (59)$$

by dehydrohalogenation of α-halogenoiminium salts with lithium dialkylamides in ether (equation 60)[73]. The starting iminium salts are readily available from carboxamides and phosgene[74].

$$R^1CH_2CONR^2_2 \xrightarrow{COCl_2} [R^1CH_2CCl{=}NR^2_2]^+Cl^- \xrightarrow[\text{ether}]{LiNR_2} R^1C{\equiv}CNR^2_2 \quad (60)$$

4. Organometallic compounds

Organolithium compounds RLi (R = Me, Et, *n*-Bu, Ph) are excellent dehydrohalogenating agents of vinyl halides in ether or THF solution under very mild conditions at temperatures below 0 °C. In a first step the acidic geminal hydrogen is replaced by lithium in a slow step to give an alkenyllithium which can be isolated at low temperature. In a second fast step lithium halide is eliminated to furnish an acetylene (equation 61). Two equivalents of alkyllithium are needed according to this equation.

$$RCH{=}CHCl \xrightarrow[-R^1H]{R^1Li} RCH{=}CClLi \xrightarrow[-R^1H,\ -LiCl]{R^1Li} RC{\equiv}CLi \quad (61)$$

Isotope effect studies indicate that the reaction proceeds by an *E*2cB mechanism[75]. In contrast to eliminations by oxygen bases and $NaNH_2$, eliminations by alkyllithium reagents are about 8 times faster with the *trans* than with the *cis* isomer[76].

The unstable acetylene **11** is obtained by dehydrohalogenation of the appropriate vinyl halide with BuLi (equation 62)[77]. A series of mono- and dihaloacetylenes are similarly obtained in low to moderate yields on using PhLi in ether at 0 °C (e.g.

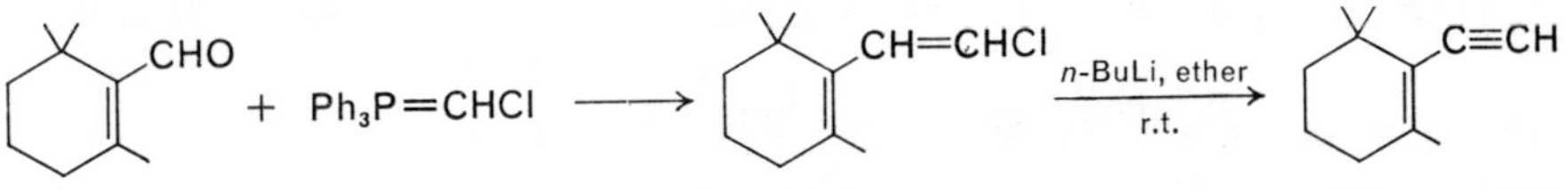

cis and *trans* **(11)** (62)

equation 63)[78]. Both Corey (Scheme 7)[79] and Villiéras (Scheme 8)[80, 69b] and their coworkers have utilized 1,1-dihaloalkenes, which are obtained from aldehydes, in

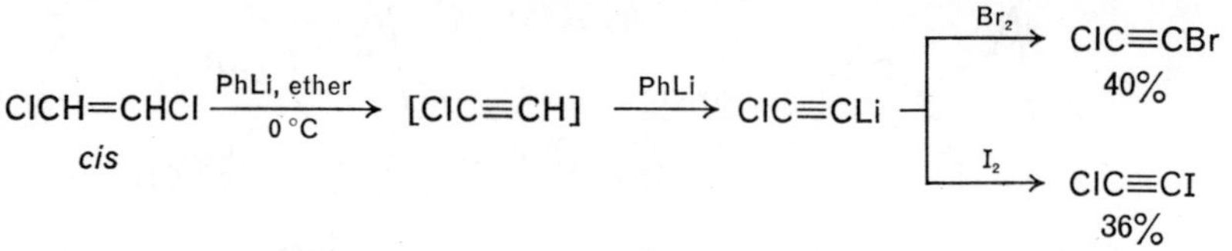

(63)

the preparation of acetylenic hydrocarbons and acids, by dehydrohalogenation with BuLi in ether or THF. Corey has prepared by this route the protected acetylenic

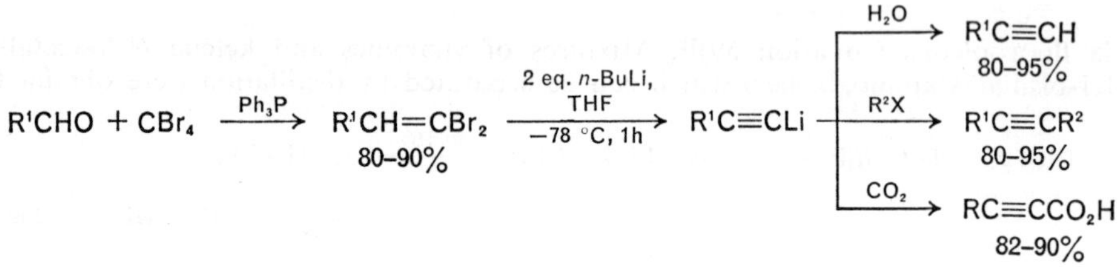

SCHEME 7

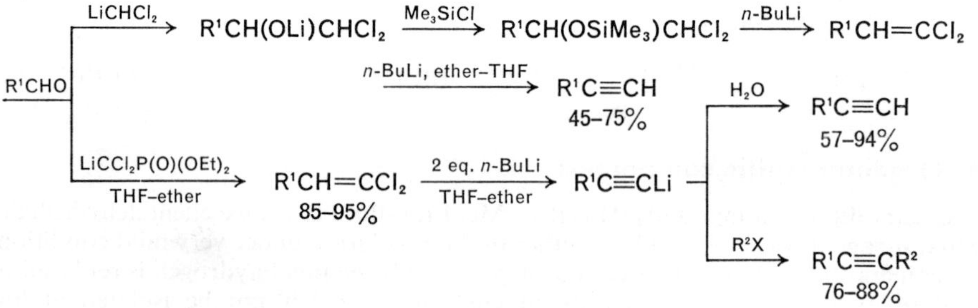

SCHEME 8

alcohol **12** from the protected α-hydroxyaldehyde **13**, in 62% overall yield, the (*S*)-antipode of the alcohol being a valuable intermediate in the synthesis of prostaglandins[79]. Acetylenic hydrocarbons and acids are obtained in similar fashion from

$HC{\equiv}CCH(OTHP)n\text{-}C_5H_{11}$ **(12)** $n\text{-}C_5H_{11}CH(OTHP)CHO$ **(13)**

1,1-dichloro-2-fluoroalkenes (Scheme 9)[81]. When the haloolefinic substrates are 1-fluoro-2-arylalkenes, dehydrohalogenation with alkyl- and phenyllithium yields

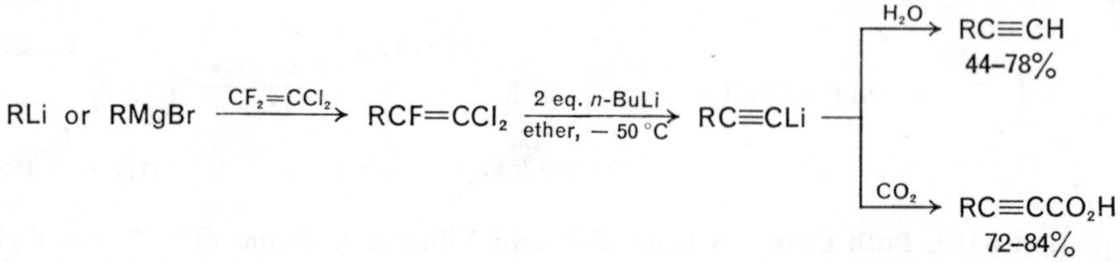

SCHEME 9

mixtures of acetylenes and substituted olefins, their ratio depending on the alkyllithium reagent used (equation 64)[82]. Dehydrohalogenation of 1,2-dichlorovinyl

$$\mathrm{ArCH{=}CHF} \xrightarrow{\mathrm{RLi}} \mathrm{ArC{\equiv}CH} + \mathrm{ArCH{=}CHR} \tag{64}$$

ethers and thioethers with BuLi yields 1-alkynyl ethers and thioethers, respectively[34, 35].

5. Metal hydrides

Sodium hydride is an effective dehydrohalogenating agent only in a strongly activating solvent, such as DMSO or HMPT. In the latter solvent sodium hydride converts β-bromostyrene to phenylacetylene in 78% yield after 20 h at 35–40 °C, and is thus comparable in efficiency to $NaNH_2$ in HMPT (Section II.A.3)[67]. It has been shown in Section II.A.3 that dehydrohalogenation of 1,2-dibromoalkanes with $NaNH_2$ in DMSO gives good yields of the corresponding 1-alkynes, although only after 9 h at 65–70 °C [46]. By contrast, treatment of the dibromoalkanes with the methylsulphinyl carbanion (generated from NaH and DMSO) gives excellent yields of the alkynes after 1 h at room temperature. Furthermore, prolonged reaction with $NaNH_2$–DMSO gives the pure 2-alkynes, whereas with the methylsulphinyl carbanion at room temperature no further reaction takes place; at 65–75 °C however, mixtures of the 2- and 3-alkynes are formed[46].

6. Organic bases and quaternary ammonium hydroxides

Amines are generally too weak to effect a double dehydrohalogenation from dihaloalkanes. If, however, the hydrogens are made highly acidic by electronegative groups, elimination does take place (e.g. equation 65)[83]. Recently it has been found

$$\underset{meso}{\mathrm{PhCOCHBrCHBrCOPh}} \xrightarrow[\text{reflux, 1 h}]{\mathrm{Et_3N,\ PhH}} \underset{82\text{–}95\%}{\mathrm{PhCOC{\equiv}CCOPh}} \tag{65}$$

that arylacetylenes can be obtained in 30–50% yields in a one-pot elimination procedure on heating a mixture of an aryl ketone, phosphorus pentachloride and pyridine in anhydrous benzene. The acetylenes formed are, however, admixed with vinyl chlorides, which are known to undergo elimination only under more vigorous conditions[84].

Quaternary ammonium hydroxides are strong bases comparable to NaOH and KOH. Recently one of them, benzyltrimethylammonium hydroxide (Triton B), has proved to be a very efficient dehydrohalogenation agent. Thus a 40% methanolic solution of Triton B in benzene affords higher yields of acetylenes than acloholic KOH or $NaNH_2$ (e.g. equations 66 and 67)[85]. In a continuation of this work it has been found that the method is not only applicable to acetylenic hydrocarbons, but

$$p\text{-}\mathrm{ClC_6H_4CCl_2Me} + p\text{-}\mathrm{ClC_6H_4CCl{=}CH_2} \xrightarrow[70\ ^\circ\mathrm{C}]{\text{Triton B, PhH}} \underset{45\text{–}50\%}{p\text{-}\mathrm{ClC_6H_4C{\equiv}CH}} \tag{66}$$

$$\mathrm{CCl_2{=}CClCH{=}CBr_2} \xrightarrow[10\ ^\circ\mathrm{C}]{\text{Triton B, PhH}} \underset{65\text{–}70\%}{\mathrm{CCl_2{=}CClC{\equiv}CBr}} \tag{67}$$

also to acetals and esters, and above all to α-acetylenic ketones, which are obtained in 30–60% yields (e.g. equation 68)[86].

$$CBr_2{=}CClCH{=}CBrCOMe \xrightarrow[-30\,^{\circ}C \text{ to } -10\,^{\circ}C]{\text{Triton B, PhH-}CH_2Cl_2} \underset{51\%}{CBr_2{=}CClC{\equiv}CCOMe} \qquad (68)$$

7. Fluorides

It has recently been observed that the fluoride ion can promote elimination from vinyl halides to yield acetylenes. Thus Et_4NF in MeCN at 25 °C converts compounds **14** to acetylenes **15** in 60–97% yields, whereas compounds **16** (R = H) under the same conditions do not react at all; compounds **16** (R = Me) afford the allenes **17**

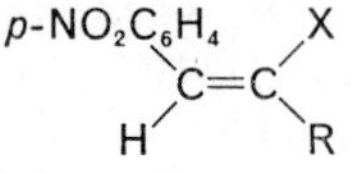

(14)

p-$NO_2C_6H_4C{\equiv}CR$

(15)

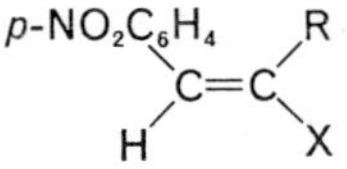

(16)

p-$NO_2C_6H_4CH{=}C{=}CH_2$

(17)

R = H, Me; X = Cl, Br

(R = Me) in 30–48% yields. With KF in DMSO even higher yields are obtained. Thus acetylenes **15** are obtained in 50–94% yields from **14** at 80–120 °C, whereas allenes **17** (R = Me) are obtained from **15** (R = Me) in 70–93% yields at 100 °C. Use of KF in the presence of a crown ether increases the rate of elimination. The above results make the fluoride ion an effective elimination agent for the preparation of acetylenes from vinyl halides in which the hydrogen and halogen are *trans*-related[87].

8. Photolysis

Aromatic vinyl halides have been recently observed to undergo photochemical dehydrohalogenation to acetylenes and by-products as illustrated in equation (69) and the adjoining table[88] (see also Section II.A.9.d).

$$\underset{\substack{cis \text{ or } trans\\ \mathbf{(18)}}}{PhCX{=}CHPh} \xrightarrow[\text{ether}]{2537\,\text{Å}} \underset{\mathbf{(19)}}{PhC{\equiv}CPh} + \underset{\substack{cis \text{ and } trans\\ \mathbf{(20)}}}{PhCH{=}CHPh} + \qquad (69)$$

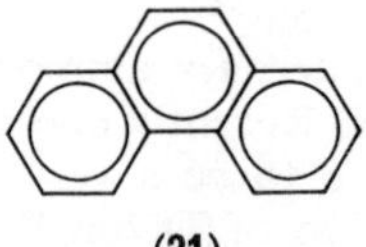

(21)

18, X	Yield (%)			
	19	*trans*-**20**	*cis*-**20**	**21**
trans, Cl	63	16	9	—
trans, Br	30	13	18	10
cis, Cl	57	18	10	—
cis, Br	25	14	19	8

9. The Fritsch–Buttenberg–Wiechell rearrangement

This route to acetylenes involves an α-elimination and a migration of an aryl group in a 1,1-diaryl-2-haloethylene (equation 70)[1k]. The reaction also proceeds when the hydrogen atom is replaced by a carboxyl group, or the halogen by an

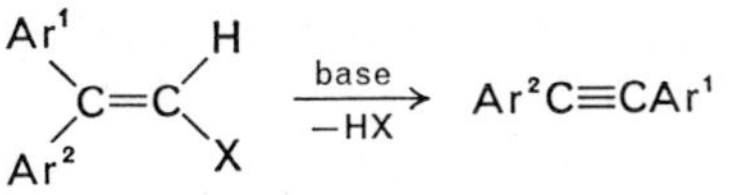

(70)

amino group. It also takes place when the substrates are 1,1-diaryl-2,2-dihaloethylenes (elimination is then induced by RLi) or 1-alkyl-1-aryl-2-haloethylenes, but does not take place with 1,1-dialkyl-2-haloethylenes. It is induced by the conventional bases discussed before and also photolytically. Reactivity is in the order $Br > I \gg Cl$ [89]. Yields vary according to the substituents on the aryl groups; electron-donating groups in the *para* position increase them[1k], whereas electron-withdrawing groups give substitution of the halogen by base[90]. Evidence has accumulated that the rearrangement does not involve a carbene but rather a highly stereoselective or even stereospecific migration of the aryl group *trans* to the halogen in the first-formed carbanion **22** (e.g. equation 71)[91, 75]. However, recent work has also implicated carbenes as the reactive intermediate (see Section II.A.9.a)[92].

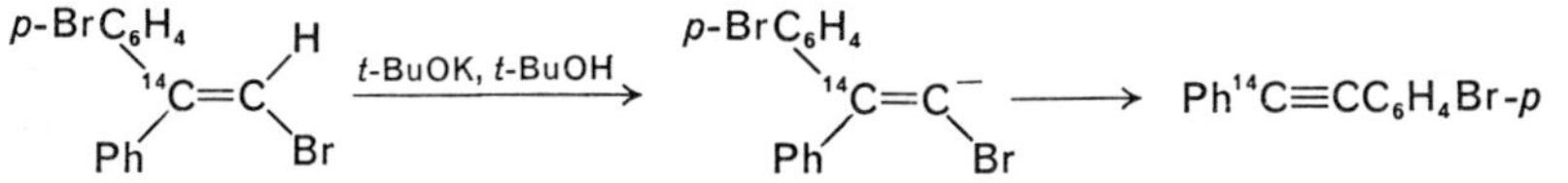

(71)

(22)

a. *Oxygen bases.* It should be stated at the outset that rearrangement with oxygen bases may take several hours whereas with $NaNH_2$ in liquid ammonia the reaction may be completed in a few minutes (see Section II.A.9.b). Molten KOH prevents the competing substitution reaction by alcoholic solvents which leads to vinyl ethers[93]. Various symmetric and asymmetric 2,2-di-*p*-alkoxyphenylvinyl chlorides and bromides are converted to the corresponding acetylenes in 90–95% yields by heating them under reflux for 5 h with sodium 2-hydroxyethoxide in ethylene glycol[94]. Similarly, refluxing 2-*p*-bromophenyl-2-phenylvinyl bromide for 3 days with *t*-BuOK in *t*-BuOH affords *p*-bromophenylphenylacetylene in 83% yield[95].

The method has found some use in the generation and trapping of strained cycloalkynes (equation 72)[92a]. Stable cycloalkynes are also obtained, but the rearrangement is accompanied by competing side-reactions, which drastically reduce the yields of the cyclic acetylenes, as illustrated for bromomethylenecyclooctane

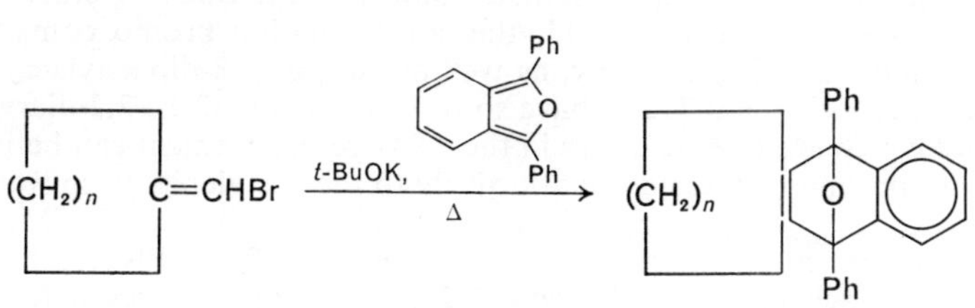

(72)

n = 4, 12·3%
n = 5, 35%

(equation 73). Bromomethylenecyclodecane and bromomethylenecyclododecane suffer similar fates[92a]. Even acyclic vinyl bromides give mixtures of acetylenes and other products (e.g. equation 74)[92]. These results have been explained by way of generation of alkylidenecarbenes (see Section II.A.9)[92].

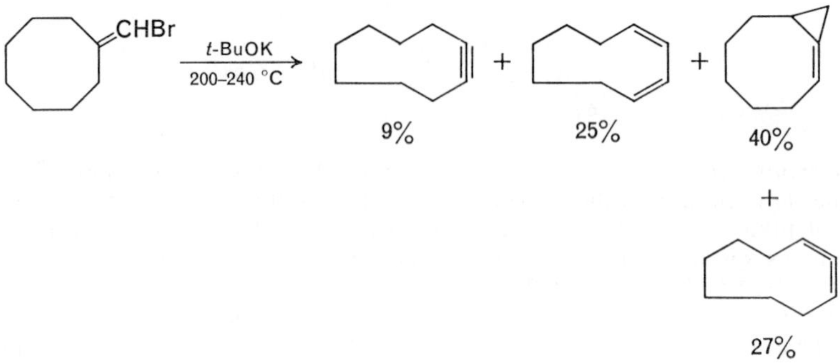

(73)

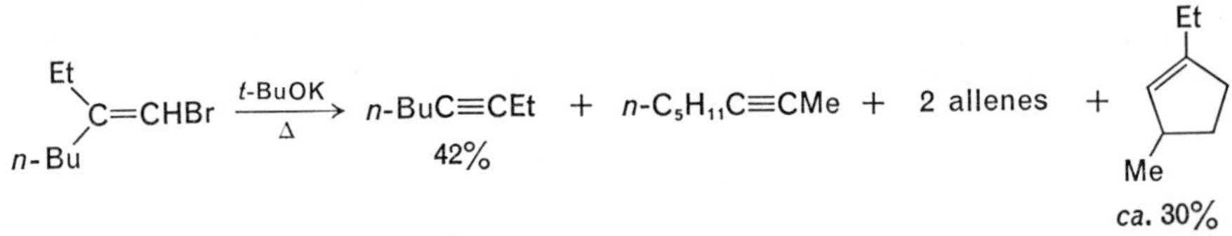

(74)

b. *Alkali metal amides*. As already indicated, these bases in liquid ammonia and in ether solvents are superior to oxygen bases and furnish the acetylenes in much higher yields. Thus 1,1-diaryl-2-chloro- and bromoethylenes with KNH_2 in liquid ammonia yield substituted diphenylacetylenes in 70–90% yields[96]. Similarly, treatment of 2,2-diphenylvinyl bromide with $NaNH_2$ in HMPT at room temperature (2 h) and at 45 °C (2 h) gives diphenylacetylene in 80% yield, but treatment of 2-methyl-2-phenylvinyl bromide under the same conditions gives only moderate yields of methylphenylacetylene[67].

c. *Organometallic compounds*. Organolithium compounds induce the rearrangement of 1,1-diarylvinyl chlorides, dichlorides and dibromides in ether solution to the corresponding acetylenes under mild conditions. 1,1-Diarylvinyl bromides, on the other hand, undergo a competing reaction as well, namely, exchange of halogen by lithium. Thus, whereas 1,1-diphenylvinyl chloride with BuLi in ether at −35 °C gives diphenylacetylene in 55% yield, the corresponding bromo compound gives diphenylacetylene in 23% yield only, as well as 30% of β,β-diphenylacrylic acid on carbonation[97]. Köbrich and Trapp have shown that many of the 2,2-diaryl-1-halo-1-lithioalkanes which are intermediates in the FBW rearrangement can be prepared at low temperatures and converted quantitatively to the acetylenes when warmed up to room temperature[98, 75]. Refluxing 1,1-diaryl-2,2-dichloroalkenes with MeLi in ether has furnished the corresponding acetylenes in 74–91% yields[99]. Dicyclopropylacetylene is obtained in 83% yield from 2,2-dicyclopropylchloroethylene on treatment with BuLi in THF at room temperature[100]; *p*,p′- and *o*,*p*′-bridged cyclic diphenylacetylenes are prepared by a FBW rearrangement (e.g. equation 75)[101].

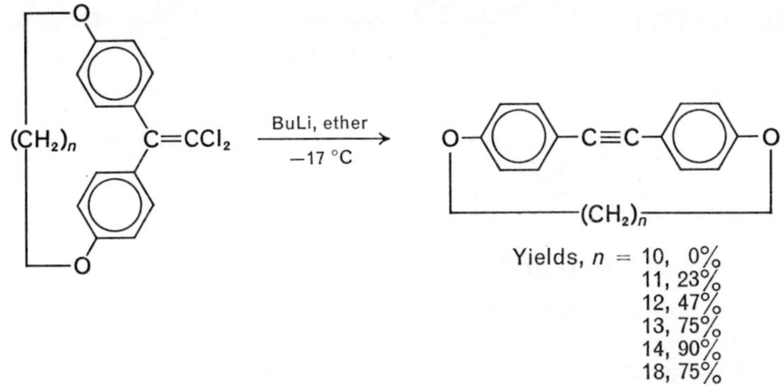

(75)

d. *Miscellaneous.* It has recently been reported that the FBW rearrangement can be induced photolytically as illustrated in equation (76)[88]. FBW-like rearrangements

$$Ph_2C{=}CHX \xrightarrow[\text{ether}]{h\nu} \underset{\mathbf{(23)}}{PhC{\equiv}CPh} + \underset{\mathbf{(24)}}{Ph_2C{=}CH_2} + \underset{\mathbf{(25)}}{Ph_2C{=}CHCH{=}CPh_2} \qquad (76)$$

	Yield (%)		
X	(23)	(24)	(25)
Cl	25	28	10
Br	23	34	10
I	19	51	6

leading to acetylenes have also been observed with substrates in which the hydrogen and halogen atoms are replaced by other substituents, as shown in equations (77)–(80).

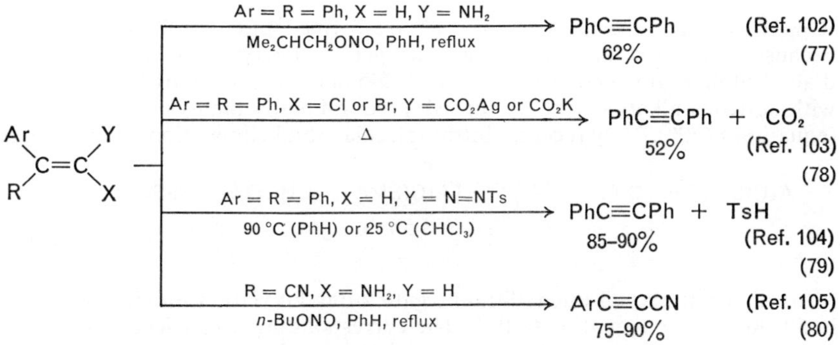

B. *Elimination of 'Acids' other than Hydrogen Halides*

1. Elimination of alcohols

Alkoxyacetylenes are obtained by alcohol and hydrogen halide elimination from dialkylacetals of α-chloroaldehydes, induced by $NaNH_2$ in liquid ammonia (equation 81)[106]. 4-Pentyn-1-ol is obtained in similar fashion (equation 82)[107].

$$ClCH_2CH(OR^1)_2 \xrightarrow{\text{3 eq. NaNH}_2\text{, liq. NH}_3} \underset{50\text{–}70\%}{NaC{\equiv}COR^1} \xrightarrow{R^2Br} R^2C{\equiv}COR^1 \quad (81)$$

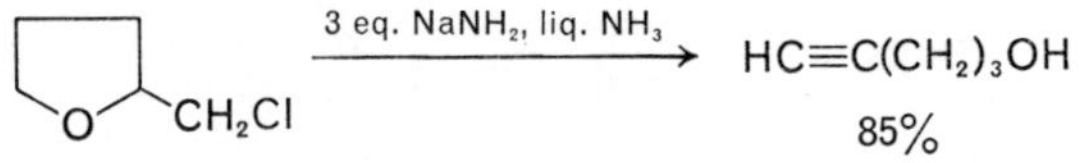
(82)

2. Elimination of thiols and sulphides

Ynamines are obtained by thiol elimination from ketene *S,N*-acetals (1-alkylthio-1-dialkylaminoalkenes) in 40–50% yields on treatment with $LiNEt_2$ at 20 °C or with $NaNH_2$ in boiling piperidine, or by leading them over solid $NaNH_2$ at 150–165 °C. In the first two procedures the formed ynamines are fractionally distilled from the reaction mixture (equation 83)[73]. When elimination is effected with KNH_2 in HMPT, aqueous work-up leads to the hydration of the ynamine. Therefore, 1,2-dibromoethane is added to the reaction mixture. It functions as a proton donor for the

$$R^1CH{=}C(SR)NR^2_2 \longrightarrow R^1C{\equiv}CNR^2_2 \quad (83)$$

$$CH_2{=}C(SBu)NMe_2 \xrightarrow{KNH_2\text{, HMPT}} KC{\equiv}CNMe_2 \xrightarrow{BrCH_2CH_2Br} \underset{55\%}{HC{\equiv}CNMe_2} + HC{\equiv}CH + H_2C{=}CHBr \quad (84)$$

potassium salt of the ynamine, which after liberation is distilled in vacuum (equation 84)[108]. In very similar fashion ynamines are obtained in 10–90% yields from thioacetamides **26** on elimination with $NaNH_2$ in boiling xylene (equation 85). With **26**,

$$\underset{(\mathbf{26})}{R^1CH_2CSNR^2_2} \longrightarrow \underset{(\mathbf{27})}{[R^1CH{=}C(SNa)NR^2_2]} \longrightarrow R^1C{\equiv}CNR^2_2 \quad (85)$$

R = H or alkyl, the thiolate salts **27** are formed but no elimination takes place because of the low acidity of the hydrogen[109]. Dialkylsulphide elimination from dialkylsulphonium methyl sulphates of β-oxocarboxylic acids[110a] proceeds readily with aqueous alkali at 0 °C to furnish predominantly high yields of alkynoic esters (equation 86)[110b]. Very recently intramolecular thiol eliminations (i.e. ring openings)

$$ArC(SR)_2CH_2CO_2Et \xrightarrow{Me_2SO_4} [ArC(SMeR){=}CHCO_2Et]^+MeSO_4^- \xrightarrow[0\,°C]{\text{aq. alkali}} ArC{\equiv}CCO_2Et \quad (86)$$

from di- or trialkyl-3-thienyllithium gave after alkylation alkylthiovinylacetylenes in 50–90% yields (equation 87)[111a]. The corresponding alkylselenovinylacetylenes[111b] and macrocyclic alkylthiovinylacetylenes[111c] were similarly obtained.

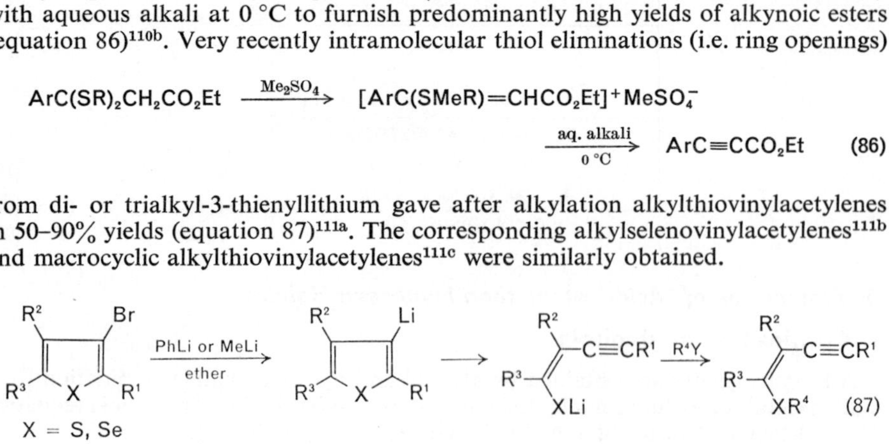
(87)

3. Elimination of sulphonic acids[10]

When the anions of these acids are good leaving groups, elimination from their enol esters can be readily induced by comparatively weak bases. This is the case for β-bromobenzenesulphonates and triflates. Thus decarboxylative elimination from enol sulphonates **28** furnishes good yields of 2-alkynoic acids (equation 88), when R

$$R^1COCH(CO_2R^2)_2 \xrightarrow[\substack{(2)\ p\text{-}BrC_6H_4SO_2Cl\ \text{or} \\ (ArSO_2)_2O}]{(1)\ EtONa\ \text{or}\ t\text{-}BuOK} \underset{\textbf{(28)}}{R^1C(OSO_2Ar){=}C(CO_2R^2)_2} \xrightarrow[\substack{\text{or}\ (1)\ TsOH \\ (2)\ \text{aq.}\ NaHCO_3}]{NaOH,\ \text{aq. dioxane}} R^1C{\equiv}CCO_2H \qquad (88)$$

is vinyl, aryl, 2-furyl, 2-thienyl or cyclopropyl; aryl groups with electron-withdrawing groups hinder the reaction[112a, b]. Overall yields were raised when the enol sulphonates were prepared by sulphonation with sulphonic anhydrides instead of sulphonyl chlorides[112c]. Deuterium labelling has shown that triflate elimination from enol triflates **29** (equation 89) proceeds by α-elimination by way of an unsaturated

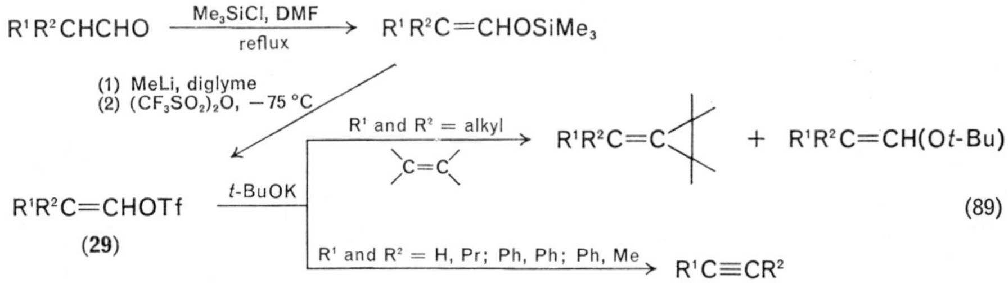

(29) (89)

carbene, and not by an *E*2 β-elimination[113a]. In fact, enol triflates **29** afford on basic elimination with *t*-BuOK in an olefinic solvent at 0 °C either cyclopropanes and vinyl ethers or acetylenes, in good yields, depending on the substituents (equation 89)[113b]. Triflate elimination also gives *t*-butylacetylene in 90% yield, starting from pinacolone (equation 90)[113c].

$$t\text{-}BuCOMe \xrightarrow{(CF_3SO_2)_2O} CH_2{=}C(t\text{-}Bu)OTf \xrightarrow[60\ ^\circ C]{\text{pyridine}} t\text{-}BuC{\equiv}CH \qquad (90)$$

4. Elimination of phosphoric acids[10]

Elimination of dialkyl phosphates from enol phosphates **30** (equation 91) with $NaNH_2$ in liquid ammonia proceeds readily and in high yields. Acetylenes are obtained when R^1 is an aryl group (equation 91); however, when R^1 is benzyl or methyl, allenes are the reaction products[114].

$$R^1COCH_2R^2 \xrightarrow{EtONa} R^1C(ONa){=}CHR^2 \xrightarrow{(EtO)_2P(O)Cl} \underset{\textbf{(30)}}{R^1C(OPO(OEt)_2){=}CHR^2} \xrightarrow{NaNH_2,\ \text{liq.}\ NH_3} R^1C{\equiv}CR^2 \qquad (91)$$

5. Elimination of trialkyltin hydrides

The formal elimination of trialkyltin hydrides from enol systems has been very recently devised by Corey and Wollenberg to prepare terminal acetylenes (equation 92)[115] and to introduce the ethynyl group into the β-position of α,β-unsaturated ketones. As equation (93) shows this may lead to the introduction of an ethynyl group into an angular position[115]. The preparation of the reagent **31** has been described[116].

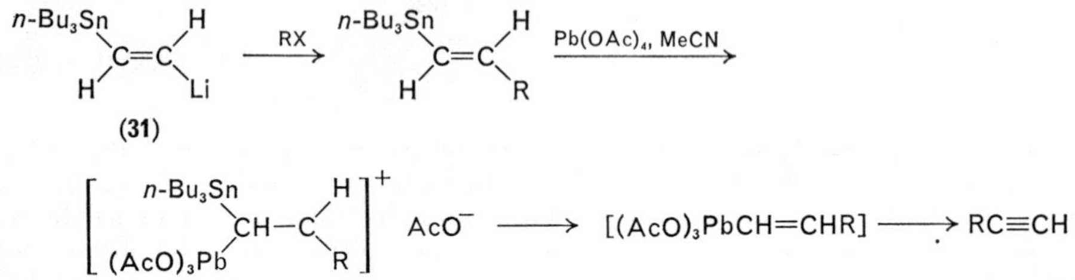

(92)

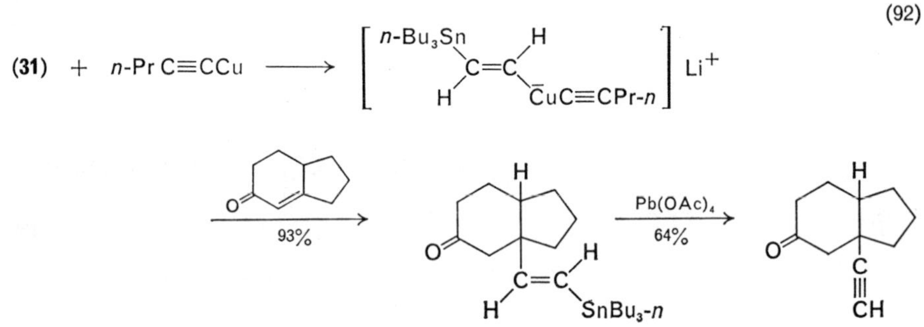

(93)

C. Hofmann Eliminations

The Hofmann degradation of quaternary ammonium hydroxides, which has been used in the preparation of olefins has also been occasionally applied to the synthesis of acetylenes, as shown in equations (94)–(96). Quaternarized enamines are also used

$$(CH_2\overset{+}{N}Me_3)_2\ 2Br^- \xrightarrow[\text{gentle warming}]{40\%\ KOH} \underset{80\%}{HC{\equiv}CH} \qquad \text{(Ref. 117)} \quad (94)$$

$$\underset{\substack{| \\ +NMe_3 \\ Br^-}}{MeCH}CH{=}CH\underset{\substack{| \\ +NMe_3 \\ Br^-}}{CHMe} \xrightarrow[\text{distillation}]{40\%\ NaOH} \underset{85\%}{MeCH{=}CHC{\equiv}CMe} \qquad \text{(Ref. 117)} \quad (95)$$

$$Ph\underset{\substack{| \\ +NMe_3 \\ I^-}}{C}{=}CHPh \xrightarrow{KNH_2,\ liq.\ NH_3} \underset{51\%}{PhC{\equiv}CPh} \qquad \text{(Ref. 118)} \quad (96)$$

(equation 96). Recent work has shown, however, that with quaternarized enamines, treatment with base may either lead to the acetylene or to the ketone form which the enamine was obtained, depending on the secondary amine used to prepare the enamine (equation 97). Thus the enamine prepared from deoxybenzoin and pyrrolidine gave, after methylation and reflux with aqueous KOH, diphenylacetylene

$$R^1COCH_2R^2 \xrightarrow{R^3_2NH} R^1C(NR^3_2){=}CHR^2 \xrightarrow{MeX} R^1C(\overset{+}{N}MeR^3_2){=}CHR^2 \; X^- \xrightarrow[\Delta]{OH^-} R^1C{\equiv}CR^2 \quad (97)$$

in 86% yield. When 3,3,4,4-tetramethylpyrrolidine was used, no acetylene was formed, and up to 40% deoxybenzoin was recovered, since the bulky tetramethylpyrrolidine prevented a *trans* elimination. Also the methylated pyrrolidine enamine from 1,3-diphenylacetone afforded on treatment with base 32% of the starting ketone and 25% of 1,3-diphenylallene. Thus it can be concluded that the method has only limited usefulness[119].

D. Dehalogenations

Dehalogenation of tetrahaloalkanes and dihaloalkenes has been effected in the past mainly by zinc and organolithium compounds and occasionally with magnesium and sodium. The solvents used in the dehalogenation should not be basic. The dehalogenation proceeds by *trans* elimination. Defluorination does not take place under the reaction conditions. Since the substrates to be dehalogenated are generally obtained by halogenation of acetylenes, the method cannot claim broad application. Equations (98) and (99) show several more recent examples of dehalogenation with

$$HC{\equiv}C(CH_2)_2C{\equiv}CH \xrightarrow[(2)\ NBS,\ h\nu]{(1)\ Br_2,\ -40\ ^\circ C} \underset{33\%}{CHBr_2CBr{=}CHCH{=}CBrCHBr_2} \xrightarrow{Zn,\ EtOH} \underset{\substack{\text{mainly } trans\\57\%}}{HC{\equiv}CCH{=}CHC{\equiv}CH} \quad \text{(Ref. 120)} \quad (98)$$

$$CF_3CCl{=}CClCF_3 \xrightarrow[reflux]{Zn\ dust,\ Ac_2O} \underset{63\%}{CF_3C{\equiv}CCF_3} \quad \text{(Ref. 121)} \quad (99)$$

zinc dust. The dehalogenation of 1,1,2-trichloro-2-pentafluorophenylethylene with zinc in ethanol or in DMF, or with magnesium powder, affords mainly pentafluorophenylacetylene with small amounts of other products[122]. Fluoroacetylene is obtained in 82% yield on dehalogenation of 1-fluoro-1,2-dichloroethylene with magnesium in THF under reflux[123]. Low-strained cycloalkynes (C_5–C_7) are generated from the 1,2-dibromocycloalkenes with magnesium in THF under reflux and trapped as their 2,5-diphenyl-3,4-benzofurane adducts[124].

Alkyl- or aryllithium compounds also occasionally induce dehalogenation with concommitant substitution (e.g. equation 100)[50]. Ynamines are similarly prepared on

$$PhCBr{=}CClF \xrightarrow{2\ eq.\ PhLi} \underset{70\%}{PhC{\equiv}CPh} \quad (100)$$

dechlorination of the corresponding dialkylamino-1,2-dichloroalkenes[125]. In like manner treatment of *gem*-dihalocyclopropanes **32** with excess MeLi in ether affords exclusively acetylenes, when R^1 = Ph, R^2 = Me and R^3 = H, mixtures of an acetylene and an allene, when R^1 = H, R^2 = Me or Ph and R^3 = H, and exclusively allenes, when R^1 = H and R^2 = R^3 = Me (equation 101). Longer reaction times and lower temperatures promote the formation of the acetylenes[126].

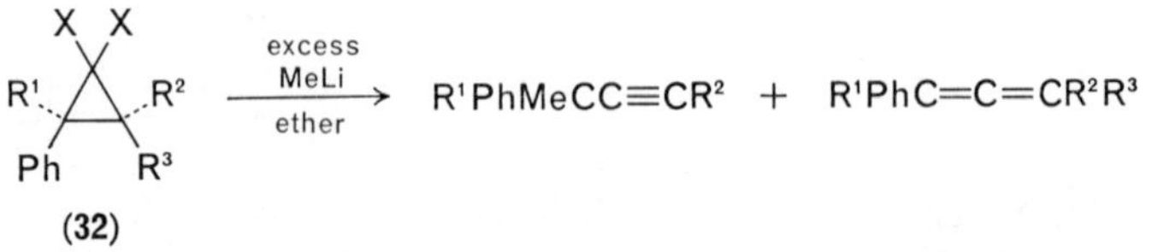

(101)

Electrolytic dechlorination of aromatic and heteroaromatic perchloroalkenes affords acetylenes (e.g. equation 102). The best conditions involve the use of a spongy lead cathode in methanol–dimethoxyethane. Yields of overreduced compounds are

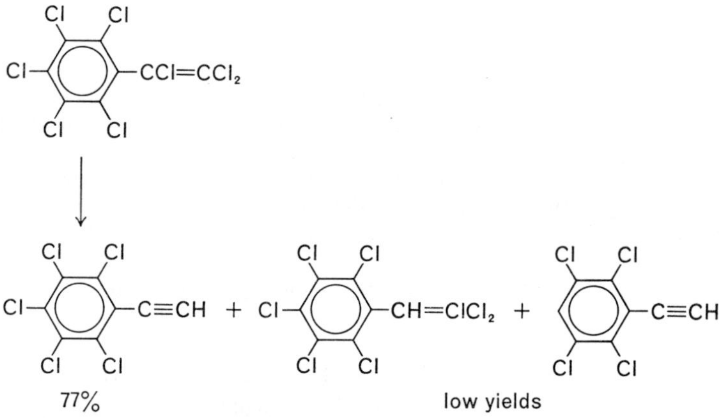

(102)

minimized by working under nearly neutral or slightly acidic conditions[127]. *cis*- and *trans*-1,2-Dichloro-3-benzenesulphonylpropenes are also electrolytically dechlorinated in DMSO at a mercury cathode to furnish 3-benzenesulphonylpropyne and 3-benzenesulphonyl-1,2-propadiene in 53 and 35% yields, respectively[128].

E. Deoxygenations

Deoxygenation of α-diketones can be effected by triethyl phosphite to furnish acetylenes and triethyl phosphate (equation 103). Either one equivalent of the α-diketone is treated with 2 equivalents of triethyl phosphite at 215 °C to furnish diaryl- or alkylarylacetylenes in 24–60% yields, or the 1 : 1 adducts **33** of the α-diketones and $(EtO)_3P$ are treated with excess reagent at 215 °C to afford the above acetylenes in 54–81% yields[129]. The reaction apparently involves a disubstituted ketene, since diphenylketene gives with $(EtO)_3P$ a 1 : 1 adduct which on pyrolysis furnishes diphenylacetylene in 40% yield (equation 103)[130]. Several diacetylenes of types **34** and **35** have been prepared in this manner[131].

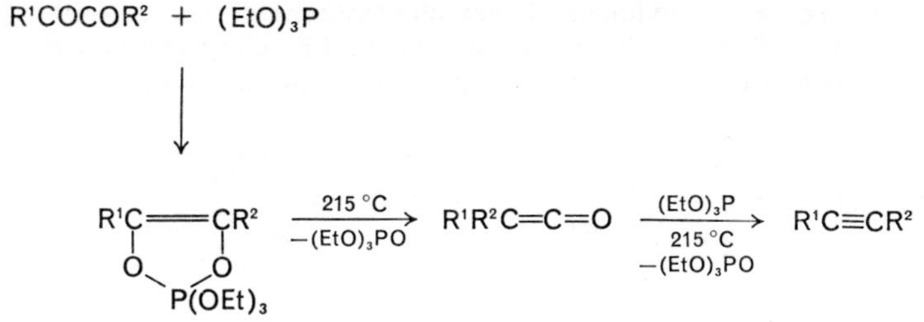

(33)

(103)

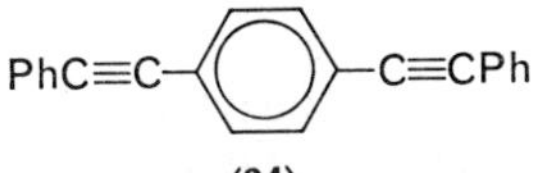

(34)

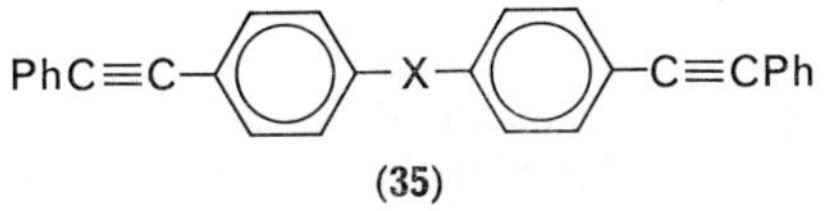

(35)

X = nil, CH_2, O, S

F. *Elimination of Triphenylphosphine Oxide*

The elimination of the stable Ph_3PO from enol phosphonium salts **36** can be induced thermally to yield acetylenes. The phosphonium salts are readily prepared on acylation of phosphoranes (equation 104). The pyrolysis of **36** is effective provided that neither R^1 nor R^2 is hydrogen, and that R^1 or R^2 is phenyl or acyl, or the

$$Ph_3P{=}CHR^1 + R^2COX \text{ or } (R^2CO)_2O \longrightarrow Ph_3P{=}CR^1COR^2$$

$$\longleftrightarrow \underset{(36)}{\underset{Ph_3P^+ \quad O^-}{R^1C{=}CR^2}} \xrightarrow{\Delta} R^1C{\equiv}CR^2 + Ph_3PO \qquad (104)$$

equivalent (e.g. CN). Thus pyrolysis of a series of acylphosphoranes at 280 °C (10 mm) furnishes acetylenic hydrocarbons and 2-alkynoic esters and nitriles in moderate to high yields; yields are improved in the presence of bases[132]. Other 2-alkynoic esters[133] and conjugated diacetylenes[134] can be similarly prepared in high and low yields, respectively. Acetylenic ketones are also obtained in high yields (equation 105)[135], as are diarylacetylenes, where the aryl groups are polycyclic rings,

$$Ph_3P{=}CHCOR^1 + (R^2CO)_2O \longrightarrow Ph_3P{=}C(COR^1)(COR^2)$$

$$\xrightarrow{250\text{–}280\ °C\ (0.01\ mm)} R^1C{\equiv}CCOR^2 + R^1COC{\equiv}CR^2 \qquad (105)$$

70–90%

such as anthryl and phenanthryl[136]. It has recently been reported that organotin halides promote the room temperature elimination of Ph_3PO from acyltriphenylphosphoranes to yield functionally substituted acetylenes[137].

G. *Eliminations of CO, SO_2 and Related Species*

Thermal extrusion at 150 °C of carbon monoxide from bis(trichlorovinyl)-cyclopropenone gives bis(trichlorovinyl)acetylene in 94% yield[138]. Flash-vacuum pyrolysis of aryl-substituted isopropylidene benzylidenemalonates at 550–600 °C gives arylacetylenes in 64–98% yields (equation 106)[139]. Labelling experiments indicate that the pyrolysis proceeds via a benzylidenecarbene intermediate[140].

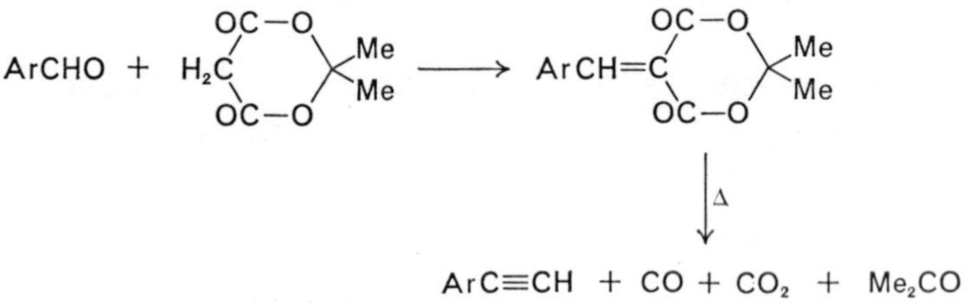

(106)

The Bamberg–Bäcklund rearrangement of α,α-dichlorodibenzyl sulphones leads to diarylthiiren 1,1-dioxides. The rearrangement is clean when induced by triethylenediamine (TED) in DMSO at ambient temperatures and furnishes the thiiren 1,1-dioxides in over 90% yields. The latter on thermal decomposition eliminate sulphur dioxide and afford diarylacetylenes in over 90% yields (equation 107)[141]. Recently it has been found that α,α-dichlorodibenzyl sulphides can be directly converted into diarylacetylenes in 62–93% yields by refluxing them with *t*-BuOK in THF [142].

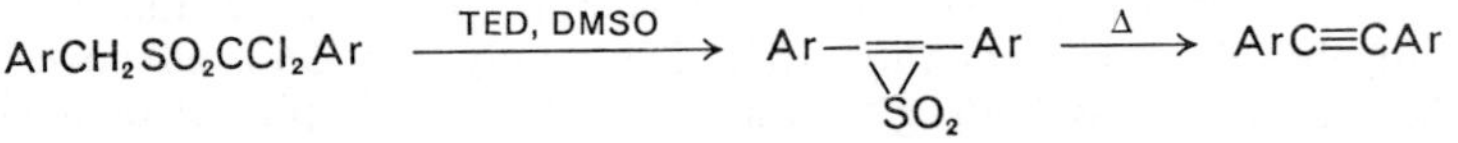

(107)

Dialkyl- and diarylacetylenes have been recently obtained in low yields (25–35%) on treating thiocarbamates with $(EtO)_3P$. The starting materials were obtained from esters or α-diketones (equation 108)[143].

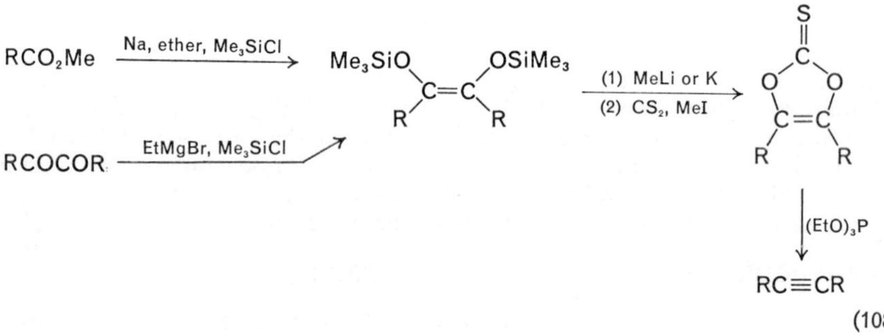

(108)

H. Elimination of Molecular Nitrogen

Molecular nitrogen, being a very stable species, is readily eliminated thermally from systems in which the two nitrogen atoms are bonded to each other. This type of elimination has the advantage of suppressing the formation of isomeric allenes and dienes, which are an accompanying feature of dehydrohalogenations and other eliminations. Of particular interest are the systems which have been developed in recent years by Eschenmoser and his coworkers, and which on heating liberate molecular nitrogen to yield acetylenes (Section II.H.10).

1. Eliminations from monohydrazones

Oxidation of hydrazones of benzyl ketones with mercurous trifluoroacetate in refluxing ether or in dioxane at 40–50 °C induces the elimination of nitrogen and formation of acetylenes in moderate yields (equation 109). Oxygenated solvents which

$$ArCH_2CR{=}NNH_2+2(CF_3CO_2)_2Hg_2 \xrightarrow{\text{ether}} ArC{\equiv}CR+4[CF_3CO_2H.\tfrac{2}{3}Et_2O]+4Hg+N_2 \quad (109)$$

form addition products with CF_3CO_2H must be used to prevent addition of the acid to the acetylene. When R = Ph or alkyl, yields amount to 60 ± 10%. Azines are the main by-products and their formation can be suppressed by adding the hydrazone solution dropwise to a slurry of the mercurous salt[144].

2. Eliminations from dihydrazones

1,2-Dihydrazones (readily available from 1,2-diketones) yield acetylenes on oxidation with a variety of oxidizing agents (equation 110). The method is applicable to aliphatic, alicyclic and aromatic dihydrazones. Recently it has been successfully applied in the preparation of cycloalkynes.

$$RC({=}NNH_2)C({=}NNH_2)+O_2 \longrightarrow RC{\equiv}CR+2N_2+2H_2O \quad (110)$$

Oxidation of benzil dihydrazone with yellow mercuric oxide in refluxing benzene gives diphenylacetylene in 67–73% yield[145]. Recently an effective and mild oxidizing agent has been developed, namely, molecular oxygen in pyridine solution, with CuCl as catalyst, and operating at room temperature. Under these conditions diphenylacetylene is obtained in 97% yield and 4-octyne in 89% yield. This reagent is superior not only to HgO, but also to CF_3CO_2Ag and $Pb(OAc)_4$ [146]. Silver trifluoroacetate is also superior to HgO and it gives diarylacetylenes in 70–85% yields on oxidation at room temperature in alcohol or acetonitrile in the presence of triethylamine[147].

Cyclodecyne[148] and cyclononyne[149] have been obtained from the corresponding cyclic dihydrazones on oxidation with HgO in refluxing benzene or toluene in 36 and 25% yields, respectively. The lower cycloalkynes (C_8–C_5) can only be trapped as adducts on their generation from their corresponding cyclic dihydrazones by oxidation with HgO in benzene, albeit in decreasing yields as the size of the ring is lowered (40, 26, 7, 0·5% of adduct)[150]. Other products are also formed in these oxidations[150, 151]. 4,4,7,7-Tetramethylcyclooctyne [152] and 3,3,7,7-tetramethylcycloheptyne[153] which are thermally stable are obtained from the corresponding

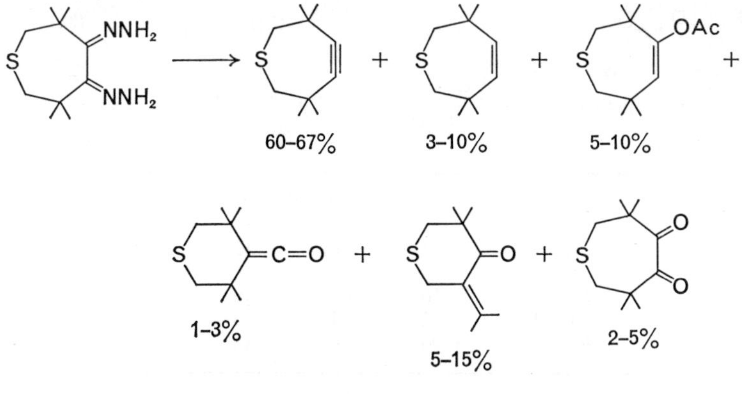

(111)

dihydrazones by $Pb(OAc)_4$ oxidation, in 28 and 25% yields, respectively. 3,3,6,6-Tetramethyl-1-thiacycloheptyne can be isolated from the corresponding dihydrazone in 5% yield on oxidation with Ag_2O in THF [154]. The yield is raised to 60–67% by carrying out the oxidation at a lower temperature in CH_2Cl_2 with $Pb(OAc)_4$, which is a more efficient oxidizing agent. Other products can also be isolated (equation 111)[155].

3. Eliminations from 3-nitroso-2-oxazolidones

Earlier work has shown that 5,5-disubstituted 3-nitroso-2-oxazolidones on treatment with aqueous KOH at room temperature give acetylenes when both substituents are aryl groups (e.g. Ph, Ph, 100% yield); aldehydes, when both of them are alkyl groups (e.g. Me, Me, 80% yield), and when one is an alkyl group and the other an aryl group, a mixture of the acetylene and a ketone is obtained (e.g. Me, Ph, 74 and 16% yields, respectively) (equation 112)[156]. A change of reagent to *n*-butylamine

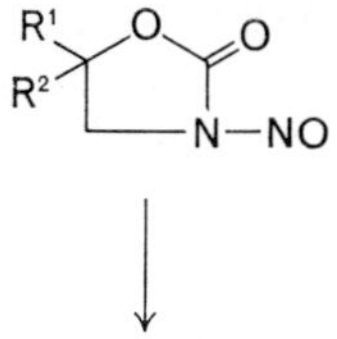

$$R^1C{\equiv}CR^2 + N_2 + CO_2 + H_2O \quad \text{and/or} \quad R^1R^2CHCHO + R^1CH_2COR^2 \qquad (112)$$

in ether at room temperature brings about a quantitative yield of arylacetylenes from 5,5-diaryl-, 5,5-arylalkyl- and 5-arylnitrosooxazolidones[157]. The same reagent gives a 78% yield of 3,5-di-*t*-butylphenylacetylene[158], and MeONa gives a 79% yield of 2-ethynylthiophene[159] from the corresponding nitrosooxazolidones.

The mechanism of this reaction has been formerly discussed in terms of a vinyl carbonium ion[156a]. It has been recently shown, however, that the reaction proceeds by a mechanism involving competition between vinyl carbonium ions and carbenes. This conclusion has been based upon the observations that on treatment of the nitrosooxazolidones with aqueous–methanolic KOH or with EtOLi in cyclohexene solution, yields of acetylenes **37** increase, and those of vinyl ethers **38** and alkylidene-bicycloheptanes **39** decrease, as R changes from methyl to cyclopropyl to phenyl (equation 113 and adjoining table)[160].

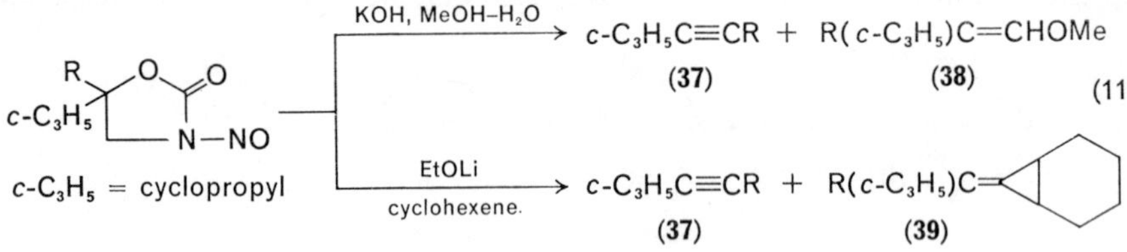

(113)

R	Yield (%) (37)	(38)	(37)	(39)
Me	16	64	26	44
c-C_3H_5	52	21	64	13
Ph	90	0	84	0

4. Eliminations from 2-pyrazolin-5-ones

Alkaline decomposition of 4,4-dihalopyrazolinones with aqueous alkali at 0–10 °C affords 2-alkynoic acids in good yields. The starting materials are obtained from the corresponding β-keto esters as illustrated in equation (114). Phenylpropiolic acid

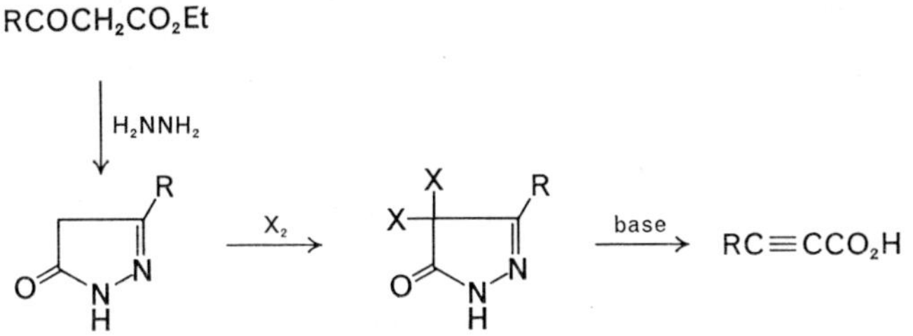

(114)

and tetrolic acid are obtained in *ca.* 75% yields[161]. Long-chain alkynoic acids are similarly obtained[162]. Under the above conditions, 5-alkyl- and 5-aryl-5-halo-pyrazolinones give *cis–trans* mixtures of 2-alkenoic acids[161b, 162, 163].

More recently 3-alkyl- and 3-aryl-5-pyrazolinones have been converted to 2-alkynoic esters by treatment with 2 equivalents of $Tl(NO_3)_3$ in MeOH under short reflux, or by direct treatment of the precursors of the pyrazolinones, namely, of the β-keto esters in methanolic solution, first with hydrazine and then with $Tl(NO_3)_3$ (equation 115). Yields amounting to 67–95% are the same for the two routes[164].

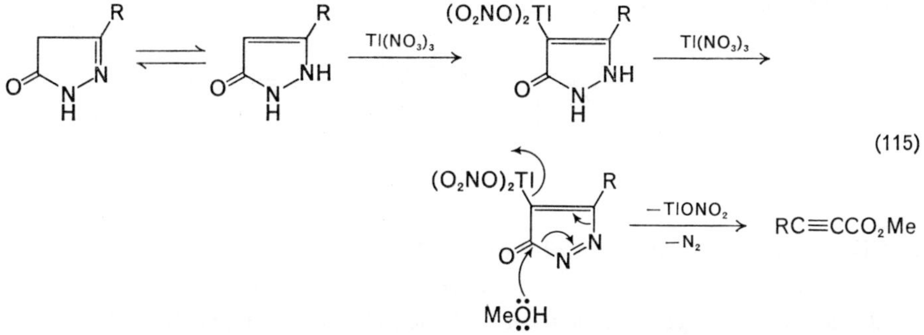

(115)

5. Eliminations from 1-tosylamino-1,2,3-triazole anions

On photolysis in dioxane or aqueous dioxane these anions liberate two equivalents of molecular nitrogen and yield acetylenes according to equation (116). Diphenyl-acetylene is obtained in 85% yield and cycloalkynes (C_6–C_8) are trapped in 54–77%

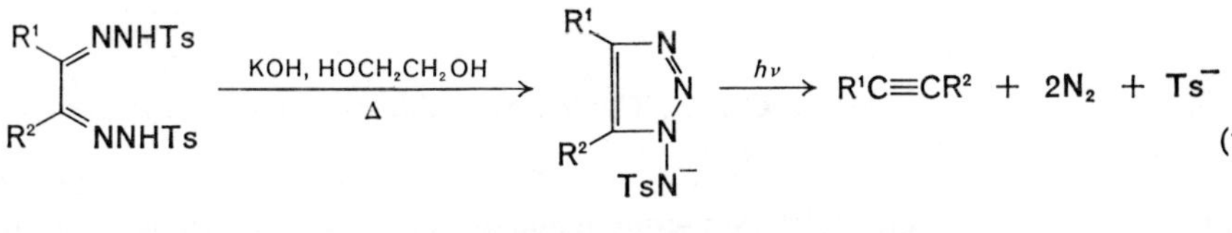

(116)

yields[165, 166]. The starting materials are readily available from the corresponding 1,2-bistosylhydrazones[166]. It has recently been shown that the latter can be directly photolysed in aqueous–methanolic NaOH to give cycloalkynes in good yields, starting from cycloalkanones (equation 117)[167].

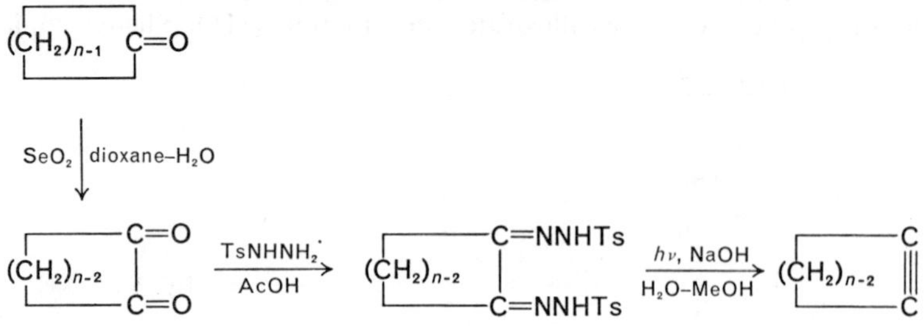

(117)

Yields, n = 8, 36%
12, 53%
15, 55%

6. Eliminations from substituted 5-halo-1*H*-tetrazoles

Substituted tetrazoles **40** (X = halogen, N_3, NH_2, OH) liberate nitrogen on heating at 110–200 °C without solvent, or in an aromatic solvent, and furnish acetylenes in 16–81% yields, depending on R^1 and R^2 (equation 118). A FBW-type of rearrangement might be involved[168].

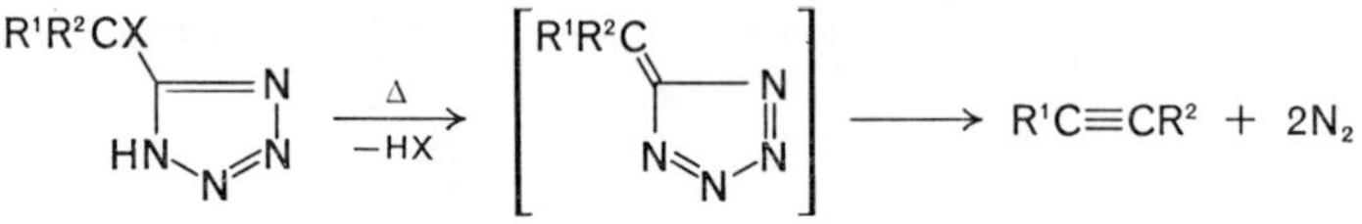

(118)

(40)

7. Eliminations from 1,2,3-selenodiazoles

Semicarbazones of aliphatic and aromatic aldehydes and ketones are converted to 1,2,3-selenodiazoles **41** on oxidation with SeO_2 [169], and on pyrolysis these afford alkynes in predominantly high yields (equation 119)[170]. R^1 and R^2 can be H, alkyl and

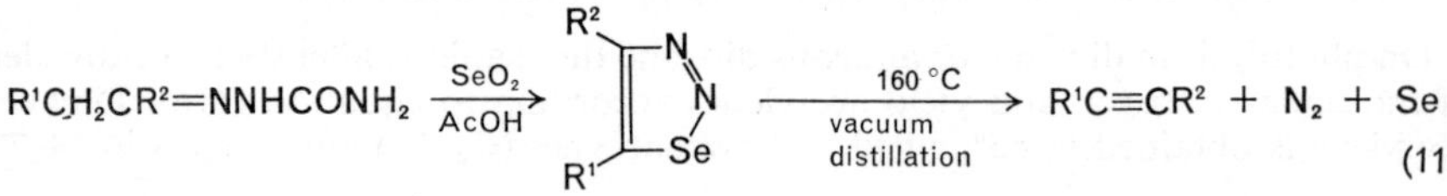

(119)

(41)

aryl, and R^1 can also be CN and CO_2Et. The lowest stable cycloalkyne, cyclooctyne, was similarly prepared by pyrolysis at 170–220 °C in 55% yield[171] and cyclododecyne was obtained in 90% yield[171b]. The lower cycloalkynes C_5 to C_8 were trapped in 0, 6, 29 and 51% yields, respectively[172]. Non-conjugated diacetylenes[173], alkynoic acids and esters[173], and acetylenic steroids and polycyclic aromatic hydrocarbons were also obtained[174].

8. Eliminations from 1,2,3-thiadiazoles

4-Alkyl- or 4-aryl-1,2,3-thiadiazoles **42** which are unsubstituted at position 5, and are obtainable from methyl ketones (equation 120), undergo ring cleavage on treatment with strong bases, such as organolithium compounds, at −60 °C, and

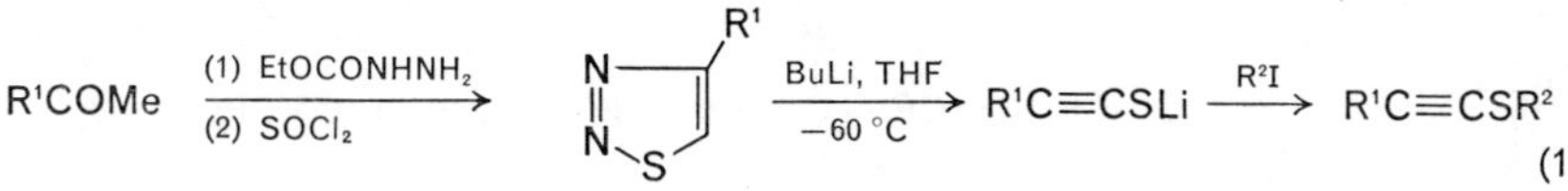

(42)

(120)

afford alkali metal alkynethiolates which can be subsequently alkylated to alkynyl thioethers (equation 120)[175]. These are obtained in good to high yields. By contrast, it has been found that, under the same conditions, a 5-substituted thiadiazole, namely 4,5-diphenyl-1,2,3-thiadiazole, gives diphenylacetylene in 78% yield[176].

9. Eliminations from vinylamines

Deamination of α-substituted-β-2-(5-nitrofuryl)vinylamines **43** (R = substituted Ph, 1-naphthyl, 2-furyl) with isoamyl nitrite in dioxane at 80 °C gives β-2-(5-nitrofuryl)acetylenes **44** in good yields[177].

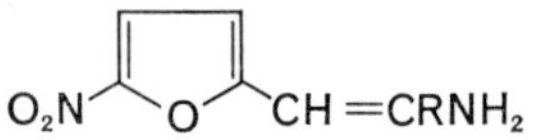

(43)

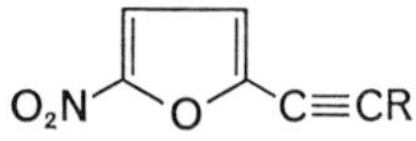

(44)

10. Eliminations with concurrent fragmentation (Eschenmoser's method)

In 1967 Eschenmoser and his coworkers devised an ingenious structure (**45**) from which molecular nitrogen was readily evolved involving neighbouring group participation from groups A and/or B. The ultimate result of this process was the formation of a carbon–carbon triple bond. The material which follows is classified according to the groups A and B used. The equations which follow should not be construed to imply statements of mechanism, but should rather serve as an illumination of the processes involved. These equations will start with a detailed version of the diazo structure **45** and the various precursors of the diazo group will be pointed

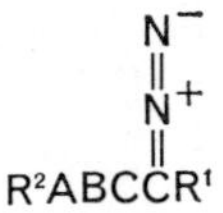

(45)

out. The precursors used have been tosylhydrazones, aminoaziridine hydrazones or the diazo compounds themselves. The diazo compounds were generated *in situ* from these precursors, affording under the reaction conditions the acetylenes concurrent with nitrogen elimination. The precursors were obtained from aldehydes and ketones on treatment with the corresponding reagents.

a. *Epoxy groups*. Eschenmoser's original system utilized the opening of an epoxy group as a leaving group[178], and the reaction can be portrayed as shown in equation (121). The diazo precursor was a tosylhydrazone. In this work[178] it was found that the

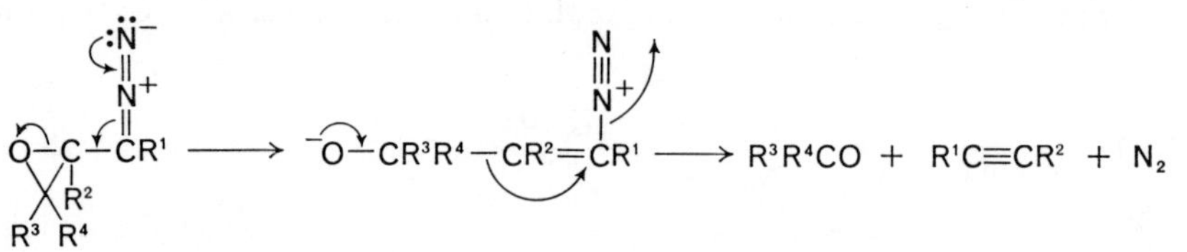

(121)

best preparative procedure is to treat the α-epoxyketone with 1·01 equivalents $TsNHNH_2$ in CH_2Cl_2–AcOH (1 : 1) for 36 h at −24 °C, 2 h at 0 °C and 4 h at room temperature. In this manner the cycloalkynone **46** (R = H) was obtained in 80–85% yield and was hydrogenated to cyclopentadecanone (exaltone) (**47**; R = H) (equation 122). Racemic muscone (**47**; R = Me) was similarly obtained. Eschenmoser and his

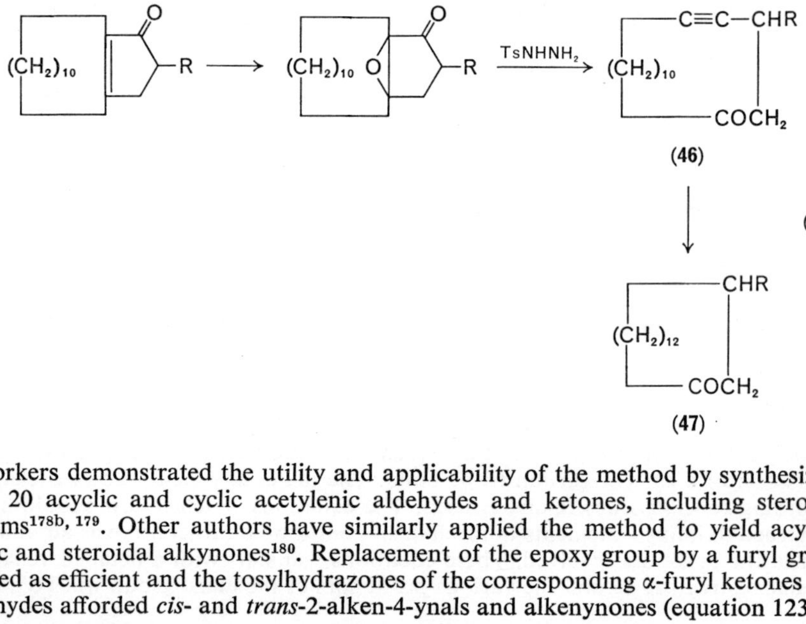

(46)

(122)

(47)

coworkers demonstrated the utility and applicability of the method by synthesizing over 20 acyclic and cyclic acetylenic aldehydes and ketones, including steroidal systems[178b, 179]. Other authors have similarly applied the method to yield acyclic, cyclic and steroidal alkynones[180]. Replacement of the epoxy group by a furyl group proved as efficient and the tosylhydrazones of the corresponding α-furyl ketones and aldehydes afforded *cis*- and *trans*-2-alken-4-ynals and alkenynones (equation 123)[181].

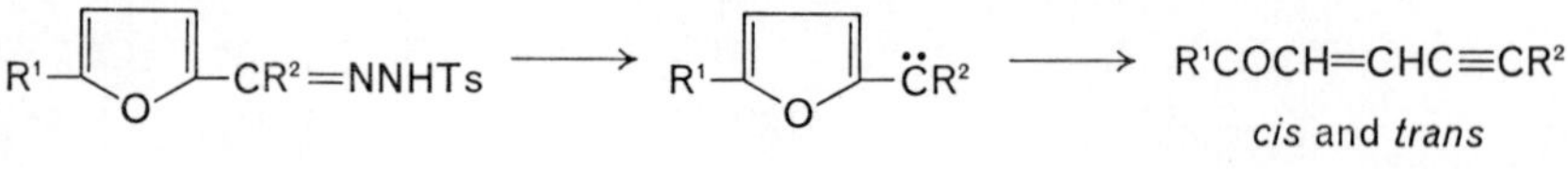

(123)

Eschenmoser and his coworkers then went on to introduce other diazo precursors. Two of them were 2-phenylaminoaziridine and *trans*-2,3-diphenylaminoaziridine[182]. The aminoaziridine hydrazones (**48**) of α-epoxyketones decompose thermally and afford acetylenes in higher yields than the tosylhydrazones[183]. Furthermore, only inert and volatile by-products are formed and the reaction takes place purely thermally in a neutral medium. Thus 5-hexynal, obtained by this route in 60–66% yield[183, 184], could not be obtained from the tosylhydrazone[180b]. Another variation

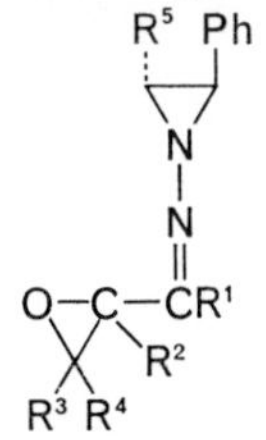

R^5 = H or Ph

(48)

uses oximes of α,β-epoxyketones to generate the diazo group. The epoxyketones were treated with hydroxylamine-*O*-sulphonic acid in alkaline solution at room temperature and afforded good yields of steroidal alkynones[185].

b. *Hydroxy, carboxylate, mesylate and fluoride groups.* Another modification of the Eschenmoser method involves the title groups as leaving groups (equation 124).

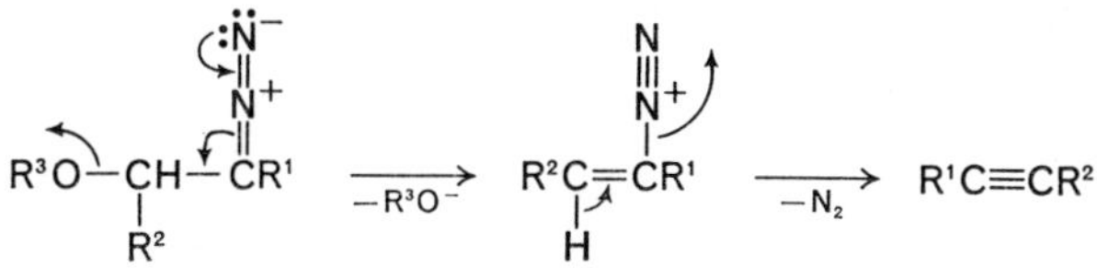

(124)

(49)

R^3 = H, F, OAc, OBz, OMs

Thus treatment of the tosylhydrazones of benzoin and its acetate and benzoate with alkoxides in protic or aprotic solvents affords increasing yields of diphenylacetylene (OH 13%, OAc 94%, OBz 98%) and decreasing yields of desoxybenzoin (OH 72%, OAc 3%, OBz 0%)[186]. 20-Oxo-21-fluoro- and -mesyloxypregnenes are similarly converted via their tosylhydrazones to pregnen-20-ynes[187]. When R^1 in **49** is an electron-withdrawing group, such as PhCO or CO_2Et, α-acylacetylenes are obtained in a single step from the unesterified α-diazo-β-hydroxycarbonyl compounds[188] as illustrated in equation (125)[189]. Very closely related is the one-step conversion of non-enolizable aldehydes and ketones to their homologous alkynes. When their mixtures

$$R^1CHO + N_2CHCOR^2 \xrightarrow{KOH} R^1CH(OH)C(=N_2)COR^2 \xrightarrow[MeCN\text{-}Et_2O]{BF_3\text{-etherate}} R^1C{\equiv}CCOR^2 \qquad (125)$$

with trimethylsilyldiazomethane or with dimethylphosphonodiazomethane are treated with *n*-BuLi in THF at −78 °C, alkynes are obtained. Thus benzophenone affords diphenylacetylene in 80% yield. Enolizable ketones give only low yields of acetylenes[190]. The method has been applied in the preparation of acetylenic sugars from aldehydosugars in about 20% yields[191].

c. *Cyclopropyl groups.* The opening of a cyclopropyl group serves here as a leaving group in still another version of Eschenmoser's method (equation 126).

Treatment of tosylhydrazones of tricyclic α-cycloketones with MeONa in MeOH at 40 °C or at room temperature gives moderate yields of cycloalkynones, occasionally intermixed with the corresponding allenes (e.g. equation 127)[192].

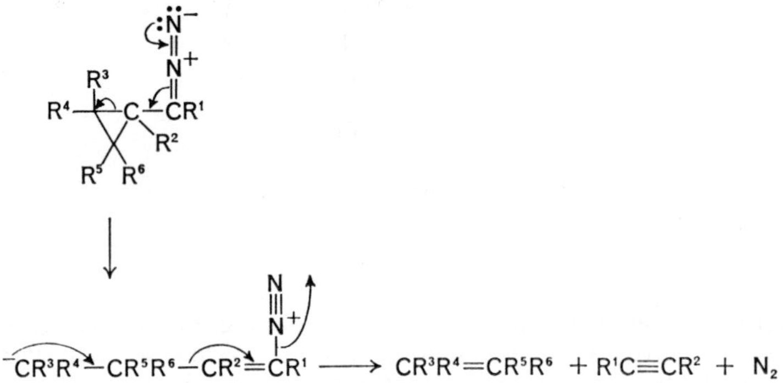

(126)

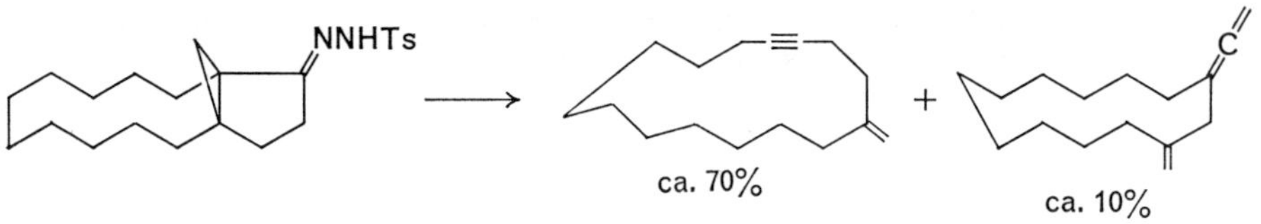

(127)

III. ACETYLENES BY SUBSTITUTION REACTIONS

Only alkylations of acetylenes (equation 128) are covered in this section, whereas the addition reactions of acetylenes to carbonyl compounds are not. Alkylation of a terminal acetylene is effected by the reaction of an alkyl or aryl halide with the

$$R^1C{\equiv}CM + R^2X \longrightarrow R^1C{\equiv}CR^2 + MX \qquad (128)$$

acetylide ion of the terminal acetylene. The acetylides are used as alkali metal acetylides, alkynylmagnesium halides (Grignard reagents), as acetylides of aluminium and copper or as complexed species with palladium and boron.

A. Alkali Metal Acetylides

The acetylide ion is a strongly basic and nucleophilic species which can induce nucleophilic substitution at positive carbon centres. Acetylene is readily converted by sodium amide in liquid ammonia to sodium acetylide. In the past alkylations were predominantly carried out in liquid ammonia. The alkylation of alkylacetylenes and arylacetylenes is carried out in similar fashion to that of acetylene. Nucleophilic substitution reactions of the alkali metal acetylides are limited to primary halides which are not branched in the β-position. Primary halides branched in the β-position as well as secondary and tertiary halides undergo elimination to olefins by the $NaNH_2$. The rate of reaction with halides is in the order I > Br > Cl, but bromides are generally preferred. In the case of α,ω-chloroiodoalkanes and α,ω-bromoiodoalkanes, metathesis at the iodo terminus is regiospecific[193]. Symmetrical dialkylacetylenes are directly prepared from the disodium salt of acetylene and two equivalents of halide, whereas unsymmetrical dialkylacetylenes are similarly obtained by adding first one halide and then the second one. Yields are in the order of 50–90% (C_3–C_{10} alkynes). 1-Hexyne is obtained in 70–77% yield from sodium acetylide and *n*-butyl bromide in liquid ammonia[194]. Because of the lower solubility of the higher halides in liquid ammonia their reaction with acetylides can still be carried out in liquid ammonia at higher temperatures and pressures in an autoclave. Aryl halides do not alkylate acetylides in liquid ammonia. Alkyl and aryl sulphates and sulphonates are more reactive than halides but their reactions are limited to the lower homologues. Alkylation of a mixture of *cis*- and *trans*-1,3-hexadien-5-ynes with methyl iodide in liquid ammonia furnishes the same *cis–trans* ratio of 1,3-heptadien-5-ynes[195]. Sodium acetylides react with alkyl halides in liquid ammonia to furnish complex mixtures of mono-, di- and tri-alkylation products (Scheme 10)[196]. Many of the products can be isolated by preparative gas chromatography and identified by n.m.r.

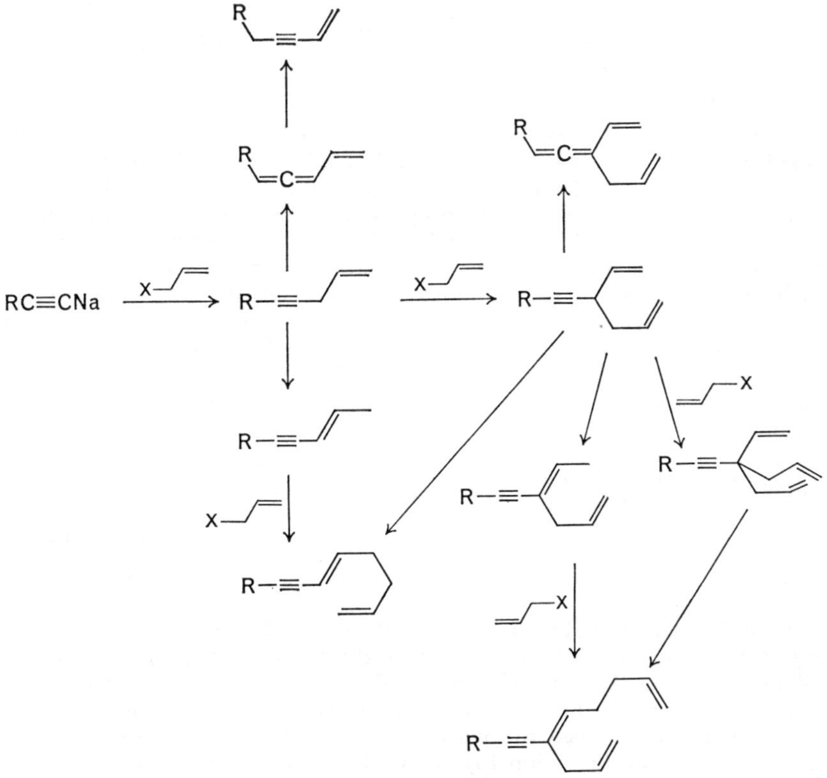

SCHEME 10

Lithium acetylide is best prepared from acetylene and $LiNH_2$ in liquid ammonia[197]. Lithium acetylides are more soluble in liquid ammonia than sodium acetylides and therefore give higher yields (50–80%) on reaction with higher halides. *trans*-2-Alken-4-ynols have been generally obtained on alkylation of sodium acetylides with epichlorohydrin in liquid ammonia, no *cis* product being formed[198], apparently because the *cis*-alkenynol formed cyclization products[198b]. However, it has recently been found that a mixture of *cis–trans*-2-penten-4-ynols (1 : 1) (47% yield) can be obtained by using lithium acetylide instead of sodium acetylide, the *cis* isomer being isolated in 15% yield by careful fractional distillation[199]. The different result is apparently due to the lesser tendency of the lithium salt of the *cis* alcohol to undergo cyclization to 2-methylfurane because of its greater covalent character compared to that of the sodium salt.

Lithium acetylide stabilized as its ethylenediamine complex[200] is a very effective reagent in reactions with alkyl halides[201]. DMSO is found to be the best polar solvent for its use (80–90% yields) but DMF is also satisfactory. These solvents have the advantage that the use of the inconvenient liquid ammonia is avoided. The reaction with iodo- and bromoalkanes requires lower temperatures (8 °C) than with chloroalkanes (25–35 °C). No internal alkynes or 1,2-dienes are formed[201]. The lithium acetylide complex has also been used in the preparation of fluoroalkynes in DMSO (e.g. equation 129)[202].

$$F(CH_2)_5Cl + LiC{\equiv}CH.EDA \xrightarrow[60\ ^\circ C]{DMSO} \underset{92\%}{F(CH_2)_5C{\equiv}CH} \qquad (129)$$

In recent years many alkylations have been carried out with lithium acetylides in polar solvents other than liquid ammonia. It has only occasionally been reported that monoalkali metal acetylides have yielded the isomerized 2-alkynes in addition to 1-alkynes. It has been found that alkylation of HC≡CLi (prepared in liquid ammonia) in HMPT or HMPT–THF with primary alkyl bromides affords increasing amounts of 2-alkynes with increase in the ratio of LiC≡CH : RBr. Thus *n*-dodecyl bromide in HMPT–THF at the ratio LiC≡CH : n-$C_{12}H_{25}Br$ 3·46, 6·09 and 10·4 gives the following percentages of 1-tetradecyne and 2-tetradecyne respectively: 88, 12; 73, 27; 32, 68. It may be inferred that with the more basic sodium and potassium acetylides the amount of 2-alkynes may be even higher at the same ratio of acetylide and halide[203]. Most lithium acetylides are now being prepared from organolithium compounds. Equation (130) illustrates a double alkylation of propyne[204]. The

$$MeC{\equiv}CH \xrightarrow[TMEDA]{2\text{ eq. BuLi, hexane–ether}} LiCH_2C{\equiv}CLi \xrightarrow{BuBr} BuCH_2C{\equiv}CLi \xrightarrow{RBr} \underset{30\text{–}80\%}{BuCH_2C{\equiv}CR} \qquad (130)$$

efficiency of HMPT or of HMPT–THF as solvents in alkylation of lithium acetylides (prepared from the terminal acetylene and BuLi) has been demonstrated in the case of medium and long-chain halides and α,ω-dihalides which give on reaction at 0 °C or at room temperature high yields of 1-alkynes and α,ω-dialkynes[205]. Acetylenic acids are similarly obtained by two routes, using HMPT and lithium acetylides (prepared from MeLi) (equation 131)[206]. Strained cyclic and macrocyclic acetylenes

$$\left.\begin{array}{l} Me(CH_2)_xC{\equiv}CH + Br(CH_2)_yCO_2H \\ Me(CH_2)_xBr + HC{\equiv}C(CH_2)_yCO_2H \end{array}\right\} \longrightarrow Me(CH_2)_xC{\equiv}C(CH_2)_yCO_2H \qquad (131)$$

(equations 132–134) as well as alkynyl-1-thiophosphonates (equation 135) are obtained in similar fashion. The sodium and lithium salts of $MeSOCH_2^-$ in DMSO are used to obtain acetylides of terminal acetylenes, and on reaction with alkyl

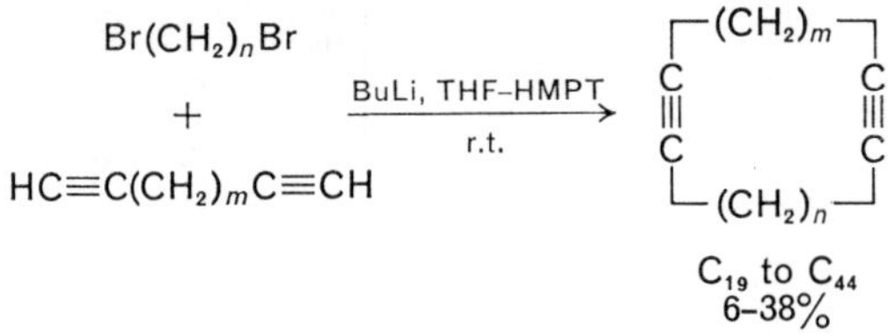

(132) (Ref. 207)

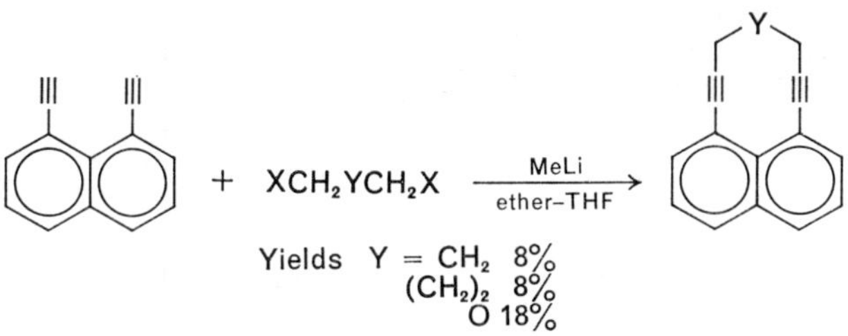

(Ref. 208) (133)

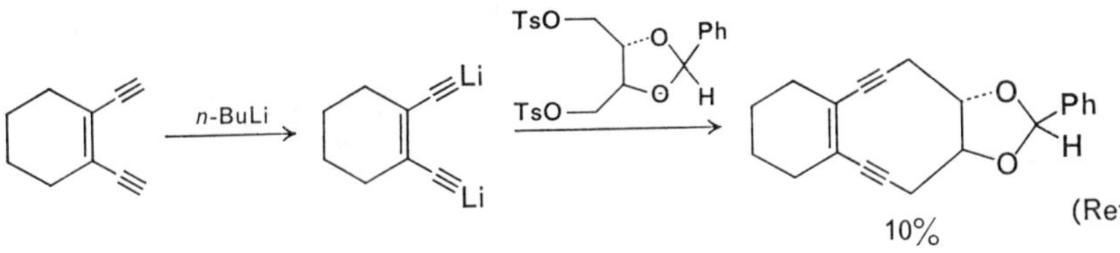

(Ref. 209) (134)

$$R^1C{\equiv}CH \xrightarrow[\text{THF–hexane}]{n\text{-BuLi}} RC{\equiv}CLi \xrightarrow[\text{r.t.}]{ClP(S)(OR^2)_2} R^1C{\equiv}CP(S)(OR^2)_2 \qquad \text{(Ref. 210)} \quad (135)$$

halides, sulphates or long-chain alkyl bromides, they give high yields of alkylation products[211]. Sodium acetylides have also been alkylated by alkyl bromides, sulphates and sulphonates in xylene[212a], xylene–DMF[212a], DMF[212b] and THF–HMPT[212c], to give alkylacetylenes in high yield.

Finally, it is important to stress the significance of carrying out alkylation experiments under a variety of conditions until satisfactory results are obtained. The situation is precisely the same as with elimination reactions, and in particular as with dehydrohalogenations. The following example is instructive. Scheme 11 shows that under four different reaction conditions no satisfactory conversions of steroid **50** to the acetylenic steroid **51** can be achieved. It is seen that the chlorosteroid **50** (X = Cl) is a major by-product. It is therefore essential to remove competing chloride ions before the metalated acetylide can react with **50** (X = OTs). The reaction is therefore carried out in dioxane where LiCl is precipitated as a LiCl–dioxane complex, furnishing **51** in 90% yield[213].

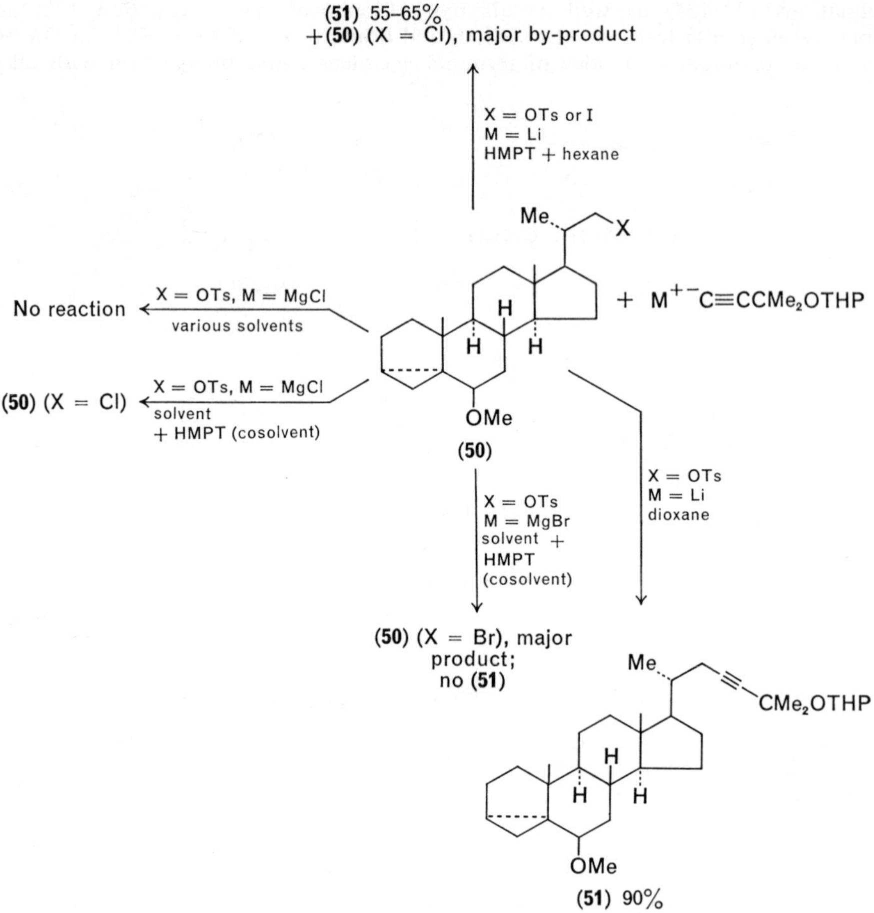

SCHEME 11

B. Alkynylmagnesium Halides (Grignard Reagents)

Alkynylmagnesium halides are less basic than the alkali metal acetylides and therefore can be applied to sensitive alkylating agents. In contrast to the alkali metal acetylides they do not react with saturated primary halides. On the other hand, they do react with allylic, propargylic and benzylic halides, but only in the presence of cuprous chloride catalysts. They also react with α-haloethers (e.g. equation 136)[214].

$$2\,O(CH_2Cl)_2 + 2\,BrMgC{\equiv}CMgBr \xrightarrow{PhH}$$ 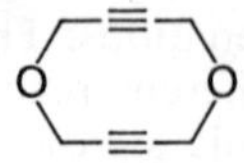 (136)

2%

With vinyl halides they react in the presence of cobaltous salts, affording low yields of alkylation products[215]. Although unreactive towards saturated alkyl halides, they are alkylated by alkyl sulphates, tosylates and mesylates[216]. The alkynyl Grignard reagents are generally prepared by reaction of the 1-alkyne with an alkylmagnesium halide, such as EtMgBr, in electron-donating solvents, such as ether, THF and higher ethers (equation 137). Occasionally diethyl ether is replaced in the Grignard solution by methylene chloride.

$$RC{\equiv}CH + EtMgBr \longrightarrow RC{\equiv}CMgBr + EtH\uparrow \qquad (137)$$

The case of ethynylmonomagnesium halide (**52**) and acetylenebismagnesium halide (**53**) deserves special comment. Reaction of acetylene with EtMgBr in ether proceeds to **53** (equation 138) because of the latter's insolubility in that solvent, and

$$\underset{}{HC{\equiv}CH} + \underset{\mathbf{(53)}}{BrMgC{\equiv}CMgBr} \overset{THF}{\rightleftharpoons} \underset{\mathbf{(52)}}{2HC{\equiv}CMgBr} \qquad (138)$$

thus **52** cannot practically be obtained. It has however been found that **53** is soluble in THF, and therefore in this solvent on reverse addition (i.e. slow addition of EtMgBr in THF to a solution containing an excess of acetylene in THF) an equilibrium mixture is obtained, containing about 85% of the mono Grignard reagent **52** [217].

A large number of 'skipped' systems, 1,4-enynes and 1,4-diynes, have been prepared by the reaction of alkynylmagnesium halides with allyl and propargyl halides, as illustrated in equations (139)–(144). Several of these products have served in the preparation of annulenes. These reactions are carried out in THF solution close to reflux temperature and in the presence of CuCl as catalyst.

$$HC{\equiv}CCH_2Br + BrMgC{\equiv}CH \longrightarrow \underset{ca.\ 70\%}{HC{\equiv}CCH_2C{\equiv}CH} \qquad \text{(Ref. 218)} \quad (139)$$

$$BrCH_2C{\equiv}CCH_2Br + BrMgC{\equiv}CH \longrightarrow HC{\equiv}C(CH_2C{\equiv}C)_nH \qquad \text{(Ref. 218)} \quad (140)$$

Yields, n = 2, 18%
4, 4%
6, 1·4%

$$HC{\equiv}CCH_2Br + BrMgC{\equiv}CPh \longrightarrow \underset{54\text{–}75\%}{HC{\equiv}CCH_2C{\equiv}CPh} \qquad \text{(Ref. 219)} \quad (141)$$

$$trans\text{-}BrCH_2CH{=}CHCH_2Br + BrMgC{\equiv}CH \longrightarrow \underset{25\%}{HC{\equiv}CCH_2CH\overset{t}{=}CHCH_2C{\equiv}CH}$$

$$+ \underset{10\%}{HC{\equiv}CCH_2CH\overset{t}{=}CHCH_2C{\equiv}CCH_2CH\overset{t}{=}CHCH_2C{\equiv}CH}$$

$$+ \underset{2\%}{HC{\equiv}CCH_2CH\overset{t}{=}CHCH_2C{\equiv}CCH_2CH\overset{t}{=}CHCH_2C{\equiv}CCH_2CH\overset{t}{=}CHCH_2C{\equiv}CH}$$

(Ref. 220) (142)

$$cis\text{-}ClCH_2CH{=}CHCH_2Cl + BrMgC{\equiv}CH \longrightarrow HC{\equiv}CCH_2CH\overset{c}{=}CHCH_2C{\equiv}CH$$

(Ref. 221) (143)

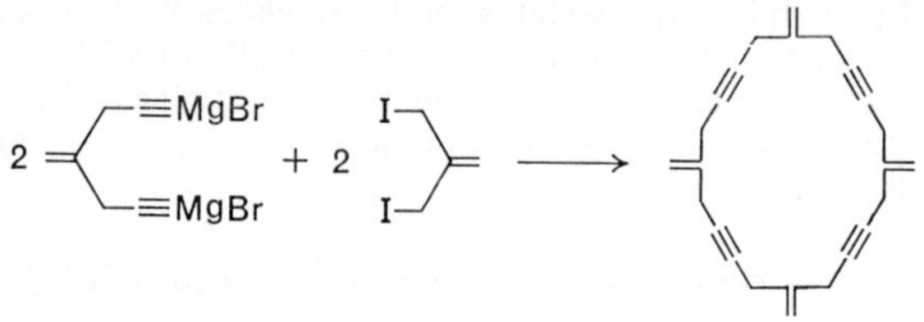

(Ref. 222)
(144)

Of other classes of acetylenic compounds which have been prepared recently by alkylation, 1-alkynylphosphonates should be mentioned. They are prepared in moderate yields from alkynylmagnesium halides and dialkyl or diaryl phosphorochloridates in ether at room temperature (equation 145)[223].

$$R^1C{\equiv}CMgBr + ClP(O)(OR^2)_2 \longrightarrow R^1C{\equiv}CP(O)(OR^2)_2 \qquad (145)$$

One of the drawbacks of this alkylation method is that under the strongly basic conditions alkylation of alkali metal acetylides and alkynyl Grignard reagents with tertiary alkyl halides and sulphonates leads to dehydrohalogenations. Since tertiary carbocations are stable under weakly basic or non-basic conditions, and as certain trisubstituted aluminium compounds not only accelerate formation of carbocations from halides, but also convert the anion residues to much weaker bases, it seemed feasible to investigate the coupling between trialkynylalanes (readily obtainable from the corresponding alkynyllithiums and anhydrous $AlCl_3$) and tertiary alkyl halides. In fact this reaction (equation 146) afforded high yields of disubstituted alkynes (e.g. equations 147 and 148)[224].

$$3R^1X + (R^2C{\equiv}C)_3Al \xrightarrow[0°\,C]{CH_2Cl_2} 3R^1C{\equiv}CR^2 \qquad (146)$$

X = Cl, Br, sulphonate

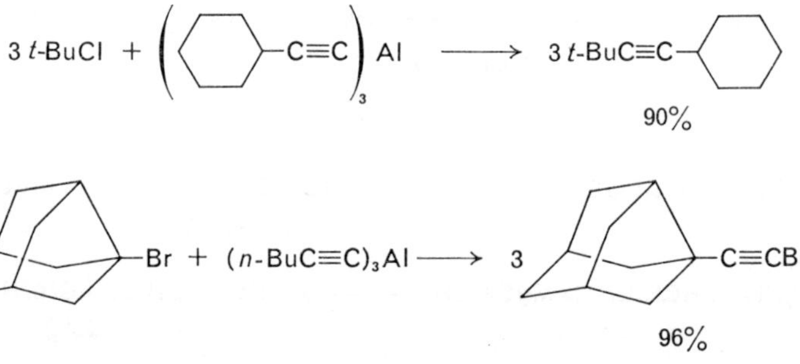

(147)

90%

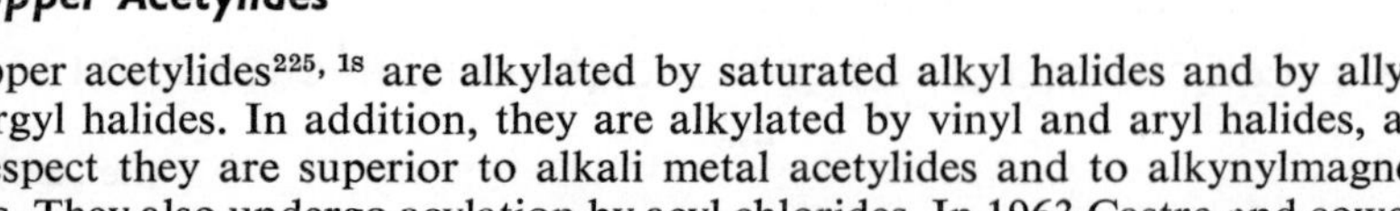

(148)

96%

C. *Copper Acetylides*

Copper acetylides[225, 1s] are alkylated by saturated alkyl halides and by allyl and propargyl halides. In addition, they are alkylated by vinyl and aryl halides, and in this respect they are superior to alkali metal acetylides and to alkynylmagnesium halides. They also undergo acylation by acyl chlorides. In 1963 Castro and coworkers reported the preparation of diarylacetylenes in good yields by treating aryl iodides with cuprous acetylides in refluxing pyridine under a nitrogen atmosphere (equation 149). Under these conditions aryl iodides bearing *ortho* nucleophilic substituents were converted exclusively to the corresponding heterocyclic compounds in high yields (equation 150)[226]. Several aspects of the stereochemistry and kinetics of cuprous acetylide substitutions have been discussed[226d].

$$Ar^1I + CuC{\equiv}CAr^2 \xrightarrow[\text{reflux}]{\text{pyridine, } N_2} Ar^1C{\equiv}CAr^2 + CuI \quad (149)$$

(150)

X = O, NH, CO_2

Since the monocuprous acetylide of acetylene is unknown, the preparation of terminal acetylenes by this method has become possible only after the development of cuprous acetylides containing readily removable substituents. Two examples are illustrated in equations (151) and (152)[227]. Polyfluorophenylacetylenes were similarly

$$PhI + CuC{\equiv}CCH(OEt)_2 \xrightarrow[\text{(2) } H^+]{\text{(1) pyridine, reflux}} PhC{\equiv}CCHO \xrightarrow[50\ ^\circ C]{4N\ NaOH} PhC{\equiv}CH \quad (151)$$

$$PhI + CuC{\equiv}CCH_2OTHP \xrightarrow[\text{(2) } 2N\ H_2SO_4]{\text{(1) pyridine, reflux}} PhC{\equiv}CCH_2OH \xrightarrow[\text{PhH, r.t.}]{NiO_2} \underset{70\%}{PhC{\equiv}CCHO}$$

$$\xrightarrow[50\ ^\circ C]{2N\ NaOH} \underset{87\%}{PhC{\equiv}CH} \quad (152)$$

synthesized[227c]. Examples of the coupling of cuprous acetylides with vinyl halides (equation 153)[228], allyl halides (equation 154)[229], and propargyl halides (equation 155)[230], as well as their acylation (equations 156[231] and 157[232]) demonstrate the

$$PhC{\equiv}CCu + CHI{=}CHCl \xrightarrow[40\text{–}100\ ^\circ C]{\text{DMF or pyridine}} \underset{90\%}{PhC{\equiv}CCH{=}CHCl} \quad (153)$$

$$n\text{-}C_5H_{11}C{\equiv}CCu + BrCH_2CH{=}CH_2 \xrightarrow[110\ ^\circ C,\ 9\ h]{Cu^+,\ K_2CO_3,\ HMPT} \underset{70\%}{n\text{-}C_5H_{11}C{\equiv}CCH_2CH{=}CH_2} \quad (154)$$

$$MeCH(OH)C{\equiv}CCu + MeCHClC{\equiv}CH \xrightarrow{50\%\ aq.\ NH_3} MeCH(OH)C{\equiv}CCH{=}C{=}CHMe \quad (155)$$

$$Me_3SiC{\equiv}CH \xrightarrow[THF]{CuI,\ t\text{-}BuOLi} Me_3SiC{\equiv}CCu \xrightarrow{n\text{-}C_5H_{11}COCl} \underset{62\%}{Me_3SiC{\equiv}CCOC_5H_{11}\text{-}n} \quad (156)$$

$$n\text{-}BuC{\equiv}CCu + n\text{-}BuCOCl \xrightarrow[\text{r.t.}]{\text{ether, LiI}} \underset{96\%}{n\text{-}BuC{\equiv}CCOBu\text{-}n} \quad (157)$$

utility of the method. The Castro coupling has been instrumental in the preparation of benzoannulenes and related macrocycles, as is shown in equations (158) and (159).

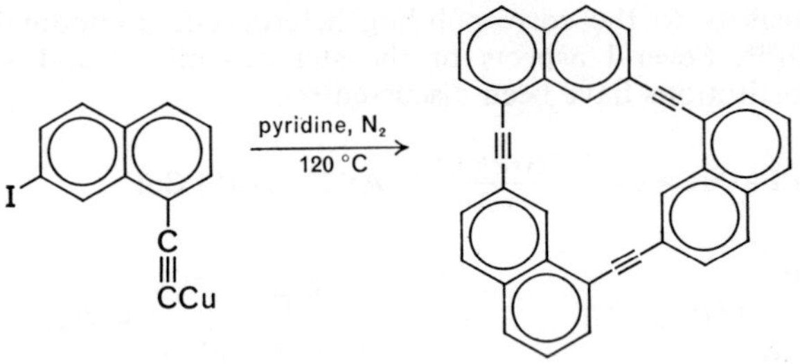

8%

(Ref. 233)
(158)

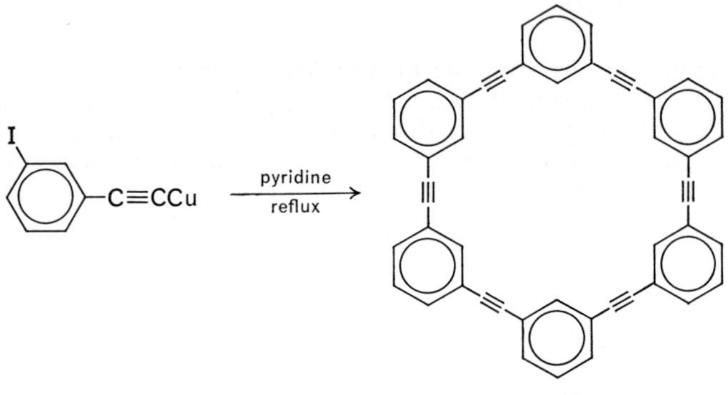

4·6%

(Ref. 234)
(159)

It has been shown that instead of using cuprous acetylides, aryl and heterocyclic halides can alkylate terminal alkyl- and arylacetylenes directly by heating the reactants in DMF or pyridine in the presence of K_2CO_3 and Cu powder to give acetylenes in high yields (e.g. equation 160)[235].

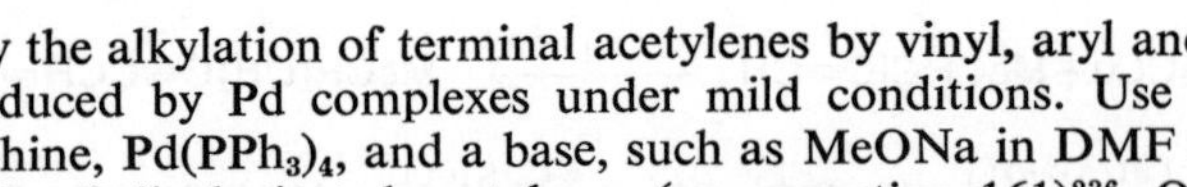

88%

(160)

D. Palladium Complexes

Very recently the alkylation of terminal acetylenes by vinyl, aryl and heterocyclic halides was induced by Pd complexes under mild conditions. Use of palladium triphenylphosphine, $Pd(PPh_3)_4$, and a base, such as MeONa in DMF at 50–100 °C, gave high yields of disubstituted acetylenes (e.g. equation 161)[236]. Other catalysts

$$\text{PhC}{\equiv}\text{CH}+\text{PhI} \xrightarrow[50\,^{\circ}\text{C}]{\text{Pd(PPh}_3)_4,\ \text{DMF}} \text{PhC}{\equiv}\text{CPh} \quad (161)$$

95%

used were diacetatobis(triphenylphosphine), $(Ph_3P)_2Pd(OAc)_2$, in the presence of an amine at 100 °C (e.g. equation 162)[237], and bis(triphenylphosphine)palladium dichloride, $(Ph_3P)_2PdCl_2$, in the presence of CuI in Et_2NH at room temperature (e.g. equation 163)[238].

$$p\text{-}NO_2C_6H_4Br + t\text{-}BuC{\equiv}CH \xrightarrow[100\ ^\circ C]{(Ph_3P)_2Pd(OAc)_2,\ Et_3N} p\text{-}NO_2C_6H_4C{\equiv}CBu\text{-}t \qquad (162)$$

$$CH_2{=}CHBr + PhC{\equiv}CH \xrightarrow[Et_3N,\ r.t.]{(Ph_3P)_2PdCl_2,\ CuI} \underset{91\%}{PhC{\equiv}CCH{=}CH_2} \qquad (163)$$

E. Boranes

Recently H. C. Brown and coworkers have developed a new convenient and general synthesis of acetylenes via the reaction of iodine with 1-alkynyltrialkylborates **54** (equation 164). The reaction of iodine with **54** takes place under very mild conditions at low temperatures and involves a migration of an alkyl group. In contrast to the

$$R^1C{\equiv}CH \xrightarrow{n\text{-BuLi, THF}} R^1C{\equiv}CLi \xrightarrow{R^2_3B} \underset{(\mathbf{54})}{Li^+[R^2_3\bar{B}C{\equiv}CR^1]}$$

$$\xrightarrow[-78\ ^\circ C\text{ to r.t.}]{I_2} R^1C{\equiv}CR^2 + R^2_2BI + LiI \qquad (164)$$

alkylation of alkali metal acetylides which proceeds only with primary alkyl groups, introduction of primary, secondary and tertiary alkyl groups and aryl groups takes place smoothly, and yields are close to quantitative[239]. The conversion of **54** to an internal acetylene is also effected by methylsulphinyl chloride, which first yields a β-methanesulphinylvinylborane **55**, followed by *cis* elimination to the acetylene (equation 165)[240]. Yields are lower (55–82%) as compared to the above method. It has also been observed in this reaction that the migration of alkyl groups is not selective, as illustrated in equation (166)[240]. Symmetric internal acetylenes cannot be

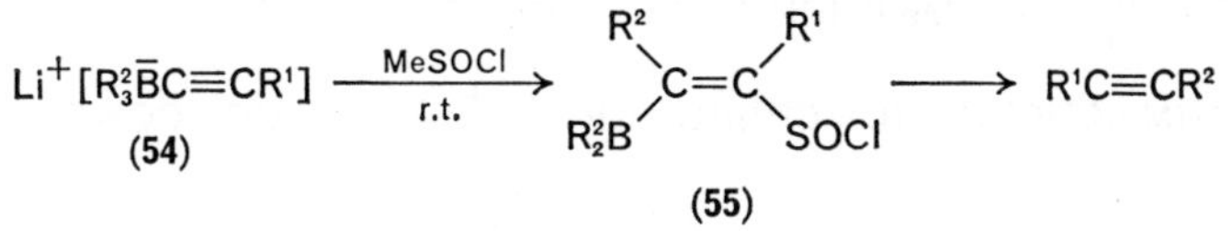

(165)

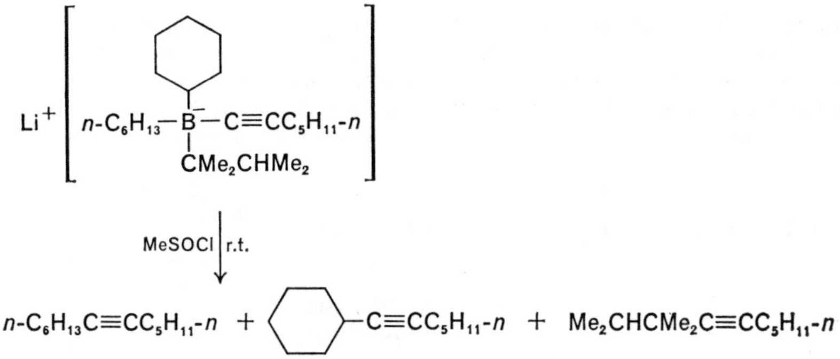

(166)

Ratio of products = 38 : 50 : 12

obtained from dilithium acetylide by the route of equation (164), but are obtained in 48–86% yields on reaction of lithium 2-chloroethynyltrialkylborates with iodine (equation 167)[241].

$$ClCH\overset{t}{=}CHCl \xrightarrow[0^\circ C]{MeLi,\ ether} LiC{\equiv}CCl \xrightarrow[r.t.]{R_3B} Li^+[R_3\bar{B}C{\equiv}CCl] \xrightarrow[-78\ ^\circ C]{I_2} RC{\equiv}CR \qquad (167)$$

It has been found that monosubstituted acetylenes can be obtained by Brown's method (equation 164) only when lithium acetylide is replaced by the lithium acetylide–ethylenediamine complex. High yields of terminal alkyl and cycloalkylacetylenes are obtained. It has also been demonstrated that the migration of an alkyl group proceeds with retention of configuration, as shown in equation (168)[242].

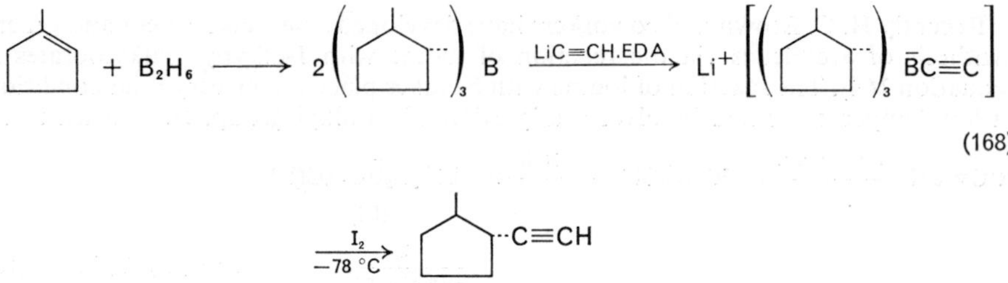

(168)

Further extension of the method has led to a 'one-pot' procedure for the synthesis of symmetrical and unsymmetrical conjugated diynes in 60–70% yields, as shown in Scheme 12[243]. The bulky 1,2-dimethylpropyl group has proved to possess a low

$$BH_3.SMe_2 + 2Me_2C{=}CHMe \xrightarrow[0\text{–}25\ ^\circ C]{THF} (Me_2CHCHMe)_2BH \xrightarrow[25\ ^\circ C]{MeOH,\ THF}$$

$$(Me_2CHCHMe)_2BOMe \xrightarrow[-78\ ^\circ C]{LiC{\equiv}CR^1,\ THF} Li^+[(Me_2CHCHMe)_2(OMe)\bar{B}C{\equiv}CR^1]$$

$$\xrightarrow[-78\text{–}25\ ^\circ C]{BF_3.OEt_2,\ THF} (Me_2CHCHMe)_2BC{\equiv}CR^1 \xrightarrow[-78\ ^\circ C]{LiC{\equiv}CR^2,\ THF}$$

$$Li^+[(Me_2CHCHMe)_2(C{\equiv}CR^2)\bar{B}C{\equiv}CR^1] \xrightarrow[-78\ ^\circ C]{I_2,\ THF} R^1C{\equiv}CC{\equiv}CR^2$$

SCHEME 12

migratory aptitude relative to the alkynyl group and thus made the synthesis of conjugated diynes possible. A related reaction leading only to symmetrical conjugated diynes has also exploited the low migratory aptitude of the 1,2-dimethylpropyl group, as well as that of the cyclohexyl group, furnishing diynes in 70–90% yields (equation 169)[244]. Primary alkyl groups are unsuitable since they show competitive migration

$$R^1_2BX + 2LiC{\equiv}CR^2 \longrightarrow Li^+[R^1_2\bar{B}(C{\equiv}CR^2)_2] \xrightarrow{I_2} \begin{cases} R^2C{\equiv}CC{\equiv}CR^2 \\ R^1C{\equiv}CR^2 \end{cases} \qquad (169)$$

R^1 = $Me_2CHCHMe$ or cyclohexyl

with respect to the alkynyl groups. The almost exclusive migration of the alkynyl group *vs* the 1,2-dimethylpropyl group has also proved useful in the synthesis of conjugated *trans*-enynes. Migration is highly stereoselective, furnishing the *trans*-enyne in over 99% isomeric purity and in 60–74% yields (equation 170)[245].

$$(Me_2CHCHMe)_2BH \xrightarrow[0^\circ C]{HC{\equiv}CR^1,\ THF} (Me_2CHCHMe)_2BCH\overset{t}{=}CHR^1 \xrightarrow[-50^\circ C]{LiC{\equiv}CR^2,\ THF}$$

$$Li^+[(Me_2CHCHMe)_2(C{\equiv}CR^2)\bar{B}CH\overset{t}{=}CHR^1] \xrightarrow[-78\text{–}25\ ^\circ C]{I_2} R^2C{\equiv}CCH\overset{t}{=}CHR^1 \qquad (170)$$

IV. MISCELLANEOUS METHODS

γ,δ-Acetylenic aldehydes have been obtained in 40–54% yields from alkenyl allenyl sulphides by a thermal [3,3]-sigmatropic rearrangement at 125–135 °C, carried out in H_2O–DMSO in the presence of calcium carbonate (equation 171)[246].

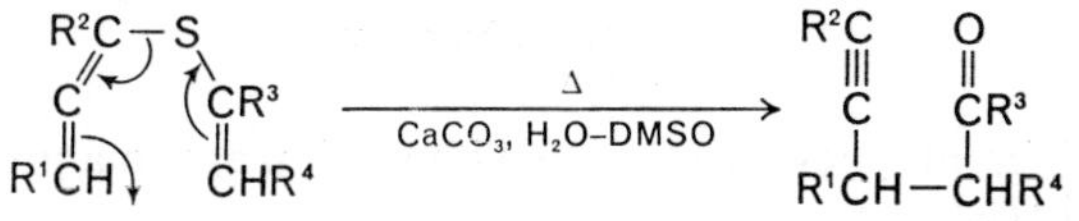

(171)

V. PROTECTION OF THE C—H AND C≡C BONDS OF ACETYLENES

Both the C—H and the C≡C bonds may require protection. The C—H bond in acetylenes is relatively highly acidic and may become involved in organometallic reactions taking place in other parts of the molecule. The C≡C bond is evidently susceptible to addition reactions and therefore may need protection. Acetylenes, and in particular terminal acetylenes, are susceptible to polymerization, which may be inhibited by protection.

A. *Protection of the C—H Bond*

The protection of the acetylenic C—H bond has recently been reviewed[247]. This reference contains a table of protecting groups, their stabilities under oxidative coupling and metalation conditions and the conditions for their removal.

In section III.C the use of an acetal and of a tetrahydropyranyl ether in the protection of monocuprous acetylide has been demonstrated (equations 151 and 152). However, the major and most popular mode of protection of the acetylenic C—H bond involves trialkylsilyl groups, mainly the trimethylsilyl and the triethylsilyl groups. These groups, as the bulky *t*-butyl group, can also inhibit the polymerization of unstable conjugated polyynes prepared in their presence. The protected acetylenes are stable to a variety of reaction conditions as detailed in this section, but can be readily cleaved by methanolic alkali, by precipitation with $AgNO_3$ and regeneration with KCN, and by *n*-Bu_4NF or KF.$2H_2O$. They are generally prepared by converting the terminal acetylene to its Grignard derivative, followed by reaction with a trialkylsilyl halide (equation 172).

$$R^1C{\equiv}CH + EtMgBr \longrightarrow R^1C{\equiv}CMgBr \xrightarrow{R^2_3SiCl} R^1C{\equiv}CSiR^2_3 \qquad (172)$$

The utility of the trialkylsilyl group for several reactions is demonstrated in the following. Internal triple bonds are selectively hydrogenated by a Lindlar catalyst to a double bond in the presence of a protected triple bond (equation 173)[248]. A Grignard reaction can take place in the presence of a protected triple bond (equation 174)[249].

$$Me_3SiC{\equiv}CCH{=}CMeCH_2C{\equiv}CCH_2CMe{=}CHC{\equiv}CSiMe_3 \xrightarrow[(2)\ AgNO_3,\ KCN]{(1)\ H_2,\ Pd\text{-}BaSO_4}$$

$$HC{\equiv}CCH{=}CMeCH_2CH\overset{c}{=}CHCH_2CMe{=}CHC{\equiv}CH \qquad (173)$$

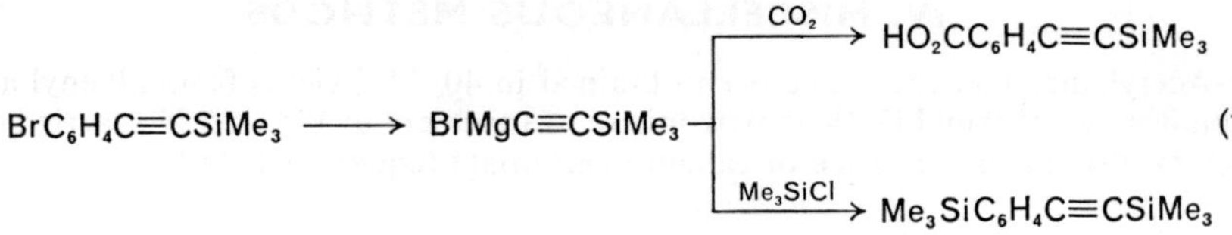

(174)

meta or *para* compounds

Alkylations of different types are effected without damage to a protected triple bond (equations 175–177).

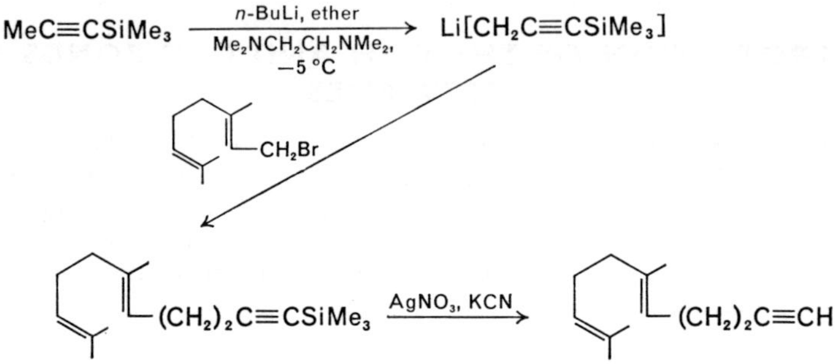

(Ref. 250) (175)

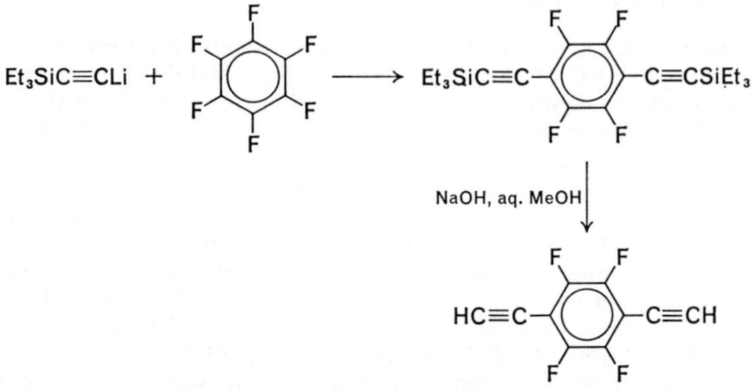

(Ref. 251) (176)

$$\text{ArMgX} + \text{CuBr} \xrightarrow{\text{THF}} \text{ArCu} \xrightarrow[0\text{–}20\,^\circ\text{C}]{\text{IC}\equiv\text{CSiMe}_3} \text{ArC}\equiv\text{CSiMe}_3 \xrightarrow{\text{dil. aq. alkali}} \text{ArC}\equiv\text{CH}$$

(Ref. 252) (177)

The trialkylsilyl group has been very useful in the preparation of unsubstituted polyynes by the Hay modification of the Glaser oxidative coupling of terminal acetylenes[253]. Thus hexadecaoctayne is obtained by the sequence shown in equation (178)[254]. This example also shows that partial cleavage of the protecting group is possible under mild conditions. An additional advantage of the trialkylsilyl group in the synthesis of conjugated polyynes is that its introduction shifts both the high and medium intensity u.v. bands of polyynes bathochromically and it thus permits all steps of the synthetic sequence to be followed quantitatively even in dilute

$$H(C{\equiv}C)_2MgBr + Et_3SiBr \longrightarrow H(C{\equiv}C)_2SiEt_3 \xrightarrow[Me_2CO,\ O_2]{CuCl,\ TMEDA} Et_3Si(C{\equiv}C)_4SiEt_3$$

$$\xrightarrow[21\ ^\circ C,\ 80\ s]{N\ NaOH,\ MeOH\text{–}hexane} Et_3Si(C{\equiv}C)_4H \xrightarrow[petroleum\ ether,\ Me_2CO]{CuCl,\ TMEDA,\ O_2} Et_3Si(C{\equiv}C)_8SiEt_3$$

$$\xrightarrow[17\ ^\circ C,\ 0.75\ min]{0.1N\ NaOH,\ MeOH\text{–}petroleum\ ether} H(C{\equiv}C)_8H \qquad (178)$$

solution[254a, b]. Protection by triethylsilyl and triethylgermyl groups in the Cadiot–Chodkiewicz couplings of arylacetylenes has also proved useful as illustrated in equation (179)[254a, b, 255].

$$ArC{\equiv}CH + BrC{\equiv}CSiEt_3 \xrightarrow{Cadiot\text{–}Chodkiewicz} Ar(C{\equiv}C)_2SiEt_3 \xrightarrow{KOH,\ aq.\ MeOH}$$

$$Ar(C{\equiv}C)_2H \xrightarrow[Cadiot\text{–}Chodkiewicz]{BrC{\equiv}CSiEt_3} Ar(C{\equiv}C)_3SiEt_3 \xrightarrow{aq.\ NAOH} Ar(C{\equiv}C)_3H \qquad (179)$$

$$Ar = mesityl$$

The preparation of Wittig reagents of compounds containing trialkylsilyl-protected acetylenic groups and their condensation with carbonyl compounds also proceeds smoothly. This is illustrated by the intermediate steps of the synthesis of a C_{18}-acid containing a 1-en-4-yne unit (equation 180)[256]. A similar example involves several steps in a synthesis of the insect juvenile hormone[257]. Conjugated *trans*-enynes are similarly obtained via a protected Wittig reagent[258].

$$Me_3SiC{\equiv}CCH_2CH_2OH \xrightarrow{(PhO)_3\overset{+}{P}MeI^-} Me_3SiC{\equiv}CCH_2CH_2I \xrightarrow[reflux]{Ph_3P,\ EtOH}$$

$$Me_3SiC{\equiv}CCH_2CH_2P^+Ph_3I^- \xrightarrow[NaH,\ THF,\ DMSO,\ 0\ ^\circ C]{OCH(CH_2)_7CO_2Me} Me_3SiC{\equiv}CCH_2CH\overset{c}{=}CH(CH_2)_7CO_2Me$$

$$\xrightarrow[(2)\ I_2,\ CH_2Cl_2]{(1)\ AgNO_3} IC{\equiv}CCH_2CH\overset{c}{=}CH(CH_2)_7CO_2Me \qquad (180)$$

Further examples of the utility of the trialkylsilyl group in the protection of terminal acetylenes are shown in reactions with hydrazine, halogenating agents and organometallic reagents (equations 181–183).

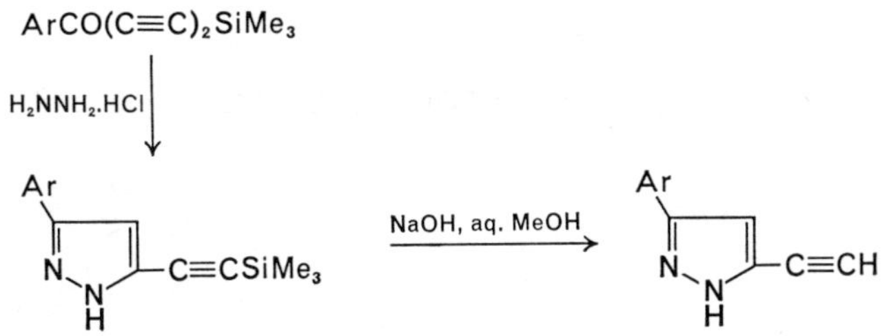

(Ref. 259) (181)

$$H(C{\equiv}C)_2CH_2OH \xrightarrow[NH_3]{Me_3SiCl} H(C{\equiv}C)_2CH_2OSiMe_3 \xrightarrow[(2)\ Me_3SiCl]{(1)\ EtMgBr} Me_3Si(C{\equiv}C)_2CH_2OSiMe_3$$

$$\xrightarrow{2N\ HCl} Me_3Si(C{\equiv}C)_2CH_2OH \xrightarrow{SOCl_2\ or\ PBr_3} Me_3Si(C{\equiv}C)_2CH_2X$$

$$\xrightarrow[X=Cl]{LiI} Me_3Si(C{\equiv}C)_2CH_2I \qquad (Ref.\ 260)\ (182)$$

$$X = Cl\ or\ Br$$

$$Co_2(CO)_8 + RC{\equiv}CSiMe_3 \longrightarrow Co_2(CO)_6[RC{\equiv}CSiMe_3]$$
$$R = Me,\ CF_3,\ Ph$$
$$\xrightarrow[\text{r.t.}]{\text{NaOH, H}_2\text{O–MeOH}} Co_2(CO)_6[RC{\equiv}CH] \qquad \text{(Ref. 261)} \quad (183)$$

B. Protection of the C≡C Bond

The carbon–carbon triple bond, as the double bond, is susceptible to many of the common addition reactions, and in some cases, such as reduction, hydroboration and acid-catalysed hydration, it is even more reactive than the double bond. An efficient protecting group for the triple bond has been developed only recently. Stirring dicobalt octacarbonyl with an alkyne at room temperature furnishes stable complexes of the alkyne in 70–90% yields. Double bonds in the protected acetylenic compound cannot be catalytically hydrogenated but reduction takes place with diimide or BH_3–AcOH. Also hydroboration proceeds exclusively at the double bonds and protected vinylacetylenes are hydrated with strong acid at the double bond. Cleavage is carried out by oxidative degradation of the complex with $Fe(NO_3)_3.9H_2O$ in 95% EtOH. Several examples are shown in equations (184)–(186)[262].

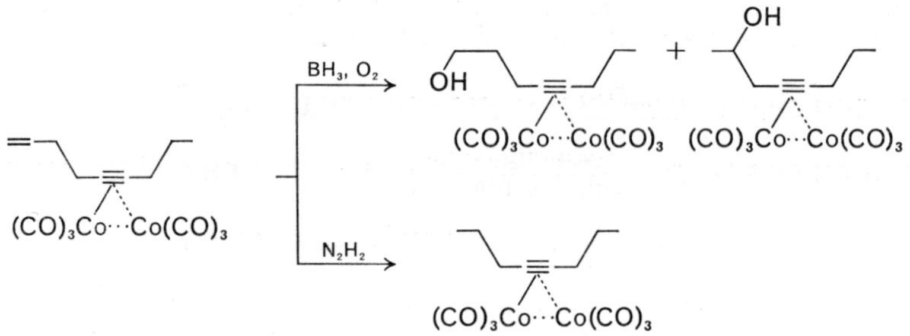

(184)

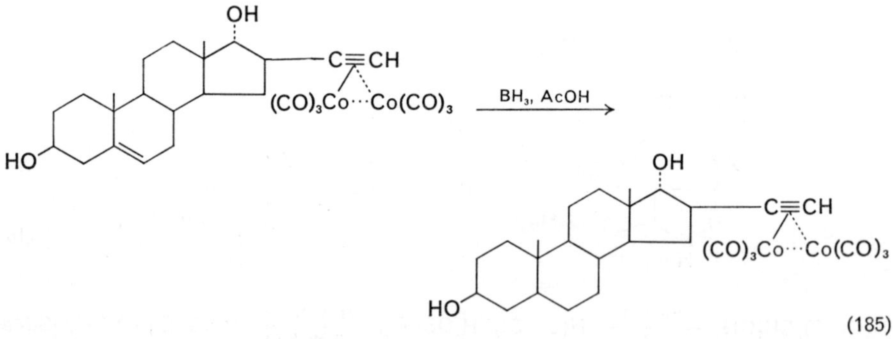

(185)

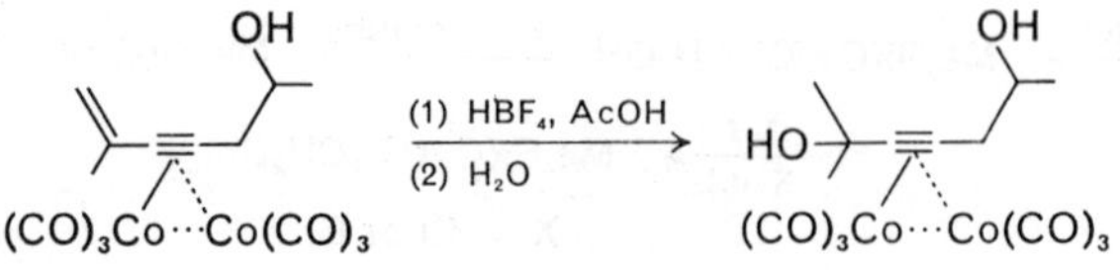

(186)

VI. REFERENCES

1. (a) J. A. Nieuwland and R. R. Vogt, *The Chemistry of Acetylene*, Reinhold, New York, 1945; (b) A. W. Johnson, *The Chemistry of the Acetylenic Compounds*, Vol. I: 'The Acetylenic Alcohols', Edward Arnold, London, 1946; Vol II: 'The Acetylenic Acids', Edward Arnold, London, 1950; (c) T. L. Jacobs in *Organic Reactions*, Vol. V (Ed. R. Adams), John Wiley and Sons, New York, 1949, Chap. 1; (d) W. Reppe, *Neue Entwicklungen auf dem Gebiet der Chemie des Acetylenes und des Kohlenoxids*, Springer, Berlin, 1949; (e) R. B. Wagner and H. D. Zook, *Synthetic Organic Chemistry*, John Wiley and Sons, New York, 1953, Chap. 3; (f) R. A. Raphael, *Acetylenic Compounds in Organic Synthesis*, Butterworths, London, 1955; (g) W. Franke, W. Ziegenbein and H. Meister in *Newer Methods of Preparative Organic Chemistry*, Vol. III (Ed. W. Foerst), Academic Press, New York, 1964, pp. 425–450; (h) W. Ziegenbein, *Einführung der Äthinyl-und Alkinyl-Gruppe in organische Verbindungen*, Verlag Chemie, Weinheim, 1963; (i) G. Eglinton and W. McCrae in *Advances in Organic Chemistry*, Vol. 4 (Eds. R. A. Raphael, E. C. Taylor and H. Wynberg), Interscience, New York, 1963, pp. 225–328; (j) W. Ried in *Newer Methods of Preparative Organic Chemistry*, Vol. IV (Ed. W. Foerst), Academic Press, New York, 1968, pp. 95–138; (k) G. Köbrich, *Angew. Chem., Int. Ed. Engl.*, **4,** 49 (1965); (l) S. R. Sandler and W. Karo, *Organic Functional Group Preparations*, Vol. I, Academic Press, New York, 1968, Chap. 3; (m) T. F. Rutledge, *Acetylenic Compounds*, Reinhold, New York, 1968; (n) H. G. Viehe (Ed.), *Chemistry of Acetylenes*, Marcel Dekker, New York, 1969; (o) J. C. Craig, M. D. Bergenthal, I. Fleming and J. Harley-Mason, *Angew. Chem., Int. Ed. Engl.*, **8,** 429 (1969); (p) C. A. Buehler and D. E. Pearson, *Survey of Organic Syntheses*, Wiley–Interscience, New York, 1970, pp. 150–173; (q) J. A. Gautier and M. Miocque, *Chim. Ther.*, **4,** 491 (1970); (r) K. E. Schulte, *Chim. Ther.*, **4,** 496 (1970); (s) J. F. Normant, *Synthesis* 63 (1972).
2. Ethynyl ethers and thioethers; (a) J. F. Arens in *Advances in Organic Chemistry*, Vol. II (Ed. R. A. Raphael, E. C. Taylor and H. Wynberg), Interscience, New York, 1960, pp. 117–212; alkynylmercaptans: (b) R. Mayer and H. Kröber, *Z. Chem.*, **15,** 91 (1975); cyclic acetylenes: (c) G. Wittig, *Angew. Chem., Int. Ed. Engl.*, **1,** 415 (1962); (d) R. Kolinskii, *Russ. Chem. Rev.* (*English Transl.*), **30,** 309 (1961); (e) R. W. Hoffman, *Dehydrobenzene and Cycloalkynes*, Academic Press, New York, 1967, Chap. 8; (f) H. Meier, *Synthesis*, 235 (1972); haloacetylenes: (g) K. M. Smirnov, A. P. Tomilov and A. I. Shchekotikhin, *Russ. Chem. Rev.* (*English Transl.*), **36,** 326 (1967); ynamines: (h) H. G. Viehe, *Angew. Chem., Int. Ed. Engl.*, **6,** 767 (1967); α-acetylenic carbonyl compounds: (i) R. I. Katkevich and L. I. Vereshchagin, *Russ. Chem. Rev.* (*English Transl.*), **38,** 900 (1969); diacetylenes: (j) M. F. Shostakovskii and A. V. Bogdanova, *The Chemistry of Diacetylenes*, John Wiley and Sons, New York, 1974; nitroacetylenes: (k) K. B. Rall, A. V. Vil'davskaya and A. A. Petrov, *Russ. Chem. Rev.* (*English Transl.*), **44,** 373 (1975).
3. L. Brandsma, *Preparative Acetylenic Chemistry*, Elsevier, Amsterdam, 1971.
4. W. Ziegenbein in *Chemistry of Acetylenes* (Ed. H. G. Viehe), Marcel Dekker, New York, 1969, Chap. 3.
5. O. M. Behr, G. Eglinton, A. R. Galbraith and R. A. Raphael, *J. Chem. Soc.*, 3614 (1960); O. M. Behr, G. Eglinton, I. A. Lardy and R. A. Raphael, *J. Chem. Soc.*, 1151 (1964).
6. C. F. H. Allen and C. O. Edens, Jr in *Organic Synthesis*, Collective Vol. 3 (Ed. E. C. Horning), John Wiley and Sons, New York, 1955, p. 731.
7. R. H. Mitchell and F. Sondheimer, *Tetrahedron*, **24,** 1397 (1968); H. A. Staab, J. Ipaktschi and A. Nissen, *Chem. Ber.*, **104,** 1182 (1971).
8. H. N. C. Wong, P. J. Garratt and F. Sondheimer, *J. Amer. Chem. Soc.*, **96,** 5604 (1974).
9. R. P. Linstead and E. G. Noble, *J. Chem. Soc.*, 933 (1937).
10. A. T. Babayan and I. Ya. Zurabov, *Zh. Obshch. Khim.*, **25,** 2445 (1955); *Chem. Abstr.*, **50,** 9280b (1956); A. T. Babayan, A. A. Grigoryan and G. T. Martirosyan, *Zh. Obshch. Khim.*, **29,** 386 (1959); *Chem. Abstr.*, **54,** 1265a (1960).
11. R. N. Haszeldine, *J. Chem. Soc.*, 2495 (1951).

12. G. Köbrich and P. Buck in *Chemistry of Acetylenes* (Ed. H. G. Viehe), Marcel Dekker, New York, 1969, Chap. 2; A. F. Cockerill in *Chemical Kinetics*, Vol. 9 (Ed. C. H. Bamford and C. F. H. Tipper), Elsevier, Amsterdam, 1973, pp. 309–316; A. Fry, *Chem. Soc. Rev.*, **1**, 163 (1972).
13. (a) W. K. Kwok, W. G. Lee and S. I. Miller, *J. Amer. Chem. Soc.*, **91**, 468 (1969); (b) S. J. Cristol and C. A. Whittemore, *J. Org. Chem.*, **34**, 705 (1969).
14. (a) A. Roedig in *Methoden der Organischen Chemie* (*Houben-Weyl*), Vol. V/4 (Ed. E. Müller), 4th ed., Georg Thieme Verlag, Stuttgart, 1960, pp. 38–152; (b) R. Stroh in *Methoden der Organischen Chemie* (*Houben-Weyl*), Vol. V/3 (Ed. E. Müller), 4th ed., Georg Thieme Verlag, Stuttgart, 1962, pp. 912–920; (c) W. Schoberth and M. Hanack, *Synthesis*, 703 (1972); (d) J. R. Sowa, E. J. Lamby, E. C. Calamai, D. A. Benko and A. Gordinier, *Org. Prep. Proced.*, **7**, 137 (1975); (e) A. Maercker in *Organic Reactions*, Vol. 14 (Ed. A. C. Cope), John Wiley and Sons, New York, 1965, Chap. 3.
15. O. Yu. Okhlobystin, V. I. Bregadze, A. A. Ponomarenko and L. I. Zakharkin, *Neftekhimiya*, **1**, 752 (1961); *Chem. Abstr.*, **57**, 14918f (1962).
16. T. L. Jacobs, R. A. Akawie and R. G. Cooper, *J. Amer. Chem. Soc.*, **73**, 1273 (1951); W. R. Moore and H. R. Ward, *J. Amer. Chem. Soc.*, **85**, 86 (1963).
17. V. L. Leighton, *Amer. Chem. J.*, **20**, 136 (1898).
18. T. W. Abbott in *Organic Synthesis*, Collective Vol. 2 (Ed. A. H. Blatt), John Wiley and Sons, New York, 1943, p. 515.
19. J. C. Hessler in *Organic Synthesis*, Collective Vol. I (Ed. A. H. Blatt), 2nd ed., John Wiley and Sons, New York, 1941, p. 438.
20. M. S. Newman and S. H. Merrill, *J. Amer. Chem. Soc.*, **77**, 5549 (1955).
21. (a) J. F. Arens, *Rec. Trav. Chim.*, **74**, 271 (1955); (b) J. R. Nooi and J. F. Arens, *Rec. Trav. Chim.*, **78**, 284 (1959).
22. A. T. Bottini, B. J. King and J. M. Lucas, *J. Org. Chem.*, **27**, 3688 (1962).
23. P. E. Eaton and C. E. Stubbs, *J. Amer. Chem. Soc.*, **89**, 5722 (1967).
24. S. J. Cristol, A. Begoon, W. P. Norris and P. S. Ramey, *J. Amer. Chem. Soc.*, **76**, 4558 (1954).
25. S. W. Staley and R. F. Doherty, *J. Chem. Soc., Chem. Commun.*, 288 (1969).
26. D. V. Banthorpe, *Elimination Reactions*, Elsevier, Amsterdam, 1963, p. 142.
27. S. I. Miller, *J. Org. Chem.*, **26**, 2619 (1961).
28. G. Marchese, F. Naso and G. Modena, *J. Chem. Soc.* (*B*), 958 (1968).
29. G. Marchese, F. Naso, N. Tangari and G. Modena, *J. Chem. Soc.* (*B*), 1196 (1970).
30. M. Schlosser and V. Ladenberger, *Chem Ber.*, **104**, 2873 (1971).
31. F. G. Bordwell, *Accounts Chem. Res.*, **5**, 374 (1972).
32. A.-H. A. Youssef and H. M. Abdel-Maksoud, *J. Org. Chem.*, **40**, 3227 (1975).
33. D. Bertin, *Compt. Rend.*, **229**, 660 (1949).
34. See also L. Brandsma, H. J. T. Bos and J. F. Arens in *Chemistry of Acetylenes* (Ed. H. G. Viehe), Marcel Dekker, New York, 1969, p. 754 and references cited therein.
35. J. F. Arens and T. Doornbos, *Rec. Trav. Chim.*, **75**, 481 (1956); see also L. Brandsma, H. J. T. Bos and J. F. Arens in *Chemistry of Acetylenes* (Ed. H. G. Viehe), Marcel Dekker, New York, 1969, p. 762 and references cited therein.
36. V. Jäger and H. G. Viehe, *Angew. Chem., Int. Ed. Engl.*, **8**, 273 (1969).
37. R. J. De Pasquale, C. D. Padgett and R. W. Rosser, *J. Org. Chem.*, **40**, 810 (1975).
38. L. F. Hatch and S. S. Nesbitt, *J. Amer. Chem. Soc.*, 72, 727 (1950).
39. C. D. Hurd and A. Tockman, *J. Org. Chem.*, **23**, 1087 (1958).
40. G. R. Newkome and D. L. Koppersmith, *J. Org. Chem.*, **38**, 4461 (1973).
41. A. M. Krubiner, N. Gottfried and E. P. Oliveto, *J. Org. Chem.*, **34**, 3502 (1969).
42. H. Adkins and R. E. Burks, Jr. in *Organic Synthesis*, Collective Vol. 3 (Ed. E. C. Horning), John Wiley and Sons, New York, 1955, p. 785.
43. G. F. Hennion, C. C. Price and T. F. McKeon, Jr. in *Organic Synthesis*, Collective Vol. 4 (Ed. N. Rabjohn), John Wiley and Sons, New York, 1963, p. 683.
44. J. B. Armitage, E. R. H. Jones and M. C. Whiting, *J. Chem. Soc.*, 44 (1951).
45. K. Bodendorf and P. Kloss, *Angew. Chem., Int. Ed. Engl.*, **2**, 98 (1963); J. Lötzbeyer and K. Bodendorf, *Chem. Ber.*, **100**, 2620 (1967).
46. J. Klein and E. Gurfinkel, *Tetrahedron*, **26**, 2127 (1970).

47. M. P. Ivitzky, *Bull. Soc. Chim. Fr.*, [4], **35,** 357 (1924); W. H. Puterbaugh and M. S. Newman, *J. Amer. Chem. Soc.*, **81,** 1611 (1959); P. J. Kocienski, *J. Org. Chem.*, **39,** 3285 (1974).
48. K. G. Untch and D. C. Wysocki, *J. Amer. Chem. Soc.*, **88,** 2608 (1966).
49. C. B. Reese and A. Shaw, *J. Chem. Soc., Chem. Commun.*, 1172 (1970).
50. H. G. Viehe, *Angew. Chem. Int. Ed. Engl.*, **2,** 477 (1963).
51. L. Meunier and E. Desparmet, *Bull. Soc. Chim. Fr*, [4], **35,** 481 (1924).
52. H. N. Miller, K. W. Greenlee, J. M. Derfer and C. E. Boord, *J. Org. Chem.*, **19,** 1882 (1954).
53. D. Caine and F. N. Tuller, *J. Org. Chem.*, **34,** 222 (1969).
54. W. J. Croxall and J. O. Van Hook, *J. Amer. Chem. Soc.*, **76,** 1700 (1954); L. Brandsma, *Preparative Acetylenic Chemistry*, Elsevier, Amsterdam, 1971, pp. 106–107, 122.
55. L. Brandsma, *Preparative Acetylenic Chemistry*, Elsevier, Amsterdam, 1971, p. 107, 122-124.
56. J. B. Armitage, C. L. Cook, E. R. H. Jones and M. C. Whiting, *J. Chem. Soc.*, 2010 (1952).
57. S. Akiyama and M. Nakagawa, *Bull. Chem. Soc. Japan*, **43,** 3561 (1970).
58. R. A. Raphael and F. Sondheimer, *J. Chem. Soc.*, 120 (1950).
59. F. Sondheimer, D. A. Ben-Efraim and R. Wolovsky, *J. Amer. Chem. Soc.*, **83** 1675 (1961).
60. F. Sondheimer and Y. Gaoni, *J. Amer. Chem. Soc.*, **84,** 3520 (1962).
61. N. A. Khan, F. E. Deatherage and J. B. Brown in *Organic Synthesis*, Collective Vol. 4 (Ed. N. Rabjohn), John Wiley and Sons, New York, 1963, p. 851.
62. P. J. Ashworth, G. H. Mansfield and M. C. Whiting in *Organic Synthesis*, Collective Vol. 4 (Ed. N. Rabjohn), John Wiley and Sons, New York, 1963, p. 128.
63. F. Fringuelli, V. Mancini and A. Taticchi, *Tetrahedron*, **25,** 4249 (1969).
64. W. Korytnyk, N. Angelino, B. Lachmann and P. G. G. Potti, *J. Med. Chem.*, **15,** 1262 (1972).
65. A. Vaitiekunas and F. F. Nord, *J. Org. Chem.*, **19,** 902 (1954).
66. H. C. Kretschmar and W. F. Erman, *Tetrahedron Letters*, 41 (1970).
67. P. Caubère, *Bull. Soc. Chim. Fr.*, 1293 (1966).
68. J. J. Van Daalen, A. Kraak and J. F. Arens, *Rec. Trav. Chim.*, **80,** 810 (1961); J. R. Nooi and J. F. Arens, *Rec. Trav. Chim.*, **81,** 517 (1962).
69. (a) D. Reisdorf and H. Normant, *Organomet. Chem. Synth.*, **1,** 393 (1972); (b) J. Villieras, P. Perriot and J. F. Normant, *Synthesis*, 458 (1975).
70. J. M. J. Tronchet, B. Baehler and A. Bonenfant, *Helv. Chim. Acta*, **59,** 941 (1976).
71. D. R. Strobach, *J. Org. Chem.*, **36,** 1438 (1971); see also Reference 50.
72. S. A. Fuqua, W. G. Duncan and R. M. Silverstein, *J. Org. Chem.*, **30,** 1027 (1965).
73. R. Buijle, A. Halleux and H. G. Viehe, *Angew. Chem., Int. Ed. Engl.*, **5,** 584 (1966).
74. H. Eilingsfeld, M. Seefelder and H. Weidinger, *Chem. Ber.*, **96,** 2671 (1963).
75. M. Schlosser and V. Ladenberger, *Chem. Ber.*, **100,** 3877 (1967); M. Schlosser and V. Ladenberger, *Chem. Ber.*, **100,** 3893 (1967); G. Köbrich, *Angew. Chem., Int. Ed. Engl.*, **6,** 41 (1967).
76. M. Schlosser and V. Ladenberger, *Chem. Ber.*, **100,** 3901 (1967).
77. G. Köbrich, H. Trapp, K. Flory and W. Drischel, *Chem. Ber.*, **99,** 689 (1966).
78. E. Kloster-Jenson, *Tetrahedron*, **27,** 33 (1971).
79. E. J. Corey and P. L. Fuchs, *Tetrahedron Letters*, 3769 (1972).
80. J. Villiéras, C. Bacquet and J. F. Normant, *Bull. Soc. Chim. Fr.*, Pt. 2, 1731 (1974).
81. K. Okuhara, *J. Org. Chem.*, **41,** 1487 (1976).
82. M. Schlosser and M. Zimmermann, *Chem. Ber.*, **104,** 2885 (1971).
83. R. L. Lutz and W. R. Smithey, Jr., *J. Org. Chem.*, **16,** 51 (1951).
84. C. M. Wong and T.-L. Ho, *Synth. Commun.*, **4,** 25 (1974).
85. A. Gorgues, *Compt. Rend. (C)*, **278,** 287 (1974).
86. A. Gorgues and A. Le Coq, *Compt. Rend. (C)*, **278,** 1153 (1974); *Bull. Soc. Chim. Fr.*, Pt. 2, 125 (1976).
87. F. Naso and L. Ronzini, *J. Chem. Soc., Perkin Trans.* 1, 340 (1974).
88. B. Šket, M. Zupan and A. Pollak, *Tetrahedron Letters*, 783 (1976).
89. J. G. Pritchard and A. A. Bothner-By, *J. Phys. Chem.*, **64,** 1271 (1960).

90. P. Beltrame, P. L. Beltrame, O. Sighinolfi and M. Simonetta, *J. Chem. Soc.* (*B*), 1103 (1967).
91. A. A. Bothner-By, *J. Amer. Chem. Soc.*, **77**, 3293 (1955); D. Y. Curtin, E. W. Flynn and R. F. Nystrom, *J. Amer. Chem. Soc.*, **80**, 4599 (1958).
92. (a) K. L. Erickson and J. Wolinsky, *J. Amer. Chem. Soc.*, **87**, 1142 (1965); (b) J. Wolinsky, G. W. Clark and P. C. Thorstenson, *J. Org. Chem.*, **41**, 745 (1976).
93. G. Köbrich and P. Buck in *Chemistry of Acetylenes* (Ed. H. G. Viehe), Marcel Dekker, New York, 1969, pp. 117-122, and references cited therein.
94. W. Tadros, A. B. Sakla and M. S. Ishak, *J. Chem. Soc.*, 4210 (1958).
95. S. Misumi, M. Kuwana, K. Murashima and M. Nakagawa, *Bull. Chem. Soc. Japan*, **34**, 1833 (1961).
96. G. H. Coleman and R. D. Maxwell, *J. Amer. Chem. Soc.*, **56**, 132 (1934); G. H. Coleman, W. H. Holst and R. D. Maxwell, *J. Amer. Chem. Soc.*, **58**, 2310 (1936).
97. D. Y. Curtin, E. W. Flynn, R. F. Nystrom and W. H. Richardson, *Chem. Ind.* (*London*), 1453 (1957); D. Y. Curtin and W. H. Richardson, *J. Amer. Chem. Soc.*, **81**, 4719 (1959).
98. G. Köbrich and H. Trapp, *Chem. Ber.*, **99**, 680 (1966).
99. H. Reimlinger, *Chem. Ind.* (*London*), 1306 (1969).
100. G. Köbrich and D. Merkel, *Angew. Chem., Int. Ed. Engl.*, **9**, 243 (1970); G. Köbrich, D. Merkel and K.-W. Thiem, *Chem. Ber.*, **105**, 1683 (1972).
101. T. Ando and M. Nakagawa, *Bull. Chem. Soc. Japan*, **44**, 172 (1971); M. Kataoka, T. Ando and M. Nakagawa, *Bull. Chem. Soc. Japan*, **44**, 177 (1971).
102. D. Y. Curtin, J. A. Kampmeier and B. R. O'Connor, *J. Amer. Chem. Soc.*, **87**, 863 (1965).
103. G. Köbrich and H. Fröhlich, *Chem. Ber.*, **98**, 3637 (1965).
104. G. Rosini and S. Cacchi, *J. Org. Chem.*, **37**, 1856 (1972).
105. M. Cariou, *Bull. Soc. Chim. Fr.*, 210 (1969).
106. E. R. H. Jones, G. Eglinton, M. C. Whiting and B. L. Shaw in *Organic Synthesis*, Collective Vol. 4 (Ed. N. Rabjohn), John Wiley and Sons, New York, 1963, p. 404; L. Brandsma, *Preparative Acetylenic Chemistry*, Elsevier, Amsterdam, 1971, p. 121; M. S. Newman, J. R. Geib and W. M. Stalick, *Org. Prep. Proced.*, **4**, 89 (1972).
107. L. Brandsma, *Preparative Acetylenic Chemistry*, Elsevier, Amsterdam, 1971, pp. 104, 122.
108. L. Brandsma, *Preparative Acetylenic Chemistry*, Elsevier, Amsterdam, 1971, pp. 105–106, 138–140; L. Brandsma, *Rec. Trav. Chim.*, **90**, 265 (1971).
109. A. Halleux, H. Reimlinger and H. G. Viehe, *Tetrahedron Letters*, 3141 (1970).
110. (a) J. Gosselck, L. Béress, H. Schenk and G. Schmidt, *Angew. Chem., Int. Ed. Engl.*, **4**, 1080 (1965); (b) J. Gosselck, L. Béress, H. Schenk and G. Schmidt, *Angew. Chem., Int. Ed. Engl.*, **4**, 1080 (1965).
111. (a) S. Gronowitz and T. Frejd, *Acta Chem. Scand.*, **B30**, 287 (1976); (b) S. Gronowitz and T. Frejd, *Acta Chem. Scand.*, **B30**, 313 (1976); (c) S. Gronowitz and T. Frejd, *Acta Chem. Scand.*, **B30**, 341 (1976).
112. (a) I. Fleming and J. Harley-Mason, *J. Chem. Soc.*, 4771 (1963); (b) E. J. D. Brown and J. Harley-Mason, *J. Chem. Soc.* (*C*), 1390 (1966); (c) I. Fleming and C. R. Owen, *J. Chem. Soc.* (*C*), 2013 (1971).
113. (a) P. J. Stang, J. Davis and D. P. Fox, *J. Chem. Soc., Chem. Commun.*, 17 (1975); (b) P. J. Stang, M. G. Mangum, D. P. Fox and P. Haak, *J. Amer. Chem. Soc.*, **96**, 4562 (1974); (c) R. J. Hargrove and P. J. Stang, *J. Org. Chem.*, **39**, 581 (1974).
114. J. C. Craig and M. Moyle, *Proc. Chem. Soc.*, 149 (1962); J. C. Craig and M. Moyle, *J. Chem. Soc.*, 3712 (1963).
115. E. J. Corey and R. H. Wollenberg, *J. Amer. Chem. Soc.*, **96**, 5581 (1974).
116. M. Fieser and L. F. Fieser, *Reagents for Organic Synthesis*, Vol. 5, John Wiley and Sons, New York, 1975, p. 43.
117. Ya. M. Slobodin and N. A. Selezneva, *Zh. Obshch. Khim.*, **26**, 691 (1956); *Chem. Abstr.*, **50**, 14502a (1956).
118. C. R. Hauser, H. T. Taylor and T. G. Ledford, *J. Amer. Chem. Soc.*, **82**, 1786 (1960).
119. J. B. Hendrickson and J. R. Sufrin, *Tetrahedron Letters*, 1513 (1973).
120. H. P. Figeys and M. Gelbcke, *Bull. Soc. Chim. Belges*, **83**, 369 (1974).
121. C. G. Krespan, R. J. Harder and J. J. Drysdale, *J. Amer. Chem. Soc.*, **83**, 3424 (1961).
122. P. L. Coe, R. G. Plevey and J. C. Tatlow, *J. Chem. Soc.* (*C*), 597 (1966).

123. H. G. Viehe and E. Franchimont, *Chem. Ber.*, **95,** 319 (1962).
124. G. Wittig and R. Pohlke, *Chem. Ber.*, **94,** 3276 (1961).
125. J. Ficini and C. Barbara, *Bull. Soc. Chim. Fr.*, 871 (1964); J. Ficini and C. Barbara, *Bull. Soc. Chim. Fr.*, 2787 (1965).
126. E. V. Dehmlow and G. C. Ezimora, *Tetrahedron Letters*, 563 (1971).
127. J. N. Seiber, *J. Org. Chem.*, **36,** 2000 (1971).
128. Y. Amiel, personal communication.
129. T. Mukaiyama, H. Nambu and T. Kumamoto, *J. Org. Chem.*, **29,** 2243 (1964).
130. T. Mukaiyama, H. Nambu and M. Okamoto, *J. Org. Chem.*, **27,** 3651 (1962).
131. J. A. Harvey and M. A. Ogliaruso, *Org. Prep. Proced.*, **8,** 37 (1976).
132. S. T. D. Gough and S. Trippett, *Proc. Chem. Soc.*, 302 (1961); S. T. D. Gough and S. Trippett, *J. Chem. Soc.*, 2333 (1962).
133. G. Märkl, *Chem. Ber.*, **94,** 3005 (1961).
134. S. T. D. Gough and S. Trippett, *J. Chem. Soc.*, 543 (1964).
135. P. A. Chopard, R. J. G. Searle and F. H. Devitt, *J. Org. Chem.*, **30,** 1015 (1965).
136. S. Akiyama, K. Nakasuji and M. Nakagawa, *Bull. Chem. Soc. Japan*, **44,** 2231 (1971).
137. J. Buckle and P. G. Harrison, *J. Organomet. Chem.*, **77,** C22 (1974).
138. C. B. Bauer and E. LeGoff, *Synthesis*, 544 (1970).
139. R. F. C. Brown, F. W. Eastwood, K. J. Harrington and G. L. McMullen, *Aust. J. Chem.*, **27,** 2393 (1974).
140. R. F. C. Brown and K. J. Harrington, *J. Chem. Soc., Chem. Commun.*, 1175 (1972); see also R. F. C. Brown, K. J. Harrington and G. L. McMullen, *J. Chem. Soc., Chem. Commun.*, 123 (1974).
141. J. C. Philips, J. V. Swisher, D. Haidukewych and O. Morales, *J. Chem. Soc., Chem. Commun.*, 22 (1971); see also L. A. Carpino, L. V. McAdams, III, R. H. Rynbrandt and J. W. Spiewak, *J. Amer. Chem. Soc.*, **93,** 476 (1971).
142. R. H. Mitchell, *J. Chem. Soc., Chem. Commun.*, 955 (1973).
143. D. P. Bauer and R. S. Macomber, *J. Org. Chem.*, **41,** 2640 (1976).
144. R. J. Theis and R. E. Dessy, *J. Org. Chem.*, **31,** 624 (1966).
145. A. C. Cope, D. S. Smith and R. J. Cotter in *Organic Synthesis*, Collective Vol. 4 (Ed. N. Rabjohn), John Wiley and Sons, New York, 1963, p. 377.
146. J. Tsuji, H. Takahashi and T. Kajimoto, *Tetrahedron Letters*, 4573 (1973).
147. M. S. Newman and D. E. Reid, *J. Org. Chem.*, **23,** 665 (1958).
148. A. T. Blomquist, R. E. Burge, Jr. and A. C. Sucsy, *J. Amer. Chem. Soc.*, **74,** 3636 (1952); V. Prelog, K. Schenker and H. H. Günthard, *Helv. Chim. Acta*, **35,** 1598 (1952).
149. A. T. Blomquist, L. H. Liu and J. C. Bohrer, *J. Amer. Chem. Soc.*, **74,** 3643 (1952).
150. G. Wittig and A. Krebs, *Chem. Ber.*, **94,** 3260 (1961); G. Wittig and H.-L. Dorsch, *Ann. Chem.*, **711,** 46 (1968).
151. E. Müller and H. Meier, *Ann. Chem.*, **716,** 11 (1968).
152. A. Krebs, *Tetrahedron Letters*, 4511 (1968).
153. A. Krebs and H. Kimling, *Angew. Chem., Int. Ed. Engl.*, **10,** 509 (1971).
154. A. Krebs and H. Kimling, *Tetrahedron Letters*, 761 (1970).
155. A. Krebs and H. Kimling, *Ann. Chem.*, 2074 (1974).
156. (a) M. S. Newman and A. Kutner, *J. Amer. Chem. Soc.*, **73,** 4199 (1951); (b) see also M. S. Newman, *J. Amer. Chem. Soc.*, **71,** 378 (1949) and M. S. Newman and W. M. Edwards, *J. Amer. Chem. Soc.*, **76,** 1840 (1954).
157. H. P. Hogan and J. Seehafer, *J. Org. Chem.*, **37,** 4466 (1972).
158. M. S. Newman and L. F. Lee, *J. Org. Chem.*, **37,** 4468 (1972).
159. T. B. Patrick, J. M. Disher and W. J. Probst, *J. Org. Chem.*, **37,** 4467 (1972).
160. M. S. Newman and S. J. Gromelski, *J. Org. Chem.*, **37,** 3220 (1972).
161. (a) L. A. Carpino, *J. Amer. Chem. Soc.*, **80,** 599 (1958); (b) L. A. Carpino, P. H. Terry and S. D. Thatte, *J. Org. Chem.*, **31,** 2867 (1966).
162. A. Silveira, Jr., T. J. Weslowski, T. A. Weil, V. Kumar and J. P. Gillespie, *J. Amer. Oil Chem. Soc.*, **48,** 661 (1971).
163. L. A. Carpino, *J. Amer. Chem. Soc.*, **80,** 601 (1958).
164. E. C. Taylor, R. L. Robey and A. McKillop, *Angew. Chem., Int. Ed. Engl.*, **11,** 48 (1972); E. C. Taylor, R. L. Robey, D. K. Johnson and A. M. McKillop in *Organic Synthesis*, Vol. 55 (Ed. S. Masamune), John Wiley and Sons, New York, 1976, p. 73.

165. F. G. Willey, *Angew. Chem., Int. Ed. Engl.*, **3**, 138 (1964).
166. W. R. Bamford amd T. S. Stevens, *J. Chem. Soc.*, 4735 (1952).
167. H. Meier and I. Menzel, *Synthesis*, 215 (1971).
168. H. Behringer and M. Matner, *Tetrahedron Letters*, 1663 (1966).
169. I. Lalezari, A. Shafiee and M. Yalpani, *Tetrahedron Letters*, 5105 (1969).
170. I. Lalezari, A. Shafiee and M. Yalpani, *Angew. Chem., Int. Ed. Engl.*, **9**, 464 (1970); I. Lalezari, A. Shafiee and M. Yalpani, *J. Org. Chem.*, **36**, 2836 (1971).
171. (a) H. Meier and I. Menzel, *J. Chem. Soc., Chem. Commun.*, 1059 (1971); (b) I. Lalezari, A. Shafiee and M. Yalpani, *J. Heterocycl. Chem.*, **9**, 1411 (1972).
172. H. Meier and E. Voigt, *Tetrahedron*, **28**, 187 (1972).
173. I. Lalezari, A. Shafiee and H. Golgolab, *J. Heterocycl. Chem.*, **10**, 655 (1973).
174. H. Golgolab and I. Lalezari, *J. Heterocycl. Chem.*, **12**, 801 (1975).
175. R. Raap and R. G. Micetich, *Can. J. Chem.*, **46**, 1057 (1968).
176. R. G. Micetich, *Org. Prep. Proced.*, **3**, 163 (1971).
177. A. Tanaka, T. Usui and S. Yoshina, *Heterocycles*, **3**, 1087 (1975).
178. (a) A. Eschenmoser, D. Felix and G. Ohloff, *Helv. Chim. Acta*, **50**, 708 (1967); (b) D. Felix, J. Schreiber, G. Ohloff and A. Eschenmoser, *Helv. Chim. Acta*, **54**, 2896 (1971).
179. J. Schreiber, D. Felix, A. Eschenmoser, M. Winter, F. Gautschi, K. H. Schulte-Elte, E. Sundt, G. Ohloff, J. Kalvoda, H. Kaufmann, P. Wieland and G. Anner, *Helv. Chim. Acta*, **50**, 2101 (1967).
180. (a) M. Tanabe, D. F. Crowe, R. L. Dehn and G. Detre, *Tetrahedron Letters*, 3739 (1967); (b) M. Tanabe, D. F. Crowe and R. L. Dehn, *Tetrahedron Letters*, 3943 (1967); (c) S. V. Sunthankar and S. D. Mehendale, *Tetrahedron Letters*, 2481 (1972); (d) G. L. Lange and T.-W. Hall, *J. Org. Chem.*, **39**, 3819 (1974); (e) F. H. Batzold and C. H. Robinson, *J. Org. Chem.*, **41**, 313 (1976).
181. R. V. Hoffman and H. Shechter, *J. Amer. Chem. Soc.*, **93**, 5940 (1971).
182. R. K. Müller, R. Joos, D. Felix, J. Schreiber, C. Wintner and A. Eschenmoser in *Organic Synthesis*, Vol. 55 (Ed. S. Masamune), John Wiley and Sons, New York, 1976, p. 114.
183. D. Felix, J. Schreiber, K. Piers, U. Horn and A. Eschenmoser, *Helv. Chim. Acta*, **51**, 1461 (1968); D. Felix, R. K. Müller, U. Horn, R. Joos, J. Schreiber and A. Eschenmoser, *Helv. Chim. Acta*, **55**, 1276 (1972); see also R. K. Müller, D. Felix, J. Schreiber and A. Eschenmoser, *Helv. Chim. Acta*, **53**, 1479 (1970).
184. D. Felix, C. Wintner and A. Eschenmoser in *Organic Synthesis*, Vol. 55 (Ed. S. Masamune), John Wiley and Sons, New York, 1976, p. 52.
185. P. Wieland, H. Kaufmann and A. Eschenmoser, *Helv. Chim. Acta*, **50**, 2108 (1967); P. Wieland and H. Kaufmann, *Helv. Chim. Acta*, **56**, 2044 (1973).
186. T. Iwadare, I. Adachi, M. Hayashi, A. Matsunaga and T. Kitai, *Tetrahedron Letters*, 4447 (1969).
187. P. Wieland, *Helv. Chim. Acta*, **53**, 171 (1970).
188. E. Wenkert and C. A. McPherson, *J. Amer. Chem. Soc.*, **94**, 8084 (1972).
189. E. Wenkert and C. A. McPherson, *Synth. Commun.*, **2**, 331 (1972).
190. E. W. Colvin and B. J. Hamill, *J. Chem. Soc., Chem. Commun.*, 151 (1973).
191. J. M. J. Tronchet, A. Gonzalez, J.-B. Zumwald and F. Perret, *Helv. Chim. Acta*, **57**, 1505 (1974).
192. G. Ohloff and W. Pickenhagen, *Helv. Chim. Acta*, **54**, 1789 (1971).
193. R. A. Raphael and F. Sondheimer, *J. Chem. Soc.*, 2100 (1950).
194. K. N. Campbell and B. K. Campbell in *Organic Synthesis*, Collective Vol. 4 (Ed. N. Rabjohn), John Wiley and Sons, New York, 1963, p. 117.
195. D. A. Ben-Efraim and F. Sondheimer, *Tetrahedron*, **25**, 2837 (1969).
196. R. Köster, A. Bussmann and G. Schroth, *Ann. Chem.*, 2130 (1975).
197. E. A. Braude in *Progress In Organic Chemistry*, Vol. 3 (Ed. J. W. Cook), Butterworths, London, 1955, p. 199.
198. (a) L. J. Haynes, I. Heilbron, E. R. H. Jones and F. Sondheimer, *J. Chem. Soc.*, 1583 (1947); (b) I. Bell, E. R. H. Jones and M. C. Whiting, *J. Chem. Soc.*, 2597 (1957); (c) S. R. Landor and E. S. Pepper, *J. Chem. Soc. (C)*, 2283 (1966).
199. J. A. Kepler and R. C. Strickland, *Org. Prep. Proced.*, **5**, 41 (1973).

200. O. F. Beumel, Jr. and R. F. Harris, *J. Org. Chem*, **28**, 2775 (1963).
201. W. N. Smith and O. F. Beumel, Jr., *Synthesis*, 441 (1974).
202. F. L. M. Pattison and R. E. A. Dear, *Can. J. Chem.*, **41**, 2600 (1963).
203. J. H. P. Tyman, *Synth. Commun.*, **5**, 21 (1975).
204. S. Bhanu and F. Scheinmann, *J. Chem. Soc., Chem. Commun.*, 817 (1975); A. J. G. Sagar and F. Scheinmann, *Synthesis*, 321 (1976).
205. D. N. Brattesani and C. H. Heathcock, *Synth. Commun.*, **3**, 245 (1973); W. Beckmann, G. Doerjer, E. Logemann, C. Merkel, G. Schill and C. Zürcher, *Synthesis*, 423 (1975).
206. N. Gilman and B. Holland, *Synth. Commun.*, **4**, 199 (1974).
207. G. Schill and U. Keller, *Synthesis*, 621 (1972); G. Schill, E. Logemann and H. Fritz, *Chem. Ber.*, **109**, 497 (1976).
208. A. Nissen and H. A. Staab, *Chem. Ber.*, **104**, 1191 (1971).
209. N. Darby, C. U. Kim, J. A. Salaün, K. W. Shelton, S. Takada and S. Masamune, *J. Chem. Soc., Chem. Commun.*, 1516 (1971).
210. M. S. Chattha and A. M. Aguiar, *J. Org. Chem.*, **36**, 2720 (1971).
211. J. Kříž, M. J. Beneš and J. Peška, *Tetrahedron Letters*, 2881 (1965); J. Gigg, R. Gigg and C. D. Warren, *J. Chem. Soc. (C)*, 1882 (1966).
212. (a) T. F. Rutledge, *J. Org. Chem.*, **24**, 840 (1959); (b) E. F. Jenny and J. Druey, *Helv. Chim. Acta*, **42**, 401 (1959); (c) J. F. Normant, *Bull. Soc. Chim. Fr.*, 859 (1965); see also H. Normant, *Angew. Chem., Int. Ed. Engl.*, **6**, 1046 (1967).
213. J. J. Partridge, S. Faber and M. R. Uskoković, *Helv. Chim. Acta.*, **57**, 764 (1974).
214. F. Sondheimer, Y. Gaoni and J. Bregman, *Tetrahedron Letters*, **26**, 25 (1960).
215. R. F. Garwood, E. Oskay and B. C. L. Weedon, *Chem. Ind. (London)*, 1684 (1962).
216. R. A. Raphael, *Acetylenic Compounds in Organic Synthesis*, Butterworths, London, 1955, p. 17, and Reference 77 cited therein.
217. E. R. H. Jones, L. Skatteböl and M. C. Whiting, *J. Chem. Soc.*, 4765 (1956); L. Skattebøl, E. R. H. Jones and M. C. Whiting in *Organic Synthesis*, Collective Vol. 4 (Ed. N. Rabjohn), John Wiley and Sons, New York, 1963, p. 792.
218. D. A. Ben-Efraim and F. Sondheimer, *Tetrahedron*, **25**, 2823 (1969).
219. H. Taniguchi, I. M. Mathai and S. I. Miller, *Tetrahedron*, **22**, 867 (1966); H. Taniguchi, I. M. Mathai and S. I. Miller in *Organic Synthesis*, Vol. 50 (Ed. R. Breslow), John Wiley and Sons, New York, 1970, p. 97.
220. Y. Gaoni, C. C. Leznoff and F. Sondheimer, *J. Amer. Chem. Soc.*, **90**, 4940 (1968).
221. D. A. Ben-Efraim and F. Sondheimer, *Tetrahedron Letters*, 313 (1963).
222. B. E. Looker and F. Sondheimer, *Tetrahedron*, **27**, 2567 (1971).
223. M. S. Chattha and A. M. Aguiar, *J. Org. Chem.*, **36**, 2719 (1971).
224. E. Negishi and S. Baba, *J. Amer. Chem. Soc.*, **97**, 7385 (1975).
225. A. M. Sladkov and L. Yu. Ukhin, *Russ. Chem. Rev. (English Transl.)*, **37**, 748 (1968).
226. (a) C. E. Castro and R. D. Stephens, *J. Org. Chem.*, **28**, 2163 (1963); (b) R. D. Stephens and C. E. Castro, *J. Org. Chem.*, **28**, 3313 (1963); (c) C. E. Castro, E. J. Gaughan and D. C. Owsley, *J. Org. Chem.*, **31**, 4071 (1966); (d) C. E. Castro, R. Havlin, V. K. Honwad, A. Malte and S. Mojé, *J. Amer. Chem. Soc.*, **91**, 6464 (1969).
227. (a) R. E. Atkinson, R. F. Curtis, D. M. Jones and J. A. Taylor, *J. Chem. Soc., Chem. Commun.*, 718 (1967); (b) R. E. Atkinson, R. F. Curtis, D. M. Jones and J. A. Taylor, *J. Chem. Soc. (C)*, 2173 (1969); (c) I. Barrow and A. E. Pedler, *Tetrahedron*, **32**, 1829 (1976).
228. J. Burdon, P. L. Coe, C. R. Marsh and J. C. Tatlow, *J. Chem. Soc., Chem. Commun.*, 1259 (1967).
229. M. Bourgain and J. F. Normant, *Bull. Soc. Chim. Fr.* 2477 (1969); J. F. Normant, M. Bourgain and A.-M. Rone, *Compt. Rend. (C)*, **270**, 354 (1970).
230. A. Sevin, W. Chodkiewicz and P. Cadiot, *Bull. Soc. Chim. Fr.*, Pt. 2, 913 (1974).
231. M. W. Logue and G. L. Moore, *J. Org. Chem.*, **40**, 131 (1975).
232. J. F. Normant and M. Bourgain, *Tetrahedron Letters*, 2659 (1970).
233. K. Endo, Y. Sakata and S. Misumi, *Bull. Chem. Soc. Japan*, **44**, 2465 (1971).
234. H. A. Staab and K. Neunhoeffer, *Synthesis*, 424 (1974).
235. M. S. Shvartsberg, I. L. Kotlyarevskii, A. N. Kozhevnikova and V. N. Andrievskii, *Bull. Acad. Sci. USSR, Chem. Div.*, 1079 (1970).
236. L. Cassar, *J. Organomet. Chem.*, **93**, 253 (1975).

237. H. A. Dieck and F. R. Heck, *J. Organomet. Chem.*, **93,** 259 (1975).
238. K. Sonogashira, Y. Tohda and N. Hagihara, *Tetrahedron Letters*, 4467 (1975).
239. A. Suzuki, N. Miyaura, S. Abiko, M. Itoh, H. C. Brown, J. A. Sinclair and M. M. Midland, *J. Amer. Chem. Soc.*, **95,** 3080 (1973).
240. M. Naruse, K. Utimoto and H. Nozaki, *Tetrahedron Letters*, 1847 (1973); M. Naruse, K. Utimoto and H. Nozaki, *Tetrahedron*, **30,** 2159 (1974).
241. K. Yamada, N. Miyaura, M. Itoh and A. Suzuki, *Tetrahedron Letters*, 1961 (1975).
242. M. M. Midland, J. A. Sinclair and H. C. Brown, *J. Org. Chem.*, **39,** 731 (1974).
243. J. A. Sinclair and H. C. Brown, *J. Org. Chem.*, **41,** 1078 (1976).
244. A. Pelter, K. Smith and M. Tabata, *J. Chem. Soc., Chem. Commun.*, 857 (1975).
245. E. Negishi, G. Lew and T. Yoshida, *J. Chem. Soc., Chem. Commun.*, 874 (1973).
246. L. Brandsma and H. D. Verkruijsse, *Rec. Trav. Chim.*, **93,** 319 (1974).
247. D. R. M. Walton in *Protective Groups in Organic Chemistry* (Ed. J. F. W. McOmie), Plenum Press, London, 1973, pp. 2–11.
248. H. M. Schmidt and J. F. Arens, *Rec. Trav. Chim.*, **86,** 1138 (1967).
249. C. Eaborn, A. R. Thompson and D. R. M. Walton, *J. Chem. Soc.* (*C*), 1364 (1967).
250. E. J. Corey and H. A. Kirst, *Tetrahedron Letters*, 5041 (1968); E. J. Corey, J. A. Katzenellenbogen, N. W. Gilman, S. A. Roman and B. W. Erickson, *J. Amer. Chem. Soc.*, **90,** 5618 (1968).
251. F. Waugh and D. R. M. Walton, *J. Organomet. Chem.*, **39,** 275 (1972).
252. R. Oliver and D. R. M. Walton, *Tetrahedron Letters*, 5209 (1972).
253. A. S. Hay, *J. Org. Chem.*, **27,** 3320 (1962).
254. (a) R. Eastmond and D. R. M. Walton, *J. Chem. Soc., Chem. Commun.*, 204 (1968); (b) R. Eastmond, T. R. Johnson and D. R. M. Walton, *Tetrahedron*, **28,** 4601 (1972); (c) T. R. Johnson and D. R. M. Walton, *Tetrahedron*, **28,** 5221 (1972).
255. (a) R. Eastmond and D. R. M. Walton, *Tetrahedron*, **28,** 4591 (1972); (b) B. N. Ghose and D. R. M. Walton, *Synthesis*, 890 (1974).
256. A. G. Fallis, E. R. H. Jones and V. Thaller, *J. Chem. Soc., Chem. Commun.*, 924 (1969).
257. J. S. Cochrane and J. R. Hanson, *J. Chem. Soc., Perkin Trans.* I, 361 (1972).
258. E. J. Corey and R. A. Ruden, *Tetrahedron Letters*, 1495 (1973).
259. M. F. Ford and D. R. M. Walton, *Synthesis*, 47 (1973).
260. B. F. Coles and D. R. M. Walton, *Synthesis*, 390 (1975); see also Reference 255b.
261. R. S. Dickson and H. P. Kirsch, *Aust. J. Chem.*, **25,** 1815 (1972).
262. K. M. Nicholas and R. Pettit, *Tetrahedron Letters*, 3475 (1971); see also D. Seyferth, M. O. Nestle and A. T. Wehman, *J. Amer. Chem. Soc.*, **97,** 7417 (1975).

CHAPTER **19**

Nucleophilic attacks on acetylenes

J. I. DICKSTEIN
College of Du Page, Glen Ellyn, Illinois, U.S.A.

S. I. MILLER
Illinois Institute of Technology, Chicago, Illinois, U.S.A.

I. INTRODUCTION

A. Scope

Nucleophilic attacks on alkynes (A) is a broad subject, some of which will be discussed here. We have divided our material somewhat arbitrarily into additions (equation 1) and substitutions (equation 2). Our intent is to compile illustrative examples, to delineate mechanistic features and to evaluate both aspects critically.

$$\underset{\textbf{(A)}}{R'C{\equiv}CR} + Nu^- \xrightarrow{E^+} R'NuC{=}CRE \quad (1)$$

$$R{-}C{\equiv}CX + Nu^- \longrightarrow R{-}C{\equiv}CNu + X^- \quad (2)$$

Besides processes (1) and (2), the reader should be aware that nucleophilic attacks on alkynes are treated in other chapters of this book, dealing with rearrangements, cyclizations, polyacetylenes, cyclic acetylenes and perhaps others. A number of publications overlap with ours in different ways and at different levels[1-4]. They treat: individual alkynes or families[5-7], e.g. acetylene, diacetylenes[8], acetylene dicarboxylic esters[9, 10], haloacetylenes[4], alkynyl ethers and thioethers[11, 12], ynamines[13], fluoroalkynes[14, 15], ethynyl ketones[16], nitroalkynes[17], etc.; synthetic targets, e.g. pyrazoles[18], *H*-1,2,3-triazoles[19], isothiazoles[20], indolizines[21], etc.; reagents, e.g. nitrones[22], lithium aluminium hydride[23], heterocyclic *N*-oxides[24, 25], azomethine ylids[25, 26], tertiary phosphorus compounds[27], miscellaneous dipolar nucleophiles[25, 28], etc. The reader will appreciate that all of these constitute alternate entries into our subject.

We shall attempt to pick up the material at the latest major survey and bring it up to the present (spring, 1976). In effect, broad introductions and background to both additions[1, 2] and substitutions[3, 4] are already at hand. In this survey our selections of factual data cover a variety of subtopics and provide a unique and useful data base. In some, the entries constitute a fraction of a large body of research; in others, subjects which have not been adequately reviewed, e.g. electron transfer to alkynes or rate data for process (2), will be considered in detail. In any case, the breadth and diversity of the subject area preclude any claim that we have seen every relevant publication. Certainly, we cannot include all we have seen. Nevertheless, this is probably the most ambitious attempt to pull together and systematize this area of acetylene chemistry.

In deciding on the boundaries of our subject we have to make several difficult decisions. Acetylene to allene conversions are usually excluded: although certain nucleophilic reductions are retained, other S_N2' attacks[29, 30] (equation 3) and base-catalysed isomerizations (equation 4) are omitted[31, 32]. We do not seek out 'concealed'

$$F_3CC{\equiv}CCF_3 + Re(CO)_5^- \xrightarrow[-78\,^\circ C]{THF} F_2C{=}C{=}C(CF_3)Re(CO)_5 \quad (3)$$

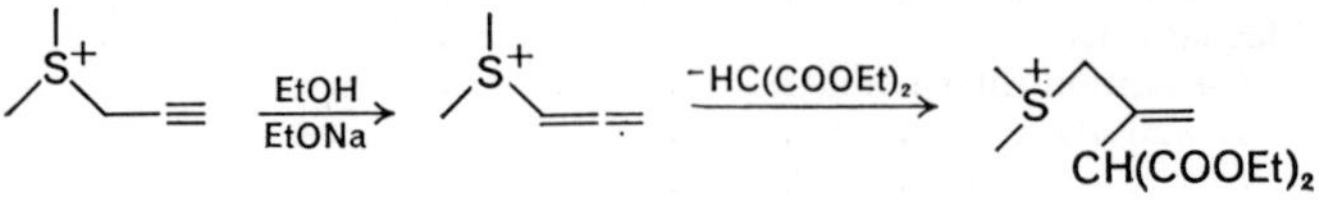

(4)

or inadvertent additions to alkynes. These may arise whenever an alkyne is generated and consumed in the course of some other intended reaction (equations 5 and 6)[33–36].

$$Z\text{- or }E\text{-MeCCl=CHCN} + \text{EtS}^- \xrightarrow{\text{EtOH}} \text{MeC}\equiv\text{CCN} \downarrow;\ \xrightarrow{?} E\text{-MeC(SEt)=CHCN} \quad (5)$$

$$\text{HC}\equiv\text{CH} + \text{RSSR} \xrightarrow[\text{12 atm}]{\text{KOH}} Z\text{-RSCH=CHSR}$$
$$\text{HC}\equiv\text{CH} \downarrow \text{HC}\equiv\text{C}^- \xrightarrow{\text{RSSR}} (\text{HC}\equiv\text{CSR} + \text{RS}^-) \uparrow Z\text{-RSCH=CHSR} \quad (6)$$

In the case of the attacks of neutral molecules on alkynes, nucleophilic attack is often difficult to distinguish from molecular cycloaddition or electrophilic initiation. Reaction (7) is typical of many which could equally as well be formulated as beginning with a dipolar cycloaddition or an acyclic zwitterion: 'Detailed mechanism of these cycloaddition–elimination reactions remains to be explored...'[37].

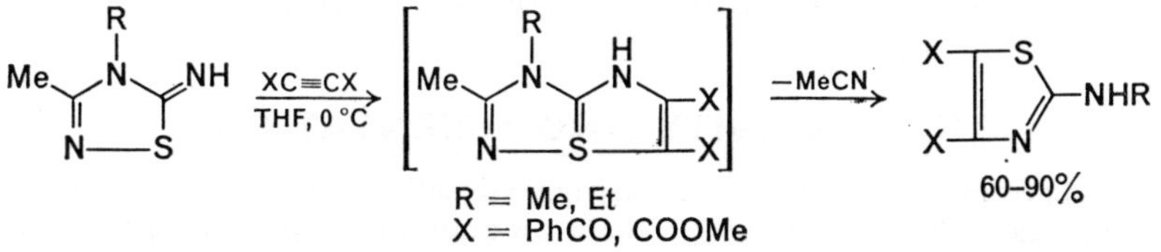

(7)

We have included certain reductions, e.g. hydride or electron transfer, but not others (equation 8) in which the nucleophilic component is absent or ambiguous[7].

$$2\text{Cr}^{2+} + 2\text{H}^+ + {-}\text{C}\equiv\text{C}{-} \longrightarrow 2\text{Cr}^{3+} + {-}\text{HC=CH}{-} \quad (8)$$

Although some reactions of organometallics, e.g. 'RMgBr', 'CuH', R_3SnH, with alkynes are admitted, it is often unclear whether one can even apply a term such as 'nucleophile' to these aggregated species[38, 39]. Likewise, if a ligand in an organometallic compound is (or can be represented as) an alkyne, we have sought those in which the triple bond is being attacked, that is, equation (9)[40], rather than equation (10)[41]. Clearly, some of our decisions to include or to exclude are not wholly satisfactory.

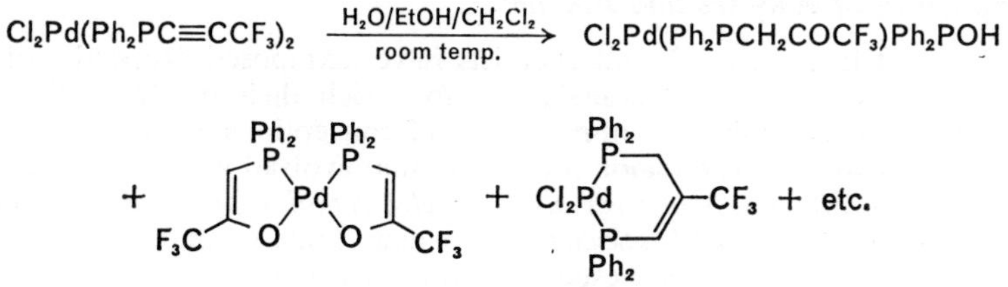

(9)

$$trans\text{-PtCl(CH}_3)\text{L}_2 + \text{RC}\equiv\text{CR}' \xrightarrow[-\text{AgCl}]{\text{Ag}^+\text{PF}_6^-} [trans\text{-Pt(CH}_3)\text{L}_2\text{RCCR}']^+\text{PF}_6^-$$
$$\xrightarrow{\text{MeOH}} \xrightarrow{-\text{CH}_4} [\text{Pt(CR=C(OCH}_3)\text{R}')\text{L}_2]^+\text{PF}_6^- \quad (10)$$

Going beyond equations (1) and (2), a nucleophilic attack on an alkyne may be one step in a *coupled* sequence. The first intermediates, anion (V^-), zwitterion ($^+V^-$) or radical anion ($A^{\dot{-}}$) are valuable synthons which may continue on in cyclization,

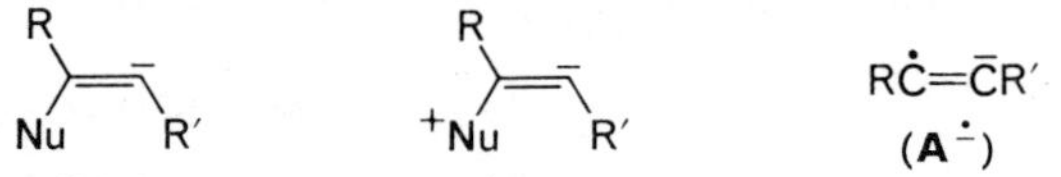

(V^-) ($^+V^-$) ($A^{\dot{-}}$)

polymerization, rearrangement, etc. Although it is doubtful that the alkylation of equation (11) could have been predicted[42], we regard the deliberate addition of a

$$2HC{\equiv}CH \xrightarrow[KOH,\,130\,^{\circ}C]{H_2O{-}Me_2SO} (H_2C{=}CHOCH{=}CH^-) \xrightarrow{Me_2SO} H_2C{=}CHOCH{=}CHMe \quad (11)$$

third reagent, a *coelectrophile* which can capture V^-, $^+V^-$ or $A^{\dot{-}}$, as one of the more interesting developments in synthetic acetylene chemistry. Here, we shall not, in general, follow a sequence beyond the first isolated products. By this limitation we necessarily lose much that is properly included under 'acetylene chemistry' or at least the *raison d'être* of a synthesis that depends on an acetylene.

II. REACTIVITY AND ORIENTATION

Recent observations bearing on reactivity have usually been scattered and of uneven quality. We can add very few kinetic data on additions (equation 1) to those of a previous review[1]; on the other hand, kinetic data for substitutions (equation 2) are available. Studies of substituent, steric and solvent effects, which influence nucleophilicity and electrophilicity orders as well as stereoselectivity, are limited and usually qualitative. For these reasons, we shall treat some of the large issues in this section and pick others up later in the context of specific nucleophiles.

There are two distinct selectivities: configurational selectivity relates to *syn vs. anti* addition; *regio* (or *directio*) selectivity is concerned with 1-Nu^-, 2-E^+ *vs.* 1-E^+, 2-Nu^- addition. The resulting orientations will be labelled 'specific' (0 or 100%) when one product or process is exclusive; otherwise they may range from highly selective to non-selective (>0 or <100%)[1, 43]. A reaction will be termed *stereoconvergent* if the same composition of product isomers is obtained on two or more reaction paths.

A. Comparison of Alkynes and Alkenes

At infrequent intervals, different chemists have juxtaposed properties of alkynes and alkenes. The purpose, of course, was to enrich their understanding of both families[3, 44–51]. In so doing, a number of misconceptions of alkyne *vs.* alkene reactivity were clarified. The major conclusion that evolved was simple: *nucleophiles react faster with alkynes than with alkenes; electrophiles (including radicals) react slower with alkynes than with alkenes.* Since new kinds of data are available, we believe that the comparisons are especially illuminating now.

The idea that an *sp* carbon is more electronegative than an sp^2 carbon is familiar. This notion can be made quantitative by examining several ionization energies. In Scheme 1, the figures are enthalpies of reaction, ΔH_r^0, in kcal/mol which were obtained or calculated from $\Delta H^0_{f,\,298^0}$ [52–58].

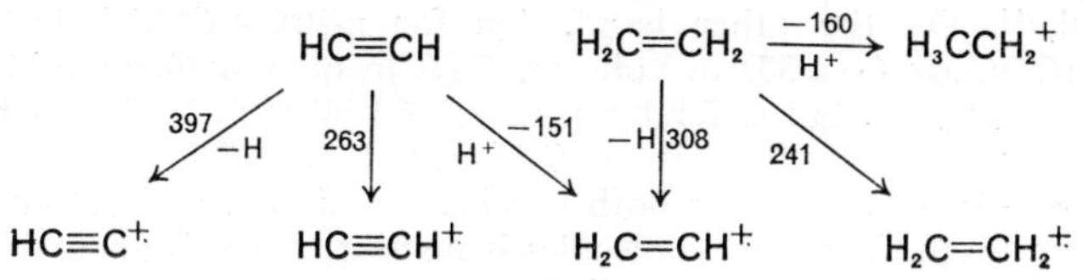

SCHEME 1

Note that ΔH_rs for proton addition to $-C{\equiv}C-$ *vs.* $>C{=}C<$ differ by only 9 kcal/mol. From a slightly different point of view, the trend in ΔH_rs of equation (12)

$$R{-}Cl \longrightarrow R^+ + Cl \qquad (12)$$

R	C_2H	C_2H_3	Ph	C_2H_5
ΔH_r(kcal/mol)	367	288	286	275

reinforces the idea that it is more difficult to form positive ions from acetylene than it is from ethylene. While these processes amount to models for charge transfer, solvolysis and electrophilic attack, they do, of course, apply only in the gas phase.

Similar models lead to the conclusion that radicals are more difficult to form by dissociation from acetylene than from ethylene[57]; but the tendency to add to these π systems is about equal (Scheme 2).

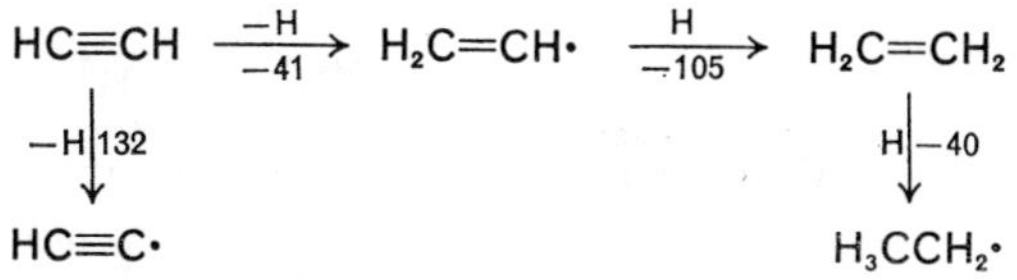

SCHEME 2

Since some gas-phase values for the affinity of these systems for electrons or other negative ions are lacking, analogous schemes for the corresponding anions are still incomplete. Preliminary vertical affinities, $EA(C_2H_4) = -1{\cdot}78 \pm 0{\cdot}1$, $EA(C_2H_2) = -2{\cdot}9 \pm 0{\cdot}2$ eV [59], $EA(C_2H) = 2{\cdot}50 \pm 0{\cdot}1$ [60, 61] and $EA(C_2H_3) > 1{\cdot}27$ eV [62, 63], as well as a useful order of proton affinities, $PA(X^-)$ of $C_2H_3^- > NH_2^- > H_2O > C_2H_2$, have become available[63]. Therefore, we can only give estimates of ΔH_r for two important processes (equations 13 and 14) in which $1{\cdot}27 > EA(C_2H_3) > EA(C_2H_5)$ eV. On the

$$C_2H_2^- \xleftarrow[e]{\sim 2{\cdot}9\text{ eV}} C_2H_2 \xrightarrow[+H^-]{-(41+EA)} C_2H_3^- \qquad (13)$$

$$C_2H_4^- \xleftarrow[e]{\sim 1{\cdot}78\text{ eV}} C_2H_4 \xrightarrow[+H^-]{-(40+EA)} C_2H_5^- \qquad (14)$$

basis of the left- and right-hand pairs of processes in equations (13) and (14), one could draw completely opposite views of the reactivities of *sp* and sp^2 systems! However, the appropriate comparison for nucleophile (H^-) addition is found in the right hand pair.

Though they are less 'fundamental', measures of *EA* in solution may be more helpful in dealing with solution phenomena. Potentiometric determinations in the solvent hexamethylphosphoramide (HMPT) gives an 'inverted' order of *EA*, i.e. 0·34 for *trans*-stilbene and 0·27 for diphenylacetylene relative to 0·0 eV for biphenyl; polarographic determinations of $E_{\frac{1}{2}}$ for these compounds in aqueous dioxane yield similar results[64]. On the other hand, $E_{\frac{1}{2}}$s for *n*-PrC≡CCOOMe (−2·26) and MeCH=CHCOOMe (−2·33) in volts *vs.* SCE in dimethylformamide (DMF) are in the 'normal' order[65], but differ by a mere 1·6 kcal/mol. It should be noted that while the energy gap in *EA* for gaseous C_2H *vs.* C_2H_3 is ~1·2, that for gaseous C_2H_2 *vs.* C_2H_4 is ~ −1·1 eV. Whether both specific substituent and solvent effects have contributed or not in the parent molecules is unclear, but it is apparent that the *EA* difference in solution can become negligible.

We shall catalogue very briefly some experimentally and theoretically derived properties, nearly all of which point to the alkyne or alkynyl being more electronegative than the alkene or alkenyl. These are: acidity (*K*, M) for HC≡CCOOH ($1·6 \times 10^{-2}$) and H_2C=CHCOOH ($5·7 \times 10^{-5}$)[66]; dipole moments (μ, Debye) for CH_3C≡CH (0·75) and CH_3CH=CH_2 (0·35)[67]; cathodic reduction waves ($E_{\frac{1}{2}}$, V) for $(HC≡C)_2$ (−2·33) and $(H_2C=CH)_2$ (−2·63)[68]; $J_{^{13}CH}$ (Hz) for C_2H_2 (250) and C_2H_4 (157)[69, 70]; ionic character of carbon to lithium bonding (%) for MeC≡CLi (38) and H_2C=CHLi (30)[71]; group electronegativities (χ) for C_2H (3·3) and C_2H_3 (2·9)[69]; Taft σ^*s for C_2H (0·25) and C_2H_3 (0·09)[72]; Hammett σ_ps for PhC≡C (0·12, 0·19) and PhCH=CH (−0·11, −0·05)[73]. Electron distributions calculated by an approximate MO theory are given in Figure 1 [74]. On the basis of their greater charge

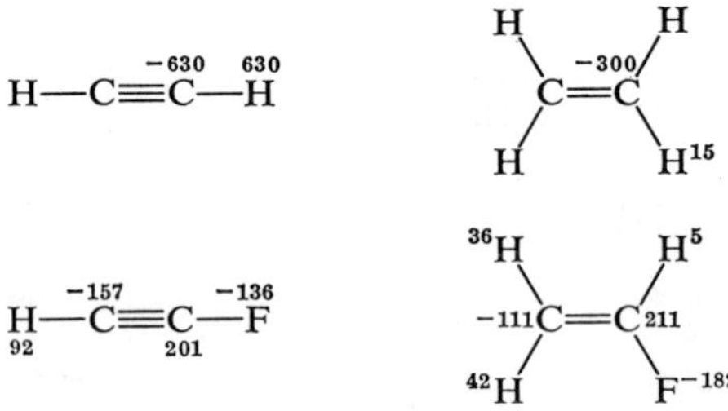

FIGURE 1. CNDO/2 charge distributions in units of 10^{-3} electrons.

density in the ground state, one might predict that alkynes would resist attacks both by electrons and by nucleophiles more strongly than do alkenes. That the latter prediction, at least, is usually reversed indicates that transition-state factors predominate, i.e. formation of sp^2 is relatively more favourable than formation of sp^3 anions (see equations 13 and 14).

Turning to kinetic comparisons of their electrophilicity, we find alkynes more reactive than alkenes, e.g. in nucleophilic additions of alkoxide[49, 75], amine[75, 76], thiolate[77, 78] and hydride[23], or substitutions of halide, e.g. by amines, thiolates, phosphines, etc.[3] (see also Section IV.B.1). The gas-phase processes (equations 15–18)

$$O^- + C_2H_2 \xrightarrow{3·5} OH^- + C_2H \tag{15}$$

$$O^- + C_2H_4 \xrightarrow{0·6} OH^- + C_2H_3 \tag{16}$$

$$O^- + C_2H_2 \xrightarrow{17} C_2H_2O + e \tag{17}$$

$$O^- + C_2H_4 \xrightarrow{4·4} C_2H_4O + e \tag{18}$$

are unusual in the present context, but their rate constants (cc/molecule s $\times 10^{10}$) conform beautifully to the notion that nucleophiles react more readily with alkynes than with alkenes[79].

We expect the reactions complementary to equations (1) and (2), namely electrophilic attacks, to be faster for alkenes than for alkynes. Thus, reactivity ratios (r_{11} and r_{22}) for corresponding alkynes and alkenes (PhC≡CH, PhCH=CH_2 and BuC≡CH, BuCH=CH_2) in radical copolymerizations favour the alkene over the alkyne[80]. Electrophilic additions of Br_2, Cl_2, ArSCl and H_3O^+ to alkenes are usually much faster than those to alkynes[45]. However, k(C=C)/k(C≡C) can vary from 10^6 to <1 for the different electrophilic processes and by 10^5 for one process (Br_2 addition) when the solvent is changed from H_2O to HOAc[45]. This unexpected trend in reactivity continues undiminished in the rates of acid-catalysed hydration k(EtOC≡CH)/k(EtOCH=CH_2) = 180 and k(>NC≡CH)/k(>NCH=CH_2) > 20,000[81]. These latter effects of substituent, electrophile and medium on rate processes are huge—they invert the 'normal' order!

Looking back at the data, we find $\Delta H_r = 9$ less favourable for addition of H^+ and probably <20 kcal/mol more favourable for addition of H^- to C_2H_2 as compared with C_2H_4. However, *equilibrium* figures are deceptive. We have seen that significant substituent and solvation effects can reduce the energy gap. In respect to electrophilic *rates*, this occurs in k(C≡C) > k(C=C), although this order is admittedly unusual. As for nucleophilic attacks, cathodic reductions may occasionally turn out to be exceptional; otherwise, the order, k(C≡C) > k(C=C), seems to be followed. A revised statement of alkyne–alkene reactivity now reads: *nucleophiles react faster with alkynes; radicals react faster with alkenes; polar electrophiles usually react faster with alkenes.*

B. *Anti* vs. *Syn* Selectivity

Scheme 3 provides a useful framework within which nucleophilic substitution and addition may be discussed. In this scheme we give only one of the possible substitution mechanisms (Section IV has others) and disregard post-isomerization of the

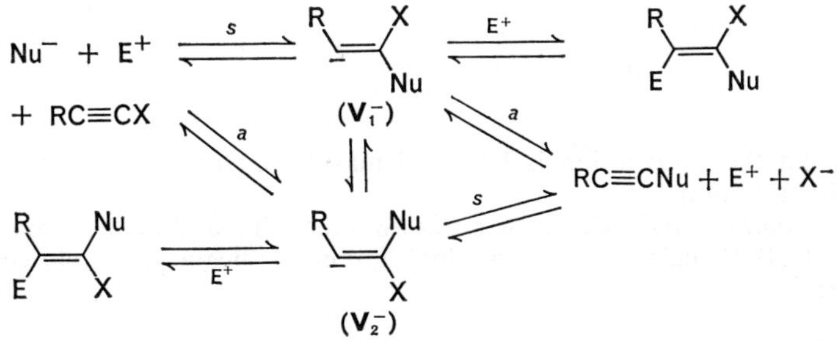

SCHEME 3

products, a complication which can usually be checked independently. If the vinyl anions are stereostable, as is often the case, overall *exchange* involves a *syn* (*s*) association of Nu^- and *anti* (*a*) dissociation of X, or *a* association of Nu^- and *s* dissociation of X: in either sequence there is one *s* step. Thus, V_1^- and V_2^- are formed competitively. On the other hand, addition, unlike exchange, may be (but need not be) *anti* stereospecific. In this unified scheme, the recognition of *syn* steps is not only interesting, but is essential for the understanding of stereoselectivity of additions which consume and eliminations which produce multiple bonds.

Michael's rules of *trans* additions were rediscovered and restated for alkynes (equation 1) in the 1950s by several groups[1, 43]. Yet there is nothing forbidden about *syn* additions. Indeed, we shall presently describe conditions under which they become favoured and even exclusive.

Qualitative bonding arguments have been produced to rationalize the *anti* preference[43, 82, 83]. Following Fukui's prescription for acyclic additions to alkenes, we separate the σ and π bonding in acetylene (Figure 2)[82]. During addition the change in

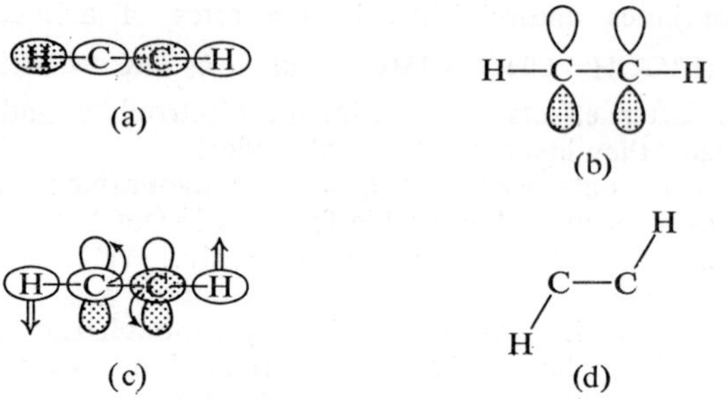

FIGURE 2. Orbital mixing and *anti* selectivity. (*a*) LU σ orbital; (b) HO π orbital; (c) AO mixing (→), nuclear direction (⇒); (d) *anti* direction.

hybridization may be regarded as an interaction of the lowest unoccupied (LU) σ orbital (a) with the highest occupied (HO) π orbital (b). Clearly, the in-plane direction of bending which facilitates optimum orbital mixing is *anti*, as in (c) and (d).

As far as we can determine, no one has computed the energies along the reaction coordinates from RC≡CX to V_1^- *vs.* V_2^- in Scheme 3. An *ab initio* SCF calculation does indicate the following relative energies for $C_2H_2^-$ [84]:

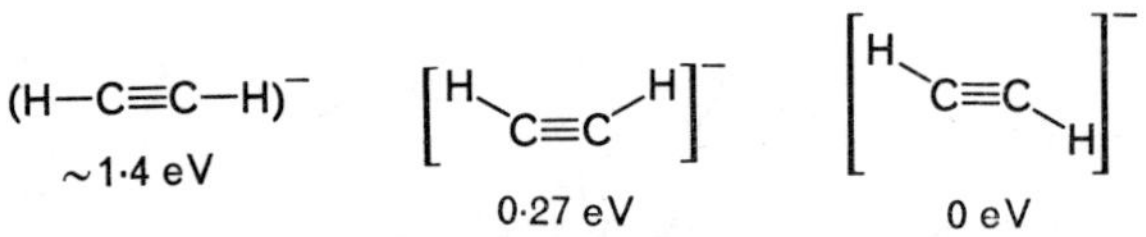

The fact that the *trans* anion is favoured may be interpreted as support for *anti* nucleophilic addition.

A rule of *anti* (or *syn*) selectivity poses an interesting and heretofore unsuspected difficulty. CNDO calculations on vinyl anions indicate the following relative energies[85]:

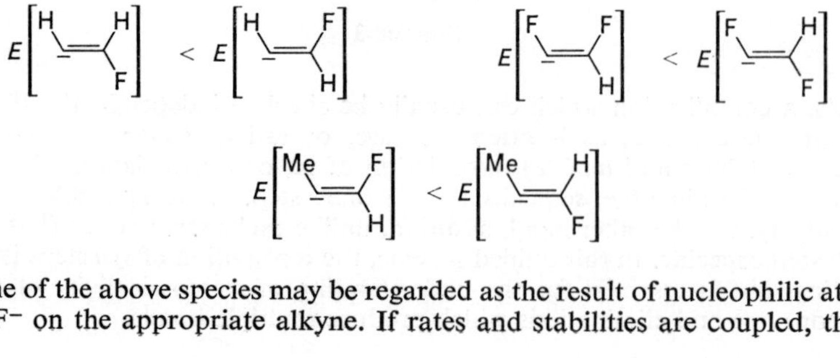

Each one of the above species may be regarded as the result of nucleophilic attack by H^- or F^- on the appropriate alkyne. If rates and stabilities are coupled, then *anti*

addition would be favoured for one and *syn* addition would be favoured for the other; moreover, these specificities interchange depending on which alkyne is the substrate. We anticipate that calculations of the energies along the two reaction coordinates from RC≡CX to V_1^- *vs.* V_2^- in Scheme 3 would not, in general, alter these relationships. The concept of a universal *anti* (or *syn*) preference is subverted by these energy considerations.

Favoured *anti* addition can be 'saved', once it is realized that it is a 'rule' that may be limited to polar solvents. In Figure 3 we have drawn activated complexes for the alternative routes for a charged and uncharged Nu: the central bond is half-formed and the separation of charge has proceeded half-way. The arcs indicate the approach distance of a solvent molecule. In a polar solvent, more effective solvation and more facile charge separation provide the necessary rationale for *anti* addition in (b) *vs.* (a). In (d) the preference is less obvious, since built-in solvation is pitted against external solvation. In non-polar solvents Nu^- is more likely to be paired with E^+ and now *syn* addition (a) may become more favourable. Likewise built-in solvation in (c) should favour the neutral Nu. By allowing the solvation energy to vary with the system and by making this a major factor, one may find the *syn* or the *anti* process or both.

There is no doubt that the solvent plays an important role in process (1). It seems to be accepted that 'activation energy for charge separation should be reduced with increasing polarity'[1, 86]. Various nucleophiles, e.g. SCN^-, I^-, MeO^-, amines, etc., do, in fact, react faster in polar than in non-polar solvents[1, 87–91]. Eliminations, the microscopic reverse of additions, take the *syn* path more often in non-polar or associating solvents[92]. All of this is supportive of the idea that polar solvents favour *anti* attack, other things being equal.

What we have just proposed is admittedly a working hypothesis whose limitations must be specified. Certainly, there are exceptions. The highest rates are usually, but not inevitably, found in the most polar solvents: in the addition of ethanol to diacetylene the rates are in the order k(dioxane–ethanol) > k(ethanol) > k(ethanol–heptane)[91]. The highest *anti*/*syn* addition ratios are usually, but not inevitably found in the most polar solvents: in equation (19), the fraction of Z isomer in the product

$$p\text{-}O_2NC_6H_4C{\equiv}CCOOEt + c\text{-}(CH_2)_4NH \longrightarrow c\text{-}(CH_2)_4NC(p\text{-}O_2NC_6H_4){=}CHCOOEt \quad (19)$$

falls in the order $CDCl_3$ (0·8), MeCN (0·66), $(CD_3)_2SO$ (rapid isomerization) and PhH (~20)[93]. It will become apparent presently that other factors besides the solvent influence *syn-anti* ratios. Unfortunately, we know of no systematic studies in which the solvent effect on stereoselectivity has been essentially isolated and studied.

Now, we return to Scheme 3 to consider V_1^- and V_2^-, whose stability may determine the stereochemical outcome of a nucleophilic attack. Based on elimination and hydrogen exchange data, an estimate of the barrier to isomerization (V_i) ⩾ 30 kcal/mol was given for $HCBr{=}CBr^-$ [94]. It has been reported the base-catalysed D for H

$Ph(2\text{-}BrC_6H_4)C{=}CHCN$

(1)

(a) *cis*, **(b)** *trans*

(2) **(a)** R = CN
(b) R = COPh

exchange of **(1)** at 30 °C and **(2a)** at temperatures > 50 °C occurs essentially without isomerization[95, 96]. However, isomerization rates may begin to approach exchange

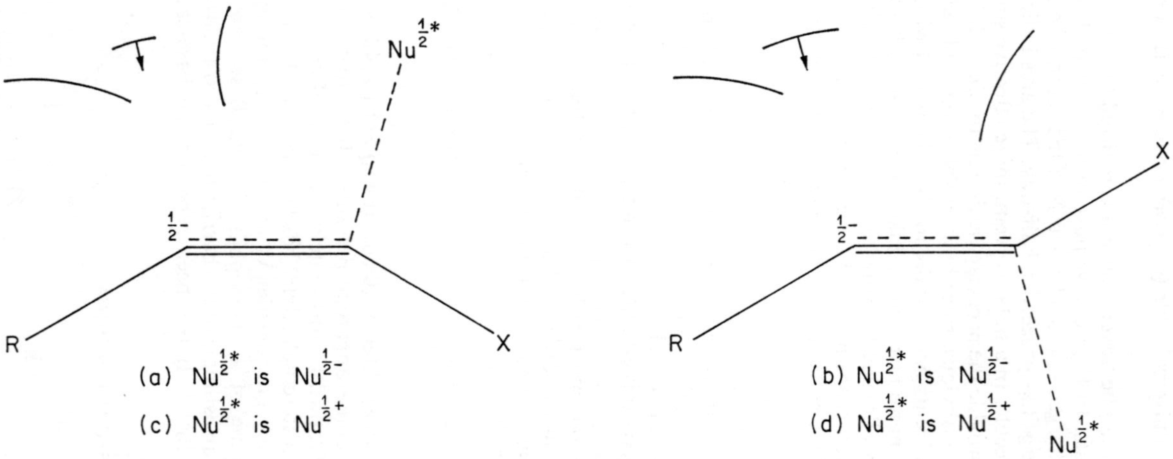

FIGURE 3. Activated complexes for *syn vs. anti* nucleophilic attack. The nucleophile (Nu) is anionic in (a) and (b) and neutral in (c) and (d). The arrows indicate the alignment of the growing orbital or the direction from which solvent or E^+ might be approaching. Distances (Å): C⩶C 1·27; R—C 1·49; C—X, C—Nu 1·35. The arcs indicate van der Waals' radii (Å): X, Nu 1·5; R 2·0; C 1·7.

rates, as in **2b**[96], or $MeCNR_2{=}CHCN$[97]. Judging by numerous data for the isoelectronic imines, high V_is are entirely reasonable[98]. Certainly, theoretical calculations support this idea[99-101]. Moreover, V^- with electron-withdrawing substituents (R = N<, O—, Hal, CF_3) should have relatively high V_i, while those in which the charge is delocalized should have relatively low V_i[99-101]. It turns out that those substituents which would be likely to lower V_i, i.e. COR, SO_2R, NO_2, Ar, CN, are important activating groups for alkynes in process (1).

In Scheme 4 we have expanded Scheme 3 to accommodate a substituent, RCO, which can delocalize the charge. Although 1,4-addition to a ketoalkyne is not a new

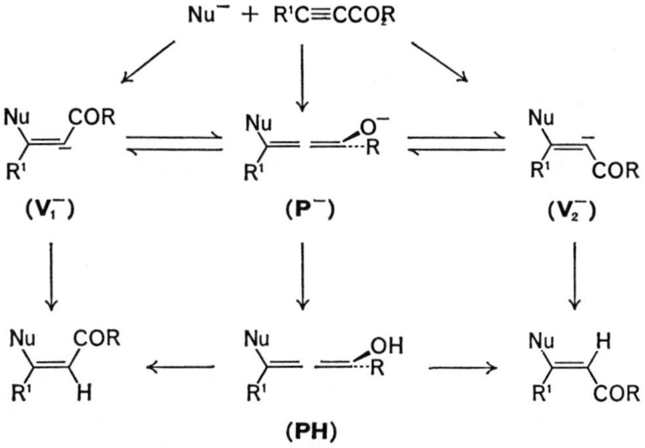

SCHEME 4

concept, the identification of an enol adduct, $MeC(OH){=}C{=}CHSMe$, by low temperature p.m.r. provides concrete support for Scheme 4[102]. If $V_i \rightarrow 0$, or the allenic intermediate (**P**⁻) forms directly, or the electrophile attacks (coordinates with) oxygen, then a stereoconvergent product becomes probable. Incidentally, this scheme allows for *Z* and *E* products without the necessity of their post-isomerization; it may be easily expanded, however, to include acid or base catalysis of the isomerization[103].

Conditions favourable to the formation of $\mathbf{P}^-$ in Scheme 4 are worth considering. Truce gives an interesting set of data for equation (20) in which this matter is probed

$$HC{\equiv}CY + p\text{-}CH_3C_6H_4S^-Na^+ \xrightarrow[0\,°C]{MeOH} p\text{-}CH_3C_6H_4SCH{=}CHY \qquad (20)$$

(Table 1)[104]. The activating groups Y were chosen to show increasing delocalization according to the ratio $\sigma_{\bar{R}}/\sigma_I$. It was established that the products were kinetically controlled[104]. In view of low %*Z* at equilibrium we are inclined to believe that here protonation of $\mathbf{V}^-$ is competitive with the formation of $\mathbf{P}^-$, which gives both isomers.

One is often compelled to monitor both the rate of protonation and interconversion of $\mathbf{V}^-$ by the stereoselectivity (%*Z*); but now the mechanistic 'explanations' may become convoluted. In equations (19) and (21), for example, a polar solvent favours greater deviation from *anti* addition. Thus, in equation (21) the %*Z* decreases in the order CCl_4 (81), PhH (80), Et_2O (74), Me_2SO (68)[103]. Considering *both* solvent polarity and the stability of the anion, we must revise our working hypothesis: when V_i is

TABLE 1. Selectivity in reaction (20)[104]

Y	Conversion (%)	Kinetic % Z, 0 °C	Equilibrium % Z, 50 °C
CN	100	100	33
p-$C_7H_7SO_2$	65	100	0
p-$O_2NC_6H_4$	98	100	0
MeOOC	~95	92	22
H_2NOC	97	87	23
MeCO	93	82	22

high, a polar solvent favours *anti* addition; when V_i is low, a polar solvent should promote stereoconvergence.

$$\text{MeC}\equiv\text{CSO}_2\text{C}_6\text{H}_4\text{CH}_3\text{-}p + \triangleright\text{NH} \longrightarrow \text{MeC(NC}_2\text{H}_4\text{-}c)\text{=CHSO}_2\text{C}_6\text{H}_4\text{CH}_3\text{-}p \qquad (21)$$

The effect of increasing the temperature on the stereochemistry resulting from Scheme 4 almost invariably leads to less selectivity[86, 105, 106]. For equation (22) under

$$\text{MeC}\equiv\text{CSOEt} + \triangleright\text{NH} \xrightarrow{\text{PhH}} \triangleright\text{NC(Me)=CHSOEt} \qquad (22)$$

kinetic control in benzene, Truce found the trend (°C, % Z): ~4 °C, 66; ~25 °C, 40; ~54 °C, 22; the increase in % E was associated with the equilibration of $\mathbf{V}_1$ and $\mathbf{V}_2$[86]. On the other hand, in the systems of equation (23), the almost inevitable

$$\text{R'COC}\equiv\text{CH+RSH} \xrightarrow{\text{EtOH}} \text{R'COCH=CHSR} \qquad (23)$$

appearance of a small proportion of *syn* addition at low temperature and the trend towards stereoconvergence at higher temperatures, may be associated with partitioning of $\mathbf{P}^-$ or **PH** in Scheme 4[102, 105, 106].

We have seen that even if V_i is low, relatively rapid capture of $\mathbf{V}^-$ by an electrophile could yield high selectivity. That is, the availability and rate of delivery of an electrophile are important. Intertwined with these factors is the question of external *vs.* internal delivery of E^+ to the anion[1]. Again we simplify by associating rapid transfer of E^+ with external-*anti* and internal-*syn*, *other things being equal.* Different types of potential internal delivery are illustrated in Figure 4. Cullen found only *anti* addition in the system given by (**3**) and concluded that rapid proton transfer from external Me_2AsH occurred[107]. In (**4**–**6**) are represented three of a wide variety of organometallic aggregates whose initial attack may be nucleophilic (these are often difficult to distinguish from electrophilic, molecular or radical processes, and may in fact be a mixed sequence of several of these)[108–110]. Of necessity, *tight* ion pairs or polymer aggregates seem to generate *syn* selectivity and this appears to apply in **4** and **5**. On the other hand, the internal coordinating O and N groups in **6** mediate an *anti* reaction[110] which is similar to a class of lithium aluminium hydride reductions which we shall discuss in detail later.

Numerous studies of amine nucleophiles have been made. If the zwitterion **7** picks up an external H^+, it *may* yield the *anti* adduct. If the proton is transferred internally from nitrogen the *syn* adduct *will be* produced. If $^+\mathbf{P}^-$ is formed first then both *syn* and *anti* adducts can be produced. Product stability is a critical factor here and manifests itself especially when internal hydrogen bonds confer stabilization as in **8** [1, 97].

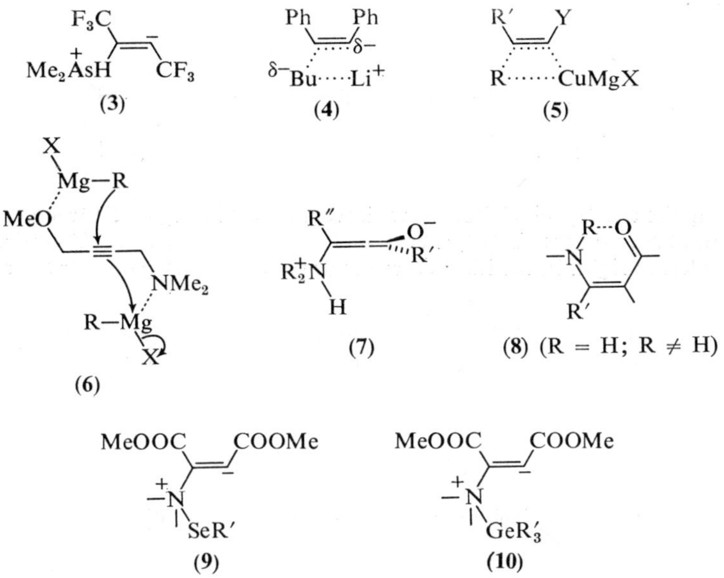

FIGURE 4. Stereoselectivity and *anti vs. syn* delivery of E^+.

A simple case of internal *syn* delivery appears to have been found in **9** in which the *Z*-$MeOOCC(NMe_2){=}C(COOMe)SePh$ is the exclusive initial product in $CHCl_3$ at -27 °C [111]. What should be an analogous example in **10** turns into 62% *Z*-$MeOOCC(NMe_2){=}C(COOMe)GeMe_3$ (and 38% *E*) whether or not ether is used as a solvent in the temperature range -70 to 20 °C; here it is assumed that $^+\mathbf{P}^-$ forms, equilibrates and 'waits' until another mole of germylamine completes the reaction[112].

There are two factors which we term 'specific' because they depend critically on the system. The first is hydrogen-bonding stabilization of the product[1]. Structures such as **8** with R=H in Figure 4 possess sufficient internal hydrogen bonding so that these may become more stable than their geometric isomers—here *anti* addition would be favoured[1, 97]. The availability of intermolecular hydrogen bonding could reverse this trend: this appears to account for the predominance of *E*-RNHCH=CHCHO in the condensed state and in solution[113]. Typically, product stability is subject to conventional polar and bulk effects, which would exert their influence most directly in **P** rather than $\mathbf{V}^-$ when R ≠ H in **8** of Scheme 4.

Since an alkyne system is so 'open', it is perhaps surprising that steric effects have been noted both in the *regio* and *syn–anti* senses. The former will be discussed in another section. In equation (24), for example, the %*Z* isomer formed under

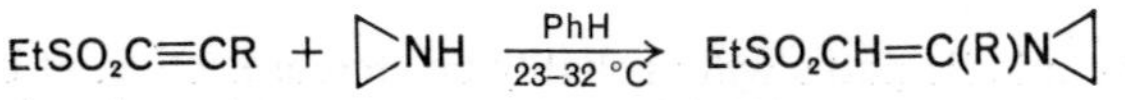

(24)

kinetic control decreases in the order H (100), Me (96), *i*-Pr (76), *t*-Bu (75)[103]. For R′COC≡CR, the rate of addition of amines decreases in the order *k*(H) > *k*(aryl) >

k(alkyl)[114]. The bulk of the nucleophile bears on process (25) in the following way (R, k_{rel}, Z/E): H, 1, 4/5; Me, 0·1, 2/3; *n*-Pr, 0·06, 0/1[115]. Since kinetic control applies, it was suggested that isomerization of one or more intermediates (V^- or P^-) is involved[115].

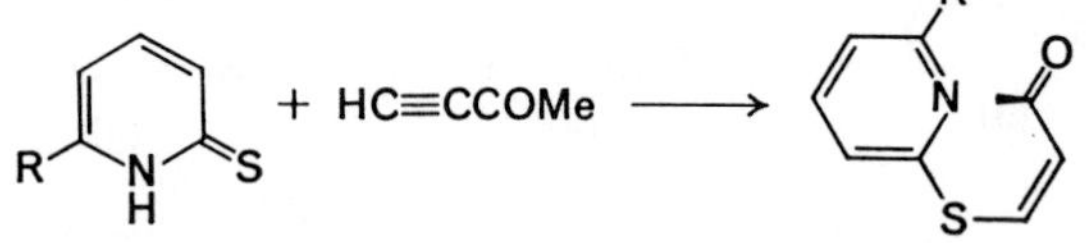

(25)

Omitting the two 'specific' factors, we have mapped in Figure 5 the stereochemical possibilities based on the three 'general' factors mentioned earlier. Consider them in turn from left to right. In the ideal cases the upper and lower paths are stereospecific,

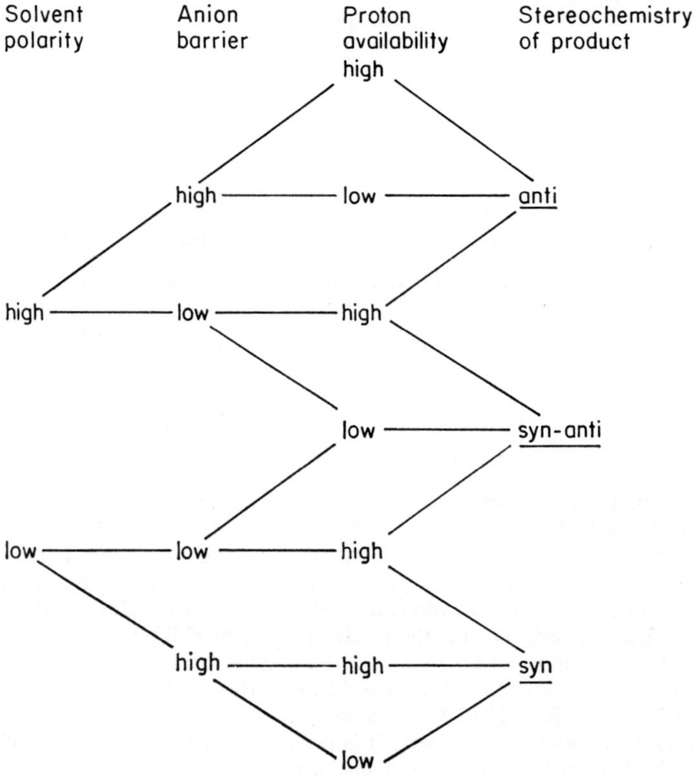

FIGURE 5. *Syn vs. anti* additions to alkynes.

while the middle paths will probably be stereoselective, at best. Both 'pure' *anti* and *syn* processes are possible! For the present, this rationale must be regarded as a working hypothesis. Even if other factors are absent, which is not usually the case, the tentative character of this hypothesis should be kept in mind.

C. Regioselectivity

1. Substitution effects

The orientation of nucleophilic attacks in equation (1) is usually regulated by the substituents on the alkyne. As a guide, molecular orbital (MO) charge distributions for several mono- and disubstituted acetylenes are given in Figure 6 [74, 116, 117]. Where

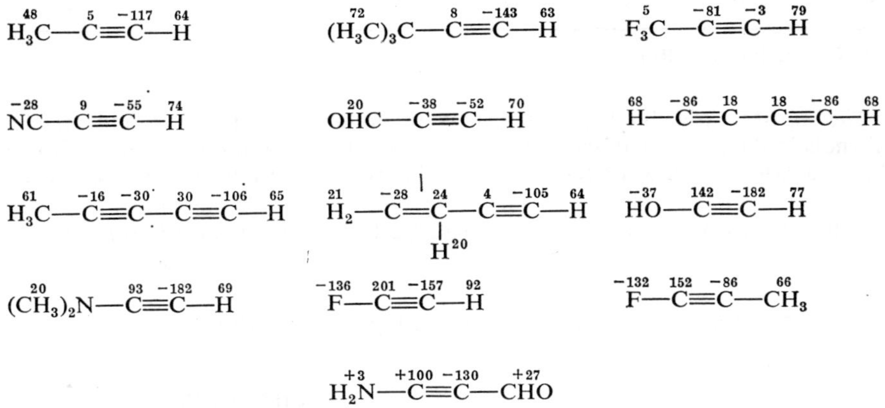

FIGURE 6. Charge densities $(q_\sigma + q_\pi) \times 10^3$ in electrons[74, 116, 117].

possible, we compare the regioselectivity that is predicted from these static CNDO/2 calculations with experimental data. Certainly, a more reliable MO treatment would include the nucleophile and examine the entire reaction surface.

In our discussion, C_α will refer to the acetylenic carbon attached to the substituent of higher priority; the other carbon is C_β. Priority assignments are made in the same manner as for *R*, *S* and *E*, *Z* specifications.

a. *Terminal acetylenes.* In methylacetylene, C_α attack is predicted (Figure 6) and this, in fact, is observed (equation 26)[118]. Here and in equation (27), nucleophilic

$$CH_3C{\equiv}CH + C_2H_5O^- \xrightarrow{C_2H_5OH} CH_3C(OC_2H_5){=}CH_2 \qquad (26)$$

$$PhSO_2CH_2C{\equiv}CH + PhSH \longrightarrow PhSO_2CH_2C(SPh){=}CH_2 \qquad (27)$$

addition in a terminal alkylacetylene follows the Markownikoff rule[119]. The charge density calculations given for *t*-butylacetylene in Figure 6 indicate that the *t*-butyl group is an α-director. Thiolates and alkoxides, however, produce substantial amounts of C_β product (equation 28)[118, 120]. Although there was a tendency to

$$t\text{-BuC}{\equiv}CH + RX^-Na^+ \longrightarrow t\text{-BuC(XR)}{=}CH_2 + t\text{-BuCH}{=}CXR \qquad (28)$$

$$R = \text{alkyl};\ X = O, S$$

minimize steric factors previously[1], the findings given in Table 2 are obviously due to such an effect. With a sufficiently bulky nucleophile, even methylacetylene undergoes C_β attack.

TABLE 2. Percentage of *anti*-Markownikoff product formed in the reaction of alkynes, $R'C{\equiv}CH$, with alkoxides, RO^-, and thiolates, RS^- [118, 120]

R′	% *anti*-Markownikoff product formed with various RO^- (RS^-)						
	R = CH_3	C_2H_5	*n*-C_3H_7	*i*-C_3H_7	*n*-C_4H_9	*i*-C_4H_9	*t*-C_4H_9
$(CH_3)_3C$	22 (39)	24 (46)	28 (46)	—[a] (52)	35 (48)	51 (54)	—[a] (92)
CH_3	0	0	0	3	0	~0	27

[a] No reaction with RO^-.

The *anti*-Markownikoff products that predominate with thiols and terminal alkynols in the presence of oxygen[121] or trace amounts of lithium[122] *may* result from the incursion of radical or radical-ion mechanisms (equations 29 and 30). In fact,

$$PhSH + HOCH_2C{\equiv}CH \xrightarrow[\text{(air)}]{KOH} \underset{75\%}{HOCH_2CH{=}CHSPh} + \underset{25\%}{HOCH_2C(SPh){=}CH_2} \quad (29)$$

$$(CH_3)_2COHC{\equiv}CH + C_2H_5SH \xrightarrow[HMPT]{Li} \underset{70\%}{(CH_3)_2COHCH{=}CHSC_2H_5} + \underset{30\%}{(CH_3)_2COHC(SC_2H_5){=}CH_2} \quad (30)$$

thiolates usually produce a higher percentage of *anti*-Markownikoff product than alkoxides of comparable size, a difference which is consistent with their greater proclivity to form radicals.

In contrast to methylacetylene, the CNDO/2 calculations for trifluoromethylacetylene indicate that the CF_3 group is a β-director (Figure 6). Supporting evidence comes from equation (31) in which the C_β product predominates[123].

$$CF_3C{\equiv}CH + CH_3OH \xrightarrow{CH_3O^-} \underset{98{\cdot}5\%}{CF_3CH{=}CHOCH_3} + \underset{1{\cdot}5\%}{CF_3C(OCH_3){=}CH_2} \quad (31)$$

The MO data for cyano-, formyl- and ethynylacetylene (Figure 6) fail to indicate the correct positional isomer. Instead of C_α attack, C_β or Michael addition products are the rule for these terminal alkynes (equations 32–34). Of course, when nucleophilic attachment occurs at C_β, the substituents are capable of stabilizing incipient

$$NCC{\equiv}CH + Nu \longrightarrow NCCH{=}CHNu \quad \text{(Refs. 104, 124)} \quad (32)$$

$$Nu = RS^-, RO^-, RR'NH$$

$$OCHC{\equiv}CH + PhNH_2 \longrightarrow OCHCH{=}CHNHPh \quad \text{(Ref. 125)} \quad (33)$$

$$(HC{\equiv}C)_2 + \underset{(o,\ m \text{ or } p)}{HOC_6H_4NH_2} \xrightarrow[Me_2SO,\ KOH]{\text{dioxane}} HC{\equiv}CCH{=}CHOC_6H_4NH_2 \quad \text{(Ref. 126)} \quad (34)$$

negative charge on C_α by inductive and/or resonance effects. Other examples of alkynes which behave similarly are given throughout this chapter.

Russian workers have observed that alkyldiynes are attacked on the terminal carbon (equation 35)[8, 127]. Avoidance of the acetylenic carbon adjacent to the methyl

$$\text{MeC}{\equiv}\text{CC}{\equiv}\text{CH} + n\text{-C}_n\text{H}_{2n+1}\text{OH} \xrightarrow[\text{KOH}]{\text{dioxane, 100 °C}} \text{MeC}{\equiv}\text{CCH}{=}\text{CHOC}_n\text{H}_{2n+1}\text{-}n \quad (35)$$

$$n = 1\text{–}5 \qquad\qquad 75\text{–}80\%$$

group is puzzling as this carbon is not particularly sterically hindered. CNDO/2 calculations (Figure 6) reveal it to be more electrophilic than the terminal carbon; attack at this site would lead to a resonance-stabilized transition state.

For vinylacetylene the static CNDO/2 data given in Figure 6 indicate that the two internal carbons are more electrophilic than the terminal ones; but terminal attack is presumably favoured because this leads to transition states which are resonance-stabilized[1, 8]. The confusing issue in the additions to vinylacetylene is the variability in the point of entry with changes in nucleophile and solvent, e.g. thiols attack primarily at the terminal *sp* carbon[128, 129], while alkoxides, phosphides and amides prefer the terminal sp^2 carbon[130]. Attack on the internal *sp* carbon may occur when the vinylacetylene contains special substituents[131], e.g. equation (36). Perhaps these matters would be clarified if equilibrium and rate studies were performed.

$$\text{Me}_2\text{NCH}{=}\text{CHC}{\equiv}\text{CH} \xrightarrow{\text{PhNH}_2} \text{Me}_2\text{NCH}{=}\text{CHC}({=}\text{NPh})\text{Me} \quad (36)$$

b. *Terminal heteroacetylenes.* Estimates of the directive powers of oxygen in terminal ethynyl ethers and nitrogen in terminal ethynyl amines are given in Figure 6. These calculations point to nucleophilic orientation on C_α. Equations (37) and (38)

$$\text{EtOC}{\equiv}\text{CH} + \text{Nu} \xrightarrow{\text{H}^+} \text{EtOC(Nu)}{=}\text{CH}_2 \quad \text{(Refs. 11, 132–134)} \quad (37)$$

$$\text{Nu} = (\text{RO})_2\text{PO}^-, \text{R}_2\text{N}^-, \text{RO}^-, \text{RS}^-, (\text{RO})_3\text{P}$$

$$\text{R}_2\text{NC}{\equiv}\text{CH} + \text{H}_2\text{O} \longrightarrow \text{R}_2\text{NCOCH}_3 \quad \text{(Refs. 135, 136)} \quad (38)$$

$$\text{R} = \text{Me, Et}$$

are in accord with the prediction. With *N,N-bis*trifluoromethylethynyl amine, however, a surprising amount of competition is shown (equation 39)[137]. The C_α

$$(\text{CF}_3)_2\text{NC}{\equiv}\text{CH} + \text{CH}_3\text{O}^- \xrightarrow{\text{CH}_3\text{OH}} (\text{CF}_3)_2\text{NC(OCH}_3){=}\text{CH}_2 + (\text{CF}_3)_2\text{NCH}{=}\text{CHOCH}_3 \quad (39)$$

$$56\% \qquad\qquad 28\%$$

product would be expected if one considers ground-state polarization of the ynamine, while the C_β product might be anticipated if there is transition-state stabilization. The unusual attack on C_β leads to a transition state in which the negative charge of C_α is stabilized by the inductive effect of the *bis*trifluoromethylamino group.

In contrast to the major course of addition in ethynyl ethers, the predominant mode in ethynyl thioethers is nucleophilic attachment to C_β, as in equation (40)[11, 138, 139]. Arens accounted for such differences in orientation by proposing that

$$\text{EtSC}{\equiv}\text{CH} + \text{Nu} \xrightarrow{\text{EtOH}} \text{EtSCH}{=}\text{CHNu}$$

$$\text{Nu} = \text{C}_2\text{H}_5\text{S}^-, \text{C}_2\text{H}_5\text{O}^-, (\text{CH}_3\text{O})_2\text{PO}^-, (\text{C}_2\text{H}_5\text{O})_3\text{P} \quad (40)$$

main-row elements in conjugation with the triple bond act as donors (as in **11a**);

heteroatoms below the main-row elements act as acceptors (as in **11b**)[11]. Whether *d*-orbital participation or some other rationalization is used[140], the facts are that

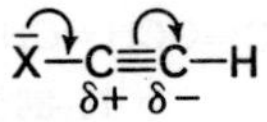

(**11a**) (X = N, O, F)

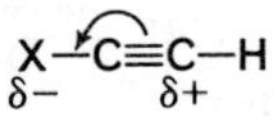

(**11b**) (X = P, S, Cl, Br, etc.)

atoms in the second and higher periods promote the formation of α-anions by C_β entry of the nucleophile.

c. *Disubstituted acetylenes.* Acetylenes with regiosynergistic groups are comprised of an α- and a β-director and those with regioantagonistic groups incorporate either two α- or two β-directors. Regiosynergism is illustrated in (41–44) equations.

$$t\text{-BuC}{\equiv}\text{CNO}_2 + \text{NuH} \longrightarrow t\text{-BuCNu}{=}\text{CHNO}_2 \quad \text{(Ref. 17)} \qquad (41)$$

$$\text{Nu} = \text{RO, RNH}$$

$$(\text{MeC}{\equiv}\text{C})_2\text{CO} + \text{NuH} \longrightarrow (\text{MeC(Nu)} = \text{CH})_2\text{CO} \quad \text{(Ref. 141)} \qquad (42)$$

$$(\text{F}_3\text{C})_2\text{NC}{\equiv}\text{CCF}_3 + \text{MeO}^- \xrightarrow{\text{MeOH}} \underset{96\%}{(\text{F}_3\text{C})_2\text{NC(OMe)}{=}\text{CHCF}_3} + \underset{4\%}{(\text{F}_3\text{C})_2\text{NCH}{=}\text{C(OMe)CF}_3} \quad \text{(Ref. 142)} \qquad (43)$$

$$\text{Et}_2\text{NC}{\equiv}\text{CCN} + (\text{H}_2\text{C}{=}\text{CHCH}_2)_2\text{NH} \longrightarrow \text{Et}_2\text{NCN(CH}_2\text{CH}{=}\text{CH}_2)_2{=}\text{CHCN} \quad \text{(Ref. 124)} \qquad (44)$$

We shall consider process (45) in some detail because it illustrates that even in alkynes containing electronically reinforcing groups, orientation can be regulated by

$$\text{Me}_2\text{NC}{\equiv}\text{CCOR} + \text{NuH} \longrightarrow \text{Me}_2\text{NC(Nu)}{=}\text{CHCOR} + \text{Me}_2\text{NCH}{=}\text{C(Nu)COR}$$

$$\text{R} = \text{H, Me, OMe} \qquad (45)$$

factors other than electronic ones[143]. The pertinent data for this reaction are given in Tables 3 and 4. On the basis of the ground-state electronic information supplied in Figure 6 for $H_2NC{\equiv}CCHO$, the predicted site of attack is C_α. Transition-state theory also indicates a definite bias for C_α attack (compare models **12a–c**). The alkoxides, in fact, yield no C_β products. The amine systems, however, show wide variations to the extent that attack on C_β may be favoured.

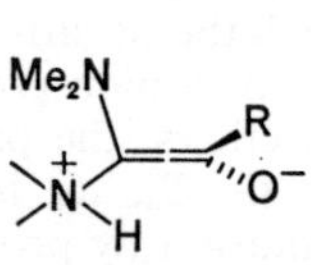

(**12a**)

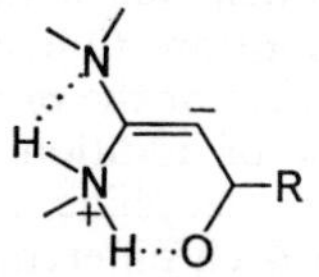

(**12b**)

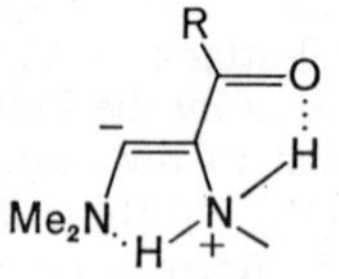

(**12c**)

The trends in Table 3 may in part be explained by steric effects, for the CHO group (R=H) is smaller in size than the Me_2N function and consequently C_β is sterically more accessible than C_α. The substituents Me_2N and COOMe are of comparable bulk and only the Michael (C_α) product is obtained. The increasing amount of *anti*-Michael product in the order for R of MeO ≪ Me < H corresponds to an increase in the C_β channel. Moreover, the C_β product should become more prevalent as the bulk of the amine becomes larger. This reasoning seems to be valid provided that the comparison is made within the same type of amine. Amine size also accounts for the difference between methyl and isopropyl amine for R=H and partly explains the data among the secondary amines and this alkyne.

TABLE 3. Regioselection in additions of amines and alcohols, NuH, to alkynes, $Me_2NC{\equiv}CCOR$, in tetrahydrofuran at 37 °C (equation 45)[143]

		Product (%)	
R	Nu	$Me_2NCNu{=}CHCOR$	$Me_2NCH{=}CNuCOR$
H	NH_2	12	88
	i-PrNH	34	66
	MeNH	52	48
	Et_2N	70	30
	Me_2N	86	14
	$C_5H_{10}N$	92	8
	c-C_2H_4N	100	0
	MeO^-	100	0
	EtO^-	100	0
Me	*i*-PrNH	74	26
	MeNH	76	24
	Et_2N	89	11
	Me_2N	94	6
	$C_5H_{10}N$	~98	~2
	c-C_2H_4N	100	0
	MeO^-	100	0
	EtO^-	100	0
OMe	All	100	0

TABLE 4. The effect of solvent on the regioselectivity of amine additions to $Me_2NC{\equiv}CCHO$ (equation 45)[143]

		Product (%)	
Amine	Solvent	$Me_2NCNu{=}CHCHO$	$Me_2NCH{=}CNuCHO$
Et_2NH	PhH	43	57
	THF	70	30
	MeCN	85	15
i-$PrNH_2$	PhH	24	76
	THF	34	66
	MeCN	70	30

The most curious feature of the additions is the switch from predominant C_β attack with ammonia to attack on C_α and C_β with primary amines and then to predominant C_α attack with secondary amines and alcohols. There seems to be no reason to suspect that the electronic character of the substituents would change as the type of amine is varied. Certainly, steric factors cannot account for this turnover since the secondary amines which are largest in size should produce the greatest amount of C_β attack. Therefore, it seems necessary to bring in additional factors.

Species **12a–c** are general models for *precursor* transition states for primary amines as nucleophiles. Each is subject to medium effects and may be involved in intra- and intermolecular hydrogen bonding[144]. In the less polar solvents, the amines will be strongly associated: hence, their steric requirements may be relatively high. There will also be relatively more hydrogen bonding with the substrate and therefore more congestion at C_α in the low polarity solvents. This may account for the dominant product (Table 3) and for the fact that the primary gives more C_β product than the secondary amine in Table 4. Of the forms with internal hydrogen bonds, the six-membered cycle (**12b**) is likely to be favoured over the smaller ring (**12c**)[144]. Since this effect is aligned with the expected electronic factor, that is, in favour of 'normal' C_α attack, it is difficult to identify in these systems; but it does seem to be present in certain *syn*-(R_2NH) and *anti*-preferred (RNH_2) additions to acetylene dicarboxylic esters (ROOCC≡CCOOR)[1].

An understanding of orientation in alkynes containing regioantagonistic substituents requires some knowledge about the relative directing powers of the substituents. The strength of an α-director is related to its electronegativity. The expected order F > O > N > C (alkyl) is supported by CNDO/2 calculations (Figure 6). But because of preparative difficulties and perhaps the reactive character of alkynes containing two highly electronegative groups, there is insufficient experimental evidence to test these expectations. In equations (46) and (47), it is the more highly

$$CH_3CH(OH)C{\equiv}COC_2H_5 \xrightarrow[H_2O]{OH^-} CH_3CH(OH)CH_2CO_2C_2H_5 \tag{46}$$

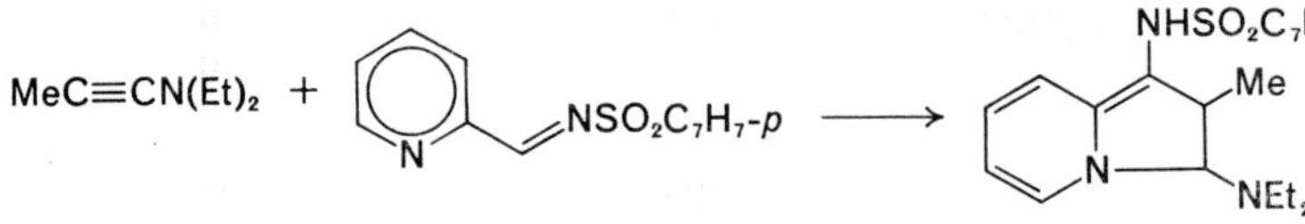
(47)

electronegative group which determines the orientation[11, 145]. Equations (48)–(50) are further examples of additions[16, 146, 147] in which the regioselectivities are in accord with the σ^- values of the substituents[148].

$$PhCOC{\equiv}CCO_2Me + RNH_2 \longrightarrow PhCOCH{=}C(NHR)CO_2Me \tag{48}$$

$$PhC{\equiv}CPO(Ph)_2 + NaCH(COOEt)_2 \longrightarrow (EtOOC)_2CHC(Ph){=}CHPO(Ph)_2 \tag{49}$$

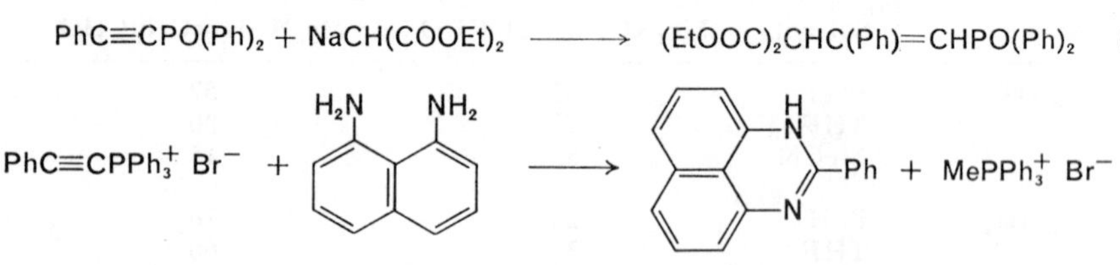
(50)

The acetylene in equation (9) contains CF_3 and Ph_2P groups whose σ^- values are 0·65 and 0·26, respectively. For the 'free' $Ph_2PC{\equiv}CCF_3$, these data call for a mode of attack *opposite* to that observed. By coordinating the diphenylphosphino group with palladium, the normal polarization is reversed, probably due to strong phosphorus to palladium back-bonding. Thus, coordination with a substituent or an acetylene may have important synthetic utility, especially when a reversal of the normal regioselectivity is desired.

In reaction (51) it might be expected that nucleophilic attack would occur at C_β [149]. Instead, the base-catalysed addition gives an equal mixture of C_α and C_β

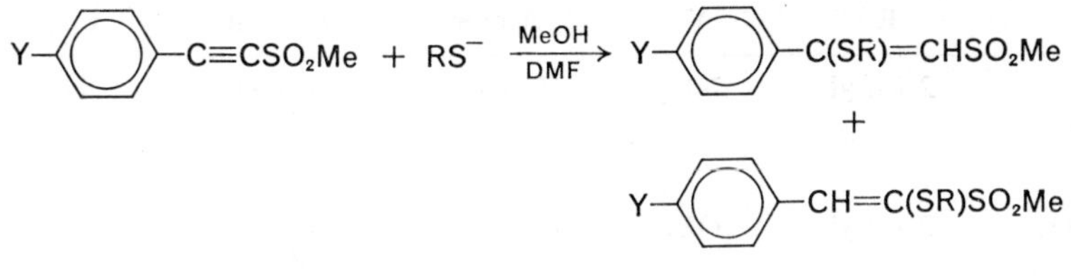

(51)

products when Y = H. The ability of the phenyl group to stabilize negative charge and to make C_β less sterically accessible may account for the unexpected amount of attack on C_α. By introducing Y = NO_2 or CH_3SO_2 on the phenyl group and using a bulky nucleophile (*n*-BuSH), the course of the reaction is steered exclusively to C_α.

We come now to two regioantagonistic alkyne systems in which the changes in product composition are often bewildering, if not inexplicable. In equation (52)

$$Et_2P(O)C{\equiv}CSMe \xrightarrow{NuH} \underset{Nu\,=\,MeO,\ EtS,\ Me_2N}{Et_2P(O)CH{=}C(Nu)SMe} + \underset{Nu\,=\,Me_2N}{Et_2P(O)C(Nu){=}CHSMe} \quad (52)$$

charged nucleophiles (MeO^-, EtS^-) attack only C_α [150]. Dimethylamine, however, gave products in the ratio $C_\alpha/C_\beta = 20/1$, when one equivalent at 20 °C in HMPT, and 5/1 when an excess at −10 °C was used[150].

As for equation (53), typical data are shown in Table 5[151]. Here some observations are: pyrollidine in ethanol at 36 °C adds to the ester of phenylpropiolic acid in *ca.*

$$X{-}R{-}C{\equiv}CCOY + R^1R^2NH \longrightarrow \underset{C_\alpha}{X{-}RCH{=}C(NR^1R^2)COY} + \underset{C_\beta}{X{-}RC(NR^1R^2){=}CHCOY} \quad (53)$$

5 h, while it takes *ca.* 5 d for comparable (100%) reaction with the amide; the amines favour C_β entry, as proton availability and solvent polarity increase and the reaction temperature decreases; electron-withdrawing groups (X) increase *anti*-Michael or C_α additions[93]. While some rationalization of trends in the above systems appears feasible, the drastic reversals in the C_β/C_α ratio do not fit neatly into our current categories.

Some years ago the *syn–anti* selectivities in amine additions were confusing—they are now largely understood! We believe that our understanding of the regioselectivities of amine additions is still primitive. It appears that we may have overrated the differences in the electronic effects exerted by substituents on the alkyne. It is also probable that we do not appreciate the extent to which the neutral amine and the substituents interact directly—this influence on orientation was mentioned years ago by Arens in regard to additions to RCOC≡CSR′ [152]. We are inclined to believe that systematic experiments in highly polar solvents in which amine associations were eliminated would be most useful in establishing 'base' substituent effects and 'base' regioselectivities.

TABLE 5. Solvent effects on regioselectivity in the additions of pyrollidine and *n*-propylamine to 2-furyl- and phenyl-propiolic acid derivatives, X—R—C≡CCOY (equation 53)[93, 151]

X	R	Y	Amine	Solvent	Product (%) C_α	Product (%) C_β
H	2-Furyl	NH_2	$(CH_2)_4NH$	EtOH	11	89
				MeCN	47	53
				Et_2O	88	12
5-NO_2	2-Furyl	OEt	$(CH_2)_4NH$	EtOH	5	95
				MeCN	13	87
				Et_2O	23	77
5-NO_2	2-Furyl	NH_2	$(CH_2)_4NH$	EtOH	55	45
				MeCN	92	8
				Et_2O	100	0
H	2-Furyl	NH_2	$PrNH_2$	EtOH	19	79
				$CHCl_3$	72	28
				Et_2O	87	13
H	Phenyl	NH_2	$(CH_2)_4NH$	EtOH	0	100
				$MeNO_2$	8	92
				Et_2O	28	72
4-NO_2	Phenyl	OEt	$(CH_2)_4NH$	EtOH	14	86
				MeCN	48	52
				Et_2O	92	8
4-NO_2	Phenyl	NH_2	$(CH_2)_4NH$	EtOH	87	13
				MeCN	97	3
				Et_2O	100	0
H	Phenyl	NH_2	$PrNH_2$	EtOH	0	100
				$CHCl_3$	20	80
				Et_2O	42	58

d. *1-Haloalkynes.* Regioselectivities in 1-halo-1-alkynes are examined separately, because of their importance in substitution reactions (see Section IV) and because they have three centres susceptible to nucleophilic attack: halogen (X), terminal carbon (C_α) and internal carbon (C_β). 1-Haloalkynes are triphilic (**13**)[3]. Occasionally,

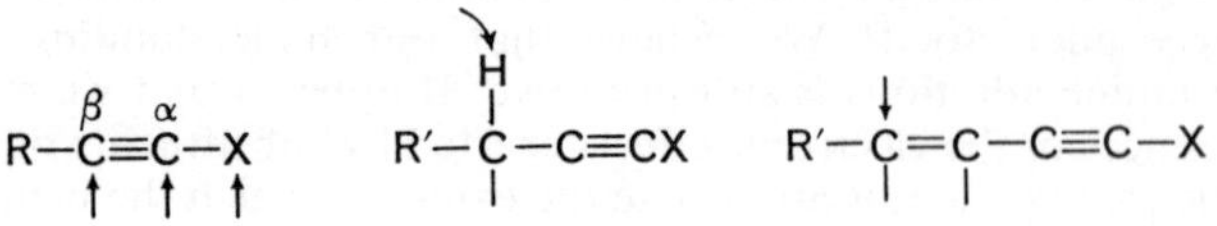

(**13a**) (**13b**) (**13c**)

three-site attack occurs in one system, i.e. methoxide ion and bromo- or chlorophenylacetylene in methanol[153]. Other examples are given in Table 6 in which the effects of substituent, nucleophile and solvent on regioselectivity are illustrated.

TABLE 6. Regioselectivity in nucleophilic attacks on 1-halo-1-alkynes, $RC_{\beta}{\equiv}C_{\alpha}{-}X$

R	X	Nu	Medium (temp., °C)	Attack (%)[a] C_{β}	Attack (%)[a] C_{α}	Attack (%)[a] X	Products (yield, %)	Reference
H	F	$(EtO)_2PSSH$	Et_3N/Et_2O	—	+	—	$H_2C{=}CFSP(S)(OEt)_2$	161
H	F	p-C_7H_7SH	ROH	—	+	—	$H_2C{=}CFSC_7H_7$-p	167
F_3C	F	F^-	H_2NCHO (20)	—	+	—	$F_3CCH{=}CF_2$ (55)	168
t-Bu	F	Me_2N^-	Et_2O	—	+	—	t-$BuC{\equiv}CNMe_2$ (48)	169
H	Cl	t-BuS^-	—	+	—	—	Z-t-$BuSCH{=}CHCl$	170
F	Cl	PhS^-	EtOH	+	—	—	$PhSFC{=}CHCl$ (70)	171
Me	Cl	RS^-	MeOH	+	—	—	$MeCSR{=}CHCl$	172
F_3C	Cl	$(EtO)_2PSSH$	—	—	+	—	$F_3CCH{=}CClSP(S)(OEt)_2$	161
Me_2COH	Cl	$(EtO)_2PO^-Na^+$	THF (106)	—	—	+	$(EtO)_2P(O)OC(Me)_2C{\equiv}CH$	159
t-Bu	Cl	PhS^-	EtOH (100)	+	—	—	t-$BuC(SPh){=}CHCl$ (76)	165
Ph	Cl	MeO^-	MeOH (78)	~1	~99	~0·2	$PhC{\equiv}CH$, $PhCH{=}CCl(OMe)$	153
Ph	Cl	$(EtO)_3P$	MeOH/THF	—	+	—	$PhC{\equiv}CPO(OEt)_2$ (40)	159, 158
Ph	Cl	$(EtO)_3P$	EtOH (b.p.)	—	+	+	$PhC{\equiv}CH$, $PhC{\equiv}CPO(OEt)_2$	163
2-C_4H_3S [b]	Cl	S^{2-}	$MeOH/H_2O$ (25)	—	67	33	$(2$-$C_4H_3S)C{\equiv}CH$	155a
2-C_4H_3S [b]	Cl	OH^-	$EtOH/H_2O$ (25)	—	+	—	— [c]	155a
H	Br	$EtNH_2$	(25)	+	—	—	$Et_2NCH{=}CHBr$	173
Me	Br	EtS^-	EtOH	—	—	+	$MeC{\equiv}CH$	162
F_3C	Br	$(EtO)_2PSSH$	Et_3N/Et_2O	—	+	—	$F_3CCH{=}CBrSPS(OEt)_2$	161
n-C_9H_{19}	Br	OH^-	EtOH	—	15	60	n-$C_9H_{19}C{\equiv}CH$, n-$C_9H_{19}CH_2COOH$	160
Ph	Br	MeO^-	MeOH (78)	11	83	5	$PhC{\equiv}CH$, $PhCH{=}CBrOMe$, $PhC(OMe){=}CBrH$, $PhC{\equiv}COMe$	153
Ph	Br	MeS^-	$MeOH/H_2O$	—	—	92	$PhC{\equiv}CH$, $(MeS)_2$ (98)	155b
Ph	Br	$(n$-$Bu)_3P$	MeOH/DMF (70)	—	—	+	$PhC{\equiv}CH$	157, 158
Ph	Br	Ph_3P	MeOH/DMF (70)	—	+	+	$PhC{\equiv}CH$, $PhC{\equiv}CPPh_3^+Br^-$, Ph_3PO	157, 158
Ph	Br	As_2O_3	H_2O	—	—	+	$PhC{\equiv}CH$	174
2-C_4H_3S [b]	Br	S^{2-}	$MeOH/H_2O$ (25)	—	—	>95	$(2$-$C_4H_3S)C{\equiv}CH$	155a
2-C_4H_3S [b]	Br	$S_2O_3^{2-}$	$MeOH/H_2O$ (25)	—	—	53	$(2$-$C_4H_3S)C{\equiv}CH$	155a
2-C_4H_3S [b]	Br	OH^-	$MeOH/H_2O$ (25)	—	—	51	$(2$-$C_4H_3S)C{\equiv}CH$	155b
1-(c-$C_6H_{10}OH$)	Br	$(i$-$PrO)_3P$	—	—	+	—	1-(c-$C_6H_{10}OH$)$CH{=}CBrPO(OPr$-$i)_2$	163a
Ph	I	$(EtOOC)_2\bar{C}HNa^+$	—	—	—	+	$PhC{\equiv}CH$, $[(EtOOC)_2CH{\text{-}}]_2$	175

TABLE 6 (*cont.*)

R	X	Nu	Medium (temp., °C)	Attack (%)[a] C_β	C_α	X	Products (yield, %)	Reference
Ph	I	RS^-	$MeOH/H_2O$ (25)	—	—	>95	PhC≡CH, $(RS)_2$	155b
$2\text{-}C_4H_3S$ [b]	I	S^{2-}	$MeOH/H_2O$	—	—	100	$(2\text{-}C_4H_3S)C{\equiv}CH$	155
Ph	I	$(EtO)_3P$	EtOH (~150)	—	—	83	PhC≡CH, $PhC{\equiv}CPO(OEt)_2$, $(EtO)_3PO$	163b
Ph	I	EtMgBr	Et_2O	—	—	96	EtI, PhC≡CMgBr	176
Ar	I	I^-	DMF	—	—	+	ArC≡CH	177

[a] A plus sign indicates that reaction was observed. Where there are propargylic hydrogens, e.g. R″CHR′C≡CX, attack at this site may yield products identical to those of C_α and C_β entry.
[b] 2-Thienyl.
[c] No product was given.

We defer (Section IV) a discussion of the mechanisms of process (2). Suffice to say that we have usually deduced the point of attack from the nature of the products, i.e. C_{α} adduct, C_{β} adduct and terminal alkyne. The latter arises from an ion–molecule pair as in equation (54) where the coproduct (NuX) may undergo subsequent reaction

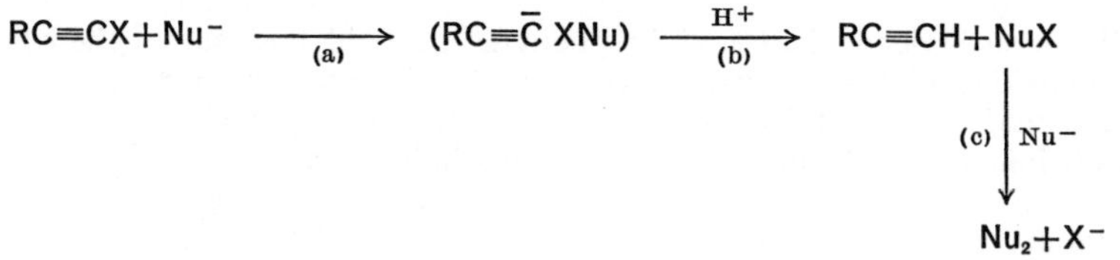

(54)

according to step (c)[154] (see also Table 6). The reader should note that *secondary* products (or their equivalents) of halogen abstraction, i.e. RC≡CH and RCOMe, will also differ from those of addition to carbon, i.e. RC≡CX and RCH_2COOH.

It is well known that the halogen in haloalkynes may be 'positive':[44] hence the expected order for nucleophilic assault on halogen is I > Br > Cl > F[3, 4, 155]. Support for this order was demonstrated in the switch in major product in equation (55)[156].

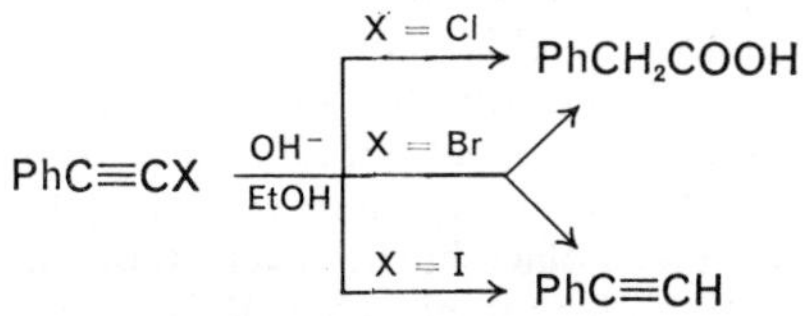

(55)

At one extreme, *positive iodine* is the exclusive site of attack in RC≡CI, regardless of the type of nucleophilic reagent and the character of the R group. At the other extreme is fluorine: its high electronegativity must render it indifferent to most nucleophilic attack. Indeed, the removal of positive fluorine strains credulity. In any case, the MO results of Figure 6 reveal that C_{α} is the most electrophilic centre in fluoroacetylenes. Experimentally, it is found that regiospecific attack occurs on this carbon regardless of the nature of the nucleophile and the R substituent (Table 6).

It is in chloro- and bromoacetylenes particularly that *sp* carbon (C_{α}, C_{β}) and halogen compete for the nucleophile (Table 6). HSAB theory has been used to rationalize the variations in the site of attack with changes in the nucleophile[155]. Bromine and chlorine are considered to be 'softer' electrophilic centres than *sp* carbon. As the softness of the nucleophile increases there is an increasing amount of attack on these halogen centres, e.g. MeS^- gives 92% attack on bromine, while hard MeO^- produces only 5% attack on bromine in phenylbromoacetylene[155], soft S^{2-} gives 33% attack on chlorine while hard OH^- attacks only *sp* carbon in 1-chloro-2-(2-thienyl) acetylene (Table 6)[155].

Attack on halogen also seems to be dependent on the basicity of the nucleophile. For instance, tributylphosphine produces substantially more halogen abstraction from bromo- and chlorophenylacetylene than the less basic triphenylphosphine (Table 6)[157, 158]. (This result seems to contradict the HSAB prediction.) Similarly, the more basic diethylphosphite anion has a greater preference for halogen in phenylhalo-acetylenes than the less basic triethylphosphite[158, 159]. In these examples the comparisons have to be made for the pairs, **14** *vs.* **15** and **16** *vs.* **17**. The changes in preferred modes of attack have not been explained convincingly.

When electron-withdrawing groups are present in a haloalkyne, there is less halogen abstraction than in halides with alkyl (electron-donating) substituents:

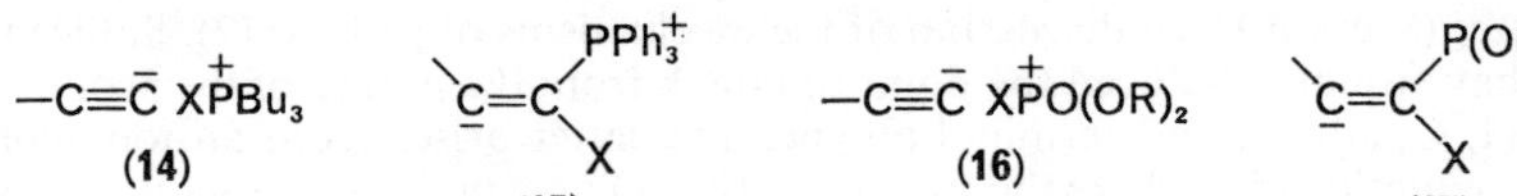

(14) (15) (16) (17)

phenylbromoacetylene gives mainly phenylacetic acid and some phenylacetylene with hydroxide while 1-bromo-1-undecyne with this base yields 60% 1-undecyne and only 15% undecanoic acid (Table 6)[156, 160]; 1-bromo-3,3,3-trifluoropropyne undergoes addition while 1-bromopropyne sustains halogen attack with sulphur nucleophiles[161, 162]. These results have been 'explained' by comparing the stabilities of the vinyl and acetylenic carbanions derived from the haloalkynes[155].

The competition for the nucleophile between *sp* carbon and halogen can also depend upon the solvent in a major way. The late development of process (2) may in part be attributed to the unfavourable solvents that were used[1]. In a protic medium halogen abstraction not only becomes visible but appears to be promoted, e.g. for PhC≡CBr + $(EtO)_3P$ [158, 163] and other examples of Table 6. In qualitative terms, the proton–solvent should favour ion–ion and ion–molecule pairs (**14** and **16**) over the larger species **15** and **17** in which the charge is more dispersed. Theoretical calculations (EHMO) tend to support this rationalization[164].

Another contest in bromo- and chloroalkynes is that between C_α and C_β. Bromo- and chloroethyne obey Arens' rule: nucleophilic attacks in these acetylenes occur on C_β. Predictably, bromo- and chloroalkylacetylenes with regiosynergistic groups undergo C_β attack. Even when there are unfavourable steric effects, Viehe demonstrated that C_β orientation takes place in the reaction of thiophenoxide and chloro-*t*-butylacetylene[165]. C_α comes under attack when R is an electron-withdrawing group and/or can delocalize incipient negative charge on C_β by resonance (Table 6).

To complete this section, we note that the possibility of other (remote) attacks, as in **13b** and **13c**, have been mentioned in connection with equation (4). They are particularly important in alkylhaloalkynes, e.g. equation (56)[32]. A propargylic

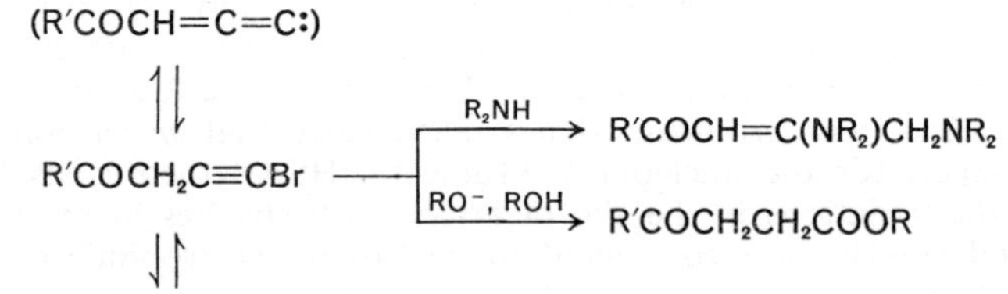

(56)

hydrogen is potentially mobile; whether its removal leads to a carbene or an allene, substitution according to equation (2) becomes improbable[166]. This accounts for some of the difficulties or 'failures' of equation (2) and the importance of taking regioselectivity factors into account when syntheses are planned.

2. Ring size

Regioselectivity in a ring closure by internal nucleophilic attack is often puzzling. While the substituent appears to make the difference in equation (57) a similar reaction of a diethynylketone with aryl amines leads to *both* 5- and 6-rings, i.e. oxo-Δ^2-pyrrolines and pyridones[141, 178, 179]. The question that arises most frequently here is illustrated by the choice given in equation (58).

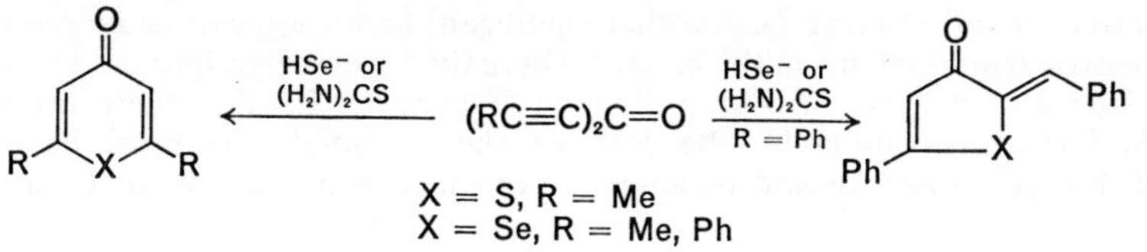

(57)

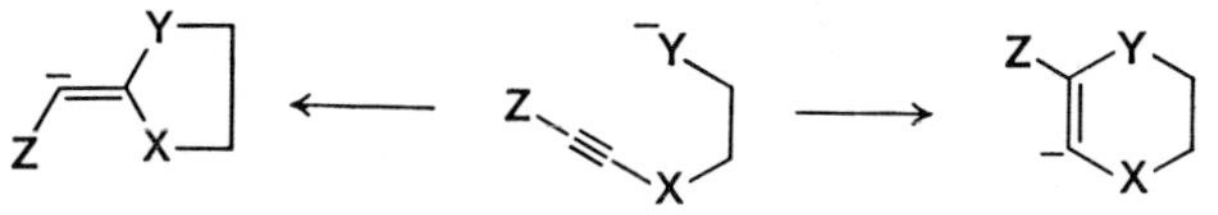

(58)

Baldwin has formulated rules for ring closure in a systematic manner[180]. For digonal systems these are: (*i*) 3- and 4-*exo-dig*, disfavoured; (*ii*) 5- to 7-*exo-dig*, favoured; (*iii*) 3 to 7-*endo-dig*, favoured. In his terminology the choice in equation (58) is between 5-*exo-dig* and 6-*endo-dig*. On the basis of his survey Baldwin concludes that *endo*-ring closures at digonal carbon predominate. For the possibilities we encountered most often, namely, 5- to 7-rings, we find that first-row nucleophilic sites, e.g. O, N, C, favour 5-*exo-dig* and 6-*exo-dig* closures. These will be illustrated here and in later sections.

Besides the regioselectivity preferences discussed previously, additional factors appear to bear on the closure process itself. The question of *exo vs. endo* double bonds has been studied. For a variety of Z groups which ranged from H to COOEt, $K(endo/exo)$ = 2–240 for equation (59) and 1140–0·2 for equation (60) at 25 °C[181].

(59)

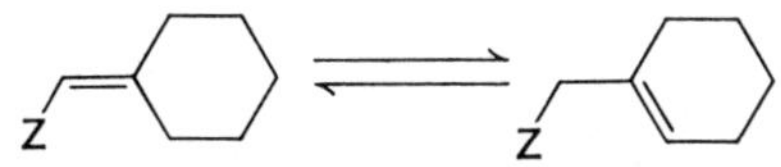

(60)

To highlight the combined factors of ring size and *endo-exo* double bonds, we have assembled heats of formation in Figure 7 (data are scanty)[54]. On thermochemical grounds we see that *endo*-6 should be favoured over *exo*-5 and *exo*-6 should be

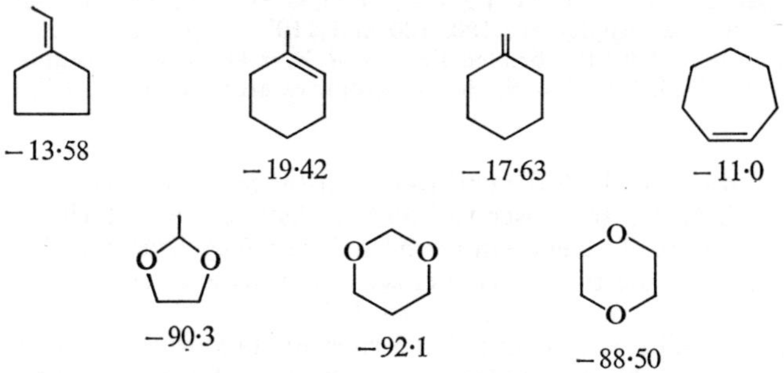

FIGURE 7. Ring size and ΔH_f^0 at 298 K for liquid compounds.

favoured over *endo*-7 (Figure 7). Another significant factor appears to be *geminal vs. vicinal* electronegative atoms in the product: here the former arrangement is favoured in dioxolane and *m*-dioxane over *p*-dioxane (Figure 7). Finally, there are stereoelectronic factors; presumably, the ends of the potential ring must be able to approach bonding distance and to align the orbitals favourably[180]. In Figure 8 we

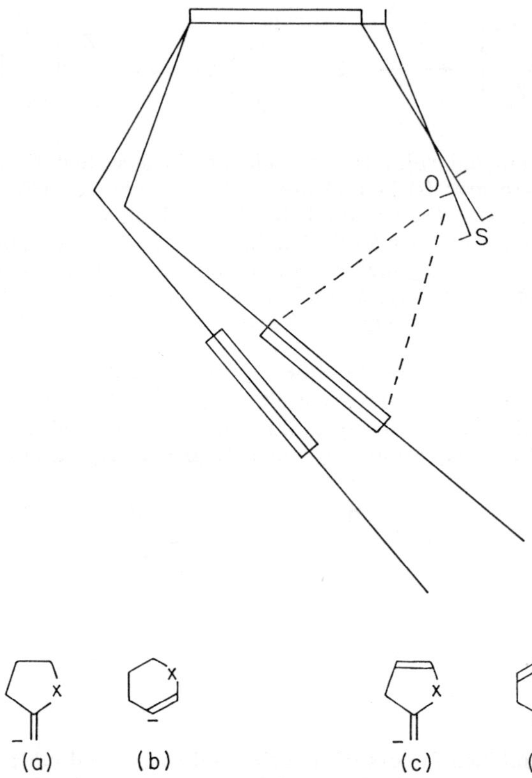

FIGURE 8. Scale diagram for alternate cyclizations to 5- *vs.* 6-membered heterocyclics. The starting compound is $MeC{\equiv}CCH_2CH_2CH_2X^-$ or $MeC{\equiv}CCH_2CH{=}CHX^-$, where X is O or S. Distances (Å) are: C≡C 1·2, C=C 1·34, C—C 1·54, ≡C—C 1·46, =C—C 1·53, C—O 1·43, C—S 1·82. Angles are 180, 120 and 110° as required. Closure distances to (a) *vs.* (b) are indicated by the broken lines. For X = O, closure distances are: (a) 1·8, (b) 1·9, (c) 1·9, (d) 2·05. For X = S, closure distances are: (a) 1·8, (b) 1·7, (c) 1·9, (d) 1·8.

give a scale diagram in which the end-to-end distances may be seen. Note that first-row elements, C, N, O, are closer to bonding distance for a 5-ring, while heavier elements such as S are close enough for both 5- and 6-cyclization. These factors and perhaps others affect the type of ring closure, but, as is often the case, kinetic and equilibrium control are not necessarily parallel.

It has been known for a long time that 5-rings are favoured kinetically: the relative *rates* for closure of ω-bromoalkylamines at 25 °C to form 5-, 6- and 7-membered rings are 5×10^4, 800 and 1 respectively; lactonization of the anions of bromo acids displays a similar trend[182]. In equations (61)–(66) we list a number of typical examples illustrating this preference in the context of process (1). In several we indicate also the route or intermediate that was *not* taken. In all cases a 5- *vs*. 6- alternative exists, although the critical step does not always involve an alkyne, e.g. equation (63) and

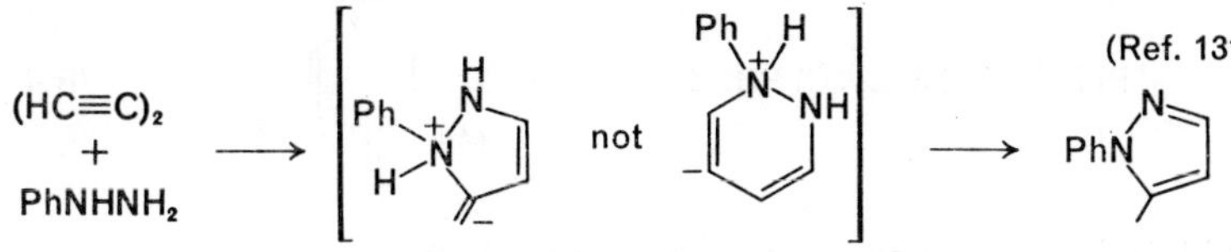

(Ref. 131) (61)

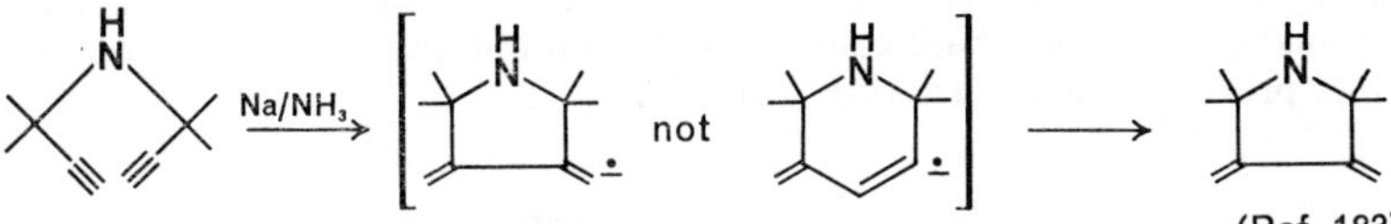

(Ref. 183) (62)

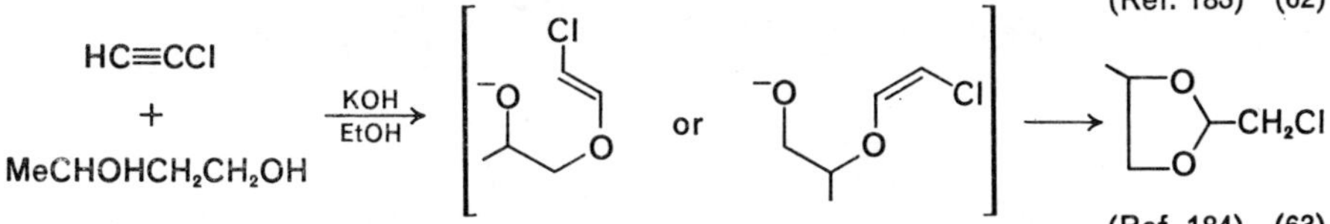

(Ref. 184) (63)

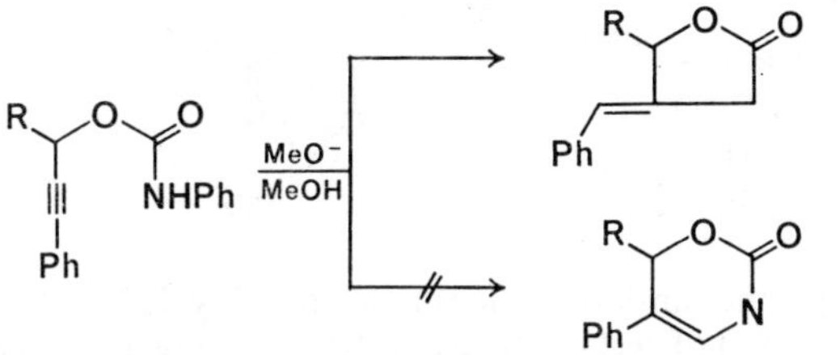

(Ref. 185) (64)

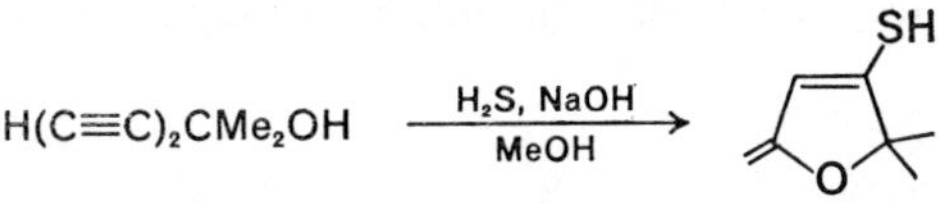

(Ref. 186) (65)

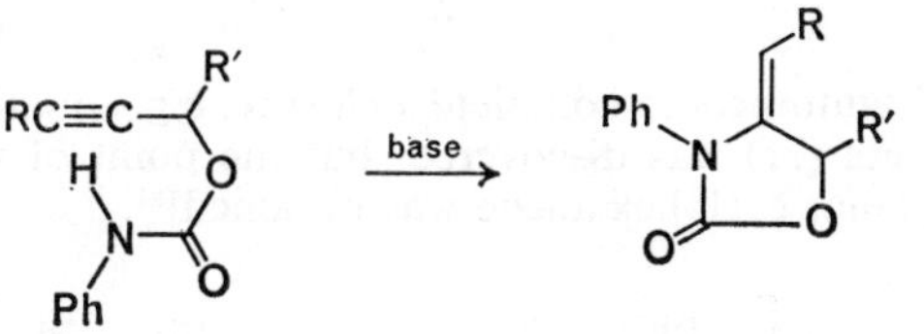

Ref. (187) (66)

perhaps (62). In equation (63) in which the geminal stability factor and Markownikoff regioselectivity are aligned, the formation of the 5-ring is easily rationalized. In other cases the simple dimensional factor may be overriding. What is quite obvious is that thermodynamic stabilities connected with ring size or *exo–endo* double bonds seem to be irrelevant here.

Equations (67) and (68) give examples of exclusive 6-ring formation. The examples may not be 'fair', since the internal S or P direct to the β-carbon, presumably by stabilizing charge most effectively in the 6-ring (see species **11**). Note, however, that in equation (57), both *anti*-Michael 5- and Michael 6-rings form.

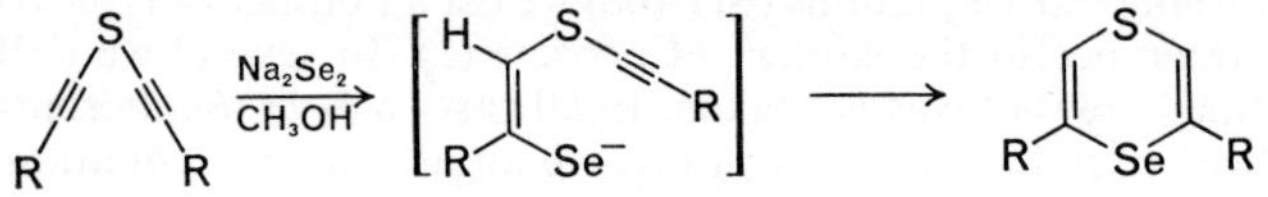

(Ref. 188) (67)

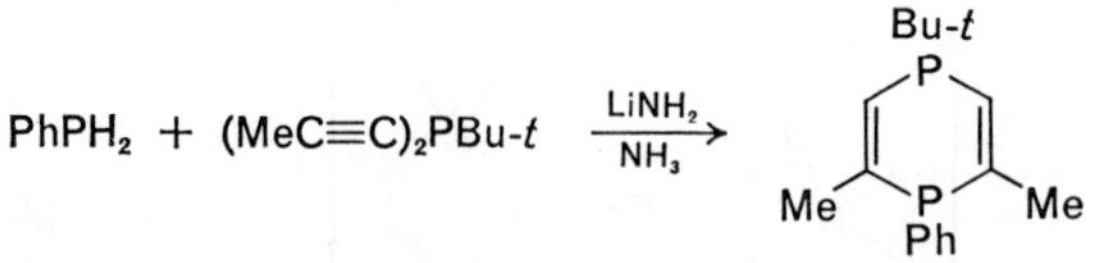

(Ref. 189) (68)

As will be indicated later, 7-rings can be made from alkynes. However, the carbanion process of (69) is a rare example in which competition between 6- and 7-ring closure is present—only the 6-ring forms[190].

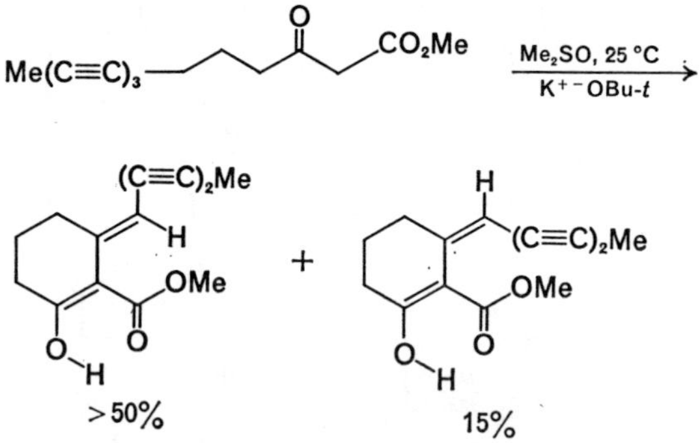

(69)

Ring aromaticity (and antiaromaticity) presumably influence the selectivity of closure. The construction of certain heterocycles, e.g. pyrazoles, isoxazoles, etc., may be rationalized in this way (equation 70). Indeed, it may be the driving force which

(70)

selects one of several tautomers to complete a closure, e.g. equation (61). It was also suggested when process (71) was discovered, but the point of view changed when evidence for a non-planar cyclohexadiene was obtained[191].

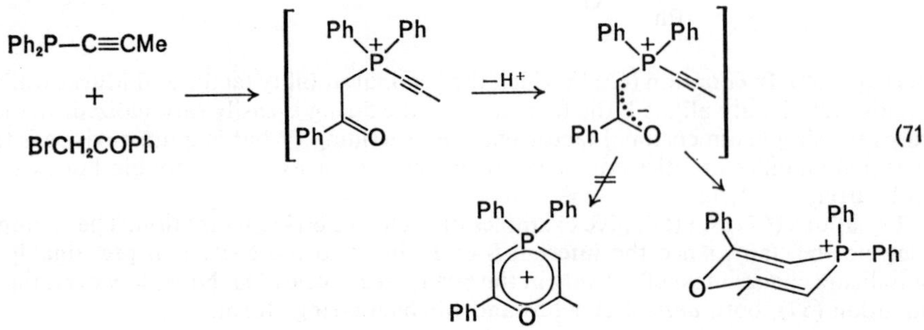

(71)

III. NUCLEOPHILIC ADDITIONS

A. Electrochemical Reduction

This is the first of several kinds of nucleophilic attack by an electron (e) on an alkyne. Here we discuss briefly two aspects of cathodic reduction, namely, polarographic measurements and controlled-potential syntheses[192]. Since the scale and the conditions often differ, the reader should be prepared for occasional apparent discrepancies resulting from the two approaches.

It is generally considered that alkylacetylenes are inert while others, especially with conjugative substituents, are active in polarographic reduction. That is, the supporting electrolyte and/or the solvent are usually reduced in preference to an alkylacetylene. It has been estimated that, relative to the standard calomel electrode (SCE), $E_{\frac{1}{2}} \leqslant -3{\cdot}1$ for cyclononyne, 5-decyne, 1-hexyne, 3-hexyne and 2,2,5,5-tetramethylhexyne and $E_{\frac{1}{2}} \simeq -3{\cdot}0$ for ethyne, where the supporting electrolyte (n-$Bu_4N^+BF_4^-$ in DMF) is already reduced at $-2{\cdot}9$ to $-3{\cdot}0$ V [193]. However, conditions for the polarographic reduction of simple alkynes have been found, e.g. dimethylacetylene in methanol containing 0·1M (n-Bu)$_4$NCl with Ag,AgCl as the reference electrode[194]. By contrast, ethynyl, vinyl, carbonyl and phenyl substituents at the triple bond facilitate electroreduction[195]. These broad trends are illustrated in Table 7 and the preparative applications are given in Table 8.

In a polarographic measurement one may, in favourable cases, determine the potentials, reversibility and electron equivalents for a given cathodic process. When combined with an analysis of products, these usually provide insights into the *gross* mechanism of reduction. By using $E_{\frac{1}{2}}$ data (Table 7) as limits, one can arrange reductions in which some group may be altered cleanly *before* the triple bond is touched, or *vice versa*. At the same time, $E_{\frac{1}{2}}$ data comprise an approximate electrophilicity scale of alkynes towards 'cathodic' electrons, that is, a sort of solution electron affinity[196]. The practical and theoretical uses of these data appear in several places in this chapter.

Cathodic reduction mechanisms follow several patterns, depending on the alkyne, the medium, the supporting electrolyte, the properties of the cathode and the applied potential[192, 197, 198] On a spongy nickel electrode in 95% ethanol and sulphuric acid, a number of alkyl- and arylalkynes are reduced in good yields in a process which resembles, but is not identical with, catalytic hydrogenation—alkenes are not reduced under these conditions[199]. In no sense does the presumed mechanism (equation 72) appear to involve nucleophilic attacks.

$$
\begin{aligned}
H^+ + e &\xrightarrow{(a)} H^{\bullet}_{ads} \\
2H^{\bullet}_{ads} &\xrightarrow{(b)} (H_2)_{ads} \\
H^{\bullet}_{ads} + RC{\equiv}CR' &\xrightarrow{(c)} R\dot{C}{=}CHR' \\
2H^{\bullet}_{ads} \text{ or } (H_2)_{ads} + RC{\equiv}CR'_{ads} &\xrightarrow{(d)} cis\text{-}RCH{=}CHR'
\end{aligned}
\qquad (72)
$$

In the second mechanism an electron makes its way from the cathode as e_{cath} to a substrate[200]. Although the detailed roles of the participating species [reactant(s), product(s), solvent, electrolyte and electrode] are often unknown, the general outline of the reduction is reasonably well established (equation 73)[197, 198]. The greatest uncertainty hinges on the timing of proton *vs.* electron delivery to the radical anion, i.e. steps (b) *vs.* (c). A similar ambiguity arises if a reagent other than the proton competes with e_{cath} [197, 201].

$$
\begin{aligned}
R C\equiv CR' + e &\xrightarrow{(a)} R\dot{C}{=}\bar{C}R' \\
R\dot{C}{=}\bar{C}R' + H^+ &\xrightarrow{(b)} R\dot{C}{=}CHR' \\
R\dot{C}{=}\bar{C}R' + e &\xrightarrow{(c)} R\bar{C}{=}\bar{C}R' \\
R\dot{C}{=}CHR' + e &\xrightarrow{(d)} R\bar{C}{=}CHR' \\
R\bar{C}{=}\bar{C}R' + H^+ &\xrightarrow{(e)} R\bar{C}{=}CHR' \\
R\bar{C}{=}CHR' + H^+ &\xrightarrow{(f)} RCH{=}CHR'
\end{aligned}
\qquad (73)
$$

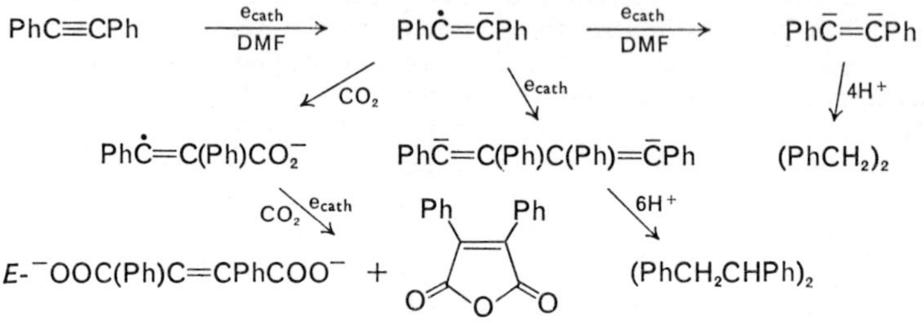

(74)

Unambiguous evidence for the cathodic generation of the radical ions from certain alkynes (equation 75) was obtained by measuring their e.s.r. spectra[202]. Other workers have generated both radical anions and dianions (equation 75) by

$$
ArC\equiv CAr' \xrightarrow{e} Ar\dot{C}{=}\bar{C}Ar' \xrightarrow{e} Ar\bar{C}{=}\bar{C}Ar' \qquad (75)
$$

electron transfer from metals or other carbanions and have measured their spectral (e.s.r., u.v.–visible, n.m.r.) properties[203, 204]. For this reason, support for the cathodic generation of the radical anion of diphenylacetylene, at least, is unequivocal. For a wider group of arylalkynes an oxidation–reduction cycle developed by oscillopolarography was a graphic indication that the first step in equations (74) and (75) is often reversible[202].

Conventional polarographic measurements are usually less definitive—one-, two-, three- or four-electron transfers have been associated with the first cathodic wave of various acetylenes (Table 7). In the case of diphenylacetylene, several polarographic studies still do not permit a clear choice between a one- or a two-electron first wave[194, 201–203]. As for the dianions of equation (75), it is probable that they are present in some cathodic processes. Since the evidence for them generally depends on the interpretation of polarographic and product data, their intermediacy is less certain.

For acetylenes with an α-H, a third mechanism, isomerization to the allene on/at the cathode, may be important. This process may be initiated by a base, e_{cath}, radical anion, etc., and the reduction may then proceed from the allene. The scheme in equation (76) is intended to represent this mechanism at a heterogeneous surface—an

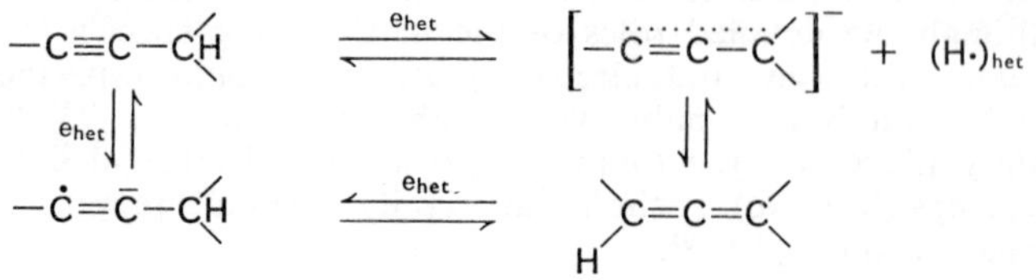

(76)

TABLE 7. Polarographic half-wave potentials of alkynes, RC≡CR′, at *ca.* 25 °C [a]

R	R′	Solvent[b]	$-E_{\frac{1}{2}}$, V[c] (n)[d]	Reference
H	H	DMF	(≃3·0)	193
n-Bu	H	DMF	(⩾3·0)	193
n-Bu	*n*-Bu	DMF	(⩾3·0)	193
Ph	H	DMF	2·37 (4)[h]	201*
2,4,6-(*t*-Bu)$_3$C$_6$H$_2$	H	G	2·7 (1 or 2)	203
Ph	*n*-Bu	DMF	2·65 (1), 2·88 (1)	210, 211*
Ph	Ph	G	2·48, ∼3·08	203
2,4,6-(Me)$_3$C$_6$H$_2$	2,4,6-(Me)$_3$C$_6$H$_2$	G	2·70 (1)	203
2,4,6-(*t*-Bu)$_3$C$_6$H$_2$	2,4,6-(*t*-Bu)$_3$C$_6$H$_2$	G	2·93 (1)	203
5-(2-MeC$_5$H$_3$N)	H	92% M	2·06	213
2-(2-C$_5$H$_4$N)	H	92% M	1·72	213
1-(1-HOC$_6$H$_{10}$)	1-(1-C$_6$H$_{10}$)	M	1·948[e]	194*
n-Pr	COMe	DMF	1·99 (∼1)	214
n-Pr	COOMe	DMF	2·26 (∼1)	214
n-Pr	COOLi	DMF	2·31 (∼1)	214
EtOOC	COOH	W, pH 2·5[f]	0·8 (∼3)	207*
EtOOC	COOEt	W, pH 2·5[f]	0·57 (∼2), 0·75 (2)	207*, 215
Ph	COOH	10% E, pH 2·5[f]	1·2, 1·4	206*
Ph	COOEt	30% E, pH 7[f]	1·32, 1·63	206*
Ph	COOEt	20% E	1·257, 1·613	208
p-ClC$_6$H$_4$	COOMe	25% E[g]	1·194 (1), 1·613 (2)	209*
p-ClC$_6$H$_4$	COOEt	20% E	1·215, 1·568	208
m-ClC$_6$H$_4$	COOEt	20% E	1·271, 1·558	208
p-MeOC$_6$H$_4$	COOEt	20% E	1·274, 1·619	208
p-O$_2$NC$_6$H$_4$	COOEt	20% E	0·212, 0·469, 1·113	208
Ph	SiMe$_3$	DMF	2·00[e]	216
p-MeC$_6$H$_4$SO$_2$	Me	M	1·46[h], 2·04	205*
p-MeC$_6$H$_4$SO$_2$CH$_2$	H	M	1·32[h], 2·08	205
Br$^-$Ph$_3$PCH$_2$$^+$	H	M	1·34	205*
Me$_2$CCl	H	DMF	1·53 (1)[e, h], 2·22[e]	217*
Ph	(CH$_2$)$_4$Br	DMF	2·35 (1)[e, h], 2·60 (2)[e], 2·80[e]	212
Ph	(CH$_2$)$_4$Cl	DMF	1·77 (1)[e, h], 2·05[e]	211
3,4-(MeO)$_2$C$_6$H$_3$CO	Ph	E–W	Variable[f]	218*, 220
Me$_3$Si	COMe	20% E	1·230 (2), 1·532	221*, 5, 6
H	CHO	E–W	0·98 (2)[f], 1·16	222*
MeSO$_2$	H	E–W	1·11 (2), 1·79 (2)	223*
PhCO	Ph	E	1·06[e], 1·62[e]	224, 218a
Ph	CHO	20% Me$_2$CO[i]	0·6[f, h], 0·78[f, h]	225, 222
PhCO	COPh	DMF	0·92 (1)[e], 1·58 (1)[e]	219

[a] This table extends (with less detail) the comprehensive compilation of Reference 192. Experimental details should be sought in the original paper. Additional examples and/or leading references are indicated by the asterisk in the last column.

[b] The solvents are: DMF, HCON(CH$_3$)$_2$; G, (CH$_3$OCH$_2$)$_2$; M, CH$_3$OH; E, C$_2$H$_5$OH; W, H$_2$O.

[c] The working electrode is mercury and the reference is the standard calomel electrode (SCE), unless otherwise indicated.

[d] The probable number of electrons transferred.

[e] The reference is not SCE.

[f] Value is dependent on pH.

[g] Value is dependent on the supporting electrolyte.

[h] This wave involves a function other than the triple bond such as Br, RSO$_2$, etc.

[i] At 16·2 °C, pH 2.

example will be given presently. Of course, the picture is simpler, if base-catalysed isomerization can be distinguished from the electrochemical process. Polarographic waves for about a dozen compounds of the type $ArSO_2CH_2C{\equiv}CH$, $ArSO_2C{\equiv}CMe$, $HC{\equiv}CCH_2PPh_3^+Br^-$, etc. at pH $\simeq 9$ have been attributed to the isomeric allenes, which are formed *before* reduction begins[205].

TABLE 8. Cathodic reduction of alkynes, RC≡CR′, on a preparative scale[a]

R	R′	Solvent[b]	Products (yield, %)	Reference
n-Bu	H	*n*-PrOH	*n*-BuCH=CH_2 (60)	194
n-Bu	Et	CH_3NH_2, LiCl[c]	*n*-BuCH=CHEt (50%; *Z*/*E* = 49)	226
Ph	H	M/W 9/1	PhC_2H_5 (38), PhC≡CH (40)	194
Ph	Me	M/W 9/1	PhC_3H_7-*n* (50)	194
Ph	*n*-Bu	DMF	PhC_6H_{13}-*n*, PhCH=C=CHC_3H_7-*n*	211
Ph	Ph	M	$(PhCH_2)_2$ (40), PhC≡CPh (40)	194
$HOCH_2$	H	M	$HOCH_2CH{=}CH_2$ (62)	194
$HOCH_2$	CH_2OH	W	$HOCH_2CH{=}CHCH_2OH$ (82)	194
Ph	CH_2OH	M/W 9/1	$Ph(CH_2)_3CH_2OH$ (70)	194
$PhCH_2$	COEt	M	$Ph(CH_2)_3CHOHMe$ (50)	194
PhC≡CCO	Ph	M	$(PhCH_2CH_2)_2CO$ (64)	194
Ph	CHO	E/W 1/9	$PhC{\equiv}CCH_2OH$ (~55)	206
HOOC	COOH	M	$(HOOCCH_2)_2$ (70)	194*
HOOC	COOH	W, HCl	*rac*-$(HOOCCHCH_3)_2$ (60)	207
HOOC	COOMe	W, HCl	*rac*-$(HOOCCHCH_3)_2$	207
EtOOC	COOEt	W, HCl	*E*-EtOOCCH=CHCOOEt (92)	207
Ph	COOH	E/W 1/9	*E*-PhCH=CHCOOH, $PhCH_2CH_2COOH$	206*
Ph	COOEt	E/W 1/9	*E*-PhCH=CHCOOEt, $PhCH_2CH_2COOEt$	206*
n-Bu	$(CH_2)_4Br$	DMF	*n*-$C_{10}H_{22}$, 1-*n*-butylcyclohexene	212
Ph	$(CH_2)_4Br$	DMF	PhC≡CBu-*n*, PhC_6H_{13}-*n*	212*
Ph	$(CH_2)_4Cl$	DMF	Benzylidenecyclopentane (81)	211
Me_2CBr	H	DMF	$(Me_2C{=}C{=}CH)_2Hg$	217
PhCO	Me	30% E	PhCOPr-*n* (35)	218b*
Ph	CHO	E	$PhCH{=}CHCH_2OH$ (20–30)	222

[a] This table extends the compilation of References 198 and 195. The cathode was mercury, except as indicated. Additional examples are indicated by the asterisk in the last column.
[b] See footnote *b* of Table 7.
[c] Pt cathode.

Concerning stereoselectivity, it is not surprising that *anti* reductions predominate[194], since *trans* are usually more stable than *cis* isomers (Table 8). This is likely to hold if any of the radical/anion species in equation (73) are *free* of the cathode[193]. Obviously, the cathode may be involved with the substrate and affect the selectivity, e.g. in equation (77)[194].

$$ROCH_2C{\equiv}CCH_2OR \xrightarrow[CH_3OH]{2e,\ 2H^+} ROCH_2CH{=}CHCH_2OR \qquad (77)$$

R = H, 100% *E*

R = Ac, 60% *E*, 40% *Z*

We conclude this section with a look at two types of reduction. The scheme of equation (78) condenses reactions of certain carboxyalkynes, among which there are

$$\text{R'C}\equiv\text{CCOOR} \xrightarrow{2e,\ 2H^+} \text{R'CH}{=}\text{CHCOOR} \xrightarrow{2e,\ 2H^+} \text{R'CH}_2\text{CH}_2\text{COOR}$$

$$\Big\downarrow\ H^+ \mid 1e \qquad (78)$$

$$(\text{R'CH}_2\text{CHCOOR})_2 \text{ or } (\text{ROOCCH}_2\text{CHR'})_2$$

considerable variations. For several compounds, e.g. PhC≡CCOOH, EtOOCC≡CCOOEt, etc.[51, 206, 207], $E_{\frac{1}{2}}$ decreases with pH, and one may consider that proton transfer is part of the rate-determining process. For others, e.g. ArC≡CCOOEt, one- or two-electron transfer to the alkyne is pH-independent and one may deduce that proton transfers are fast[208].

The $E_{\frac{1}{2}}$ of the one-electron wave indicated in equation (78) may fall between the other two in the case of PhC≡CCOOH or essentially coincide with the first one in HOOCC≡CCOOR (R = H, Me); under controlled hydrolysis the respective products are the alkene, the dimer, $(PhCHCH_2COOH)_2$, and a decarboxylated reduced product, *rac*-$(ROOCCHCH_3)_2$ [51, 206, 207]. Apart from the *p*-NO_2 compound, Krishnamurthy found that the typical polarography of $XC_6H_4C{\equiv}CCOOEt$ in 20% ethanol consists of a pH-independent two-electron wave followed by a pH-dependent two-electron wave[208]. This contrasts with the compounds mentioned previously, all of which have pH-dependent $E_{\frac{1}{2}}$s. A rather different dimension was uncovered by Missan and coworkers: although *m*-$ClC_6H_4C{\equiv}CCOOEt$ shows the 'normal' two two-electron waves, these merge at the lower $E_{\frac{1}{2}}$ when the supporting electrolyte, $Me_4N^+Cl^-$, is changed to *n*-$Bu_4N^+Cl^-$; this effect was tentatively ascribed to adsorption of the alkylammonium ions on the cathode[197, 209].

Our second example concerns the cathodic reductions of 6-X-1-phenyl-1-hexynes in DMF [210–212]. 1-Phenyl-1-hexyne (>0·002M) first isomerizes to 1-phenyl-1,2-hexadiene which is reduced in steps, or (<0·002M) is reduced in one 4e-step to 1-phenylhexane[211]. At the higher concentrations, a radical anion could be detected by cyclic voltametry (100 V/s^{-1}), although the usual measurements indicate irreversibility[210]. Here the concentration dependence is attributed to competition between further reduction (low M) and isomerization (high M) of the starting material.

For 6-chloro-1-phenyl-1-hexyne (X = Cl, >0·002M), isomerization to the allene is promoted at the cathode and subsequent stepwise reduction (equation 79) to three

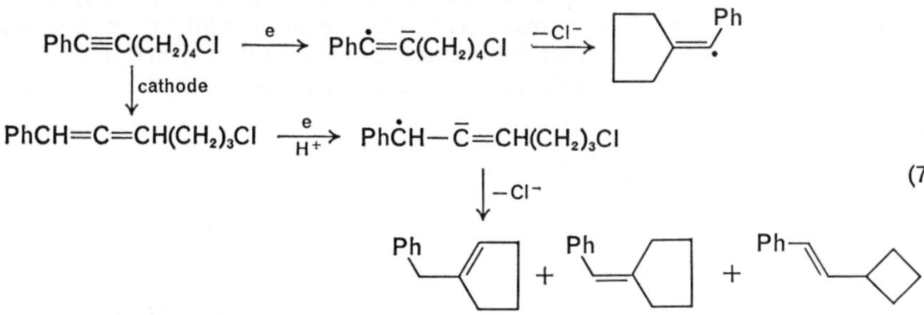

(79)

major and *ca.* seven minor products is observed (−1·75 V). At low concentrations ($2{\cdot}5 \times 10^{-4}$M), this hexyne shows waves of $E_{\frac{1}{2}} = -1{\cdot}77$ and $-2{\cdot}05$ V; controlled electrolysis at $-1{\cdot}75$ V indicates that no allene, but benzilidenecyclopentane (81%) and at least seven other products are formed. When X = Br or I, the most facile cathodic process becomes removal of halogen (for Br, $E_{\frac{1}{2}} = -2{\cdot}35$ V) to produce 1-phenyl-1-hexyne, which is then reduced in its characteristic way[210, 212].

B. Electron Transfer

Chemical, as opposed to cathodic, delivery of electrons to the triple bond has been accomplished in various ways. The conventional *anti*-selective reductions of alkynes by dissolved metals is still important[227], but new reagents and solvents have widened the scope of nucleophilic reductions (Table 9). Indeed, an understanding of the mechanistic options has made for greater flexibility: different initiation and entry/departure of participants is now possible.

Clean metallic mirrors (Na, K, etc.) in aprotic solvents are especially useful for producing anions whose spectra or reactivities may be measured (equation 80)[64, 204, 216, 228–236]. Whether the electron leaps from the metal surface directly or

$$\mathrm{PhC{\equiv}C{-}Bu\text{-}}t + \mathrm{K^{\bullet}} \xrightarrow[-70\,^{\circ}\mathrm{C}]{\mathrm{THF}} \mathrm{PhCCBu\text{-}}t^{\dot{-}}, \mathrm{K^+} \qquad (80a)$$

$$\mathrm{M} + \mathrm{solvent} \rightleftharpoons \mathrm{M^+_{solv}} + \mathrm{e_{solv}} \qquad (80b)$$

travels through the solvent may be difficult to distinguish in many cases. In others, solutions of alkali and alkaline earth metals in ammonia, amines, ethers, hexamethylphosphortriamide, etc., are known and do, in fact, store solvated electrons (equation 80b). The 'structure' of the solvated electron is probably variable, from an electron in a solvent sheath, perhaps e_{NH_3}, to a solvated ion, e.g. $[((Me_2)N)_3PO]^{\dot{-}}_{HMPT}$ or $(HC{\equiv}CCOOEt)^{\dot{-}}_{NH_3}$ [64, 232, 233]. Since the standard electrode potentials of e_{solv} are $E^0(H_2O, 25\,^{\circ}C) = 2{\cdot}58$ V and $E^0(NH_3, -36\,^{\circ}C) = 1{\cdot}89$ V [237, 238] and its reactivity is high, such solutions are powerful reducing media (Table 9)[64, 237].

Ions and ion clusters may also donate electrons to the triple bond. Typical of aryl mono- and dianions are those of naphthalene and biphenyl (equation 81)[64, 235]. Not

$$\mathrm{Ar^{\dot{-}}, Na^+ + PhCCPh^{\dot{-}}, Na^+} \rightleftharpoons \mathrm{2Na^+, PhCCPh^{2-} + Ar} \qquad (81)$$

so typical is the finding that KCNS or KI interact with MeOOCC≡CCOOMe and HC≡COOR in DMF to give paramagnetic species; if the formation of a radical anion is valid, this would appear to be one of the simplest methods of generating it[239].

At first glance it appears that other reductants of alkynes such as Cr(II) might also function by nucleophilic electron transfer. In fact, an electrophilic process appears to be involved[7b]. On the other hand, there is some evidence that 'CuH' does start with electron transfer (equation 82)[240]. The fact that the *Z* isomer is the major product

$$\begin{array}{c}\mathrm{RC{\equiv}CR'}\\ + \\ [\mathrm{LiAlH(OCH_3)_3 + CuBr}]\end{array} \xrightarrow{\mathrm{THF}} (\mathrm{R\dot{C}{=}\bar{C}R'}) \xrightarrow{\mathrm{CuH}} (\mathrm{R\bar{C}{=}CHR')CuX} \xrightarrow{\mathrm{D_2O}} Z\text{-}\mathrm{RCD{=}CHR'} \qquad (82)$$

suggests predominant intramolecular control and delivery on one side of the face of the alkyne by the copper cluster.

TABLE 9. Electron transfer (non-electrochemical) to alkynes, RC≡CR′ [a]

R	R′	Reductant, solvent (°C) [b]	Product (yield, %) [c]	Reference
H	H	Na, C_2H_2–560 psig (~25)	NaC_2H (75), C_2H_4 (< 10), C_2H_6 (< 2), H_2 (< 1)	248*
Et	Et	Na, HMPT/THF (−33)	*E*-*n*-PrCH=CHMe (44), *E*-EtCH=CHEt (13), *Z*-*n*-PrCH=CHMe (17)	193*
Et	Et	Na, HMPT/THF–*t*-BuOH (25)	*E*-*n*-PrCH=CHMe (2), *E*-EtCH=CHEt (95)	193
Et	Et	Na, HMPT/THF–*t*-BuOH (25)	*E*-EtCH=CHEt (29), *Z*-EtCH=CHEt (5), *n*-C_6H_{14} (47)	193, 242*
n-$C_5H_{11}CT_2$	H	Na, NH_3, NH_4^+ (−33)	*n*-$C_5H_{11}CT_2CH{=}CH_2$ (89)	249
n-C_7H_{15}	C_2H_5	Na, HMPT/THF (r.t.)	*n*-$C_9H_{19}C{\equiv}CH$ (55)	243*
n-C_5H_{11}	*n*-Bu	Na, HMPT/PhH 2/15 (r.t.)	*n*-C_5H_{11}CH=CHBu-*n* (64)	243
t-Bu	Ph	K, THF (−70)	*E*-*t*-BuCH=CHPh (15), 4-(*t*-BuCH=CPh)C_6H_4CH=CHBu-*t*, and 4-(PhCH=CHBu-*t*)C_6H_4CH=CHBu-*t* (85)	236*
Ph	Ph	($B^{\bullet -}$, Na^+)[d] HMPT (< −78)	$A^{\bullet -}$, Na^+ (λ_{max} 450, 860 nm)	204*, 230
Ph	Ph	Na, THF (< −78)	A^{2-}, $2Na^+$ (λ_{max} 580 nm)	204, 230
Ph	Ph	Li, ether (r.t.)	$(PhC(Li){=}CPh)_2$	250
p-MeC_6H_4	Ph	Li, ether (0)	$(PhC(Li){=}CC_7H_7\text{-}p)_2$	251*, 228*
PhC≡C	Ph	Na, G (−70)	$A^{\bullet -}$, Na^+	231*, 229
2,4,6-(*t*-Bu)$_3C_6H_2$	2,4,6-(*t*-Bu)$_3C_6H_2$	K, THF/G (−80)	$A^{\bullet -}$, K^+ (λ_{max} 414, 470, 554 nm) A^{2-}, $2K^+$ (645 nm)	203*
Ph	$SiMe_3$	Na, THF (−80)	$A^{\bullet -}$, Na^+; $(PhC{=}CSiMe_3)_2^{2-}$, $2Na^+$	216
$MeNHCMe_2$	H	Na, NH_3/NH_4^+ (−33)	$MeNHCMe_2CH{=}CH_2$ (48)	185*
n-C_4H_9	$(CH_2)_6OH$	Li, NH_3/THF–EtOH (−33)	*E*-*n*-$C_4H_9CH{=}CH(CH_2)_6OH$	252
(*n*-Pr)$_2C{=}CH(CH_2)_2$	$(CH_2)_4OH$	Na, NH_3 (−33)	*E*-(*n*-Pr)$_2C{=}CH(CH_2)_2CH{=}CH(CH_2)_4OH$ (84)	253
$C_{10}H_{21}$	$(CH_2)_4$Pr-*i*	Na, NH_3, ether (−33)	*E*-$C_{10}H_{21}$CH=CHPr-*i* (82)	254
Ph	CN	($N^{\bullet -}$, Na^+)[e], THF (−78)	N(PhCCCN)$_n$H (M.W. 630–850)	255*
$HO(CH_2)_4CHNEt_2$	Et	Ca, NH_3/THF (60)	*E*-$HO(CH_2)_4CHN(Et)_2CH{=}CHEt$ (90)	256*
$MeOCH_2$	CH_2NMe_2	Na, NH_3/NH_4^+ (−33)	*E*-$MeOCH_2CH{=}CHNMe_2$	257*
n-$C_{12}H_{25}$	CH_2OH	Na, E/W 19/1 (~80)	*E*-*n*-$C_{12}H_{25}CH{=}CHCH_2OH$	258
n-Bu	$CH(OEt)_2$	Na, HMPT/PhH 3/1 (r.t.)	*n*-BuCH=$CHCH(OEt)_2$ (20), $HC{\equiv}C(CH_2)_4CH(OEt)_2$ (5)	243*
n-BuCHOH	$CH(OEt)_2$	Na, NH_3 (−33)	*E*-*n*-BuCHOHCH=$CHCH(OEt)_2$ (72)	259

TABLE 9 (*cont.*)

R	R′	Reductant, solvent (°C) [b]	Product (yield, %) [c]	Reference
$H_2C{=}CMe$	SMe	'CuH' ($NaCuAlHBr(OMe)_3$) (−60)	*Z*-$CH_2{=}CMeCH{=}CHSMe$ (95)	39
Me	COOMe	'CuH' ($NaCuAlH_2Br(OR)_2$) (−20)	$MeCH{=}CHCOOMe$ (*Z*/*E* = 29/25), *n*-PrCOOMe (18)	240*
EtOOC	H	e_{NH_3}, NH_3/Na^+ (−63)	$\dot{A}^-$, $H_2C{=}CH\dot{C}OOEt^-$	232*
^-OOC	COO^-	e_{H_2O}, H_2O (pH 14) (r.t.)	*E*-$^-OOCCH{=}\dot{C}COO^-$	260
Ph	COOEt	e_{NH_3}, NH_3/Na^+ (−63)	(*E*-$PhCH{=}CHCO_2^{2-}$)$^\cdot$	232*
Me	$(CH_2)_7COOH$	Na, NH_3 (−33)	*E*-$MeCH{=}CH(CH_2)_7COOH$ (75)	261
4-methyl-4,5-secocholest-3-yn-5-one		($\dot{N}^-$, Na^+), THF (r.t.)	(*Z*,*E*-3-ethylidene)-A-norcholestan-5β-ol (100)	246
$HOCH_2CH_2$	$CH_2CH_2CMe{=}CH_2$	Na, Et_2O/NH_3	*E*-$HOCH_2CH_2CH{=}CHCH_2CH_2CMe{=}CH_2$ (97)	262
t-$BuOCH_2$	CH_2OBu-*t*	Li, BuOH, NH_3/THF (−33)	*E*-*t*-$BuOCH_2CH{=}CHCH_2OBu$-*t* (70)	263

[a] The entries in this table are representative. Additional experimental details and/or leading references are indicated by the asterisk in the final column.

[b] See Table 7 (p. 845) for the solvent code. HMPT is $[(CH_3)_2N]_3PO$; r.t. is room temperature.

[c] $\dot{A}^-$ indicates that physical measurements were made on the anion.

[d] $\dot{B}^-$, Na^+ is $Ph_2^{\dot{-}}$, Na^+.

[e] $\dot{N}^-$, Na^+ is $C_{10}H_8^{\dot{-}}$, Na^+.

All of the above processes may be formally represented by Scheme 5 with the proviso that MC$\lessgtr$ or MC, the 'covalent' representation, stands for contact ions as

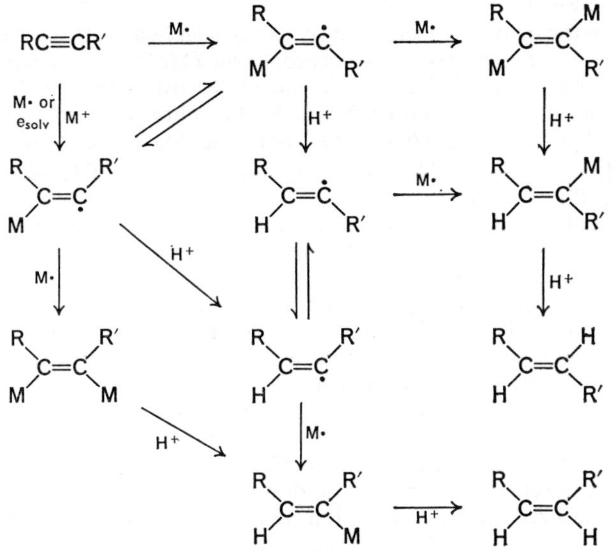

SCHEME 5

well as polymeric aggregates (equation 83)[64, 193]. That these species coexist is affirmed

$$(MC)_s \rightleftharpoons (MC)_{n,\text{solv}} \rightleftharpoons \ldots M^+ + C^- \quad (83)$$

by fairly detailed information about a few acetylenic mono- and dianions (equation 84), namely, $E_{\frac{1}{2}}$, ionization energy, lifetime, configuration, spectra (e.s.r., u.v.–visible), solubility and state of aggregation[64]. Some of the physical data are included in Table 9.

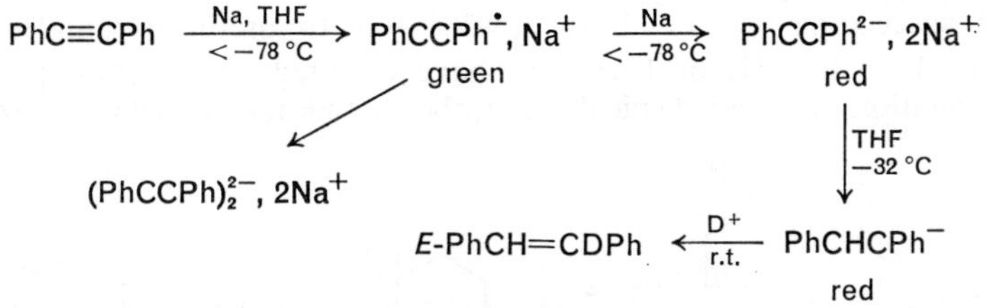

(84)

Most of the electron attacks eventually lead to *anti* addition of hydrogen to the triple bond. Therefore, it has often been assumed that a *trans* dianion is a necessary intermediate. Because both free dianions and dimetal adducts may be isolated or detected, either or both may be important in different systems.

In equation (84), for example, we note that these ions lead quite independent lives and that the monoreduced anion, which is relatively stable at ~25 °C, is the immediate precursor of the *trans*-stilbene[228]. Likewise, *t*-butylphenylacetylene leads to *t*-BuCCPh$^{\dot{-}}$, Ph$\bar{C}$=CHBu-*t*, Ph$\bar{C}$HCH$_2$Bu-*t*, PhCH$_2\bar{C}$HBu-*t*, four dimers, as well as *t*-BuCCPh^{2-} [236]. Furthermore, the reduction potentials of alkylacetylenes (see

Table 7) are probably lower than those for a solution of sodium, which is −2·96 V (*vs.* SCE at 28 °C) in hexamethylphosphoramide or *ca.* −2·3 V in ammonia (−33 °C); reduction presumably occurs because both Na^+, e_{solv} and/or $Na^{\bullet}$ add to the triple bond in such cases[193].

The necessity for involving metalated species is seen in a pretty case of specificities: Levin and coworkers observed that protonation (MeOH) of a slurry of reddish $PhCCPh^{2-}$, $2Li^+$ at −77 °C gives *cis*-stilbene while protonation of THF solutions of this material or of the disodium salt yields the more accessible *trans*-stilbene[204]. Under somewhat different conditions, both analogous and different species, $PhCCPh^{\dot{-}}$, M^+, $PhCCPh^{2-}$, $2M^+$, $(PhCH_2)_2$ and *cis,cis*-$(PhCH{=}CPh)_2$ have been prepared from PhC≡CPh [204].

The practice of selective reduction is still an art[227, 241]. Note the different products for EtC≡CEt and PhC≡CPh in Table 9 produced under different conditions. Now, Scheme 5 was devised to provide a broad rationale for diverse results in both named (Birch, Benkeser, Normant) and unnamed reductions. One learns that the presence of acids, i.e. NH_4^+ in Na–NH_3 or *t*-BuOH in Na–HMPT, and low reaction temperatures favour *anti* reduction[193, 227, 241, 242]. Proton donors usually preclude alkyne–allene rearrangement (equation 73) and/or rearrangement to terminal alkynes (acetylides) presumably by intercepting newly formed radical anions or metalated species[193, 243]. But if these intermediates do form, they appear to equilibrate rapidly in solution, favouring the more stable *trans* (in equation 85) and *cis* products (in equation 86)[193, 244]. In the absence of strong proton donors and at low temperatures, dimerization of the stabilized (aryl) radical ion may be competitive with formation of

$$Z\text{-EtCCl=CHEt} \xrightarrow[\text{HMPT/THF}]{\text{Na, 10 °C}} \text{EtCH=CHEt} \quad E/Z = 4/1 \qquad (85)$$

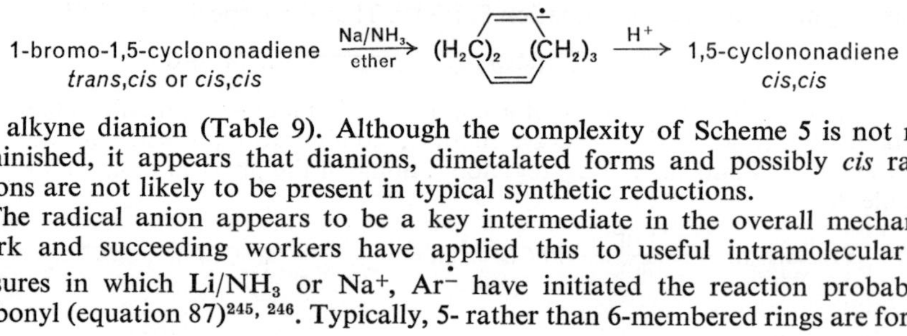

(86)

the alkyne dianion (Table 9). Although the complexity of Scheme 5 is not much diminished, it appears that dianions, dimetalated forms and possibly *cis* radical anions are not likely to be present in typical synthetic reductions.

The radical anion appears to be a key intermediate in the overall mechanism. Stork and succeeding workers have applied this to useful intramolecular ring closures in which Li/NH_3 or Na^+, $Ar^{\dot{-}}$ have initiated the reaction probably at carbonyl (equation 87)[245, 246]. Typically, 5- rather than 6-membered rings are formed,

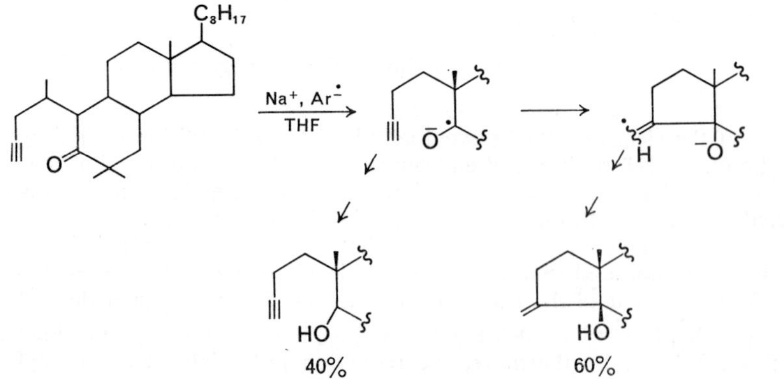

(87)

although the latter does happen[247]. Further, the A,B rings of the product are *cis* in about 10 examples; the *syn–anti* ratio across the triple bond varies (when suitable labelling substituents are present) with the metal (Na, K, Li), the solvent (THF, DME) and the temperature and appears to be determined by the interconversions of the cyclic radical anions of the type indicated in equation (87)[246].

C. Hydride Attack

In a comprehensive survey, main group hydrides of aluminium and possibly tin were regarded as reagents which could add H^- to an alkyne[264]. ('CuH' is treated as an electron donor in equation 82.) Thus, R_2AlH [265], R_2BH [266], R_3GeH [264], $LiAlH_4$–$Ni[(MeCO)_2CH_2]_2$ [267], etc., which are or appear to be non-nucleophilic, are barred. Tin hydrides, which often engage in radical processes, may, however, deliver H^- to an alkyne[1, 38, 264, 268]. Obviously, conventional catalytic reductions and the curious reductions of C_2H_2 by dispersed Fe [269] or by powdered metals (Cr, Fe, Zn) (equation 88) do not appear to be appropriate here[270]. Essentially, this leaves us with lithium

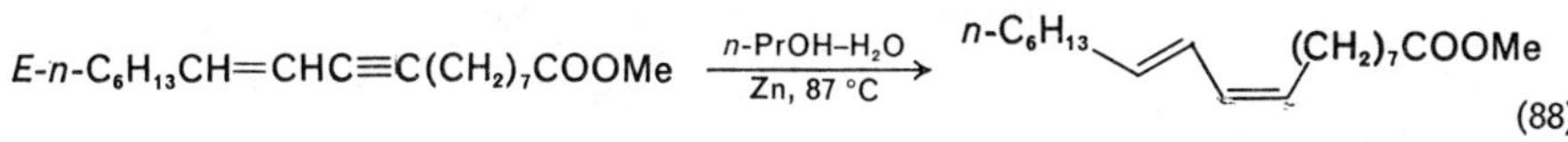

(88)

aluminium hydride (LAH) and some of its modified forms $[LiAl(OR)_3H]$ as donors of H^-.

In a thorough review, Pizey has covered the reactions of $LiAlH_4$ up to about 1970[23]. While it appears that most alkynes can be reduced, the presence of α-OH,

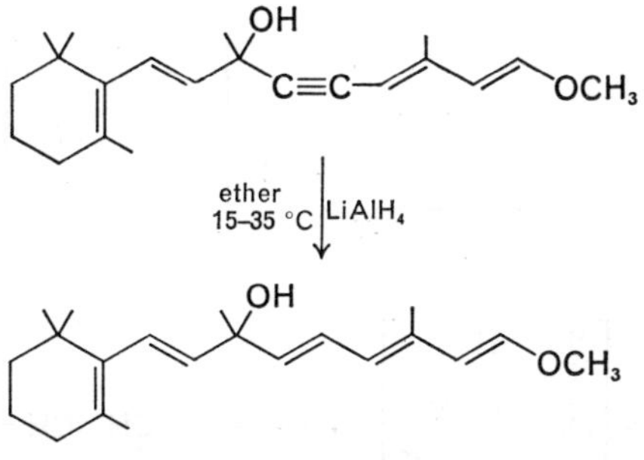

(89)

—OR, —COOR, etc. facilitates the process, e.g., equation (89). Kinetic data in Table 10 illustrate this activation and, incidentally, indicate how simple and convenient this reduction can be[271]. In Table 10 the first three compounds provide data for a rudimentary Hammett plot: $\rho \approx 1{\cdot}4$, that is, electron-withdrawing substituents on the ring increase the rate. The remaining entries show a fall-off in rate as the OH group becomes more distant from the triple bond: a β-OH and even a γ-OH appear to be rate-enhancing compared with no OH [271]. From Bohlmann's data, some of which are in Table 10, one can deduce the following order of activating substituents on the triple bond: PhC≡C > MeC≡C ≫ Ph > RCH=CH and $HOCH_2 > HO(CH_2)_2 > HO(CH_2)_3$ [271]. In a 4-phenylbut-3-yn-1-ol, the two substituents are not sufficiently activating to provide a measurable rate at 20 °C.

The mechanisms appropriate to the normal and α-activated processes are illustrated in equations (90) and (91)[272–274]. In these equations the transfer of H^- from LAH is regarded as rate-determining. The special role of an α-group, which can

TABLE 10. The reduction reaction in ether at 20 °C of equimolar lithium aluminium hydride and acetylenic alcohol (M = $7{\cdot}55 \times 10^{-3}$) [271]

Alkynol	$t_{\frac{1}{2}}$ (min)
$4\text{-MeC}_6H_4C{\equiv}CCH_2OH$	32
$PhC{\equiv}CCH_2OH$	5
$4\text{-ClC}_6H_4C{\equiv}CCH_2OH$	1·1
$C_6H_5C{\equiv}CCH_2CH_2OH$	∞
$C_6H_5(C{\equiv}C)_2CH_2OH$	0·022
$C_6H_5(C{\equiv}C)_2(CH_2)_2OH$	7
$C_6H_5(C{\equiv}C)_2(CH_2)_3OH$	392
$C_6H_5(C{\equiv}C)_2(CH_2)_2OTHP$ [a]	∞

[a] THP is 2-tetrahydropyranyl.

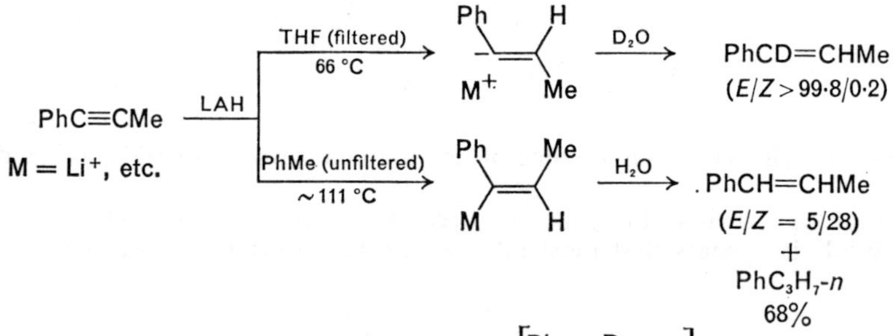

(90)

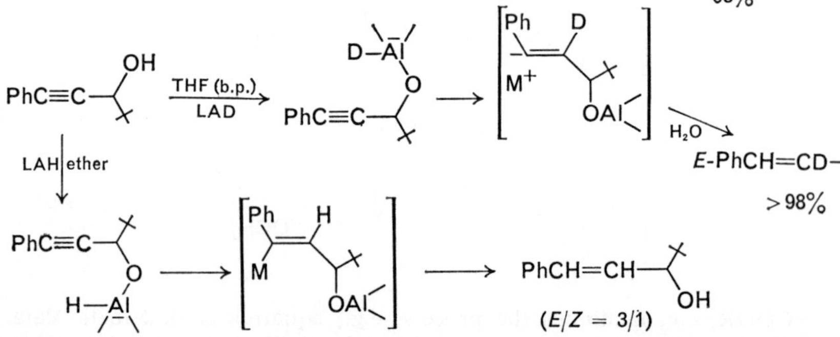

(91)

coordinate with aluminium, is indicated in equation (91). For 1-heptyn-3-ol, at least, the rate of reduction does not increase once the stoichiometric amount of LAH has been added; this is consistent with observed formation of an Al to ynol complex followed by slow intramolecular H^- transfer[273]. The organometallic complex seems to survive right to the final work-up stage in which H, I or AcO [257] may be introduced (Table 11). In an interesting case of complete reduction (equation 92), protonation of an intermediate containing a carbon–metal double bond has been proposed[275].

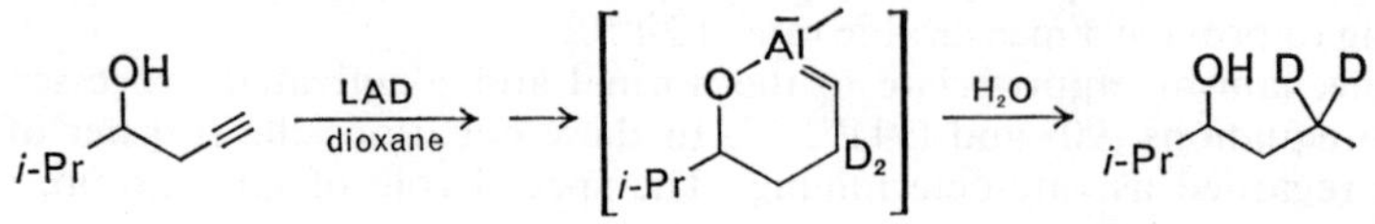

(92)

TABLE 11. Syntheses of alkenes and allenes by hydride transfer to alkynes, RC≡CR′ [a]

R	R′	Hydride[b]	Medium (°C)[c]	Product (yield, %)	Reference
Me	Et	LAH	THF (~130)	*E*-MeCH=CHEt (~100)	272
Me	Et	LAH	$PhCH_3$ (125)	MeCH=CHEt (*E*/*Z* = 1/1)[d], C_5H_{12} (97·6)	272
Me	Ph	LAH	THF (66)	*E*-MeCH=CHPh (>99·8)	272
Me	Ph	LAH	$PhCH_3$ (~111)	MeCH=CHPh (*E*/*Z* = 4·7/27·7)[d], PhC_3H_7-*n* (67·6)	272
Me	$CH_2CH_2C≡CMe$	$NaAlH_3OMe$	G (b.p.)	$MeCI=CHCH_2CH_2CI=CHCH_2OH$ [e]	287*
Me	CH_2OMe	LAH/$AlCl_3$	—	*E*-$MeCH=CHCH_2OMe$	243*, 279
H	$CH_2CHOHPr$-*i*	LAH	Dioxane (b.p.)	*i*-$PrCHOHCH_2CH_2CD_2H$ [f]	275
Me	17-(4-estren-17β-ol)	LAH	THF	17α-(*trans*)propenyl-4-estren-17β-ol (56), 21-methyl-19-nor-4,17,20-pregnatriene (32)	286
Ph	COOEt	LAD	Ether (r.t.)	*E*-$PhCH=CDCD_2OH$ (71)	288
t-Bu	SMe	LAH	THF (~60)	*E*-*t*-BuCH=CHSMe (92)	39
CH_2=CH	CH_2NEt_2	LAH/$AlCl_3$	Ether (36)	$CH_2=CHCH=CHCH_2NEt_2$ (70)	286*, 276
$HOCH_2$	CHOMePr-*n*	LAH	Et_2O (40)	*n*-$PrCH(OMe)CH=CHCH_2OH$	281
$HOCH_2$	$CH_2CH_2CMe=CHCH_2O$—THP[g]	LAH	THF (~55)	$HOCH_2CI=CHCH_2CH_2CMe=CHCH_2O$—THP(100)[g]	289
CH_2=CH	H	(*n*-Pr)$_3$SnH	THF (b.p.)	(*n*-Pr)$_3$SnCH=C=CHMe (38)	285
H	CH=CHCHClMe	LAH	Ether	$MeCH=CHCH=C=CH_2$ (85)	282*
Me_2C=C=CH	MeC=CMeCHOAcMe	$LiAlH_3OMe$	THF (r.t.)	MeCHOHCHMeCMe=C=CHCH=C=CMe_2(50)	284*, 290
R	Br	LAH/$AlCl_3$	—	*E*-RCH=CHBr (30–60)	291*
NC	CN	Bu_3SnH	MeOH (65)	*E*-NCCH=CHCN (93)	268*
H	Glucofuranosyl	LAH	THF (b.p.)	Glucofuranosylethene (83)	292–294*
n-C_5H_{11}	CH_2OH	LAH	—	*E*-$HOCH_2CH=CHC_5H_{11\text{-}n}$	295
Z-*n*-PrCH=CH	CH_2OH	LAH	—	*Z*,*E*-*n*-Pr$(CH=CH)_2CH_2OH$ (96)	296*
t-Bu	NO_2	LAH	Ether	*t*-$BuCH_2CH_2NH_2$	297
PhCO	COPh	LAH	THF (25)	PhCOCH=CHCOPh (18), $(PhCH(OH)CH_2)_2$ (21), PhCH(OH)C≡CCHOHPh (29)	298

[a] The entries in this table are representative. Additional experimental examples and/or leading examples are indicated by the asterisk in the last column.

[b] LAH is $LAlH_4$; LAD is $LAlD_4$.

[c] At the boiling point is b.p., room temperature is r.t. See Table 7 (p. 845) for solvent code.

[d] The figures making up the ratio are yields.

[e] Work-up with I_2.

[f] Work-up with D_2O.

[g] THP is 2-tetrahydropyranyl.

The labelling, LAH then D_2O or LAD then H_2O, establishes the points of hydride attack, metalation and proton–metal interchange. What does not easily lend itself to simplified mechanistic description are the effects of additives on LAH, e.g. $AlCl_3$ or MOR, which appear to alter—often invert—expected stereo- and regio-selectivities[276–278].

Although there are precedents for regioselectivity, rationalization of the results is often lacking. In equation (90), for example, the phenyl group obviously controls the direction of H^- entry[272]. Again, *anti* reductions of RC≡CSMe with LAH in THF are regiospecific, hydride entering α to MeS when R = Ph and β to MeS when R = alkyl[39]. However, one set of reducing conditions may yield low selectivity while another may be specific (equations 93 and 94)[277, 279]. Where an α-OĀl⋜ complex is

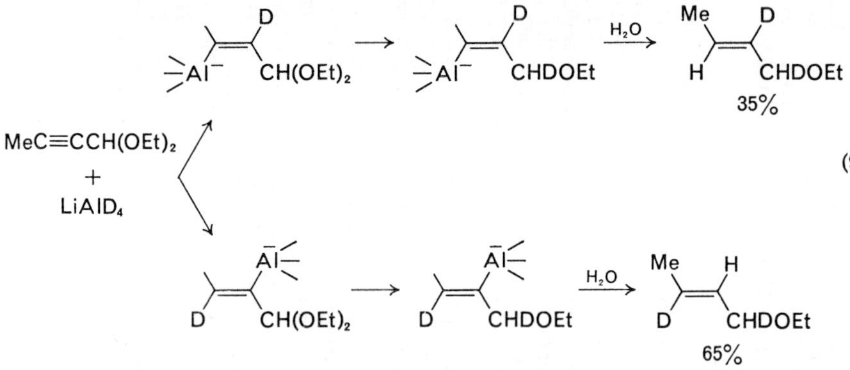

(93)

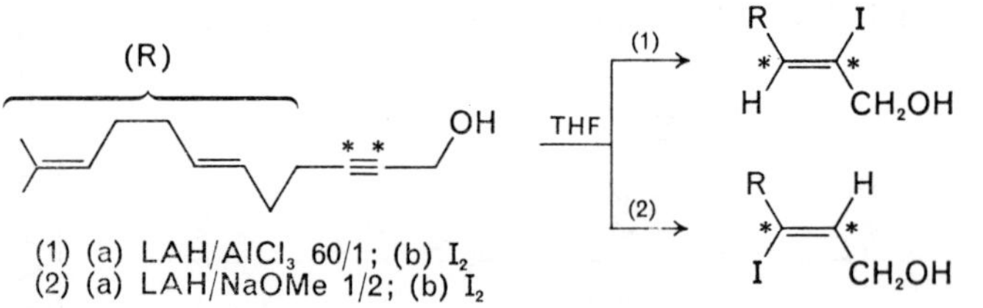

(94)

possible, it does appear that H^- *usually* enters at the nearest acetylenic carbon, e.g. equations (93) and (95)[274].

$$n\text{-}C_4H_9\text{—}C\equiv CCH_2OH \xrightarrow[(2)\ D_2O]{(1)\ LAH} E\text{- and/or } Z\text{-}n\text{-}C_4H_9CD{=}CHCH_2OH \qquad (95)$$

The stereoselectivity in equations (90) and (91) indicates, albeit in oversimplified form, the possible difference in *syn* and *anti* reductions. Process (90) is stereospecific in THF—only the *E*-alkene is produced; in toluene, both alkenes and the alkane are produced (see Table 11)[272]. Process (91) is highly selective, yielding 98% of the *E*-alkene in THF but yielding some of the *Z*-alkene, i.e. $E/Z = 3/1$, in ether[273]. Our interpretation of these results is that in the more polar solvent, THF, in which LAH is probably somewhat dissociated[264], normal ionic addition occurs; a coordinating metal ion (if any) and a final proton come in *anti* from the medium—hence the *ionic representation.* In the less polar solvents, ether and toluene, H^- and then metal (M) are delivered *syn* from associated LAH to one side of the alkyne—hence the *aggregate representation.*

The correlation of these mechanisms with solvent polarity (Lewis basicity) is strongly supported by a study of the solvent effect (% *E*) on the *E*/*Z* product ratio of equation (95) at 25 °C: dioxane (100), THF (100), THF + $AlCl_3$ (100), 2,5-dimethyltetrahydrofuran (55), Et_2O (60), $Et_2O + AlCl_3$ (60), $(i\text{-}Pr)_2O$ (25)[274]. The addition of a crown ether raised the yield in $i\text{-}Pr_2O$ to 70% *E*, presumably by facilitating dissociation and the ionic route; a drop in reaction temperature to −25 °C lowered the yield in Et_2O to 45% *E*, presumably by facilitating association and the aggregate route[274].

It is worth noting that LAH reduction of alkynes is only partly function-selective (Table 11). Generally, the alkene products or isolated carbon double bonds in the molecule are not reduced. On the other hand, carbonyl- or hydroxyl-related functions, e.g. ester, OAc, alkyl halide, acetal, etc., tend to end up as —OH or —OR, e.g. equation (93)[279].

Another diversion is to allenes by reductive 'displacement', often an S_N2' reaction in which an α-function is changed. These allene syntheses, of which there are many examples, come in two main forms. Equations (96) and (97) are variants of an overall

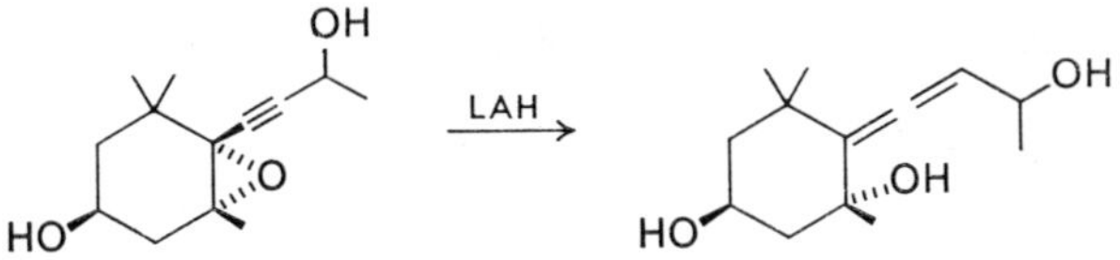

(96)

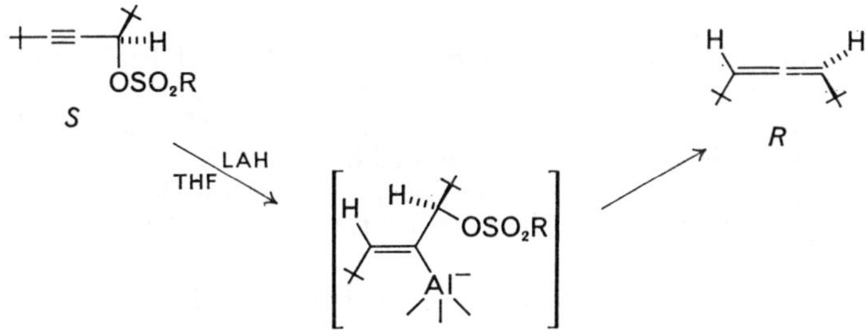

(97)

S_N2' reaction in which the leaving group may be halogen, RCOO, RO or R_3N [265, 276, 280–283].

In equation (96), H^- entry and RO^- departure are 1,3-*syn*, in accord with theoretical predictions and some experimental precedents[280]. Reaction (97) may proceed *via* an intermediate analogous to those written for equation (93); there is 1,2-*anti* addition followed by intramolecular elimination, giving an overall *anti* entry/departure.

The second route to allenes is from enynes (equation 98)[7b, 261, 282, 284–286]. Variations in the placement of the electronegative group, the triple and double bonds as well as

$$R'C{\equiv}CCH{=}CHCH_2OTHP \xrightarrow[\text{(2) HCl, 38 °C}]{\text{(1) LiAlH}_4\text{, ether}} R'CH{=}C{=}CHCH_2CH_2OH \qquad (98)$$

THP = 2-tetrahydropyranyl

$R' = n\text{-}C_{11}H_{23}$ (63%)

$R' = MeCH{=}CH(CH_2)_8$ (81%)

in their structural environment have been examined, e.g. equation (99). Among these are examples in which the hydride may be delivered either to the double or the triple bond[282, 284].

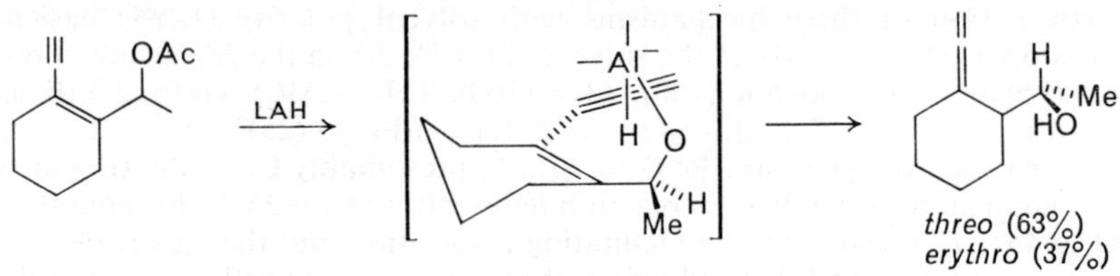

(99)

D. Halide

As indicated in Table 3 of the previous review[1], most halide attacks required highly activated alkynes and were usually associated with kinetic studies of addition (equation 100). The vinyl anion intermediate (V^-) then reacts rapidly with a protic

$$\underset{(\mathbf{A})}{R'C{\equiv}CR+X^-} \longrightarrow \underset{(\mathbf{V}^-)}{R'XC{=}CR^-} \xrightarrow{E^+} R'XC{=}CRE \qquad (100)$$

$$\begin{array}{ccccccc} (\mathbf{V}^-) & \xrightarrow{A} & X(R'C{=}CR)_2^- & \xrightarrow{A} & X(R'C{=}CR)_3^- & \longrightarrow \;\longrightarrow & X(R'C{=}CR)_n^- \\ & & \downarrow E^+ & & \downarrow E^+ & & \downarrow E^+ \end{array}$$

acid or some other electrophile. In this manner, HX (X = Cl, Br, I) and Cl_2 have been added across the triple bond (Table 12). In the newer developments, V^- has been recognized as a potential nucleophile and has been captured in novel and useful ways. Much of the work in this area is represented in Table 12.

The Kiev group is still one of the few providing kinetic data on nucleophilic addition to alkynes[87, 88, 299–303]. Their general rate law (equation 101) is adaptable to

$$-d[A]/dt = (k_2+k_3[E])\,[A]\,[X^-] \qquad (101)$$

several situations (**A** is the alkyne, X^- is halide ion and **E** is an electrophile, e.g. halogen or H^+)[88]. In other sections we cite analogous work with $X^- = AcO^-$ or SCN^- and E = another acetylene. In practice, equation (101) usually has to be modified, e.g. for non-polar solvents, when the participants are involved in complexation, equilibria, etc.[87, 88] One of these variants is discussed in connection with AcO^- [304].

An example of equation (101) with $X^- = I^-$ and E = HI was found for methanol–DMF solutions; both second- and third-order terms were retained[88]. Dvorko's group also found that when $X^- = Cl^-$, a weak nucleophile, the formation of V^- in equation (100) was slow and subsequent protonation was fast, leading to second-order kinetics ($k_3 \simeq 0$)[88, 302]; when $X^- = I^-$ or SCN^-, the formation of V^- was often fast and the protonation slow, leading to third-order kinetics ($k_2 = 0$)[300]. Examples of solvent effects superimposed on the basic rate laws are found in Figure 9[88], and even more complex effects are found when chlorobenzene, toluene, chlorobenzene–THF mixtures, etc. were used[87, 299].

The promotion of polymerization of certain alkynes by nucleophiles is another Kiev contribution of the late 1960s: polypropiolanhydride of molecular weight 12000 could be produced[304]; elsewhere (see Section III.F) are mentioned other anions that promote both trimerization and polymerization of methyl propiolate[305].

TABLE 12. Halide-initiated attacks on alkynes, RC≡CR′ [a]

R	R′	Nu⁻	Medium (temp., °C); coreactant(s)	Product(s) (yield, %); comments[b]	Reference
H	COOMe	HI	PhCl (25)	IHC=CHCOOMe; $k \simeq 0{\cdot}06$ [c]	87*
MeOOC	COOMe	HI	$PhCH_3$ (30)	MeOOCIC=CHCOOMe; $k \simeq 0{\cdot}48$ [c]	299*
H	COOMe	I^-	DMF (60); HOAc	IHC=CHCOOMe; $k = 9 \times 10^{-5}$ [d]	300*
EtOOC	COOEt	I^-	EtOH, ArCOOH	EtOOCIC=CHCOOEt; $E_a = 12{\cdot}7$, $\Delta S^{\ddagger} = -36{\cdot}9$	301*
EtOOC	COOEt	Cl^-	MeOH (50); 2,4-$(O_2N)_2C_6H_3COOH$	EtOOCClC=CHCOOEt; $k = 9 \times 10^{-5}$	88*
$ClCH_2CH_2OOC$	$COOCH_2CH_2Cl$	Cl^-	2-PrOH (50); 2,4-$(O_2N)_2C_6H_3COOH$	RClC=CHCOOR; $k = 1{\cdot}3 \times 10^{-3}$	302*
MeOOC	COOMe	I^-	PhCl (~25); I_2	MeOOCCI=CICOOMe	299
H	COOCOC≡CH	Br^-	DMF (70)	Polypropiolanhydride (65); M.W. = 12 000	304*
H	$CH_2(n\text{-}Pr)_2NH^+$	HCl_2^-	None (80)	$(n\text{-}Pr)_2\overset{+}{N}HCH_2CCl{=}CH_2$ (~25), Z-$(n\text{-}Pr)_2\overset{+}{N}HCH_2CH{=}CHCl$ (~50)[e]	311*
$BF_4^-PhI^+$	Ph	Cl^-	H_2O (r.t.)	Z-PhCCl=CHIPh$^+BF_4^-$ (50)	312
F_3C	CF_3	CsF	CH_3CN (25)	E-$F_3CCCs{=}CFCF_3$, polyhexafluorobutyne	308, 306*
F_3C	CF_3	CsF	F_5C_6CN	$F_3CCF{=}CCF_3(C_6F_4CN\text{-}p)$, polyhexafluorobutyne	309*
EtOOC	COOEt	CsF	$(CH_2)_4SO_2$ (100)	EtOOCCF=CHCOOEt	307*
F_3C	CF_3	CsF	$(CH_2)_4SO_2$ (20); 1,3-$C_4N_2F_4$	4-(1,3-$C_4N_2F_4$)$CCF_3{=}CFCF_3$ (70, $E/Z \simeq 9$)	310*
PhCO	H	HCl	Ether; $CHCl_3$ (−40)	Z-ArCOCH=CHCl (40)	313

[a] This table updates the entries of Tables 3 and 4 of Reference 1. The availability of additional results in a cited paper is indicated by an asterisk in the last column.

[b] For the rate data, the units are: for k, $M^{-1}\,s^{-1}$, unless otherwise noted; for E_a, kcal/mol; for $\Delta S^{\ddagger}$, e.u.

[c] Rate law is complex.

[d] k in $M^{-2}\,s^{-1}$.

[e] The authors suggest an electrophilic mechanism[8].

In the same period, fluorine chemists discovered such processes independently and have also been able to obtain both low and high molecular weight products. First,

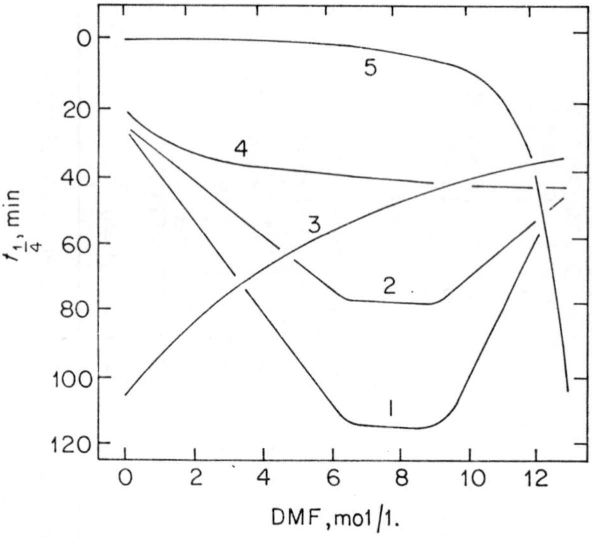

FIGURE 9. The effect of added dimethylformamide on the rates of addition of HX to MeOOCC≡CCOOMe in the presence of 0·5M added ArCOOH: $t_{\frac{1}{4}}$ is a quarter-life, Q is quinoline. The rate data derive from equation (101)[88]. Curve 1: I^-(KI), PhCOOH in MeOH at 40 °C. Curve 2: I^-(HI), 2,4-$(O_2N)_2C_6H_3COOH$ in MeOH at 40 °C. Curve 3: Cl^-(QHCl), PhCOOH in 2-PrOH at 50 °C. Curve 4: I^-(QHI), 2,4-$(O_2N)_2C_6H_3COOH$ in 2-PrOH at 40 °C. Curve 5: CNS^-(HCNS), 2,4-$(O_2N)_2C_6H_3COOH$ in 2-PrOH at 20 °C.

there was the surprising (at the time) addition of equation (102)[306]. Hexafluorobut-2-yne (A) is a strong electrophile and polymerizes in stages according to equation (100); up to three vinyl anions were trapped and the products isolated in subsequent

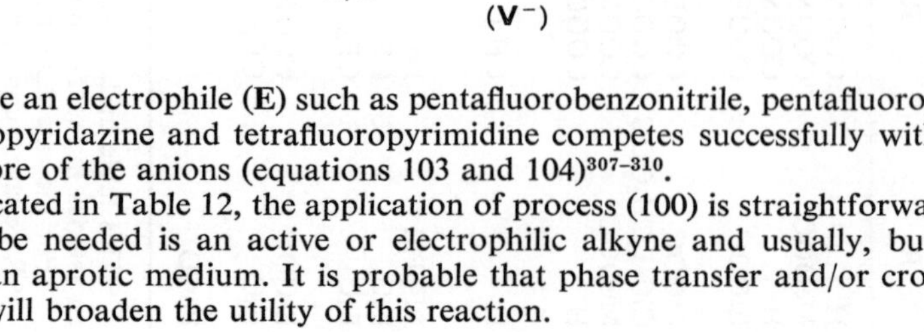

(102)

work. Here an electrophile (E) such as pentafluorobenzonitrile, pentafluoropyridine, tetrafluoropyridazine and tetrafluoropyrimidine competes successfully with V^- for one or more of the anions (equations 103 and 104)[307–310].

As indicated in Table 12, the application of process (100) is straightforward; what seems to be needed is an active or electrophilic alkyne and usually, but not inevitably, an aprotic medium. It is probable that phase transfer and/or crown ether catalysis will broaden the utility of this reaction.

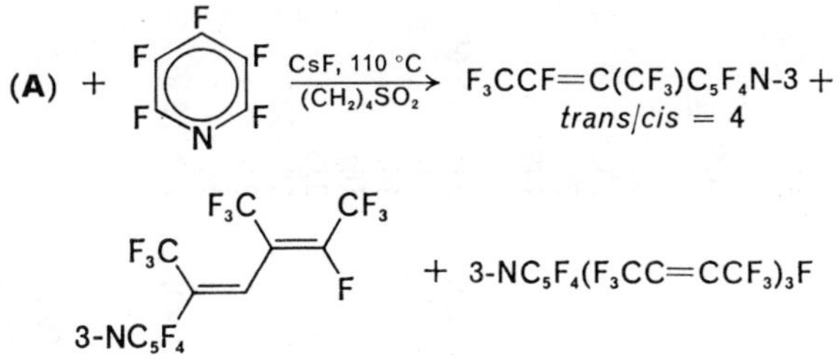

(103)

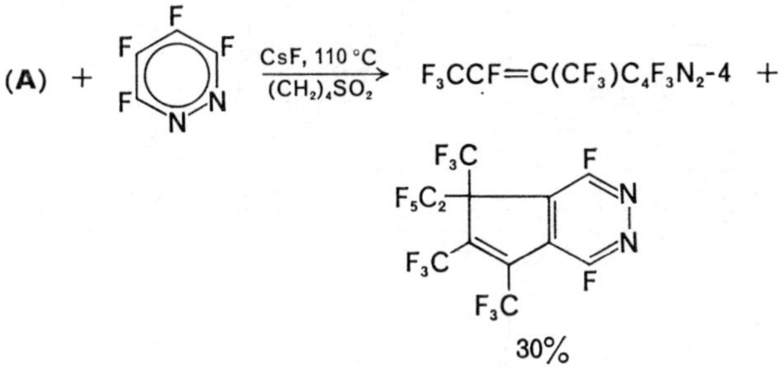

(104)

E. Oxygen

1. Acyclic additions

Perhaps the most important applications of equation (1) involve C_2H_2 and alcohols in Reppe vinylation[314, 315]. It is not surprising, therefore, that such systems and their products have been studied repeatedly[1, 5, 6, 42, 316–318]. For accessibility, ease of reaction and synthetic exploration, however, the acetylenedicarboxylic esters have been the favourite electrophiles[9, 10].

Elsewhere in this chapter we have used oxygen nucleophilic processes to illustrate stereoselectivity (Sections II.B and II.C), coelectrophiles (Section I.A), etc. and shall not repeat these here. The literature of oxygen nucleophile attacks on alkynes includes trends in substituent effects but the data are often qualitative and usually scattered. Numerous examples indicate that the reactions of alkoxides with C_2H_2 are relatively slow and that most substituents facilitate the addition (Tables 13 and 14):

TABLE 13. Rate data for the addition of oxygen nucleophiles to alkynes

Alkyne	RO^-	Solvent[a]	Temp. (°C)	k ($M^{-1}\,s^{-1}$)	$\Delta H^{\ddagger}$ (kcal/mol)	$\Delta S^{\ddagger}$	Ref.
HC≡CH	KOH	EtOH	~155	—	36	—	327
$(HC{\equiv}C)_2$	EtO^-	E/D	50·0	$9{\cdot}3\times10^{-3}$	25·2	1	91
$(HC{\equiv}C)_2$	EtO^-	EtOH	50·0	$5{\cdot}5\times10^{-5}$	24·9	0	91
$(HC{\equiv}C)_2$	EtO^-	E/H	50·0	$3{\cdot}7\times10^{-5}$	15·7	−30	91
PhC≡CCl	MeO^-	MeOH	78	1×10^{-4}	—	—	153
PhC≡CH	MeO^-	MeOH	126	$3{\cdot}7\times10^{-4}$	28·5	−8	328
MeOOCC≡CCOOMe	LiOAc	DMF[b]	100	$\sim2\times10^{-4}$ [c]	—	—	305

[a] E = EtOH, D = dioxane, H = *n*-heptane.
[b] HOAc is present.
[c] This $k = k'$[LiOAc]/[HOAc].

TABLE 14. Additions of oxygen nucleophiles to alkynes, RC≡CR′ to give acyclic products[a]

R	R′	Nu	Medium (temp., °C)	Product(s) (yield, %)[b]	Reference
MeOOC	COOMe	*o*-$C_6H_4(OH)CONH_2$	MeO^-, MeOH (b.p.)	*Z*-2-$H_2NCOC_6H_4O(MeOOC)C=CHCOOMe$ (>46)	330*
H	CF_3	MeO^-	MeOH (r.t.)	*E*-$CF_3CH=CHOMe$ (1·8), $CH_2=CCF_3(OMe)$ (1·5)	123*
H	COOMe	AcO^-	DMF, HOAc (100)	*Z*-MeOOCCH=$CHOOCCH_3$ (7), 1,3,5-$(MeOOC)_3C_6H_3$ (93), *k*	303*
H	COOMe	PhO^-	DMF, HOAc (0–100)	*Z*-MeOOCCH=CHOPh, $H(MeOOCC=CH)_nOPh$	305*
MeCO	*n*-Bu	MeO^-	MeOH (5)	*n*-BuC(OMe)=CHCOMe (72)	331*
$(CH_3)_2N$	Me	H_2O	CH_2Cl_2 (b.p.)	$MeCOCH_2CON(Me)_2$ (88)	332*
H	COOMe	MeOH	CH_3OH, Bu_3SnOMe	MeOCH=CHCOOMe (*Z*/*E* > 1)	268*
p-$O_2NC_6H_4$	SO_2Me	EtO^-, MeS^-	EtONa, EtOH/DMF 1/1 (r.t.)	*E*-ArCSMe=CHSMe (30), $ArCH_2COOEt$ (79)	149*
H	H	$(HOCH_2CH_2)_2S$	5% KOH, 16 atm (140)	$H_2C=CHOCH_2CH_2SCH_2CH_2OH$, etc.	318*, 5*, 6*
H	H	$C_{18}H_{37}OH$	85% KOH, MeOH (180)	$CH_2=CHOMe$ (70), $C_{18}H_{37}OCH=CH_2$	316
Me_2N	COMe	MeOH	*t*-BuOK (b.p.)	$Me_2N(MeO)C=CHCOMe$ (81)	143*
t-Bu	Cl	PhO^-Na^+	DMF, MeOH (180)	PhO(*t*-Bu)C=CHCl (32)	165
MeOOC	COOMe	HpNOH[c]	DMSO (r.t.)	HpNOC(COOMe)=CHCOOMe (~20)[c]	333*
H	COOMe	MeOH	Et_3N	MeOCH=CHCOOMe (*E*/*Z* = 7/3)	323–325
Me	COC≡CMe	EtO^-Na^+	EtOH (b.p.)	$(MeC(OEt)=CH)_2CO$	334*
Ph	CN	$C_5H_5N^+O^-$	$(ClCH_2)_2$ (b.p.)	$(2-C_5H_4N)CH(CN)COPh$, $(3-C_5H_4NH^+)C(CN)=C(Ph)O^-$ (56), *Z,E*-$(3-C_5H_4NH^+)C(CN)=C(Ph)C(CN)=C(Ph)O^-$ (~12)	141*, 335*, 336
p-$C_7H_7SO_2$	Ph	*t*-BuNHOH	EtOH (r.t.)	*p*-$C_7H_7SO_2CH_2C(Ph)=\overset{+}{N}(t\text{-Bu})\overset{-}{O}$ (56), *Z*-*p*-$C_7H_7SO_2CH=C(Ph)ONHBu$-*t* (38)	323
p-C_7H_7CO	H	PhOH	Et_3N, PhH (b.p.)	*E*-PhOCH=$CHCOC_7H_7$-*p* (80)	337*
$(n\text{-Bu})_2P(O)$	H	MeO^-	MeOH (20)	$(n\text{-Bu})_2P(O)CH=CHOMe$ (*Z*, *E*)	338*
Ph	Br	MeO^-	MeOH	PhC(OMe)=CHBr, PhCH=CBr(OMe), *k*	154*
R	H	MeOH	KOH, NH_3 (−33)	2-(*N*-methylbenzimidazolyl)vinyl methyl ether (72)	339*

MeSe	HC=CH_2	MeO^-Na^+	MeOH (120)	MeSeCH=C(OMe)CH=CH_2 (12), MeSe(MeO)C=CHCH=CH_2 (24)	340*
H	H	H_2O	KOH, DMSO–H_2O (130)	*Z*-H_2C=CHOCH=$CHCH_3$ (2)	42
H	H	$(4\text{-}HOC_6H_4)_2CH_2$	KOH (180)	$CH_2(C_6H_4(OCH=CH_2)\text{-}p)_2$ (60)	317*, 5
HC≡C	H	*o*-$HOC_6H_4NH_2$	KOH, dioxane, Me_2SO (100)	*o*-$H_2NC_6H_4OCH$=CHC≡CH (59, Z/E = 79/21)	126*
PhCO	Br	PhOH	K_2CO_3, Me_2CO (r.t.)	PhCOCH=$C(OPh)_2$ (94)	337*, 341, 32
HC≡C	H	EtOH	Various (50–75)	*k*	91, 342
PhC≡CCO	Ph	PhO^-Na^+	EtOH (r.t.)	PhC(OPh)=CHCOC≡CPH (45)	75*, 343*
i-Bu	H	*t*-BuOK	*t*-BuOH (225)	*i*-BuC(OBu-*t*)=CH_2 (51), *i*-BuCH=CHOBu-*t* (<1)	118*
PhC≡C	H	EtO^-	KOH, EtOH (70)	PhC≡CCH=CHOEt (70)	344*, 5
Me	H	*n*-$C_{18}H_{37}OH$	KOH, 520 p.s.i. (145)	*n*-$C_{18}H_{37}OC(Me)$=CH_2	345*
MeOOC	COOMe	ArOH	Me_2CO (r.t.)	2,6-$ClC_6H_3OC(COOMe)$=CHCOOMe (70, Z/E = 49/51)	321*, 5, 346
$(CF_3)_2N$	H	MeONa	MeOH (100)	$(CF_3)_2NCOMe$=CH_2 (56), *E*-$(CF_3)_2NCH$=CHOMe (28)	137
$(CF_3)_2N$	CF_3	MeONa	MeOH (r.t.)	*Z*-$(CF_3)_2NCOMe$=$CHCF_3$ (85), *Z*-$(CF_3)_2NCH$=$CCF_3(OMe)$ (4)	142
1-(*c*-$C_6H_{10}OH$)	Br	$(i\text{-}PrO)_3P$	EtOH	$(i\text{-}PrO)_2POCBr$=$CH(1\text{-}C_6H_{10}OH\text{-}c)$	163a
Et_2N	R	$HOCR^1R^2C≡CR^3$	—	$Et_2NCOCHRC(R^3)$=CR^1R^2	329*

[a] Additional examples of a similar type and leading references are indicated by an asterisk in the last column.
[b] *k* indicates that rate data are available.
[c] Hp is dihydroisoxazole.

of these, electron-withdrawing substituents, e.g. Cl_5C_6, CF_3, RCO, RC≡C, etc. are most effective (Table 3; equations 43 and 105)[1, 10, 14, 15]. For example, the relative rates of addition of MeOH in THF at 37 °C are given for R in $Me_2NC{\equiv}CCOR$ as H ≫ Me > OMe[143]. In fact, with a strong activating substituent, the addition may become a complication in the synthesis of an alkyne (equation 105)[319]. Finally, the

$$(m\text{-}CF_3C_6H_4C{\equiv}CH) \xrightarrow[\text{EtOH}]{\text{KOH}} m\text{-}CF_3C_6H_4CH{=}CHOEt \qquad (105)$$

reader should recall that a dipolar aprotic solvent can enhance the reactivity of alkoxide, e.g. $RO^- + ArC{\equiv}CCl$ in DMSO or HMPT *vs.* ROH (see Section IV.B.1) or $EtO^- + (HC{\equiv}C)_2$ in dioxane (Table 13)[91].

Alkynes may be reactive because they are strained; this arises in connection with cycloalkynes (equation 106)[320] which are dealt with in another chapter of this

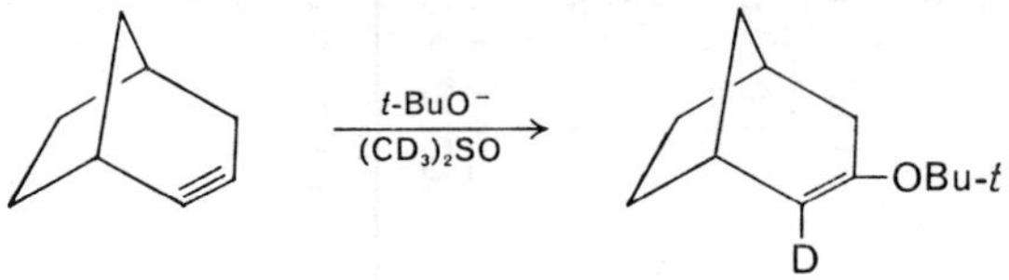

(106)

volume. However, normal steric factors are apparent in more typical situations. The changing regioselectivity in the addition of ROH (R = *t*-Bu, Me) to R′C≡CH (R′ = alkyl) has been taken up in relation to Table 2. Note the influence of electronic and steric effects in equation (107) where *t*-BuOH is unreactive under the conditions

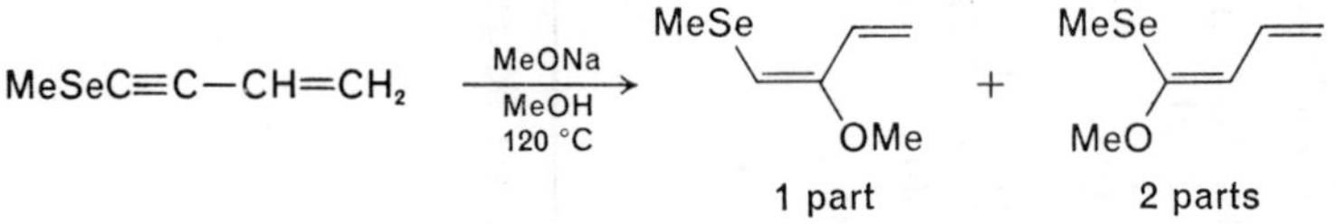

(107)

given and *i*-PrOH and *i*-BuOH give adducts at the 3- and 4-carbons. As for *syn vs. anti* selectivity, a study of process (108) indicated that exclusive *anti* addition prevails

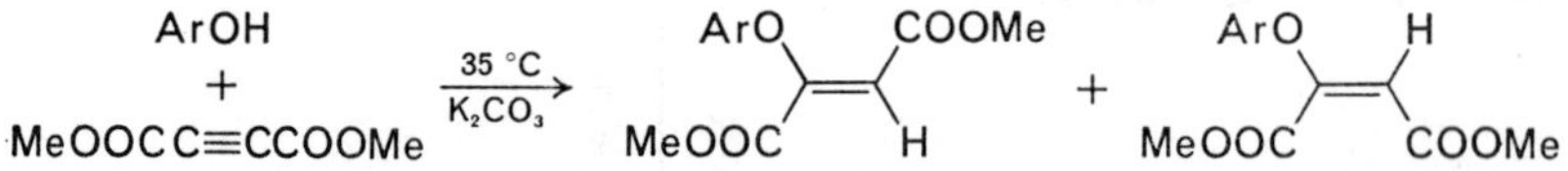

(108)

for $p\text{-}BrC_6H_4OH$ in Me_2CO and MeCN[321]. It is not just that the fumarate is more stable, since the amount of fumarate in other solvents does drop: PhH (80%), THF (65%) and dioxane (35%)[321]. Slower proton delivery may account for the isomerization of V^- prior to product formation. In this process, steric effects become evident in the *ortho*-substituted phenols in which the fumarate is the dominant but not exclusive product: *o*-Me (79%), *o*-(*i*-Pr) (73%) and 2,6-Cl_2 (49%)[75, 319].

Diethynylketones appear to follow the chemistry set out in equation (109). The relative amounts of mono- or diadduct and γ-pyrone may, however, vary with reagents (and perhaps the presence of water and acid). While thiolates give comparable products, H_2S (from thiourea), surprisingly, produces a thiocyclopentenone[178].

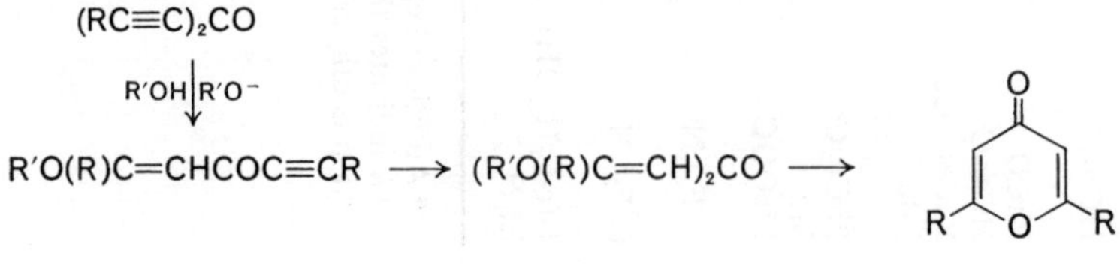

(109)

Examples of furan synthesis by carbon attack from diacyl sulphonium methylides are given elsewhere (see Section III.H)[322]. Here we illustrate an unusual pattern in which the cyclic diacyl sulphonium methylide appears to lead with O^-, e.g. equation (110)[322].

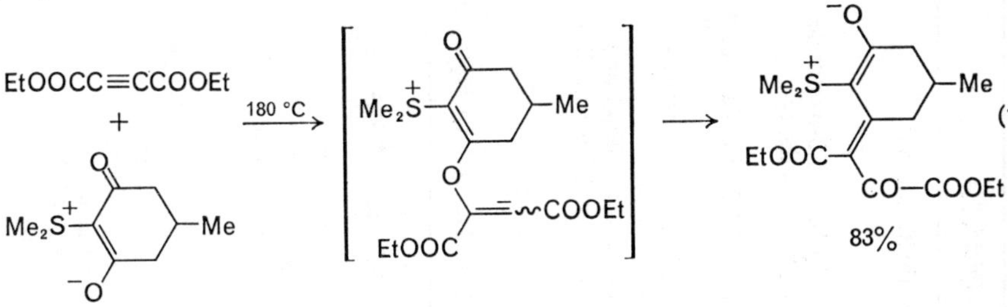

(110)

Hydroxylamines are also binucleophiles with respect to alkynes[323]. Nitrogen attack is usually favoured[10, 324, 325], although oxygen attack is also found (equation 111)[323]. The conditions in equation (111) are sufficiently mild that Winterfeldt's isomerization

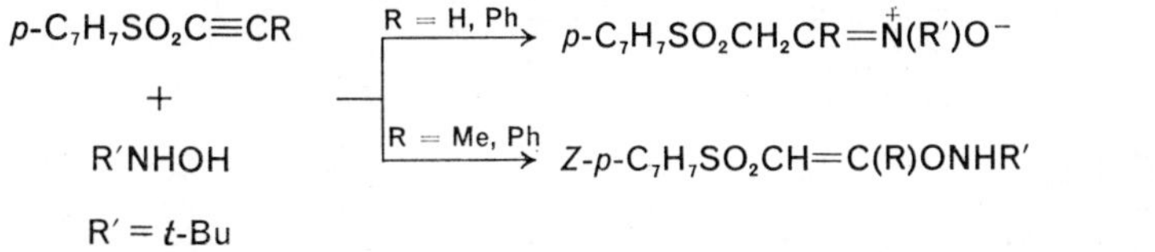

(111)

(equation 112) is improbable[324]. Again, oxygen attack on alkyne seems certain to give one of the products of process (113)[326].

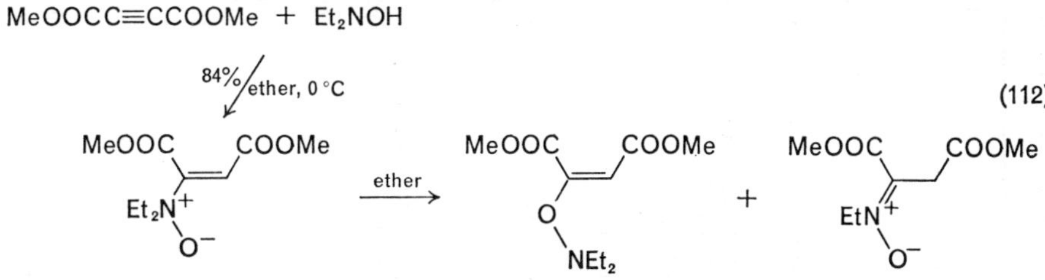

(112)

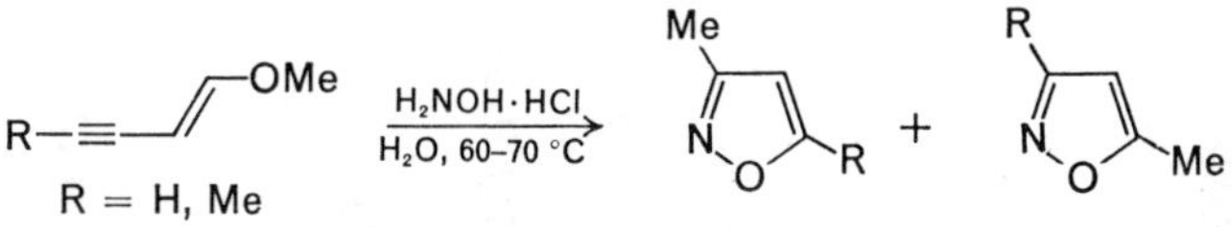

(113)

2. Cyclic adducts

It is intended that the heterocyclic compounds that are collected here should derive from products of process (1). Those of interest are variations of equation (1) which lead to cyclic products in at least two steps rather than from cycloadditions. As was evident in the last section, both acyclic and cyclic examples are often conveniently discussed together. Only a few representatives of the possible structures, i.e. with different combination of atoms of oxygen, sulphur, nitrogen, etc., can be mentioned.

Furan and related cycles (methylenetetrahydrofuran, methylenedihydrofuran) are usually formed in preference to pyrans from alkynes (Table 15)[6a, 347–349]. Processes

TABLE 15. Additions of oxygen nucleophiles to alkynes, $RC{\equiv}CR'$, to give cyclic products[a]

R	R′	Nu	Medium (temp., °C)	Product(s) (yield, %)	Ref.
$HOCH_2CH{=}CMe$	HCMeOMe	—	*t*-BuOH, *t*-BuOK, Me_2SO (35°, 0·5 h)	2-(1′-Propenyl)-3-methylfuran (47)	348*, 349, 186
CH_3COS	*t*-Bu	H_2O	MeOH (r.t.)	2-Methyl-2-methoxy-4-*t*-butyl-1,3-oxathiole	358*
EtOOC	COOEt	*o*-$C_6H_4(OH)_2$	MeONa/MeOH (b.p.)	2-Carboethoxymethyl-2-carboethoxy-benzodioxole (78)	352*, 359
H	Cl	$MeCHOHCH_2$-CH_2OH	KOH/EtOH, N_2, 20 atm (120)	2-Chloromethyl-4-methyl-1,3-dioxolane (47)	184, 331*
CH_3OOC	$COOCH_3$	2-$HOC_6H_4CONHPh$	NaOMe/MeOH (r.t.)	2-Carbomethoxy-2-carbomethoxymethyl-*N*-phenyl-benzoxazinone (93)	360*, 330
Ph	COOEt	$C_6H_{11}NOH$	NaOH/EtOH (hot)	2-Cyclohexyl-5-phenyl-4-isoxazolin-3-one (19), 2-Cyclohexyl-3-phenyl-3-isoxazolin-5-one (17)	361*
(2-C_4H_3O)CO	COOEt	H_2NOH	NaOEt, EtOH (r.t.)	5-(2′-Furyl)-3-carboethoxyisoxazole (72)	362*, 363, 343
$H(C{\equiv}C)_2$	$(CHOH)_2CH_2OH$	—	OH^-, H_2O, N_2 (20)	$H(C{\equiv}C)_2CH{=}CCH(OH)CH(OH)CH_2O$ (ring)	349, 353*
$PhC{\equiv}C$	COPh	$HN(CH_2CH_2OH)_2$	96% EtOH (r.t.)	2,6-Diphenyl-γ-pyrone	364*, 75
H	COOMe	PhCONHOH	NaH, Me_2SO (r.t.)	2-Carboethoxymethyl-5-phenyl-1,3,4-dioxazole (45)	355
Ph_2P	Ph	$BrCH_2COMe$	HMPT	2-Methyl-4,4,6-triphenyl-1-oxa-4-phosphonia-2,5-cyclohexadiene (76)	191
$PhCOCH_2P(O)Ph$	Ph	—	EtOH, MeO^-Na^+ (r.t.)	4-Oxo-2,4,6-triphenyl-1-oxa-4-phosphorin (86)	365
o-HOC_6H_4	$C{\equiv}CC(OH)R_2$	—	Base	2-($R_2C(OH)C{\equiv}C$)benzofuran	6a
MeOOC	COOMe	*o*-HOC_6H_4CHO	PhH, K_2CO_3 (r.t.)	2,3-Dicarbomethoxychrom-2-en-4-ol (22), dimethyl *o*-formylphenoxyfumarate (14)	346
H	CH_2Br	$MeN(CH_2CH_2OH)_2$	(0)	2-Methylene-4-methylmorpholine (9), 2,4-dimethyl-1,2-oxazine (12)	366
$MeOCH{=}CH$	H	$H_2NOH \cdot HCl$	H_2O (~70)	(3- and 5-)Methylisoxazoles (60)	326
PhCO	COPh	PhCHOHCOPh	Me_2CO, K_2CO_3 (b.p.)	2,3-Dibenzoyl-4,5-dihydro-diphenyl-4-hydroxyfuran (39)	367

$MeCO(CH_2)_2$	H	$MeOOC(CH_2)_2$-CN^+O^-	Et_3N, PhH, PhNC	3,5-Dicarboalkoxyisoxazole (80), $(PhNH)_2CO$	357*
Ph	$PPh_3^+Br^-$	o-$H_2NC_6H_4OH$[b]	$CHCl_3$ (b.p.)	2-Phenylbenzoxazole (14)	147, 368
MeOOC	COOMe	$MeOOCCH{=}\overset{+}{N}(Me)O^-$	Ether (r.t.)	2-Methyl-3,4,5-tricarbomethoxydihydroisoxazole (70)	325
MeCH—CHCH$_2$ O	H	—	$K^{+-}OBu$-t, Me_2SO (25)	2,5-Dimethylfuran (50), $HC{\equiv}CCH{=}CHCH(Me)OH$ (20)	369*, 347
Me	Me	Oxadiazin-6-one-4-oxide	Heptane (b.p.)	3-Acetyl-3,4-dimethyl-4-phenylbutenolide (25), 3,4-Dimethyl-5-acetyl-5-phenylbutenolide (8)	370*

[a] Additional examples of a similar type as well as leading references are indicated by an asterisk in the last column; cyclic products are also given in Table 16.

[b] It is probable that nitrogen leads the nucleophilic attack.

(114) and (115) are typical; the authors give no indication of the role of $CuSO_4$ in equation (115), a reagent which is not usually used in such cyclization[350].

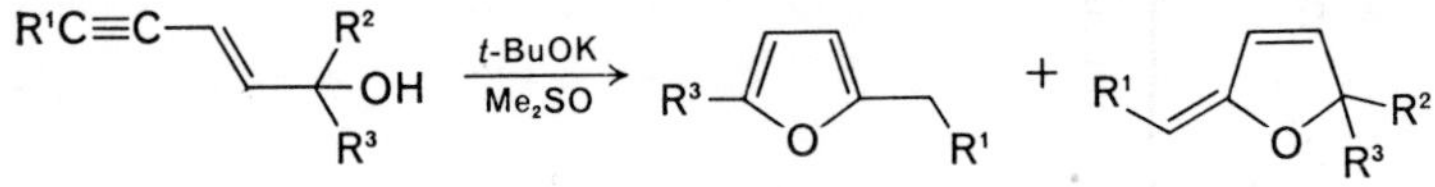
(Ref. 347) (114)

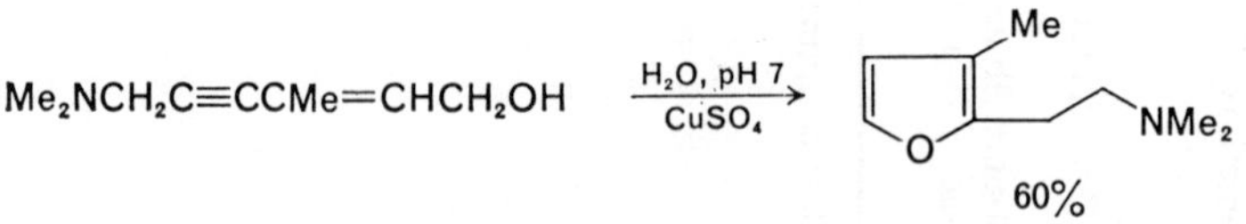
(115)

Numerous variations on the reactions of *o*-substituted phenols with alkynes have been attempted (equations 116, 117; Table 15)[1, 10]. Although amines are normally

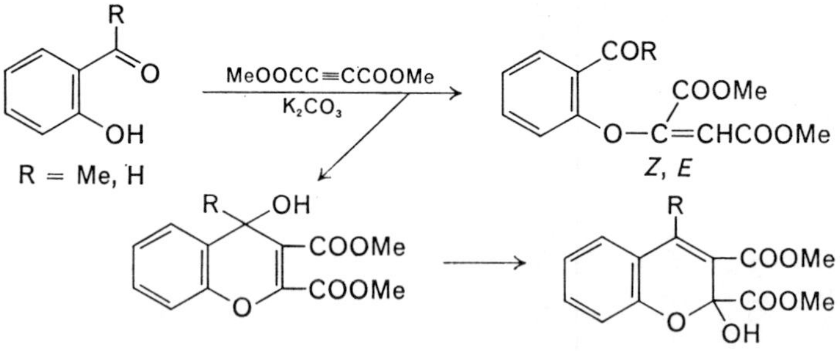
(116)

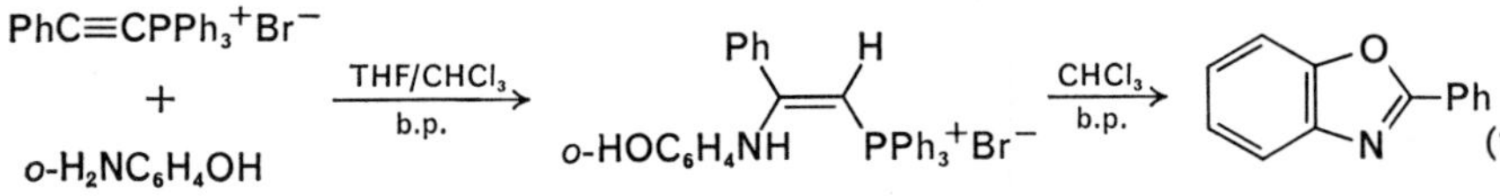
(117)

more nucleophilic than alcohols (see Section III.G), oxygen appears to lead the attack in equation (34). This is contrary to what appears to be similar additions in equations (117)[147] and (118)[351].

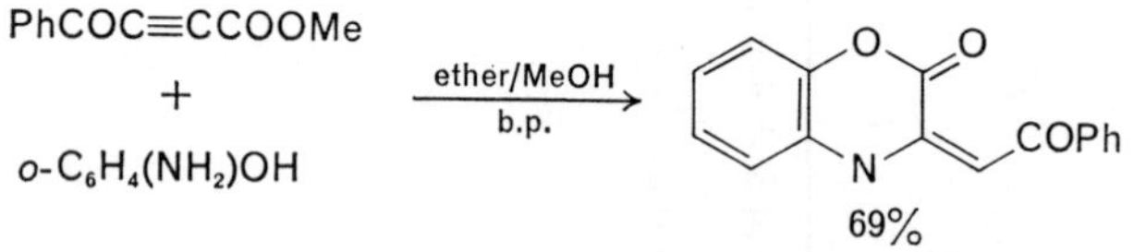
(118)

1,3-Dioxo rings may arise in several ways. In equation (119) the chloro group directs to C-2 and the second oxygen follows the first[184]. This geminal addition is

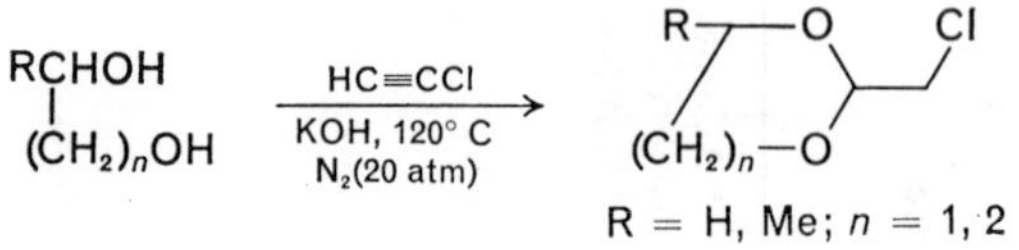
(119)

found repeatedly[314, 331, 345, 352]. A less usual route to dioxoles is given in equation (120)[353]. Bottini and Maroski treat the complex problem of ring formation from propargyloxyethanols (equation 121) in some detail[354]. In Table 16 are given some trends in their data covering substituents, base and solvent. It is interesting that high yields of the 7-membered ring **(19)** are obtained and that the methylenedioxole is essentially absent. In the light of this paper, predictions of regioselectivity in ring closure should be made cautiously and with suitable qualifications[354].

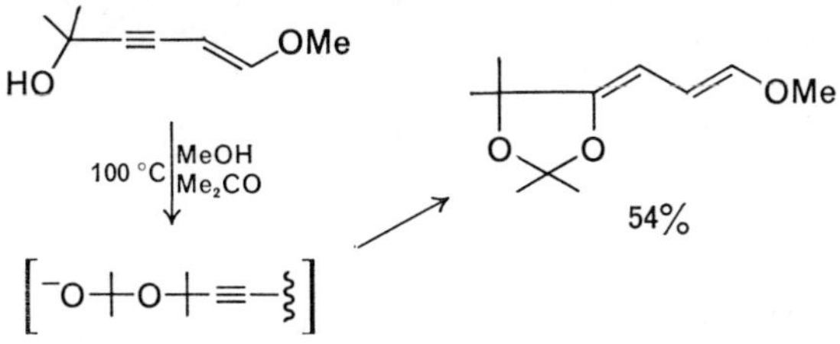

(120)

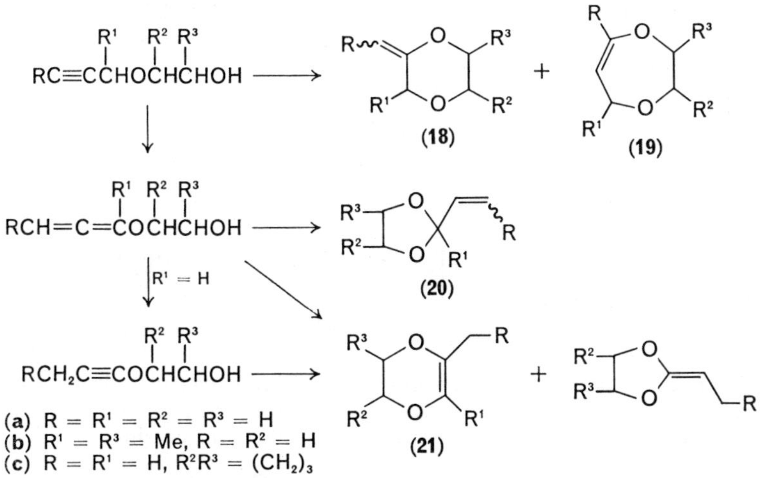

(121)

TABLE 16. Products from the reactions of propargyloxyethanols, $HC{\equiv}CCH(R^1)OCH(R^2)CH(OH)R^3$, with base (equation 121)[354]

Compounds[a] 18–21	Medium (h, °C)	Yield (%)	Composition (%) (18)	(19)	(20)	(21)
(a)	KOH, H_2O (12, b.p.)	54	36	44	20	—
(b)	KOH, H_2O (12, b.p.)	72	87	12	<1	<1
(c)	KOH, H_2O (12, b.p.)	52	<1	~95	<1	~5
(a)	KOH, Me_2SO (0·7, 100)	33	4	7	18	71
(b)	KOH, Me_2SO (0·1, 100)	80	32	7	23	38
(c)	KOH, Me_2SO (12, 100)	78	—	35	—	65
(a)	KOH, *t*-BuOH (12, 100)	61	5	8	65	22
(b)	KOH, *t*-BuOH (12, 100)	58	35	7	25	33
(c)	KOH, *t*-BuOH (12, 100)	61	—	48	—	52

[a] (a) $R^1 = R^2 = R^3 = H$; (b) $R^1 = R^3 = Me$, $R^2 = H$; (c) $R^1 = H$, $R^2R^3 = (CH_2)_3$.

It is generally considered that nitrones and alkynes react by cycloaddition (rather than process 1)[10, 325]. Since the distinction is usually difficult to prove and some acyclic products are found, we include some examples (equations 122 and 123)[325]: presumably, the anion of hydroxamic acid leads the attack[355]. Similarly, forced

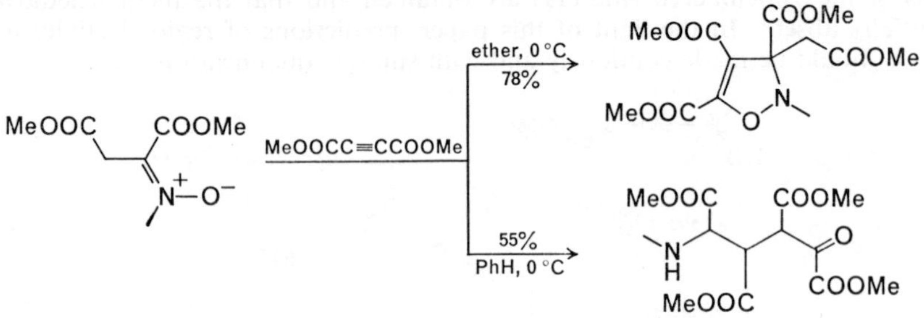

(122)

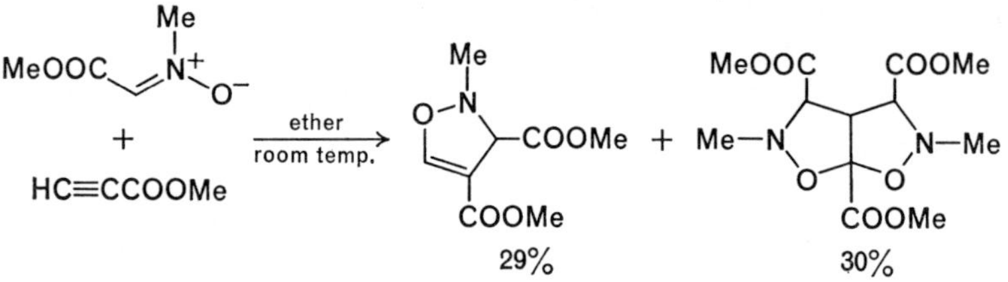

(123)

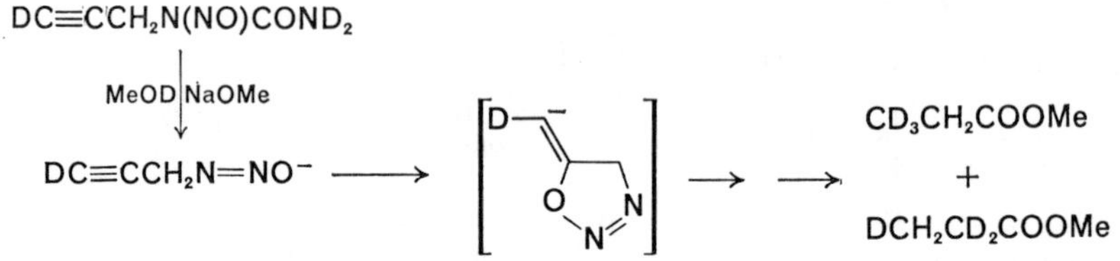

(124)

cyclization of diazotates of the type in equation (124) during (following) the expulsion of nitrogen leads to rearranged products[356]. A related case may be that of the nitrile oxides in which O⁻ attack is found in a potential route to the corrins (equation 125)[357].

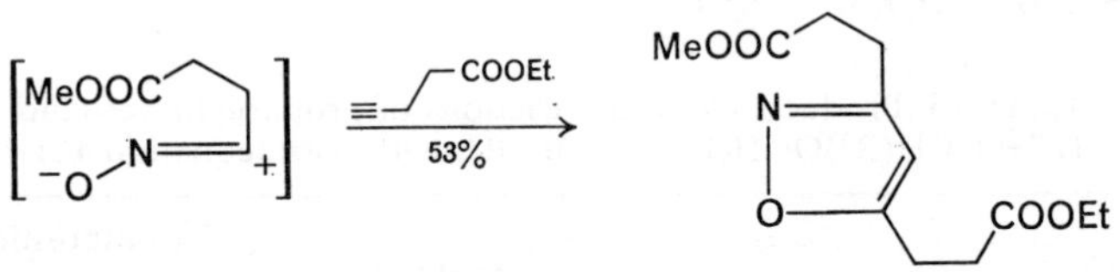

(125)

The attacks of heterocyclic *N*-oxides, e.g. of pyridine, quinoline, isoquinoline, phenanthridine, etc., on activated alkynes (RC≡CR′: R = R′ = COOMe; R = Ph, R′ = COOEt; R = Ph, R′ = CN) pose similar problems[24, 335]. An acyclic intermediate has been postulated but is rarely detected. Some of the possibilities are illustrated in equation (126)[335]. If the open intermediate is formed, then the paths to the ylid and the 2-substituted quinoline in equation (126) seem simple enough, but several possible mechanisms can lead to the 3-substituted products[24]. Other workers regard the reaction of the nitrone (or azomethine oxide) with alkyne as simple cycloadditions[21, 22] which yield 2,3-dihydro-1,2-oxazoles; since these are often unstable, only decomposition products may be found (equation 127)[22]. The construction of the indolizine skeleton initiated by a similar process has been reviewed (equation 128)[21].

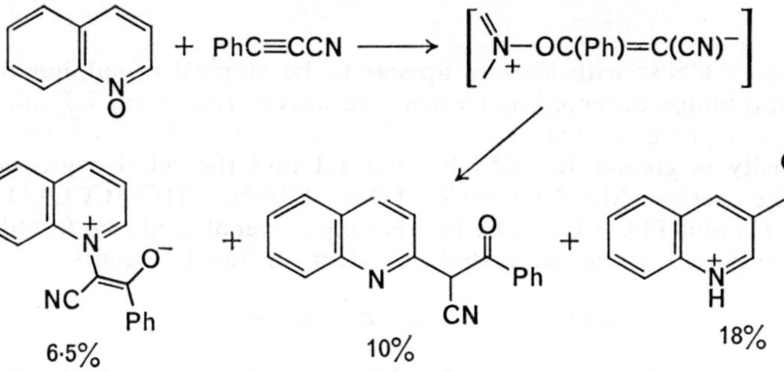

+ PhC≡CCN
N—OC(Ph)=C(CN)⁻
CN
Ph
O⁻
NC
Ph
6·5%
+
O
N
Ph
CN
10%
+
N
H
O⁻
18%

(126)

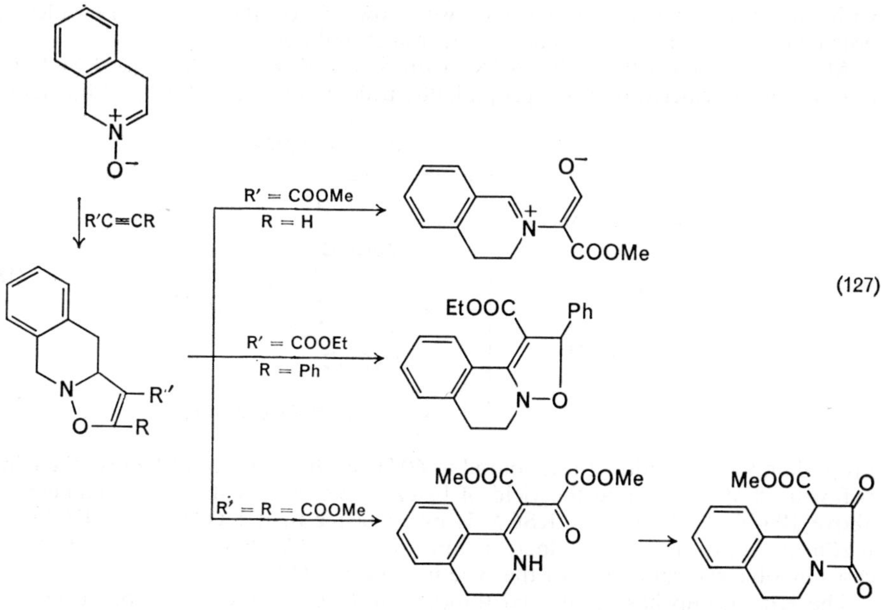

N
O⁻
R′C≡CR
R′ = COOMe
R = H
O⁻
COOMe
R′ = COOEt
R = Ph
EtOOC
Ph
N—O
N
O
R′
R
R′ = R = COOMe
MeOOC
COOMe
O
NH
MeOOC
O
N
O

(127)

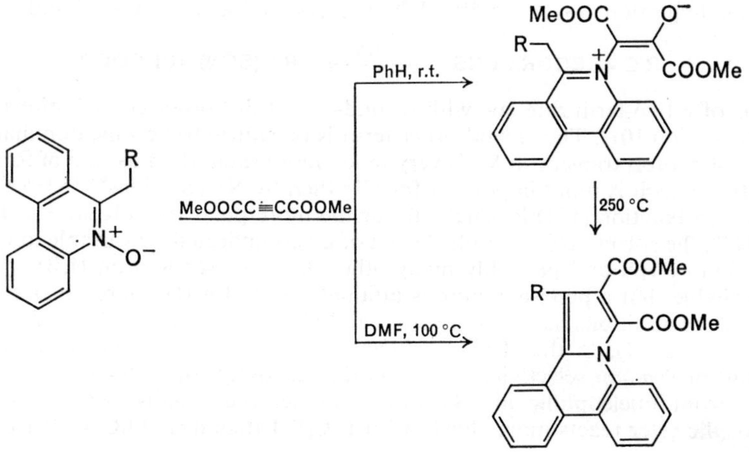

R
N
O⁻
MeOOCC≡CCOOMe
PhH, r.t.
MeOOC
O⁻
R
N
COOMe
250 °C
DMF, 100 °C
COOMe
R
COOMe
N

(128)

F. Sulphur

Some reactions of CNS^- with alkynes appear to be atypical of sulphur nucleophiles. Dvorko and Shilov observed that when solutions of KCNS (or KI) are mixed with electrophilic acetylenes in DMF at 60–80 °C an e.p.r. signal begins to grow in[239]. The signal intensity is greater for KSCN than KI and the relative growth rate decreases in the series MeOOCC≡CCOOMe (2000), $(HC{\equiv}CCO)_2O$ (30), HC≡CCOOMe (1) and PhC≡CCOOR (0). No e.p.r. signal could be found when protic solvents were used. It was suggested that electron transfer occurs:

$$\mathrm{A + CNS^- \longrightarrow A^{\dot{-}} + CNS^{\bullet}} \tag{129}$$

The fact that polymerization of the acetylene diester was also observed is consistent with this initiating reaction. To date, we know of no other laboratory which has noted electron transfer to an alkyne under such mild conditions.

Along with a few other salts (K^+X^- with X = CN, I, Br, Cl, OAc), KSCN promotes the polymerization of propiolanhydride (equation 130)[304]. With methyl

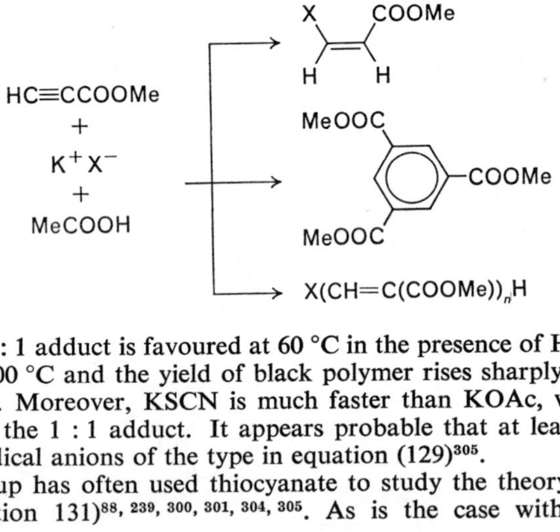

(130)

propiolate the 1 : 1 adduct is favoured at 60 °C in the presence of HOAc, the trimer is favoured at 100 °C and the yield of black polymer rises sharply at temperatures above 100 °C [305]. Moreover, KSCN is much faster than KOAc, while KOPh and KSPh give only the 1 : 1 adduct. It appears probable that at least some of these cases involve radical anions of the type in equation (129)[305].

The Kiev group has often used thiocyanate to study the theory of nucleophilic additions (equation 131)[88, 239, 300, 301, 304, 305]. As is the case with halide attacks,

$$\mathrm{R'C{\equiv}CCOOR + CNS^- \xrightarrow{HX} R'C(SCN){=}CHCOOR} \tag{131}$$

versions of a two-term rate law with second- and third-order contributions usually apply (equation 101). The second-order term is presumed to become dominant when the rate of proton transfer to V^- is very much more rapid than its rate of formation, a situation which is more important for Cl^- then SCN^- additions[87, 88, 300, 301].

Although reaction (131) is usually faster the more polar the solvent, e.g. DMF in *n*-PrOH [88], the solvent effects studied were often complicated by variable ion pairing, association of HX and probably many other 'factors' (see Section II.B)[87, 88, 299–301]. It is probable that a precise picture is attainable only for the more polar solvents.

Since thiol additions may be initiated readily by radicals this is always a complication to process (1). In the absence of controlled reaction conditions, the resulting *regio* and/or *syn-anti* selectivity as well as the reactivity could differ from what one expects from nucleophilic attack. For example, the usually reactive acetylenedicarboxylic ester reacts more slowly with F_5C_6SH than does PhC≡CR (R = H or

Ph); one would have to conclude that radical additions are involved in these systems[371]. Likewise, the additions of the type in equation (132) may be radical,

$$(EtO)_2P(S)SH + RC{\equiv}CH \xrightarrow{110\,°C} (EtO)_2P(S)SCR{=}CH_2 \quad (132)$$

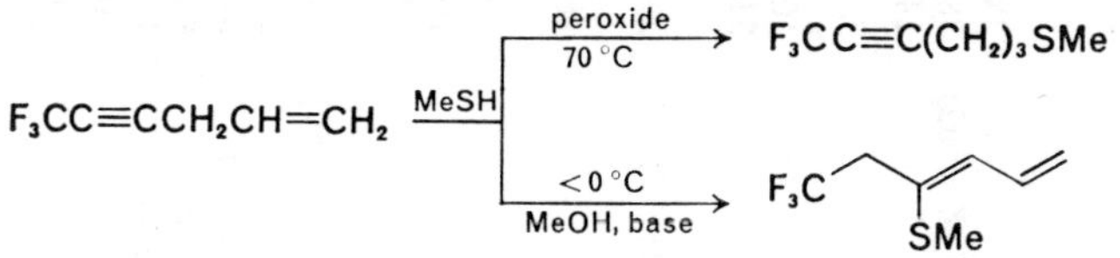

(133)

since the regioselectivity is 'wrong' for R = phenyl [372]. No speculation on the mechanism of equation (133) is necessary, since the radical (upper) and polar (lower) branches are clear from the reaction conditions[373].

The simple nucleophilic additions of thiolates (equation 1) have been noted previously[1, 5] and are updated in Table 17. Although post-isomerization is often facile, the *anti* product can usually be obtained (see Section II.B). For example, ketoalkynes, in which V_i for $\mathbf{V}^-$ is low, yield *anti* adducts in alcohols in which protons are abundant and tend to yield isomer mixtures in HMPT and Me_2SO in which $\mathbf{P}^-$ is stabilized with respect to $\mathbf{V}^-$ (Scheme 4) and proton delivery is slow[6a]. The conditions for these reactions range from forcing to facile, depending, of course, on the presence of activating substituents. (Engineering data on the vinylation of thiols and industrial uses of the products have been described in detail[314].) There is an impressive enhancement in the rate of reaction of alkanethiols with the usually sluggish C_2H_2 when an aprotic solvent is used (equation 134)[374].

$$C_2H_2 + RSH + KOH \xrightarrow[20\text{–}40\,°C]{\text{HMPT or DMSO}} RSCH{=}CH_2 \quad (134)$$

Kinetic data for process (135) in $CDCl_3$ at 0 °C were used to establish the following trends: for R′, k(COMe) > k(COOEt) ≫ k(CONHEt); for R, k(H) : k(Me) : k(n-Pr) = 1 : 0·1 : 0·6, when R′ = CONHMe; *anti* addition dominates when R′ is amide, *syn* addition dominates when R′ is ketone, and *anti* is only slightly favoured when R′ is ester[115].

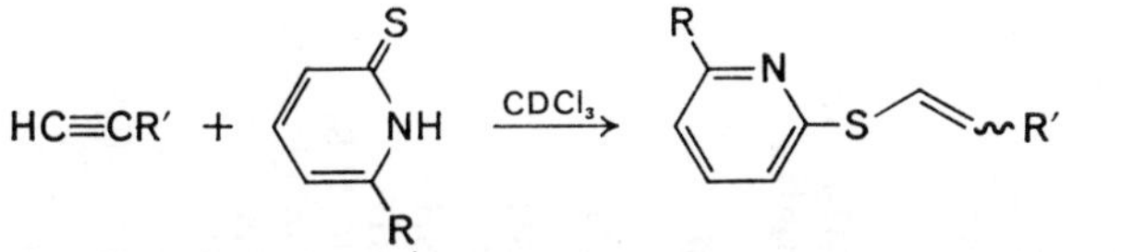

(135)

Russian workers have devoted much attention to conjugated acetylenes. With respect to regioselectivity, nucleophiles appear to favour unhindered sites (equation 136)[5, 344]; and in a fair competition in an enyne, the thiolate attacks the triple bond (equation 137)[375, 376].

$$PhC{\equiv}CC{\equiv}CR \xrightarrow[\text{KOH, MeOH}]{n\text{-}C_6H_{13}SH} \begin{cases} \xrightarrow{R\,=\,H} PhC{\equiv}CCH{=}CHSC_6H_{13}\text{-}n \\ \xrightarrow[R\,=\,Me_2COH]{} PhCH{=}CS(C_6H_{13}\text{-}n)CH{=}CHC(OH)Me_2 \end{cases} \quad (136)$$

$$E\text{-}PhCH{=}CHCOC{\equiv}CPh + p\text{-}C_7H_7S^-Na^+ \xrightarrow{\text{MeOH}} E,Z\text{-}PhCH{=}CHCOCH{=}C(Ph)SC_7H_7\text{-}p \quad (137)$$

TABLE 17 Additions of sulphur nucleophiles to alkynes, RC≡CR′ [a]

R	R′	Nu⁻	Medium (temp., °C)	Product(s) (yield, %)[d]	Reference
EtOOC	COOEt	SCN⁻	2-PrOH, 0·5M $(O_2N)_2C_6H_3COOH$ (20)	*k*	301, 308, 88
Me	SC≡CMe	Na_2S	MeOH	2,6-Dimethyl-1,4-dithiin (55)	381*, 382*
Ph	COC_6H_4OMe-*p*	$SC(NH_2)_2$	Et_3N,MeOH (r.t.)	$(ArCOCH{=}CPh)_2S$ (*Z*, *Z* = 20, *E*, *Z* = 62)	383*, 385
Ph	SO_2Me	MeS⁻	EtO⁻Na⁺, EtOH/DMF 1 : 1 (r.t.)	*Z*-PhCSMe=$CHSO_2Me$ (14), *E*-PhCH=C(SMe)SO_2Me(50), *E*-PhCSMe=CHSMe (36)	149*, 104
H	COMe	*p*-$C_7H_7S^-Na^+$	MeOH (0)	ArSCH=CHCOMe (*Z*/*E* = 82/18)	104*
H	H	$(CH_2SH)_2$	*t*-BuO⁻K⁺, *t*-BuOH, pyrogallol, 10 atm	$(H_2C{=}CHSCH_2)_2$	318*
5-(2-RC_4-H_2Se)CO	R¹	R²SH	—	5-(2-RC_4H_2Se)COCH=CR¹SR² (23–92)	398*
MeC≡CPO(Ph)	Me	Na_2S_2	NH_3/MeOH (−33)	2,6-Dimethyl-4-oxo-4-phenyl-1,4-thiaphosphorin (69)	188*
MeCO	H	MeSH	Triton B [b], CD_3OH (−75)	MeSCH=C=CMe(OH)	102*, 400*, 105*
H	SO_2Et	MeSH	Triton B [b], EtOH (−10)	*Z*-MeSCH=$CHSO_2Et$ (83)	401*, 104*
H	H	EtSH	KOH, HMPT (20)	EtSCH=CH (85)	374*
MeOOC	COOMe	4-ClC_6H_4SH	MeOH (b.p.)	*Z*-(4-ClC_6H_4S)CCOOMe=CHCOOMe (59)	380*
n-Bu	H	(*n*-BuO)$_2$P(S)SH	(110)	(2-BuO)$_2$P(S)SC(Bu-*n*)=CH_2 (78)	372*, 376
H	COOMe	ArSH	MeOH (b.p.)	(4-I,2-$CONH_2$)C_6H_3SCH=COOMe (77)	378*
(2-C_4H_3S)CO	H	PhSH	MeOH, Et_3N (20)	1-(2-Thienyl)-3-phenylthiopropenone	402*
t-Bu	Cl	PhS⁻Na⁺	EtOH (100)	PhSC(Bu-*t*)=CClH (76, *Z*/*E* = 93/7)	165
H	$CH_2PPh_3^+$,Br⁻	*o*-$H_2NC_6H_4SH$	MeCN (r.t.)	2-Methylbenzothiazole (62)	403, 368*
Me	COC≡CMe	$(H_2N)_2CS$	DMF (r.t.)	2,6-Dimethyl-4*H*-thiapyran-4-one	141
H	CONHEt	(2-thione)	$CDCl_3$ (0)	*E*-(2-C_5H_4N)SCH=CHCONHEt (87)	115*, 404*
H	COOEt	$C_6F_5S^-Li^+$	THF (−70)	C_6F_5SCH=CHCOOEt (~80, *Z*/*E* = 7/1)	379*
(*n*-Bu)$_2$P(O)	H	EtS⁻Na⁺	NH_3 (−33)	*Z*-(*n*-Bu)$_2$P(O)CH=CHSEt (82)	338
E-PhCH=CHCO	Ph	*p*-MeC_6H_4SH	MeO⁻Na⁺, MeOH (r.t.)	*E*, *Z*-PhCH=CHCOCH=C(Ph)SC_7H_7-*p*	185
H	H	$SC(NH_2)_2$	—	$(H_2C{=}CH)_2S$ (83)	384*, 405*

PhCO	C≡CH	MeSH	Triton B [b], MeOH (−10)	$PhCOCH_2C(SMe)_2CH{=}CHSMe$ (70, $Z/E = 4/1$)	375*, 376*
H	H	Na_2S_2	Moist DMSO (100)	$H_2C{=}CHSCH{=}CHSCH{=}CH_2$ (10, $Z/E = 41/59$)	406
$HOCMe_2$	$C{\equiv}CC(OH)Me_2$	H_2S	NaOH, MeOH (60)	2,5-di[2-(2-Hydroxypropyl)] thiophene (50)	186, 407*
PhCHOH	CHOHPh	PhSH	KOH (r.t.)	$PhCH(OH)C(SPh){=}CHCH(OH)Ph$ (30)	408*
H	COOMe	$(H_2N)_2CS$	H_2O, H^+ (r.t.)	*Z*-$MeOOCCH{=}CHSC(NH_2)_2^+Cl^-$, *k*	409, 386*
$Me_2C(OH)$	$C(OH)Me_2$	PhSH	KOH (170)	$Me_2C(OH)C(SPh){=}CHC(OH)Me_2$	106*, 410*, 121*
H	H	$(i\text{-}BuS)_2$	KOH, 12 atm (120)	*Z*-*i*-$BuSCH{=}CHSBu$-*i* (90)	5, 409
CF_3	Cl	EtS^-	EtO^-Na^+, EtOH	$EtSCCl{=}CHCF_3$	411*
PhC≡C	H	*n*-BuS^-	KOH, MeOH (70)	$PhC{\equiv}CCH{=}CHSBu$-*n* (75)	344*
CF_3	$CH_2CH{=}CH_2$	EtS^-Na^+	MeOH (<0)	$CF_3CH_2C(SEt){=}CHCH{=}CH_2$ (~85)	373*
$Me_2C(OH)$	H	EtSH	Li, HMPT (r.t.)	$Me_2C(OH)CH{=}CHSEt$ (70), $Me_2C(OH)CSEt{=}CH_2$ (30)	122
Me	NEt_2	$HC{\equiv}CCH_2SH$	Base	$MeCH{=}C(NEt_2)SCH_2C{\equiv}CH$	412*
$Me_2C(NH_2)$	H	PhSH	—	*Z*-$Me_2C(NH_2)CH{=}CHSPh$ (65)	6a, 399*
PhCO	Ph	$PhCH_2SH$	Et_3N, MeOH (r.t.)	*Z*-$PhCH_2SCPh{=}CHCOPh$ (85)	413*
H	H	$EtOCS_2Bu$-*n*	Me_2SO/H_2O, KOH (135)	$BuSCH{=}CH_2$, $EtSCH{=}CH_2$	414*
H	Na [c]	—	S, CSe_2 or Se, CS_2, Et_2O (r.t.)	2-Thio(or seleno)-1,3-dithia(or selena)-cyclopentene	390*

[a] Additional examples of a similar type and leading references are indicated by an asterisk in the last column.
[b] $PhCH_2NMe_3^+OH^-$.
[c] $HC{\equiv}\bar{C}Na^+$.
[d] *k* indicates that rate data are available.

In those instances in which two nucleophilic sites compete for the acetylenic carbon, RS^- usually wins over RO^- or RNH_2. As indicated in Table 17, the usual synthetic technique is to generate a thiolate by means of the alkoxide. When the nucleophile is bifunctional, thiolate usually leads (equation 138)[147]. It does, in fact,

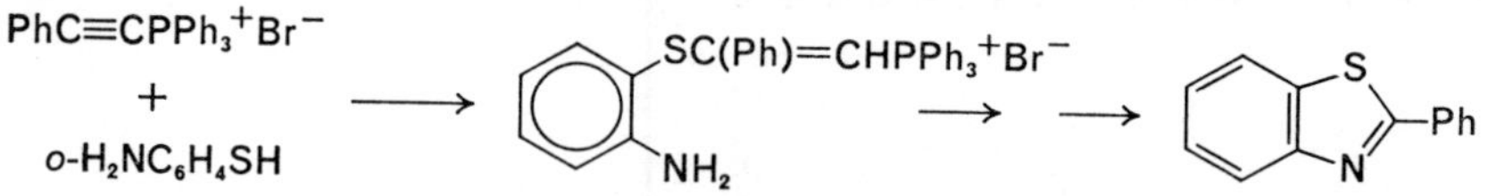

(138)

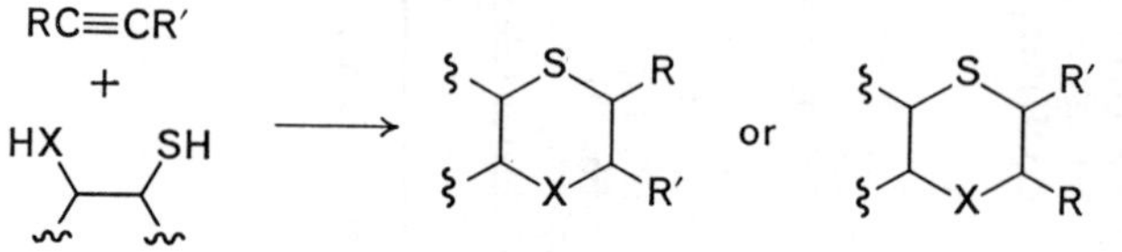

(139)

make an important difference in some products, particularly when the real goal is to carry the reaction through to the heterocyclic, say in process (139).

Acetylenes provide a convenient starting point for the syntheses of several heterocycles. The patterns in ring closure vary somewhat so that structural assignments have to be made critically. The similar result in equations (138) and (140)[368] differs

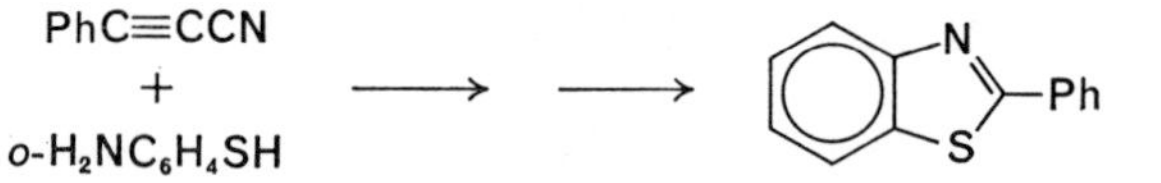

(140)

from that in equation (141)[377]. The cyclization in equation (142) did have the potential of forming a 7-membered ring but took the standard course[378]. In a few

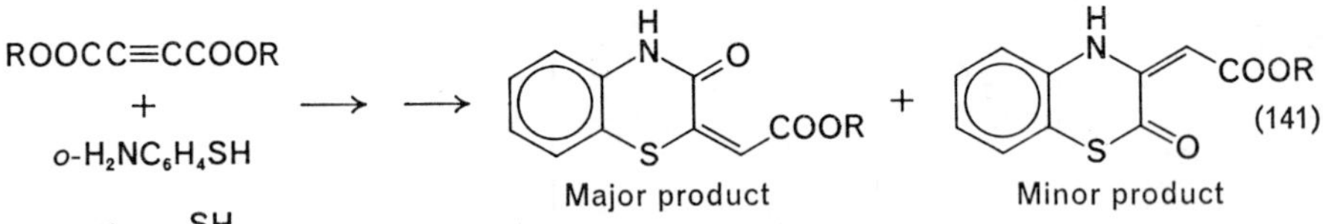

(141)

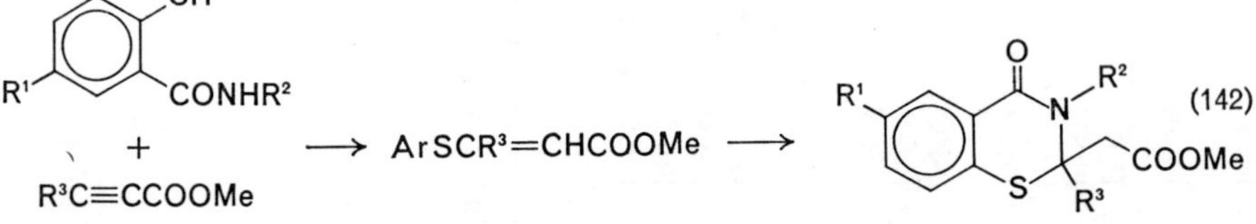

(142)

examples the fact that cyclizations occur at all is perhaps surprising (equations 143 and 144)[379]. Certainly, process (143) is analogous to equation (104) discussed in Section III.D but the oxidative cyclization in equation (144) is novel[380]. Since both fumaric and maleic esters were detected in the product mixture, it appears that

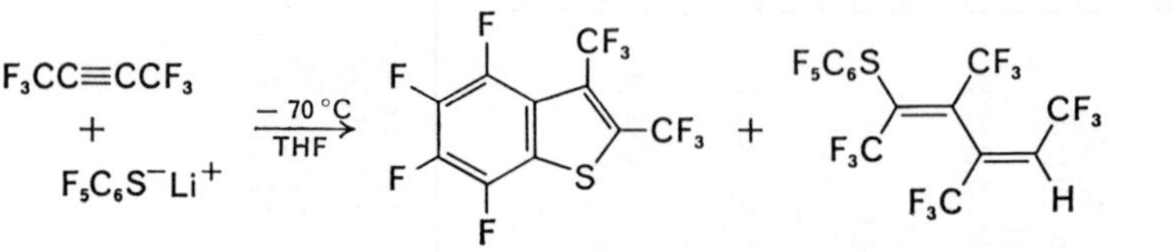

(143)

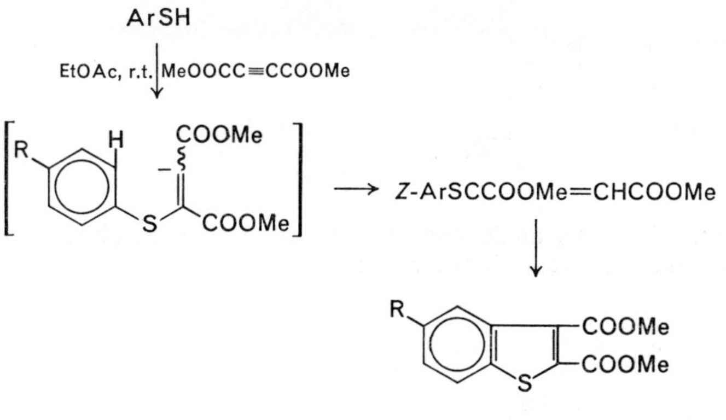

(144)

MeOOCC≡CCOOMe accepts H^- from the anion pictured in equation (144). Incidentally, the reaction goes more slowly when R is electron-withdrawing (R = Cl, NO_2) and gives benzo[b]thiophene for R = H, Me, Cl (not NO_2), when R′ = Me and H in R′OOCC≡CCOOR′ [380].

The formation of thiophenes from conjugated diacetylenes and H_2S (or its equivalent) is fairly standard[1, 5]. Cyclization in the skipped diacetylene is also becoming familiar (equation 146)[5, 188, 381, 382].

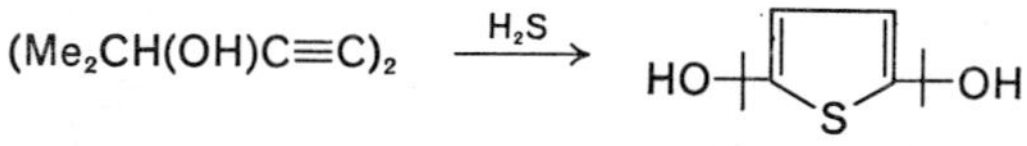

(145)

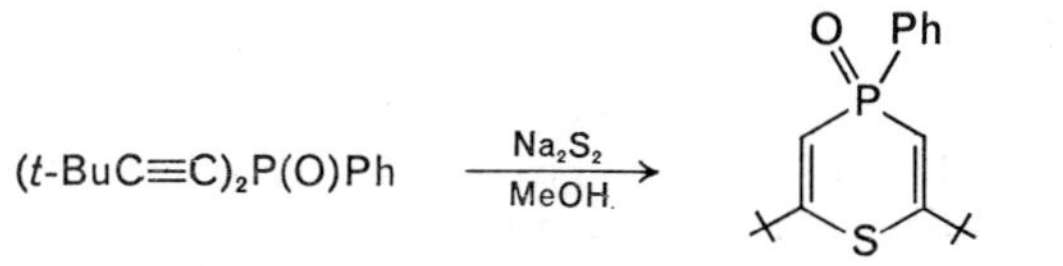

(Ref. 381) (146)

Thiourea, diphenylthiourea and ammonium dithiocarbamate are often simply masked versions of H_2S in single (equation 147)[383–385] and double additions (equation 148)[141, 178]. Although acyclic thiourea intermediates have sometimes been

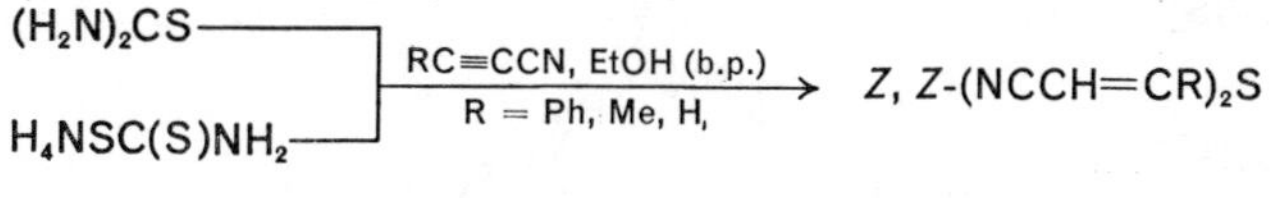

(Ref. 385) (147)

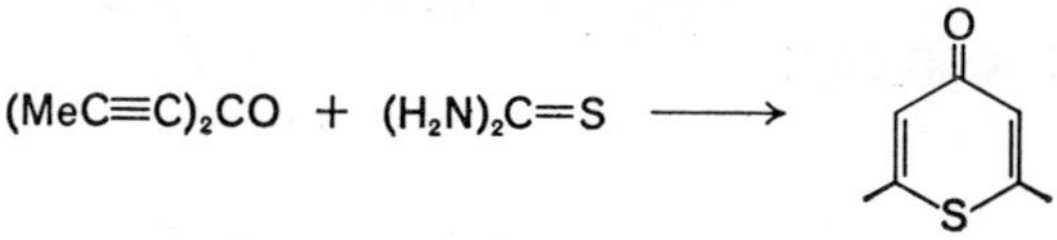

(148)

noted, they were not characterized[178]. It does seem to be possible to capture the first adduct of acetylene mono- and dicarboxylic acids and their esters or proceed to the thiazine without losing ammonia (equation 149)[385, 386]. Preformed acetylenic

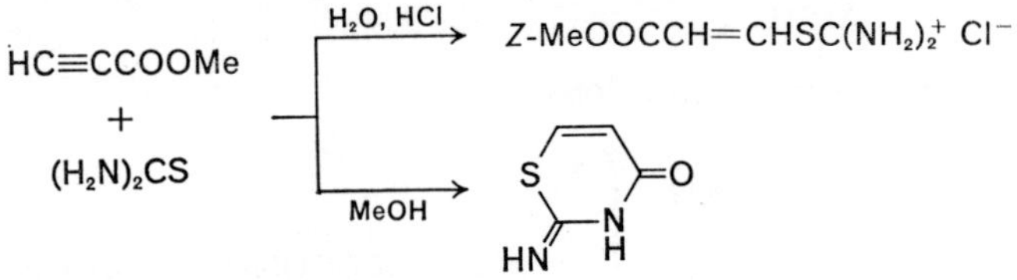

(149)

thioamides are, of course, analogous to intermediates in the above reaction and might be expected to behave in similar ways:

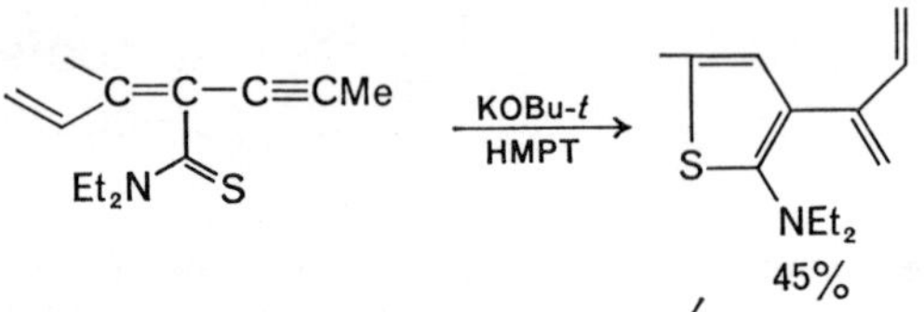

(Ref. 387a) (150)

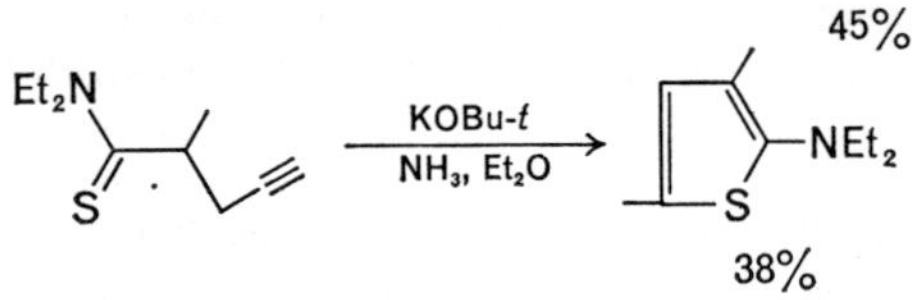

(Ref. 387b) (151)

An interesting example which may also fall into this group is given in equation (152); the published mechanism is rather different and more complex[388].

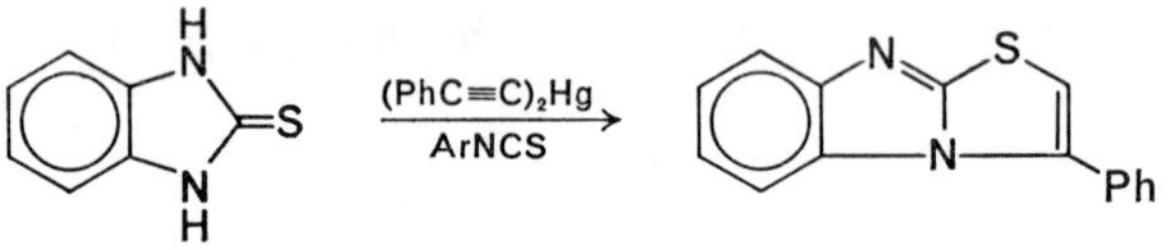

(152)

Dithiocarboxylic acids and related compounds yield both acyclic and cyclic adducts. A number of reactions of the type given in equation (153)[389], which may continue on to tetrathiafulvalenes[390], have been regarded as cycloadditions. Nevertheless equation (153) may be initiated, as equations (154)–(156) seem to be, by nucleophilic attacks.

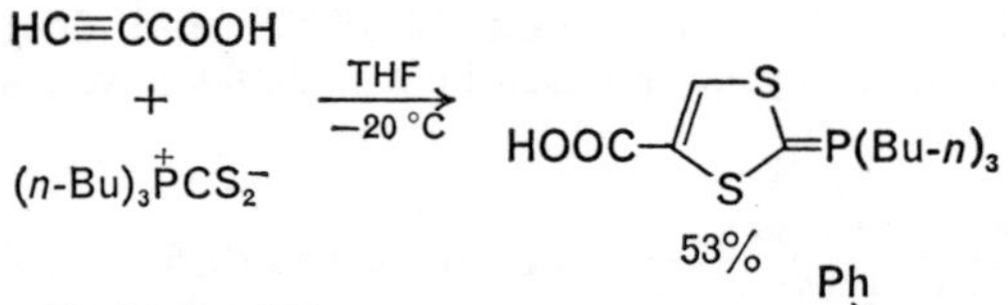

(153)

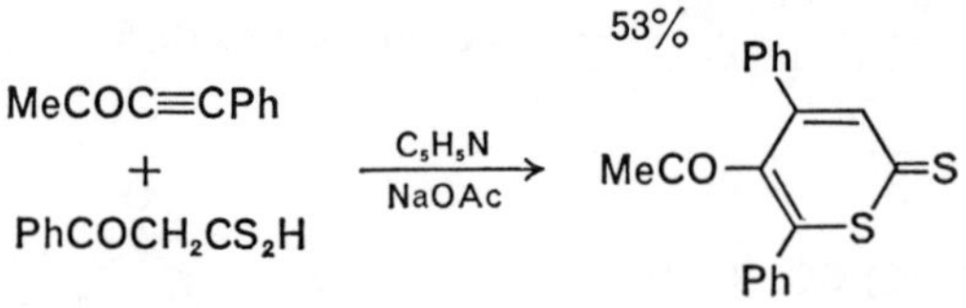

(Ref. 391) (154)

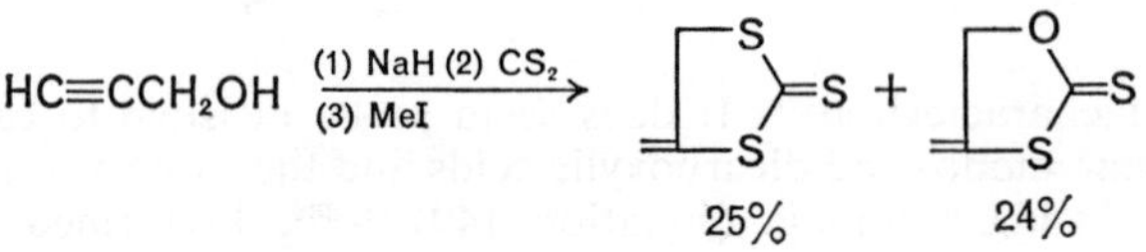

(Ref. 392) (155)

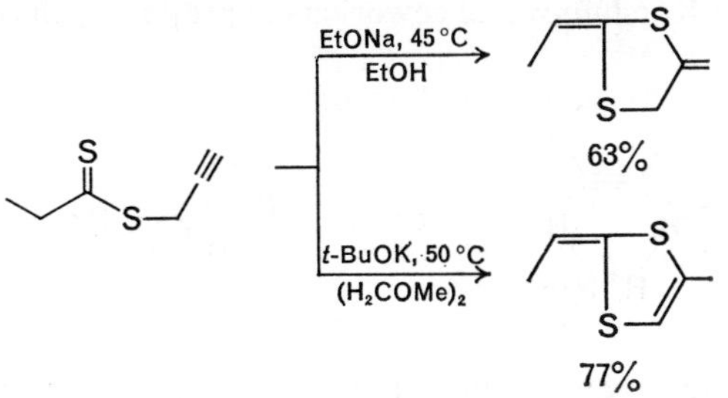

(Ref. 393) (156)

Processes such as equation (157) have been effected in boiling CCl_4 or xylene; the alkyne is usually activated with R and R^3 = COOMe or COPh [394–397]. Although their

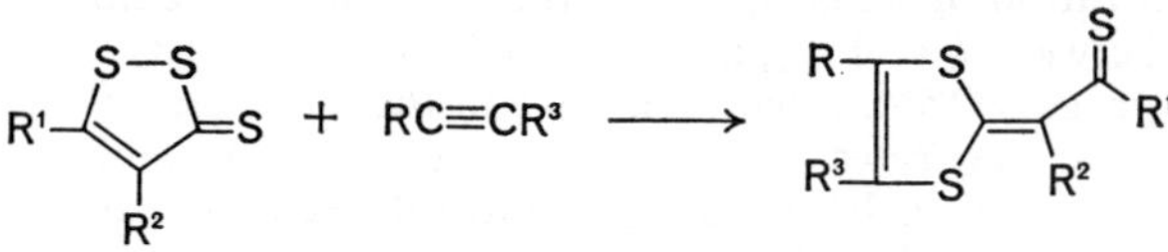

(157)

mechanisms could involve initial nucleophilic attack, these reactions have been regarded as dipolar cycloadditions and will not be considered further.

G. Nitrogen

In their various forms nitrogen nucleophiles probably comprise the largest family of reagents in process (1)[1, 8, 10, 314]. Although C_2H_2 and amines constitute sluggish systems, modern practice makes possible the formation of a large array of vinylation products[5, 314, 415, 416], one of which is given in equation (158)[5a, 115]. Even omitting this chemistry, we have had to compress our material considerably.

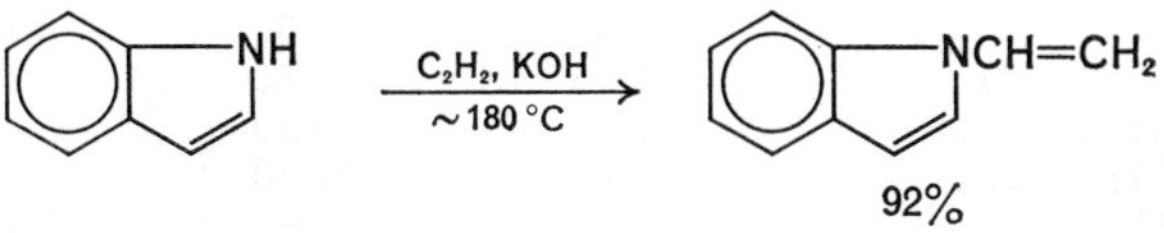

(158)

I. Mechanistic data

Consider reaction (159) which has been studied in depth. With morpholine, the process is first order in alkyne and first order in amine in ethanol: $\Delta H^{\ddagger} = 9{\cdot}7$ kcal/

$$\text{PhCOC}{\equiv}\text{CPh} + \text{R}_2\text{NH} \longrightarrow \text{PhCOCH}{=}\text{C(NR}_2\text{)Ph} \qquad (159)$$

mol and $\Delta S^{\ddagger} = -36$ e.u. at 20 °C [224]. In an aprotic solvent, dioxane, the reaction is first order in alkyne but second order in amine: $\Delta H^{\ddagger} = 4{\cdot}0$ kcal/mol and $\Delta S^{\ddagger} = -59$ e.u. at 20 °C; the addition of triethylamine retards the reaction and methanol increases the rate[417]. Specifically, with piperidine in equation (159), the effects on k ($M^{-1}\ s^{-1}$, 30 °C) are: C_6H_{12} (2·67), C_6H_6 (4·43), t-BuOH (14·4)[90]. These observations and the high $\Delta S^{\ddagger}$ led Korshunov and coworkers to propose cyclic transition states of the type

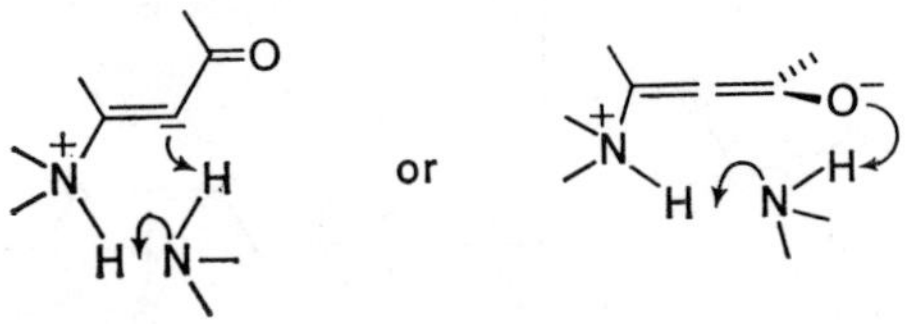

These polymolecular forms are presumably most appropriate for the aprotic solvents. They illustrate well the notion that polar solvents with high proton availability increase the rates of addition.

The effect of solvent polarity on rates has been delineated for equation (45) in which R = H and Nu = c-C_2H_4N [89]. First the specific rate constant increased as the aziridine concentration increased, when the solvent was benzene. Then a fair proportionality between k and E_T, a measure of solvent polarity, was demonstrated.

Returning to process (159), large variations in the structures of the reactants and the solvents indicate that rates of addition also correlate roughly with amine base strength. Vereshchagin gives lifetimes (t) for PhCOC≡CH in n-BuOH with primary and secondary amines ($pK_b \simeq 4$–6) of 30–700 min, anilines ($pK_b \simeq 10$) of 4000–6400 min and Ph_2NH ($pK_b \simeq 13$) of $\geqslant 10^4$ min [114]. The quantitative data cover a smaller range[114, 418]: for eight amines in 95% ethanol (Table 18), Korshunov finds a reasonable correlation with steric parameters (E_N)[418].

TABLE 18. Addition of amines to PhCOC≡CPh [114, 418]

	k ($M^{-1}\ s^{-1}$) at 20·8 °C in EtOH	pK_a	k ($M^{-1}\ s^{-1}$) at 30 °C in C_6H_6
c-$C_6H_{11}NH_2$	15·4	11·22	4·43
$O(CH_2CH_2)_2NH$	5·5	8·36	0·79
$(Me_2CHCH_2CH_2)_2NH$	4·7	—	—
Et_2NH	4·23	10·93	0·21
$(n\text{-}Bu)_2NH$	4·0	11·3	0·116
$(HOCH_2CH_2)_2NH$	1·80	8·88	—
$(CH_2{=}CHCH_2)_2NH$	1·72	9·29	—
$(Me_2CHCH_2)_2NH$	1·19	10·50	—
c-C_2H_4NH	—	—	0·205

The importance of both electronic and steric factors is borne out by the reactions of piperidine in equation (159) at 30 °C in benzene (k, $M^{-1}\ s^{-1}$): PhCOC≡CPh (4·4), PhCOC≡CBu-n (1·9), PhCOC≡CBu-t (0·012), MeCOC≡CPh (1·15)[114]. If the steric requirements are made constant, as in a series of anilines in process (159) in 95% ethanol, the resulting rate data yield a satisfactory Brønsted plot and Hammett correlations: $\beta \simeq 0{\cdot}93$ and $\rho = -(2{\cdot}12$ to $2{\cdot}18)$ in the range 20–50 °C [419]. Further structure–reactivity variations were made with morpholine as nucleophile and $XC_6H_4C{\equiv}CCOC_6H_4Y$ as electrophile: in 95% ethanol, the Hammett $\rho = 1{\cdot}13$ to 0·97 for the X series and 1·42 to 1·20 for the Y series from 20–50 °C [420, 421]; in t-BuOH at 40 °C, these ρ values are 1·2 and 1·6 respectively[422]. Our interpretation of these near-equal ρ-values is that there is considerable (> 50%) delocalization of negative charge into the carbonyl centre in the activated complexes of reaction (159).

Relative rates of additions to $Me_2NC{\equiv}CCOR$ also indicate an interplay of steric and polar effects (equation 45). With aziridine in THF at 37 °C, the relative second order rates are k(H) 14·5, k(Me) 1·9 and k(OMe) 1: the trend is consistent with a polar effect. It will be recalled, however, that regioselectivity in these ynamines was governed by both steric and polar effects (Tables 3–5, see Section II.B,C). Indeed, the relative rates for amine additions to $Me_2NC{\equiv}CCOOMe$ are Me_2NH (24), piperidine (17), $MeNH_2$ (8), ▷NH (2·2), i-$PrNH_2$ (1·5), Et_2NH (1) and NH_3 (~0·1)[89, 143].

2. Additions

Because typical additions have already been treated, we merely point out here that Table 19 provides further acyclic products. Likewise, Table 20 catalogues the formation of some heterocyclic families. What follows, therefore, are several nitrogen examples which are in some sense special.

Tertiary amines are known to interact with activated alkynes. Presumably a zwitterion of the type analogous to **9**, **10** or **15** forms first[1]. Although the chemistry of such zwitterions from aliphatic amines has not yet been greatly developed, their reactions are theoretically interesting and should become synthetically useful. Trapping of the zwitterion with H^+ or CO_2 seems straightforward[1], but occasionally the adduct loses the amine so that its mediating role is invisible (equation 160)[334].

$$\mathrm{HC{\equiv}CCOOEt} + \mathrm{Et_3N} \xrightarrow{\text{inert solvent}} \begin{cases} E\text{-}\mathrm{EtOOCC{\equiv}CCH{=}CHCOOEt} \\ \xrightarrow{\text{little } H_2O} (E\text{-}\mathrm{EtOOCCH{=}CH})_2\mathrm{O} \\ \xrightarrow{\text{much } H_2O} E\text{-}\mathrm{Et_3\overset{+}{N}CH{=}CHCOO^-} \end{cases} \qquad (160)$$

Just as a secondary amine can transfer its proton internally once the zwitterion has formed (7), so too can tertiary amines deliver suitable groups. In Table 19 are examples of SnR_3, GeR_3 and SbR_2 making 1,3-shifts from nitrogen analogous to those of SPh and SePh in equation (161)[111]. Process (162) is again similar to (161)

$$\mathrm{R'OOCC{\equiv}CCOOR'} + \mathrm{Me_2NX} \longrightarrow Z\text{-}\mathrm{R'OOCC(NMe_2){=}CXCOOR'} \qquad (161)$$

$$\mathrm{X = SePh, SPh}$$

in that the tertiary nitrogen initiates the addition and a group attached to it 'departs', breaking the aziridine ring[423]. A related (vinylogous) example is found in equation (163)[424].

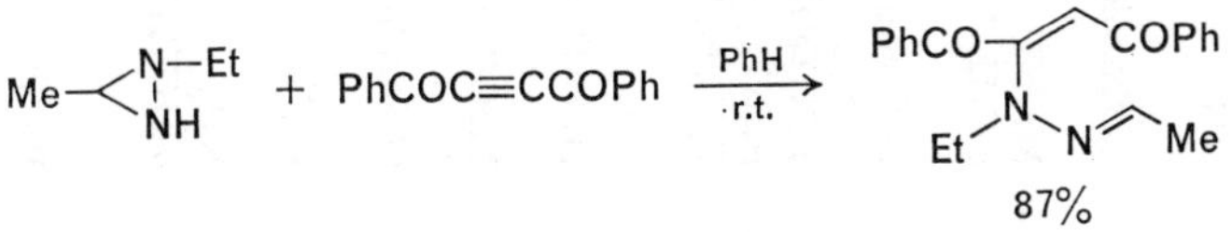

(162)

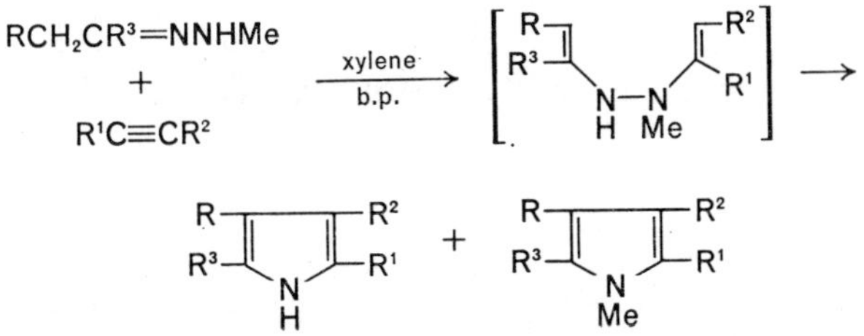

(163)

TABLE 19. Additions of nitrogen nucleophiles to alkynes, RC≡CR′, to form acyclic adducts[a, b]

R	R′	Nu	Medium (temp., °C)	Products (yield, %)	Reference
F_3C	CF_3	NH_3	Et_2O (0)	$F_3CCH_2C(CF_3){=}NH$ (25), *Z*-$F_3CC(NH_2){=}CHCF_3$ (37)	444
H	H	Carbazole	NaH, Me_2NCOMe, 1 atm (160)	Vinylcarbazole (96)	415, 416, 314*, 5*
c-C_5H_5	H	Me_2NH	EtOH (−30)	*c*-$C_5H_4{=}CMeNMe_2$	445
HC≡C	H	*c*-$C_6H_{11}NH_2$	(r.t.)	*c*-$C_6H_{11}NHCH{=}CHC{\equiv}CH$ (80)	131*, 446–448, 91, 8, 5*
Ph	H	2-Piperidone	NaH, Me_2SO (60)	*N*-Styryl-2-piperidone(*E*, *Z*)	449*
PhC≡C	H	Et_2NH	(70)	$PhC{\equiv}CCH{=}CHNEt_2$ (70)	344, 450*, 451*
NC	H	$TsNHNH_2$	$(HOCH_2)_2$ (60)	$(TsNHNH)_2CHCH_2CN$ (80)	452*
NC	Me	Et_2NH	EtOH (20)	*E*-$Me(NEt_2)C{=}CHCN$	97*, 453*, 454
OCH	NMe_2	$C_5H_{10}NH$	THF (20)	$Me_2NC(NC_5H_{10}){=}CHCHO$ (82), $Me_2NCH{=}C(NC_5H_{10})CHO$ (7)	143*, 113, 125, 332*
O_2N	Bu-*t*	$C_5H_{10}NH$	Et_2O (r.t.)	$C_5H_{10}NC(t\text{-}Bu){=}CHNO_2$	455
EtCO	H	$HC{\equiv}CCH_2NH_2$	MeCN (40)	*Z*-$EtCOCH{=}CHNHCH_2C{\equiv}CH$ (60)	456*, 90*, 114*, 141, 104*
PhCO	H	$MeOOCCH_2NH_2$	KOAc, MeOH (r.t.)	$PhCOCH{=}CHNHCH_2COOMe$ (60)	457–462*, 364*, 351*, 141, 5, 75*
Ph	CHO	(Triazolide)	Et_3N, THF (−20)	β-(4,5-Dicarbomethoxy-1,2,3-triazol-2-yl) cinnamaldehyde (71, *Z*/*E* = 27/44)	19*, 432*
Me_2COH	Br	Et_2NH	Bu_2O (b.p.)	$Me_2C(OH)COCH_2NEt_2$ (85), $Me_2C(OH)CH_2CONEt_2$	463*, 5*, 32*
Me_2COH	COMe	NH_3	PhH (40)	$Me_2C(OH)C(NH_2){=}CHCOMe$ (33)	464*, 461*, 465*
PhCO	COPh	*o*-$HOC_6H_4NH_2$	MeOH	$PhCOC(HNC_6H_4OH\text{-}o){=}CHCOPh$ (84)	351*, 364*, 363*, 466–469*, 37
PhCO	COPh	$Ph_2S^+NH^-$	$HCCl_3$ (r.t.)	$PhCOCH{=}C(COPh)N^-SPh_2^+$ (90)	433*, 440*, 470*, 439
MeOOC	H	5-NH_2-tetrazole	THF/Me_2SO 2/1 (b.p.)	5-[2-Methoxycarbonyl vinylamino]-2-[2-methoxycarbonyl vinyl]-tetrazole (16), 5-[2-methoxycarbonyl vinylamino]-1-[2-methoxycarbonyl vinyl]-tetrazole (8)	429, 423*

EtOOC	H	$(H_2NCH_2)_2$	MeCN (~0)	*E*-EtOOCCH=CHNHCH$_2$CH$_2$NH$_2$ (~20), (*E*-EtOOCCH=CHNHCH$_2$)$_2$ (~20)	471–474*, 429*, 368*
MeOOC	H	MeCH=NNHMe	Xylene (r.t.)	*E*-MeCH=NN(Me)CH=CHCOOMe (75)	542*, 475*, 424
MeOOC	H	Benzimidazole	MeOH	Methyl benzimidazol-1-ylacrylate (15)	476*
MeOOC	H	C_5H_5N	MeNO$_2$, Et$_2$O (r.t.)	4-O$_2$NCH$_2$-1-(MeOOCCH=CH)C$_5$H$_5$N	477*
EtOOC	H	2-MeC$_5$H$_4\overset{+}{N}\overset{-}{N}$CH=C(Me)COOMe	PhH (r.t.)	2-Me-3-[C(COOMe)=CHNHCH=C(COOEt)-Me]C$_5$H$_3$N (51), 2-MeC$_5$H$_4$N$^+$N$^-$CH=C(Me)CH=C(COOMe)(COOEt) (9)	441*
MeOOC	COOMe	α-Naphthyl-NH$_2$	EtOH (25)	Dimethyl *N*-(α-naphthyl)aminofumarate (100)	478–485*, 471, 368*
MeOOC	COOMe	PhCONHNH$_2$	MeOH (r.t.)	PhCONHN=C(COOMe)CH$_2$COOMe (72)	486–492*, 479–481*
MeOOC	COOMe	EtNHN=CHMe	EtOH, HOAc (0)	*E*-MeOOCCH=C(COOMe)N(Et)N=CHMe (72)	493–495*, 423
MeOOC	COOMe	Et$_2$NOH	Et$_2$O (0)	*E*-Et$_2$N(O)C(COOMe)=CHCOOMe (80)	323–325
MeOOC	COOMe	Sb(NMe$_2$)$_3$	(70)	(*Z*-MeOOC(NMe$_2$)=CCOOMe)$_3$Sb (38)	496*
MeOOC	COOMe	(Pyrrole)	*i*-PrCH(COOH)-NHCOPr-*i*, Ac$_2$O (130)	1-[*cis*-1,2-bis-Methoxycarbonyl vinyl]-2,5-diisopropyl-3,4-dicarbomethoxypyrrole (91)	497–499*, 476
MeOOC	COOMe	(Pyrazolone)	MeCN (b.p.)	2-Phenyl-1-(1′,2′-dicarboxyvinyl)-5-methyl-pyrazolidin-3-one, dimethyl 1,2,6,7-tetrahydro-7-methyl-5-oxo-2-phenyl-1,2-diazepine-3,4-dicarboxylate (56)	499*
MeOOC	COOMe	*o*-C$_6$H$_4$($\overset{-}{N}\overset{+}{P}Ph_3$)$_2$	—	*o*-C$_6$H$_4$[N=C(COOMe)C(COOMe)=PPh$_3$]$_2$, 2-methoxy-3-(carbomethoxytriphenyl-phosphoniummethyl)quinoxaline	500*
EtOOC	COOEt	Me$_2$NSnMe$_3$	Petroleum ether (r.t.)	*Z*-EtOOCC(NMe$_2$)=C(COOEt)SnMe$_3$ (98)	501*
EtOOC	COOEt	Me$_2$NGeEt$_3$	Et$_2$O (20)	EtOOCC(NMe$_2$)=C(COOEt)GeEt$_3$ (57, *Z*/*E* = 62/38)	112*
MeOOC	COOMe	Ph$_3\overset{+}{As}\overset{-}{N}$Ph	PhH	Ph$_3$As=C(COOMe)CC(COOMe)=NPh	502
MeOOC	COOMe	(2-Me-1-PhCO$\overset{-}{N}$)-C$_5$H$_4\overset{+}{N}$	PhH (r.t.)	2-Me-(3 or 5)-[*Z*-MeOOCC=CCOOMe-(NHCOPh)]C$_5$H$_3$N [9, (3-)/(5-) = 7/2]	434*, 505*
(*n*-Bu)$_2$PO	H	Et$_2$NH	NH$_3$ (−33)	*Z*-(*n*-Bu)$_2$POCH=CHNEt$_2$ (91)	338*, 365*, 503*
Br$^-$Ph$_3$P$^+$CH$_2$	H	*o*-H$_2$NC$_6$H$_4$CONH$_2$	MeCN	*E*-*o*-H$_2$NCOC$_6$H$_4$NHC(Me)=CHPPh$_3^+$Br$^-$ (96)	403, 504*

TABLE 19 (*cont.*)

R	R′	Nu	Medium (temp., °C)	Products (yield, %)	Reference
$Br^-Ph_3P^+CH_2$	H	$Ph_3\overset{+}{P}\overset{-}{N}Ph$	—	$Ph_3P{=}CHC(NPh)CH_2PPh_3^+Br^-$ (~100)	504*
$PhSO_2$	Me	c-C_2H_4NH	PhH (r.t.)	$PhSO_2CH{=}C(Me)NC_2H_4$-c (~100, $Z/E = 81/19$)	103*, 37*, 368, 453
MeOOC	Ph	(Thiazoline)	—	2-(2′-Carbomethoxy-1′-phenylvinyl)imino-3,4-diphenylthiazoline (46)	37b

[a] Additional examples of a similar type and leading references are indicated by an asterisk in the last column.
[b] Incomplete names or structures are indicated by parentheses.

TABLE 20. Additions of nitrogen nucleophiles to alkynes, RC≡CR′, to form cyclic adducts[a, b]

R	R′	Nu	Medium (temp., °C)	Products (yield, %)	Reference
PhCO	COPh	$Me_2S^+C(MeOOC)=C(COOMe)NAr^-$	PhMe (b.p.)	1-(*p*-Chlorophenyl)-2,3-dicarbomethoxy-4,5-dibenzoylpyrrole (50)	439*
$ClCH_2C(OH)$-Me	H	$(H_2NCH_2)_2$	EtOH (b.p.)	1-(2-Aminoethyl)-3-methylpyrrole (47)	506*, 5, 507*, 508*
MeOOC	COOMe	$PhCOCH_2NHAr$	$MeOH/CHCl_3$, 1/1 (b.p.)	Dimethyl-1-(*p*-bromophenyl)-4-phenylpyrrole-2,3-carboxylate (52), dimethyl phenacyl (*p*-bromoanilino) maleate (20)	487*, 480*, 462
MeOOC	H	$(Et)_2C=NNHMe$	Xylene (b.p.)	1-*R*-2-ethyl-3-methyl-4-carbomethoxypyrroles [20, (R = H)/(R = Me) = 1/1]	424*
MeOCH=CH	H	$MeC(NH_2)=CHCOOMe$	NH_4OAc, HOAc (~100)	2,4-(or 2,6)-Dimethyl-3-carbomethoxypyridine [20, (2,4-)/(2,6-) = 89/11]	509*, 508*, 510*
MeOOC	COOMe	ArCH=NR	—	1-(2-Phenylethyl)-2-aryl-3,4,5,6-tetracarbomethoxy-1,2-dihydropyridine	511*, 425, 495*, 431*
PhC≡C	Ph	*p*-$C_7H_7CH_2NH_2$	Me_2SO (145)	2-*p*-Tolyl-3,6-diphenylpyridine (51)	450*
MeOOC	COOMe	PhN=C(Ph)CH-(Me)C(Ph)=NH	THF (60)	2,3-Dicarbomethoxy-5-methyl-4,6-diphenylpyridine (61)	512*
MeOOC	H	*o*-$H_2NC_6H_4COPh$	MeOH (b.p.)	3-Carbomethoxy-4-phenylquinoline (31)	473*, 481, 367, 485
EtOOC	H	3,4-$(MeO)_2C_6H_3NH_2$	(r.t.)	Ethyl 6,7-dimethoxy-4-ethoxycarbonylmethyl-1,4-dihydroquinoline-3-carboxylate/ethyl 6,7-dimethoxy-4-ethoxycarbonylmethylquinoline-3-carboxylate sp = 5·6/1	513*
MeC≡CCO	Me	3,4-$Me_2C_6H_3NH_2$	(1) Ethanol (b.p.); (2) xylene (b.p.)	*N*-(3,4-$Me_2C_6H_3$)-2,6-dimethyl-4-pyridone (~90)	141, 514, 515
HC≡CPO(Ph)	H	$EtNH_2$	$MeOH–H_2O$ (r.t.)	1-Ethyl-4-oxo-4-phenyl-1,4-azaphosphorin (36)	188
MeOOC	H	C_5H_5N	Et_2O	1-Carbomethoxymethyl-2-carbomethoxyindolizine (0·6)	21*
MeOOC	COOMe	(2-Me-1-PhCO$\bar{N}$)-$C_5H_4N^+$	PhH (r.t.)	1-Benzoyl-2,3-dicarbomethoxy-3a-methyl-1,3a-dihydro-pyrazolo[1,5a]-pyridine (5), other products	505, 516, 434

TABLE 20 (*cont.*)

R	R′	Nu	Medium (temp., °C)	Products (yield, %)	Reference
MeOOC	H	2-($PhCOCH_2$)C_5H_4N	MeCN (r.t.)	Methyl *trans*-3(1-benzoyl-4-oxo-4H-quinolizin-3-yl)acrylate (~10)	426*, 517–519
MeOOC	COOMe	(Pyrazolinone)	MeCN (b.p.)	(Tetrahydrodiazepine) (25)	499
HC≡C	H	$PhNHNH_2$	—	1-Phenylpyrazole	131*, 520*,
t-Bu	C(C≡CBu-*t*)=NNLi(Tos)	—	EtOH (50)	1-*p*-Toluenesulphonyl-3-(*t*-butylethynyl)-4-*t*-butylpyrazole (~100)	521*,
Me_2N	CHO	N_2H_4	THF (25)	3-Dimethylaminopyrazole (56)	332*, 362*, 141*, 514*, 343*, 364*, 112*, 423*, 368* 363*, 492, 522–524
MeOOC	COOMe	$PhNHNH_2$	Pyridine (b.p.)	3-Carbomethoxy-1-phenylpyrazolin-5-one	481, 491*, 525
MeOOC	COOMe	PhCH=NNHPh	(145)	Dimethyl-1,3-diphenylpyrazole-4,5-dicarboxylate (16)	526*, 527*, 493, 475
Me_2COH	CH=CHOMe	N_2H_4	(160)	3-($Me_2C(OH)CH_2CH_2$)pyrazole (75)	528*, 529*
PhCO	COPh	$PhC\overset{+}{N}\overset{-}{N}PH$	MeCN, Et_3N (r.t.)	1,3-Diphenyl-4,5-dibenzoylpyrazole (~100)	507
PhCO	COPh	$NH_2OH \cdot HCl$	(b.p.)	Phenyl 3-phenyl-5-isoxazolylketone	363*, 343*, 325, 151, 362
MeOCH=CH	H	$NH_2OH \cdot HCl$	H_2O (70)	5-(or 3-)Methylisoxazole [60, (5-)/(3-) = 3/2]	456*
PhCO	COPh	Ph_2SNH	$CHCl_3$ (b.p.)	3-Benzoyl-5-phenylisoxazole (75)	433*
p-C_7H_7-NHCOOC-(Me_2)	H	—	C_5H_5N (b.p.)	3-*p*-Tolyl-4-methylene-5,5-dimethyl-2-oxazolidinone (86)	187*
MeOOC	H	3,4-Dihydroisoquinoline *N*-oxide	Me_2NCOH (80)	Methyl 6,10b-dihydro-5*H*-isoxazolo [3,2-a]isoquinoline-1-carboxylate (83)	22*
MeOOC	COOMe	ArCH=NCH$_2$CH$_2$OH	MeOH	(Oxazolidine) (82)	511*
MeOOC	COOMe	$(H_2N)_2C$=NH	—	Methyl 2-amino-5-oxo-$\Delta^{4,a}$-2-imidazolinylacetate	541*

HC≡C	H	$H_2NC(NHCN)=NH$	NaOMe	2-Cyanamino-4-methylpyrimidine (62)	530*
MeOOC	COOMe	PhCH=NCPh=NH	PhH (b.p.)	2,4-Diphenyl-5,6-dicarbomethoxy-pyrimidine (40)	531
MeOOC	COOMe	*E*-(2-H_2N,3-Cl)-$C_6H_3C(Ph)=NOH$	MeOH (r.t.)	2-Carboxy-6-chloro-1,2-dihydro-4-phenyl-2-quinazolineacetic acid dimethyl ester (72)	484
MeOOC	Ph	3-H_2N-1,2,4-triazole	*n*-BuOH (b.p.)	7-Oxo-5-phenyl-7,8-dihydro-*s*-triazolo-[4.3-a]pyrimidine (10%)	428*
(2-Benzimid-azole)CH_2S	H	—	EtOH, EtO^-Na^+ (b.p.)	3-Methylthiazolo[3,2-a]benzimidazole (60)	532
PhCO	COPh	(Thiazolidine)	$CHCl_3$ (r.t.)	2-Anilino-4,5-dibenzoylthiazole (74)	37*
MeOOC	COOMe	(Azoimine)	$HCONMe_2$ (r.t.)	(Dibenzoimidazolinodihydrodiazepine) (80)	437*
MeOOC	COOMe	$(H_2NCH_2)_2$	MeCN (0)	*Z, E*-2-Oxo-3-carbomethoxymethylene-piperazine	471*, 351
PhCO	COOMe	*o*-$HOC_6H_4NH_2$	Ether/MeOH (b.p.)	3-Carbomethoxymethylene-3,4-dihydro-1,4-benzoxazin-2-one	351, 6a
MeOOC	COOMe	*o*-$C_6H_4(NH_2)$	MeCN (0)	*Z, E*-2-Oxo-3-carbomethoxymethylene-1,2,3,4-tetrahydroquinoxaline	471, 351, 6a, 500*, 533
MeOOC	COOMe	*o*-$H_2NC_6H_4\overset{-}{N}\overset{+}{P}Ph_3$	—	2-Hydroxy-3-(carbomethoxytriphenyl-phosphonium methyl)quinoxaline, 2-methoxy-3-carbomethoxymethyl-quinoxaline	500*
MeOOC	COOMe	2,4,5,6-$(H_2N)_4$-pyrimidine sulphate	H_2O/EtOH, NaOAc (r.t.)	Methyl 2,4-diamino-7(8*H*)-pteridinone-6-acetate (~50)	535*
Ph	$PPh_3^+Br^-$	*o*-$C_6H_4(NH_2)_2$	$CHCl_3$ (b.p.)	2-Phenylbenzimidazole (27), $MePPh_3^+Br^-$	147*
Ph	$PPh_3^+Br^-$	$Na^+N_3^-$	$HCONMe_2$ (60)	4-Phenyl-5-triphenylphosphonium-1,2,3-triazolyl ylid (93)	432a*
Ph	Ph	$Na^+N_3^-$	Me_2SO (140)	4,5-Diphenyl-1,2,3-triazole (16)	536*, 19*, 185
Ph	$Ph_3P^+Br^-$	1,8-$(H_2N)_2C_{10}H_6$	—	(Naphthopyrimidine)	147*
MeOOC	Ph	3-H_2N-*s*-triazole	*n*-BuOH (b.p.)	7-Oxo-5-phenyl-7,8-dihydro-*s*-triazolo-[4,3-a]pyrimidine (10)	428a*
MeOOC	Ph	3-H_2N-*s*-triazole	H_2O, KOH (b.p.)	5-Oxo-7-phenyl-4,5-dihydro-*s*-triazolo-[1,5-a]pyrimidine (30)	428b*

TABLE 20 (*cont.*)

R	R′	Nu	Medium (temp., °C)	Products (yield, %)	Reference
MeOOC	H	2-PhCH$_2$-benzimidazole	(120–180)	Methyl *E*-(2-benzylbenzimidazol-1-yl) acrylate (23), methyl 3-(4-benzyl-2,3-bismethoxycarbonyl-4,5-pyrrolo-[1,2-a]dihydroquinoxalin-5-yl)-acrylate (6)	476*
MeOOC	COOMe	5-NH$_2$-tetrazole	THF/Me$_2$SO 2/1 (r.t.)	2-Azido-6-methoxy-4-methoxycarbonyl-pyrimidine (5), (5- or 7)-oxo (7- or 5)-carbomethoxytetrazolo[1,5-a]-dihydropyrimidine (31)	429
EtOOC	Ph	1-H$_2$N-isoquinoline	EtOH (b.p.)	2-Oxo-4-phenyl-2*H*-pyrimido[2,1-a]-isoquinoline (55)	430*
MeOOC	H	3-H$_2$N-benzisoxazole	EtOH (b.p.)	(2 or 4)-Oxo-(2 or 4)*H*-pyrimido [1,2-b]benzisoxazole (15 + 8)	537*
MeOOC	Me	3-H$_2$N-pyrazole	EtOH (b.p.)	(5 or 7)-Oxo-(7 or 5)-methyl pyrazolo-[1,5-a]dihydropyrimidine (28 + 12)	427*, 538, 539
MeOOC	COOMe	PhNNC(Me)NNHPh	PhH (b.p.)	(1,4-Dihydro-1,2,4-triazine) (13), MeOOCRC=CHCOOMe (21)	540

[a] Additional examples of a similar type and leading references are indicated by an asterisk in the last column.
[b] Incomplete names or structures are indicated by parentheses.

The reactions of tertiary imino nitrogen and activated alkynes (equation 164) have received a great deal of attention[1, 9, 10]. Again there are possibilities in the zwitterion

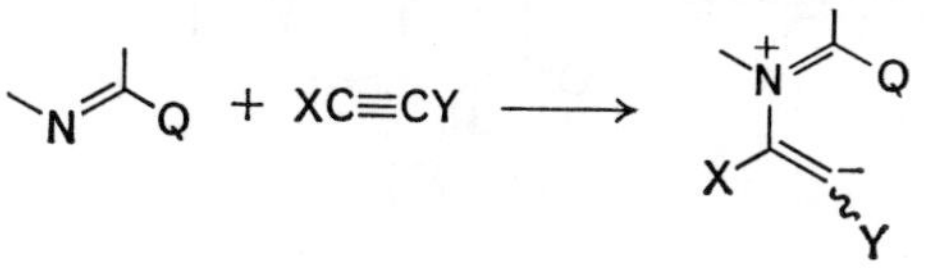

(164)

for internal, e.g. at Q, or external attacks. These are implicit in several syntheses in Table 20 of pyridines, quinolizines, indolizines, pyrimidines, etc. and are illustrated in equations 165–167. A variety of bicyclic heterocyclics have been prepared from a

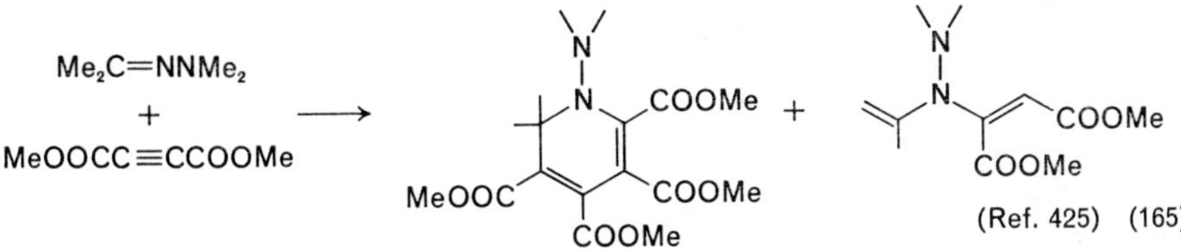

(Ref. 425) (165)

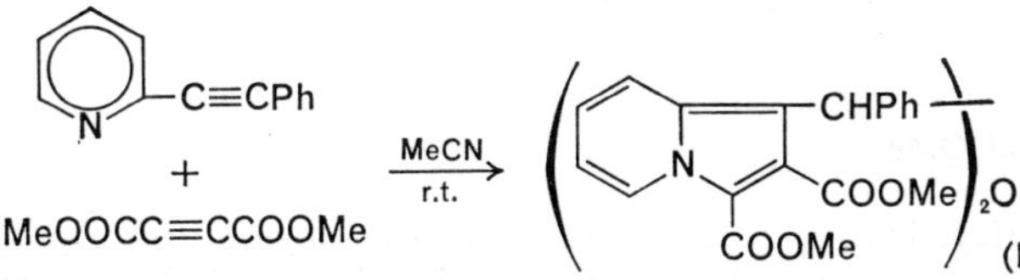

(Ref. 426) (166)

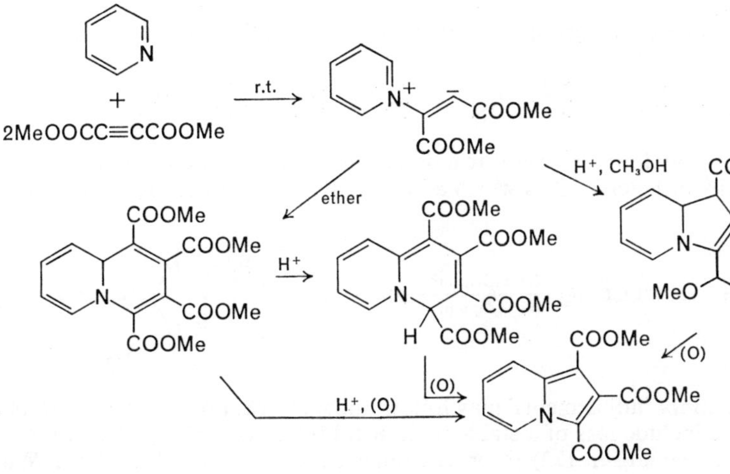

(Ref. 21) (167)

1,3-binucleophile of the type in equation (168)[427–430]. It appears that either the primary or nuclear (possibly tertiary) nitrogen may attack the triple bond. Similar reactions of 3-amino-*s*-triazole[428], 5-aminotetrazole[429], 1-aminoisoquinoline[430], etc. are given in Table 20.

The sensitivity of the zwitterion in these reactions is seen in equation (169) in which the favoured product depends on the medium: in anhydrous methanol a yield

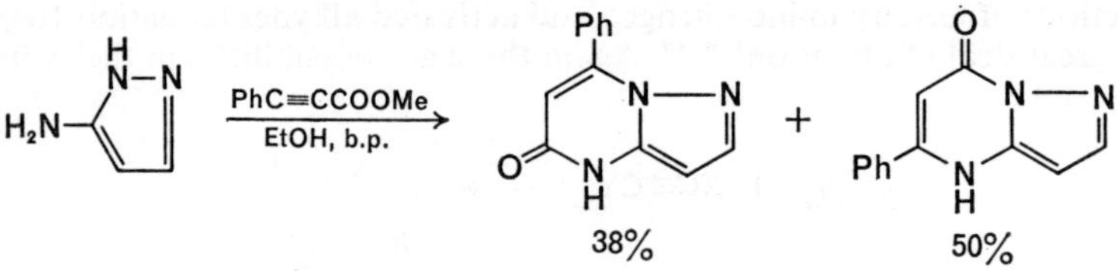
(168)

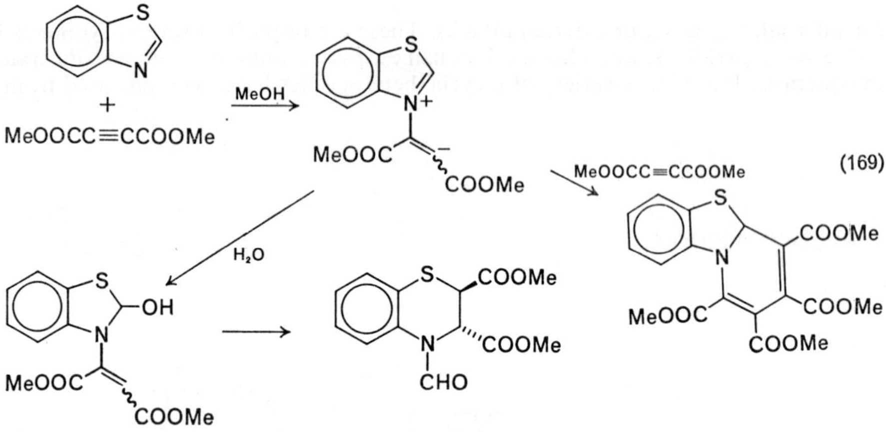
(169)

of 5% of the tricyclic (upper path) and 0% of the bicyclic (lower path) product are obtained; in water/methanol 1/6, 0% of the tricyclic and 85% of the bicyclic product are obtained[431].

Ylids form an interesting group. Beginning with diverse 'imido-onium' sites, their

anionic nitrogen may initiate several reaction sequences. The ylid of equation (170) is one of a family of heterocyclics which adds simply to an alkyne (Table 19)[432]. On

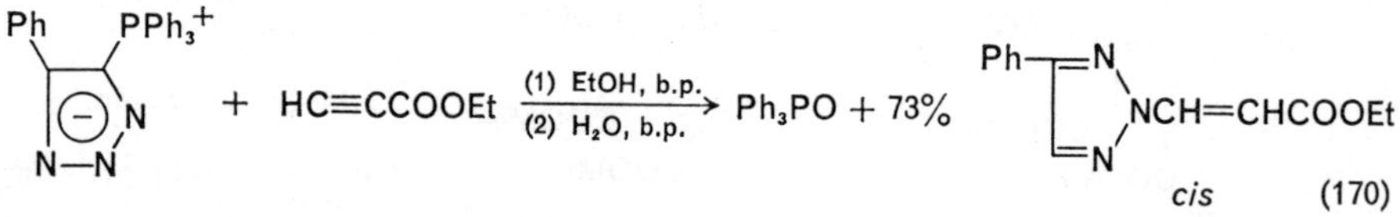
(170)

the other hand, minor 'adjustments' may follow initial attack, providing a π system is accessible. These include loss of a stable molecule, Ph_2S in equation (171)[433], proton (group) transfer (equation 172)[434] or rearrangement (equation 173)[435–438]. With activated alkynes the sulphonium imidoylides yield 1,2-adducts first, which in turn

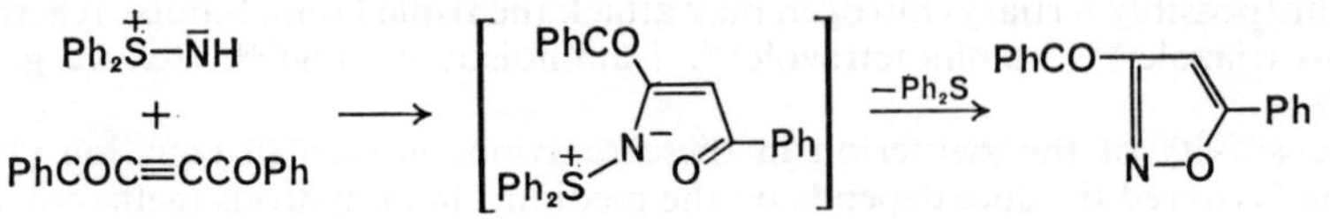
(171)

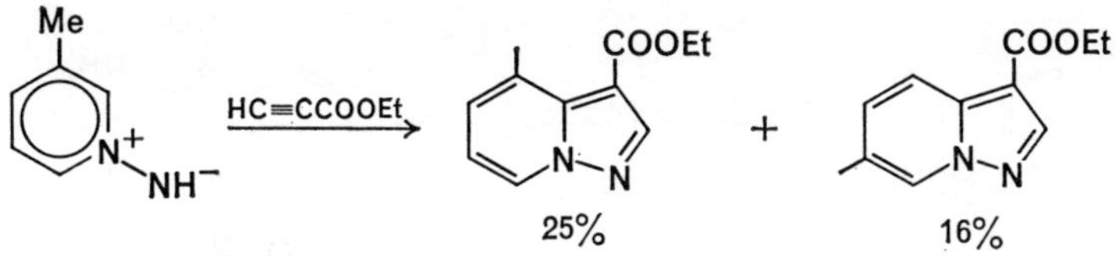

(172)

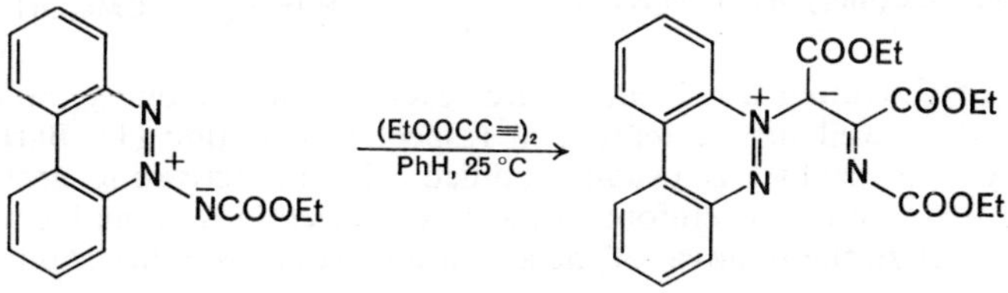

(173)

may add to alkynes to yield pyrroles (equation 174)[439, 440]. Similar first products are obtained from phosphonium and arsonium ylids (Table 19). An analogue to the 1,3-migration, which presumably gives the first adduct in equation (174), also holds

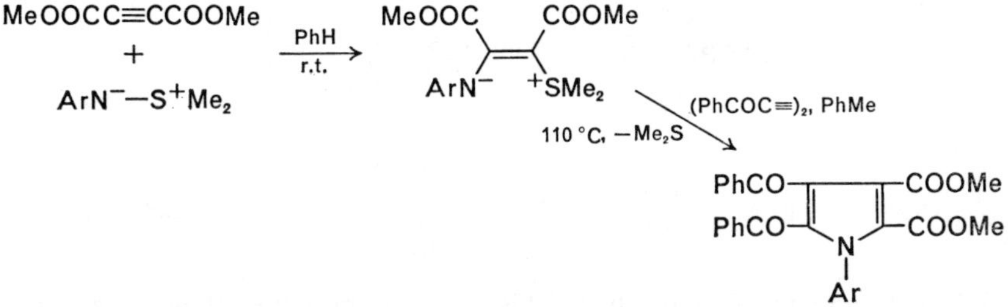

(174)

in equation (173). This same theme is seen in equation (175) except that there is superimposed an interesting carboalkoxy transfer for one product and probably a 3,3-shift to the other[441].

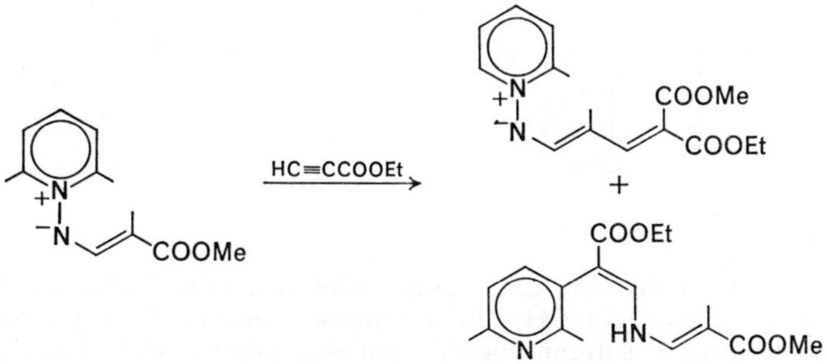

(175)

In Section III.E we noted that alkynes may be attacked by hydroxylamines at oxygen or nitrogen and illustrated nitrone formation in equation (111). The possibility of rearrangement of the nitrogen and oxygen products was also indicated. Here we give an example of the former (equation 176)[442]. By comparison, isoxazole formation from keto alkynes is routine (equation 177)[343].

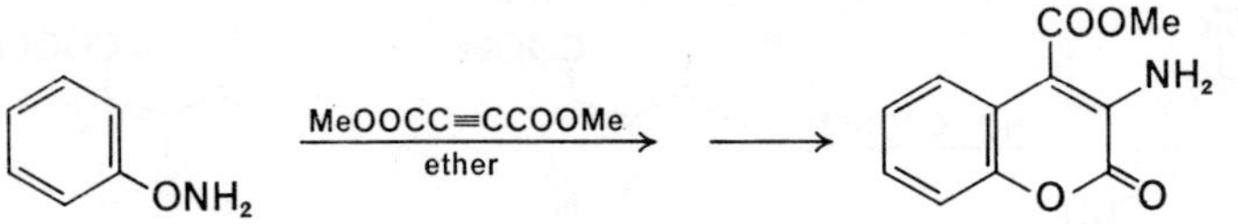

(176)

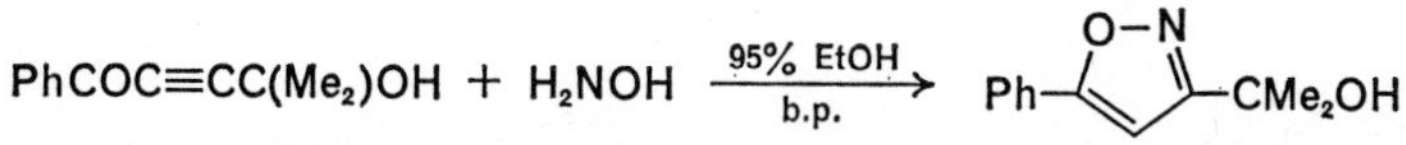

(177)

The notion of activating a reagent before reaction and removing the activating group later has received little attention in respect to equation (1). Barring fairly drastic surgery, the product is usually saddled with its activating group (RCO, ROCO, NO_2, RSO_2, etc.). Therefore, some novel exceptions should be noted: in equations (178)–(180), the removal of the activating groups is either spontaneous or

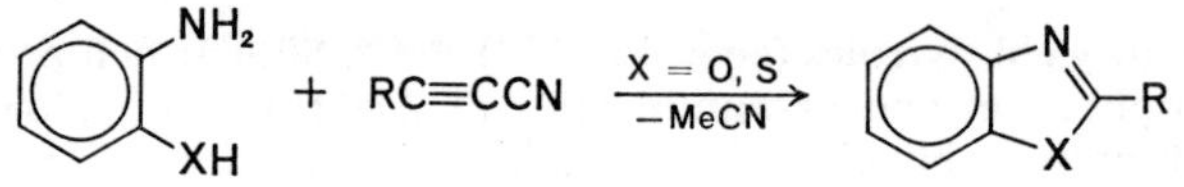

(Ref. 368) (178)

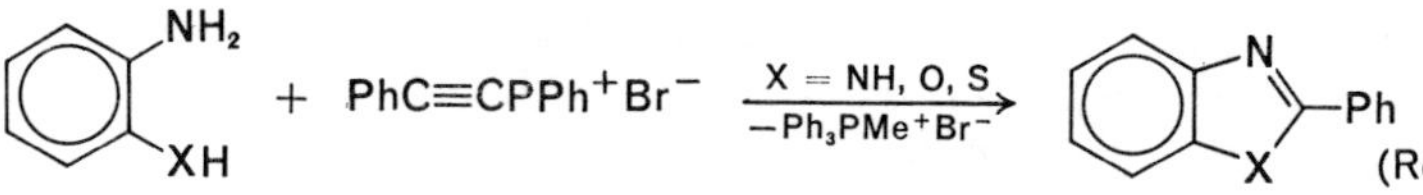

(Ref. 147) (179)

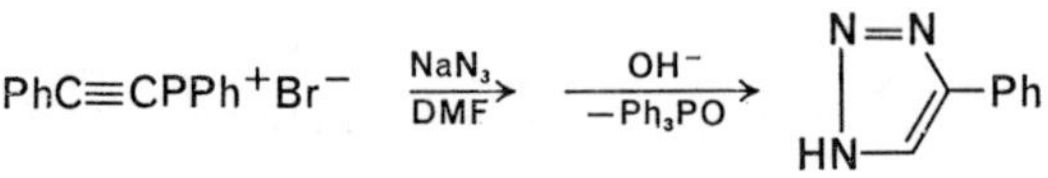

(Ref. 432) (180)

readily effected from the first product. This synthetic principle of *activate–react–remove* is a familiar notion in molecular cycloadditions and others which are or may be 1,3-dipolar[25], e.g. equations (7) and (181)[443].

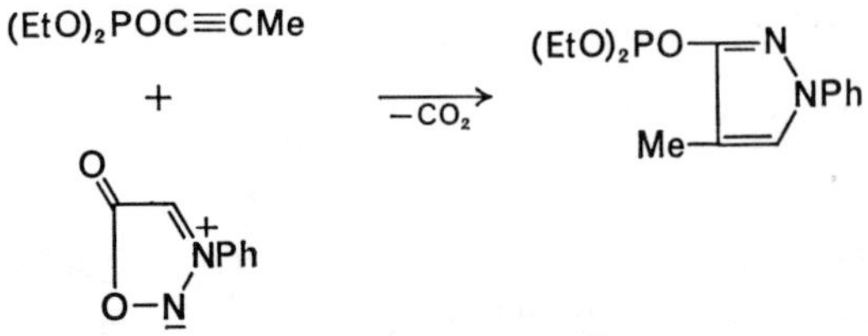

(181)

H. Carbon

Carbanions (C^-) do not react readily with unactivated alkynes. Thus most carbanionic or carbon acid (CH) additions possess some facilitating feature, be it an activating substituent, solvent, coordination site, catalyst, etc[1]. Among all of the nucleophilic sites, carbon provides the greatest mechanistic variety. Inevitably, this often takes one to an ill-defined border region between unequivocal additions according to equation (1) and distinctly non-nucleophilic additions. Our plan is to treat carbon acids and organometallics, then pick up methylides and dipolar species, in all cases continuing as long as the carbanionic character appears to dominate the reagent which attacks the alkyne.

We note again that allenic species frequently intrude in this chemistry. Abstraction of the propargylic proton, e.g. BuLi+R′C≡CCH$_2$OR [543], or attacks on a conjugated enyne or diyne, e.g. EtCaX+RC≡C—CH=CH$_2$ [544], which lead to the anion of equation (182), will be excluded. In one of our examples (equation 183),

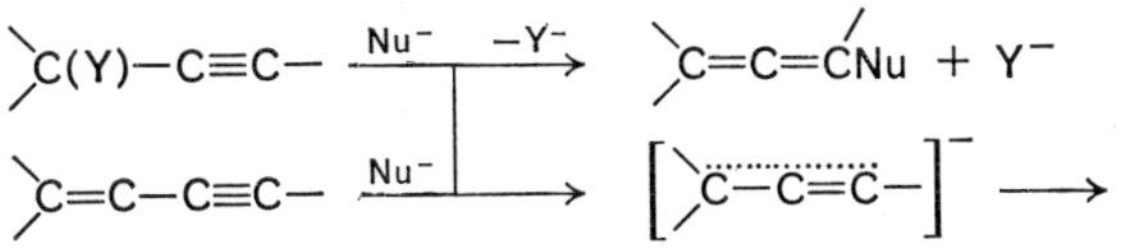

(182)

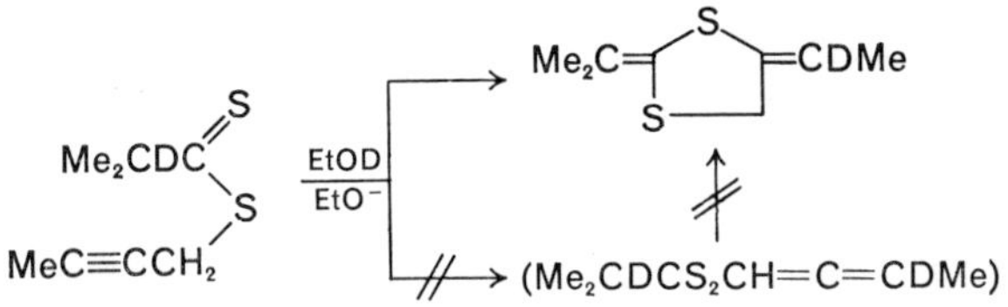

(183)

the path of deuterium from reactant to medium to product precludes the propargyl anion as an intermediate[393]. Otherwise, attacks on the triple bond which produce allenes (S_N') will usually be omitted.

Anions from the stronger carbon acids are accessible and give expected products (see Table 22, p. 897). Invariably, the alkyne carries an activating substituent. Cyclic products form when possible (the additions of the azallyl anions of equation (184)

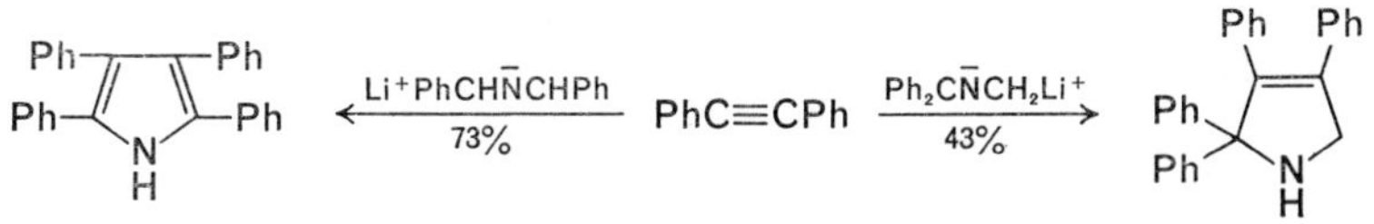

(184)

are perhaps better regarded as cycloadditions[545]). There are a fair number of examples in which a second mole of alkyne behaves as a coelectrophile (equation 185)[546]. Some

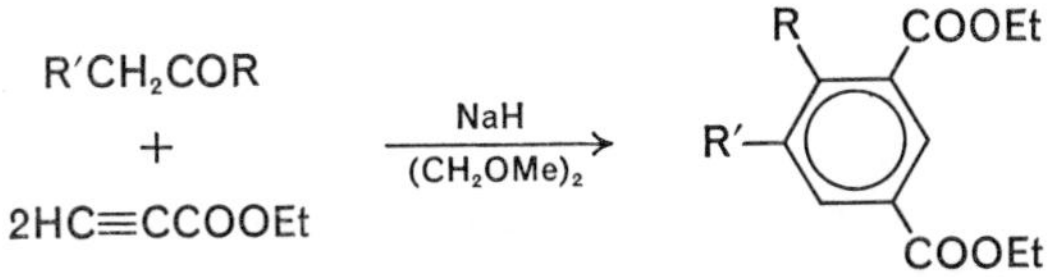

(185)

R′ = 1-H$_3$C$_3$N$_2$, R = Ph; R′ = Ph, R = H or Ph

of these 'typical' features are indicated in the syntheses of heterocyclics in equations (186)[547] and (187)[548], although such examples often involve enamine or amine attacks (see below). With respect to regioclosure, the absence of a cyclohexadienone product in equation (188) indicates that *exo–dig* has won over *endo–dig* closure[141, 179]. This preference is confirmed in the base-catalysed cyclizations of equation (69) and of HC≡C(CH$_2$)$_3$CH(COOEt)$_2$ [549]; on the other hand, the *exo–dig* to *endo–dig* ratio is 1/1·4 in equation (189)[549].

Carbon nucleophiles from organometallics containing Li, Mg, Ca, Zn and Cu are undoubtedly delivered from associated species. By writing them as RLi, RMgX, etc. we are, of course, indicating the simplest of the associated forms which might better be represented as R$_n$Li$_n$·(solvent)$_x$, etc. In some cases, at least, there is no

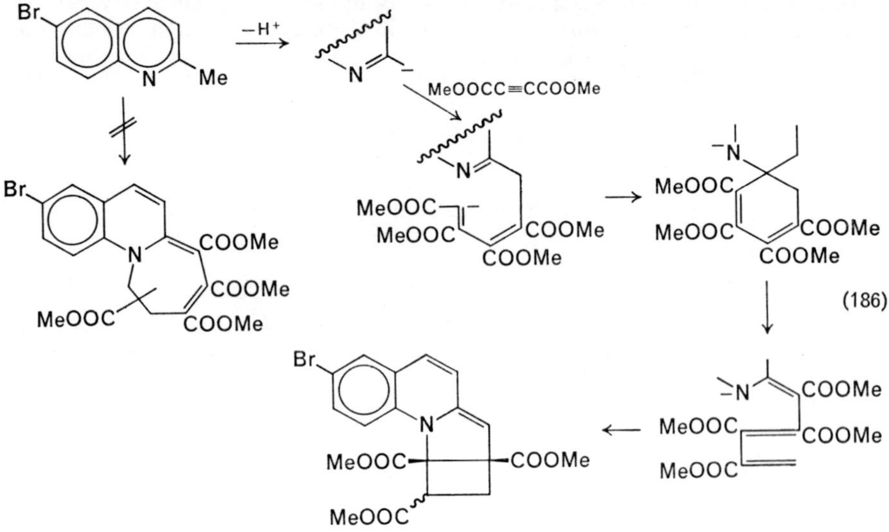

(186)

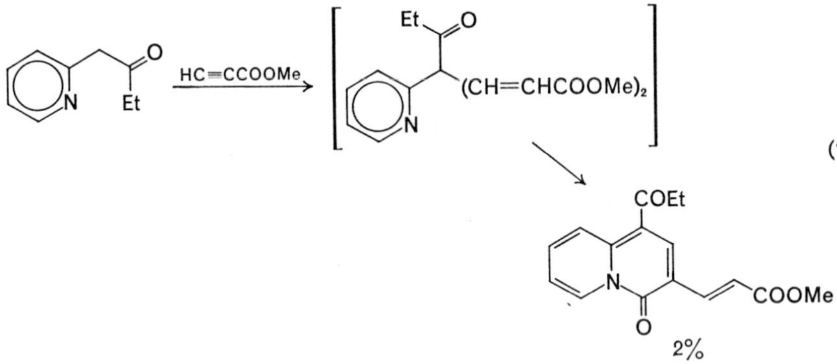

(187)

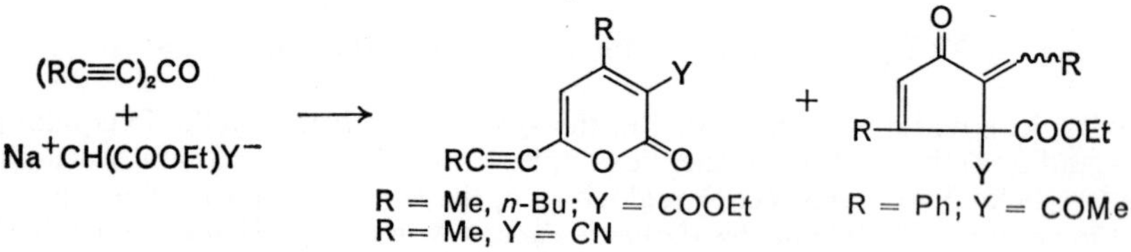

(188)

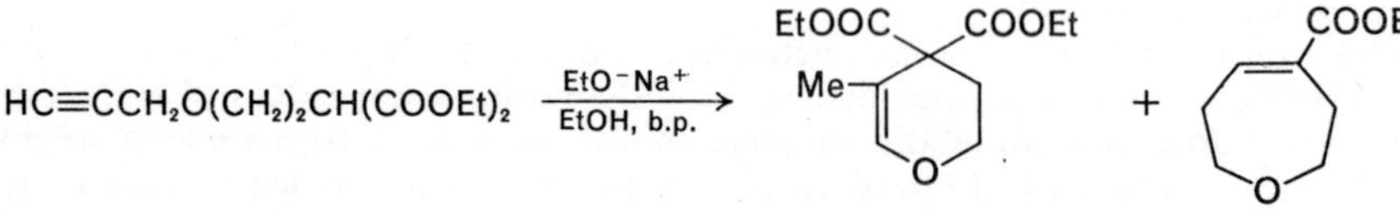

(189)

question that an anionic intermediate has been produced, e.g. equation (190). The added amine in equation (190) may promote the reaction by coordinating Li and

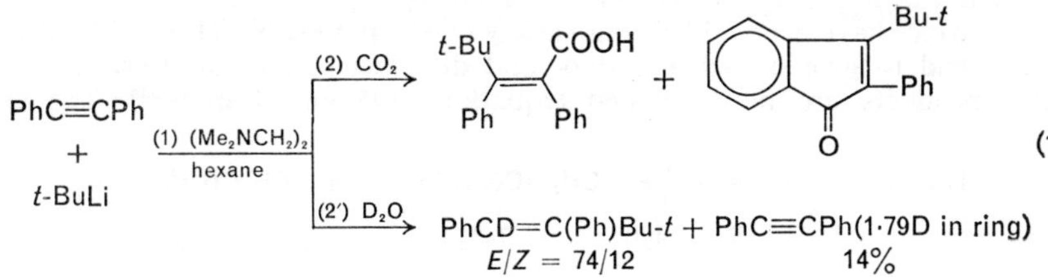

(190)

thereby reduce the size of polymeric aggregates and/or 'liberate' the carbanion[108]. A similar role may be assigned to a site within the alkyne molecule, as in equation (191)[550]. It must be admitted that all of these are *ad hoc* rationalizations: confronted

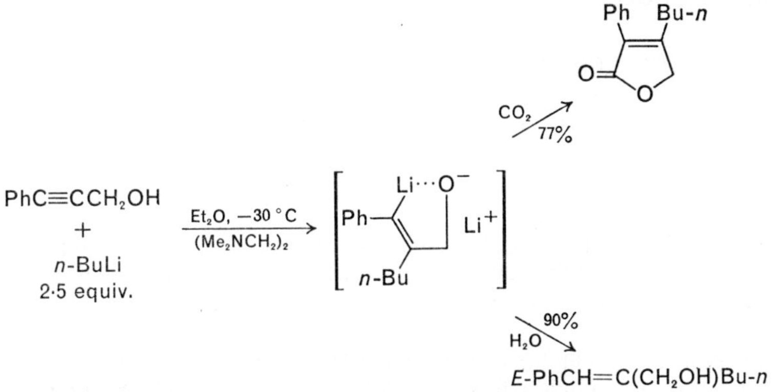

(191)

with a 'new' complex in equation (192), one might be hard pressed to predict the *syn* specificity of the first (Grignard) product and its probable fate in the work-up[551].

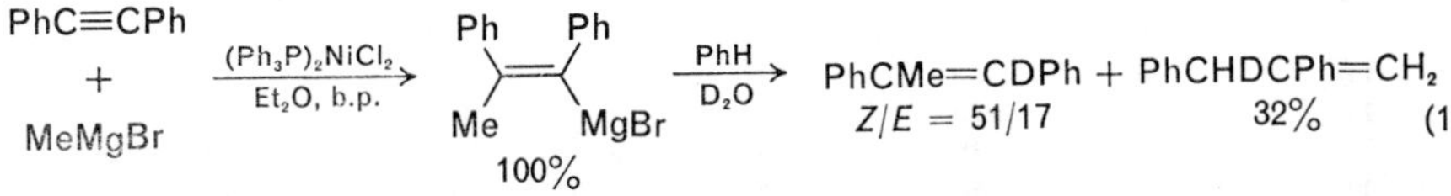

(192)

It has been pointed out in connection with **4–6** that selectivities are variable. Organolithiums appear to favour *syn* addition, e.g. equation (190); when internal coordination is available, e.g. equation (191), the *anti* adducts are preferred. The latter effect appears to prevail for Grignard reagents too (see Table 22, equation 193)[110, 552, 553]. That is, an intermediate such as that pictured in equation (191) is

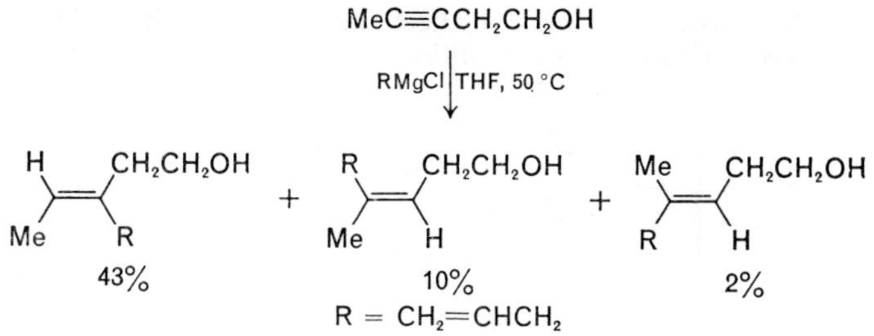

(193)

formed and is 'held' until it is destroyed by acid, e.g. D_2O. When the binding site is moved one atom further from the triple bond, the *anti* preference applies, although the reaction does become selective rather than regiospecific[553].

Organozincs have been added to a variety of alkynes (Table 21 and 22). Terminal alkynes tend to give branched mono- and diadducts (equation 194)[554], although 'linear' products are not excluded (equations 195 and 196)[555, 556]. The special

$$HC{\equiv}CCH_2OH \xrightarrow[RZnBr]{20\,^{\circ}C,\ 3h} \underset{74\%}{CH_2{=}C(CH_2OH)R} + \underset{4\%}{CH_3C(CH_2OH)R_2} \quad (R = \text{allyl}) \qquad (194)$$

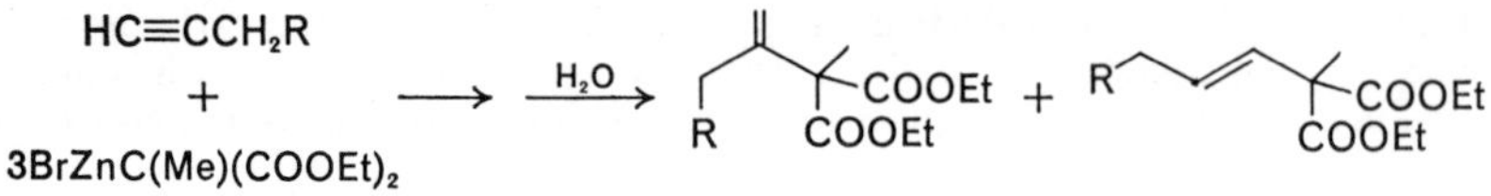
(195)

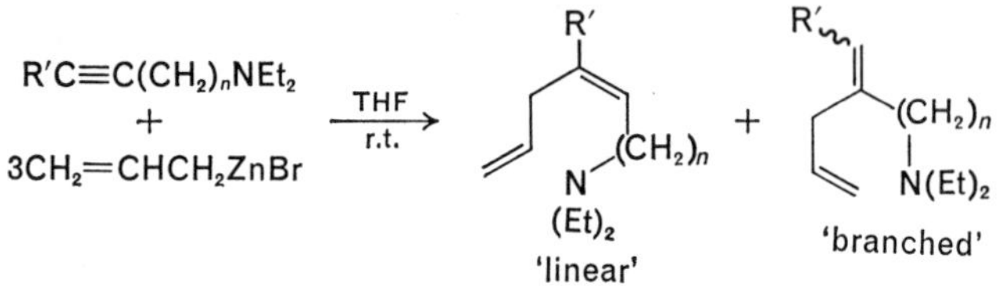
(196)

character of the terminal alkynes is seen in the regioselective products of process (196) given in Table 21. Trapping evidence indicates that the terminal alkyne actually forms the intermediate $(BrZn)_2C{=}C(CH_2CH{=}CH_2)CH_2N(Et)_2$, when $R' = H$ in

TABLE 21. Products of addition of $H_2C{=}CH_2CH_2ZnBr$ to alkynes $R'C{\equiv}C(CH_2)_nNEt_2$ in tetrahydrofuran at reflux temperature according to equation (196)[556]

R′	n	Time (h)	Yield	'Linear' (%)	'Branched' (%)
H	1	23	70	0	100
H	2	23	60	0	100
H	3	46	16	28	72
H	4	46	25	0	100
Me	1	96	trace	—	—
Me	2	23	65	100	0
Me	3	46	25	100	0
Me	4	46	trace	—	—

equation (196)[556]. (It would be simple to check for mono- and dizinc intermediates in equation 197[554].) Surprisingly, these additions are facile not only with many α- and

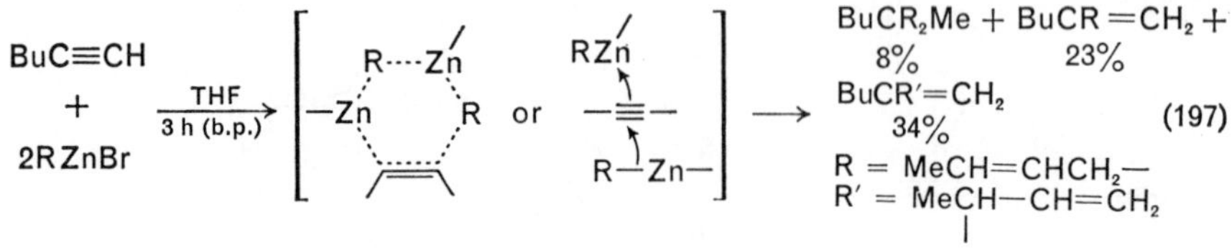
(197)

TABLE 22. Additions of carbon nucleophiles to alkynes, RC≡CR′, to form acyclic and cyclic adducts[a, b, c]

R	R′	Nu	Medium (temp., °C)	Product(s) (yield, %)	Ref.
PhNHC(S)	Ph	$BrCH_2NO_2$	Et_3N, Me_2CO (b.p.)	2-Nitro-3-phenyl-5-anilinothiophene (50)	596*
9-Acridinyl	H	$H_2C(CN)_2$	NaOH, EtOH (r.t.)	3-(9-Acridinylidene)-1,1-dicyanopropene (91)	597*
$Ph_2P(O)$	Ph	$Na^{+-}CH(COOEt)Ph$	THF (r.t.)	$Ph_2POCH{=}C(Ph)CH(Ph)COOEt$ (75)	146*
MeC≡CCO	Me	$Na^{+-}CH(CN)COOEt$	DMF/PhH (r.t.)	3-Cyano-4-methyl-6-propynyl-2-*H*-pyran-2-one	141*
EtOOC	H	—	$C_5H_{10}NMe$, PhH	*E*-EtOOCC≡CCH=CHCOOEt (60)	334*
MeOOC	H	$(Pz)COCH_2R$	NaH, $(CH_2OMe)_2$ (0)	1-(1-Pyrazolyl)-2-phenyl-3,5-dicarbomethoxy-benzene (40)	546
EtOOC	H	ArCH(CN)Et	Triton B, dioxane/MeOH (80)	EtOOCCH=CHC(CN)Et(C_6H_4-*p*) (48)	599
$Ph_2C(OMe)$	COPh	$PhCOCH_2Ph$	$K^{+-}OBu$-*t*, THF (r.t.)	$PhCOCPh{=}C(CH_2COPh)COMePh_2$ (84)	466*, 600*
$(EtOOC)_2$-$CHCH_2CH_2$-OCH_2	H	—	NaOEt, EtOH (b.p.)	3-Methylene-4,4-dicarbomethoxypyran (32), 2,3,6,7-tetrahydro-5-carbomethoxyoxapin (46)	549*
Me_2CHCS_2-CH_2	Me	—	NaOEt, HOEt	2-Isopropylidene-4-ethylidene-1,3-dithiolane (75), 2-isopropylidene-4-ethyl-1,3-dithiole (25)	393*
MeOOC	COOMe	(α-Cyanolactone)	Et_3N, THF (r.t.)	α-Cyano-α(1,2-dicarbomethoxyethylene)-γ-butyrolactone	602, 598*
PhCO	Ph	9-EtOOC-fluorene	NaOEt, Et_2O (r.t.)	Ph(9-carboethoxyfluorenyl-9)C=CHCOPh	601
MeOOC	COOMe	Oxindole	NaH, dioxane (r.t.)	(2-Oxo-3-indolylidene)=C(COOMe)-CH_2COOMe (80)	483
MeOOC	COOMe	2-Methylindole	PhH (b.p.)	Dimethyl 3-(2-methylindolyl) maleate (53)	497*
MeOOC	MeOOC	(Pyrrole)	Et_2O (−70)	3,4-Dimethyl-5-ethoxy-2-pyrrolyl-CCOOMe=CHCOOMe (83)	603
Ph	Ph	*n*-BuLi	C_6H_{14}, $(Me_2NCH_2)_2$ (r.t.)	*E*-Ph(Bu-*n*)C=CHPh (69)	108*
Ph	CH_2OH	*n*-BuLi	Et_2O, $(Me_2NCH_2)_2$ (−30)	*E*-PhCH=C(CH_2OH)Bu-*n* (77)	550
Ph	$CONMe_2$	MeLi	Et_2O (r.t.)	*E*-PhC(Me)=$CHNMe_2$ (86)	552*
Ph	$CONH_2$	MeMgBr	Et_2O	*E*-PhC(Me)=CHCOMe (68), PhC≡CCOMe (17)	552*
Me	CH_2OH	$CH_2{=}CHCH_2MgCl$	Et_2O (50)	*E*-MeCH=C(CH_2OH)CH=CH_2 (85)	553*
$MeOCH_2$	CH_2N-(Pr-*n*)$_2$	EtMgBr	(*n*-Bu)$_2$O (80)	*E*-$MeOCH_2$C(Et)=CHN(Pr-*n*)$_2$ (43)	110*
Bu	H	$PhCH{=}CHCH_2ZnBr$	THF (b.p.)	BuC(CH_2CH=CHPh)=CH_2 (55)	554*

TABLE 22 (*cont.*)

R	R′	Nu	Medium (temp., °C)	Product(s) (yield, %)	Ref.
EtNHCH-(Ph)CH_2	H	$H_2C{=}CHCH_2ZnBr$	THF (b.p.)	$EtNHCHPhCH_2C(CH_2CH{=}CH_2){=}CH_2$ (52)	556*
$EtOCH_2CH_2$	H	$(EtOOC)_2C(Me)ZnBr$	(42)	$(EtOOC)_2C(Me)C(CH_2CH_2OEt){=}CH_2$ (56)	555*
$HOCH_2$	H	$(EtOOC)_2CMeZnBr$	(42)	α-Carboethoxy-α-methyl-β-methylene-γ-butyrolactone (60)	557*
H	H	$(n\text{-}C_7H_{15})_2CuLi$	(1) Et_2O (−40), (2) I_2	$Z\text{-}n\text{-}C_7H_{15}CH{=}CHI$ (80)	613*
H	H	EtCu, $MgBr_2$	Et_2O/C_5H_{12} 1/1 (−20)	*Z,Z*,-$(EtCH{=}CH)_2$ (20), *Z,Z,Z*-$(EtCH{=}CHCH_2$ (10), *Z,Z,Z,Z*-$(EtCH{=}CHCH{=}CH)_2$ (25)	561*, 563*
Ph	COOMe	MeCu	Et_2O (r.t.)	$PhCMe{=}CHCOOMe$ (*E*/*Z* = 35/63)	552*
Ph	COMe	Me_2CuLi	THF (−80, 20)	$PhCMe{=}CHOMe$ (*E*/*Z* = 8/92)	567*, 559*
$MeOCH{=}CH$	H	*n*-BuCu, $MgBr_2$	Et_2O (−15)	$BuCH{=}CHCH{=}CHOMe$ (54), *Z*-$BuCH{=}C(Bu)CH{=}CHOMe$ (12)	565*
$MeSO_2$	H	EtMgX/CuBr 1/3	THF (−70)	$EtCH{=}CHSO_2Me$ (~80, *E*/*Z* = 99/1)	605*
EtOOC	H	(*Z*-$MeCH{=}CH)_2CuLi$	Et_2O (−20)	*Z,E*-$MeCH{=}CHCH{=}CHCOOEt$ [77, (*Z,E*)/(*Z,Z*) = 95/5]	606–608*, 560
MeOOC	H	$(H_2C{=}CH$—△—$)_2CuLi$	THF/Et_2O (−78)	*E,E*-$H_2C{=}CH$—△—$CH{=}CHCOOMe$ (80), 1-carbomethoxy-2,5-cycloheptadiene (15)	614
$(Ph)_2P$	Me	*t*-$Bu_2CuMgCl$	THF (−60, 20)	*E*-$Ph_2PCH{=}C(Me)Bu$-*t* (100)	609*
$(EtO)_2CH$	H	*n*-$Bu_2CuMgBr$	Et_2O (−45)	*n*-$BuCH{=}CHCH(OEt)_2$ (65), *n*-$BuCH{=}CHCHO$ (20)	610*
$EtSCH_2CH_2$	H	*n*-BuCu,$MgBr_2$	Et_2O (−20)	$H_2C{=}C(Bu)CH_2CH_2SEt$ (54), *E*-$BuCH{=}CHCH_2CH_2SEt$ (36)	565*
MeS	*i*-Pr	MeMgBr,CuBr	THF (30)	*Z*-Me(*i*-Pr)$C{=}CHSMe$ (80)	611*, 604*
EtSO	*n*-Bu	Me_2CuLi	THF (−78)	*Z*-*n*-$BuCMe{=}CHS(O)Et$ (~100)	612*, 604*
Ph	$(CH_2)_4Br$	Bu_2CuLi	C_5H_{12}/Et_2O 10/1 (−30; b.p.)	$\overline{(CH_2)_4C}{=}CHPh$ (79), $\overline{(CH_2)_4C}(Ph)Bu$ (13)	564*
NC	H	$(o\text{-}C_4N_2H_4)^+{-}^-C(CN)_2$	MeCN (b.p.)	7-Cyano-pyrrolo[1,2-b]pyridazine (30)	518*, 573

MeOOC	COOMe	$(MeOOC)_2\overset{-}{C}\overset{+}{N}\langle$	MeOH	(Dihydropyrrolo[2,1-a]isoquinoline), (Pyrrolo[2,1-a]isoquinoline)	572*, 573, 615*, 616
MeOOC	COOMe	$(EtOOC)_2\overset{-}{C}\overset{+}{N}\langle$	$HCONMe_2$ (r.t.)	(Pyrazolopyridazine)	617*
MeOOC	COOMe	2-Me-thiazoline	$HCONMe_2$ (r.t.)	(Tetrahydropyrrolothiazole)	618*
EtOOC	COOEt	$Me_2\overset{+}{S}—C_5H_4^-$	CH_2Cl_2 (5)	2-$Me_2\overset{+}{S}$—($C_5H_3^-$)-C(COOEt)=CHCOOEt (89)	569
EtOOC	COOEt	$Ph_3\overset{+}{P}—C_5H_4^-$	CH_2Cl_2 (r.t.)	2-$Ph_3\overset{+}{P}$—C_5H_4—C(COOEt)=CHCOOEt (~100)	619
EtOOC	H	PhCH=C(Ph)-$CO\overset{-}{C}H—\overset{+}{S}Me_2$	EtOH (r.t.)	PhCH=C(Ph)$CO\overset{-}{C}(\overset{+}{S}Me_2)$CH=CHCOOEt (~80)	620*
PhCO	H	$Et_2\overset{+}{S}—\overset{-}{C}HCN$	$HCCl_3$, NaOH	*E*-PhCOCH=$CH\overset{-}{C}(CN)\overset{+}{S}Et_2$	568*, 621
EtOOC	H	$Me_2\overset{+}{S}—\overset{-}{C}(COMe)_2$	(160)	2-Methyl-3-acetyl-5-carbomethoxyfuran (70)	322*, 621*
PhCO	Ph	$Me_2\overset{+}{S}—\overset{-}{C}CO(CH_2)_4$ (ring)	Me_2SO (18)	$Me_2\overset{+}{S}$-2-Ph-3-PhCO-4-oxo-2-cyclooctenylide (82)	622
t-BuCO	Ph	$Me_2\overset{+}{S}(O)\overset{-}{C}H_2$	Me_2SO	1-Methyl-3-phenyl-5-*t*-butyl-thiabenzene-1-oxide	623*
PhCO	Ph	$Me_2\overset{+}{S}(O)\overset{-}{C}H_2$	THF/Me_2SO (−8)	$Me_2\overset{+}{S}(O)—\overset{-}{C}HC(Ph)$=CHCOPh (66)	623*, 570*

[a] For additions of enamines, enol ethers and isonitriles see text.
[b] Additional examples of a similar type and leading references are indicated by an asterisk in the last column.
[c] Incomplete names or structures are indicated by parentheses.

β-acetylenic alcohols, ethers and amines but also with alkylacetylenes[554]. With suitable reagents ring syntheses become possible, e.g. equation (198)[556, 557].

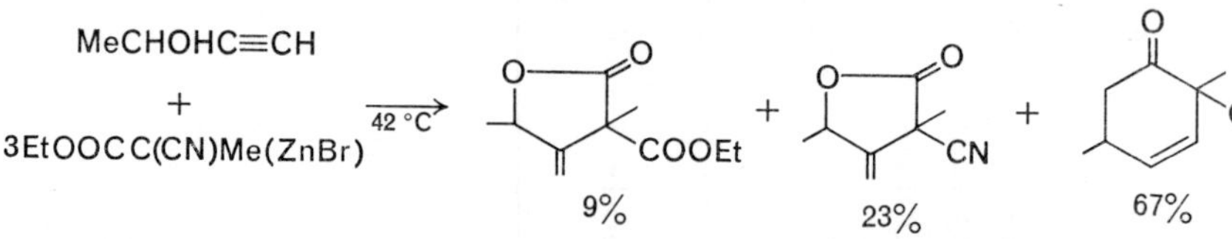

(198)

Although the mechanism of organozinc additions has been considered, only its coarse features are as yet apparent. As is the case with Grignards, e.g. **6**, we believe that a strong *electrophilic* component is present—this seems to be particularly important with compounds such as 1-hexyne in which the qualitative rates of addition of one or two organozincs are roughly similar (equation 197)[554, 556]. It is *a propos* to mention that zinc (cadmium, mercury) salts promote additions to alkynes and that organolithiums and -magnesiums do not add easily to simple alkenes[1]. Stereochemical studies in the additions of equation (199) yield convergent products, a result

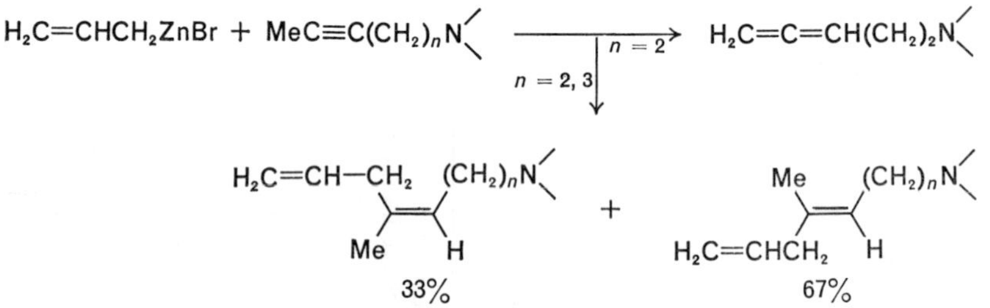

(199)

which is consistent with equilibrium control of precursors to the products. Whether this involves final adducts containing ZnBr or 'first' complexes which lead to inter- or intramolecular attacks, that is, **22** *vs*. **23**, has not been clarified[556].

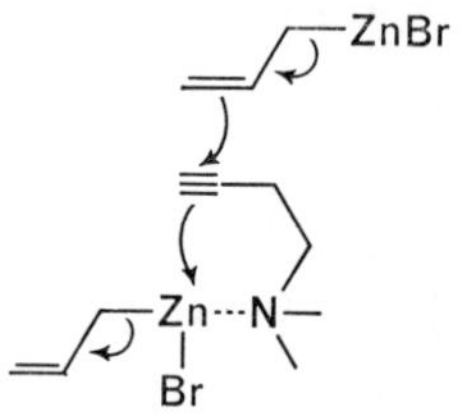

(22)

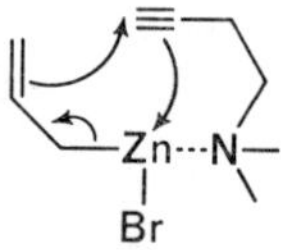

(23)

In the past few years, organocoppers have probably become the most useful of the organometallics with respect to addition to alkynes[558, 559]. They include compounds which are insoluble $(RCu)_n$, or soluble $(R_2CuLi)_n$, $(RCuMgX_2)_n$ and which have been used in small (catalytic) or large (⩾equivalent) amounts in the presence of potential copper ligands such as Cl^-, Br^-, $P(OR)_3$, etc. Understandably, these reagents cover a fair range of reactivity and selectivity: additions are usually predominantly *syn* and the acetylene substituents may be H, alkyl, aryl, COOH, COOR, $CONH_2$, SR, SOR″, SO_2R'', $CH(OR)_2$, CH_2OR, PR_2.

The conjugate additions of organocoppers to α,β-unsaturated carbonyl compounds, among them alkynes, have been reviewed[559]. These generally show a *syn* preference and are directed in the Michael sense (equation 200)[560]. Similar alkyl

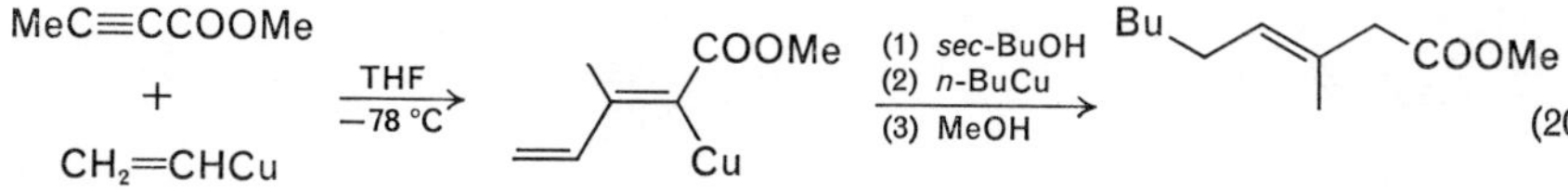

(200)

additions take place with terminal alkynes, e.g. equation (201)[561]; here too coupling reactions may become important competing processes (equation 202)[562, 563]. One of

$$n\text{-BuC}{\equiv}\text{CH} + n\text{-BuCu} \xrightarrow[-15\,^{\circ}\text{C}]{\text{Et}_2\text{O, MgBr}_2} (n\text{-Bu})_2\text{C}{=}\text{CH}_2 \quad 85\% \tag{201}$$

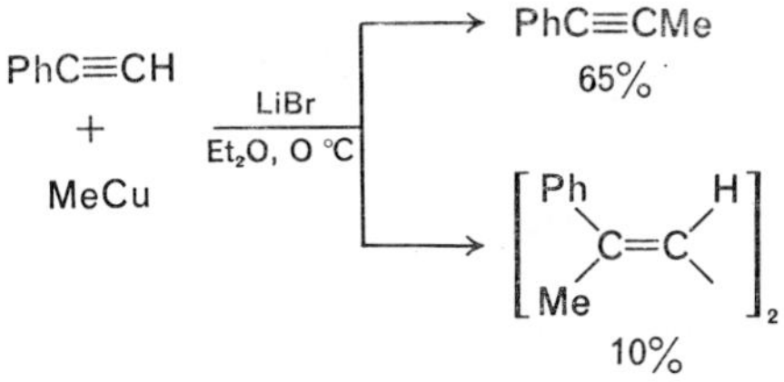

(202)

the more complex cases is given in equation (203) where the kinds of products from addition, coupling and disproportionation depend on the solvent composition (pentane–ether)[564].

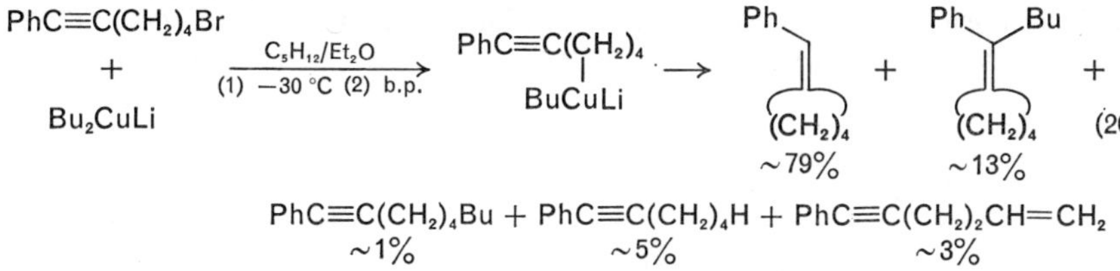

(203)

It must be emphasized that the conditions for carrying out organocopper additions must be optimized for the alkynes involved. The yields and selectivities in process (201), for example, depend on the presence of MgX_2; other variables such as solvent, added ligands, temperature and structure of both major reagents have been examined[561]. Although these conditions can be manipulated so that the addition may be *followed* by further couplings of the organocopper, as in equation (204), these often become competitive with additions (equations 202, 205 and 206)[565].

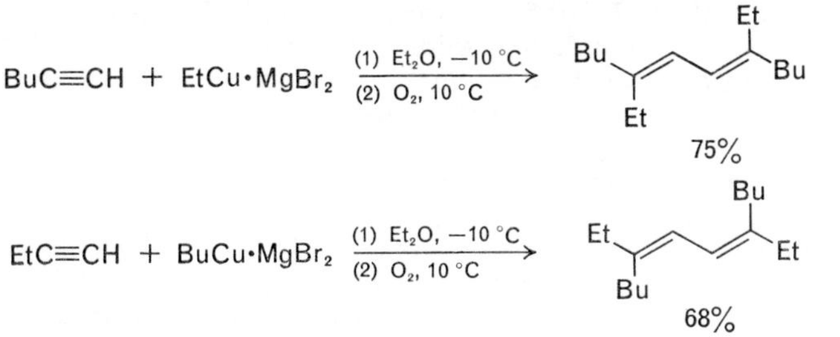

(204)

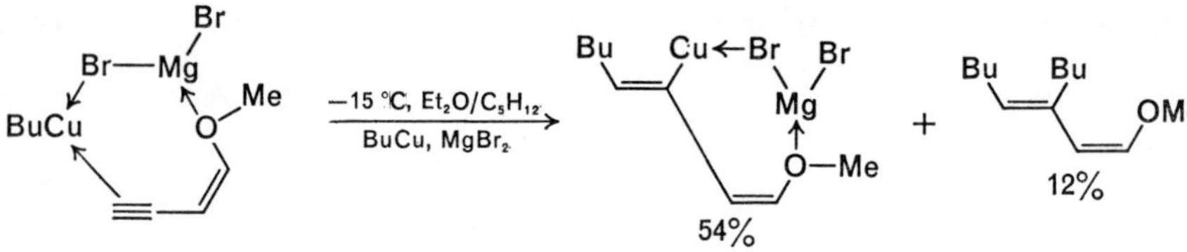

(205)

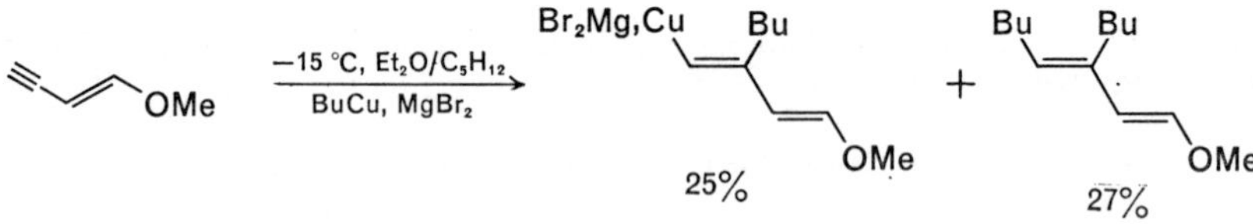

(206)

The mechanism(s) of organocopper reactions are by no means settled[65, 559, 566]. The exchange of equation (207) has been discussed as an oxidative addition followed

$$Li_2Cu_2Me_4 + MeI \longrightarrow Li_2CuMe_3I + C_2H_6 \qquad (207)$$

by a reductive elimination[565a]. Basically different mechanisms, namely, nucleophile *vs.* electron transfer, have been considered for additions of $LiCuR_2$ to unsaturated carbonyl compounds[65]. In our view, it is not at all certain that a single term applies to the copper reagents or indeed whether such a label is even useful. However they are represented, e.g. RCu, R_2CuLi, etc., the structures of the organocoppers are probably polymeric, e.g. tetramers or higher[558]. Structural evidence for $R_4Cu_2Li_2$ in solution has been considered to favour an almost planar (**24**) over a tetrahedral (**25**) metal skeleton[556a]. As a class, the ethynylcoppers are particularly interesting in that X-ray

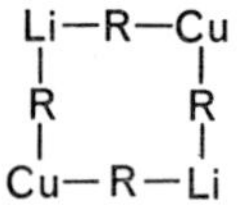

(**24**)

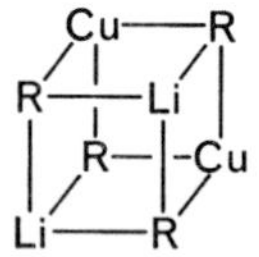

(**25**)

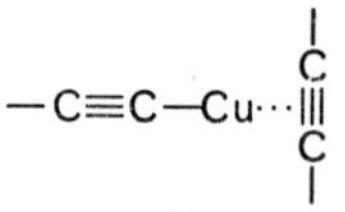

(**26**)

and i.r. spectroscopic data indicate that the copper atom interacts (coordinates) with the triple bonds of other monomeric units (**26**)[558]. If analogous coordination occurs in the addition[561], one would like to know whether Cu(I) is an electron donor, or acceptor or both in the activated complex.

Organocopper reactions can be highly stereoselective both with respect to the configuration of R (from RCu)[558] and the substrate, be it alkyne or diene (equation 200). On this basis, we exclude *free* anions, radicals or cations during the addition. [Our chief concern here is with addition; coupling mechanisms which may complete several processes such as equation (204) will not be considered.] Now, Me_2CuLi appears to be less basic (towards toluene) and less nucleophilic than MeLi (towards RCOR′)[559]. On the other hand, the mixed organometallics, Me_2CuLi or Me_2CuMgX, usually add much more cleanly and rapidly than organolithiums and -magnesiums add to triple bonds. While organocoppers and alkynes usually yield adducts of the 'correct' regioselectivity for nucleophiles and electrophiles—this perhaps justifies our inclusion of these reactions—the details and timing of the transfer are not clear.

Within the polymer aggregate these may be electrophilic attack (* is + in **27**), oxidative addition (* is − in **27**), nucleophilic attack (* is − in **28**), 1,2-cycloaddition (**29, 30**) and for dienes, 1,4-cycloaddition (**31**), etc.

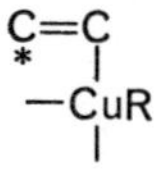

(27)

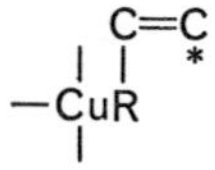

(28)

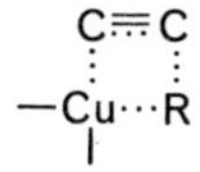

(29)

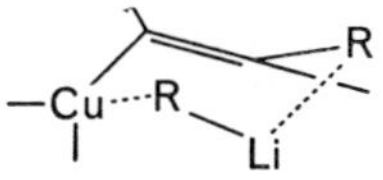

(30)

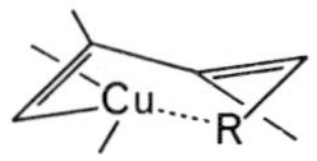

(31)

We have implied that the high *syn* selectivity in alkynes requires relatively tight association or bonding from beginning to end, whether the participants are covalent, ion pair, radicaloid, etc. By 'participants' we mean the four reaction centres pictured in **27** to **30**. Other ligands to copper, e.g. Br or $(EtO)_3P$, may be labile and are effectively replaced by the triple bond as the addition occurs[566b]. At higher temperatures, when *syn* selectivity often decreases[567], one would have to allow the copper adduct to dissociate, isomerize and recombine.

Ylids of various types, e.g. $>\overset{+}{N}-\overset{-}{C}<$, $>\overset{+}{P}-\overset{-}{C}<$, $>\overset{+}{S}-\overset{-}{C}<$, etc., may also provide anionic carbon. Mechanistically these may be simple versions of equation (1) (equations 208–210).

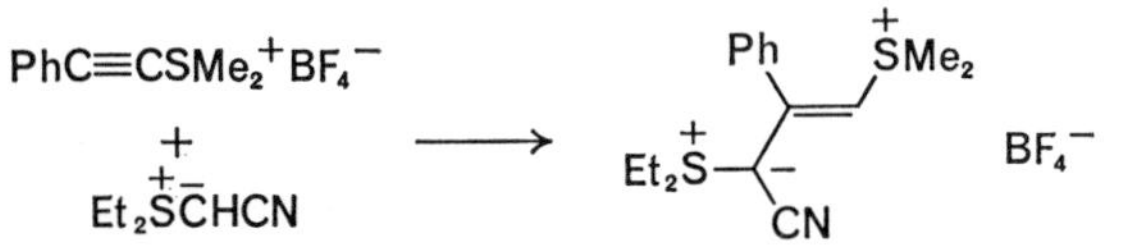

(Ref. 568) (208)

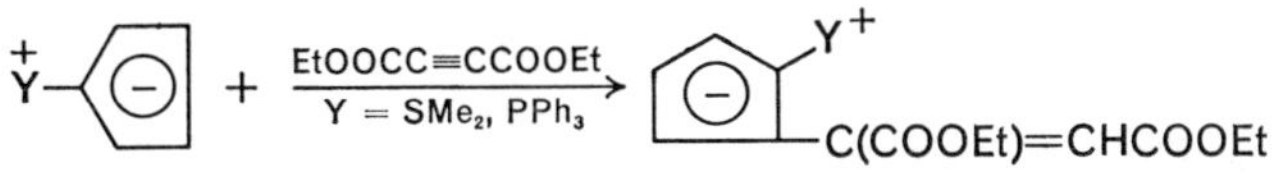

(Ref. 569) (209)

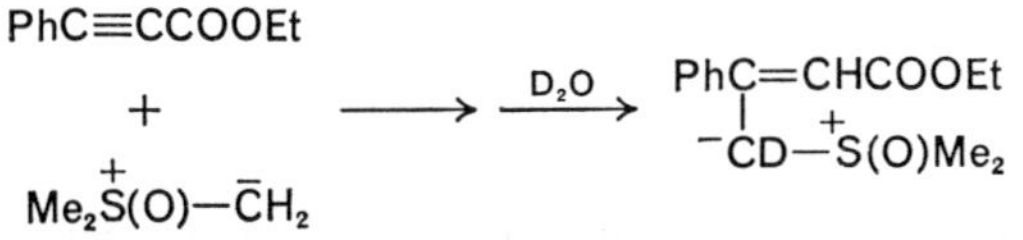

(Ref. 570) (210)

In equation (211), we see the potential choice (or ambiguity) between addition as in equation (1) and cycloaddition[571]. While one cannot regard a cyclic product as a proof of cycloaddition, one might, for purposes of convenience or simplicity and the

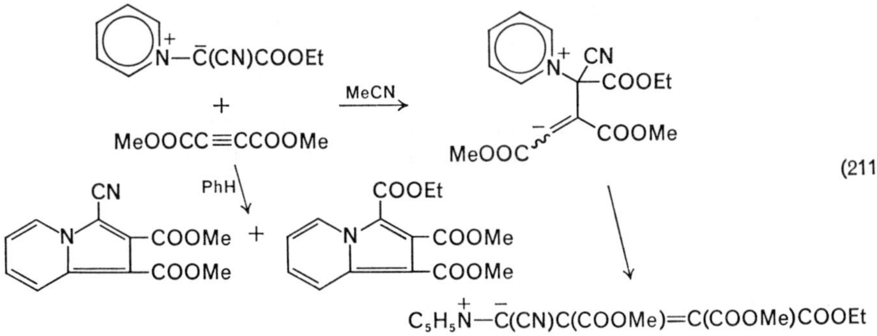

(211)

absence of other evidence, adopt this criterion. We have included several cyclic products in equations (212)–(215) and in Table 22 to emphasize the point that there

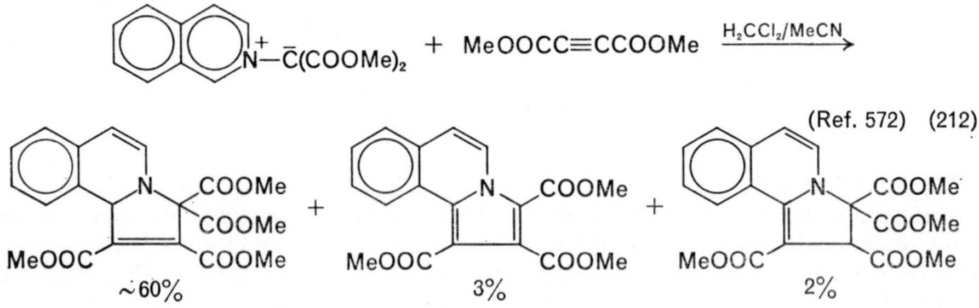

(Ref. 572) (212)

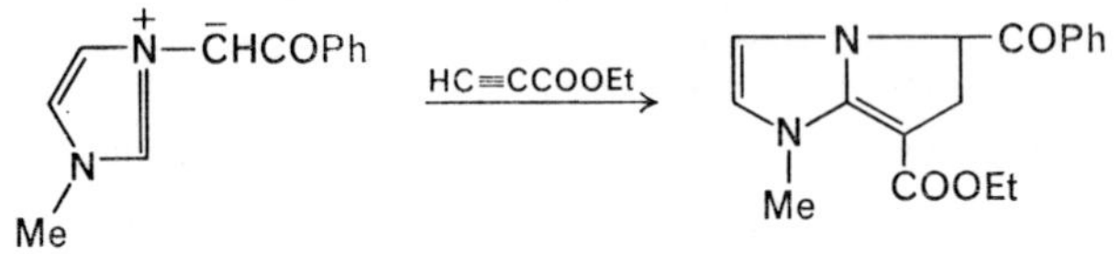

(Ref. 573) (213)

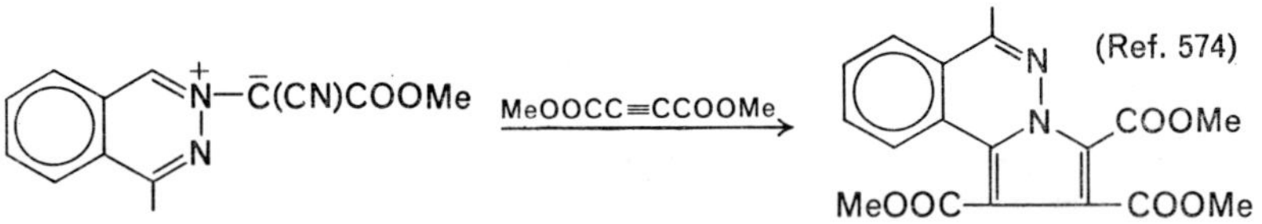

(Ref. 574) (214)

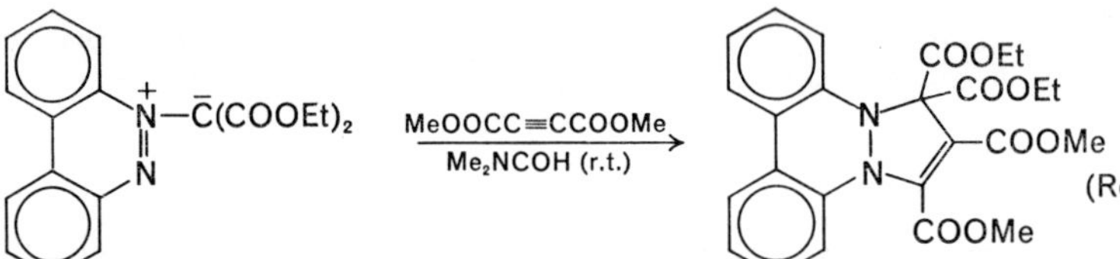

(Ref. 426) (215)

is a mechanistic issue here which is usually unsettled. On the other hand, it must be conceded that there is compelling evidence that certain zwitterions, e.g. **32–36**, are

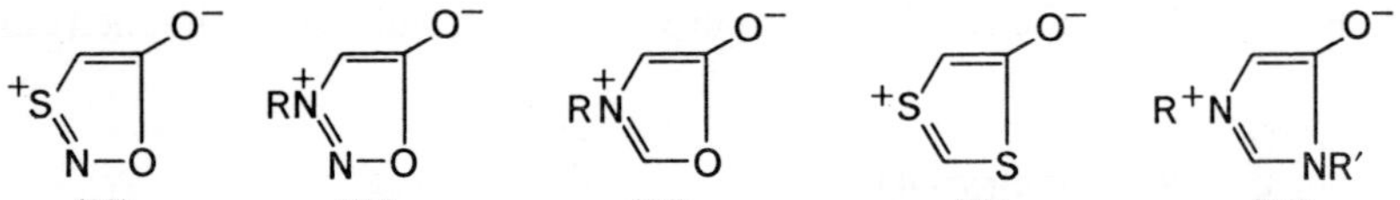

(32) **(33)** **(34)** **(35)** **(36)**

dipolarophiles which normally enter into cycloadditions with simultaneous cycloelimination of a stable molecule such as CO_2, COS, PhNCO, etc.,[575–580] e.g. equation (216).

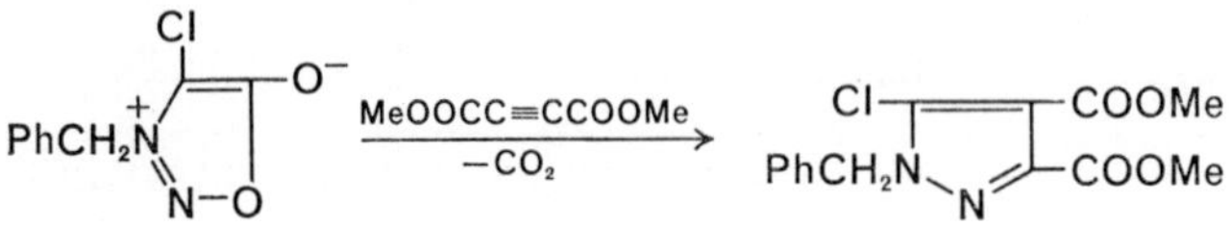

(Ref. 581) (216)

Enamines, enol ethers and related compounds again encompass one-site nucleophiles and two-site dipolarophiles. Where the adducts are acyclic or where a co-electrophile is involved these appear to be examples of equation (1). In equations (217)–(219), the paths to the major products seem straightforward. While a competing

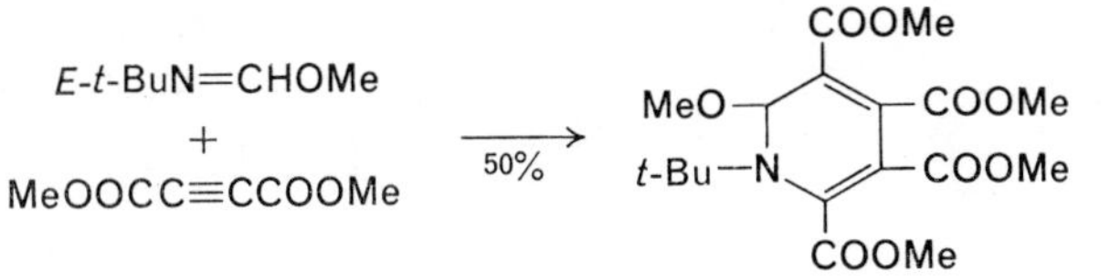

(Ref. 582) (217)

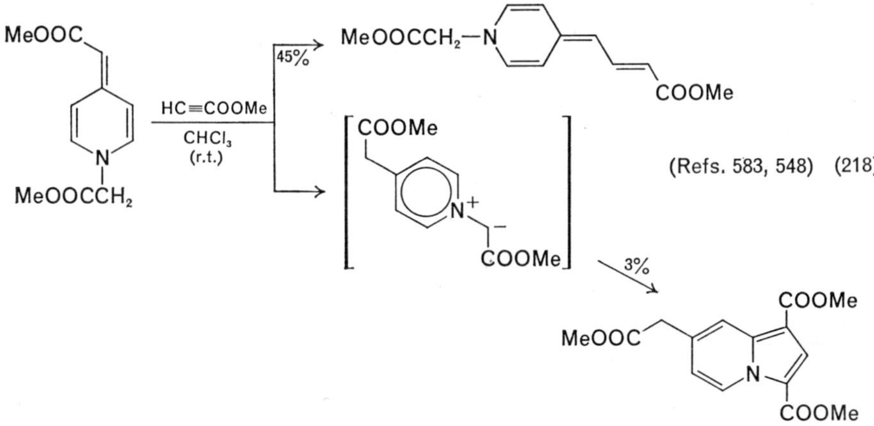

(Refs. 583, 548) (218)

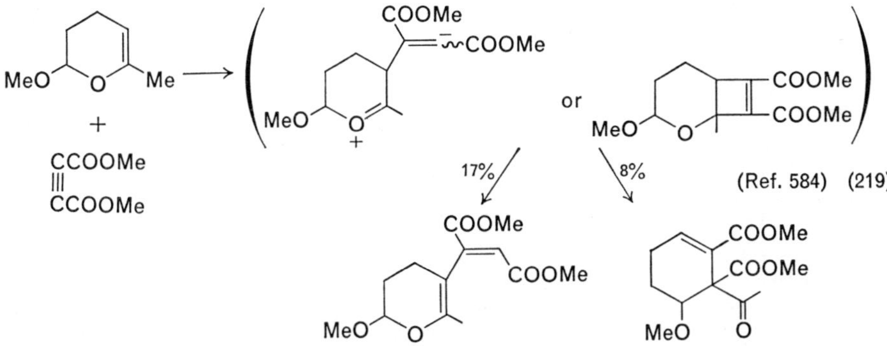

(Ref. 584) (219)

[2+2] cycloaddition had to be considered in equation (219), this appears to be the major path in equations (220) and (221). Process (222) may be initiated by nucleophilic attack to give the first product which undergoes a further cycloaddition[17]. It is probable, therefore, that there will be examples in which one or both mechanisms are present without, however, allowing one to choose between them, e.g. equation (223)[587].

At first glance, reaction (224) seems to be similar to the enol ether additions[588], although one could well have expected oxygen transfer processes, e.g. equation (110) or (126) to take precedence. That is, the αβ-product of cycloaddition or the corresponding acyclic zwitterion is unexpected, but each is readily related to the final

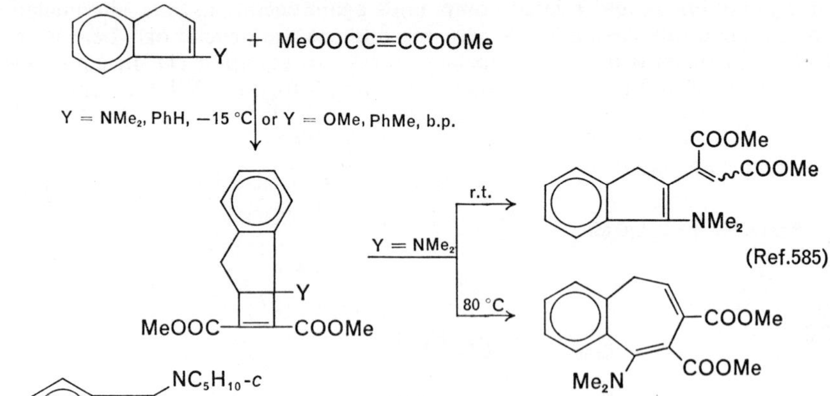

(Ref.585) (220)

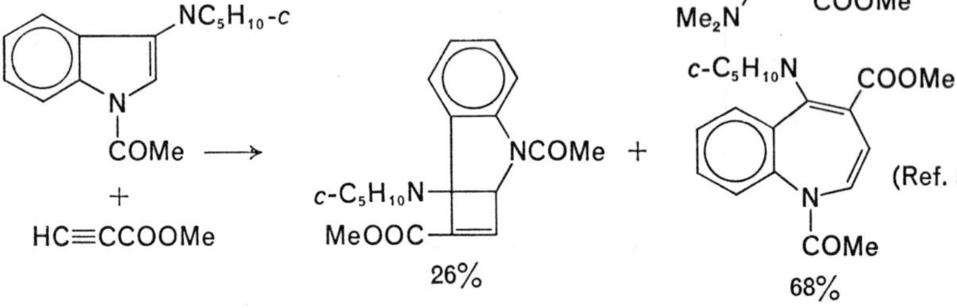

(Ref. 586) (221)

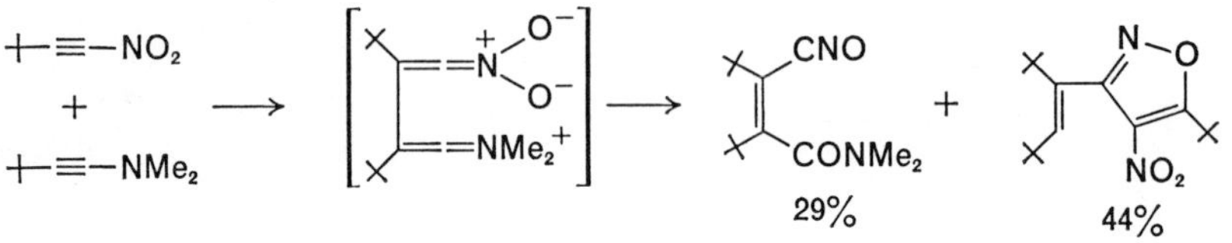

(222)

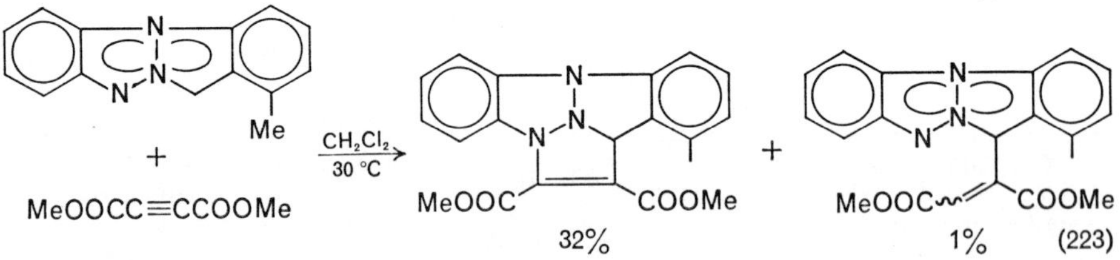

(223)

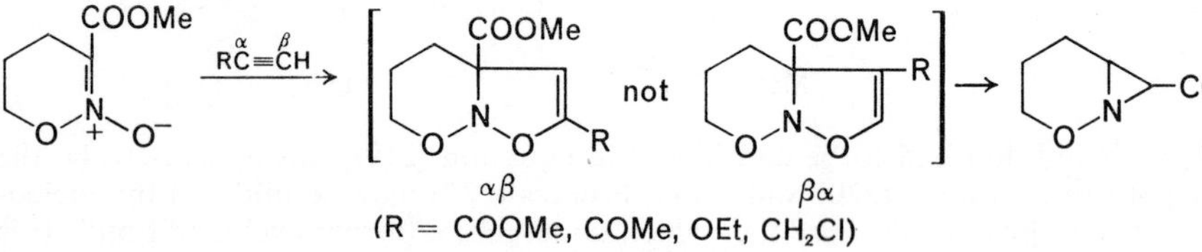

(224)

product. The more plausible βα-cycloadduct or its acyclic isomers are not easily related to the isolated product. We regard the mechanism of (224) as a curious unsettled matter.

Although isonitriles (and CO) are electron donors both as bases and nucleophiles, these properties are more in evidence in transition metal than in organic reactions. Admittedly, isonitriles often react as carbenes; but there are additions to activated alkynes in which the most plausible *beginning* is with RNC as a nucleophile, e.g. equations (225)–(227)[589]. Several groups have postulated that the betaine, $-N{=}\overset{+}{C}{-}\overset{|}{C}{=}\overset{-}{C}-$, forms and that attacks by electrophiles, nucleophiles, dipolarophiles, etc. then follow. Clearly, the ensuing possibilities are numerous and they are often complex as has been demonstrated in the literature[593–595].

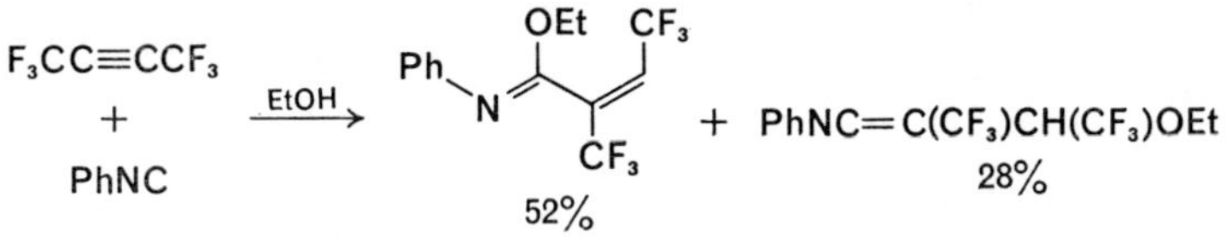

(Ref. 590) (225)

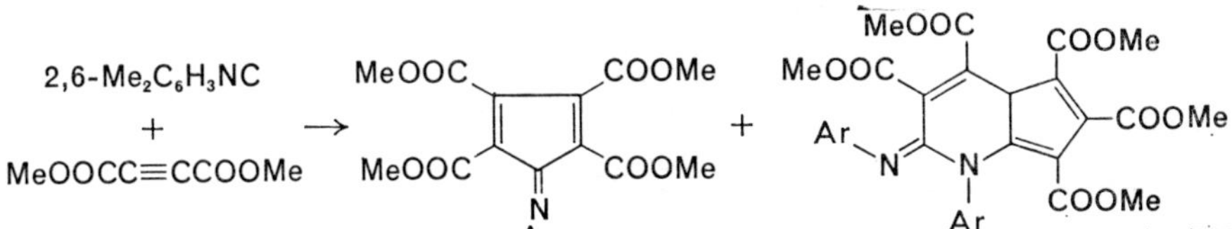

(Ref. 591) (226)

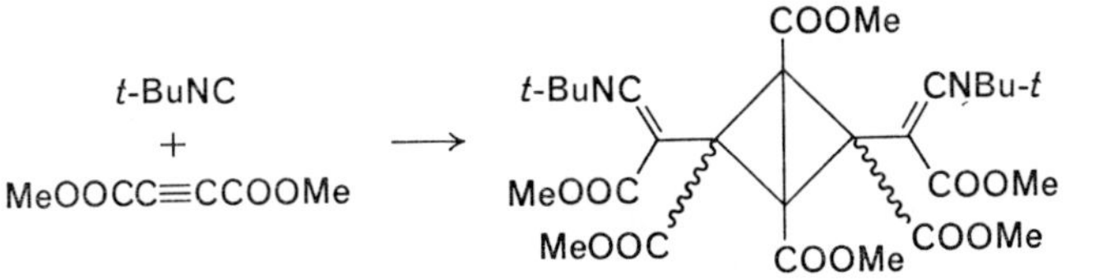

(Ref. 592) (227)

I. Miscellaneous Nucleophiles

Because there are fewer of them we have collected all the remaining nucleophiles in Table 23. These turn out to be 'heavy' atoms. Normally, these elements have widely different chemistries. It is interesting that their hydrides, or the anions formed from them, i.e. Et_3GeLi, Et_3SnNa, $[C_5H_5Fe(CO)_2]Na$, RSe^-, RTe^-, R_2P^-, $(RO)_2PO^-$, $HMn(CO)_5$, $HRh(CO)(PPh_3)_3$ and Me_2AsH behave conventionally: a 1 : 1 adduct forms (equations 228–233) and, if the possibility exists, ring closure may follow

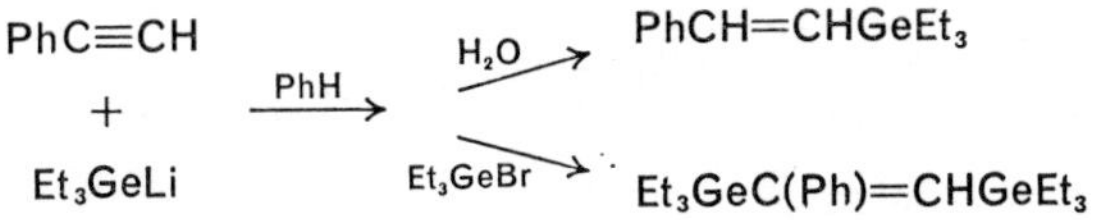

(Ref. 624) (228)

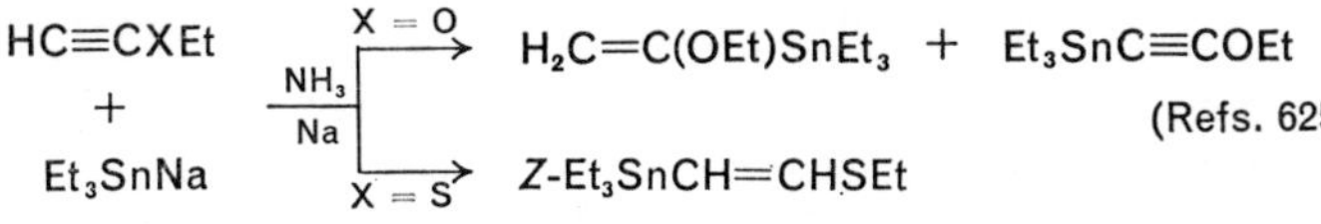

(Refs. 625 and 626) (229)

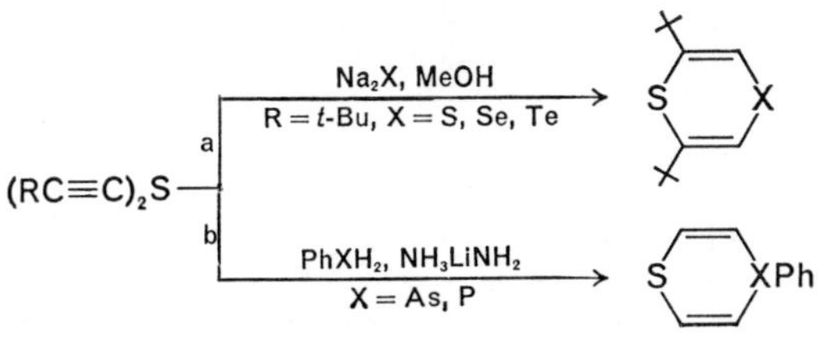

(Ref. 382)

(230)

(Ref. 627)

TABLE 23. Additions of heavy atom nucleophiles to alkynes, RC≡CR′ [a]

R	R′	Nu or Nu⁻	Medium (temp., °C)	Product(s) (yield, %)	Ref.
H	SC≡CH	$PhPH_2$	$LiNH_2$, NH_3 (b.p.)	4-Phenyl-1,4-thiaphosphorin (34)	627*
HC≡C	P(Bu-*t*)C≡CH	$PhPH_2$	$LiNH_2$, NH_3 (b.p.)	1-*t*-Butyl-1,4-dihydro-1,4-diphosphabenzene (57)	189*
MeOOC	COOMe	$(Ph_2P)_2CH_2$	Ether (r.t.)	Dimethyl-1,1,3,3-tetraphenyl-5*H*-diphosphole-4,5-dicarboxylate (75)	636
PhCO	COPh	Ph_3P	Ether (0)	$(Ph_3P{=}CCOPh)_2$ (97)	637
Ph	$PPh_3^+Br^-$	$(n\text{-Bu})_3P$	MeCN (b.p.)	$[(n\text{-Bu})_3PC(Ph){=}CHP(Bu\text{-}n)_3]^{2+}2Br^-$	157
p-$O_2NC_6H_4$	H	Ph_3P	$(HOCH_2)_2$ (b.p.)	$Ph_2P(O)CH(Ph)CH_2C_6H_4NO_2$-*p* (50)	633
MeOOC	COOMe	$Ph_2PC{=}CH_2$	Moist ether (r.t.)	$Ph_2POCH_2CH{=}C(COOMe)CH_2COOMe$ (5), 1,2,3-tricarbomethoxy-3-carbomethoxy-methyl-4-diphenylphosphorylcyclopentene (5)	638
PhCO	Ph	Ph_3P	THF, H_2O	PhCOCH=CHPh (62)	632*
PhCO	COPh	Ph_3P	Et_2O/MeOH (<0)	$Ph_3P{=}C(COPh)CH(OMe)COPh$	630*
EtO	H	$(MeO)_2PO^-Na^+$	Na^+OMe^-, MeOH (r.t.)	$H_2C{=}C(OEt)PO(OMe)_2$ (~55)	133
EtS	H	$(MeO)_2PO^-Na^+$	K^+OMe^-, MeOH (10)	*Z*-EtCH=$CHPO(OMe)_2$ (~70)	139, 138*
MeO	Me	$(i\text{-PrO})_3P$	$HgCl_2$, THF (r.t.)	$(i\text{-PrO})_2POC(OMe){=}C(Me)HgCl$ (~100)	631*
R′O	H	$(RO)_3P$	—	$H_2C{=}C(OR')PO(OR)_2$ (10–98%)	639*
Et	SC≡CEt	$PhAsH_2$	$LiNH_2$, NH_3 (b.p.)	4-Phenyl-1,4-thiarsenin (72)	627*
CF_3	CF_3	Me_2AsH	Et_2O (−30 to 22)	$F_3CC(AsMe_2){=}CCF_3H$ [89, *Z*/*E*~(10–20)/1], *k* [b]	107
MeC≡C	P(Bu-*t*)C≡CMe	$PhAsH_2$	$LiNH_2$, NH_3 (b.p.)	1-*t*-Butyl-2,6-dimethyl-1,4-dihydro-1-phospha-4-arsabenzene (98%, *cis*/*trans* = 3/2)	189*
EtO	H	$(t\text{-BuO})_2PH$	(20)	$H_2C{=}C(OEt)P(OBu\text{-}t)_2$ (50)	132
PhC≡CCO	Ph	H_2Se	Et_3N, MeOH (35)	2-Benzylidene-3-oxo-5-phenyl-2,3-dihydro-selenophene (28), 2,6-diphenyl-1-seleno-γ-pyrone (17)	179
MeC≡CCO	Me	H_2Se	Et_3N, MeOH (35)	2,6-Dimethyl-1-seleno-γ-pyrone (57)	179
MeCH=CH	SC≡CCH=CHMe	Na_2Se	NH_3/MeOH (−33)	2,6-Di(1-propenyl)-1,4-thiaselenin (71)	382*
t-BuC≡CPO-(*c*-C_6H_{11})	Bu-*t*	Na_2Se_2	NH_3/MeOH (−33)	2,6-Di-*t*-butyl-4-oxo,4-cyclohexyl-1,4-selenaphosphorin (62)	188*
R	CR^1R^2OH	PhSeH	—	$PhSeC(R){=}CHC(OH)R^1R^2$	640*, 6a

H	COOH	$(H_2N)_2CSe$	H_2O, H^+ (r.t.)	*Z*-HOOCCH=CHSe$(NH_2)_2^+Cl^-$ (~90)	386*
$HOCH_2C\equiv C$	CH_2OH	Na_2Te	MeOH (20)	2,5-Dihydroxymethyltellurophen	641*
Me	SC≡CMe	Na_2Te	NH_3/MeOH 2/1 (−33)	2,6-Dimethyl-1,4-thiatellurin (78)	382*
HC≡CPO(Ph)	H	Na_2Te	NH_3/MeOH (−33)	4-Oxo,4-phenyl-1,4-telluraphosphorin (70)	188*
Ph	Ph	Et_3GeLi	PhH	PhCH=C(Ph)Ge$(Et)_3$ (98)	624*
H	OCH_3	$Et_3Sn^-Na^+$	NH_3 (b.p.)	H_2C=C(OEt)$SnEt_3$ (68)	625*, 626
H	CF_3	$[cpFe(CO)_2]^-$	THF (−78)	*Z*-CF_3CH=CHFe$(CO)_2$cp (13)	642*, 30
HOOC	COOH	HRh(CO)$(PPh_3)_3$	Ether (r.t.)	*Z*-HOOCCH=C(COOH)Rh(CO)$(PPh_3)_3$ (72)	628*
Ph	COMe	$HMn(CO)_5$	(r.t.)	*Z*-$(OC)_4$MnCPh=CHCOMe (11)	643
H	Se^-	—	CSe_2, Et_2O (r.t.)	2-Seleno-1,3-diselenacyclopentene	390*

[a] Additional examples of a similar type and leading references are indicated by an asterisk in the last column.
[b] Rate measurements were taken.

$$HC{\equiv}CXEt + (MeO)_2PO^- \xrightarrow{MeOH} \begin{cases} \xrightarrow{X = O} H_2C{=}C(OEt)PO(OMe)_2 \\ \xrightarrow{X = S} Z\text{-}(MeO)_2P(O)CH{=}CHSEt \end{cases}$$ (Refs. 133 and 139) (231)

$$HC{\equiv}CCF_3 + Re(CO)_5^- \xrightarrow{THF} Z\text{-}(CO)_5ReCH{=}CHCF_3$$ (Ref. 30) (232)

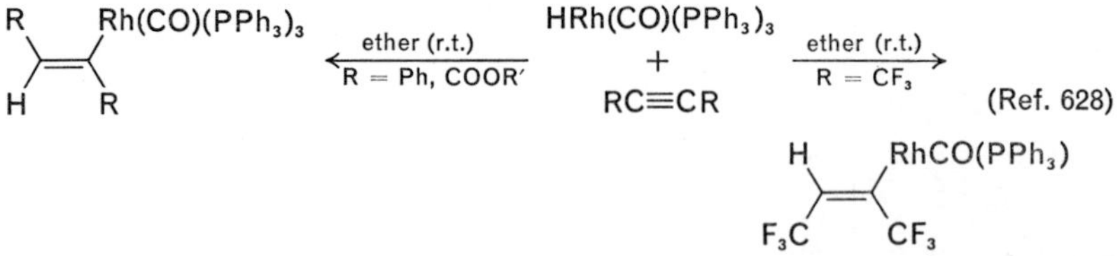
(Ref. 628) (233)

(equation 230)[1]. The mechanisms of addition of neutral hydride rather than anion, as in the examples of As, Mn and Rh, are, in fact, not obvious. While *anti* selectivity is indicative, kinetic and D-labelling studies (**3**) in the case of Me_2AsH provide strong support for process (**1**)[107]. Examples of 'normal' *regio*- and *anti*-selectivity are found in equations (229) and (231)[625, 626]. The 'undesired' acetylenic product in equation (229) probably results from a competing one-electron transfer, since Et_6Sn_2 is a coproduct.

Our remaining discussion is concerned chiefly with tertiary phosphorus and similar nucleophiles. Although the products are often complex—1 : 1, 2 : 1, 1 : 2, 2 : 2 and 3 : 2 adducts have been found[27]—we believe that the initial steps are straightforward. Evidently an ylid ($^+$V$^-$) forms readily as in equation (234)[629, 630].

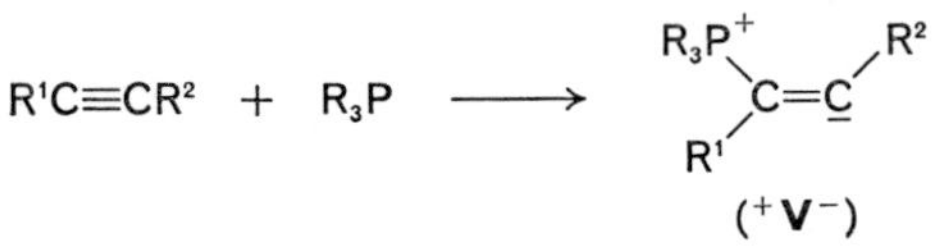
(234)

The reactive ylid may then undergo electrophilic (H^+, Hg^{2+}, CO_2, RC≡CR′) or nucleophilic [R_3P, $(RO)_3P$, RO^-] attack and/or rearrangement (equation 235). Note the amazing variety of products which arise in equations (235–240) from formally similar ylids.

$$[(i\text{-}PrO)_3\overset{+}{P}C(OMe){=}CMe^-] \xrightarrow[THF]{HgCl_2} (i\text{-}PrO)_2P(O)C(OMe){=}C(Me)HgCl \quad (235)$$
(Ref. 631)

$$[Ph_3\overset{+}{P}C(Ph){=}\bar{C}COPh] \xrightarrow{H_2O} PhCH{=}CHCOPh \quad (236)$$
(Refs. 632 and 629)

$$[Ph_3P^+C(COOMe){=}\bar{C}COOMe] \xrightarrow{MeOH} Ph_3P{=}C(COOMe)CH(OMe)COOMe \quad (237)$$
(Ref. 630)

$$[Bu_3P^+C(Ph){=}C{=}PPh_3, Br^-] \xrightarrow[Bu_3P]{CH_3CN} Bu_3\overset{+}{P}C(Ph){=}CHPBu_3^+, 2Br^- \quad (238)$$
(Ref. 30)

$$[(MeO)_3\overset{+}{P}C(COOMe){=}\bar{C}COOMe] \xrightarrow{(MeO)_3P} \longrightarrow [(MeO)_2P(O)(MeOOC)MeC]_2 \quad (239)$$
(Ref. 27)

$$Ph_3\overset{+}{P}CH{=}\overset{-}{C}C_6H_4NO_2\text{-}4 \xrightarrow{(HOCH_2)_2} Ph_2P(O)PhCHCH_2C_6H_4NO_2\text{-}4 \qquad (240)$$

(Ref. 633)

The reactions of the initial ylid vary depending on the stability of the product(s) and the coreactants—compare equation (237) with equation (241)[1, 27, 630]. If a

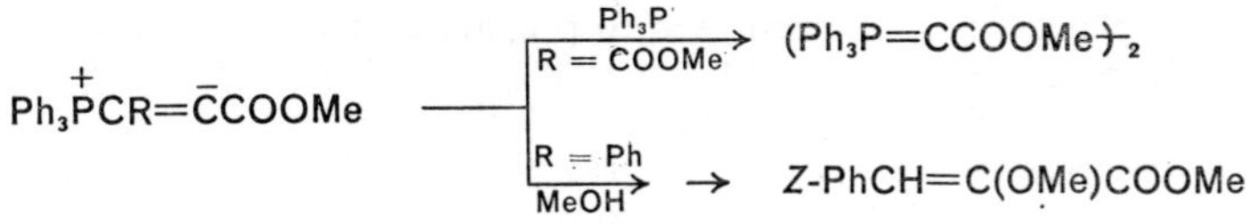

(241)

complex structure is produced and then identified—the latter seems to be the most challenging problem—one can usually provide a mechanism for its formation. The paths to (**37**)–(**39**), for example, seem simple enough[27, 634]. In equation (242) we give

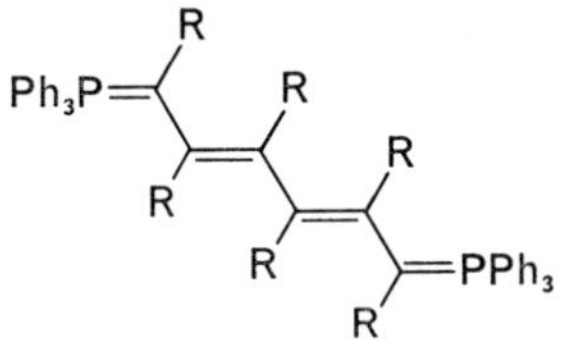

(**37**) R = CN

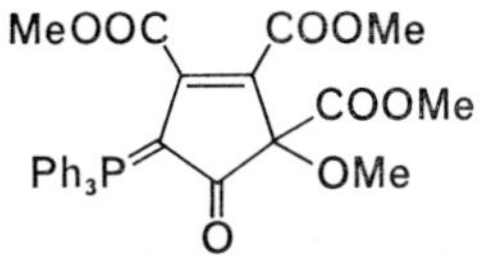

(**38**)

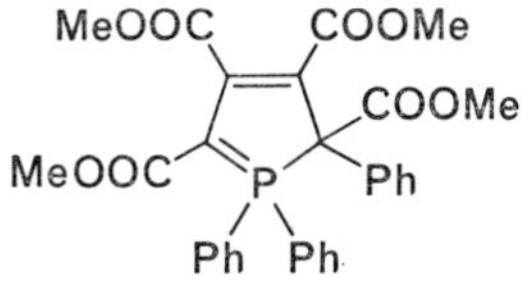

(**39**)

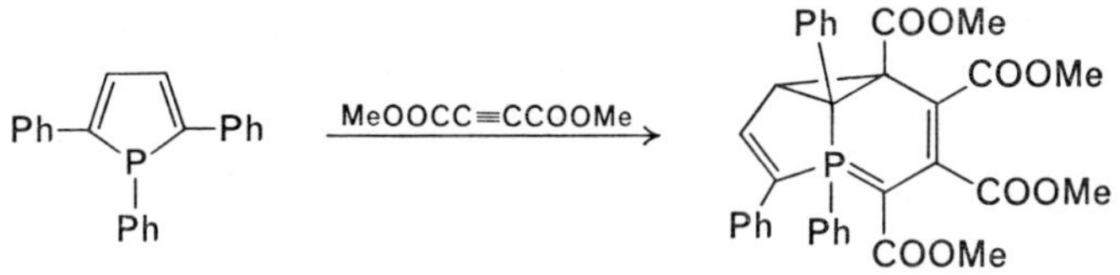

(242)

the structure of a 1 : 2 adduct which was proposed by the second research group to study the reaction[635]. Since it is complex, it would be reassuring to have an X-ray determination to validate this interesting structure.

IV. NUCLEOPHILIC SUBSTITUTIONS

A. Introduction

Nucleophilic substitution reactions with aliphatic[644], carbonyl[645], aromatic[646] and vinylic[647] substrates were well developed, before a single example of process (2) was observed. This late beginning with the most unsaturated centre, acetylenic carbon, was by no means due to a lack of effort. For about seventy years after the first recorded failure in 1892[648], attempts to find an acetylenic substitution product were generally unsuccessful[4, 160, 649]. The resistance to substitution at an acetylenic carbon appeared to be demonstrated in a study of relative reactivities of various organic chlorides towards sodium iodide in absolute acetone at 60 °C: $PhCOCH_2Cl$ (1×10^5), $PhC{\equiv}CCH_2Cl$ (780), PhCOCl (700), *n*-BuCl (1·0), $PhC{\equiv}CCl$ (0)[650]. Nevertheless, if one allows an organometallic reagent to be the 'nucleophile', then Ott must be credited with the first example of equation (2)[651]:

$$ClC{\equiv}CCl + Na^{+}\,{}^{-}CEt(COOEt)_2 \longrightarrow ClC{\equiv}CCEt(COOEt)_2 \qquad (243)$$

In 1962 several groups published successful syntheses with conventional nucleophiles[652–657], e.g. equation (244)[652]. Surprisingly, a haloalkyne could be more reactive

$$p\text{-}C_7H_7S^- + C_6H_5C{\equiv}CX \xrightarrow[X=Br,\ Cl]{DMF,\ -25\ °C} C_6H_5C{\equiv}CSC_7H_7\text{-}p + X^- \qquad (244)$$

than a haloalkane: with p-$C_7H_7S^-$ in DMF at -25 °C, $k(PhC{\equiv}CCl)/k(n\text{-}BuCl) \simeq 60$ [657]. To date, process (2) has been used to prepare either for the first time or most directly several important acetylenic families, e.g. ynamines[13], ethers[658], thioethers[659], phosphonium salts[157, 652, 654], phosphines[653] and phosphites[159, 655, 660].

If controversy over the mechanism of a reaction may be taken as a measure of its importance, then process (2) has attained considerable status. In this section we shall consider the mechanistic data and proposals and synthetic facts of process (2). Since the latter subject was last reviewed in 1969[4], our emphasis will be on the more recent synthetic findings. We shall give a more detailed description of the mechanistic aspects of process (2) than the brief account published in 1976[3].

B. *Kinetics and Mechanism*

1. Rate data

A listing of alkyne–nucleophile systems whose substitution kinetics have been studied is given in Table 24 for each of these systems. Rate constants and enthalpies and entropies of activation, if available, are tabulated. In order to compare the reactivity of haloalkynes with other organic halides we have also included in Table 24 related rate data for vinylic, aromatic and alkyl halides.

Several features of this table stand out. All of the acetylenic systems exhibit second-order kinetics, first order in nucleophile and first order in haloalkyne. When a haloalkyne is coupled with a neutral nucleophile, a large negative value for $\Delta S^{\ddagger}$ is observed. This is consistent with other molecule–molecule reactions in which ions are formed[661]. Among comparable halounsaturates, the reactivity order is alkynyl $\gg$ alkenyl $>$ aryl. Only when the vinyl and aryl halides are substituted with strongly activating groups can they match the reactivity of unsubstituted alkynyl halides. In some instances, the reactivity of a haloacetylene is even greater than a haloalkane. Certainly, these data should dispel any impression that 1-halo-1-alkynes need be inert towards nucleophilic substitution.

We shall also use element effects (Table 25) and ρ-values (Table 26) as diagnostic probes for mechanism. Here, too, we have compiled data for other organic series so that comparisons can be made between these systems and haloalkynes. More specific use of the data in Tables 24–26 together with regioselectivity material in Section II.C.1.d and Table 6 will be made in the following section.

2. Mechanisms

Of the seven mechanisms to be mentioned for process (2), we can evaluate three in fair detail. Two others are admittedly hypothetical. Only recently recognized, the last two are probably widely applicable but still poorly characterized.

a. *Carbanion intermediates*. In Section II.C.1.d we cited evidence for nucleophilic attack on RC≡CX at X, C_α and C_β (see **13**). By considering one or other of these centres as the principal site of attack, three different groups proposed three different

TABLE 24. Selected rate data for nucleophilic displacement reactions of haloalkynes and other halides

Reactants	Solvent	Temp. (°C)	k ($M^{-1}\,s^{-1}$)	$\Delta H^{\ddagger}$ (kcal/mol)	$-\Delta S^{\ddagger}$ (e.u.)	Reference
$HC{\equiv}CBr + (C_2H_5)_3N$	DMF	81	$6{\cdot}14 \times 10^{-5}$	11·8	43	173
$H_2C{=}CHBr + C_5H_{10}NH$	$C_6H_5NO_2$	100	~0 (100 h)	—	—	662
$C_6H_5Br + C_5H_{10}NH$	C_6H_6	130	0 (200 h)	—	—	663
$C_2H_5Br + (C_2H_5)_3N$	$(CH_3)_2CO$	100	55×10^{-5}	—	—	664
$C_6H_5C{\equiv}CCl + TED$[a]	CH_3CN	60	$10{\cdot}6 \times 10^{-4}$	10·7	40	665
$C_6H_5C{\equiv}CBr + TED$	CH_3CN	60	$7{\cdot}95 \times 10^{-4}$	14·2	30	665
$2,4\text{-}(O_2N)_2C_6H_3Cl + TED$	CH_3CN	51	$1{\cdot}13 \times 10^{-4}$	—	—	666
$C_6H_5CH_2CH_2Cl + TED$	CH_3CN	55	$6{\cdot}21 \times 10^{-5}$	—	—	667
$C_6H_5CH_2CH_2Br + TED$	CH_3CN	55	$6{\cdot}16 \times 10^{-3}$	—	—	667
$n\text{-}C_4H_9Cl + TED$	CH_3CN	60	$1{\cdot}82 \times 10^{-4}$	13·8	34	665
$C_6H_5C{\equiv}CCl + CH_3O^-$	CH_3OH	78	$1{\cdot}6 \times 10^{-4}$	—	—	153
$C_6H_5C{\equiv}CBr + CH_3O^-$	CH_3OH	78	$1{\cdot}0 \times 10^{-4}$	—	—	153
$(p\text{-}O_2NC_6H_4)_2C{=}CHCl + C_2H_5O^-$	C_2H_5OH	50	$3{\cdot}8 \times 10^{-3}$	—	—	668
$C_6H_5Cl + CH_3O^-$	CH_3OH	232	$\sim 6{\cdot}6 \times 10^{-7}$	—	—	669
$n\text{-}C_4H_9Cl + C_2H_5O^-$	C_2H_5OH	77	$1{\cdot}0 \times 10^{-4}$	—	—	670
$C_6H_5C{\equiv}CCl + (n\text{-}C_4H_9)_3P$	DMF	36	$5{\cdot}92 \times 10^{-2}$ [l]	11·5[b]	27[b]	157
$C_6H_5C{\equiv}CBr + (n\text{-}C_4H_9)_3P$	DMF	36	$2{\cdot}20 \times 10^{-1}$ [l]	5·4[c]	44[c]	157
$C_6H_5C{\equiv}CCl + (C_6H_5)_3P$	DMF	36	$1{\cdot}75 \times 10^{-4}$ [l]	14·5[d]	29[d]	157
$C_6H_5C{\equiv}CBr + (C_6H_5)_3P$	DMF	36	$8{\cdot}45 \times 10^{-5}$ [l]	16·8[e]	23[e]	157
$CH_3Br + (C_6H_5)_3P$	DMF	36	$2{\cdot}88 \times 10^{-3}$	11·8[f]	31[f]	157
$n\text{-}C_3H_7Br + (n\text{-}C_4H_9)_3P$	$(CH_3)_2CO$	35	$6{\cdot}0 \times 10^{-5}$	—	—	671
$C_6H_5C{\equiv}CCl + (C_2H_5O)_3P$	THF	60	$3{\cdot}86 \times 10^{-5}$	17·6	26	159
$C_6H_5C{\equiv}CBr + (C_2H_5O)_3P$	THF	60	$2{\cdot}96 \times 10^{-5}$	18·0	25	159
$C_6H_5CH{=}CHBr + (C_2H_5O)_3P$	—	200	~0	—	—	672
$C_2H_5I + (C_2H_5O)_3P$	CH_3CN	60	$2{\cdot}0 \times 10^{-5}$	—	—	673
$C_6H_5C{\equiv}CCl + p\text{-}C_7H_7S^-$	DMF	−25	$6{\cdot}12 \times 10^{-2}$	12·3	16·8	657
$C_6H_5C{\equiv}CBr + p\text{-}C_7H_7S^-$	DMF	−25	$2{\cdot}15 \times 10^{-2}$	—	—	164
$(C_6H_5)_2C{=}CHCl + p\text{-}C_7H_7S^-$	DMF	−25	$1{\cdot}74 \times 10^{-7}$ [l]	16·8[g]	23·9[h]	674
$n\text{-}C_4H_9Cl + p\text{-}C_7H_7S^-$	DMF	−12	$4{\cdot}13 \times 10^{-3}$	14·5[g]	—	657
$C_6H_5C{\equiv}CBr + C_2H_5S^-$	CH_3OH	26	$8{\cdot}6 \times 10^{-1}$	15·2	8	155
$2\text{-}C_4H_3S{-}C{\equiv}CCl$[i] $+ S^{2-}$	$CH_3OH\text{-}H_2O$[j]	26	$2{\cdot}3 \times 10^{-4}$	—	—	155
$2\text{-}C_4H_3S{-}C{\equiv}CBr + S^{2-}$	$CH_3OH\text{-}H_2O$[j]	26	47·7	—	—	155

TABLE 24 (*cont.*)

Reactants	Solvent	Temp. (°C)	k ($M^{-1} s^{-1}$)	$\Delta H^{\ddagger}$ (kcal/mol)	$-\Delta S^{\ddagger}$ (e.u.)	Reference
$2\text{-}C_4H_3S{-}C{\equiv}CI + S^{2-}$	$CH_3OH\text{-}H_2O$[j]	26	9×10^4	—	—	155
$2\text{-}C_4H_3S{-}C{\equiv}CCl + C_2H_5S^-$	$CH_3OH\text{-}H_2O$[j]	26	9×10^{-3}	—	—	155
$2\text{-}C_4H_3S{-}C{\equiv}CBr + C_2H_5S^-$	$CH_3OH\text{-}H_2O$[j]	26	23·5	—	—	155
$2\text{-}C_4H_3S{-}C{\equiv}CI + C_2H_5S^-$	$CH_3OH\text{-}H_2O$[j]	26	3×10^4	—	—	155
$C_6H_5C{\equiv}CBr + I^-$	Dioxane–H_2O[k]	127	~0·5	—	—	675
$4\text{-}O_2NC_6H_4CH{=}CHBr + I^-$	$n\text{-}C_4H_9OCH_2CH_2OH$	174	$1{\cdot}04 \times 10^{-4}$	—	—	676
$2,4\text{-}(O_2N)_2C_6H_3Br + I^-$	CH_3OH	100	$1{\cdot}5 \times 10^{-1}$	—	—	677, 678
$CH_3Br + I^-$	CH_3OH	25	1×10^{-3}	—	—	678

[a] TED = triethylenediamine, $N(CH_2CH_2)_3N$. [b] At 12·50 °C. [c] At −15·00 °C. [d] At 60·40 °C. [e] At 72·20 °C. [f] At 36·30 °C. [g] Arrhenius E_A. [h] At 50 °C. [i] 2-Thienylchloroacetylene. [j] $v/v = 1/1$. [k] $v/v = 9/1$. [l] Extrapolated value.

TABLE 25. Element effects in nucleophilic substitution reactions of haloalkynes and other halides

Nucleophile	Halide	Solvent	Temp. (°C)	X	$k(F) : k(Cl) : k(Br) : k(I)$	Reference
CN^-	PhC≡CX	$MeOH/H_2O$ 1/1	26	Br, I	1 : 664	155[b]
TED[a]	PhC≡CX	MeCN	60	Cl, Br	1·3 : 1	665
TED	$PhCH_2CH_2X$	MeCN	55	Cl, Br	1 : 99	666
MeO^-	PhC≡CX	MeOH	78	Cl, Br	1·9 : 1	153
EtO^-	$(p\text{-}O_2NC_6H_4)_2C{=}CHX$	EtOH	50	Cl, Br	1·3 : 1	667
EtO^-	$(C_6H_5)_2C{=}CHX$	EtOH	100	F, Cl	290 : 1	647
EtO^-	$p\text{-}NO_2C_6H_4X$	EtOH	91	F, Cl, Br, I	3·0 : 14 : 12 : 1	679
$(n\text{-}Bu)_3P$	PhC≡CX	DMF	36	Cl, Br	1 : 3·7	157
$(n\text{-}Bu)_3P$	$n\text{-}C_3H_7X$	Me_2CO	35	Cl, Br	1 : 25	671
Ph_3P	PhC≡CX	DMF	36	Cl, Br	2·1 : 1	157
Ph_3P	HC≡CX	Et_2O	r.t.	F, Cl	> 200 : 1[b]	654
$(EtO)_3P$	PhC≡CX	THF	60	Cl, Br	1·3 : 1	159
$(EtO)_2P(O)^-$	PhC≡CX	THF	0[c], −70[d]	Cl, Br	$k(Cl) \ll k(Br)$[b]	159, 660
$(EtO)_2P(O)^-$	$Me_2C(OH)C$≡CX	THF	106[c], −70[d]	Cl, Br	$k(Cl) \ll k(Br)$[b]	169, 660
$p\text{-}C_7H_7S^-$	PhC≡CX	DMF	−25	Cl, Br	3·1 : 1	164
$p\text{-}C_7H_7S^-$	$E\text{-}O_2NC_6H_4C(Ph){=}CHX$	DMF	24	Cl, Br	1·7 : 1	680
$C_6H_5S^-$	$2,4\text{-}(O_2N)_2C_6H_3X$	MeOH	0	F, Cl, Br, I	25 : 0·8 : 1·29 : 1	646
S^{2-}	$2\text{-}C_4H_3SC$≡CX[e]	$MeOH/H_2O$ 1/1	26	Cl, Br, I	$1 : 2{\cdot}1 \times 10^5 : 4 \times 10^8$	155
$C_2H_5S^-$	$2\text{-}C_4H_3SC$≡CX	$MeOH/H_2O$ 1/1	26	Cl, Br, I	$1 : 2{\cdot}6 \times 10^3 : 3{\cdot}3 \times 10^6$	155

[a] TED = triethylenediamine, $N(CH_2CH_2)_3N$.
[b] Estimated from reaction conditions.
[c] Reaction temperature for PhC≡CCl.
[d] Reaction temperature for PhC≡CBr.
[e] $2\text{-}C_4H_3S$ is 2-thienyl.

TABLE 26. Hammett ρ-values for nucleophilic attack on alkynes and related unsaturated halides

Reaction	Solvent	Temp. (°C)	ρ	Site of attack[a]	Reference
m,p-$YC_6H_4C{\equiv}CBr + C_2H_5S^-$	MeOH	26	1·15	Br	155
p-$YC_6H_4C{\equiv}CBr + t$-$C_4H_9S^-$	MeOH	26	1·25	Br	155
p-$YC_6H_4C{\equiv}CBr + p$-$C_7H_7S^-$	DMF	−25	3·9	C_α	164
p-$YC_6H_4C{\equiv}CCl + p$-$C_7H_7S^-$	DMF	−25	3·4	C_α	164
$(p$-$YC_6H_4)_2C{=}CHX^b + p$-$C_7H_7S^-$	DMF	24	2·2[c]	C_α	680, 681
p-$Y(2$-$O_2NC_6H_3)Cl + C_6H_5S^-$	MeOH	35	5·1	C_α	682
p-$YC_6H_4C{\equiv}CCl + (C_2H_5O)_3P$	THF	25	~2·3	C_α	159
p-$YC_6H_4C{\equiv}CBr + (C_2H_5O)_3P$	THF	25	~2·0	Br, C_α	159
p-$YC_6H_4C{\equiv}CT + OH^-$	MeOH/H_2O 1/4	25	0·77	T	683
p-$YC_6H_4C{\equiv}CH + OH^- + BrO^-$	H_2O	25	0·76	H	174
$(p$-$YC_6H_4)_2C{=}CHX^d + C_2H_5O^-$	EtOH	50	2·1[c]	C_α	668
p-$Y(2$-$O_2NC_6H_3)X^d + C_5H_{10}NH$	Various	25	>4	C_α	646

[a] C_α refers to the carbon containing the halogen.
[b] X = F, Cl.
[c] ρ-Values were corrected for the two phenyl groups.
[d] X = Cl, Br.

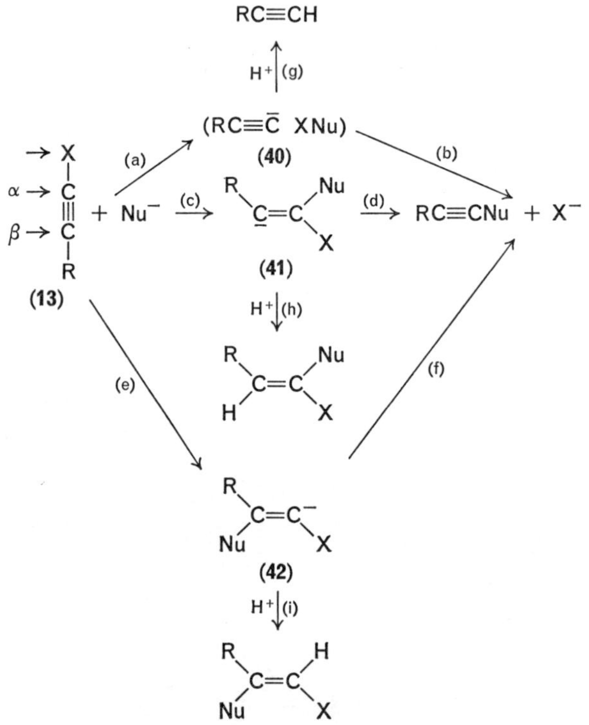

SCHEME 6

channels for process (2) as laid out in Scheme 6. Attack at X, C_α and C_β results in the formation of carbanions **40, 41** and **42**, respectively. Evidence for these species comes from experiments in which proton traps (solvent, nucleophile or alkyne) are present, i.e. steps (i), (g) and (h) (see Table 6 for examples of such experiments). Of course, it can only be assumed that one or other of these carbanion intermediates lies on the reaction pathway to the displacement product. In the absence of diversionary reagents, steps (b), (d) and (f) are followed and the substitution product is formed. The three mechanisms are kinetically indistinguishable, since each could obey second-order kinetics.

The first mechanism assigned to process (2) was proposed by the IIT (Illinois Institute of Technology) group (1962)[652]. The principal target for nucleophilic attack is C_α of the haloalkyne. The substitution product arises via an association–dissociation sequence, steps (c) and (d) of Scheme 6. These steps are analogous to those given in the replacement mechanism of other unsaturated halides, e.g. vinyl[647], carbonyl[645] and aromatic halides[646]. The pattern of reactivity, $F > Cl \sim Br > I$, for these halounsaturates (Table 25) is expected to hold with the haloacetylenes provided the first step of the mechanism is rate-determining.

Several other lines of evidence support the IIT mechanism. First, the carbanion (alkenide) intermediate is real—it turns up in other reactions. Without it how would one rationalize most of the standard *anti* additions to alkynes exemplified by equation (1)? Then, base-induced deprotonations of alkenes, which have been demonstrated by proton labelling, lead directly to this anion[94, 135, 684]. In Scheme 3 (see Section II.B) we showed how addition (equation 1), substitution (equation 2) and elimination may be mechanistically interrelated by this anion. Later we shall show that for certain systems, product distributions, element effects and Hammett ρ-values are in accord with the IIT pathway.

An early challenge to the IIT mechanism came from Arens[154]. It seemed important to emphasize that 'nucleophilic substitution at atoms other than carbon may occur especially when rather stable carbanions can be expelled'. Besides the illustrations given in Table 6, the haloalkyne synthesis[4] and its reversion (equation 245) which involve attacks on hydrogen and halogen are examples supporting his contention[4, 174].

$$RC{\equiv}CH + OH^- + X_2 \underset{}{\overset{X=Br,\ Cl}{\rightleftharpoons}} RC{\equiv}CX + X^- + H_2O \tag{245}$$

Steps (a) and (b) of Scheme 6 constitute the Arens mechanism. Therefore, equation (244) was simply a case of attack on Cl in which the ion-molecule ($PhC{\equiv}C^-\ ClSC_7H_7$-*p*) was the key intermediate. Attack of acetylide on the sulphur of the sulphenyl chloride leads to the product given in equation (244). Support for this step is the well-known reaction of sulphenyl chloride with carbanions to yield sulphides[685]. The disulphides which sometimes turn up in the haloalkyne–thiolate processes (see Table 6) are easily explained by the sulphenyl halide reacting with the thiolates.

A nice demonstration of the Arens process and its several consequences emerged from a clever experiment[4]. $R^1C{\equiv}CBr$, $R^2C{\equiv}CH$ and RS^- were dissolved in DMF; after a few minutes at -35 °C $R^1C{\equiv}CH$, $R^2C{\equiv}CH$ and RSSR could be isolated; at -10 °C both products $R^1C{\equiv}CSR$ and $R^2C{\equiv}CSR$ were found. Steps (246)–(250)

$$R^1C{\equiv}CBr + RS^- \longrightarrow R^1C{\equiv}C^- + RSBr \tag{246}$$

$$R^1C{\equiv}C^- + RSBr \longrightarrow R^1C{\equiv}CSR \tag{247}$$

$$RS^- + RSBr \longrightarrow RSSR \tag{248}$$

$$R^2C{\equiv}C^- + RSSR \longrightarrow R^2C{\equiv}CSR \tag{249}$$

$$R^1C{\equiv}C^- + R^2C{\equiv}CH \rightleftharpoons R^1C{\equiv}CH + R^2C{\equiv}C^- \tag{250}$$

presumably apply here. Analogues of these steps are also required to rationalize processes such as equations (251)[36] and (252)[686], which might seem obscure.

$$HC{\equiv}CH + (i\text{-}Bu)_2S \xrightarrow[110-120\,^{\circ}C]{KOH} i\text{-}BuSCH{=}CHSBu\text{-}i,\quad Z/E > 1 \tag{251}$$

$$HIC{=}CIH + MeO^- \xrightarrow{MeOH} (HC{\equiv}CI) \longrightarrow IC{\equiv}CI + C_2H_2 \tag{252}$$

In contrast to the IIT mechanism, the element effect on the Arens path is expected to be $k(I) > k(Br) > k(Cl)$ as is also the case with S_N2 reactions of alkyl halides (Table 25)[644]. This reactivity order has been observed in the reactions of 1-halo-2-(2-thienyl)acetylenes with sulphides and thiolates in methanol–water[155] in which the C—X bond is presumably broken in the rate-determining step by attack on halogen (see Section II.C.1.d). In both S_N2 and halogen abstraction reactions, the magnitude of the element effect is quite large (Table 25).

The third carbanion mechanism for process (2) was proposed by Viehe[4]. While it may well be the most interesting, it is the least established by precedent and example. Here, C_β is the principal site of nucleophilic attack in RC≡CX (see Table 6 for examples), except when X = F or when attack on X is facile. In this mechanism, the substitution product forms *via* carbanion **42** in which Nu slides from C_β over to C_α with ejection of X^-, i.e. steps (e) and (f) of Scheme 6. Viehe labels this an onium rearrangement through species **43** and **44** in equation (253). As precedent for this

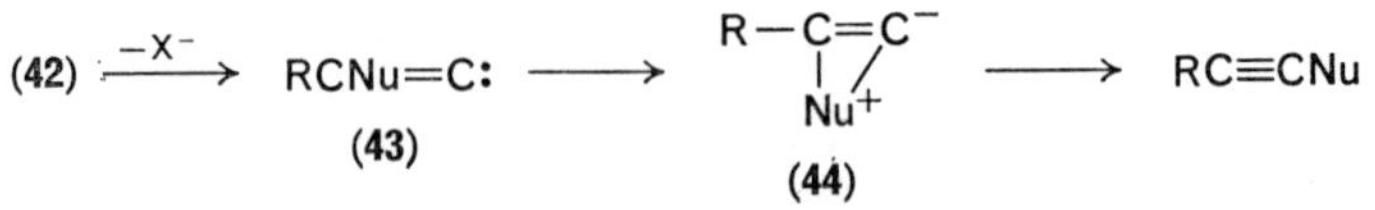

(253)

unusual transformation there is the Fritsch–Buttenberg–Wiechell (FBW) rearrangement (equation 254)[687] and a few model reactions (e.g. equation 255)[688]. By

$$(C_6H_5)_2C{=}CHCl \xrightarrow{base} C_6H_5C{\equiv}CC_6H_5 \tag{254}$$

$$(R_2N)_2C{=}CHCl \xrightarrow{LiNR_2} R_2NC{\equiv}CNR_2 \tag{255}$$

process (256), Viehe demonstrated that even *t*-butylchloroacetylene is attacked by nucleophiles at C_β (see Section II.C.1.d)[165]. Next, he rearranged the adduct from this

$$\left.\begin{array}{r} Z,E\text{-}t\text{-}BuC(SC_6H_5){=}CHCl \\ t\text{-}BuC{\equiv}CCl + C_6H_5S^- \xrightarrow{DMF} \end{array}\right\} \longrightarrow t\text{-}BuC{\equiv}CSC_6H_5 \tag{256}$$

chloroalkyne and thiophenol to the same acetylenic thioether as obtained from *t*-butylchloroacetylene and thiophenoxide in DMF. Therefore, Viehe concluded that the same carbanion intermediate, $t\text{-}BuC(SC_6H_5){=}\bar{C}Cl$, is formed in both branches of equation (256). Once generated, the anion rearranges to the acetylenic thioether *via* the sequence given in equation (253). It is interesting that the FBW process, a rearrangement–elimination, and its reverse are competitive with normal 1,2-addition in equation (257)[405].

Although model reactions such as equation (255) and the upper branch of equation (256) are pictured as occurring by intramolecular rearrangements, there is the

$$\mathrm{PhC{\equiv}CPh + (H_2N)_2CS \xrightarrow[120-140\ ^\circ C]{KOH,\ Me_2SO} PhCOCH_2Ph + (Ph_2C{=}CH)_2S} \quad (257)$$

$$\text{10\%} \qquad \text{16·5\%}$$

$$\mathrm{(Ph_2C{=}CH)_2S \xrightarrow[(H_2O)\ {\sim}140\ ^\circ C]{KOH,\ Me_2SO} PhC{\equiv}CPh}\ \ 48\%$$

distinct possibility that these processes are due instead to intermolecular paths. The conditions employed in these processes are similar to those which lead to elimination[11, 689], e.g.:

$$\mathrm{Me_2N(\mathit{n}\text{-}BuS)C{=}CH_2 \xrightarrow[NH_3]{KNH_2} Me_2NC{\equiv}CH + \mathit{n}\text{-}BuS^-} \quad (258)$$

In equation (256), haloacetylene and nucleophile could be eliminated and these could form the substitution products by addition–elimination steps, the IIT mechanism or the Arens process. By using $LiNR'_2$ as a base in equation (255) and looking for two ynamines, or by adding ArSH to reaction (256) and looking for two thioether products, one can seek evidence of the elimination process. Until such tests are performed, the applicability of the Viehe substitution mechanism remains uncertain.

For some time we have recommended that the prudent approach in this field is to consider any new system within the framework of Scheme 6. One should, in fact, also consider the mechanisms to be discussed following this section. Several systems involving different nucleophiles are discussed in detail to illustrate the 'flexible' approach.

In the reaction of methoxide ion with bromo- and chlorophenylacetylene in methanol (equation 259)[153], the initially detectable products result from the three-site attack **(13)** of methoxide ion on the phenylhaloalkyne. Graphic evidence for the

$$\mathrm{PhC{\equiv}CX + MeO^-Na^+ \xrightarrow[X=Br,\ Cl]{MeOH} PhC{\equiv}COMe + PhC{\equiv}CH}$$
$$\mathrm{+ PhCH{=}CXOMe + PhCOMe{=}CHX} \quad (259)$$

three competing modes, i.e. attack on X, C_α and C_β, is given in the time–products profile (Figure 10). To assign the initial points of attack, note the rates of growth of the several products (Table 27). Since the ethynyl ether forms faster than most of the other products and all of these are relatively stable under the reaction conditions, none of them is considered to be a plausible precursor to the ethynyl ether. We consider this to be strong evidence for three independent channels to the observed products.

Neither the Viehe nor the Arens routes to the ethynyl ether are plausible for this system. The 1-phenyl-1-methoxy-2-haloalkenes of equation (259), for example, may be recovered intact when treated in MeOH with 4M NaOMe at 155 °C[153]. Though these conditions are presumably suitable for the generation of C_β or C_α vinyl anions, no onium process (equation 253) seems to have occurred. Further, it seems improbable that phenylacetylide could be a precursor of the ethynylether (Arens mechanism), since this ion abstracts protons from protic solvents ($k \simeq 10^8 \mathrm{M^{-1}\,s^{-1}}$ at 25 °C in water)[690] and halogen from hypohalite (OX^-) [k(Cl) $= 2{\cdot}3 \times 10^{-4} \mathrm{M^{-2}\,s^{-1}}$ at 25 °C in water][691, 174]. Thus, the possibility that there is an Arens ion-molecule intermediate, which can survive long enough in methanol to rearrange and form the alkynyl ether by attack on oxygen, appears to be highly improbable. Besides, sodium phenylacetylide is known to attack chlorine in ethyl hypochlorite to give chlorophenylacetylene and not ethylphenylethynyl ether[692].

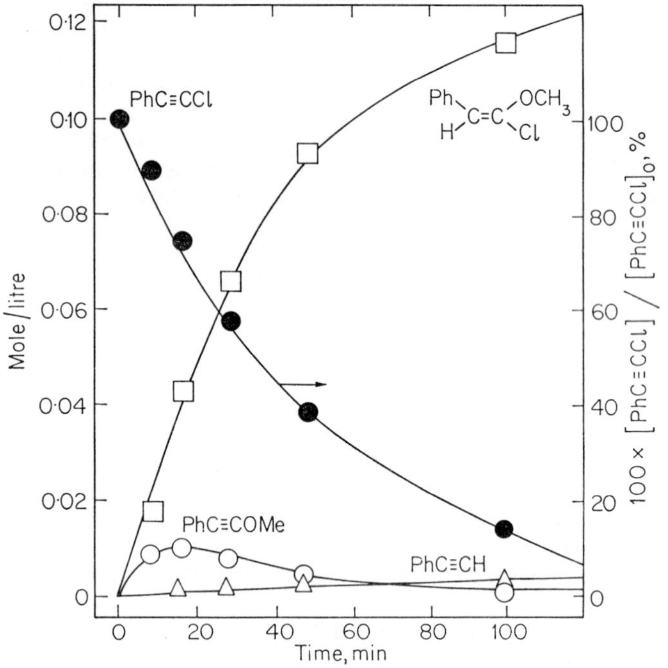

FIGURE 10. Reaction of PhC≡CCl (0·313M) with NaOMe (1·95M) in methanol at 78 °C.

TABLE 27. Rate constants ($M^{-1}\,s^{-1} \times 10^4$) for the systems PhC≡CX–CH_3ONa in methanol at 78 °C [153]

	Process[a]	k(Cl)	k(Br)	k(Cl)/k(Br)
PhC≡CX	k_d	1·63	1·03	1·6
PhC≡$COCH_3$	k_f	0·60	0·51	1·2
E-PhCH=$CXOCH_3$	k_f	1·03	0·34	3·0
PhC≡CH	k_f	0·02	0·05	0·4
C_α (total)	k_f	1·63	0·85	1·9
Z-$PhCOCH_3$=CHX	k_f	0·004	0·11	0·036

[a] Rate constants for disappearance (k_d) and formation (k_f) of the species are given.

Another reason to support the IIT mechanism and discount the other two routes to the ethynyl ether is the element effect observed in the formation of this product, $k(Cl)/k(Br) = 1{\cdot}2$ (Table 27). Attacks at X or C_β do, in fact, yield element effects <1 while attack at C_α gives a value >1.

Turning to a second case, the Arbuzov reaction of substituted phenylbromo- and phenylchloroacetylene with triethylphosphite in THF, we have used the parts of Scheme 6 applicable to attack at C_α and X (equation 260)[159]. Rate data for this

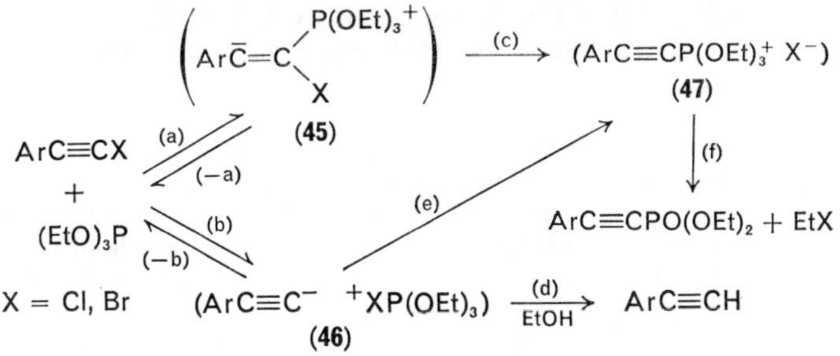

(260)

system are given in Table 24. Since quasi-phosphonium salts of the type generated in Arbuzov reactions are known to react rapidly with weak nucleophiles[693], it is assumed that the ejection of diethyl phenylethynylphosphonate from its quasi-salt, **47**, is a fast process, i.e. k_f is large. The reactivity comparison of phenylchloro- and phenylbromoacetylene toward triethylphosphite is $k(\text{Cl})/k(\text{Br}) = 1{\cdot}3$, entirely consistent with a bond-making rate-determining step. On the other hand, when halogen attack is observed with a phosphite, e.g. 1-halo-3-methyl-butyn-3-ol with sodium diethylphosphite, $k(\text{Cl})/k(\text{Br}) \ll 1$[660].

Hammett ρ-values were found to be 2·3 and 2·0 at 25 °C for the arylchloro- and arylbromoethynes, respectively, with triethylphosphite. It has been noted that for related reactions there is a decrease in the reaction constant, ρ, as the distance between the reaction centre and aromatic ring increases[694, 695]:

$$\text{Ar—Z—Y—X—}$$
$$\uparrow \quad \uparrow \quad \uparrow$$

Judging from the ρ-values of the related systems listed in Table 26, a ρ-value of $\geqslant 2$ indicates attack at C_α while a ρ-value of *ca.* 1 is a sign of abstraction from an alkyne, e.g. PhC≡CT with hydroxide[683] or p-$YC_6H_4C{\equiv}CBr$ with $C_2H_5S^-$ in MeOH (Table 26)[155].

If ρ and $k(\text{Cl})/k(\text{Br})$ were the only mechanistic tests performed on the ArC≡CX–$(EtO)_3P$–THF systems, then it would have been concluded that substitution occurs *via* intermediate **45**. The addition of ethanol to the phenylchloroacetylene reaction produced no phenylacetylene, confirming the supposition that steps (a) (rate-determining), (c) and (f) occur in the process (equation 260). When phenylbromoacetylene was treated with triethylphosphite in THF with added ethanol, phenylacetylene was produced. This fact indicates that step (b) of equation (260) occurs in the phenylbromoacetylene reaction.

To determine whether phenylbromoacetylene follows step (b) exclusively or whether step (a) competes, the product ratios (PR) of diethylphenylethynylphosphonate to phenylacetylene have been studied. For reaction (260), the PR is given by equation (261a). This relation is valid provided that the steady-state assumption

$$\text{PR} = \frac{k_e}{k_d[\text{ROH}]} + \frac{k_a k_e}{k_b(k_{-a}+k_c)}\left(\frac{k_{-b}+k_e}{k_d[\text{ROH}]}+1\right) \tag{261}$$

$$\text{PR} = \frac{[\text{PhC}{\equiv}\text{CPO(OEt)}_2]}{[\text{PhC}{\equiv}\text{CH}]};\quad [\text{ROH}] = [\text{EtOH}] \tag{261a}$$

$$\text{PR} = \frac{[\text{PhC}{\equiv}\text{C}\overset{+}{\text{P}}\text{R}'_3\ \text{X}^-]}{[\text{PhC}{\equiv}\text{CH}]};\quad [\text{ROH}] = [\text{MeOH}] \tag{261b}$$

applies to intermediates **(45)**–**(47)**. A plot of PR *vs.* 1/[EtOH] for the phenylbromoacetylene–triethylphosphite reaction in THF–EtOH was linear and had an intercept of $\leqslant 0{\cdot}25$ and a slope of *ca.* 6. If attack by triethylphosphite on the bromine of phenylbromoacetylene is exclusive, i.e. $k_a = 0$ or the above plot has a zero intercept, then only the first term on the right-hand side of equation (261a) is retained. In this PR equation, the slope which is *ca.* 6 is represented by k_e/k_d. This means that proton transfer (step d) is slower than step (e) which involves rearrangement and collapse of ion pair **46**. Such a situation is highly improbable since the rate of proton transfer to the phenylacetylide ion is extremely fast (see above)[690]. Step (a) must compete with step (b), i.e. $k_a \neq 0$! Phenylbromoacetylene with triethylphosphite uses both (a) and (b) as parallel rate-determining steps to form diethyl phenylethynylphosphonate.

Mechanistic problems similar to those we have just described were also found in the study of bromo- and chlorophenylacetylene with tributyl- and triphenylphosphine in DMF (equation 262)[157]. Rate data for the production of the ethynylphosphonium

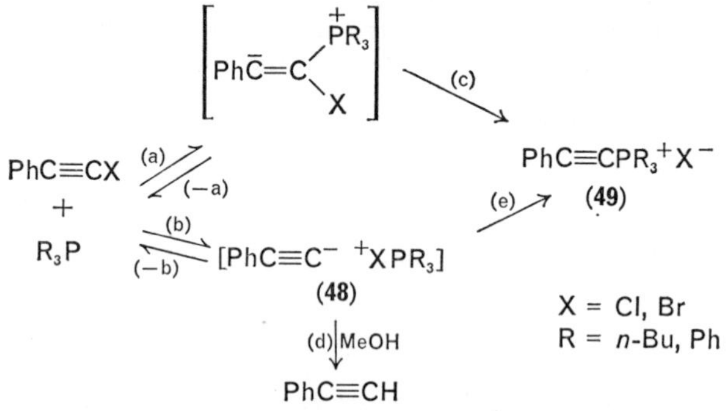

(262)

salts are given in Table 24. In comparison to the other phosphine reactions, the $\Delta H^{\ddagger}$ and $\Delta S^{\ddagger}$ of the tributylphosphine–phenylbromoacetylene reaction are unusually low. These results were taken as an indication that the mechanism of this system differed from the other haloalkyne–phosphine processes. Another clue to a mechanistic change came from element effect data (Table 25). For the phenylbromoacetylene–tributylphosphine system, the element effect $[k(\mathrm{Cl})/k(\mathrm{Br}) = 1/3{\cdot}7]$ parallels those of processes in which the carbon–halogen bond is broken in the rate-determining step, e.g. S_N2 on carbon or halogen. The other haloalkyne–phosphine systems exhibit element effects, $k(\mathrm{Cl})/k(\mathrm{Br}) > 1$, which are like those of other unsaturated organic halides. While the other systems in DMF–MeOH produced MeX, R_3PO, PhC≡CH and the substitution product **49**, tributylphosphine with phenylbromoacetylene gave all of the above products *except* the last one. It is unlikely that these results could be attributed to differences in the partitioning of **48** with changes in structure. Based on nucleophilic substitution rates of other halophosphorus compounds[696], the ion-pair derived from the bromoalkyne should be diverted to the substitution product more rapidly than that formed from the chloroalkyne. These arguments lead to the conclusion that step (e) of equation (262) is insignificant for all systems except when $[\mathrm{MeOH}] \rightarrow 0$.

Additional insight into the mechanisms of these systems was gained by determining their product ratios. As we have indicated above, $k_e = 0$ and thus only the second term on the right-hand side of equation (261b) need be considered. For phenylbromoacetylene with tributylphosphine, $k_a = 0$, since none of the substitution product was formed when methanol was present (PR = 0). For the other systems, $k_a \neq 0$ since the observed PR values are significant. In summary, phenylbromoacetylene with tributylphosphine takes only one path to **49**, that is, steps (b) (rate-determining) and (e) in equation (262); the other systems form the substitution product by a mechanism which has both steps (a) and (b) of equation (262) as competitive and rate-determining.

The two branches of Scheme 6 which were initially considered in the reaction of triethylenediamine (TED) with bromo- and chlorophenylacetylene in MeCN (equation 263) were those which were initiated by attacks at X and C_α (see Table 24

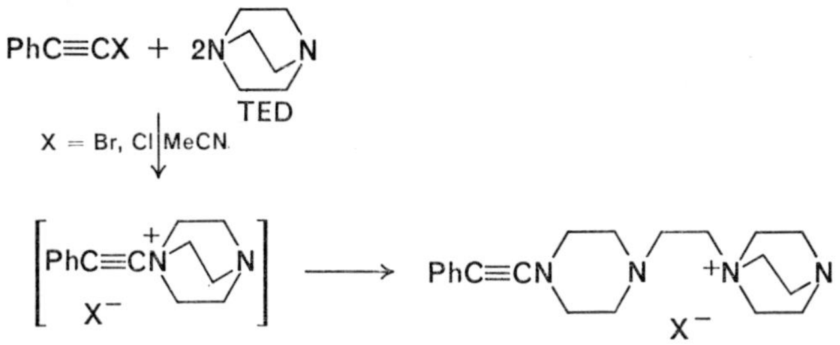

(263)

(50)

for rate data)[665]. These mechanistic choices seemed consistent with the observed element effect (Table 25) and the fact that phenylacetylene along with **50** were formed when methanol was added to the system. Although the latter result is compelling evidence for attack on halogen, it is improbable that the ion-pair which is generated in this process is a precursor to **50**. The formation of the ynamine **50** *via* ion-pair **51** presumably requires backside attack of the acetylide ion on nitrogen. Since the nitrogen in **51** is a bridgehead atom, such a process cannot take place. Because the bicyclic ring is opened in these reactions, another possible arrangement

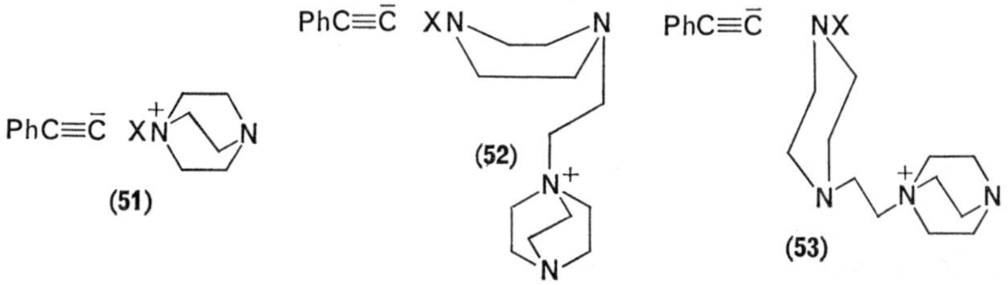

(51) **(52)** **(53)**

for the ion pair is **52**. Backside attack could take place by having **52** rearrange to **53** within the solvent cage. However, formation of **52** seems unlikely because it requires that TED molecules penetrate the solvent cage in preference to the smaller and more abundant methanol molecules. Moreover, the phenylacetylide ion would probably move from the N—X centre to the new charged site. With all of this movement, the acetylide could hardly survive in the presence of a proton source to give **50**. All of this suggests that halogen abstraction and the formation of an ion-pair is a dead-end process. For equation (263) the only branch of Scheme 6 that is applicable is the one which involves C_α attack (IIT mechanism).

In the last three studies (equations 260, 262 and 263) that we have discussed, the Viehe mechanism was not considered because there was no evidence to show that attack occurs on C_β of the haloalkynes. In the case of bromoacetylene with triethylamine in DMF[173], this mechanism cannot be excluded *a priori*, the more so since bromoacetylene is known to orient nucleophiles to the β-carbon (see Section II.C.1.d).

Rate data for this system are given in Table 24 and a general mechanistic scheme is shown in equation (264).

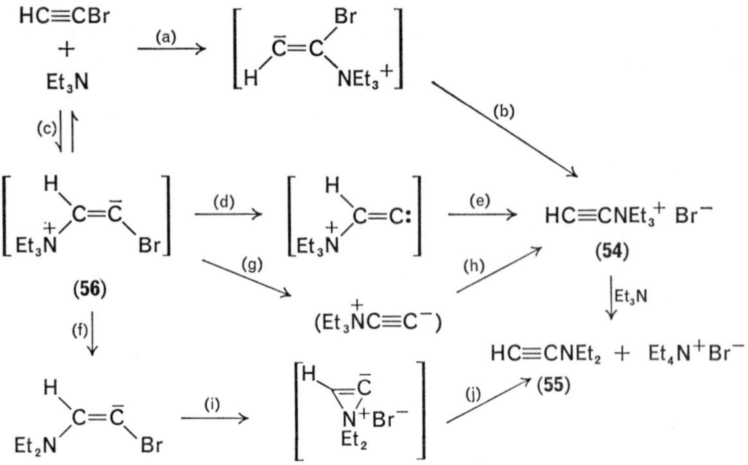

(264)

Consider the Viehe mechanism, i.e. steps (c), (f), (i) and (j) of equation (264); the problem with this route is that it does not reach product **54**. This is because triethylamine is a stronger base (nucleophile) than **55** and hence the direction shown in equation (264), namely **54** → **55**, is preferred rather than **55** → **54**. Attack on bromine (Arens mechanism) is not shown in the above scheme because no acetylene was produced. If the ion-pair [HC≡C⁻ BrNEt$_3^+$] had formed, it would almost certainly have yielded acetylene in the presence of the proton source, HC≡CBr.

Two mechanisms which cannot be ruled out involve intermediate **56**. This species could shed bromide ion (step d) to give a carbene which then could go on to produce **54** by a 1,2-hydride shift (step e). The *anti*-dehydrobromination of **56** (step g), followed by proton uptake (step h), also leads to **54**. The third mechanistic alternative which cannot be discounted involves steps a and b (IIT mechanism).

We close this section with a final comment on equation (244). A 'neutral' party[164, 697] reinvestigated this system in DMF and found that $k(\mathrm{Cl})/k(\mathrm{Br}) = 3{\cdot}1$ and that the Hammett ρ-values for chloro- and bromophenylacetylene were 3·41 and 3·94, respectively. Moreover, when 1 M ethanol was added to this system no phenylacetylene was produced. These results are in agreement with the mechanism that was suggested first, i.e. the IIT mechanism.

b. *Reactivity*. Reactivities of haloalkynes in equation (2) were often comparable with those of alkyl halides and exceeded those of vinyl and aryl halides: alkynyl ≫ alkenyl > aryl (Table 24). This ranking of unsaturated halides is attributed to the acetylenic carbon having the greatest *s*-character and hence the greatest electrophilicity so that $k(sp) > k(sp^2) > k(\mathrm{aryl})$.

Although most of our information about substituent effects on RC≡CX in process (2) comes from qualitative experiments (Table 6), there are quantitative data for X = Cl, Br, I (Tables 24 and 25). Because of the very limited number of examples[165, 688], there is no information on how R and X affect the reactivity of a haloalkyne when C_β is the exclusive nucleophilic target. We have noted previously that where C_α is attacked, the reactivity increases as the electronegativity of X increases (see Section II.C.1.d). When X is the primary site of attack, abstraction becomes more facile as the electronegativity of X diminishes. Whether attack is at C_α or X, the reactivity of RC≡CX is enhanced as the electron-withdrawing ability of R increases. This is understandable, since in either case incipient anions are formed in the transition state. A second issue is stereochemical. If *anti* association and dissociation are favoured in solution in the sense of Scheme 3, the IIT intermediate V_1^- must isomerize to V_2^- to facilitate departure of X (see Section II.B). Groups such as aryl or carbonyl delocalize the negative charge, lower the barrier (V_i) and enhance reactivity. Alkyl groups, if anything, retard substitution. Thus, alkyl- are often weaker electrophiles than arylhaloacetylenes. For example[665], phenylbromoethyne with triethylenediamine in ether at 25 °C affords an 85% yield of the substitution product in several days; 2′-(3-chloro-1,1-dimethyl-2-propynyloxy)tetrahydropyran with this nucleophile in ether requires *ca.* 1 month at 35 °C to produce a 40% yield of the displacement product. Worse than being slow, too often alkyl—C≡CX have a predilection for avoiding the substitution process (2), since nucleophilic attacks on X, C_β or propargylic hydrogen (**13a**) usually lead to other products (see also Section II.C.2.d).

Except for one report, there have been no systematic studies on how the rate of process (2) is affected by the character of the nucleophile. In the reaction of 1-halo-2-(2-thienyl)acetylenes with alkyl thiolates in MeOH–H_2O, it was established that the rate of halogen attack increases as the basicity of the thiolate increases. Specifically, a good Taft correlation with $\rho^* \simeq -1{\cdot}7$ was obtained[155].

There are some general indications that in a series of nucleophiles of the same type, basicity is also an important factor when C_α is under attack: phenylbromoacetylene in DMF yields the substitution product with PhS^- at −30 °C and with $Cl_5C_6S^-$ at *ca.* 100 °C [659].

Whether one looks at C_α or X attack, the data given in Table 24 suggest that polarizability or the softness factor is important in determining nucleophilicity. Phosphorus nucleophiles appear to be more reactive than structurally similar nitrogen nucleophiles, e.g. with chlorophenylacetylene, $k(Bu_3P, DMF)/k(TED, MeCN) = 260$ at 30 °C [157, 685]. As compared with oxygen, sulphur compounds are more potent nucleophiles in MeOH, e.g. with PhC≡CBr, $k(EtS^-)/k(MeO^-) > 5 \times 10^3$ [153, 155], but it is difficult to generalize from these specific instances. However, judging from the trend in reactivity with PhC≡CBr in the series $Ph_3P \gg Ph_3Sb \sim Ph_3Bi \sim 0$ [157], and from the miscellaneous data mentioned, one might *tentatively* conclude that the changing factors of proton basicity and polarizability yield the highest rates at P and S.

Verploegh and coworkers have attempted to be more critical. Using a variety of nucleophiles with 1-iodo or 1-bromo-(2-thienyl)acetylene they have shown that good correlations may be obtained with the oxibase scale of Edwards, $\log k(Nu)/k(H_2O) = \alpha E + \beta H$. Both polarizability ($E$) and basicity ($H$)-related terms enhance the rate[155].

Finally, we shall consider the effect of the solvent on process (2). As expected, the rates of ion-molecule examples of equation (2) are greater in aprotic than in protic solvents (see also Section II.C.1.d). For instance, PhC≡COMe forms from $NaOCH_3$ and PhC≡CCl in DMSO at 25 °C at roughly the same rate that it forms in MeOH at 80 °C [153, 658]. A similar comparison can be made for the reaction of PhC≡CBr with p-$C_7H_7S^-$ at −25 °C: $k_{Br}(MeOH) \simeq 10^{-4}$ *vs.* $k_{C_\alpha}(DMF) \simeq 2 \times 10^{-2} M^{-1} s^{-1}$ in which the point of attack in PhC≡CBr is indicated[155, 164]. Since PhC≡CSAr is not produced in methanol, $k_{C_\alpha}(MeOH) \leqslant 10^{-6} M^{-1} s^{-1}$. Hence there is a rate factor of *at least* 10^4 favouring C_α attack in DMF over MeOH.

Despite basic differences in their mechanisms, it is nevertheless instructive to compare the effect of solvent on sp^3 with sp sites. We compare the PhC≡CBr–p-$C_7H_7S^-$ system given above with n-BuBr–PhS^- reaction. The rate factor for the sp^3

system at −25 °C is $k(\text{DMF})/k(\text{MeOH}) \simeq 10^4$ [698]. It would appear that as large as rate enhancements are for sp^3, they are even larger for sp systems. We believe this must be ascribed to differential solvation favouring the sp-activated complex, solvation energy which accrues to the 'softer' species in the more polarizable aprotic solvent as compared to the more compact sp^3 activated complex. This differential solvation appears to amount to a rate factor of at least 10^2.

There is another somewhat puzzling differential effect relating to attack at the sites C_α *vs.* X in process (2). As mentioned above, k_{Br}/k_{C_α} appears to increase as the solvent becomes more protic. This is evident, for example, in the reactions of $(EtO)_3P$ or Ph_3P with haloalkynes in alcohol–THF mixtures[157, 158, 163]. It would appear that species **41** (Scheme 6) is not as well solvated as species **40** in a protic solvent. It should be noted, however, that Simpson and Burt have studied the variation of products in similar systems, e.g. $(EtO)_3P{-}PhC{\equiv}CCl$, and found a maximum in [$PhC{\equiv}CH$] and a minimum in $PhC{\equiv}CPO(OEt)_2$ between 0 and 5M added ethanol[163].

c. S_N1. Gas-phase heats of reaction given in Section II.A show that the ethynyl cation is by far the most difficult to form from the parent hydrocarbon. In contrast to vinyl cations, it is impossible for ethynyl carbocations to have stabilizing groups on the carbon bearing the positive charge. All of this indicates that ethynyl cations may be inaccessible. So far, no one has dared to consider even the fleeting existence of these species in solution. However, attempts have been made to generate ethynyl cations in solution (equations 265 and 266)[699–701]. Although the type of chemistry shown in equation (267) 'works' for suitably activated alkyl[702] and vinyl halides[703],

$$Ag^+ + PhC{\equiv}CBr \not\longrightarrow PhC{\equiv}C^+ \tag{265}$$

$$\xrightarrow{?} (RC{\equiv}CN_2^+) \not\longrightarrow RC{\equiv}C^+ + N_2 \tag{266}$$

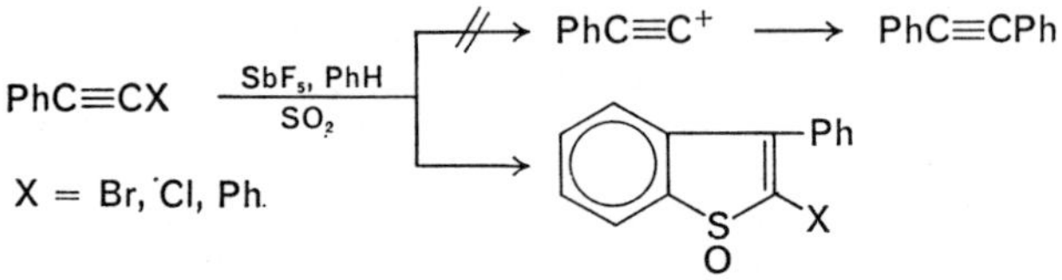

(267)

it failed for ethynyl halide. Instead of forming the phenylethynyl cation and trapping it with benzene in equation (267), thiophene oxides were obtained[701]. It may be that antimony pentafluoride has insufficient 'pulling power' on a halogen linked to an sp carbon to promote ionization and/or the triple bond effectively competes for this Lewis acid. This latter suggestion is supported by the fact that X need not be a halogen for the lower branch of equation (267) to occur. If the ethynyl cation is to be prepared in solution, then some other approach is obviously required.

d. S_N2. Transition state **57**, which is an analogy to that of the Walden inversion, has justifiably been called, 'patently absurd' [173]. The arrangement in transition state **58**

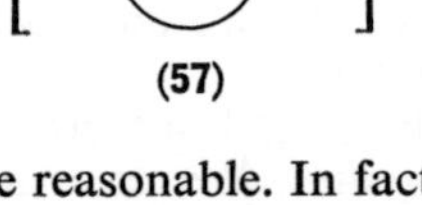

(57)

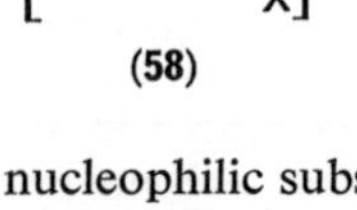

(58)

appears to be more reasonable. In fact, one-step nucleophilic substitution involving front-side attack and retention of configuration has recently been considered[704] and rejected for alkenes[685]. Although absolute arguments cannot be presented, there are a number of objections to such a one-step process taking place in acetylenic substrates: there are few examples of nucleophilic displacement at other unsaturated centres which proceed by a single step; species **58** appears to be geometrically close to the vinyl anion which is normally at a potential minimum on the energy surface[705, 706].

e. *Aggregate*. In reactions such as (268) and (269), the nucleophiles are probably polymeric species[701, 707]. We place these processes in a category of mechanisms which

$$PhC{\equiv}CX + CuCN \longrightarrow PhC{\equiv}CCN \qquad (268)$$

$$t\text{-}BuC{\equiv}CBr + HC{\equiv}CCMe_2OH \xrightarrow[EtNH_2]{CuCl} t\text{-}Bu(C{\equiv}C)_2CMe_2OH \qquad (269)$$

is termed 'aggregate'. We mean by this that the rate-determining step involves an ion pair, dimer or higher clusters along with RC≡CX. Indeed, for most organometallic processes, e.g. equations (243) and (270)[654], aggregate mechanisms probably predominate over those given in Scheme 6. Although organometallic coupling processes

$$HC{\equiv}CF + n\text{-}BuLi \xrightarrow{\text{ether}} HC{\equiv}CBu\text{-}n \qquad (270)$$

are synthetically useful (see next section) and may involve a variety of metals, e.g. Li, Na, Mg, Cu, Sn, relatively little is known about their mechanisms. Undoubtedly, these are likely to be varied and complex.

Consider one aggregate process, the Cadiot–Chodkiewicz reaction[708]. A terminal alkyne and iodo-, bromo- or chloroacetylene are coupled in the presence of Cu(I) and an amine, as illustrated in equation (269). (Variations from this recipe yield Glaser, Straus and Eglinton couplings[708].) Studies on the coupling of $Me_2C(OH)C{\equiv}CCl$ with $Me_2C(OH)C{\equiv}CH$ in aqueous methanol in the presence of NH_2OH and $EtNH_2$ indicate that the reaction is first order in copper species $[CuCl]_0$, first order in chloroalkyne, first order in amine and of uncertain order in terminal alkyne[709]. The rates (v_0) clearly indicate that, more important than simply neutralizing HCl, the amine probably coordinates with the copper (v_0, pK): $MeNH_2$ (13·5, 10·6), i-$PrNH_2$ (2·8, 10·6), t-Bu (0·25, 10·4), Et_2NH (0·13, 11), C_5H_5N (0·7, 5·2), Et_3N (0, 10·7). Under conditions favourable to dissociation of $[Cu(H_2NEt)_nX]_m$, that is, high dilution, the coupling is retarded[709]. In addition, bromoalkynes react much faster than chloroalkynes and water-soluble alkynes react more rapidly than hydrophobic compounds, e.g. PhC≡CH. To couple the latter, one must go to solvents in which the RC≡CCu dissolves, e.g. Me_2SO, $(Me_2N)_3PO$ or morpholine[710].

On the basis of these observations and the known structures of a few polynuclear copper complexes, we suppose that CuC≡CR is a part of a soluble complex, e.g. structure **26** [558]. The coordination number of the copper is probably 4 and the ligands (L) may be amine, alkyne, etc. In the coupling reaction, the slow step is the release or transfer of X^- from carbon. These speculations lead to something like the progression in equation (271).

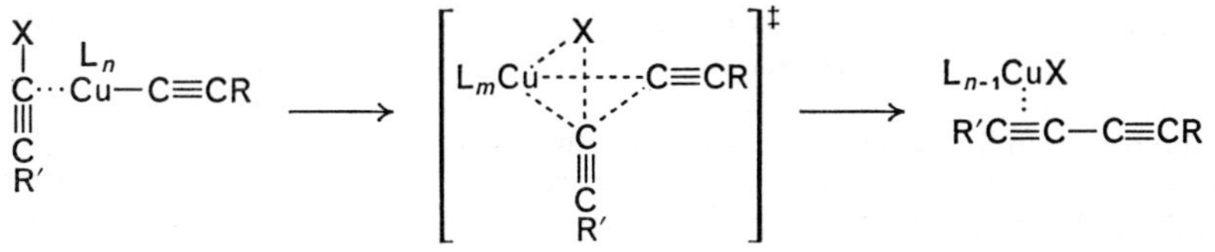

(271)

f. *Radical anion* (S_RN). The seventh mechanism for process (2) involves radical anion intermediates. Its steps, which are presented in Scheme 7, have been adapted

$$RC{\equiv}CX + Nu^- \longrightarrow RC{\equiv}CX^{\bullet -} + Nu^{\bullet}$$

$$RC{\equiv}CX^{\bullet -} \longrightarrow RC{\equiv}C^{\bullet} + X^-$$

$$RC{\equiv}C^{\bullet} + Nu^- \longrightarrow RC{\equiv}CNu^{\bullet -}$$

$$RC{\equiv}CNu^{\bullet -} + RC{\equiv}CX \longrightarrow RC{\equiv}CNu + RC{\equiv}CX^{\bullet -}$$

(Nu = nucleophile)

SCHEME 7

from other systems[711, 712]. This mechanism has been recognized in nucleophilic substitutions at aromatic[711] and at vinylic carbon sites[712]. It was suggested as a possibility by Verploegh[155] but Glaser coupling (equation 272)[688] and equation

$$2RC{\equiv}CH \xrightarrow[RNH_2]{Cu^+,\ O_2} (RC{\equiv}C)_2 \qquad (272)$$

$$PhC{\equiv}CH + R_2NH + Cu(OAc)_2.H_2O \xrightarrow[0\,°C]{O_2,\ PhH} PhC{\equiv}CNR_2 + (PhC{\equiv}C)_2 \qquad (273)$$

(273)[713] seem to be more definite examples of S_RN. Eventually, process (274) was deliberately developed according to Scheme 7[712].

$$PhC{\equiv}CI + NH_4SPh \xrightarrow[NH_3,\ -33\,°C]{h\nu} PhC{\equiv}CSPh + PhC{\equiv}CH + (PhS)_2 \qquad (274)$$

An important feature of Scheme 7 is the presence of radicals. In equation (273) they were generated in a redox process; in equation (274) they were produced by photons. Once formed, radicals may couple with the wrong partners, as is evident in both equations (273) and (274). Indeed, whenever process (2) is attempted and 'wrong' couplings turn up, one should consider the possibility that radicals were formed. Although simple halogen metal interchange may have occurred in equations (275)[714] and (276)[715], it is conceivable that the S_RN mechanism applied.

$$PhC{\equiv}CBr + Na^+\bar{C}H(CO_2Et)_2 \xrightarrow{DMF} [(EtO_2C)_2CH]_2 \qquad (275)$$

$$PhC{\equiv}CCl + Ph_2CH_2 \xrightarrow[DMSO]{KOH} (PhC{\equiv}C)_2 \qquad (276)$$

Glaser coupling has found many applications[708]. As for other 'heterocouplings' the wide utility of Scheme 7 for synthesis has yet to be developed. The chief obstacle seems to be the preference for self-coupling of ethynyl radicals.

C. Synthesis

Synthetic applications of process (2) are given in Table 28. This compilation is intended to be representative rather than exhaustive; examples of process (2) prior to 1968 can be found in Viehe's excellent book[4]. Instead of surveying each class of nucleophile found in Table 28, we shall limit our discussion to some of the complications and successful strategies associated with process (2). In this manner, we hope to provide the reader with insight concerning the scope and limitations of this reaction.

TABLE 28. Synthesis *via* nucleophilic substitution at an acetylenic carbon[a]

$$Nu^- + RC{\equiv}CX \longrightarrow RC{\equiv}CNu + X^-$$

Nucleophile	R	X	Conditions: Medium	Conditions: Time (h)	Conditions: Temp. (°C)[b]	Products (yield, %)	Ref.
B_{12s}('CoH')	H	Br	H_2O	0·5	r.t.	(Ethynylcobalamin)	727
$LiAlH_4$	Ph	Br	Et_2O	5	b.p.	PhC≡CH (50), PhCH=CH_2 (13), PhEt (31), PhCH=CHBr (5), PhC≡CBr (1)	291
LiCuHBr·$Al(OCH_3)$ ('CuH')	Ph	SMe	THF	1	50–60	PhC≡CH, *E*-PhCH=CHSEt	39
R^1Li	*n*-Bu	OEt	Et_2O	0·5	b.p.	*n*-BuC≡CR^1; e.g. R^1 = Me (34), *n*-Bu (70), *t*-Bu (50), Ph (60), 2-C_4H_3S (33), *n*-BuC≡C (24)[c]	728
R^1Li	$(CH_3)_2(OR^2)$	OR^3	Et_2O, Et_2O/THF, Et_2O/DME[d]	2	—	$(CH_3)_2C(OR_2)C{\equiv}CR^1$, $(CH_3)_2C{=}C{=}(OR^3)R^1$, $(CH_3)_2C(OR^2)C{\equiv}COR^3$, e.g. R^1, R^2, R^3 = Me, Me, Me (39), (37), (24)[e]; Ph, Me, Et (40), (14), (46)[e]	729
C_6F_5Li	C_6F_5	F[f]	Et_2O	120	−40	$C_6F_5C{\equiv}CC_6F_5$	731
Sodio-2-phenyl-1,3-indandione	Ph	IPh^+Cl^-	*t*-BuOH	0·17	r.t.	2-Phenyl-2-phenylethynyl-1,3-indandione (73)	312
Ph_3CK[g]	Ph	Cl	Me_2SO	2	r.t.	PhC≡$CCPh_3$ (12)	3, 715
$Ph_2C(CN)K$[g]	Ph	Cl	Me_2SO	36	r.t.	PhC≡$CC(CN)Ph_2$ (61)	3, 715
$R^1C_5H_4K$[g]	Ph	Cl	Me_2SO	24	r.t.	R^1 = H: phenylethynylferrocene (21), 1,1′-di(phenylethynyl)ferrocene (10). R^1 = Me: 1,1′-dimethyl-3-phenylethynylferrocene (55); 1,1′-dimethyl-3,3′-di(phenylethynyl)-ferrocene (13)	3, 715
CuCN	Ph	Cl	DMF	3	b.p.	PhC≡CCN (30)	701
ArCu[h, i]	Me_3Si	I	THF	6	20	$Me_3SiC{\equiv}CAr$; e.g. Ar=Ph (64), 3-MeC_6H_4 (80), 3-$MeOC_6H_4$ (33), 2-$F_3CC_6H_4$ (51), 2-C_4H_3S (90)[j]	725

TABLE 28 (*cont.*)

Nucleophile	R	X	Conditions: Medium	Conditions: Time (h)	Conditions: Temp. (°C)[b]	Products (yield, %)	Ref.
C_6H_5Cu	Et_3Si	Br	THF	10	b.p.	$Et_3SiC{\equiv}CC_6F_5$ (85)	732
$R^1C(Et){=}CHCu$	R^2	I, Br, Cl	Et_2O, THF, TMEDA[k]	1–2	−15	$R^1C(Et){=}CHC{\equiv}CR^2$; e.g. R^1, R^2 = *n*-Bu, *n*-Bu (77)[l]; *n*-Bu, Me_3Si (82)[m]; Me, CO_2Me (78); Me, EtS (82)	296
$R^1C{\equiv}CCu$	R^2	Cl	HMPT, $MeOH{-}H_2O$	24	r.t.	$R^1(C{\equiv}C)_2R^2$; e.g. R^1, R^2 = Me_2COH, Me_2COH (70); Ph, Me_2COH (30); Ph, Ph (30)	710
$R^1C{\equiv}CCu$	CH_2OH	Br	C_5H_5N	—	35	$R^1(C{\equiv}C)_2CH_2OH$; e.g. R^1 = Me, Ph, $MeC{\equiv}C$	724
$PhC{\equiv}CCu$	$Et_3Si(C{\equiv}C)_2$	Br, I	DMF	2	25	$Ph(C{\equiv}C)_3SiEt_3$ (30)	733
m,p-$XC_6H_4C{\equiv}CCu$	Et_3M	Br	DMF, $EtNH_2$	0·5	25	*m,p*-$XC_6H_4(C{\equiv}C)_2MEt_3$; e.g. X, M = H, Si (50); H, Ge (40); *m*-Me, Si (35); *p*-F, Si (60); *p*-NO_2, Si (40)	734
2-$HO_2CC_6H_4CO_2CH_2C{\equiv}CCu$	*t*-Bu	Br	$EtNH_2$	—	—	2-$HO_2CC_6H_4CO_2CH_2(C{\equiv}C)_2Bu$-*t* (65)	707
3-Furylcopper	*E*-THPOCMe=CH[n]	I		—	—	3-Furyl$C{\equiv}CCH{=}C(Me)OTHP$-*E*	735
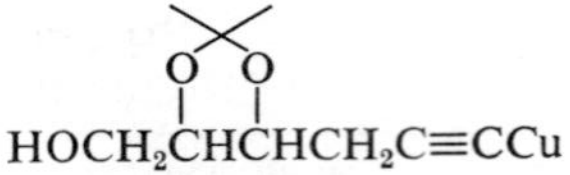 $HOCH_2CHCHCH_2C{\equiv}CCu$	$H_3C(C{\equiv}C)_3$	Br	$H_2O{-}MeOH$, $EtNH_2$	1	20	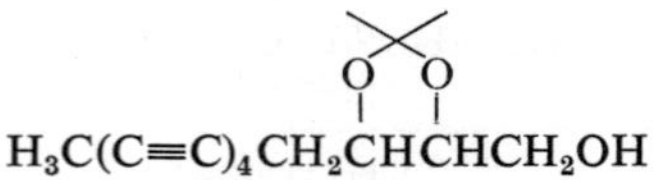 $H_3C(C{\equiv}C)_4CH_2CHCHCH_2OH$	736
$BrZnCMe(CO_2Et)_2$	Ph	Br	Xylene	—	b.p.	$PhC{\equiv}CCMe(Co_2Et)_2$, $PhC(CMe(CO_2Et)_2){=}CH_2$	737
NH_3	CN	Cl	Various	0·33	30	$CH_2(CN)_2$ (60–87)	718
Me_2NH	Ph	H	PhH[o]	0·5	0	$PhC{\equiv}CNMe_2$ (>40), $(PhC{\equiv}C)_2$	713
Et_2NH, PhMgBr	Cl	Cl	Et_2O	—	0	$PhC{\equiv}CNEt_2$ (70)	688
$(H_2C{=}CHCH_2)_2NH$	CN	Cl	Et_2O	1·5	−60	$NCC{\equiv}CN(CH_2CH{=}CH_2)_2$ (68)	124

R^1_2NH	$R^2C{\equiv}C$	Br	Et_2O	5	r.t.	$R^2(C{\equiv}C)_2NR^1_2$; R^1, R^2 = Me, $(Me)_2COH$ (69); 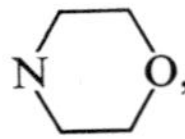, Me_2COH (72); Me, n-C_4H_9 (92); Me, Me (91)	738
Et_3N	H	Br	Et_2O	48	−78	$HC{\equiv}CN(Et)_3^+Br^-$, $H_3CCONEt_2$	173
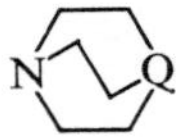	R	Cl, Br	Et_2O	>48	r.t.	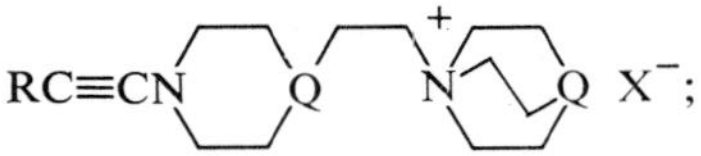; R,Q,X = Ph,N,Cl (70); Ph,N,Br (85); Ph,C,Br (64); $(CH_3)_2COTHP$,N,Cl (40)[n,p]	665
$LiNR^1R^2$	F	Cl	Et_2O	3	−78	$R^1R^2NC{\equiv}CY$; R^1, R^2, Y = Et, Et, NEt_2 (57); CH_3, Ph, Cl (20)	171
$LiNEt_2$	R^1	F[q]	C_6H_{14}	~12	0–r.t.	$R^1C{\equiv}CNEt_2$; e.g. R^1 = Ph (73), p-$CH_3C_6H_4$ (77), o-ClC_6H_4 (84), C_5H_{11} (74), C_6H_{13} (45)	739
$LiNEt_2$	R^1	Cl	HMPT	—	—	$R^1C{\equiv}CNEt_2$; R^1 = t-Bu (50), Ph (60)	740
$(n\text{-}Pr)_2NLi$	C_5H_{11}	OEt	Et_2O	0·5	110–120	$C_5H_{11}C{\equiv}CN(Pr\text{-}n)_2$	145
R^1O^-	Ph	Cl	Me_2SO	1	0–25	$PhC{\equiv}COR$; R^1 = Me (53), Et (42), i-Pr (32)[r], t-Bu (46)	658 658
$(MeO)_3P$	Cl	Cl	Et_2O	~12	0–r.t.	$(MeO)_2POC{\equiv}CY$; Y = Cl (88), $PO(OMe)_2$ (72)[s]	741
$(R^1O)_3P$	Cl	Cl	Et_2O	1	−5 to 35	$(R^1O)_2POC{\equiv}CY$; R^1, Y = Et, Cl (81); Et, $PO(OEt)_2$ (48)[t]; i-Pr, Cl (100); i-Pr, $(i\text{-}PrO)_2PO$ (9)[u]	742
$(R^{1,2}O)_3P$	Me_3Si	Cl	—	5	b.p.	$Me_3SiC{\equiv}CPO(OR^{1,2})_2$; R^1, R^2 = Me, Me (70); Et, Et (82); i-Pr, i-Pr (98); i-Pr, Me_3Si (80)[v]	717
$(R^1O)_3P$	R^2	Br, Cl, I	THF	2–4	80–110	$R^2C{\equiv}CPO(OR^1)_2$; R^1, R^2 = Me, Ph (81); Et, Ph (75);[w] Et, $(CH_3)_2COH$; Et, H[x]	159, 163b

TABLE 28 (*cont.*)

Nucleophile	R	X	Medium	Conditions: Time (h)	Conditions: Temp. (°C)[b]	Products (yield, %)	Ref.
$(R^1O)_3P$	R^2	Br	—	—	120	$R^2C{\equiv}CPO(OR^1)_2$; e.g. R^1R^2 = Et, $H_2C{=}CEt$ (60); Et, $CH_3CH{=}CCH_3$ (52); Me, $CH_3CH{=}CCH_3$ (52)	730
R^1_3P	R^2	Br, Cl	Et_2O	70–100	r.t.	$R^2C{\equiv}CPR^1_3{}^+X^-$; R^1R^2, X^- = Ph, Ph, Br (90); Ph, Ph, Cl (86); *n*-Bu, Ph, Br (92); *n*-Bu, Ph, Cl (83); Ph, H, Cl; Ph, H, Br	157
$(EtO)_2PONa$	$(EtO)_2CH$	Br	THF	16	−70	$(EtO)_2P(O)C{\equiv}CCH(OEt)_2$ (64)	743
PhS^-	*t*-Bu	Cl	DMF	—	r.t.	*t*-BuC≡CSPh (63)	165
ArS^-	Ar′	Br, Cl	DMF	<0·2	−25	Ar′C≡CSAr; Ar, Ar′ = *p*-C_7H_7, *p*-C_7H_7; *p*-C_7H_7, *p*-ClC_6H_4; *p*-C_7H_7, *p*-$O_2NC_6H_4$; *p*-C_7H_7, *p*-$CH_3OC_6H_4$	164
Cl^-[y]	Ph	Br	Me_2SO	240	90	PhC≡CCl (30)	726

[a] The examples listed in this table are intended to update those given in Reference 4.
[b] B.p. is boiling point of the solvent.
[c] Reaction in dioxane at 100 °C for 30 h.
[d] Dimethoxyethane.
[e] Mole % of (alkynylether), (allenylether), (starting product).
[f] Prepared from $HCF{=}CF_2$ or $BrCF{=}CF_2$ and C_6F_5Li.
[g] Prepared from KOH. The sodium salt was prepared in DME with Na metal.
[h] Other examples of coupling between alkynyl halides and organocopper compounds can be found in Reference 559.
[i] The organocopper reagents may be aggregates.
[j] Prepared from $Me_3SiC{\equiv}CBr$.
[k] Tetramethylenediamine.
[l] Prepared from *n*-BuC≡CI.
[m] Prepared from $Me_3SiC{\equiv}CI$.
[n] THP is 2-Tetrahydropyranyl.

[o] Solution contains $Cu(OAc)_2.H_2O$.
[p] Isolated as an amide.
[q] Prepared from $R^1CH{=}CF_2$ and $LiNEt_2$.
[r] Reaction time 20 h.
[s] Reaction kept at 10–15 °C for 15 h.
[t] Reaction time 12 h.
[u] Reaction kept at 80 °C for 12 h.
[v] Prepared from $Me_3SiC{\equiv}CBr$.
[w] Prepared also from $PhC{\equiv}CCl$ and $(EtO)_2PO^-$ in 61% yield and $(EtO)_3P$ in 86% yield.
[x] Obtained in the reaction of 1-chloro-3-methylbutyn-3-ol with triethylphosphite.
[y] NaCl or $(CH_3)_4N^+Cl^-$.

Difficulties arise when process (2) is carried out in the presence of mobile protons. This stems from the fact that carbanionic intermediates such as those shown in Scheme 6 are the precursors to the substitution product. Interception of these intermediates by proton traps may prevent the formation of the desired product or at least diminish its yield. Moreover, process (2) proceeds more rapidly in the presence of an aprotic solvent than a protic solvent. This is well illustrated in the synthesis of ethynyl ethers[658]. When methanol is used as the solvent in equation (277), the

$$PhC{\equiv}CCl + RO^- \longrightarrow PhC{\equiv}COR + Cl^- \qquad (277)$$

$$R = Me, Et, i\text{-}Pr, t\text{-}Bu$$

substitution product could not be isolated (Table 6). On the other hand, substantial quantities of the ethynylethers were obtained when the synthesis was conducted in an aprotic solvent (Me_2SO, HMPT).

Yields of the substitution product are also suppressed when the available proton is a part of the alkyne structure. The alkynes used in equation (278)[159] (upper branch)

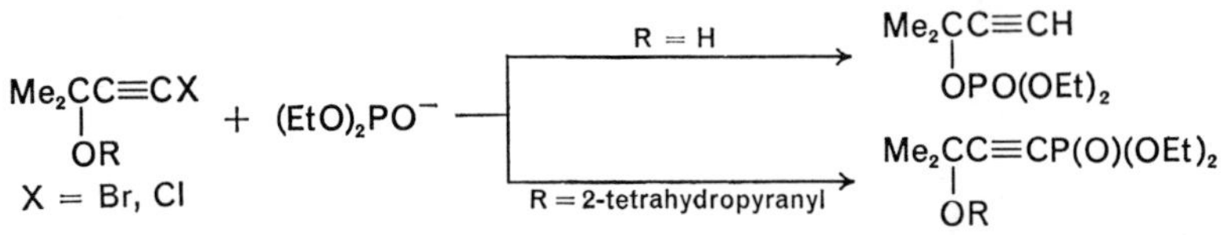

(278)

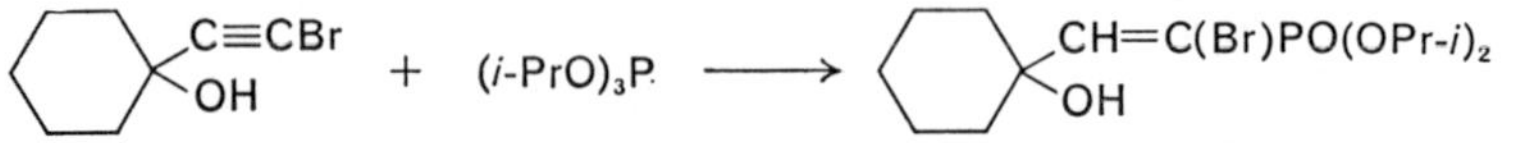

(279)

and reaction (279)[636] contain a 'built-in' trap for the carbanion intermediates. Some other complications which arise when the acetylene has a mobile proton are shown by equations (280)[659] and (281)[716]. One solution to this problem is to replace the troublesome proton in the alkyne with a blocking group. In several cases where this

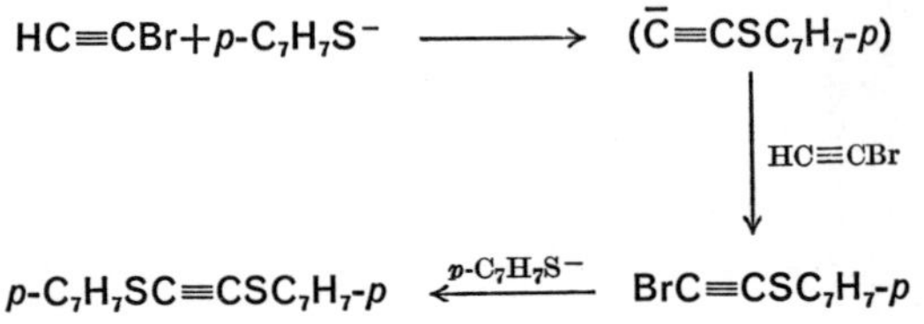

(280)

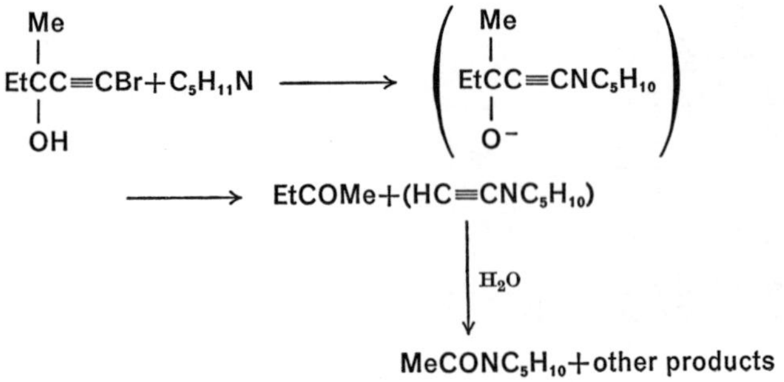

(281)

tactic was employed, successful syntheses were achieved, e.g. equations (278) (lower branch), (282)[717] and (283)[715].

$$Me_3SiC{\equiv}CCl + (MeO)_3P \longrightarrow Me_3SiC{\equiv}CP(O)(OMe)_2 \xrightarrow{Na_2CO_3,\ H_2O} HC{\equiv}CP(O)(OMe)_2 \qquad (282)$$

$$Hg(C{\equiv}CCl)_2 + Ph_2\bar{C}CN \longrightarrow Hg(C{\equiv}CCPh_2CN)_2 \qquad (283)$$

A different kind of problem is caused by protons on the nucleophile[651]. In equation (284), tautomerization accounts for the failure to isolate the substitution product[718].

$$NCC{\equiv}CCl \xrightarrow{NH_3} (NCC{\equiv}CNH_2) \longrightarrow CH_2(CN)_2 \qquad (284)$$

Another side-reaction of process (2) is due to the susceptibility of the substitution product to undergo attack by the nucleophile, e.g. equations (285)[715] and (286)[164].

FlH

$$FlH + PhC{\equiv}CCl \xrightarrow[DMSO]{KOH} (PhC{\equiv}CFl) \longrightarrow (PhCH{-}C(Fl)_2 \qquad (285)$$

$$ArC{\equiv}CX + p\text{-}C_7H_7S^- \xrightarrow{DMF} ArC{\equiv}CSC_7H_7\text{-}p \longrightarrow ArC(SC_7H_7\text{-}p){=}CHSC_7H_7\text{-}p \qquad (286)$$

X = Br, Cl

These undesirable reactions are facilitated by excess nucleophile and proton availability.

Some of the difficulties in obtaining a substitution product may be due to its inherent instability. The apparent order of chemical stability, $RC{\equiv}CF < RC{\equiv}COR < RC{\equiv}CNR < RC{\equiv}CCR_3$, coincides with the decreasing electronegativity of the substituent attached to the electronegative *sp* carbon[4]. The ease of electrophilic, nucleophilic and radical additions to the triple bond as well as isolation difficulties will be expected to follow the order, $RC{\equiv}CF > RC{\equiv}COR > RC{\equiv}CNR_2 > RC{\equiv}CCR_3$. Equations (287)–(289) are examples of some of these complicating side-reactions[11, 13, 665, 719, 720].

$$3\ t\text{-}BuC{\equiv}CF \longrightarrow \qquad (287)$$

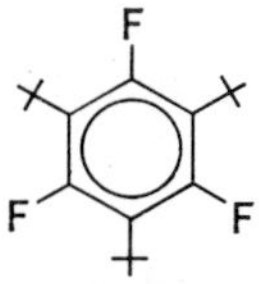

$$RC{\equiv}COR' \xrightarrow{H^+/H_2O} RCH_2CO_2R' \qquad (288)$$

$$RC{\equiv}CNR'_2 \xrightarrow{H_2O} RCH_2CONR'_2 \qquad (289)$$

The chance of success of process (2) appears to be enhanced when the alkyne contains a leaving group which is both a powerful activator of the triple bond and a

C_α director, regardless of the grouping at the other terminal position of the acetylenic function. Ideally such groups should obviate orientation and reactivity problems found in alkyl and unsubstituted acetylenes, prevent metal–halogen exchange processes from occurring and possibly allow for conditions mild enough for the survival of most heteroacetylenic products. Triflates, sulphonates and acetates might serve in this capacity but acetylenes substituted with these groups are unknown. Fluorides might appear to be ideal but they are not easily prepared and they react in other ways (e.g. equation 287). One way to circumvent some of these problems is to generate highly reactive RC≡CX *in situ*. In some cases products may be obtained in high yields, even with fluoroalkylalkynes, which are unobtainable with chloroacetylenes, e.g. Table 28 and equation (290)[169]. When fluorine is present in an alkylacetylene, high yields of the substitution product are obtained (Table 28).

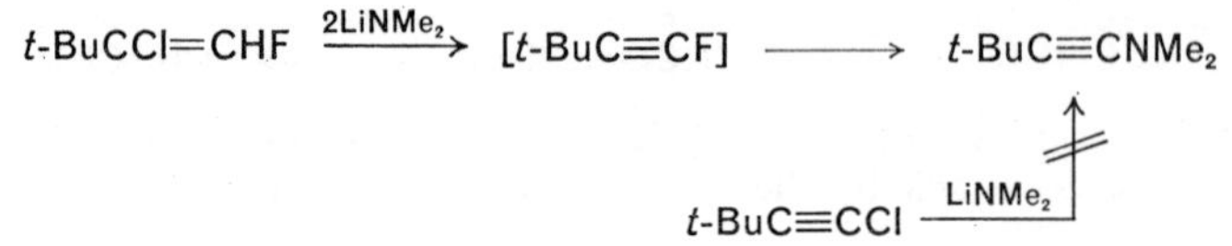

(290)

An alkoxy group is another leaving group which is capable of overcoming the deactivating and directing influences of the alkyl substituent (Table 28). The disadvantage with alkoxy groups is that they seem to require strong bases as nucleophiles to replace them (equations 291–293)[11, 145]. However, ethynyl ethers are

$$RC{\equiv}COR'+LiNR''_2 \longrightarrow RC{\equiv}CNR''_2+LiOR' \qquad (291)$$

$$RC{\equiv}COR'+LiR''_2 \longrightarrow RC{\equiv}CR''+LiOR' \qquad (292)$$

$$R'C{\equiv}COCH_3+HAlR_2 \longrightarrow R'C{=}CH+CH_3OAlR_2 \qquad (293)$$

available[11]. Since it is not yet clear what limits there may be on the leaving ability of alkoxides, these ethers deserve wider testing as possible substrates in equation (2).

The phenyliodonium group is a special group which may be displaced from an acetylenic carbon by nucleophilic attack[312]. Unfortunately, onium leaving groups

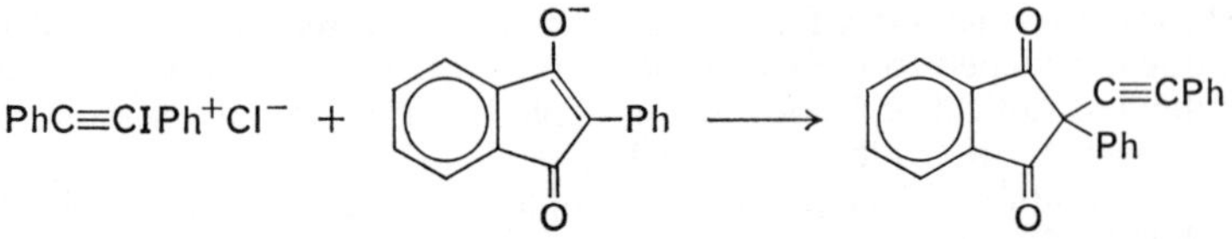

(294)

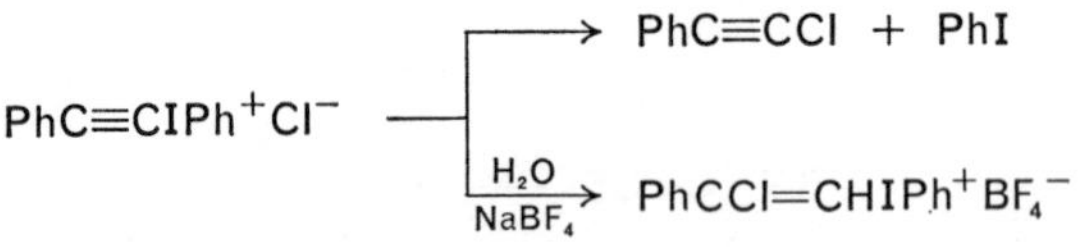

(295)

(296)

promote C_β addition[157, 432a, 721], e.g. equation (296), and may attract direct attack on the positive heteroatom, with the resulting loss of the ethynyl group[157, 722], e.g. equation (297). If the scope of process (2) is to be expanded then alkynes with leaving groups other than those we have mentioned need to be designed and investigated.

For various reasons touched on previously, bromo- and chloroalkylacetylenes have proven to be poor substrates for process (2). Alkyl groups are not activating substituents and the propargylic proton, halogen or C_β are usually more attractive to the

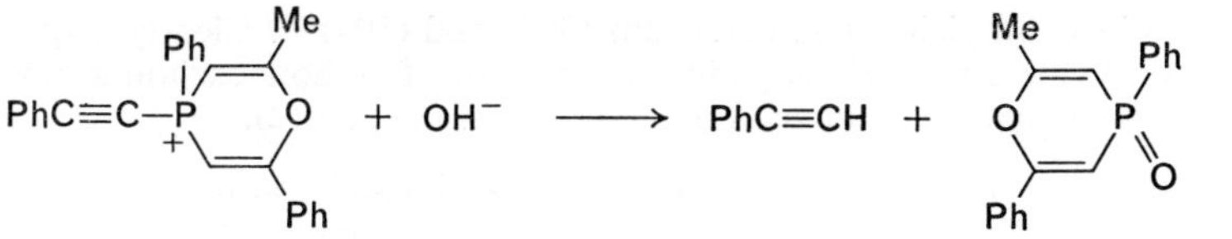

(297)

nucleophile than the C_α site (equations 56, 278 and 281 and Tables 6 and 28)[32, 341, 463]. It is conceivable, however, that if halogen abstraction can be favoured, for example with RC≡CI, and if strictly aprotic conditions prevail, the Arens substitution path (Scheme 6) might predominate as in equation (278) (lower branch). Halide ion, which is a coproduct, might have to be removed in some way, since it may begin to compete as a halogen abstractor[177]. Undoubtedly there are a few 'inadvertent' examples in which the preceding methodology is illustrated, but the one (equations 247–249) originating with the Arens group is perhaps the best.

It is worth special mention that until recently there have been few examples in which carbanions have been successfully used in process (2)[3, 13]. Organoalkali reagents derived from carbon acids frequently enter competing side-reactions such as metal-halogen exchange or induce radical processes ($S_{R}N$). Such reactions may cause the formation of the 'wrong' product or at least limit the yield of the 'right' one, e.g. (275)[714] and (276)[715]. These kinds of problems are accentuated with acetylenic iodides and bromides, attenuated with chlorides and non-existent with fluorides (see Section II.C.1.d). A few successful syntheses have been achieved with organo-alkali metal reagents derived from highly stabilized carbanions and phenylchloro-acetylene[3, 715]:

$$\text{PhC}\equiv\text{CCl} + \text{Ph}_3\text{CH} \xrightarrow[\text{DMSO}]{\text{KOH}} \text{PhC}\equiv\text{CCPh}_3 \qquad (298)$$

$$\text{PhC}\equiv\text{CCl} + \text{Ph}_2\text{C(CN)Na} \xrightarrow[\text{Glyme}]{} \text{PhC}\equiv\text{CC(CN)Ph}_2 \qquad (299)$$

The number of examples where Grignard reagents have been successfully used in process (2) is also small, e.g.[723]

$$\text{C}_6\text{H}_5\text{C}\equiv\text{CBr} + n\text{-C}_5\text{H}_{11}\text{MgBr} \xrightarrow{\text{HMPT}} \underset{55\%}{\text{C}_6\text{H}_5\text{C}\equiv\text{CC}_5\text{H}_{11}} \qquad (300)$$

The most effective carbon nucleophiles in process (2) appear to be organo-copper reagents. Table 28 lists several examples of organocopper 'nucleophiles' coupling with alkynyl halides. Such reagents can be used in conjunction with functional groups (e.g. OH, COOH) where Grignard and organoalkali compounds would not survive, e.g.[724, 707]

$$\text{H}_2\text{C(OH)C}\equiv\text{CBr} + \text{MeC}\equiv\text{CCu} \xrightarrow{\text{Pyridine}} \text{H}_2\text{C(OH)(C}\equiv\text{C)}_2\text{Me} \qquad (301)$$

$$2\text{-HO}_2\text{CC}_6\text{H}_4\text{CO}_2\text{CH}_2\text{C}\equiv\text{CCu} + t\text{-BuC}\equiv\text{CBr} \xrightarrow{\text{EtNH}_2} 2\text{-HO}_2\text{CC}_6\text{H}_4\text{CO}_2\text{CH}_2\text{(C}\equiv\text{C)}_2\text{Bu-}t \qquad (302)$$

Another advantage of organocopper compounds is that unlike other organometallic reagents, they couple reasonably well with alkylhaloacetylenes, e.g. equations (302) and (303)[296]. Furthermore, iodoalkynes, which generally fail to give the desired product with other organometallic reagents[4, 155b], condense with organocopper

$$n\text{-BuC(Et)}=\text{CHCu} + n\text{-BuC}\equiv\text{CI} \longrightarrow \underset{77\%}{n\text{-BuC(Et)}=\text{CHC}\equiv\text{CBu-}n} \qquad (303)$$

compounds in good yields, e.g. equations (303) and (304)[725]. Clearly, organocopper reagents constitute a breakthrough in the synthesis of carbon–carbon bonds: further work should enlarge their range and versatility in process (2).

$$PhCu + Me_3SiC{\equiv}CI \longrightarrow \underset{64\%}{Me_3SiC{\equiv}CPh} \quad (304)$$

In this section we have touched on synthetic applications of equation (2) and associated problems. Compared to process (1), the substitution (equation 2) is in its infancy. There is obviously room for improved understanding on the mechanistic front, especially of the aggregate and S_RN processes. The enhanced and specific nucleophilicities brought out by crown ethers and phase-transfer conditions have yet to be applied synthetically. Thus, there is ample reason to predict substantial development of process (2) in the near future.

Acknowledgement is made to the donors of the Petroleum Research Fund, administered by the American Chemical Society, for current support of our research.

V. REFERENCES

1. S. I. Miller and R. Tanaka in *Selective Organic Transformations*, Vol. 1 (Ed. B. S. Thyagarajan), Wiley–Interscience, New York, 1970, p. 143.
2. E. Winterfeldt, *Newer Methods of Preparative Organic Chemistry*, **6**, 243 (1971).
3. S. I. Miller and J. I. Dickstein, *Accounts Chem. Res.*, **9**, 358 (1976).
4. S. Y. Delavarenne and H. G. Viehe in *The Chemistry of Acetylenes*, (Ed. H. G. Viehe), Marcel Dekker, New York, 1969, Chap. 10.
5. A. A. Petrov (Ed.), *Khim. Atsetilena Tr. Vses. Konf.* 3rd *Dushanbe Tadzh. SSR*, 1968 (Pub. 1972).
6. (a) *Dokl. Vses. Konf. Khim. Atsetilena* 4*th*, 1972; (b) M. F. Shostakovskii (Ed.), *Khim. Atsetilena*, M. F. Izd. 'Nauka', Moscow, USSR, 1968. These publications were unavailable to the authors.
7. T. F. Rutledge, *Acetylenic Compounds*, Reinhold, New York, 1968.
8. M. F. Shostakovskii and A. V. Bogdanova, *The Chemistry of Diacetylenes*, translated from Russian by IPST Staff, Wiley, New York, 1974.
9. M. Baumgarth, *Chem. Ztg.*, **96**, 361 (1972).
10. M. V. George, S. K. Khetan and R. K. Gupta in *Advan. Heterocylic Chem.*, **19**, 279 (1976).
11. L. Brandsma, H. J. T. Bos and J. F. Arens in *The Chemistry of Acetylenes*, (Ed. H. G. Viehe), Marcel Dekker, New York, 1968, Chap. 11.
12. L. Brandsma and J. F. Arens in *Chemistry of the Ether Linkage*, (Ed. S. Patai), John Wiley and Sons, London, 1967, Chap. 13.
13. H. G. Viehe in *Chemistry of Acetylenes*, (Ed. H. G. Viehe), Marcel Dekker, New York, 1969, Chap. 12.
14. M. I. Bruce and W. R. Cullen, *Fluorine Chem. Rev.*, **4**, 79 (1969).
15. R. D. Chambers, *Fluorine in Organic Chemistry*, Wiley-Interscience, New York, 1973, pp. 196, 303.
16. R. L. Bol'shedvorskaya and L. I. Vereshchagin, *Usp. Khim.*, **42**, 511 (1973).
17. K. B. Rall', A. I. Vil'davskaya and A. A. Petrov, *Usp. Khim.*, **44**, 744 (1975).
18. G. Coispeau, J. Elguero and R. Jacquier, *Bull. Soc. Chim. Fr.*, **689**, 2717 (1970).
19. Y. Tanaka, S. R. Velen and S. I. Miller, *Tetrahedron*, **29**, 3271 (1973).
20. K. R. H. Wooldridge, *Advan. Heterocyclic Chem.*, **14**, 1 (1972).
21. T. Uchida and K. Matsumoto, *Synthesis*, 209 (1976).
22. D. St. C. Black, R. F. Crozier and V. C. Davis, *Synthesis*, 205 (1975).
23. J. S. Pizey, *Synthetic Reagents*, Vol. I, Halsted Press, New York, 1974, pp. 232–236.
24. R. A. Abramovitch and I. Shinkai, *Accounts Chem. Res.*, **9**, 192 (1976).
25. J. Bastide, J. Hamelin, F. Texier and Y. Vo Quang, *Bull. Soc. Chim. Fr.*, 2555, 2871 (1973).

26. C. G. Stuckwisch, *Synthesis*, 469 (1973).
27. M. A. Shaw and R. S. Ward, *Topics in Phosphorus Chem.*, **7,** 1 (1972).
28. R. S. Atkinson in *Aliphatic Chemistry*, Vol. 3, A Specialist Periodical Report, The Chemical Society, London, 1975, p. 1.
29. G. F. Hennion and C. V. DiGiovanna, *J. Org. Chem.*, **31,** 1977 (1966).
30. R. J. Goodfellow, M. Green, N. Mayne, A. F. Rest and F. G. A. Stone, *J. Chem. Soc. (A)*, 177 (1968).
31. G. Griffiths, P. D. Howes and C. J. M. Stirling, *J. Chem. Soc., Chem. Comm.*, 296 (1976).
32. L. I. Vereshchagin, L. D. Gavrilov and R. L. Bol'shedvorskaya, *Zh. Org. Khim.*, **11,** 2311 (1975).
33. J. C. Chalchat, F. Théron and R. Vessière, *Bull. Soc. Chim. Fr.*, 2501 (1973).
34. G. Modena, *Accounts Chem. Res.*, **4,** 73 (1971).
35. J. Chanet-Ray and R. Vessière, *Bull. Soc. Chim. Fr.*, 1661 (1974).
36. B. A. Trofimov, N. K. Gusarova and S. V. Amosova, *Zh. Org. Khim.*, **8,** 272 (1972).
37. (a) K. Akiba, T. Tsuchiya and N. Inamoto, *Tetrahedron Letters*, 1877 (1976); (b) K. Akiba, M. Ochiumi, T. Tsuchiya and N. Inamoto, *Tetrahedron Letters*, 459 (1975).
38. P. G. Harrison, *J. Organometal. Chem.*, **79,** 17 (1974).
39. P. Vermeer, J. Meijer, C. Eylander and L. Brandsma, *Recl. Trav. Chim. Pays-Bas*, **95,** 25 (1976).
40. A. J. Carty, S. E. Jacobson, R. T. Simpson and N. J. Taylor, *J. Amer. Chem. Soc.*, **97,** 7254 (1975).
41. M. H. Chisholm and H. C. Clark, *Accounts Chem. Res.*, **6,** 202 (1973).
42. B. A. Trofimov, S. V. Amosova, O. A. Tarasova, V. V. Keiko and V. K. Voronov, *Zh. Org. Khim.*, **10,** 127 (1974).
43. S. I. Miller, *Advan. Phys. Org. Chem.*, **6,** 185 (1968).
44. J. Dale in *Chemistry of Acetylenes*, (Ed. H. G. Viehe), Marcel Dekker, New York, 1969, Chap 1.
45. K. Yates, G. H. Schmid, T. W. Regulski, D. G. Garratt, H.-W. Leung and R. McDonald, *J. Amer. Chem. Soc.*, **95,** 160 (1973).
46. I. P. Bachkovskii, *Khim, Shk.*, **81** (1973); *Chem. Abstr.*, **80,** 58981 (1974).
47. F. Bohlmann, *Angew. Chem.*, **69,** 82 (1957).
48. R. Daniels and L. Bauer, *J. Chem. Educ.*, **35,** 444 (1958).
49. S. I. Miller, *J. Org. Chem.*, **21,** 247 (1956).
50. R. A. Raphael, *Acetylenic Compounds in Organic Synthesis*, Academic Press, London, 1955, pp. 21, 35.
51. C. H. Prévost, P. Souchay and J. Chauvelier, *Bull. Soc. Chim. Fr.*, **18,** 714 (1951).
52. NSRDS–NBS 26, *Ionization Potentials, Appearance Potentials, and Heats of Formation of Gaseous Positive Ions*, National Bureau of Standards, U.S. Govt. Printing Office, Washington, 1969.
53. F. P. Lossing and G. P. Semeluk, *Can. J. Chem.*, **48,** 955 (1970); F. P. Lossing, *Can. J. Chem.*, **49,** 357 (1971).
54. D. R. Stull, E. F. Westrum, Jr and G. C. Sinke, *The Chemical Thermodynamics of Organic Compounds*, Wiley, New York, 1969, Chap. 14.
55. E. Kloster-Jensen, C. Pascual and J. Vogt, *Helv. Chim. Acta*, **53,** 2109 (1970).
56. D. K. Sen Sharma and J. L. Franklin, *J. Amer. Chem. Soc.*, **95,** 6562 (1973).
57. H. Okabe and V. H. Dibeler, *J. Chem. Phys.*, **59,** 2430 (1973).
58. J. R. Wyatt and F. E. Stafford, *J. Phys. Chem.*, **76,** 1913 (1972).
59. P. D. Burrow and K. D. Jordan, *Chem. Phys. Letters*, **36,** 594 (1975); P. D. Burrow, private communication.
60. B. M. Hughes, C. Lifshitz and T. O. Tiernan, *J. Chem. Phys.*, **59,** 3162 (1973).
61. D. K. Bohme, G. I. Mackay, H. I. Schiff and R. S. Hemsworth, *J. Chem. Phys.*, **61,** 2175 (1974).
62. D. K. Bohme and L. B. Young, *J. Amer. Chem. Soc.*, **92,** 3301 (1970).
63. D. K. Bohme, E. Lee-Ruff and L. B. Young, *J. Amer. Chem. Soc.*, **94,** 5153 (1972).
64. M. Szwarc and J. Jagur-Grodzinski in *Ions and Ion Pairs in Organic Reactions*, Vol. 2 (Ed. M. Szwarc), Wiley, New York, 1974, Chap. 1.
65. H. O. House, *Accounts Chem. Res.*, **9,** 59 (1976).

66. J. Guilleme and B. Wojtkowiak, *Bull. Soc. Chim. Fr.*, 3007 (1969).
67. V. I. Minkin, O. A. Osipov and Y. A. Zhdanov, *Dipole Moments in Organic Chemistry*, Plenum, New York, 1970, p. 89.
68. M. V. Stackelberg and V. Stracke, *Z. Elektrochem.*, **53,** 118 (1949).
69. P. R. Wells, *Progr. Phys. Org. Chem.*, **6,** 111 (1968).
70. L. M. Jackman and S. Sternhell, *Application of Nuclear Magnetic Resonance Spectroscopy in Organic Chemistry*, 2nd Ed., Pergamon, Oxford, 1969, p. 346.
71. H. F. Ebel, *Tetrahedron*, **24,** 459 (1968).
72. M. Charton, *Prog. Phys. Org. Chem.*, **10,** 81 (1973); *J. Org. Chem.*, **29,** 1222 (1964).
73. A. R. Katritzky, D. J. Short and A. J. Boulton, *J. Chem. Soc.*, 1516 (1960).
74. J. A. Pople and D. L. Beveridge, *Approximate Molecular Orbital Theory*, McGraw-Hill, New York, 1970, pp. 118,119.
75. L. I. Vereshchagin, N. C. Sushkova and L. P. Vologdina, *Zh. Org. Khim.*, **8,** 1376 (1972); *Chem. Abstr.*, **77,** 113974 (1972).
76. S. T. McDowell and C. J. M. Stirling, *J. Chem. Soc.* (*B*), 351 (1967).
77. G. S. Krishnamurthy and S. I. Miller, *J. Amer. Chem. Soc.*, **83,** 3961 (1961).
78. C. E. Lough, D. J. Currie and H. L. Holmes, *Can. J. Chem.*, **46,** 771 (1968).
79. J. L. Franklin and P. W. Harland, *Ann. Rev. Phys. Chem.*, **25,** 485 (1974).
80. Y. Yukawa, *Handbook of Organic Structural Analysis*, W. A. Benjamin, New York, 1965.
81. W. F. Verhelst and W. Drenth, *J. Amer. Chem. Soc.*, **96,** 6692 (1974).
82. K. Fukui, *Topics in Current Chem.*, **15,** 1 (1970).
83. G. Klopman, *Chemical Reactivity and Reaction Paths*, Wiley–Interscience, New York, 1974, pp. 86–89.
84. S. Y. Chu and L. Goodman, *J. Amer. Chem. Soc.*, **97,** 7 (1975).
85. S. I. Miller, unpublished results.
86. W. E. Truce and L. D. Markley, *J. Org. Chem.*, **35,** 3275 (1970).
87. G. F. Dvorko, T. F. Karpenko and E. A. Shilov, *Kinet. Katal.*, **6,** 809 (1965); *Chem. Abstr.*, **64,** 1918 (1966).
88. G. F. Dvorko and T. P. Travchuk, *Reakts. Sposobnost. Org. Soedin.*, **5,** 995 (1968); *Chem. Abstr.*, **71,** 21409 (1969).
89. M. Neuenschwander and P. Bigler, *Helv. Chim. Acta*, **56,** 959 (1973).
90. L. I. Vereshchagin, L. P. Kirillova, S. R. Buzilova, R. L. Bol'shedvorskaya and G. V. Chernysheva, *Zh. Org. Khim.*, **11,** 531 (1975); *Chem. Abstr.*, **83,** 28136 (1975).
91. A. A. Ul'yanov and I. A. Maretina, *Zh. Org. Khim.*, **9,** 228 (1973).
92. W. F. Saunders and A. F. Cockerill, *Mechanisms of Elimination Reactions*, Wiley, New York, 1973.
93. E. Åkerblom, *Chem. Scripta*, **3,** 232 (1973).
94. S. I. Miller and W. G. Lee, *J. Amer. Chem. Soc.*, **81,** 6313 (1959).
95. B. A. Feit, L. Bohor and S. Rubinraut, *J. Chem. Soc.*, *Perkin II*, 253 (1976).
96. J. F. Arnett and H. M. Walborsky, *J. Org. Chem.*, **37,** 3678 (1972).
97. J. C. Chalchat, F. Théron and R. Vessière, *Bull. Soc. Chim. Fr.*, 711 (1970).
98. H. O. Kalinowski and H. Kessler, *Topics in Stereochem.*, **7,** 295 (1973).
99. G. W. Koeppl, D. S. Sagatys, G. S. Krishnamurthy and S. I. Miller, *J. Amer. Chem. Soc.*, **89,** 3396 (1967).
100. J. M. Lehn, B. Munsch and P. Millie, *Theoret. Chim. Acta*, **16,** 351 (1970).
101. F. Kerek, G. Ostrogovich and Z. Simon, *J. Chem. Soc.* (*B*), 541 (1971).
102. V. S. Bogdanov, I. L. Mikhelashvili and E. N. Prilezhaeva, *Izv. Akad. Nauk SSSR, Ser. Khim.*, 2374 (1972); *Chem., Abstr.*, **78,** 28756 (1973).
103. W. E. Truce and D. W. Onken, *J. Org. Chem.*, **40,** 3200 (1975).
104. W. E. Truce and G. J. W. Tichenor, *J. Org. Chem.*, **37,** 2391 (1972).
105. E. N. Prilezhaeva, I. L. Mikhelashvili and V. S. Bogdanov, *Zh. Org. Khim.*, **8,** 1505 (1972).
106. E. N. Prilezhaeva, G. S. Vasil'ev, I. L. Mikhelashvili and V. S. Bogdanov, *Zh. Org. Khim.*, **7,** 1349 (1971).
107. W. R. Cullen and W. R. Leeder, *Can. J. Chem.*, **47,** 2137 (1969).
108. J. E. Mulvaney and D. J. Newton, *J. Org. Chem.*, **34,** 1936 (1969).
109. A. Alexakis, A. Commerçon, J. Villiéras and J. F. Normant, *Tetrahedron Letters*, 2313 (1976).

110. R. Mornet and L. Gouin, *J. Organometal. Chem.*, **86**, 57 (1975).
111. H. J. Reich, J. M. Renga and J. E. Trend, *Tetrahedron Letters*, 2217 (1976).
112. M. Rivière-Baudet and J. Satgé, *J. Organometal. Chem.*, **56**, 159 (1973).
113. S. M. Makin, A. A. Ishmael and V. V. Yastrebov, *Zh. Org. Khim.*, **7**, 2120 (1971); *Chem. Abstr.*, **76**, 13712 (1972).
114. L. I. Vereshchagin, R. L. Bol'shedvorskaya, G. A. Pavlova, A. A. Vereschagina and N. V. Alekseeva, *Zh. Org. Khim.*, **11**, 99 (1975).
115. K. Undheim and L. A. Riege, *J. Chem. Soc., Perkin I*, 1493 (1975).
116. (a) Yu. I. Porfir'eva, B. Yu. Simkin, V. I. Minkin and A. A. Petrov, *Zh. Org. Khim.*, **11**, 496 (1975); (b) R. T. C. Brownlee and R. W. Taft, *J. Amer. Chem. Soc.*, **92**, 7007 (1970).
117. A. Niederhauser and M. Neuenschwander, *Helv. Chim. Acta*, **56**, 1331 (1973).
118. V. I. Laba, A. A. Kron, S. P. Sitnikova and E. N. Prilezhaeva, *Izv. Akad. Nauk SSSR, Ser. Khim.*, 2129 (1972); *Bull. Acad. Sci. USSR, Div. Chem. Sci.*, **21**, 2077 (1972).
119. J. W. Batty, P. D. Howes and C. J. M. Stirling, *J. Chem. Soc., Perkin I*, 59 (1973).
120. V. I. Laba, A. V. Sviridova, A. A. Kron, V. N. Dudnikova and E. N. Prilezhaeva, *Izv. Akad. Nauk SSSR, Ser. Khim.*, 2874 (1974).
121. A. Behzadi and L. N. Owen, *J. Chem. Soc., Perkin I*, 25 (1974).
122. R. Mantione and H. Normant, *Bull. Soc. Chim. Fr.*, 2261 (1973).
123. E. K. Raunio and T. G. Frey, *J. Org. Chem.*, **36**, 345 (1971).
124. T. Sasaki, A. Kojima and M. Ohta, *J. Chem. Soc. (C)*, 196 (1971).
125. J. S. Walia, A. S. Walia, and L. J. Guillot, *Chem. Ind. (London)*, 933 (1972).
126. A. N. Volkov, L. V. Sokolyanskaya, E. O. Tsetlina and R. Ya. Mushii, *Zh. Org. Khim.*, **9**, 64 (1973); *Chem. Abstr.*, **78**, 97246 (1973).
127. B. P. Gusev and V. F. Kucherov, *Izv. Akad. Nauk SSSR, Ser. Khim.*, 208 (1974); *Chem. Abstr.*, **80**, 107930 (1974).
128. V. N. Petrov, G. M. Andrianova and E. N. Prilezhaeva, *Izv. Akad. Nauk SSSR, Ser. Khim.*, 2180 (1966); *Bull. Acad. Sci. USSR, Div. of Chem. Sci.*, 2109 (1966).
129. E. N. Prilezhaeva, G. S. Vasil'ev and V. N. Petrov, *Izv. Akad. Nauk SSSR, Ser. Khim.*, 2217 (1967); *Bull. Acad. Sci. USSR, Div. of. Chem Sci.*, 2128 (1967).
130. A. A. Petrov, *Usp. Khim.*, **29**, 1049 (1960).
131. W. Schroth, J. Peschel and A. Zschunke, *Zeit. Chem.*, **9**, 108, 110 (1969).
132. I. F. Lutsenko, M. V. Proskurnina and N. B. Karlstedt, *Phosphorus*, **3**, 55 (1973).
133. M. L. Petrov and A. A. Petrov, *Zh. Obshch. Khim.*, **42**, 1863 (1972); *Chem. Abstr.*, **78**, 29931 (1973).
134. M. L. Petrov, V. G. Salishchev and A. A. Petrov, *Zh. Obshch. Khim.*, **44**, 2107 (1974); *Chem. Abstr.*, **82**, 43533 (1975).
135. W. K. Kwok, W. G. Lee and S. I. Miller, *J. Amer. Chem. Soc.*, **91**, 468 (1969).
136. L. Brandsma, *Preparative Acetylenic Chemistry*, Elsevier, New York, 1971, p. 139.
137. J. Freear and A. E. Tipping, *J. Chem. Soc. (C)*, 411 (1969).
138. M. L. Petrov and A. A. Petrov, *Zh. Obshch. Khim.*, **43**, 691 (1973).
139. M. L. Petrov and A. A. Petrov, *Zh. Obshch. Khim.*, **42**, 2345 (1972).
140. N. D. Epiotis, R. L. Yates, F. Bernardi and S. Wolfe, *J. Amer. Chem. Soc.*, **98**, 5435 (1976).
141. K. G. Migliorese and S. I. Miller, *J. Org. Chem.*, **39**, 843 (1974).
142. J. Freear and A. E. Tipping, *J. Chem. Soc. (C)*, 1848 (1969).
143. A. Niederhauser, A. Frey and M. Neuenschwander, *Helv. Chim. Acta*, **56**, 944 (1973).
144. M. D. Joesten and L. J. Schaad, *Hydrogen Bonding*, Marcel Dekker, New York, 1974.
145. M. E. Kuehne and P. J. Sheeran, *J. Org. Chem.*, **33**, 4406 (1968).
146. M. Dupré, *C. R. Acad. Sci., Paris, Sér. C*, **277**, 891 (1973).
147. N. Morita, J. I. Dickstein and S. I. Miller, unpublished results.
148. O. Exner in *Advances in Linear Free Energy Relationships* (Ed. N. B. Chapman and J. Shorter), Plenum Press, New York, 1972, Chap. 1.
149. H. A. Selling, *Tetrahedron*, **31**, 2387 (1975).
150. W. Hagens, H. J. T. Bos and J. F. Arens, *Rec. Trav. Chim. Pays-Bas*, **92**, 762 (1973).
151. E. Åkerblom and Å. Norrman, *Chem. Scripta*, **2**, 183 (1972).
152. J. Bonnema, H. J. Alkema and J. F. Arens, *Rec. Trav. Chim. Pays-Bas*, **79**, 937 (1960).

153. R. Tanaka, M. Rodgers, R. Simonatis and S. I. Miller, *Tetrahedron*, **27**, 2651 (1971).
154. J. F. Arens, *Rec. Trav. Chim. Pays-Bas*, **82**, 183 (1963).
155. (a) M. C. Verploegh, L. Donk, H. J. T. Bos and W. Drenth, *Rec. Trav. Chim. Pays-Bas*, **90**, 765 (1971); (b) M. C. Verploegh, *Ph.D. Thesis*, '*Nucleofiele Substitutie op Halogeen in* 1-*Halogeen*-1-*alkynen*', University of Utrecht, 1971.
156. R. Truchet, *Ann. Chim.*, (*Paris*) [10] **16**, 309 (1931).
157. J. I. Dickstein and S. I. Miller, *J. Org. Chem.*, **37**, 2168 (1972).
158. A. Fujii, J. I. Dickstein and S. I. Miller, *Tetrahedron Letters*, 3435 (1970).
159. A. Fujii and S. I. Miller, *J. Amer. Chem. Soc.*, **93**, 3694 (1971).
160. V. Grignard and H. Perrichon, *Ann. Chim.*, (*Paris*) [10] **5**, 5 (1926).
161. N. I. Gazieva, L. V. Abramova, A. I. Schekotikhin and V. A. Ginsburg, *Zh. Obshch. Khim.*, **38**, 1198 (1968); *Chem. Abstr.* **69**, 86259 (1968).
162. J. Loevenich, J. Losen and A. Dierichs, *Chem. Ber.*, **60**, 950 (1927).
163. (a) P. Simpson and D. W. Burt, *Tetrahedron Letters*, 4799 (1970); (b) D. W. Burt and P. Simpson, *J. Chem. Soc.* (*C*), 2872 (1971).
164. P. Beltrame, P. L. Beltrame, M. G. Cattania and M. Simonetta, *J. Chem. Soc.*, *Perkin II*, 63 (1973).
165. H. G. Viehe and S. Y. Delavarenne, *Chem. Ber.*, **103**, 1216 (1970).
166. C. D. Beard, J. C. Craig and M. D. Solomon, *J. Amer. Chem. Soc.*, **96**, 7944 (1974).
167. V. Concialini, G. Modena and F. Taddei, *Boll. Sci. Fac. Chim. Ind. Bologna*, **21**, 207 (1963).
168. R. E. Banks, M. G. Barlow, W. D. Davies, R. N. Haszeldine and D. R. Taylor, *J. Chem. Soc.* (*C*), 1104 (1969).
169. H. G. Viehe and M. Reinstein, *Angew. Chem.*, **76**, 537 (1964).
170. J. J. Flynn, *Diss. Abstr.*, **27**, 406-B (1966).
171. S. Y. Delavarenne and H. G. Viehe, *Chem. Ber.*, **103**, 1198 (1970).
172. L. Maioli, G. Modena and F. Taddei, *Boll. Sci. Fac. Chim. Ind. Bologna*, **18**, 58 (1960).
173. R. Tanaka and S. I. Miller, *J. Org. Chem.*, **36**, 3856 (1971).
174. R. R. Lii and S. I. Miller, *J. Amer Chem. Soc.*, **95**, 1602 (1973).
175. J. U. Nef, *Justus Liebigs Ann. Chem.*, **308**, 264 (1899).
176. V. Franzen, *Chem. Ber.*, **87**, 1148 (1954).
177. G. F. Dvorko, Y. P. Filinov and I. M. Vasilkevich, *Reakt. Sposobn. Organ. Soedin.* (*Org. Reactivity*), **11**, 847 (1975); S. I. Miller, *Reakt. Sposobn. Organ. Soedin.* **13**, 85 (1976).
178. T. Metler, A. Uchida and S. I. Miller, *Tetrahedron*, **24**, 4285 (1968).
179. A. I. Tolmachev and M. A. Kudinova, *Khim. Geterotsikl. Soedin.*, 274 (1974); *Chem. Abstr.*, **81**, 13365 (1974).
180. J. E. Baldwin, *J. Chem. Soc.*, *Chem. Comm.*, **734** (1976).
181. V. Sváta and M. Procházka, *Coll. Czech. Chem. Comm.*, **40**, 2315 (1975).
182. J. Hine, *Physical Organic Chemistry*, 2nd ed., McGraw-Hill., New York, 1962, p. 181.
183. G. F. Hennion and C. V. DiGiovanna, *J. Org. Chem.*, **30**, 2645 (1965).
184. W. Schroth and J. Peschel, *Zeit. Chem.*, **9**, 141 (1969).
185. K. G. Migliorese, Y. Tanaka and S. I. Miller, *J. Org. Chem.*, **39**, 739 (1974).
186. A. N. Volkov, Yu. M. Skvortsov, Yu. V. Kind and M. G. Voronkov, *Zh. Org. Khim.*, **10**, 174 (1974).
187. T. Francis and M. P. Thorne, *Can. J. Chem.*, **54**, 24 (1976).
188. A. Naakgeboren, J. Meijer, P. Vermeer and L. Brandsma, *Rec. Trav. Chim. Pays-Bas*, **94**, 92 (1975).
189. G. Märkl, D. Matthes, A. Donanbauer and H. Baier, *Tetrahedron Letters*, 3171 (1975).
190. F. Bohlmann and H. Mönch, *Chem. Ber.*, **102**, 3298 (1969).
191. (a) M. Simalty and H. Chahine, *Bull. Soc. Chim. Fr.*, 4938 (1968); (b) M. Simalty and M. H. Mebazaa, *Bull. Soc. Chim. Fr.*, 3532 (1972).
192. A. P. Tomilov, *Russ. Chem. Rev.*, **31**, 569 (1962).
193. H. O. House and E. F. Kinloch, *J. Org. Chem.*, **39**, 747 (1974).
194. L. Horner, and H. Röder, *Justus Liebigs Ann. Chem.*, **723**, 11 (1969).
195. H. Siegerman in *Techniques of Electroorganic Synthesis*, (Ed. N. L. Weinberg), Vol. V, Part II, Wiley–Interscience, New York, 1975, pp. 667 ff.
196. V. D. Parker, *J. Amer. Chem. Soc.*, **98**, 98 (1976).

197. A. J. Fry, *Synthetic Organic Electrochemistry*, Harper and Row, New York, 1972, pp. 143, 163, 238.
198. M. R. Rifi in *Techniques of Electroorganic Synthesis*, Vol. V, Part II (Ed. N. L. Weinberg), Wiley–Interscience, New York, 1975, Chap. VIII.
199. K. N. Campbell and E. E. Young, *J. Amer. Chem. Soc.*, **65**, 965 (1943).
200. R. K. Sen, E. Yeager and W. E. O'Grady, *Ann. Rev. Phys. Chem.*, **26**, 287 (1975).
201. S. Wawzonek and D. Wearring, *J. Amer. Chem. Soc.*, **81**, 2067 (1959).
202. R. E. Sioda, D. O. Cowan and W. S. Koski, *J. Amer. Chem. Soc.*, **89**, 230 (1967).
203. H. E. Zimmerman and J. R. Dodd, *J. Amer. Chem. Soc.*, **92**, 6507 (1970).
204. G. Levin, J. Jagur-Grodzinski and M. Szwarc, *J. Amer. Chem. Soc.*, **92**, 2268 (1970); *J. Org. Chem.*, **35**, 1702 (1970).
205. R. W. Howsam and C. J. M. Stirling, *J. Chem. Soc., Perkin II*, 847 (1972).
206. S. Ono, *Nippon Kagaku Zasshi*, **77**, 665 (1956); *Chem. Abstr.*, **52**, 9020, 9066 (1958).
207. I. Rosenthal, J. R. Hayes, A. J. Martin and P. J. Elving, *J. Amer. Chem. Soc.*, **80**, 3050 (1958).
208. G. S. Krishnamurthy, *Ph.D. Thesis*, Illinois Institute of Technology, Chicago, 1960.
209. S. R. Missan, E. I. Becker and L. Meites, *J. Amer. Chem. Soc.*, **83**, 58 (1961).
210. W. M. Moore and D. G. Peters, *J. Amer. Chem. Soc.*, **97**, 139 (1975).
211. W. M. Moore, A. Salajegheh and D. G. Peters, *J. Amer. Chem. Soc.*, **97**, 4954 (1975).
212. W. M. Moore and D. G. Peters, *Tetrahedron Letters*, 453 (1972).
213. V. D. Bezuglyi and T. A. Alekseeva, *Zh. Anal. Khim.*, **20**, 244 (1965).
214. H. O. House, L. E. Huber and M. J. Umen, *J. Amer. Chem. Soc.*, **94**, 8471 (1972).
215. V. P. Gul'tyai, S. G. Mairanovskii, N. K. Lisitsina and M. V. Mavrov, *Izv. Akad. Nauk SSSR, Ser. Khim.*, 1655 (1968).
216. A. G. Evans, J. C. Evans, P. J. Emes and T. J. Phelan, *J. Chem. Soc.* (*B*), 315 (1971).
217. J. Simonet, H. Doupeux, P. Martinet and D. Bretelle, *Bull. Soc. Chim. Fr.*, 3930 (1970).
218. (a) C. Degrand, B. Grosjean and P.-L. Compagnon, *Bull. Soc. Chim. Fr.*, 281 (1973); (b) C. Degrand and M. Lacour, *Bull. Soc. Chim. Fr.*, 817 (1976).
219. R. Breslow, D. Murayama, R. Drury and F. Sondheimer, *J. Amer. Chem. Soc.*, **96**, 249 (1974).
220. S. G. Mairanovskii, V. P. Gul'tyai, L. A. Spiridonova, I. L. Mikhelashvili and E. N. Prilezhaeva, *Elektrokhim.*, **9**, 1607 (1973).
221. B. V. Prokop'ev, N. M. Deriglazov and V. B. Pukhnarevich, *Izv. Sib. Otd. Akad. Nauk SSSR, Ser. Khim. Nauk*, **6**, 89 (1970); *Chem. Abstr.*, **74**, 93811 (1971). See also References 5 and 6.
222. G. Capobianco, E. Vianello and G. Giacometti, *Gazz. Chim. Ital.*, **97**, 243 (1967).
223. V. I. Laba, M. K. Polievktov, E. N. Prilezhaeva and S. G. Mairanovskii, *Izv. Akad. Nauk SSSR, Ser. Khim.*, 2149 (1969).
224. S. P. Korshunov, N. A. Kudryavtseva, V. S. Pisareva and A. A. Nazarova, *Zh. Org. Khim.*, **7**, 2125 (1971).
225. S. G. Mairanovskii and L. A. Spiridonova, *Elektrokhim.*, **5**, 739 (1969).
226. R. A. Benkeser and C. A. Tincher, *J. Org. Chem.*, **33**, 2727 (1968).
227. E. M. Kaiser, *Synthesis*, 391 (1972).
228. G. Levin, J. Jagur-Grodzinski and M. Szwarc, *Trans. Faraday Soc.*, **67**, 768 (1971).
229. J. G. Broadhurst and E. Warhurst, *J. Chem. Soc.* (*A*), 351 (1966).
230. D. A. Dadley and A. G. Evans, *J. Chem. Soc.* (*B*), 107 (1968); 418 (1967).
231. I. B. Goldberg and A. J. Bard, *Chem. Phys. Letters*, **7**, 139 (1970).
232. I. H. Elson, T. J. Kemp, D. Greatorex and H. D. B. Jenkins, *J. Chem. Soc., Faraday II*, 1402 (1973).
233. J. J. Lagowski in *Electrons in Fluids*, (Ed. J. Jortner and N. R. Kestner), Springer-Verlag, New York, 1973, p. 29.
234. A. G. Evans, *Proc. Roy. Soc.* (*A*), **302**, 331 (1968).
235. N. L. Holy, *Chem. Rev.*, **74**, 243 (1974).
236. Z. Csürös, P. Caluwe and M. Szwarc, *J. Amer. Chem. Soc.*, **95**, 6171 (1973).
237. E. J. Hart, *Survey of Progress in Chem.*, **5**, 129 (1969).
238. J. J. Lagowski and G. A. Moczygemba in *The Chemistry of Non-Aqueous Solvents*, Vol. II, (Ed. J. J. Lagowski), Academic Press, New York, 1967, Chap. 7.

239. G. F. Dvorko and E. A. Shilov, *Teor. Eksp. Khim.*, **3**, 606 (1967); *Chem. Abstr.*, **68**, 95200 (1967).
240. M. F. Semmelhack and R. D. Stauffer, *J. Org. Chem.*, **40**, 3619 (1975).
241. A. J. Birch and G. Subba Rao, *Advan. Org. Chem.*, **8**, 1 (1972).
242. G. M. Whitesides and W. J. Ehmann, *J. Org. Chem.*, **35**, 3565 (1970).
243. M. Larchevêque, *Ann. Chim. (Paris)*, **5**, 129 (1970).
244. A. Kumar, A. Singh and D. Devaprabhakara, *Tetrahedron Letters*, 3343 (1975).
245. G. Stork, S. Malhotra, H. Thompson and M. Uchibayashi, *J. Amer. Chem. Soc.*, **87**, 1148 (1965).
246. S. K. Pradhan, T. V. Radhakrishnan and R. Subramanian, *J. Org. Chem.*, **41**, 1943 (1976).
247. R. J. Balf, B. Rao and L. Weiler, *Can. J. Chem.*, **49**, 3135 (1971).
248. R. J. Tedeschi and G. L. Moore, *J. Org. Chem.*, **34**, 435 (1969).
249. A. Saus and P. Krings, *Tetrahedron Letters*, 947 (1967).
250. Y. Nakadaira and H. Sakurai, *J. Organometal. Chem.*, **47**, 61 (1973).
251. J. C. Brunet and C. Lemahieu, *J. Organometal. Chem.*, **49**, 381 (1973).
252. B. G. Kovalev and R. I. Ishchenko, *Zh. Org. Khim.*, **10**, 463 (1974).
253. K. Utimoto, M. Kitai, M. Naruse and H. Nozaki, *Tetrahedron Letters*, 4233 (1975).
254. B. G. Kovalev, R. I. Ishchenko, V. A. Marchenko and M. P. Filippova, *Zh. Org. Khim.*, **9**, 4 (1973).
255. M. Janić, M. J. Beneš and J. Peška, *Makromol. Chem.*, **138**, 99 (1970).
256. D. Couturier, *Ann. Chim. (Paris)*, **7**, 145 (1972).
257. R. I. Kruglikova and Z. F. Abramova, *Zh. Org. Khim.*, **10**, 433 (1974).
258. W. G. Knipprath and R. A. Stein, *Lipids*, **1**, 81 (1966).
259. H. E. Seyfarth, *Chem. Ber.*, **101**, 2283 (1968).
260. P. Neta and R. W. Fessenden, *J. Phys. Chem.*, **76**, 1957 (1972).
261. J. S. Cowie, P. D. Landor, S. R. Landor and N. Punja, *J. Chem. Soc., Perkin I*, 2197 (1972).
262. G. D. Abrams, W. R. Bartlett, V. A. Funs and W. S. Johnson, *Bioorg. Chem.*, **1**, 243 (1971).
263. R. H. Everhardus, A. Peterse, P. Vermeer, L. Brandsma and J. F. Arens, *Rec. Trav. Chim. Pays-Bas*, **93**, 90 (1974).
264. E. Wiberg and E. Amberger, *Hydrides of the Elements of Main Groups I–IV*, Elsevier, Amsterdam, 1971, Chaps. V and VIII.
265. J. J. Eisch and S.-G. Rhee, *J. Amer. Chem. Soc.*, **96**, 7276 (1974).
266. H. C. Brown and N. Ravindran, *J. Amer. Chem. Soc.*, **98**, 1785 (1976).
267. M. F. Lappert and S. Takahashi, *J. Chem. Soc., Chem. Comm.*, 1272 (1972).
268. J. P. Quintard and M. Pereyre, *J. Organometal. Chem.*, **42**, 75 (1972).
269. G. N. Schrauzer and T. D. Guth, *J. Amer. Chem. Soc.*, **98**, 3508 (1976).
270. S. G. Morris, P. Magidman and S. F. Herb, *J. Amer. Oil Chem. Soc.*, **49**, 505 (1972).
271. F. Bohlmann, R. Enkelmann and W. Plettner, *Chem. Ber.*, **97**, 2118 (1964).
272. E. F. Magoon and L. H. Slaugh, *Tetrahedron*, **23**, 4509 (1967).
273. W. T. Borden, *J. Amer. Chem. Soc.*, **92**, 4898 (1970).
274. B. Grant and C. Djerassi, *J. Org. Chem.*, **39**, 968 (1974).
275. J. Gore and R. Baudouy, *Tetrahedron Letters*, 3743 (1974).
276. R. I. Kruglikova, L. G. Babaeva, N. A. Polteva and B. V. Unkovskii, *Zh. Org. Khim.*, **10**, 956 (1974); *Chem. Abstr.*, **81**, 49179 (1974).
277. E. J. Corey, J. A. Katzenellenbogen and G. H. Posner, *J. Amer. Chem. Soc.*, **89**, 4245 (1967).
278. B. B. Molloy and K. L. Hauser, *Chem. Comm.*, 1017 (1968).
279. W. W. Zajac, Jr and K. J. Byrne, *J. Org. Chem.*, **40**, 530 (1975).
280. T. E. DeVille, M. B. Hursthouse, S. W. Russell and B. C. L. Weedon, *Chem. Comm.*, 754 (1969).
281. A. Claesson, *Acta Chem. Scand.*, **29B**, 609 (1975).
282. M. Santelli and M. Bertrand, *Bull. Soc. Chim. Fr.*, 2331, 2335 (1973).
283. J. K. Crandall, D. J. Keyton and J. Kohne, *J. Org. Chem.*, **33**, 3655 (1968).
284. R. Baudouy and J. Gore, *Tetrahedron Letters*, 1593 (1974).
285. E. C. Juenge, S. J. Hawkes and T. E. Snider, *J. Organometal. Chem.*, **51**, 189 (1973).

286. L. A. Van Dijck, K. H. Schönemann and F. J. Zeelen, *Rec. Trav. Chim. Pays-Bas*, **88**, 254 (1969).
287. E. J. Corey, J. A. Katzenellenbogen, S. A. Roman and N. W. Gilman, *Tetrahedron Letters*, 1821 (1971).
288. D. G. Lee and J. R. Brownridge, *Can. J. Chem.*, **51**, 2102 (1973).
289. E. J. Corey, K. Achiwa and J. A. Katzenellenbogen, *J. Amer. Chem. Soc.*, **91**, 4318 (1969).
290. R. Baudouy and J. Gore, *Synthesis*, 573 (1974).
291. R. I. Kruglikova, L. P. Kravets and B. V. Unkovskii, *Zh. Org. Khim.*, **11**, 263 (1975).
292. J. G. Buchanan, A. D. Dunn and A. R. Edgar, *J. Chem. Soc.*, *Perkin I*, 68 (1976).
293. W. S. Chilton, W. C. Lontz, R. B. Roy and C. Yoda, *J. Org. Chem.*, **36**, 3222 (1971).
294. R. Hems, D. Horton and M. Nakadate, *Carbohyd. Res.*, **25**, 205 (1972).
295. E. J. Corey, G. W. J. Fleet and M. Kato, *Tetrahedron Letters*, 3963 (1973).
296. J. F. Normant, A. Commerçon and J. Villiéras, *Tetrahedron Letters*, 1465 (1975).
297. V. Jäger and H. G. Viehe, *Angew. Chem. Int. Ed.*, **9**, 795 (1970).
298. L. I. Vereshchagin, E. I. Titova, N. V. Sushkova and L. D. Gavrilov, *Zh. Org. Khim.*, **7**, 912 (1971).
299. G. F. Dvorko and E. A. Shilov, *Kinet. Katal.*, **6**, 37 (1965).
300. G. V. Dvorko and T. F. Karpenko, *Ukr. Khim. Zh.*, **37**, 41 (1971); *Chem. Abstr.*, **74**, 140445 (1971).
301. G. V. Dvorko and T. P. Travchuk in *Khim. Str.*, *Svoistva Reaktivnost. Org. Soedin.* (Ed. A. I. Izd. Kiprianov), 'Naukova Dumka', Kiev, USSR, 1969, p. 19; *Chem. Abstr.*, **73**, 44555 (1970).
302. G. F. Dvorko and T. P. Travchuk, *Dopov. Akad, Nauk Ukr. RSR. Ser. B*, **30**, 1105 (1968); *Chem. Abstr.*, **70**, 56903 (1969).
303. G. F. Dvorko and N. M. Soboleva, *Kinet. Katal.*, **13**, 854 (1972); *Chem. Abstr.*, **78**, 71013 (1973).
304. R. I. Yakhimovich, E. A. Shilov and G. F. Dvorko, *Dokl. Akad. Nauk SSSR*, **166**, 388 (1966); *Chem. Abstr.*, **64**, 12800 (1966).
305. G. F. Dvorko, N. M. Soboleva and T. F. Karpenko, *Dokl. Akad. Nauk SSSR*, **184**, 850 (1969); *Chem. Abstr.*, **70**, 96122 (1969).
306. W. T. Miller, R. H. Snider and R. J. Hummel, *J. Amer. Chem. Soc.*, **91**, 6532 (1969).
307. R. D. Chambers, W. K. R. Musgrave and S. Partington, *Chem. Comm.*, 1050 (1970).
308. J. A. Jackson, *J. Polymer Sci.*, *Part A*-1, **10**, 2935 (1972).
309. W. T. Flowers, R. N. Haszeldine and P. G. Marshall, *Chem. Comm.*, 371 (1970).
310. R. D. Chambers, S. Partington and D. B. Speight, *J. Chem. Soc.*, *Perkin I*, 2673 (1974).
311. J. Cousseau and L. Gouin, *Bull. Soc. Chim. Fr.*, 244 (1976).
312. F. M. Beringer and S. A. Galton, *J. Org. Chem.*, **30**, 1930 (1965).
313. B. Cavalchi, D. Landini and F. Montanari, *J. Chem. Soc.* (*C*), 1204 (1969).
314. S. A. Miller, *Acetylene*, Vol. II, Academic Press, New York, 1966, Chaps. 4 and 6.
315. M. Sittig, *Acetylene Processes and Products*, Noyes Development Corp., Park Ridge, N.J., 1968, p. 124.
316. J. F. Vitcha, *U.S. Patent, No.* 3,370,095; *Chem. Abstr.*, **68**, 77735 (1968).
317. V. K. Voinova, B. Kh. Kimsanov and B. I. Minkhant'ev, *Dokl. Akad. Nauk Tadzh. SSR*, **14**, 27 (1971).
318. S. V. Amosova, A. S. Atavin and B. A. Trofimov, *Izv. Akad. Nauk SSSR*, *Ser. Khim.*, 619 (1967); *Chem. Abstr.*, **68**, 48975 (1968).
319. I. J. Solomon, *Ph.D. Thesis*, Illinois Institute of Technology, 1962.
320. P. K. Freeman and T. A. Hardy, *Tetrahedron Letters*, 3317 (1973).
321. M. N. Gudi, J. G. Hiriyakkanavar and M. V. George, *Indian J. Chem.*, **7**, 971 (1969).
322. Y. Hayasi, M. Kobayasi and H. Nozaki, *Tetrahedron*, **26**, 4353 (1970).
323. J. A. Sanders, K. Hovius and J. B. F. N. Engberts, *J. Org. Chem.*, **39**, 2641 (1974).
324. E. Winterfeldt and W. Krohn, *Chem. Ber.*, **102**, 2336 (1969).
325. E. Winterfeldt, W. Krohn and H.-U. Stracke, *Chem. Ber.*, **102**, 2346 (1969).
326. A. N. Belyaeva, I. A. Maretina, A. A. Petrov and E. A. Lisitsyn, *Zh. Org. Khim.*, **6**, 2371 (1970).
327. E. S. Wassermann and A. B. Bedrintzeva, *Compt. Rend. Acad. Sci. USSR*, **33**, 34 (1941); *Chem. Abstr.*, **37**, 6539 (1943).
328. S. I. Miller and G. Shkapenko, *J. Amer. Chem. Soc.*, **77**, 5038 (1955).

329. J. Ficini, *Tetrahedron*, **32**, 1449 (1976).
330. N. D. Heindel and L. A. Schaeffer, *J. Org. Chem.*, **35**, 2445 (1970).
331. L. I. Vereshchagin, N. V. Sushkova, S. R. Buzilova, L. P. Kirillova and S. I. Demina, *Zh. Org. Khim.*, **11**, 286 (1975); *Chem. Abstr.*, **82**, 155211 (1975).
332. H. J. Gais, K. Hafner and M. Neuenschwander, *Helv. Chim. Acta*, **52**, 2641 (1969).
333. M. P. Collins, D. Gardiner and W. G. Overend, *Carbohyd. Res.*, **35**, 251 (1974).
334. A. W. McCulloch and A. G. McInnes, *Can. J. Chem.*, **52**, 3569 (1974).
335. (a) R. A. Abramovitch, G. Grins, R. B. Rogers, J. L. Atwood, M. D. Williams and S. Crider, *J. Org. Chem.*, **37**, 3383 (1972); (b) R. A. Abramovitch, G. Grins, R. B. Rogers and I. Shinkai, *J. Amer. Chem. Soc.*, **98**, 5671 (1976).
336. R. Huisgen, H. Seidl and J. Wulff, *Chem. Ber.*, **102**, 915 (1969).
337. P. S. Venkataramani, N. K. Saxena, V. K. Tripathi and G. Mehta, *Indian J. Chem.*, **13**, 852 (1975).
338. W. Hagens, H. J. T. Bos, W. Voskuil and J. F. Arens, *Rec. Trav. Chim. Pays-Bas*, **88**, 71 (1969).
339. I. I. Popov, A. M. Simonov and A. A. Zubenko, *Zh. Org. Khim.*, **11**, 1139 (1975).
340. S. I. Radchenko, *Zh. Org. Khim.*, **11**, 652 (1975).
341. L. I. Vereshchagin, L. D. Gavrilov, R. L. Bol'shedvorskaya, L. P. Kirillova and L. L. Okhapkina, *Zh. Org. Khim.*, **9**, 506 (1973).
342. A. A. Ul'yanov and I. A. Maretina, *Zh. Org. Khim.*, **8**, 1560 (1971).
343. L. I. Vereshchagin, L. D. Gavrilov, E. I. Titova, S. R. Buzilova, N. V. Sushkova and A. V. Maksikova, *Zh. Org. Khim.*, **11**, 47 (1975).
344. A. N. Volkov, Yu. M. Skvortsov, I. I. Danda and M. F. Shostakovskii, *Zh. Org. Khim.*, **6**, 897 (1970).
345. E. S. Rothman, G. G. Moore, J. M. Chirinko and S. Serota, *J. Amer. Oil Chem. Soc.*, **49**, 376 (1972).
346. R. K. Gupta and M. V. George, *Tetrahedron*, **31**, 1263 (1975).
347. (a) P. H. M. Schreurs, W. G. Galesloot and L. Brandsma, *Rec. Trav. Chim. Pays-Bas*, **94**, 70 (1975); (b) P. H. M. Schreurs, A. J. deJong and L. Brandsma, *Rec. Trav. Chim. Pays-Bas*, **95**, 75 (1976).
348. A. J. deJong, L. Brandsma and J. F. Arens, *Rec. Trav. Chim. Pays-Bas*, **93**, 15 (1974).
349. E. R. H. Jones and J. S. Stephenson, *J. Chem. Soc.*, 2197 (1959).
350. R. I. Kruglikova and L. A. Kundryutskova, *Zh. Org. Khim.*, **9**, 2477 (1973).
351. L. I. Vereshchagin, L. D. Gavrilov, R. L. Bol'shedvorskaya, E. I. Titova, S. R. Buzilova, A. V. Maksikova and G. A. Kalabin, *Zh. Org. Khim.*, **10**, 2059 (1974).
352. K. Nagarajan, V. R. Rao and R. K. Shah, *Indian J. Chem.*, **9**, 532 (1971).
353. A. N. Volkov, A. N. Nikol'skaya, E. P. Levanova and K. A. Volkova, *Zh. Org. Khim.*, **9**, 1594 (1973).
354. A. T. Bottini and J. G. Maroski, *J. Org. Chem.*, **38**, 1455 (1973).
355. F. M. F. Chen and T. P. Forrest, *Can. J. Chem.*, **51**, 1368 (1973).
356. W. Kirmse, A. Engelmann and J. Heese, *Chem. Ber.*, **106**, 3073 (1973); W. Kirmse and A. Engelmann, *Chem. Ber.*, **106**, 3086 (1973).
357. (a) R. V. Stevens, *Tetrahedron*, **32**, 1599 (1976); (b) R. V. Stevens, L. E. DuPree, Jr., W. L. Edmonson, L. L. Magid and M. P. Wentland, *J. Amer. Chem. Soc.*, **93**, 6637 (1971).
358. W. Drenth and G. H. E. Nieuwdorp, *Rec. Trav. Chim. Pays-Bas*, **88**, 307 (1969).
359. V. Rosnati, F. Sannicolò and G. Zeechi, *Gazz. Chim. Ital.*, **100**, 3 (1970).
360. N. D. Heindel and L. A. Schaeffer, *J. Med. Chem.*, **13**, 981 (1970).
361. N. Nakamura, Y. Kishida and N. Ishida, *Chem. Pharm. Bull. (Tokyo)*, **19**, 1389 (1971).
362. H. Saikachi and T. Kitagawa, *Chem. Pharm. Bull. (Tokyo)*, **19**, 1562 (1971).
363. E. I. Titova, L. D. Gavrilov, R. L. Bol'shedvorskaya and L. I. Vereshchagin, *Zh. Org. Khim.*, **5**, 2113 (1969).
364. L. I. Vereshchagin, S. R. Gainulina, R. L. Bol'shedvorskaya and L. P. Kirillova, *Izv. Vyssh. Ucheb. Zaved. Khim. Khim. Tekhnol.*, **12**, 1527 (1969); *Chem. Abstr.*, **72**, 111351 (1970).
365. M. Maumy, *Bull. Soc. Chim. Fr.*, 1600 (1972).
366. M. F. Shostakovskii, A. S. Atavin, Z. T. Dmitrieva, T. V. Kashik and B. A. Trofimov, *Zh. Org. Khim.*, **6**, 902 (1970).
367. K. T. Potts and A. J. Elliott, *J. Org. Chem.*, **38**, 1769 (1973).

368. Z. T. Fomum, P. D. Landor, S. R. Landor and G. B. Mpango, *Tetrahedron Letters*, 1101 (1975).
369. P. H. M. Schreurs, J. Meijer, P. Vermeer and L. Brandsma, *Tetrahedron Letters*, 2387 (1976).
370. J. P. Freeman, J. A. Kassner and R. C. Grabiak, *J. Org. Chem.*, **40**, 3402 (1975).
371. T. S. Leong and M. E. Peach, *J. Fluorine Chem.*, **6**, 145 (1975).
372. A. N. Pudovik and O. S. Shulyndina, *Zh. Obshch. Khim.*, **44**, 221 (1974).
373. Yu. I. Porfir'eva, V. A. Konotopov and A. A. Petrov, *Zh. Org. Khim.*, **5**, 1914 (1969).
374. E. N. Prilezhaeva, N. P. Petukhova, L. I. Shmonina and I. A. D'yakonova, *Izv. Akad. Nauk SSSR, Ser. Khim.*, 956 (1972); *Chem. Abstr.*, **77**, 87746 (1972).
375. I. L. Mikhelashvili and E. N. Prilezhaeva, *Zh. Org. Khim.*, **10**, 2524 (1974).
376. E. N. Prilezhaeva and I. L. Mikhelashvili, *Zh. Org. Khim.*, **9**, 1129 (1973).
377. M. D. Nair, *Indian J. Chem.*, **7**, 229 (1969).
378. N. D. Heindel and C. C. Ho Ko, *J. Heterocycl. Chem.*, **7**, 1007 (1970).
379. G. M. Brooke and Md. A. Quasem, *J. Chem. Soc., Perkin I*, 429 (1973).
380. K. Undheim and R. Lie, *Acta Chem. Scand.*, **27**, 595 (1973).
381. A. Zilverschoon, J. Meijer, P. Vermeer and L. Brandsma, *Rec. Trav. Chim. Pays-Bas*, **94**, 163 (1975).
382. J. Meijer, P. Vermeer, H. D. Verkruijsse and L. Brandsma, *Rec. Trav. Chim. Pays-Bas*, **92**, 1326 (1973).
383. M. N. Basyouni and M. T. Omar, *Aust. J. Chem.*, **27**, 1585 (1974).
384. B. A. Trofimov, S. V. Amosova, O. A. Tarasova, G. A. Kalabin and V. M. Bzhezovskii, *Zh. Org. Khim.*, **10**, 1988 (1974).
385. Y. Kishida and A. Terada, *Chem. Pharm. Bull. (Tokyo)*, **16**, 1351 (1968).
386. E. G. Kataev, L. K. Konovalova and E. G. Yarkova, *Zh. Org. Khim.*, **5**, 621 (1969).
387. (a) J. Meijer and L. Brandsma, *Rec. Trav. Chim. Pays-Bas*, **92**, 1331 (1973); (b) *Rec. Trav. Chim. Pays-Bas*, **91**, 578 (1972).
388. W. Ried, W. Merkel and S.-W. Park. *Justus Liebigs Ann. Chem.*, 79 (1975).
389. C. U. Pittman, Jr and M. Narita, *J. Chem. Soc., Chem. Comm.*, 960 (1975).
390. M. Narita and C. U. Pittman, Jr, *Synthesis*, 489 (1976).
391. J.-P. Pradère, A. Guenec, G. Duguay, J.-P. Guemas and H. Quiniou, *C. R. Acad. Sci., Paris, Sér. C*, **269**, 929 (1969).
392. K. Tomita and M. Nagano, *Chem. Pharm. Bull. (Tokyo)*, **20**, 2302 (1972).
393. P. Vermeer, J. Meijer, H. J. T. Bos and L. Brandsma, *Rec. Trav. Chim. Pays-Bas*, **93**, 51 (1974).
394. H. Davy and J. Vialle, *Bull. Soc. Chim. Fr.*, 1435 (1975).
395. H. Davy and J. M. Decrouen, *Bull. Soc. Chim. Fr.*, 115 (1976).
396. D. B. J. Easton, D. Leaver, and T. J. Rawlings, *J. Chem. Soc., Perkin I*, 41 (1972).
397. D. M. McKinnon and J. M. Buchshriber, *Can. J. Chem.*, **49**, 3299 (1971).
398. V. N. Elokhina and A. S. Nakhmanovich, *Khim. Geterotsikl. Soedin.*, 1637 (1971); *Chem. Abstr.*, **76**, 153485 (1971).
399. I. N. Azerbaev, A. B. Asmonova, L. A. Tsoi and G. P. Von, *Khim. Atsetilena Tekhnol. Karbida Kal'tsiya*, 100 (1973); *Chem. Abstr.*, **79**, 115279 (1973).
400. E. N. Prilezhaeva, G. S. Vasil'ev, I. L. Mikhelashvili and V. S. Bogdanov, *Izv. Akad. Nauk SSSR, Ser. Khim.*, 1922 (1970); *Chem. Abstr.*, **74**, 41828 (1970).
401. E. N. Prilezhaeva, V. I. Laba, V. I. Snegotskii and R. I. Shekhtman, *Izv. Akad. Nauk SSSR, Ser. Khim.*, 1602 (1970); *Chem. Abstr.*, **74**, 3285 (1970).
402. A. S. Nakhmanovich, E. N. Deryagina and V. N. Elokhina, *Khim. Geterotsikl. Soedin.*, 45 (1971).
403. E. E. Schweizer and S. V. DeVoe, *J. Org. Chem.*, **40**, 144 (1975).
404. N. A. Korchevin, V. N. Elokhina, V. A. Usov and M. G. Voronkov, *Zh. Org. Khim.*, **11**, 1532 (1975).
405. S. V. Amosova, O. A. Tarasova, B. A. Trofimov, G. A. Kalabin, V. M. Bzhezovskii and V. B. Modonov, *Zh. Org. Khim.*, **11**, 2026 (1975).
406. B. A. Trofimov, N. K. Gusarova and S. V. Amosova, *Zh. Org. Khim.*, **10**, 128 (1974).
407. A. G. Makhsumov, A. Safaev and E. A. Mirabaev, *Zh. Org. Khim.*, **5**, 1510 (1969).
408. A. I. Nogaideli, R. Sh. Tkeshelashvili and N. P. Makharashvili, *Zh. Org. Khim.*, **9**, 2481 (1973).

409. A. I. Konovalov, L. K. Konovalova and E. G. Kataev, *Zh. Org. Khim.*, **7**, 69 (1971); **6**, 2158 (1970); **5**, 1538 (1969).
410. A. I. Nogaideli, R. Sh. Tkeshelashvili and N. P. Makharashvili, *Zh. Org. Khim.*, **8**, 2472 (1972).
411. N. I. Gazieva, A. I. Shchekotikhin and V. A. Ginsburg, *Zh. Org. Khim.*, **7**, 1815 (1971).
412. L. Brandsma, P. J. W. Schuijl, D. Schuijl-Laros, J. Meijer and H. E. Wijers, *Int. J. Sulfur Chem. (B)*, **6**, 85 (1971).
413. M. T. Omar and M. N. Basyouni, *Bull. Chem. Soc. Japan*, **47**, 2325 (1974).
414. B. A. Trofimov, S. V. Amosova and O. A. Tarasova, *Zh. Org. Khim.*, **11**, 2224 (1975).
415. S. R. Sandler, *Chem. Ind. (London)*, 134 (1973).
416. G. G. Skvortsova, M. F. Shostakovskii and G. N. Kurov, *Zh. Org. Khim.*, **8**, 382 (1972); *Chem. Abstr.*, **76**, 126896 (1972).
417. S. P. Korshunov, N. A. Kudryavtseva, G. I. Shalyapina, A. V. Zemlyanskaya and N. V. Korzhova, *Zh. Org. Khim.*, **8**, 1913 (1972).
418. N. V. Korzhova, V. S. Pisareva, S. P. Korshunov and L. L. Kucherenko, *Zh. Org. Khim.*, **10**, 2542 (1974).
419. N. V. Korzhova, V. S. Pisareva and S. P. Korshunov, *Zh. Org. Khim.*, **11**, 1030 (1975).
420. V. S. Pisareva, N. V. Korzhova, V. M. Kazantseva and S. P. Korshunov, *Zh. Org. Khim.*, **11**, 1034 (1975).
421. V. S. Pisareva, N. V. Korzhova, V. A. Minaeva, V. M. Kazantseva and S. P. Korshunov, *Zh. Org. Khim.*, **10**, 1900 (1974).
422. N. V. Korzhova, V. S. Pisareva, O. M. Slyusareva, V. M. Kazantseva and S. P. Korshunov, *Zh. Org. Khim.*, **11**, 2255 (1975).
423. H. W. Heine, T. R. Hoye, P. G. Williard and R. C. Hoye, *J. Org. Chem.*, **38**, 2984 (1973).
424. R. Baumes, R. Jacquier and G. Tarrago, *Bull. Soc. Chim. Fr.*, 1147 (1974).
425. S. F. Nelsen, *J. Org. Chem.*, **34**, 2248 (1969).
426. R. M. Acheson and J. N. Bridson, *J. Chem. Soc. (C)*, 1143 (1969).
427. H. Reimlinger, M. A. Peiren and R. Merényi, *Chem. Ber.*, **103**, 3252 (1970).
428. (a) H. Reimlinger, R. Jacquier and J. Daunis, *Chem. Ber.*, **104**, 2702 (1971); (b) H. Reimlinger and M. A. Peiren, *Chem. Ber.*, **103**, 3266 (1970).
429. H. Reimlinger, M. A. Peiren and R. Merényi, *Chem. Ber.*, **105**, 103 (1972).
430. H. Reimlinger, F. Billiau, M. A. Peiren and R. Merényi, *Chem. Ber.*, **105**, 108 (1972).
431. A. McKillop and T. S. B. Sayer, *Tetrahedron Letters*, 3081 (1975).
432. (a) Y. Tanaka and S. I. Miller, *J. Org. Chem.*, **38**, 2708 (1973); (b) Y. Tanaka and S. I. Miller, *Tetrahedron*, **29**, 3285 (1973).
433. Y. Tamura, K. Sumoto, H. Matsushima, H. Taniguchi and M. Ikeda, *J. Org. Chem.*, **38**, 4324 (1973).
434. T. Sasaki, K. Kanematsu, A. Kakehi and G. Ito, *Bull. Chem. Soc. Japan*, **45**, 2050 (1972).
435. S. F. Gait, M. J. Rance, C. W. Rees and R. C. Storr, *Chem. Comm.*, 688 (1972).
436. S. R. Challand, S. F. Gait, M. J. Rance, C. W. Rees and R. C. Storr, *J. Chem. Soc., Perkin I*, 26 (1975).
437. M. J. Rance, C. W. Rees, P. Spagnolo and R. C. Storr, *J. Chem. Soc., Chem. Comm.*, 658 (1974).
438. Y. Tamura, H. Hayashi, Y. Nishimura and M. Ikeda, *J. Heterocycl. Chem.*, **12**, 225 (1975).
439. P. Barraclough, M. Edwards, T. L. Gilchrist and C. J. Harris, *J. Chem. Soc., Perkin I*, 716 (1976).
440. Y. Hayashi, Y. Iwagami, A. Kadoi, T. Shono and D. Swern, *Tetrahedron Letters*, 1071 (1974).
441. T. Sasaki, K. Kanematsu and A. Kakehi, *Tetrahedron Letters*, 5245 (1972).
442. T. Sheradsky and S. Lewinter, *Tetrahedron Letters*, 3941 (1972).
443. A. N. Pudovik and N. G. Khusainova, *Zh. Obshch. Khim.*, **40**, 697 (1970).
444. C. G. Krespan, *J. Org. Chem.*, **34**, 42 (1969).
445. W. D. Crow and M. N. Paddon-Row, *J. Amer. Chem. Soc.*, **94**, 4746 (1972).
446. A. N. Volkov, L. V. Sokolyanskaya and R. Ya. Mushii, *Zh. Org. Khim.*, **9**, 1599 (1973).
447. A. A. Ul'yanov and I. A. Maretina, *Zh. Org. Khim.*, **7**, 2052 (1971).

448. S. Holand and R. Epsztein, *Tetrahedron Letters*, 1897 (1974).
449. H. Möhrle and R. Kilian, *Tetrahedron*, **25**, 5745 (1969).
450. A. J. Chalk, *Tetrahedron Letters*, 3487 (1972); *Tetrahedron*, **30**, 1387 (1974).
451. R. Mestres, *J. Chem. Soc., Perkin I*, 805 (1972).
452. M. Franck-Neumann and G. LeClerc, *Bull. Soc. Chim. Fr.*, 247 (1975).
453. W. E. Truce and M. L. Gorbaty, *J. Org. Chem.*, **35**, 2113 (1970).
454. T. Sasaki and A. Kojima, *J. Chem. Soc. (C)*, 476 (1970).
455. V. Jäger and H. G. Viehe, *Angew. Chem., Int. Ed.*, **8**, 273 (1969).
456. E. V. Vasil'eva, E. M. Auvinen, I. A. Favorskaya, *Zh. Org. Khim.*, **11**, 313 (1975).
457. L. I. Vereshchagin, E. I. Titova, L. G. Tikhonova, S. R. Buzilova, L. D. Gavrilov and G. A. Kalabin, *Zh. Org. Khim.*, **10**, 978 (1974).
458. R. L. Bol'shedvorskaya, G. A. Pavlova, L. D. Gavrilov, N. V. Alekseeva and L. I. Vereshchagin, *Zh. Org. Khim.*, **8**, 1879 (1972).
459. Y. Kishida, T. Hiraoka and M. Yoshimoto, *Chem. Pharm. Bull. (Tokyo)*, **17**, 2126 (1969).
460. E. Suzuki, H. Sekizaki and S. Inoue, *J. Chem. Soc., Chem. Comm.*, 568 (1973).
461. L. I. Vereshchagin, E. I. Titova, L. G. Tikhonova, S. I. Demina and L. D. Gavrilov, *Zh. Org. Khim.*, **11**, 955 (1975).
462. L. G. Tikhonova, E. I. Titova, L. D. Gavrilov, L. I. Vereshchagin and A. G. Praidakov, *Zh. Org. Khim.*, **11**, 2510 (1975).
463. L. D. Gavrilov, G. G. Mordvina, E. I. Titova and L. I. Vereshchagin, *Zh. Org. Khim.*, **8**, 2009 (1972).
464. L. P. Safronova and A. S. Medvedeva, *Zh. Org. Khim.*, **8**, 1373 (1972).
465. L. I. Vereshchagin, S. R. Gainulina, L. P. Kirillova and T. V. Lipovich, *Zh. Org. Khim.*, **5**, 1557 (1969).
466. W. Ried and E. König, *Justus Liebigs Ann. Chem.*, 375 (1973).
467. Z. Alaune, Z. Talaikyte and G. Dienys, *Leit. TSR Mokslu Akad. Darb., Ser. B*, 65 (1968); *Chem. Abstr.*, **70**, 28569 (1969).
468. N. D. Heindel, *J. Org. Chem.*, **35**, 3138 (1970).
469. L. I. Vereshchagin, R. L. Bol'shedvorskaya and L. L. Okhapkina, *Zh. Org. Khim.*, **6**, 32 (1970).
470. Y. Tamura, H. Matsushima, M. Ikeda and K. Sumoto, *Tetrahedron*, **32**, 431 (1976).
471. P. Blanc and J. Bellan, *Ann. Pharm. Fr.*, **32**, 139 (1974).
472. N. D. Heindel, P. D. Kennewell and V. B. Fish, *J. Heterocycl. Chem.*, **6**, 77 (1969).
473. N. D. Heindel, P. D. Kennewell and C. J. Ohnmacht, *J. Org. Chem.*, **34**, 1168 (1969).
474. Z. B. Papanastassiou, R. J. Bruni and E. White, V, *J. Med. Chem.*, **10**, 701 (1967).
475. W. Sucrow and M. Slopianka, *Chem. Ber.*, **105**, 3807 (1972).
476. R. M. Acheson and M. S. Verlander, *J. Chem. Soc., Perkin I*, 2348 (1973).
477. R. M. Acheson and J. Woollard, *J. Chem. Soc., Perkin I*, 446 (1975).
478. Y. Iwanami, *Bull. Chem. Soc. Japan*, **48**, 1657 (1975).
479. N. D. Heindel and C. C. Ho Ko, *J. Heterocycl. Chem.*, **11**, 1109 (1974).
480. S. K. Khetan, J. G. Hiriyakkanavar and M. V. George, *Tetrahedron*, **24**, 1567 (1968).
481. M. D. Nair, *Indian J. Chem.*, **9**, 104 (1971).
482. G. Schmidt and E. Winterfeldt, *Chem. Ber.*, **104**, 2483 (1971).
483. E. Winterfeldt and J. M. Nelke, *Chem. Ber.*, **103**, 1183 (1970).
484. J. B. Hester, Jr, *J. Org. Chem.*, **39**, 2137 (1974).
485. S. K. Khetan and M. V. George, *Can. J. Chem.*, **47**, 3545 (1969).
486. M. N. Gudi, J. G. Hiriyakkanavar and M. V. George, *Indian J. Chem.*, **9**, 743 (1971).
487. S. K. Khetan and M. V. George, *Tetrahedron*, **25**, 527 (1969).
488. M. D. Nair, *Indian J. Chem.*, **11**, 109 (1973).
489. W. Sucrow and K.-P. Grosz, *Chem. Ber.*, **109**, 261 (1976).
490. N. D. Heindel, P. D. Kennewell and M. A. Pfau, *Chem. Comm.*, 757 (1969).
491. N. D. Heindel, P. D. Kennewell and M. A. Pfau, *J. Org. Chem.*, **35**, 80 (1970).
492. A. Fehlauer, K.-P. Grosz, M. Slopianka, W. Sucrow, W. J. S. Lockley and W. Lwowski, *Chem. Ber.*, **109**, 253 (1976).
493. W. Sucrow, C. Mentzel and M. Slopianka, *Chem. Ber.*, **107**, 1318 (1974).
494. V. Bardakos, W. Sucrow and A. Fehlauer, *Chem. Ber.*, **108**, 2161 (1975).
495. A. DeSavignac and A. Lattes, *Bull. Soc. Chim. Fr.*, 4476 (1970).

496. A. Kiennemann and R. Kieffer, *Bull. Soc. Chim. Fr.*, 1185 (1975).
497. (a) H. O. Bayer, H. Gotthardt and R. Huisgen, *Chem. Ber.*, **103,** 2356 (1970); (b) R. Huisgen, H. Gotthardt, H. O. Bayer and F. C. Schaefer, *Chem. Ber.*, **103,** 2611 (1970).
498. M. N. Sharma, *Curr. Sci.*, **42,** 201 (1973).
499. S. N. Egé, M. L. C. Carter, R. L. Spencer, C. E. Nordman and H. Z. Friedman, *J. Chem. Soc., Perkin I*, 868 (1976).
500. E. M. Briggs, G. W. Brown, W. T. Dawson and J. Jiricny, *J. Chem. Soc., Chem. Comm.*, 641 (1975).
501. T. A. George and M. F. Lappert, *J. Organometal. Chem.*, **14,** 327 (1968).
502. P. Frøyen, *Acta Chem. Scand.*, **27,** 141 (1973).
503. N. A. Portnoy, C. J. Morrow, M. S. Chattha, J. C. Williams, Jr and A. M. Aguiar, *Tetrahedron Letters*, 1397 (1971).
504. E. E. Schweizer, C. S. Kim, C. S. Labaw and W. P. Murray, *J. Chem. Soc., Chem. Comm.*, 7 (1973).
505. T. Sasaki, K. Kanematsu and A. Kakehi, *J. Org. Chem.*, **36,** 2978 (1971).
506. B. P. Curwain, C. S. Fake, J. L. Gordon, M. D. Mehta, D. Miller, R. H. Poyser and E. A. Watts, *J. Med. Chem.*, **14,** 737 (1971).
507. (a) N. A. Ampilogova and F. Ya. Perveev, *Zh. Obshch. Khim.*, 42, 1873 (1972); (b) N. A. Ampilogova and F. Ya. Perveev, *Zh. Obshch. Khim.*, **42,** 1874 (1972).
508. J. M. Patterson, *Synthesis*, 281 (1976).
509. M. A. Kirillova, Y. A. Zaichenko, I. A. Maretina and A. A. Petrov, *Zh. Org. Khim.*, **8,** 1575 (1972).
510. Yu. A. Zaichenko, I. A. Maretina and A. A. Petrov, *Zh. Org. Khim.*, **11,** 708 (1975).
511. M. Sakamoto and Y. Tominatsu, *Yakugaku Zasshi*, **90,** 1339 (1970); *Chem. Abstr.*, **74,** 53468 (1971).
512. J. Barluenga, S. Fustero and V. Gotor, *Synthesis*, 191 (1975).
513. N. D. Harris, *Synthesis*, 48 (1973).
514. V. N. Yandovskii and T. K. Klindukhova, *Zh. Org. Khim.*, **10,** 730 (1974).
515. F. Ya. Perveev and N. V. Koshmina, *Zh. Org. Khim.*, **5,** 1334 (1969).
516. Y. Tamura, Y. Sumida, Y. Miki and M. Ikeda, *J. Chem. Soc., Perkin I*, 406 (1975).
517. M. Petrovanu, A. Sauciuc, I. Gabe and I. Zugrăvescu, *Rev. Roum. Chim.*, **13,** 513 (1968).
518. T. Sasaki, K. Kanematsu, Y. Yukimoto and S. Ochiai, *J. Org. Chem.*, **36,** 813 (1971).
519. M. D. Nair, *Indian J. Chem.*, **6,** 630 (1968).
520. S. G. Matsoyan, E. G. Darbinyan and Yu. B. Mitardzhyan, *Arm. Khim. Zh.*, **21,** 998 (1968); *Chem. Abstr.*, **70,** 96693 (1969).
521. H. Hauptmann, *Tetrahedron*, **32,** 1293 (1976).
522. V. N. Yandovskii and T. K. Klindukhova, *Zh. Org. Khim.*, **10,** 136 (1974).
523. A. S. Nakhmanovich and E. N. Deryagina, *Khim. Geterotsikl. Soedin.*, 33 (1971).
524. M. F. Ford and D. R. M. Walton, *Synthesis*, 47 (1973).
525. W. J. S. Lockley and W. Lwowski, *Tetrahedron Letters*, 4263 (1974).
526. M. K. Saxena, M. N. Gudi and M. V. George, *Tetrahedron*, **29,** 101 (1973).
527. W. Sucrow, M. Slopianka and V. Bardakos, *Angew. Chem., Int. Ed.*, **14,** 560 (1975).
528. E. P. Levanova, A. N. Volkov, A. N. Nikol'skaya, K. A. Volkova and E. O. Tsetlina, *Zh. Org. Khim.*, **10,** 1835 (1974).
529. E. A. Lisitsyn, A. N. Belyaeva, A. K. Maleeva, I. A. Maretina and A. A. Petrov, *Zh. Org. Khim.*, **6,** 439 (1970).
530. M. A. Kirillova, I. A. Maretina and A. A. Petrov, *Zh. Org. Khim.*, **6,** 1347 (1970).
531. V. M. Cherkasov, N. A. Kapran and V. N. Zavatskii, *Ukr. Khim. Zh.*, **35,** 1057 (1969); *Chem. Abstr.*, **72,** 43596 (1970).
532. K. K. Balasubramanian and R. Nagarajan, *Synthesis*, 189 (1976).
533. M. D. Nair and S. R. Mehta, *Indian J. Chem.*, **6,** 490 (1968).
534. H. Suschitzky, B. J. Wakefield and R. A. Whittaker, *J. Chem. Soc., Perkin I*, 401 (1975).
535. Y. Iwanami, *Bull. Chem. Soc. Japan*, **44,** 1314 (1971).
536. F. P. Woerner and H. Reimlinger, *Chem. Ber.*, **103,** 1908 (1970).
537. H. Reimlinger, M. A. Peiren and R. Merényi, *Chem. Ber.*, **105,** 794 (1972).
538. B. Gavrilenko and S. I. Miller, unpublished results.

539. V. Sprio and S. Plescia, *J. Heterocycl. Chem.*, **9**, 951 (1972).
540. V. M. Cherkasov, I. A. Nasyr and V. T. Tsyba, *Khim. Geterotsikl. Soedin.*, 1704 (1970); *Chem. Abstr.*, **74**, 100003 (1971).
541. A. S. Katner and E. A. Ziege, *Chem. Comm.*, 864 (1971).
542. R. Baumes, R. Jacquier and G. Tarrago, *Bull. Soc. Chim. Fr.*, 260 (1976).
543. Y. Leroux and C. Roman, *Tetrahedron Letters*, 2585 (1973).
544. L. N. Cherkasov, *Izv. Vyssh. Ucheb. Zaved., Khim. Khim. Tekhnol.*, **14**, 1117 (1971); *Chem. Abstr.*, **75**, 140173 (1971).
545. T. Kauffmann, *Angew. Chem., Int. Ed.*, **13**, 627 (1974).
546. V. Boekelheide and J. E. Nottke, *J. Org. Chem.*, **34**, 4134 (1969).
547. R. M. Acheson, G. Procter and S. R. Critchley, *J. Chem. Soc., Chem. Comm.*, 692 (1976).
548. (a) R. M. Acheson and J. Woollard, *J. Chem. Soc., Perkin I*, 740 (1975); (b) R. M. Acheson and J. Woollard, *J. Chem. Soc., Perkin I*, 744 (1975).
549. A. T. Bottini, J. G. Maroski and V. Dev, *J. Org. Chem.*, **38**, 1767 (1973).
550. L.-I. Olsson and A. Claesson, *Tetrahedron Letters*, 2161 (1974).
551. J. G. Duboudin and B. Jousseaume, *J. Organometal. Chem.*, **96**, C 47 (1975).
552. J. Klein and N. Aminadav, *J. Chem. Soc. (C)*, 1380 (1970).
553. F. W. Von Rein and H. G. Richey, Jr, *Tetrahedron Letters*, 3777 (1971).
554. F. Bernadou and L. Miginiac, *C. R. Acad. Sci., Paris, Sér. C*, **280**, 1473 (1975); *Tetrahedron Letters*, 3083 (1976).
555. M. T. Bertrand, G. Courtois and L. Miginiac, *C. R. Acad. Sci., Paris, Sér. C*, **280**, 999 (1975).
556. B. Mauzé, C. Nivert and L. Miginiac, *J. Organometal. Chem.*, **44**, 69 (1972).
557. M. T. Bertrand, G. Courtois and L. Miginiac, *Tetrahedron Letters*, 1945, 3147 (1975).
558. A. E. Jukes, *Advan. Organometal. Chem.*, **12**, 215 (1974).
559. G. H. Posner, *Org. Reactions*, **19**, 1 (1972); **22**, 253 (1975).
560. E. J. Corey and R. H. K. Chen, *Tetrahedron Letters*, 1611 (1973).
561. J. F. Normant, G. Cahiez, M. Bourgain, C. Chuit and J. Villiéras, *Bull. Soc. Chim. Fr.*, 1656 (1974).
562. R. Levene, J. Y. Becker and J. Klein, *J. Organometal. Chem.*, **67**, 467 (1974).
563. J. F. Normant and M. Bourgain, *Tetrahedron Letters*, 2583 (1971).
564. J. K. Crandall, P. Battioni, J. T. Wehlacz and R. Bindra, *J. Amer. Chem. Soc.*, **97**, 7171 (1975).
565. (a) A. Alexakis, J. F. Normant and J. Villiéras, *J. Organometal. Chem.*, **96**, 471 (1975); (b) A. Alexakis, J. F. Normant and J. Villiéras, *J. Mol. Catal.*, **1**, 43 (1975); (c) J. F. Normant, A. Alexakis and J. Villiéras, *J. Organometal. Chem.*, **57**, C 99 (1973).
566. (a) R. G. Pearson and C. D. Gregory, *J. Amer. Chem. Soc.*, **98**, 4098 (1976); (b) C. Jallabert, H. Riviere and P. W. Tang, *J. Organometal. Chem.*, **104**, 1 (1976).
567. J. Klein and R. Levene, *J. Chem. Soc., Perkin II*, 1971 (1973).
568. D. Jeckel and J. Gosselck, *Tetrahedron Letters*, 2101 (1972).
569. Z. Yoshida, S. Yoneda and M. Hazama, *J. Org. Chem.*, **37**, 1364 (1972).
570. J. Ide and Y. Kishida, *Chem. Pharm. Bull. (Tokyo)*, **16**, 784 (1968).
571. C. Leonte and I. Zugrăvescu, *Tetrahedron Letters*, 2029 (1972).
572. (a) N. S. Basketter and A. O. Plunkett, *J. Chem. Soc., Chem. Comm.*, 594 (1975); (b) T. Kutsuma, K. Fujiyama, Y. Sekine and Y. Kobayashi, *Chem. Pharm. Bull. (Tokyo)*, **20**, 1558 (1972).
573. H. Ogura and K. Kikuchi, *J. Org. Chem.*, **37**, 2679 (1972).
574. M. Caprosu, M. Petrovanu and I. Zugrăvescu, *Bul. Inst. Politeh. Iasi Sect.* 2, **21**, 61 (1975); *Chem. Abstr.*, **84**, 43964 (1976).
575. H. Gotthardt and F. Reiter, *Tetrahedron Letters*, 2163 (1976).
576. H. Gotthardt, M. C. Weisshuhn and B. Christl, *Chem. Ber.*, **109**, 753 (1976).
577. H. Gotthardt, *Chem. Ber.*, **105**, 196 (1972).
578. R. Knorr, R. Huisgen and G. K. Staudinger, *Chem. Ber.*, **103**, 2639 (1970).
579. H. Gotthardt, R. Huisgen and H. O. Bayer, *J. Amer. Chem. Soc.*, **92**, 4340 (1970).
580. G. Singh and P. S. Pande, *Tetrahedron Letters*, 2169 (1974).
581. B. V. Badami and G. S. Puranik, *Rev. Roum. Chim.*, **20**, 981 (1975).
582. D. H. Aue and D. Thomas, *J. Org. Chem.*, **40**, 2360 (1975).

583. R. M. Acheson, N. D. Wright and P. A. Tasker, *J. Chem. Soc., Perkin I*, 2918 (1972).
584. S. S. Hall and A. J. Duggan, *J. Org. Chem.*, **39**, 3432 (1974).
585. T. W. Doyle, *Can. J. Chem.*, **48**, 1629, 1633 (1970).
586. M.-S. Lin and V. Snieckus, *J. Org. Chem.*, **36**, 645 (1971).
587. O. Tsuge and H. Samura, *Tetrahedron Letters*, 597 (1973).
588. I. E. Chlenov, I. L. Sokolova, S. S. Novikov and V. A. Tartakovsky, *Izv. Akad. Nauk SSSR, Ser. Khim.*, 473 (1973); *Chem. Abstr.*, **79**, 5304 (1973).
589. E. Winterfeldt, D. Schumann and H. J. Dillinger, *Chem. Ber.*, **102**, 1656 (1969).
590. T. R. Oakes and D. J. Donovan, *J. Org. Chem.*, **38**, 1319 (1973).
591. Y. Suzuki, N. Obata and T. Takizawa, *Tetrahedron Letters*, 2667 (1970).
592. H. J. Dillinger, G. Fengler, D. Schumann and E. Winterfeldt, *Tetrahedron*, **30**, 2553 (1974).
593. J. Z. Gougoutas and W. Saenger, *J. Org. Chem.*, **36**, 3632 (1971).
594. T. Saegusa, Y. Ito, S. Tomita, H. Kinoshita and N. Taka-ishi, *Tetrahedron*, **27**, 27 (1971).
595. T. Takizawa, N. Obata, Y. Suzuki and T. Yanagida, *Tetrahedron Letters*, 3407 (1969).
596. W. Ried and L. Kaiser, *Synthesis*, 120 (1976).
597. O. Tsuge and A. Torii, *Bull. Chem. Soc. Japan*, **46**, 283 (1973).
598. S. F. Tan and T.-H. Tjia, *J. Chem. Soc., Perkin I*, 2137 (1976); 2405 (1975).
599. R. Paul, R. P. Williams and E. Cohen, *J. Med. Chem.*, **17**, 539 (1974).
600. W. Ried and E. König, *Justus Liebigs Ann. Chem.*, **757**, 153 (1972).
601. K. Buggle, G. P. Hughes and E. M. Philbin, *Proc. Roy. Irish Acad. (B)*, 71 (1971).
602. H. Wamhoff and J. Hartlapp, *Chem. Ber.*, **109**, 1269 (1976).
603. S. A. Khan, H. Plieninger, D. Wild and A. Siddiqui, *Chem. Ber.*, **106**, 812 (1973).
604. P. Vermeer, J. Meijer and C. Eylander, *Rec. Trav. Chim. Pays-Bas*, **93**, 240 (1974).
605. J. Meijer and P. Vermeer, *Rec. Trav. Chim. Pays-Bas*, **94**, 14 (1975).
606. F. Näf and P. Degen, *Helv. Chim. Acta*, **54**, 1939 (1971).
607. E. J. Corey and D. J. Beames, *J. Amer. Chem. Soc.*, **94**, 7210 (1972).
608. E. J. Corey, C. U. Kim, R. H. K. Chen and M. Takeda, *J. Amer. Chem. Soc.*, **94**, 4395 (1972).
609. J. Meijer, H. Westmijze and P. Vermeer, *Rec. Trav. Chim.*, **95**, 102 (1976).
610. G. Tadema, P. Vermeer, J. Meijer and L. Brandsma, *Rec. Trav. Chim. Pays-Bas*, **95**, 66 (1976).
611. P. Vermeer, C. D. Graaf and J. Meijer, *Rec. Trav. Chim. Pays-Bas*, **93**, 24 (1974).
612. W. E. Truce and M. J. Lusch, *J. Org. Chem.*, **39**, 3174 (1974).
613. A. Alexakis, J. Normant and J. Villiéras, *Tetrahedron Letters*, 3461 (1976).
614. J. P. Marino and L. J. Browne, *Tetrahedron Letters*, 3245 (1976); *J. Org. Chem.*, **41**, 3629 (1976).
615. T. Kutsuma, K. Fujiyama and Y. Kobayashi, *Chem. Pharm. Bull. (Tokyo)*, **20**, 1809 (1972).
616. T. Sasaki, K. Kanematsu and Y. Yukimoto, *J. Chem. Soc. (C)*, 481 (1970).
617. S. F. Gait, M. J. Rance, C. W. Rees, R. W. Stephenson and R. C. Storr, *J. Chem. Soc., Perkin I*, 556 (1975).
618. J. Roggero and C. Divorne, *C. R. Acad. Sci., Paris, Sér. C*, **268**, 870 (1969); **271**, 763 (1970).
619. Z. Yoshida, S. Yoneda and Y. Murata, *J. Org. Chem.*, **38**, 3537 (1973).
620. A. Terada and Y. Kishida, *Chem. Pharm. Bull. (Tokyo)*, **18**, 505 (1970).
621. M. Higo and T. Mukaiyama, *Tetrahedron Letters*, 2565 (1970).
622. M. Higo, T. Sakashita, M. Toyoda and T. Mukaiyama, *Bull. Chem. Soc. Japan*, **45**, 250 (1972).
623. A. G. Hortmann and R. L. Harris, *J. Amer. Chem. Soc.*, **93**, 2471 (1971).
624. N. S. Vyazankin, E. N. Gladyshev, E. A. Arkhangel'skaya, G. A. Razuvaev and S. P. Korneva, *Izv. Akad. Nauk SSSR, Ser. Khim.*, 2081 (1968).
625. V. S. Zavgorodnii, T. P. Zubova and A. A. Petrov, *Zh. Obshch. Khim.*, **43**, 115 (1973); *Chem. Abstr.*, **78**, 111455 (1973).
626. T. P. Zubova, V. S. Zavgorodnii and A. A. Petrov, *Zh. Obshch. Khim.*, **43**, 120 (1973); *Chem. Abstr.*, **78**, 111458 (1973).
627. M. Schoufs, J. Meijer, P. Vermeer and L. Brandsma, *Rec. Trav. Chim. Pays-Bas*, **93**, 241 (1974).

628. B. L. Booth and A. D. Lloyd, *J. Organometal. Chem.*, **35**, 195 (1972).
629. K. Akiba, T. Yoneyama and N. Inamoto, *Bull. Chem. Soc. Japan*, **49**, 2023 (1976).
630. I. F. Wilson and J. C. Tebby, *J. Chem. Soc., Perkin I*, 2830 (1972).
631. M. A. Kazankova, T. Y. Satina and I. F. Lutsenko, *Zh. Obshch. Khim.*, **45**, 1195 (1975).
632. E. M. Richards, J. C. Tebby, R. S. Ward and D. H. Williams, *J. Chem. Soc. (C)*, 1542 (1969).
633. E. M. Richards and J. C. Tebby, *J. Chem. Soc. (C)*, 1059 (1971).
634. M. A. Shaw, J. C. Tebby, R. S. Ward and D. H. Williams, *J. Chem. Soc. (C)*, 1609 (1968).
635. N. E. Waite and J. C. Tebby, *J. Chem. Soc. (C)*, 386 (1970).
636. M. A. Shaw, J. C. Tebby, R. S. Ward and D. H. Williams, *J. Chem. Soc. (C)*, 504 (1970).
637. M. A. Shaw and J. C. Tebby, *J. Chem. Soc. (C)*, 5 (1970).
638. A. N. Hughes and M. Davies, *Chem. Ind. (London)*, 138 (1969).
639. V. G. Salishchev, M. L. Petrov and A. A. Petrov, *Khim. Elementoorg. Soedin.*, (Ed. S. N. Danilov), 'Nauka', Leningr. Otd., Leningrad, USSR, 1976, p. 91; *Chem. Abstr.*, **85**, 192821 (1976).
640. I. N. Azerbaev, A. B. Asmanova, L. A. Tsoi and V. S. Bazalitskaya, *Izv. Akad. Nauk Kaz. SSR, Ser. Khim.* **20**, 53 (1970); *Chem. Abstr.*, **73**, 76788 (1970).
641. W. Mack, *Angew. Chem. Int. Ed.*, **5**, 896 (1966).
642. M. I. Bruce and F. G. A. Stone, *Angew. Chem., Int. Ed.*, **7**, 747 (1968).
643. B. L. Booth and R. G. Hargreaves, *J. Chem. Soc. (A)*, 308 (1970).
644. C. A. Bunton, *Nucleophilic Substitution at a Saturated Carbon Atom*, Elsevier, New York, 1963.
645. M. L. Bender, *Chem. Rev.*, **60**, 53 (1960).
646. J. Miller, *Aromatic Nucleophilic Substitution*, Elsevier, New York, 1968.
647. Z. Rappoport, *Advan. Phys. Org. Chem.*, **7**, 1 (1969).
648. J. Bode, *Justus Liebigs Ann. Chem.*, **267**, 268 (1892).
649. J. A. Nieuwland and R. R. Vogt, *The Chemistry of Acetylene*, Reinhold, New York, 1945, p. 71.
650. M. J. Murray, *J. Amer. Chem. Soc.*, **60**, 2662 (1938).
651. E. Ott and G. Dittus, *Chem. Ber.*, **76**, 80 (1943).
652. S. I. Miller, C. E. Orzech, C. A. Welch, G. R. Ziegler and J. I. Dickstein, *J. Amer. Chem. Soc.*, **84**, 2020 (1962).
653. K. Isslieb and G. Harzfeld, *Chem. Ber.*, **95**, 268 (1962).
654. H. G. Viehe and E. Franchimont, *Chem. Ber.*, **95**, 319 (1962).
655. B. I. Ionin and A. A. Petrov, *Zh. Obshch. Khim.*, **32**, 2387 (1962); **33**, 2863 (1963).
656. V. S. Zavgorodnii and A. A. Petrov, *Dokl. Akad. Nauk SSSR*, **143**, 855 (1962); V. S. Zavgorodnii and A. A. Petrov, *Zh. Obshch. Khim.*, **32**, 3527 (1962).
657. A. K. Kuriakose and S. I. Miller, *Tetrahedron Letters*, 905 (1962).
658. R. Tanaka and S. I. Miller, *Tetrahedron Letters*, 1753 (1971).
659. G. R. Ziegler, C. A. Welch, C. E. Orzech, S. Kikkawa and S. I. Miller, *J. Amer. Chem. Soc.*, **85**, 1648 (1963).
660. G. Sturtz, C. Charrier and H. Normant, *Bull. Soc. Chim. Fr.*, 1707 (1966).
661. A. A. Frost and R. G. Pearson, *Kinetics and Mechanism*, 2nd ed., Wiley, New York, 1961, p. 138.
662. G. Salomon and A. J. Ultée, Sr, *Rec. Trav. Chim. Pays-Bas*, **69**, 95 (1950).
663. F. Kalberer, *Bull. Soc., Fribourg Sci. Nat.*, **44**, 225 (1954); *Chem. Abstr.*, **50**, 16718 (1956).
664. N. Menschutkin, *Z. Phys. Chem.*, **5**, 589 (1890); **6**, 41 (1890).
665. J. I. Dickstein and S. I. Miller, *J. Org. Chem.*, **37**, 2175 (1972).
666. S. D. Ross, J. J. Bruno and R. C. Petersen, *J. Amer. Chem. Soc.*, **85**, 3999 (1963).
667. M. Auriel and E. de Hoffmann, *J. Amer. Chem. Soc.*, **97**, 7433 (1975).
668. P. Beltrame, P. L. Beltrame, O. Sighinolfi and M. Simonetta, *J. Chem. Soc. (B)*, 1103 (1967).
669. J. Miller and W. Kai-Yan, *J. Chem. Soc.*, 3492 (1963).
670. C. A. Vernon, *J. Chem. Soc.*, 4462 (1954).
671. W. A. Henderson, Jr and S. A. Buckler, *J. Amer. Chem. Soc.*, **82**, 5794 (1960).
672. B. A. Arbuzov and B. P. Lugovkin, *Zh. Obshch. Khim.*, **21**, 99 (1951).

673. G. Aksnes and D. Aksnes, *Acta Chem. Scand.*, **18,** 38 (1964); **20,** 2463 (1966).
674. P. Beltrame, D. Pitea and M. Simonetta, *J. Chem. Soc.* (*B*), 1108 (1967).
675. P. K. Yonan, *M.S. Thesis*, Illinois Institute of Technology, 1956.
676. S. I. Miller and P. K. Yonan, *J. Amer. Chem. Soc.*, **79,** 5931 (1957).
677. J. Miller and A. J. Parker, *J. Amer. Chem. Soc.*, **83,** 117 (1961).
678. A. J. Parker, *Chem. Rev.*, **69,** 1 (1969).
679. C. W. L. Bevan, *J. Chem. Soc.*, 2340 (1951).
680. P. Beltrame, P. L. Beltrame and L. Bellotti, *J. Chem. Soc.* (*B*), 932 (1969).
681. P. Beltrame, P. L. Beltrame, M. L. Cereda and G. Lazzerini, *J. Chem. Soc.* (*B*), 1100 (1969).
682. A. M. Porto, L. Altieri, A. J. Castro and J. A. Brieux, *J. Chem. Soc.* (*B*), 963 (1966).
683. C. Eaborn, G. A. Skinner and D. R. M. Walton, *J. Chem. Soc.* (*B*), 922 (1966).
684. D. Daloze, H. G. Viehe and G. Chiurdoglu, *Tetrahedron Letters*, 3925 (1969).
685. N. Kharasch, *Organic Sulfur Compounds*, Vol. 1, Pergamon Press, New York, 1961, p. 390.
686. S. I. Miller and R. M. Noyes, *J. Amer. Chem. Soc.*, **74,** 629 (1952).
687. G. Köbrich and P. Buck in *The Chemistry of Acetylenes* (Ed. H. G. Viehe) Marcel Dekker, New York, 1969, Chap. 2.
688. S. Y. Delavarenne and H. G. Viehe, *Chem. Ber.*, **103,** 1209 (1970).
689. L. Brandsma, *Rec. Trav. Chim. Pays-Bas*, **90,** 265 (1971).
690. E. A. Halevi and F. A. Long, *J. Amer. Chem. Soc.*, **83,** 2809 (1961).
691. R. R. Lii and S. I. Miller, *J. Amer. Chem. Soc.*, **91,** 7524 (1969).
692. M. M. Mousseron and P. Froger, *Bull. Soc. Chim. Fr.*, **12,** 69 (1945).
693. J. H. Finley, D. Z. Denney and D. B. Denney, *J. Amer. Chem. Soc.*, **91,** 5826 (1969).
694. C. D. Ritchie and W. F. Sager, *Progr. Phys. Org. Chem.*, **2,** 323 (1964).
695. W. K. Kwok, R. A. More O'Ferrall and S. I. Miller, *Tetrahedron*, **20,** 1913 (1964); R. A. More O'Ferrall and S. I. Miller, *J. Amer. Chem. Soc.*, **85,** 2440 (1963).
696. A. J. Kirby and S. G. Warren, *The Organic Chemistry of Phosphorus*, Elsevier, New York, 1967, p. 274.
697. P. Beltrame, P. L. Beltrame, M. G. Cattania and M. Simonetta, *Gazz. Chim. Ital.*, **102,** 164 (1972).
698. E. C. F. Ko and A. J. Parker, *J. Amer. Chem. Soc.*, **90,** 6447 (1968).
699. Z. Rappoport, private communication.
700. M. Hanack, private communication.
701. J. I. Dickstein and S. I. Miller, unpublished results.
702. G. A. Olah, J. R. De Member, R. H. Schlosberg and Y. Halpern, *J. Amer. Chem. Soc.*, **94,** 156 (1972).
703. M. Hanack, *Accounts Chem. Res.*, **9,** 364 (1976).
704. W. D. Stohrer, *Tetrahedron Letters*, 207 (1975); W. D. Stohrer and K. R. Schmieder, *Chem. Ber.*, **109,** 285 (1976).
705. S. I. Miller, *Tetrahedron*, **33,** 1211 (1977).
706. P. Beltrame, A. Gavezzotti and M. Simonetta, *J. Chem. Soc.*, *Perkin II*, 502 (1974).
707. M. Rösner and G. Köbrich, *Angew. Chem.*, *Int. Ed.*, **14,** 708 (1975).
708. P. Cadiot and W. Chodkiewicz in *The Chemistry of Acetylenes*, (Ed. H. G. Viehe), Marcel Dekker, New York, 1969, Chap. 9.
709. J. L. Philippe, W. Chodkiewicz and P. Cadiot, *C. R. Acad. Sci.*, *Paris*, *Sér. C.*, **270,** 1419 (1970).
710. J. L. Philippe, W. Chodkiewicz and P. Cadiot, *Tetrahedron Letters*, 1795 (1970).
711. J. K. Kim and J. F. Bunnett, *J. Amer. Chem. Soc.*, **92,** 7463 (1970); **92,** 7464 (1970).
712. J. F. Bunnett, X. Creary and J. E. Sundberg, *J. Org. Chem.*, **41,** 1707 (1976).
713. L. I. Peterson, *Tetrahedron Letters*, 5357 (1968); *U.S. Patent*, No. 3,499,928; *Chem. Abstr.*, **73,** 14430 (1970); *U.S. Patent*. No. 3,499,904; *Chem. Abstr.*, **73,** 45145 (1970); *U.S. Patent*, No. 3,657,342; *Chem. Abstr.*, **77,** 48068 (1972).
714. G. R. Ziegler and S. I. Miller, unpublished results.
715. T. Izumi and S. I. Miller, *J. Org. Chem.*, in press.
716. V. Wolf, W. Block and H. Piater, *Justus Liebigs Ann. Chem.*, **682,** 112 (1965).
717. D. W. Burt and P. Simpson, *J. Chem. Soc.* (*C*), 2273 (1969).

718. N. Hashimoto, K. Matsumura, T. Saraie, Y. Kawano and K. Morita, *J. Org. Chem.*, **35**, 675 (1970).
719. H. G. Viehe, R. Merényi, J. F. M. Oth and P. Valange, *Angew. Chem., Int. Ed.*, **3**, 746 (1964).
720. H. G. Viehe, R. Merényi, J. F. M. Oth, J. R. Senders and P. Valange, *Angew. Chem., Int. Ed.*, **3**, 755 (1964).
721. H. Hoffmann and H. Förster, *Tetrahedron Letters*, 983 (1964).
722. M. Simalty and M. H. Mebazaa, *Tetrahedron*, **28**, 3343 (1972).
723. J. Normant, *Bull. Soc. Chim. Fr.*, 1888 (1963).
724. R. F. Curtis and J. A. Taylor, *J. Chem. Soc. (C)*, 186 (1971).
725. R. Oliver and D. R. M. Walton, *Tetrahedron Letters*, 5209 (1972).
726. R. Tanaka (Kyushu University), private communication.
727. R. H. Abeles and D. Dolphin, *Accounts Chem. Res.*, **9**, 114 (1976).
728. J. G. A. Kooyman, H. P. G. Hendriks, P. P. Montijn, L. Brandsma and J. F. Arens, *Rec. Trav. Chim. Pays-Bas*, **87**, 69 (1968).
729. J. F. H. Braams, P. P. Montijn and H. J. T. Bos, *Rec. Trav. Chim. Pays-Bas*, **91**, 700 (1972).
730. G. Peiffer, A. Guillemonat, J. C. Traynard and M. Faure, *C. R. Acad. Sci., Paris, Sér. C*, **268**, 358 (1969).
731. D. D. Callander, P. L. Coe, J. C. Tatlow and R. C. Terrell, *J. Chem. Soc. (C)*, 1542 (1971).
732. F. Waugh and D. R. M. Walton, *J. Organometal Chem.*, **39**, 275 (1972).
733. B. N. Ghose and D. R. M. Walton, *Synthesis*, 890 (1974).
734. R. Eastmond and D. R. M. Walton, *Tetrahedron*, **28**, 4591 (1972).
735. D. W. Knight and G. Pattenden, *Chem. Comm.*, 188 (1974).
736. F. Bohlmann and R. Weber, *Chem. Ber.*, **105**, 3036 (1972).
737. K. E. Schulte, G. Rücker and J. Feldkamp, *Chem. Ber.*, **105**, 24 (1972).
738. B. P. Gusev, L. A. Tsurgozen and V, F. Kucherov, *Izv. Akad. Nauk SSSR, Ser. Khim.*, 1098 (1972); *Chem. Abstr.*, **77**, 125815 (1972).
739. D. R. Strobach, *J. Org. Chem.*, **36**, 1438 (1971).
740. T. Cuvigny and H. Normant, *C. R. Acad. Sci., Paris, Sér. C.*, **268**, 834 (1969).
741. D. Seyferth and J. D. H. Paetsch, *J. Org. Chem.*, **34**, 1483 (1969).
742. L. Maier, *Phosphorus*, **2**, 229 (1973); *U.S. Patent*, No. 3,789,092 (1974); *Chem. Abstr.*, **80**, 83232 (1974).
743. A. J. Rudinskas and T. L. Hullar, *J. Med. Chem.*, **19**, 1367 (1976).

CHAPTER **20**

Synthesis and uses of isotopically labelled acetylenes

J. C. LAVALLEY and J. SAUSSEY

U.E.R. de Sciences, Université de Caen, 14032 Caen Cedex, France

I. INTRODUCTION

Preparations of labelled acetylenes have been compiled first in Murray and Williams' work[1] and then in a paper relating to some deuterated compounds[2], but they have not been the subject of a recent review. Excellent publications by Shatenshtein[3] and Thomas[4] are more general, but no review has been uniquely dedicated to labelled acetylenes. In this chapter we have tried to classify the different ways of preparing labelled acetylenes, emphasizing particularly the recent ones ('the isomerization-exchange' method for instance). We have selected the preparations of some of the simplest compounds which are often used as starting products for the preparation of more sophisticated acetylenes. We shall complete this part by briefly considering the different methods generally used in determining the extent of labelling and isotopic effects.

The main use of isotopically labelled compounds is in spectrometry (i.r., u.v., n.m.r., microwave,...). We have reported a few examples of such studies relative to acetylenes, principally those oriented towards structure determination. However, chemists are more interested in using labelled compounds to investigate reaction mechanisms. It is obvious that knowledge of the labelled state (labelled positions and percentage of labelling) provides vital information for determining complicated mechanisms, the starting products being well-known labelled compounds. Another possibility is the comparison of reaction rates, e.g. the study of kinetic isotope effects. We have added to these a third one, based on the determination of the structure of species strongly held on a surface, which may be reaction intermediates. Here, the labelled compounds are used, as we shall show, to determine the precise structure of the species. On the other hand, we have not dealt with different studies using labelled acetylenes as starting materials or as intermediates in preparations of more complex labelled molecules.

II. PREPARATION

A. Acetylenes Containing the DC≡C or TC≡C Group

1. The DC≡C group

a. Hydrolysis. Hydrolysis by D_2O of alkaline salts of 1-alkynes is a method very often used. In the case of the sodium salt, prepared from sodamide in liquid ammonia, it has been shown that the yield is about 85%[5], but the isotopic purity does not exceed more than 95%[5,6], even if traces of ammonia are exactly neutralized by deuterated phosphoric acid[5]. Hydrolysis of lithium salt has been used with success to prepare, for instance, $C_6H_5C{\equiv}CD$[7] and $CH_3(CH_2)_3C{\equiv}CD$[8] with high isotopic purity, the lithium salt being obtained from phenyllithium[9] or *n*-butyllithium[8]. To avoid the formation of polylithium compounds, it is necessary to employ the same quantities of 1-alkyne and *n*-butyllithium. Otherwise, as shown by Eberly and Adams[10], dilithiated alkynes may be formed and so, after hydrolysis, the deuterium content of the final compound may be higher than expected.

Hydrolysis of alkynylmagnesium bromides with D_2O has been used to prepare $CH_3CH_2C{\equiv}CD$[11], $DC{\equiv}CCH_2OH(D)$[12] and $CH_3(CH_2)_3C{\equiv}CD$[13]. The yield is not very high (<70%). Romanet and Wojtkowiak[5], testing the method, obtained a slightly higher yield but found a deuterium content of only 95%. On the other hand, reaction of deuteriochloric acid, DCl, with the mercury derivative of propyne gave 1-propyne(D-1) of much better purity[14].

Acetylene(D_2) has been prepared on numerous occasions from calcium carbide and D_2O[1]. Generally it is necessary to enrich the acetylene obtained by repeated exchanges with alkaline heavy water. However, it has been shown that C_2D_2 with 99·1–99·5% isotopic abundance may be prepared directly, without additional exchanges, if commercial calcium carbide is broken up in dry air and baked at 650 °C for 2 days[15]. The author believes that C_2D_2 of higher abundance would be obtained if the carbide was baked for an even longer period. Chemical impurities are often present; it is not necessary to remove them if the acetylene is used in a later step. If not, the C_2D_2 must be purified, for instance by means of gas chromatography or trap-to-trap distillation[16].

HC≡CD of 90% isotopic purity has recently been prepared from pure monolithium acetylenide and acetic acid, CH_3COOD [17]:

$$\mathrm{HC{\equiv}CH} \xrightarrow[-80\,^{\circ}\mathrm{C}]{n\text{-}\mathrm{C_4H_9Li}} \mathrm{HC{\equiv}CLi} \xrightarrow{\mathrm{CH_3COOD}} \mathrm{HC{\equiv}CD}$$

If a purer compound is needed, HC≡CD may be separated from C_2H_2 and C_2D_2 by gas chromatography[18]. The best results are obtained with triethylamine supported on 60/80 mesh Chromosorb P at −78 °C, employing helium as a carrier gas. After three runs, the product contains 1·3% of C_2H_2 and less than 1% of C_2D_2. C_2HD, stored at room temperature over mercury, is stable to disproportionation for several weeks at least.

b. Isotopic exchange. Due to the *s*-bond character of the acetylenic C—H bond, the acid strength of acetylenes is notable [p$K(C_2H_2) \sim 25$][4, 19] and there is exchange of acetylenic hydrogen with D_2O in the presence of a base. The exchange is rather slow[5]: 5h of thorough stirring are necessary for equilibrium to be reached. At room temperature, contrary to higher temperatures, the triple bond is stable (see Section II.B.3). Using 0.25 mole of D_2O for 0·12 mole of 1-alkyne and 0·20 g of sodium, it was found that, after three exchanges, the isotopic purity was >98%; after four exchanges, it was >99%. This seems to be the preferred method for the preparation of 1-alkynes (D-1). However, a larger amount of heavy water is consumed in this preparation than in preparations involving hydrolysis.

Samples of $DC{\equiv}CCH_2X$ (X = Cl, Br) were prepared by passing $HC{\equiv}CCH_2X$ through a preparative gas–liquid chromatographic column packed with Chromosorb W (30–60 mesh) coated with 15% E 20 M and 10% KOH which had been pretreated with deuterium oxide[20]. Deuterium-exchanged samples of 90% purity were collected in a trap. The same technique has been used to prepare $DC{\equiv}CCH_2OD$ [21]. Since exchange between acetylenic hydrogen and D_2O occurs under basic conditions, there is no ≡CH → ≡CD substitution when $HC{\equiv}CCH_2OH$ or HC≡CCOOH are mixed with D_2O ($HC{\equiv}CCH_2OD$ or HC≡CCOOD are obtained). To prepare DC≡CCOOH, the pH of a solution of sodium propiolate dissolved in D_2O was adjusted to 12 by adding a few drops of a solution of sodium deuteroxide in D_2O. After standing for 2 days and removing the heavy water, the residue was dissolved in H_2O, then cooled. A few drops of concentrated hydrochloric acid were added and DC≡CCOOH was extracted with ether[22].

2. The TC≡C group

Both methods involving hydrolysis or exchange have been used. C_2T_2 has been prepared by the reaction of T_2O and C_2Ca [23] or C_2Li_2 [24]. It was found that polymerization of C_2T_2 occurred at the rate of about 25% per day. Some phenylacetylenes were tritiated by treating the arylethynylmagnesium bromide with T_2O [25]. On the other hand, the exchange method has been used to prepare tritiated propargyl halides[25], using a weak alkaline mixture of tritiated water (~50 mc/ml) and alcoholic alkali.

B. General Methods of Preparing Other Deuterated or Tritiated Alkynes

1. Hydrolysis of inorganic carbides

As pointed out above (Section II.A.1.a), C_2D_2 may be prepared from calcium carbide and D_2O. The method is also applicable to propyne; propyne(D_4) may be obtained from magnesium carbide, C_3Mg_2 and D_2O [26]. Magnesium carbide is prepared by passing pentane (or butane) vapour over magnesium at 650 °C [2]. It has been found that by raising the temperature to 740 °C the yield of carbide increases from 25 to 90% [15]. Hydrolysis with D_2O is possible directly in the reaction oven[27] and this renders the preparation easier. The isotope content of $CD_3C{\equiv}CD$ is not less than 99% [2]. However, there is some allene(D_4) admixed with it. The reaction is exothermic, and the higher the temperature of hydrolysis, the greater the amount of allene in the product (this amount becomes appreciable if the hydrolysis temperature is higher than 150 °C [27]). Therefore, the method is suitable if perdeuterated propyne is used in a later step in which allene(D_4) is eliminated. If not, it is necessary to purify it, either by precipitation of the silver acetylide–silver nitrate complex and regeneration of propyne (by treatment of the complex with aqueous sodium cyanide or with ammonium thiocyanate) or via the mercury salt.

Some attempts have been made to prepare deuterated alkynes from their lithium salts. It has been shown that polylithiated 1-alkynes bearing α-hydrogen atoms are readily prepared using alkyllithiums. For instance, propyne reacts with *n*-butyllithium in hexane to form the tetralithium compound, C_3Li_4, perhaps mixed with some C_3Li_3H [28]. Hydrolysis with D_2O of this compound would lead to perdeuterated propyne, but certainly admixed with some deuterated allene. It has been reported that when one mole of 1-phenylpropyne is treated with six moles of *n*-butyllithium in hexane under reflux, followed by addition of D_2O to the reaction mixture, isomerization takes place and deuterated 3-phenylpropyne is obtained[29]. Its n.m.r. spectrum shows no aliphatic protons, and deuterium analysis reveals the presence of 3·22 atoms of deuterium/molecule. The authors suggest that the product is essentially C_6H_5-$CD_2C{\equiv}CD$, probably admixed with a small amount of $C_6H_4DCD_2C{\equiv}CD$. In the same way, from 1,3-pentadiyne, three isomers (1,3-pentadiyne, 1,4-pentadiyne and 1,2-pentadien-4-yne) were obtained[30]; they are mainly tetradeuterated but they are not isotopically pure. Hence it appears that hydrolysis by D_2O of polylithiated alkynes is a difficult method to apply to the preparation of specifically deuterated acetylenes.

2. Reaction of alkyl halides or alkynyl halides with alkaline acetylides

The reaction between sodium acetylide and alkyl halides in liquid ammonia is the most general method of preparing substituted acetylenes. It has very often been used to prepare labelled alkynes[2]. Appropriately labelled alkyl halides have to be prepared first (a review of the preparation of the most common ones has already been given[2]), but more and more are now commercially available. The yield of the condensation in liquid ammonia is about 75% and, under experimental conditions, no isotopic exchange has been found; the isotopic purity of the final alkynes is the same as that of the starting alkyl halides.

A majority of labelled acetylenes have been prepared according to this method as, for instance, $HC{\equiv}CCD_3$ [31], $HC{\equiv}CCD_2CH_3$, $HC{\equiv}CCH_2CD_3$, $HC{\equiv}CCH_2CH_2D$, $HC{\equiv}CCHDCH_2D$, $HC{\equiv}CCHDCD_2H$, $HC{\equiv}CCD_2CD_2H$, $HC{\equiv}CCD_2CD_3$, $HC{\equiv}CCD_2H$ [32] and $HC{\equiv}CCT_2(CH_2)_4CH_3$ [33].

A variant of this type of synthesis is the reaction between $CD_3C{\equiv}CNa$ and RX in liquid ammonia to prepare compounds with the $CD_3C{\equiv}C$— group. This method is less expensive than the one involving the use of perdeuterated methyl halide, as $CD_3C{\equiv}CD$ may be prepared from C_3Mg_2 and D_2O. $CD_3C{\equiv}CCH_3$ [34] and $CD_3C{\equiv}CC_2H_5$ [35] have also been obtained according to this process.

Salts other than the sodium one may be used: $CH_3C{\equiv}CCD_3$ has been prepared by reaction between ICD_3 and $CH_3C{\equiv}CLi$ [36] (made from propyne and *n*-butyllithium in hexane). The yield is excellent (88%) but the isotopic purity of the final product is not very high: there was about 1–3% of 2-butyne(D_6) (presumably originating in acetylene present in the starting propyne) in the final product $CD_3C{\equiv}CCH_3$. The Grignard salt $C_6H_5C{\equiv}CMgBr$ has been used to prepare $(C_6H_5)_2CDC{\equiv}CC_6H_5$ [37]. Generally it is the cuprous salt which is employed to prepare diarylacetylenes; a series of deuterated diphenylacetylenes have been prepared from iodobenzene(D_5) and the cuprous derivative of several phenylacetylenes[38]. Chodkiewicz's method[39] also involves the cuprous salt for the preparation of α-diyne; this method has recently been used to make $CD_3C{\equiv}CC{\equiv}CCD_3$ [40] (from $CD_3C{\equiv}CBr$ and $CD_3C{\equiv}CD$), $CH_3C{\equiv}CC{\equiv}CCD_3$ [40] and $CD_3C{\equiv}CC{\equiv}CH$ [41]. Concerning the latter, it is not possible to use acetylene itself but its carbinol form, $HC{\equiv}CC(OH)(CH_3)_2$, which necessitates a further thermal decomposition[41]. The carbinol obtained, $CD_3C{\equiv}CC{\equiv}CC(OH)(CH_3)_2$, is a good product for storing since 1,3-diynes are not very stable, even at low temperature. No isotopic dilution has been observed when the cuprous salt is involved.

3. Isomerization and isomerization-exchange methods

For a long time[42], it has been known that there is rearrangement of 1-alkynes to 2-alkynes under basic conditions (for instance by alcoholic potassium hydroxide or powdered potassium hydroxide) at ~175 °C. It has been shown that in fact an equilibrium mixture is obtained[43]. It contains isomeric compounds, mainly the starting 1-alkyne and isomeric 1,2-alkadiene and 2-alkyne, but the latter, more stable from thermodynamic considerations, is predominant. Therefore the method could be used to prepare 2-alkynes from corresponding 1-alkynes. The reverse reaction is possible: disubstituted acetylenes can be converted to sodium derivatives of 1-alkynes by sodium[44] or sodamide[45]. 1-Alkynes are recovered by hydrolysis. We have tried to apply both reactions to prepare acetylenes deuterated in defined positions with high isotopic purity.

a. 2-Alkyne → 1-alkyne isomerization[46]. Isomerization of $CH_3(CH_2)_2C{\equiv}CCD_3$ (prepared from $CD_3C{\equiv}CD$) under the effect of an excess of sodamide in heptane at 150 °C, followed by hydrolysis, gives $CH_3(CH_2)_2CD_2C{\equiv}CH$. However, from the n.m.r. spectrum analysis, it appears that the deuterium purity is only 70%, which indicates deuterium exchanges with NH_2^- hydrogen atoms during the propargylic rearrangement. Another experiment, using $CD_3CD_2C{\equiv}CCH_3$ (made from CD_3-CD_2Br and $NaC{\equiv}CCH_3$, 99% isotopic purity) as a starting compound, leads to $CD_3CD_2CH_2C{\equiv}CH$. However, the isotopic purity of the methyl group is 92%, the methylene purity being only 88%. We can explain this result by the intermediate formation of $CD_3CD{=}C{=}CHCH_3$, a symmetrical product if isotopic effects are excepted, leading to a variety of deuterated 1-pentynes. We conclude that the isomerization method involving NH_2Na gives impure deuterated compounds (except of course if perdeuterated 2-alkynes and ND_2Na are used), and cannot be applied to prepare specifically labelled acetylenes.

On the other hand, use of sodium as an isomerization agent gives isotopically pure 1-alkynes. Starting from $CD_3CD_2C{\equiv}CCH_3$ and an excess of sodium in heptane at 130 °C, we obtained isotopically pure (99%) $CD_3CD_2CH_2C{\equiv}CH$. Unfortunately the yield was rather low (~20%), which was mainly due to a partial reduction of the triple bond to a double one (formation of sodium 1-alkynide involves the evolution of hydrogen). Another difficulty was the destruction of sodium excess before hydrolysis (we did this by transforming the excess into sodium amide in liquid ammonia[46]). We conclude that the poor yield and the difficulties inherent in the use of sodium limit the utilization of this method, which could however be useful in some particular cases, since at 130 °C, no isotopic dilution is observed. An attempt at a higher temperature (170 °C) showed that the CD_3 group was partly exchanged.

b. 1-Alkyne → 2-alkyne isomerization. Base-catalysed rearrangement of acetylenes is prototropic[47]. We suspected that the use of deuterium reservoirs, the most common and inexpensive being deuterium oxide, could induce an H → D exchange between the hydrogen atoms of the hydrocarbon and the deuterium atoms of the reservoir during the isomerization process and so lead to partially deuterated compounds. We undertook this study, using NaOD or LiOD (from Na or Li in D_2O) in heavy water as isomerization agents. We indeed found that a mixture of deuterated 2-butynes was obtained by shaking 1-butyne and LiOD in D_2O at 180 °C for 3 days[48]. In fact a mixture of three isomers (94% 2-butyne, 4·5% 1,2-butadiene and 1·5% 1-butyne) was obtained. The overall yield was ~75%. Attempts at higher temperatures lead to lower yields due to the formation of 2-butanone. Repeated 'isomerization-exchange' reactions give well-deuterated compounds; 2-butyne is separated from its isomers by gas chromatography only after the last exchange. After four exchanges, the isotopic purity of $CD_3C{\equiv}CCD_3$ is close to 99·5%. This is certainly the most convenient method of preparing 2-butyne(D_6). The same results are obtained if 2-butyne itself is the starting product. The only difference is in the isotopic composition of the different compounds present at the beginning of the reaction.

The same reactions, involving propyne and 1-pentyne, are in process of investigation. From the first attempts, it appears that after three exchanges $CD_3C{\equiv}CD$ of high isotopic purity is formed, but containing allene(D_4) and also acetone(D_6); this could indicate that a lower temperature than 170 °C would be more convenient to avoid acetone formation. From 1-pentyne, it appears that only $CH_3CH_2C{\equiv}CCD_3$ is formed. However, in this case, the isotopic purity is not very high. Further exchanges increase the purity, but it is then difficult to determine whether small quantities of deuterium are present in the methylene group. Nevertheless, it appears that the method gives good results in the cases of 2-butyne and propyne and that it could be convenient to prepare compounds tritiated in the $CH_3C{\equiv}C$ group from the corresponding 1-alkynes and T_2O.

C. ^{13}C- or ^{14}C-labelled Acetylenes

Hydrolysis of labelled acetylides (made from labelled $BaCO_3$ or CO_2) has been used to prepare ^{13}C- or ^{14}C-labelled acetylenes. For instance, mole quantities of acetylene ($^{13}C_2$) have recently been obtained, according to the reactions[49]:

$$2\ {}^{13}CO_2 + 10\ Li \xrightarrow{650\ ^\circ C} Li_2\ {}^{13}C_2 + 4\ Li_2O$$

$$Li_2\ {}^{13}C_2 + 2\ H_2O \longrightarrow {}^{13}C_2H_2 + 2\ LiOH$$

The yield of labelled acetylene is 90–100%. A variant, using labelled barium carbonate and excess of barium to make labelled BaC_2, then hydrolysis, has been used to prepare $^{14}C_2H_2$ [50] and $^{13}C_2H_2$ [51]. The isotopic yield is excellent if all traces of paraffin have been removed from the barium metal[51].

Substituted labelled acetylenes, such as $CH_3{}^{13}C{\equiv}{}^{13}CH$ [52], can be prepared from labelled sodium acetylide and the alkyl halide in liquid ammonia, following the general method (see Section II.B.2). The dehydrohalogenation method is however preferable for the preparation of labelled phenylacetylene $C_6H_5{}^{13}C{\equiv}CH$, from either $C_6H_5\,{}^{13}CH{=}CHBr$ [53] or $C_6H_5\,{}^{13}CCl{=}CH_2$ [54]. Renaud and Leitch[55] have discussed the different means of dehydrobromination of $C_6H_5CHBr^{13}CH_2Br$ and find that phenylacetylene might best be obtained in good yield (~90%) by dehydrobromination with an excess of sodamide in anhydrous ether (or tetrahydrofuran) in the presence of traces of ammonia.

$^{13}CH_3C{\equiv}CH$ has been prepared from $^{13}CH_3I$ and sodium acetylide[52]. Myers and Schmidt-Bleek[56], rather than employ a similar method to make $^{14}CH_3C{\equiv}CH$, preferred to use the lithium acetylide complexed with 1,2-ethanediamine and $^{14}CH_3I$. After purification, the yield is only 20%. Some other labelled acetylenes have been prepared by the usual methods: for instance, $C_6H_5C{\equiv}{}^{13}C^{13}CH_3$ [55], $HC{\equiv}C^{14}COOH$, $HC{\equiv}C^{14}CH_2OH$ and $HC{\equiv}C^{14}CH_2Br$ [57].

D. Selected Methods of Preparing Common Labelled Acetylenes

From the various studies reported above, it appears that the best methods of preparing specifically labelled acetylenes vary from one compound to another. The preferred methods are presented in Table 1.

TABLE 1. Preferred methods of preparing some simple labelled acetylenes[a]

Compound	Preferred method	References	Comments
$DC{\equiv}CD$	C_2Ca+D_2O	15	Eventually followed by isotopic exchange[1]
$RC{\equiv}CD$	$RC{\equiv}CH$, D_2O exchange	5	Basic conditions, 25 °C, four exchanges are necessary
$HC{\equiv}CD$	$HC{\equiv}CLi+CH_3COOD$	17	Pure compound necessitates separation by gas chromatography[18]
$CD_3C{\equiv}CD$	$C_3Mg_2+D_2O$	15	$CD_2{=}C{=}CD_2$ is also present
$CD_3C{\equiv}CR$ ($R\neq CD_3$)	$CD_3C{\equiv}CNa+RX$	2	$CD_3C{\equiv}CD$ is prepared from C_3Mg_2
$CD_3C{\equiv}CCD_3$	$CH_3CH_2C{\equiv}CH$, D_2O 'isomerization-exchange'	48	Basic conditions, 180 °C, four exchanges are necessary
$^*C_2H_2$	$^*C_2Li_2+H_2O$	49	$^*C_2Li_2$ from $^*CO_2$
$C_6H_5{}^*C{\equiv}CH$	Dehydrohalogenation	53–55	—

[a] $^*C = {}^{13}C$ or ^{14}C.

Other labelled acetylenes are generally prepared from labelled alkyl halides and sodium acetylide[2]. The following two methods seem to be quite specific: $CD_3CH_2C{\equiv}CCH_3$ has been made[2] from deuterated methyl iodide and $BrMgCH_2C{\equiv}CCH_3$ (Grignard–Würtz method), in ~50% yield; $(CH_3)_2CDC{\equiv}CH$ has been prepared as follows[58]:

$$(CH_3)_2C{=}C{=}CHBr \xrightarrow{LiAlD_4} (CH_3)_2CDC{\equiv}CH$$

This method is very convenient in the case of branched alkynes, and leads to almost chemically pure compounds[58].

III. ISOTOPE PURITY DETERMINATION AND ISOTOPE EFFECTS

Before using labelled acetylenes, it is necessary to check their isotopic purity. In the following, we shall summarize some separation and spectrometric methods which are most often employed and also consider isotope effects. Since general accounts of isotope effects have already been given, for instance by Halevi[59] or Laszlo and Welvart[60], and brought up to date by Thomas[4], in this chapter we shall mainly consider the effects related to the presence of the triple bond.

A. Gas Chromatography

The separation of C_2HD from C_2D_2 and C_2H_2 by gas chromatography has been reported as the only method allowing the preparation of C_2HD with an isotopic purity of higher than 95%[18]. In our laboratory, we have found that disubstituted acetylenes, differing only in their isotopic content, can also be separated by gas chromatography; squalane in a capillary column (length 100 m, temperature −20 °C) separates various deuterated 2-butynes[48]. The retention times decrease with increasing H → D substitution, as already observed in the case of a mixture of deuterated

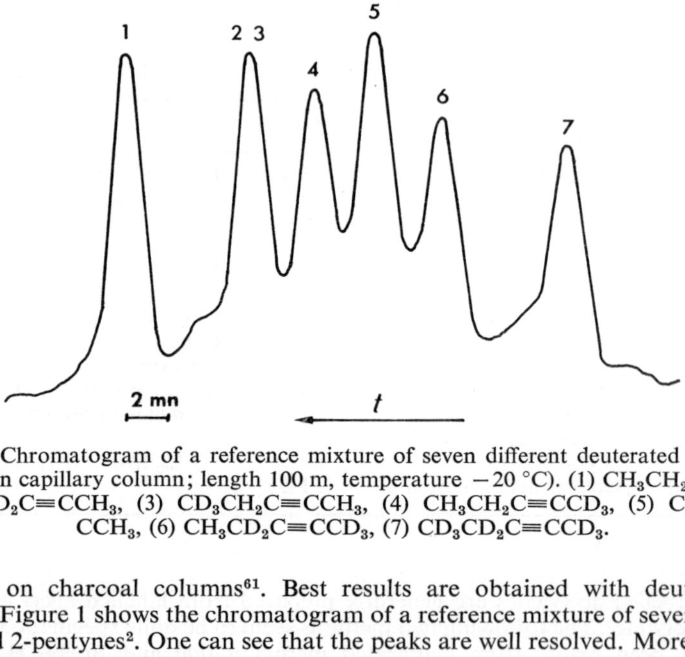

FIGURE 1. Chromatogram of a reference mixture of seven different deuterated 2-pentynes (squalane in capillary column; length 100 m, temperature −20 °C). (1) $CH_3CH_2C{\equiv}CCH_3$, (2) $CH_3CD_2C{\equiv}CCH_3$, (3) $CD_3CH_2C{\equiv}CCH_3$, (4) $CH_3CH_2C{\equiv}CCD_3$, (5) $CD_3CD_2C{\equiv}CCH_3$, (6) $CH_3CD_2C{\equiv}CCD_3$, (7) $CD_3CD_2C{\equiv}CCD_3$.

methanes on charcoal columns[61]. Best results are obtained with deuterated 2-pentynes. Figure 1 shows the chromatogram of a reference mixture of seven different deuterated 2-pentynes[2]. One can see that the peaks are well resolved. Moreover, two very weak extra peaks appear, which are due to two other deuterated 2-pentynes (impurities of the main compounds). The isotope effect due to the $CH_3 \rightarrow CD_3$ substitution is different depending on whether the methyl group is in the α or the β position in relation to the triple bond, being larger in the α position.

The application of these effects in the determination of the isotopic purity of disubstituted alkynes is obvious and we have used it to follow the H → D exchange when we prepared 2-butyne(D_6)[48]. The method is very convenient since it does not need any prior purification of the mixture of chemical isomers which are obtained, contrary to mass spectrometry which necessitates the previous elimination of 1-butynes and 1,2-butadienes. Unfortunately, under the same conditions, attempts to separate various deuterated 1-alkynes (such as e.g. propynes) failed.

B. Mass Spectrometry

Mass spectrometry is certainly the most common method of measuring the total amount of deuterium or ^{13}C contained in labelled compounds and of determining their isotopic distribution. For instance, Whitesides and Ehmann[36] have determined the isotopic composition of $CH_3C{\equiv}CCD_3$ samples from intensity data (corrected to ^{13}C), the ionizing voltage being taken such that the intensity of the M-1 peaks is negligible. However, as stated above, it is first necessary to separate chemical isomers which could be present, such as allene in propyne. This may be realized by coupling a gas chromatograph with the mass spectrometer. Further, when $RC{\equiv}CD$ compounds are analysed, it is necessary to pay attention to possible exchanges involving the acetylenic deuterium. It has been found for instance that an unexpected result observed in $CD_3C{\equiv}CD$ analysis[62] is due to the exchange of the acetylenic deuterium with the water adsorbed on the gas injection system of the GC–MS or on the walls of the ion source of the mass spectrometer. This gives rise to $CD_3C{\equiv}CH$ and explains why an intense peak is found at m/e 43 and not, as expected, at m/e 44.

The determination of the position of the deuterium atoms in the molecule by mass spectrometry is a difficult problem and it is necessary to take into account the possibilities of scrambling[4]. It has been shown, for instance, that in the 1-phenylpropyne(D_3) case there is H/D randomization in the molecular ion prior to fragmentation[63]. The same H/D scrambling was observed when Safe[64] studied the mass spectra of $C_6H_5C{\equiv}CD$. He found too, when analysing the spectra of bromo- or chlorophenylacetylenes, that 100% H/D scrambling occurred in the $[M{-}X]^+$ ion prior to expulsion of the acetylene fragment[64]. The same phenomenon occurred when substituted diphenylacetylenes were analysed[65]. Hence it is impossible to deduce the position of the deuterium atoms in the molecules from analysis of the fragment ions. It does not seem that such scramblings have been studied in the case of aliphatic acetylenes.

Important isotope effects have been observed by mass spectrometry; it was found, for instance, that in $HC{\equiv}CD$, the breaking of the CH bond is 1·9 times more probable than the breaking of the CD bond[66]. Such effects are not peculiar to acetylene and have also been observed in the case of methane, ethylene and biphenyl[4, 67].

C. Infrared Spectroscopy

Infrared spectroscopy has not very often been used to check the deuterium purity of alkynes, with the exception of the case of compounds with the $C{\equiv}CD$ group. For these compounds a quantitative method, analogous to the one used to determine the deuterium content of deuterium oxide by n.m.r., has been described[5]. Known quantities of the non-deuterated hydrocarbon $RC{\equiv}CH$ are successively added to the corresponding $RC{\equiv}CD$ compound of unknown deuterium purity, and the transmittance of the $\nu({\equiv}CH)$ or $\nu({\equiv}CD)$ band is followed *vs.* the added amounts of $RC{\equiv}CH$. The method is long but it does not require precise knowledge of the coefficient of molar extinction, ε, of the $\nu({\equiv}CH)$ band. Its main advantage is the elimination of errors due to molecular associations. It is not applicable to volatile compounds.

From the analysis of residual ν(CH) bands of the deuterated methyl or methylene groups, it is possible to determine the nature of isotopic impurities. The comparison of their intensity to the ones relative to the pure compound indicates their percentage. So, from $CD_2HCHDC{\equiv}CH$ spectrum analysis, we have found that a sample of $CD_3CD_2C{\equiv}CH$ contained 4% of 1-butyne(D_4-3,4,4,4) and 6% of 1-butyne(D_4-3,3,4,4) [68]. In the same manner, we have found that the deuterium impurities of $CD_3CH_2C{\equiv}CH$ and $CH_3CD_2C{\equiv}CH$ were respectively $CHD_2CH_2C{\equiv}CH$ and $CH_3CHDC{\equiv}CH$ [69]. The method is now used to determine the deuterium purity of propyne(D_4) prepared by exchange in NaOD/D_2O at high temperatures[46]. The method is very convenient in this case since the bands due to allene do not overlap those of the propynes.

D. Nuclear Magnetic Resonance Spectroscopy

1. Proton magnetic resonance

Lavalley, Thiault and Braillon[70] have studied the possibility of estimating the deuterium content of an organic molecule by p.m.r. They give a survey of the theory and illustrate it by some examples relative to deuterated alkynes [$CH_3(CH_2)_3$-$C{\equiv}CD$ and $C_6H_{11}C{\equiv}CD$], using an internal standard. They have found that the method requires a certain quantity of alkyne and that its precision is not very high (limited generally by the noise). However, the method does not necessitate any preliminary purification of samples and very often allows location of the positions of the deuterium atoms in the molecule. A good example is given by the analysis of spectra of 1,2-butadienes which are formed during the 'isomerization-exchange' of 1-butyne to 2-butyne[48]. The presence of $CH_3CD{=}C{=}CD_2$, $CH_3CH{=}C{=}CD_2$, $CH_3CH{=}C{=}CHD$ and $CD_3CD{=}C{=}CH_2$ is shown, which confirms the proposed mechanism of isomerization. Unfortunately, the spectra of some 2-butynes, such as $CH_3C{\equiv}CCHD_2$ and $CD_2HC{\equiv}CCH_2D$, are not directly accessible and the method is not easily applicable to 2-butynes. Another striking application of p.m.r. spectroscopy is the determination of the purity of a $^{13}C_2H_2$ sample. This spectrum consists of the superposition of three spectra: the one-line spectrum of $H^{12}C{\equiv}^{12}CH$, the eight-line spectrum of $H^{13}C{\equiv}^{12}CH$ (ABX system) and the ten-line spectrum of $H^{13}C{\equiv}^{13}CH$ (AA'XX' system). It was found that it contained about 1% $H^{12}C{\equiv}^{12}CH$, 18% $H^{13}C{\equiv}^{12}CH$ and 81% $H^{13}C{\equiv}^{13}CH$ [49].

Lavalley[68] studied the p.m.r. spectra of several deuterated alkynes and observed the existence of isotope effects on chemical shifts (Table 2). It was found that

(*i*) $\sigma_H(D) \geqslant \sigma_H(H)$ always occurs (the substitution causes upfield shifts),
(*ii*) the effects are additive, and
(*iii*) their value depends upon the number of bonds between the substituted site and the hydrogen atom, whose signal is observed. For one substitution, the effect is:

$$\Delta\sigma \times 10^6 = 0{\cdot}015 \quad \text{(geminal effect)}$$
$$= 0{\cdot}008 \quad \text{(vicinal effect)}$$
$$= 0{\cdot}004 \quad \text{(distant effect, through four bonds, one of which is the triple one)}$$

It seems that a correlation exists between the value of the isotopic effect and the coupling constant[68]. However, the distant effect is smaller than expected in the case of propargyl halides. This could perhaps be explained by the presence of the electronegative atom, as was already postulated for other halides[71].

From these results it appears that the triple bond transmits isotope effects on chemical shifts very well. This result has to be compared with the observed effect on the solvolysis rate[72] of 4-chloro-4-methyl-2-pentyne(D_0) and (D_3-1,1,1). The rate

TABLE 2. Isotope effects on chemical shifts in some acetylenes[68]

Reference molecule	Substituted molecule	$[\sigma_H(D)-\sigma_H(H)]\times 10^6$
$\underline{CH_3}CD_2C\equiv CH$	$\underline{CHD_2}CD_2C\equiv CH$	0·031 ±0·01
$\underline{CH_3}CH_2C\equiv CH$	$\underline{CH_3}CD_2C\equiv CH$	0·015 ±0·003
$\underline{CH_3}CH_2C\equiv CCH_3$	$\underline{CH_3}CD_2C\equiv CCH_3$	0·017 ±0·003
$CH_3\underline{CH_2}C\equiv CD$	$CD_3\underline{CH_2}C\equiv CH$	0·025 ±0·003
$CH_3\underline{CH_2}C\equiv CCD_3$	$CD_3\underline{CH_2}C\equiv CCH_3$	0·022 ±0·003
$CD_2HCDHC\equiv \underline{CH}$	$CD_2HCD_2C\equiv \underline{CH}$	0·0042 ±0·0008
$CD_2HCDHC\equiv C\underline{H}$	$CD_3CD_2C\equiv C\underline{H}$	0·005 ±0·0008
$CD_3CH_2C\equiv C\underline{H}$	$CD_2HCDHC\equiv C\underline{H}$	0·002 ±0·0007
$CD_3CH_2C\equiv C\underline{H}$	$CD_3CD_2C\equiv C\underline{H}$	0·008 ±0·0017
$Br\underline{CH_2}C\equiv CH$	$Br\underline{CH_2}C\equiv CD$	0·001 ±0·0005
$Cl\underline{CH_2}C\equiv CH$	$Cl\underline{CH_2}C\equiv CD$	0·001 ±0·0005

retardation, $k_H/k_D = 1{\cdot}09$, due to deuterium in the latter, is another example of transmission of the effect across the triple bond. They both demonstrate that the deuterium substitution affects the electronic properties of the molecule and that the triple bond is a very efficient conductor of electronic effects.

2. ^{13}C nuclear magnetic resonance

Deuterium isotope effects on ^{13}C chemical shifts are of a different magnitude to those on proton chemical shifts. We have used this property to determine the deuterium content of 2-butynes prepared from the 'isomerization-exchange' method[48]. $\underline{C}H_3$, $\underline{C}HD_2$, $\underline{C}H_2D$ and $\underline{C}D_3$ group signals are easily located using the proton-decoupling technique (Figure 2). However, due to the Overhauser effect, it is not possible to deduce from the relative intensity of the peaks the proportion of the different groups. Using the same method, we found in the 2-pentyne case that the signal of the methylene ^{13}C is single, which shows that there are no deuterium atoms, or a very small quantity of them, present in the methylene group[46].

From Figure 2, we deduce that the isotope effect for one $H \rightarrow D$ substitution is

$$[\sigma_{^{13}C}(D)-\sigma_{^{13}C}(H)]\times 10^6 = 0{\cdot}22$$

which has a value quite similar to those already observed[73]. A striking result, on the other hand, arises from an analysis of $RC\equiv CD$ compounds: the β deuterium isotope shift (+0·50 p.p.m.) is more than twice the α one (+0·22 p.p.m.)[73]. There is no doubt about this result: we have obtained the same result using a mixture of 1-pentyne(D_0) and (D-1). Doddrell and Burfitt explain it by changes in $C\equiv C$ bond length when acetylenic hydrogen is substituted by deuterium[73].

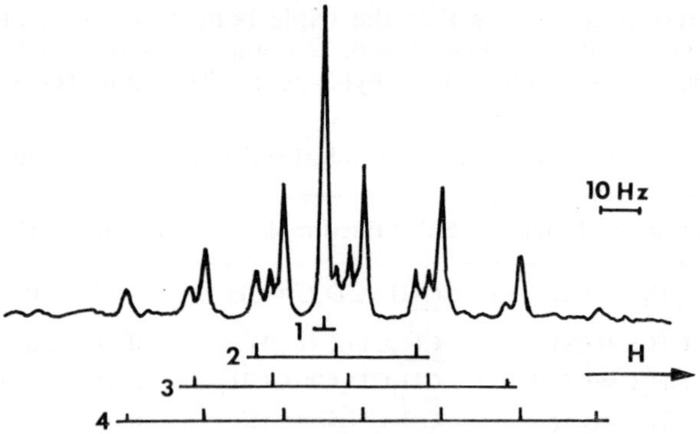

FIGURE 2. Proton-decoupled ^{13}C n.m.r. spectrum of a mixture of deuterated 2-butynes prepared from 1-butyne and D_2O (one 'isomerization-exchange' at 180 °C). (1) $\equiv\underline{C}H_3$ group, (2) $\equiv\underline{C}H_2D$ group, (3) $\equiv\underline{C}HD_2$ group, (4) $\equiv\underline{C}D_3$ group.

IV. EXAMPLES OF USES OF LABELLED ACETYLENES

A. Spectroscopic Studies

It appears that a large proportion of labelled acetylenes are prepared for use in molecular spectroscopy. In infrared spectroscopy for instance, it is well known that the substitution of hydrogen by deuterium changes the vibration frequencies of the bonds without changing the force constants. The assignment of the spectra of labelled (deuterated and ^{13}C-labelled) compounds and normal molecules leads to the experimentally determined fundamentals[41, 74] and then specifies the force field. For instance, this has been used to determine the harmonic force field of methylacetylene, taking into account corrections for anharmonicity which are unfortunately only approximate[52]. Without these corrections, it was found that the anharmonic k(CD) force constants (calculated by ignoring anharmonicity) are significantly higher than for CH, when, for instance, normal coordinate analysis of acetylenes C_2H_2, C_2HD and C_2D_2 has been carried out[75]. In ultraviolet spectroscopy, labelled acetylenes are mainly used to analyse the fine structure of absorption systems[40, 76]. In microwave spectroscopy, isotopically substituted molecules are used to determine the bond lengths and angles, with the aid for instance of Kraitchman's equations[77].

A more recent method of determining CH bond lengths with accuracy has been developed by McKean and coworkers[78] in infrared spectroscopy, using 'isolated' CH frequencies, derived from the spectra of compounds in which every hydrogen atom is deuterated except one. The method, first used by Lavalley[68], demonstrates the asymmetry of the methyl group of 1-butyne. Correlations between ν(CH) frequencies and bond lengths and dissociation energies enable this kind of information to yield accurate values for the latter quantities. In the case of dissociation energies, the method assumes no resonance stabilization of the radicals. In the 1-butyne case, from ν(CH) wave numbers of $CHD_2CD_2C{\equiv}CH$ and $CD_3CHDC{\equiv}CH$, it was found that

$$r_0\underline{CH}H_2 = 1{\cdot}095\ \text{Å} \quad D_0^{298} = 100{\cdot}7\ \text{kcal/mole}$$

$$r_0\underline{CHH} = 1{\cdot}099\ \text{Å} \quad D_0^{298} = 97{\cdot}7\ \text{kcal/mole}$$

Another example relates to propargyl alcohol, which is mainly in the *gauche* form in the gas phase[79]. Study of the $HC{\equiv}CCHDOH$ spectrum[80] showed that the CH bond *trans* to an OH bond is a little shorter ($\Delta r_0 \sim 0{\cdot}005$ Å) than the other, *trans* to an oxygen lone pair. An application of 'isolated' CH frequencies is the determination of molecular conformations. A good example is the study of the infrared spectrum of 2-ethynyl-1,3-dioxolane(D_4-4,4,5,5)[81]. The spectrum showed two ν(CH) bands, situated at 2863 and 2956 cm^{-1}. We assigned them to isomers (**1**) and (**2**) respectively, taking into account the specific effect of oxygen lone pairs on the *trans* CH vibrator:

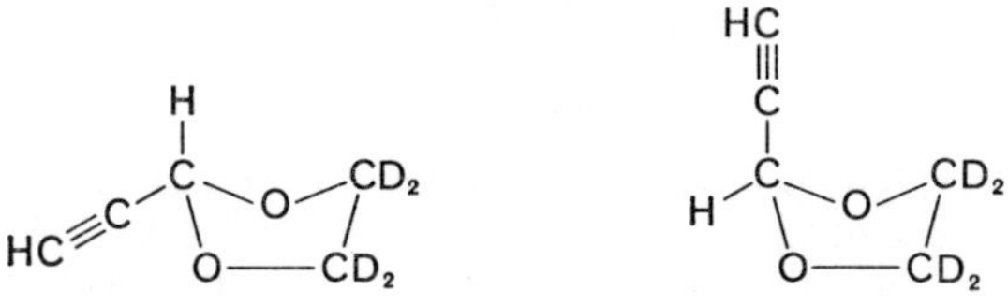

(**1**) (**2**)

From intensity considerations, isomer **2** predominates in the CCl_4 solution. On the other hand, isomer **1** is relatively more abundant in CD_3CN than in CCl_4 solution. These results agree with those obtained from n.m.r. analysis[82].

B. Studies of Mechanism

1. From the isotope composition of the products obtained

a. Isomerization. We have used the isomerization of acetylenes to prepare deuterated acetylenes (see Section II.B.3). Analysis of the deuterium content of the different compounds which are formed, i.e. 1-butynes, 1,2-butadienes and 2-butynes, is in accord with the mechanism proposed by Jacobs and collaborators[43] involving the removal of a proton by the base and rearrangement of the resultant carbanion.

Cram and coworkers[37], using $(C_6H_5)_2CDC{\equiv}CC_6H_5$ to study the 1,3-intramolecular proton transfer, found that the intramolecularity ranged from 88% in dimethylsulphoxide–methanol–triethylenediamine to 19% in methanol–potassium methoxide at 30 °C. To explain the 1,3-intramolecular proton transfer they propose the 'conducted tour' mechanism. On the other hand, Wotiz and coworkers[83] propose a concerted mechanism involving the $[NHCH_2CH_2NH_2]^-$ ion and the propargyl group in a nine-membered ring transition state in order to explain the relatively fast rearrangement occurring when 3-hexyne is mixed with a NH_2Na/ethylenediamine solution.

Isotopic labelling of the reactant and examination of the isotopic distribution in the products have been used to study whether the isomerization, $CH_2{=}C{=}CH_2 \rightleftharpoons CH_3C{\equiv}CH$, in a single-pulse shock-tube is intramolecular or whether it proceeds via another mechanism[62]. A 50–50 mixture of $CH_2{=}C{=}CH_2$ and $CD_2{=}C{=}CD_2$ was shocked at 1030–1220 K, the gas being highly diluted in argon. If the formed propynes are unscrambled, i.e. if only $CH_3C{\equiv}CH$ and $CD_3C{\equiv}CD$ are present, the reaction is clearly intramolecular. On the other hand, if C_3D_3H and C_3H_3D are observed, either an isotope exchange reaction has occurred or the reaction is not intramolecular. Taking account of the D → H exchange between propyne(D_4) and water in the mass spectrometer, it was found that the isomerization is first order with respect to allene. As only a few decomposition products were found (0·17% after 30% of the allene had isomerized) it is concluded that the following mechanism proceeds to a large extent[62]:

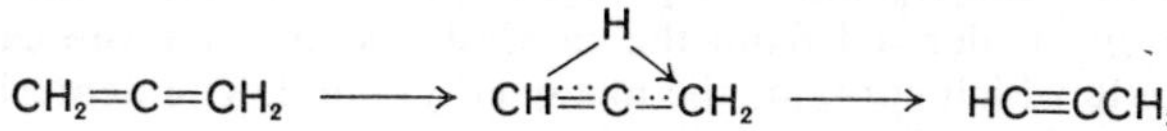

b. Metathesis. It has been found that metathesis of $CH_3(CH_2)_2C{\equiv}C^{14}CH_3$ over MoO_3–SiO_2 at 350 °C leads to labelled 2-butyne and unlabelled 4-octyne[84]. This suggests the following mechanism:

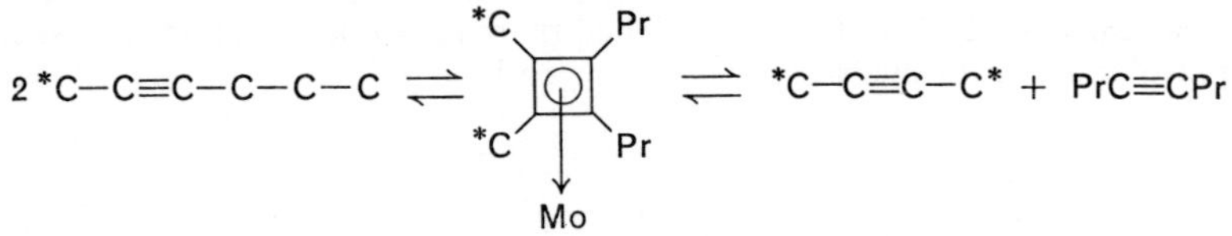

Over the same catalyst, propyne does not form products of metathesis but yields small quantities of cyclotrimerization products, one of them being 1,2,3-trimethylbenzene (4%), which may indicate that a similar mechanism could occur.

c. Cyclotrimerization. Whitesides and Ehmann[36] have carried out a series of studies on cyclotrimerization by transition metal catalysts, using $CH_3C{\equiv}CCD_3$. The idea was to determine whether 1,2,3-trimethyl-4,5,6-trimethyl(D_3)-benzene (**3**) is a product of the reaction. If not, this excludes a cyclobutadiene intermediate (**4**),

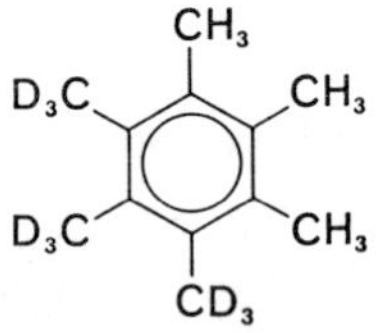

(3)

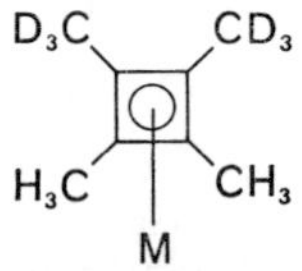

(4)

formed by dimerization of 2-butyne(D_3-1,1,1). The assumption has been made that the four ring carbon atoms and carbon–carbon bonds of the metal cyclobutadiene are chemically equivalent. With supplementary assumptions, particularly if deuterium kinetic isotope effects are neglected, it was found that, within the limit of detection of the involved experimental procedure ($\sim \pm 0{\cdot}5\%$), no compound (3) is formed in the reactions involving triphenyltris(tetrahydrofuran)chromium(III), dimesitylcobalt(II), dicobalt octacarbonyl, bis(acrylonitrile)nickel(0) and the Ziegler catalyst $TiCl_4 + (iso\text{-}C_4H_9)_3Al$. These five catalysts represent the majority of the structural types commonly associated with transition metal reagents displaying activity in acetylene cyclotrimerization reactions. On the other hand, results involving the use of $AlCl_3$ as a catalyst indicate that an intermediate having the same effective symmetry as a tetramethylcyclobutadiene may be involved. The yield of (3) obtained from the cyclization catalysed by bis(benzonitrile)palladium(II) dichloride is intermediate between the two extremes. A similar study has been carried out with triphenyltris(tetrahydrofuran)chromium(III) and $CH_3C{\equiv}CCD_3$ [85]. A free or metal-complexed tetramethylcyclobutadiene has been excluded as an intermediate in this reaction because it has been observed that 1,2-dimethyl-3,4-di(methyl-D_3)-naphthalene is not a product of the reaction.

In the same manner, reactions involving acetylene and acetylene(D_2) have been studied[86]. The catalyst system used was a Ziegler–Natta catalyst, tris(acetylacetonato)titanium(III) and diethylaluminium chloride. With C_2H_2, it leads to formation of benzene, a trace of ethylbenzene and a small amount of polyacetylene[86]. Reaction with C_2D_2 gives benzene(D_6) and $C_6D_5CH_2CH_3$, which suggests that the ethyl group in the ethylbenzene is derived from the catalyst system, or more exactly from the active intermediate which contains the metal–ethyl bond. The deuterium position in the deuterated benzenes, obtained from an equimolar mixture of acetylene and acetylene(D_2), has been investigated. The results suggest that the mechanism involving the cyclobutadiene intermediate may be discarded, as benzene(D_2-1,4) and benzene(D_4-1,2,4,5) are not formed.

2. From kinetic isotope effects

Deuterium substitution may give rise to important effects on the reaction kinetics. The maximum kinetic isotope effect is obtained when the bond is broken in the transition state (primary isotope effect). In reactions involving several stages, isotope effects can naturally be observed only if the bond to the isotope is broken in the rate-determining step[87]. In this case, deuterium substitution would be expected to depress the reaction rate. We give below two examples relative to acetylenes, showing how the effect, k_H/k_D, may specify the reaction process or, on the contrary, allow the rejection of a possible mechanism.

The mechanism of 1,2,3,4-tetramethylnaphthalene formation from 2-butyne and triphenyltris(tetrahydrofuran) chromium(III) has been clarified by using deuterated compounds[85]. For instance, the reaction of (**5**) with 2-butyne leads to a mixture of 1,2,3,4-tetramethylnaphthalenes (**6** and **7**). The ratio k_H/k_D is $2{\cdot}7 \pm 0{\cdot}1$. This result,

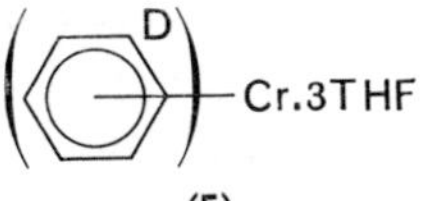

(**5**)

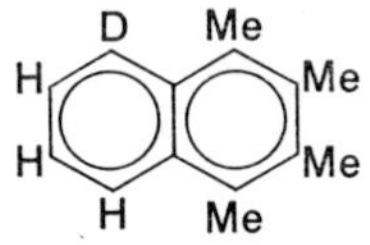

(**6**)

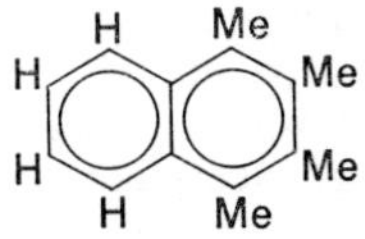

(**7**)

added to that obtained by the reaction of 2-butyne with a mixture of triphenyl(D_0)- and triphenyl(D_5)tris(tetrahydro)chromium(III) ($k_H/k_D = 0{\cdot}97 \pm 0{\cdot}02$) discredits benzyne complexes as intermediates in the formation of final products[85].

Another example concerns the trimerization of acetylene over $[(C_6H_5)_3P]_4Ni$. This could be represented by the following scheme:

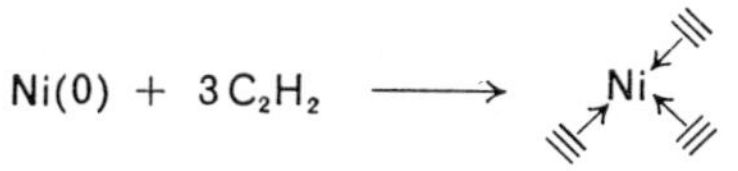

However, a study[88] of the comparative kinetics of the transformation of C_2H_2 and C_2D_2 over this catalyst showed that the rate of adsorption of acetylene is very much higher than the rate of adsorption of C_2D_2 ($k_H/k_D = 2{\cdot}3$). This effect implies the rupture of the H(D)—C bond in the process of isomerization, and confirms the first step of Meriwether's scheme[89]:

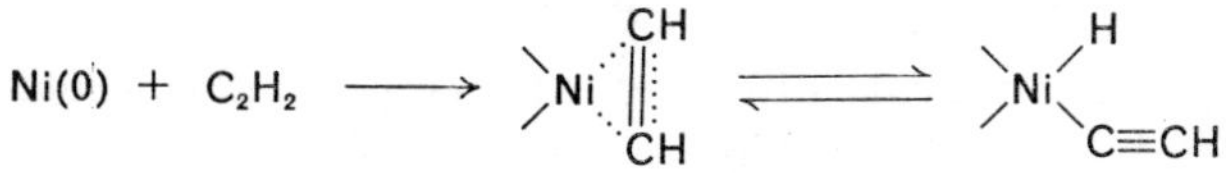

3. From the structure of adsorbed species

A way to study reaction intermediates and therefore the reaction mechanisms is the determination by spectroscopy of the structure of adsorbed species formed on the surface of catalysts. The assumption is that the observed species take part in the chemical process which is being investigated. Infrared spectroscopy seems a powerful technique of investigation of such structures and we report some recent results from our laboratory on this topic. It is necessary, however, to take account of the occurrence of species which are due to side-reactions. They seem to appear often and should first be characterized. Use of deuterated compounds make their determination easier. The other species may be intermediates of the reaction studied. It is important to determine their structure exactly. Here again, use of labelled compounds, or more precisely the study of the shifts of the characteristic bands due to $H \rightarrow D$ substitution, is of great assistance. Examples are given below in relation to the adsorption of acetylenes on metal oxides.

a. Elimination of species due to side-reactions. Infrared study of adsorbed acetylenes on alumina showed that strongly held species are attached to the surface by the acetylenic end, giving rise to RC≡C···Al(surface) species[90]. Moreover, on alumina, there is isomerization, which explains the nature of strongly held species given by 2-butyne adsorbed on Al_2O_3 [91]. A similar study with ZnO as the oxide did not take into account the possibility of the presence of a great deal of acetylenic $RC{\equiv}C^-$ carbanion species[92]. Use of 3,3-dimethyl-1-butyne and 3,3-dimethyl-1-butyne(D-1) proved that this assumption was untrue. The spectra given by these particular two compounds, adsorbed on ZnO, give rise to strong bands which are closely similar if we except the shift of the band at 3300 cm^{-1} to 2460 cm^{-1} due to the formation of the OH(OD) groups on the surface[93]. This proves that dissociative adsorption occurs with formation of $(CH_3)_3CC{\equiv}C\cdots Zn$(surface) species. Moreover, use of $(CH_3)_3CC{\equiv}CD$ allows us to assign the two bands observed between 2050 and 2150 cm^{-1} to two kinds of 'acetylide' species.

b. Study of the structure of 'intermediate' species. Contrary to the alumina case, for which no species due to isomerization intermediates has been found[90], Chang and Kokes[92] have shown on ZnO the formation of a propargyl species, which is a likely intermediate in the allene–methylacetylene isomerization reaction. It was important to define the structure of the species as accurately as possible, which we did by using deuterated acetylenes[93]. The two extreme resonance forms for these species derived from methyl acetylene are:

$$^{-}CH_2C{\equiv}CH \quad \text{and} \quad CH_2{=}C{=}CH^{-}$$

(8) **(9)**

Both of these imply a linear three-carbon skeleton, but different angular dispositions of the CH bonds, i.e. corresponding approximately to sp^3 and sp hybridization of the terminal carbons in the first case, and $2 \times sp^2$ in the second case. Intermediate resonance forms such as

$$\tfrac{1}{2}^{-}CH_2\overset{\cdots}{-}C\overset{\cdots}{\equiv}CH\tfrac{1}{2}^{-}$$

(10)

are also possible, and the actual disposition taken up by the ion would be expected to depend on the details of the interaction (e.g. one-ended or not) between the carbanion and the surface zinc ion. Our choice between the various alternatives were based on frequency shifts due to $H \rightarrow D$ substitution on the characteristic band near 1850 cm^{-1}. The frequencies are reported in Table 3. In the propyne case, the extreme structure **(9)** leads to the expectation that substitution of the three CH bonds by CD should each lead to a lowering of the asymmetrical $\nu(C{=}C{=}C)$ frequency (in allene itself near 1980 cm^{-1})[94]. In fact (Table 3), the substitution of only one CH bond by CD affects the frequency of the band in the 1900–1800 cm^{-1} region. This latter observation is supported qualitatively by the extreme structure (**8**), but the observed isotopic shift ($\sim 30\ cm^{-1}$) is much smaller than that observed[14] for the 'C≡C' frequencies of the parent gas-phase molecules ($133\ cm^{-1}$). Hence a structure between (**8**) and (**10**) is to be preferred, with the CC bond to which the lone CH is attached retaining a relatively high bond order and the CH bond being at an angle between 120° and 180° in relation to it.

TABLE 3. Position (in cm^{-1}) of the bands near $1850\ cm^{-1}$ for different acetylenes adsorbed on ZnO [93]

Acetylene	Position of bands (cm^{-1})
$CH_3C{\equiv}CH$	1880 (sh) [a] 1865
$CH_3C{\equiv}CD$	(1880) (sh) [c] 1834 (1865) [c]
$CD_3C{\equiv}CH$	1880 (sh) 1863 (1830) [c]
$CD_3C{\equiv}CD$ [b]	1829
$CH_3CH_2C{\equiv}CH$	1880 1865
$CD_3CD_2C{\equiv}CD$ [b]	1880 1839

[a] sh = shoulder.
[b] Adsorbed on deuterated ZnO (containing —OD groups).
[c] The intensity of the bands in brackets increases with time; they are due to isotope exchanges.

It may be noted that the order of appearance and intensities of the *ca.* 1865 and $1835\ cm^{-1}$ bands agree with the propargylic formulation. Initially, with $CH_3C{\equiv}CD$ and $CD_3C{\equiv}CH$, the respective frequencies corresponding to the C⩦CD or C⩦CH groups of the initially adsorbed molecule are observed. After a period of time the two bands are both present in the ratio expected if there is easy (but not instantaneous) equilibrium between the three H or D atoms in the species.

From comparison of frequencies and frequency shifts due to H → D substitution (Table 3) with ethylacetylenes it is deduced[92] that two types of propargyl species are observed:

$$[HC\overset{\cdots}{\equiv}C\overset{\cdots}{-}CHCH_3]^- \qquad [CH_3C\overset{\cdots}{\equiv}C\overset{\cdots}{-}CH_2]^-$$

As might be expected, the former species predominates at first when 1-butyne is the adsorbate and the latter when 2-butyne is the adsorbate. From the results with 2,3-pentadiene it appears that, as is the case with the parent hydrocarbons, adsorbed species with an internal high-order CC bond seem to be the most stable. Finally, the fact that 2,4-dimethyl-2,3-pentadiene does not give strongly held species on ZnO is consistent with the overall conclusion that either a CH group α to C≡C, or an allene-type CH group, is necessary if propargylic surface species are to be formed.

Since similar propargylic spectra are obtained, starting from isomeric substituted acetylenes and allenes, we conclude that this surface species does act as a catalytic intermediate. This clarifies the mechanism: isomerization apparently occurs via a 1,3-hydrogen shift involving the propargyl species as an intermediate. Chang and Kokes suggest three steps[92]: the first represents dissociative adsorption, the second represents the surface rearrangement and the third represents readdition of the hydrogen atom and desorption of the product. However, we think that some further experiments are necessary before coming to definite conclusions; without deuterated compounds such a study would have been almost impossible, as the main information on the structure of the species was deduced from isotope shifts of characteristic bands.

V. ACKNOWLEDGEMENTS

We are grateful to Professor Romanet and Professor Sheppard for their interest in this work. Thanks are also due to A. Janin, A. Vallet, J. Lamotte, O. Saur and J. Travert for their very kind collaboration.

VI. REFERENCES

1. A. Murray, III and D. L. Williams, *Organic Syntheses with Isotopes*, Interscience, New York, 1958.
2. J. C. Lavalley, A. Janin and R. Romanet, *Proceedings of the Second International Conference on Methods of Preparing and Storing Labelled Compounds*, Brussels, Nov. 1966, p. 631.
3. A. I. Shatenshtein, *Isotopic Exchange and the Replacement of Hydrogen in Organic Compounds*, translated from the Russian, Consultants Bureau, New York, 1962.
4. A. F. Thomas, *Deuterium Labeling in Organic Chemistry*, Appleton–Century–Crofts, Educational Division, Meredith Corporation, New York, 1971.
5. R. Romanet and B. Wojtkowiak, *Bull. Soc. Chim. Fr.*, 1500 (1960).
6. B. S. Rabinovitch and F. S. Looney, *J. Amer. Chem. Soc.*, **75**, 2652 (1953).
7. G. Szeimies and J. Harnisch, *Chem. Comm.*, 739 (1973).
8. G. Zweifel and H. Arzoumanian, *J. Amer. Chem. Soc.*, **89**, 291 (1967).
9. J. Jonas and T. M. Digennaro, *J. Chem. Phys.*, **50**, 2392 (1969).
10. K. C. Eberly and H. E. Adams, *J. Organomet. Chem.*, **3**, 165 (1965).
11. A. Streitwieser, Jr, L. Verbit and R. Bittman, *J. Org. Chem.*, **32**, 1530 (1967).
12. R. Couffignal and M. Gaudemar, *Bull. Soc. Chim. Fr.*, 3218 (1969).
13. R. H. Summerville and P. v. R. Schleyer, *J. Amer. Chem. Soc.*, **96**, 1110 (1974).
14. R. J. Grisenthwaite and H. W. Thompson, *Trans. Farad. Soc.*, **50**, 212 (1954).
15. B. C. Patterson, *J. Chem. Soc. (C)*, 942 (1966).
16. I. Yasumori and K. Hirabayashi, *Trans. Farad. Soc.*, **67**, 3283 (1971).
17. L. Lompa-Krzymien and L. C. Leitch, *Synthesis*, 124 (1976).
18. W. A. Van Hook and J. T. Phillips, *J. Chromatog.*, **30**, 211 (1967).
19. D. J. Cram, *Fundamentals of Carbanion Chemistry*, Academic Press, New York, 1965.
20. R. A. Nyquist, T. L. Reder, F. F. Stec and G. J. Kallos, *Spectrochim. Acta*, **27A**, 897 (1971); R. A. Nyquist, T. L. Reder, G. R. Ward and G. J. Kallos, *Spectrochim. Acta*, **27A**, 541 (1971).
21. R. A. Nyquist, *Spectrochim. Acta*, **27A**, 2513 (1971).
22. D. G. Lister and J. K. Tyler, *Spectrochim. Acta*, **28A**, 1423 (1972).
23. L. H. Jones, M. Goldblatt, R. S. McDowell and D. E. Armstrong, *J. Mol. Spectrosc.*, **23**, 9 (1967).
24. V. D. Nefedov, E. N. Sinotova, M. V. Korsakov and E. G. Alekseev, *Radiokhimiya*, **15**, 635 (1973); *Chem. Abstr.*, **79**, 136635z (1973).
25. C. Eaborn, G. A. Skinner and D. R. M. Walton, *J. Chem. Soc. (B)*, 922 (1966); 989 (1966).

26. L. C. Leitch and R. Renaud, *Can. J. Chem.*, **30,** 79 (1952).
27. H. Hunziker, W. Good, R. Meyer and H. H. Günthard, *Helv. Chim. Acta*, **45,** 59 (1962).
28. R. West and P. C. Jones, *J. Amer. Chem. Soc.*, **91,** 6156 (1969).
29. J. E. Mulvaney, T. L. Folk and D. J. Newton, *J. Org. Chem.*, **32,** 1674 (1967).
30. T. L. Chwang and R. West, *J. Amer. Chem. Soc.*, **95,** 3324 (1973).
31. B. Bak, D. Christensen, L. Hansen-Nygaard and E. Tannenbaum, *J. Chem. Phys.*, **26,** 241 (1957).
32. J. L. Duncan, D. C. McKean, P. D. Mallinson and R. D. McCulloch, *J. Mol. Spectrosc.*, **46,** 232 (1973).
33. A. Saus and P. Krings, *Tetrahedron Letters*, 947 (1967).
34. J. Saussey, J. Lamotte and J. C. Lavalley, Spectrochim. Acta, to be published.
35. L. C. Leitch and A. T. Morse, *Can. J. Chem.*, **31,** 785 (1953).
36. G. M. Whitesides and W. J. Ehmann, *J. Amer. Chem. Soc.*, **91,** 3800 (1969).
37. D. J. Cram, F. Willey, H. P. Fisher, H. M. Relles and D. A. Scott, *J. Amer. Chem. Soc.*, **88,** 2759 (1966).
38. S. Safe, *J. Lab. Compds, VI,* 197 (1970).
39. W. Chodkiewicz, *Ann. Chim.*, **12,** 852 (1957).
40. J. Lamotte, *Thesis*, Caen (1976).
41. J. Lamotte, J. C. Lavalley and R. Romanet, *J. Chim. Phys.*, **70,** 1077 (1973).
42. A. Favorskii, *J. Prakt. Chem.*, **37,** 382 (1888).
43. T. L. Jacobs, R. Akawie and R. G. Cooper, *J. Amer. Chem. Soc.*, **73,** 1273 (1951).
44. M. Behal, *Ann. Chim.*, [6], **15,** 408 (1888).
45. M. Bourguel, *Ann. Chim.*, [10], **3,** 190 (1925).
46. J. Saussey, J. C. Lavalley, A. Janin and A. Vallet, to be published.
47. J. H. Wotiz in *Chemistry of Acetylenes* (Ed. H. G. Viehe), Marcel Dekker, New York, 1969, pp. 365–424.
48. J. Saussey, J. C. Lavalley, A. Janin and A. Vallet, *J. Lab. Compds.*, XIII, 417 (1977).
49. T. W. Whaley and D. G. Ott, *J. Lab. Compds*, **X,** 461 (1974).
50. J. D. Cox and R. J. Warne, *J. Chem. Soc.*, 1893 (1951).
51. R. M. Lynden-Bell and N. Sheppard, *Proc. Roy. Soc.* (*London*), *A*, **269,** 385 (1962).
52. J. L. Duncan, D. C. McKean and G. D. Nivellini, *J. Mol. Struct.*, **32,** 255 (1976).
53. R. F. C. Brown, F. W. Eastwood, K. J. Harrington and G. L. McMullen, *Aust. J. Chem.*, **27,** 2393 (1974).
54. J. Casanova, M. Geisel and R. N. Morris, *Org. Prep. Proced.*, **1,** 81 (1969).
55. R. N. Renaud and L. C. Leitch, *Can. J. Chem.*, **42,** 2089 (1964).
56. D. C. Myers and F. Schmidt-Bleek, *J. Lab. Compds*, **III,** 62 (1967).
57. C. George, E. W. Gill and J. A. Hudson, *J. Chem. Soc.* (*C*), 74 (1970).
58. J. K. Crandall, D. J. Keyton and J. Kohne, *J. Org. Chem.*, **33,** 3655 (1968).
59. E. A. Halevi, *Prog. Phys. Org. Chem.*, **1,** 109 (1963).
60. P. Laszlo and Z. Welvart, *Bull. Soc. Chim. Fr.*, 2412 (1966).
61. P. L. Gant and K. Yang, *J. Amer. Chem. Soc.*, **86,** 5063 (1964).
62. A. Lifshitz, M. Frenklach and A. Burcat, *J. Phys. Chem.*, **79,** 1148 (1975).
63. S. Safe, W. D. Jamieson, W. R. Pilgrim and S. Wolfe, *Can. J. Chem.*, **48,** 1171 (1970).
64. S. Safe, *J. Chem. Soc.* (*B*), 962 (1971).
65. S. Safe, *Chem. Comm.*, 534 (1969); *Org. Mass Spectrom.*, **3,** 239 (1970); **7,** 1329 (1973).
66. P. M. Guyon and F. Fiquet-Fayard, *J. Chim. Phys.*, **66,** 32 (1969).
67. M. I. Gorfinkel, Y. A. Koptyug and K. M. Salikhov, *Izv. Sib. Otd. Akad. Nauk. SSSR, Ser. Khim. Nauk.*, 103 (1967).
68. J. C. Lavalley, *Thesis*, Caen, 1969.
69. J. C. Lavalley and R. Romanet, *J. Chim. Phys.*, **69,** 1431 (1972).
70. J. C. Lavalley, B. Thiault and B. Braillon, *Bull. Soc. Chim. Fr.*, 1875 (1967).
71. H. Batiz-Hernandez, *Thesis*, The Pennsylvania State University, 1965; R. A. Bernheim and H. Batiz-Hernandez, *J. Chem. Phys.*, **45,** 2261 (1966).
72. V. J. Shiner and G. S. Kriz, *J. Amer. Chem. Soc.*, **86,** 2643 (1964).
73. D. Doddrell and I. Burfitt, *Aust. J. Chem.*, **25,** 2239 (1972).
74. J. Saussey, J. Lamotte and J. C. Lavalley, *Spectrochim. Acta*, **32A,** 763 (1976).
75. M. K. Phibbs and D. G. Robinson, *Can. J. Spectrosc.*, **19,** 29 (1974).
76. G. W. King and S. P. So, *J. Mol. Spectrosc.*, **37,** 543 (1971).

77. L. B. Szalanski and R. G. Ford, *J. Mol. Spectrosc.*, **54**, 148 (1975).
78. D. C. McKean, J. L. Duncan and L. Batt, *Spectrochim. Acta*, **29A**, 1037 (1973).
79. E. Hirota, *J. Mol. Spectrosc.*, **26**, 335 (1968).
80. J. Travert, O. Saur, A. Janin and J. C. Lavalley, *J. Mol. Struct.*, **33**, 265 (1976).
81. O. Saur, A. Janin, A. Vallet and J. C. Lavalley, *J. Mol. Struct.*, **34**, 171 (1976).
82. W. F. Wailey and E. L. Eliel, *J. Amer. Chem. Soc.*, **96**, 1798 (1974).
83. J. H. Wotiz, P. M. Barelski and D. F. Koster, *J. Org. Chem.*, **38**, 489 (1973).
84. A. Mortreux and M. Blanchard, *Bull. Soc. Chim. Fr.*, 1641 (1972).
85. G. M. Whitesides and W. J. Ehmann, *J. Amer. Chem. Soc.*, **92**, 5625 (1970).
86. H. Shirakawa and S. Ikeda, *J. Polymer. Sci., Polym. Chem. Ed.*, **12**, 929 (1974).
87. H. Simon and D. Palm, *Angew. Chem., Int. Ed.*, **5**, 920 (1966).
88. G. A. Chukhadzhyan, Z. I. Abramyan and G. A. Gevorkyan, *J. Gen. Chem. (USSR)*, **43**, 1998 (1974).
89. L. S. Meriwether, M. F. Leto, E. C. Colthup and G. W. Kennerly, *J. Org. Chem.*, **27**, 3930 (1962).
90. D. J. Yates and P. J. Lucchesi, *J. Chem. Phys.*, **35**, 243 (1961).
91. J. Saussey, J. Lamotte, J. C. Lavalley and N. Sheppard, *J. Chim. Phys.*, **72**, 818 (1975).
92. C. C. Chang and R. J. Kokes, *J. Catal.*, **28**, 92 (1973).
93. J. Saussey, J. C. Lavalley and N. Sheppard, *J. Chim. phys.*, **74**, 329 (1977).
94. L. M. Sverdlov and M. G. Borisov, *Opt. Spectrosc.*, **9**, 227 (1960).

Author Index

This author index is designed to enable the reader to locate an author's name and work with the aid of the reference numbers appearing in the text. The page numbers are printed in normal type in ascending numerical order, followed by the reference numbers in parentheses. The numbers in *italics* refer to the pages on which the references are actually listed.

Beetz, T. 431, 433 (362), *445*
Beggs, D. P. 189 (155), *197*
Begoon, A. 762 (24), *806*
Bégué, J. P. 171 (69), *195*, 244 (492), *269*
Behal, M. 961 (44), *975*
Behr, O. M. 696 (239), *712*, 729 (55a, b), *737*, 757 (5), *805*
Behringer, H. 472 (82), *518*, 786 (168), *810*
Behzadi, A. 828, 875 (121), *941*
Beissner, C. 122 (380), *135*
Beisswenger, U. 254, 255 (596), *272*
Beletskaya, I. P. 76 (19), *127*
Bell, A. P. 232 (329), *266*
Bell, I. 792 (198b), *810*
Bell, N. A. 186 (112), *196*
Bell, R. A. 125 (413), *136*
Bellamy, L. J. 91, 92 (91), *129*, 143 (32), *154*
Bellan, J. 883, 887 (471), *949*
Bellotti, L. 915, 916 (680), *954*
Belluco, U. 40 (196), *55*
Beltrame, A. 5, 50 (3), *50*
Beltrame, P. 490 (169), 491 (173), *520*, 773 (90), *808*, 838 (164), 913 (164, 668, 674), 915 (164, 680), 916 (164, 668, 680, 681), 924 (164, 697), 925 (164), 927 (706), 932, 935 (164), *942*, *953*, *954*
Beltrame, P. L. 773 (90), *808*, 838 (164), 913 (164, 668), 915 (164, 680), 916 (164, 668, 680, 681), 924 (164, 697), 925, 932, 935 (164), *942*, *953*, *954*
Belyaeva, A. N. 578 (114), *617*, 865, 866 (326), 886 (529), *945*, *950*
Benaim, J. 361 (49), *378*, 399, 401 (161), *440*
Bender, C. F. 21 (70), *52*
Bender, M. L. 911, 917 (645), *953*
Beneche, H. P. 487 (153), *520*
Benedikt, G. 216, 217 (171), *262*
Ben-Efraim, D. A. 389 (76), 392 (116), 394 (125, 126), 395 (76, 125, 136), *439*, *440*, 529 (33), *550*, 572 (85), 579, 603, 604 (121), 612 (277), *616*, *617*, *620*, 663 (132), 666 (138), *710*, 767 (59), 791 (195), 795 (218, 221), *807*, *810*, *811*
Beneš, M. J. 364 (60), *378*, 793 (211), *811*, 849 (255), *944*
Benghiat, V. 11 (15), 12 (19, 20), *51*
Benkeser, R. A. 227, (271–273, 275, 276), *264*, 366 (70), 371 (103), *378*, *379*, 400 (176), *441*, 749, 750 (26, 27), *753*, 846 (226), *943*
Benko, D. A. 760, 765 (14d), *806*
Benn, W. R. 243 (467), *269*, 403 (208), *441*
Bennett, M. A. 14 (37), 39 (175), *51*, *54*, 111 (275–278), 115 (316), 119 (278), 121 (378), *133*, *135*
Benson, S. W. 69 (4), 70 (11), 71 (4, 11, 13), *73*, 382 (15), *437*
Benson, W. R. 247 (535, 541), *270*, 302 (101), *339*
Bentley, R. K. 186 (120), *196*, 595 (195), *619*, 631 (58), *633*
Berbalk, H. 96 (113), *129*
Berdnikov, V. M. 567 (61–63), 568 (61, 62), *616*
Béress, L. 776 (110a,b), *808*
Berg, M. H. 391 (108), *439*
Bergel'son, L. D. 325 (186), *340*
Bergenthal, M. D. 756, 797 (1o), *805*
Bergman, R. G. 26 (102), *53*, 428 (338), 429 (338, 341), *444*, 565 (51), 600, 601 (230), 607 (252), 609 (263–266), 611 (273), *615*, *619*, *620*
Bergmark, T. 23 (81), *52*
Beringer, F. M. 859, 929, 936 (312), *945*
Berkowitz, J. 161 (9), 169 (55), *194*, *195*
Berlan, J. 431, 432 (355), *445*
Berliner, E. 327, 328 (189, 190), 329 (190, 191), 330 (191), *341*
Bernadou, F. 202 (77), *260*, 896, 897, 900 (554), *951*
Bernady, K. F. 221, 222 (222), 225, 226 (222, 259), *263*, *264*
Bernardi, F. 830 (140), *941*
Berngruber, W. 525, 537 (16), *549*
Bernhardt, J. C. 203, 240 (84), *261*
Bernheim, R. A. 966 (71), *975*
Bernstein, H. J. 92 (94), *129*
Berry, D. J. 367 (81), *378*
Berry, R. S. 161 (10), *194*
Berteli, D. J. 672, 681 (161), *710*
Berthier, G. 33 (143), *54*
Bertin, D. 763 (33), *806*
Bertolini, J. C. 49 (231), *56*
Bertrand, M. 389 (80, 96, 100), 392 (111, 112), 397 (143, 144), 398 (80), 400 (80, 169), 417 (284), *439*, *440*, *443*, 855, 857 (282), *944*
Bertrand, M. T. 896 (555), 898 (555, 557) 900 (557), *951*
Besseyre, J. 452 (15), *517*
Bethke, H. 176, 177 (83), 178 (84, 88), 185 (83, 84), *195*, *196*, 564 (48), *615*
Betts, T. C. 30 (126), *53*
Beumel, D. F. Jr 692, 694, 704 (218), *711*
Beumel, O. F. Jr 792 (200, 201), *811*
Bevan, C. W. L. 915 (679), *954*
Beveridge, D. L. 60 (35), *66*, 818, 827 (74), *940*
Bey, P. 241, 242 (454), *269*
Beynon, J. H. 170 (64), 184 (100), *195*, *196*
Bezman, S. A. 39, 40 (188), *55*, 97, 117 (134), *130*
Bezuglyi, V. D. 845 (213), *943*
Bhacca, N. S. 527 (25), *549*
Bhanu, B. 257, 258 (636), *272*

Subject Index